The Science and Technology of Rubber

Fourth Edition

The Science and Technology of Rubber

Fourth Edition

Edited by

Burak Erman
Department of Chemical and Biological Engineering
Koc University
Rumeli Feneri Yolu 34450 Istanbul, Turkey

James E. Mark
Department of Chemistry
University of Cincinnati
Cincinnati, OH 45221-0172, USA

C. Michael Roland
Naval Research Laboratory
Chemistry Division, Code 6120
Washington, DC, USA

ELSEVIER

AMSTERDAM • BOSTON • HEIDELBERG • LONDON
NEW YORK • OXFORD • PARIS • SAN DIEGO
SAN FRANCISCO • SINGAPORE • SYDNEY • TOKYO

Academic Press is an Imprint of Elsevier

Academic Press is an imprint of Elsevier
225 Wyman Street, Waltham, MA 02451, USA
The Boulevard, Langford Lane, Kidlington, Oxford, OX5 1GB, UK

Notices
Knowledge and best practice in this field are constantly changing. As new research and experience broaden our understanding, changes in research methods, professional practices, or medical treatment may become necessary.

Practitioners and researchers must always rely on their own experience and knowledge in evaluating and using any information, methods, compounds, or experiments described herein. In using such information or methods they should be mindful of their own safety and the safety of others, including parties for whom they have a professional responsibility.

To the fullest extent of the law, neither the Publisher nor the authors, contributors, or editors, assume any liability for any injury and/or damage to persons or property as a matter of products liability, negligence or otherwise, or from any use or operation of any methods, products, instructions, or ideas contained in the material herein.

Library of Congress Cataloging-in-Publication Data
A catalog record for this book is available from the Library of Congress

British Library Cataloguing-in-Publication Data
A catalogue record for this book is available from the British Library.

ISBN: 978-0-12-394584-6

Printed and bound by CPI Group (UK) Ltd, Croydon, CR0 4YY
13 14 15 9 8 7 6 5 4 3 2 1

CONTENTS

Chapter 1

Rubber Elasticity: Basic Concepts and Behavior

A.N. Gent
The University of Akron, Akron, OH, USA

1.1 INTRODUCTION

The single most important property of elastomers—that from which their name derives—is their ability to undergo large elastic deformations; that is, to stretch and return to their original shape in a reversible way. Theories to account for this characteristic high elasticity have passed through three distinct phases: (1) the early development of a molecular model relating experimental observations to the known molecular features of rubbery polymers; (2) generalization of this approach by means of symmetry considerations taken from continuum mechanics that are independent of the molecular structure; and (3) a critical reassessment of the basic premises on which these two quantitative theories are founded. In this chapter, the theoretical treatment is briefly outlined and shown to account quite successfully for the observed elastic behavior of rubbery materials. The special case of small elastic deformations is then discussed in some detail because of its technical importance. Finally, attention is drawn to some aspects of rubber elasticity that are still little understood.

1.2 ELASTICITY OF A SINGLE MOLECULE

The essential requirement for a substance to be rubbery is that it consist of long flexible chainlike molecules. The molecules themselves must therefore have a "backbone" of many noncollinear single valence bonds, about which rapid rotation is possible as a result of thermal agitation. Some representative molecular subunits of rubbery polymers are shown in Figure 1.1; thousands of these units linked together into a chain constitute a typical molecule of the elastomers listed in Figure 1.1. Such molecules change their shape readily and continuously at normal temperatures by Brownian motion. They take up random conformations in a stress-free state but assume somewhat oriented

The Science and Technology of Rubber. http://dx.doi.org/10.1016/B978-0-12-394584-6.00001-7

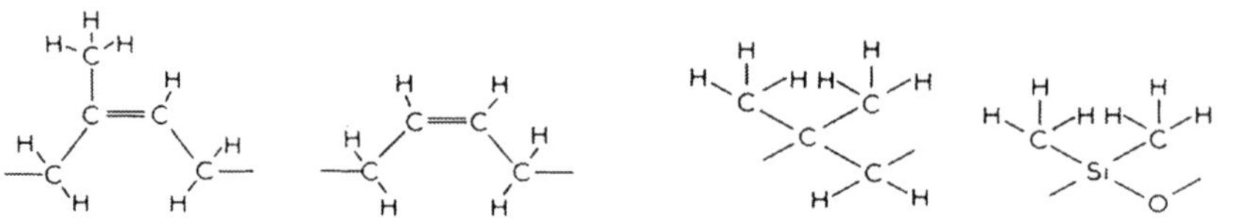

cis-1,4-polyisoprene cis-1,4-polybutadiene poly(iso-butylene) poly(dimethylsiloxane)

FIGURE 1.1 Repeat units for some common elastomer molecules.

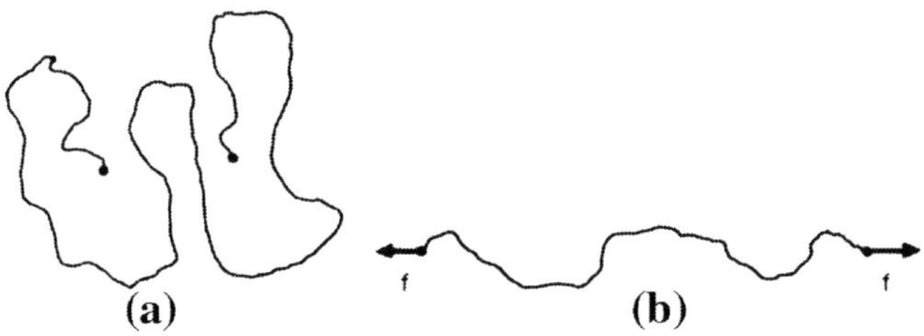

FIGURE 1.2 (a) Random chain and (b) oriented chain (from Gent, 1958).

conformations if tensile forces are applied at their ends (Figure 1.2). One of the first questions to consider, then, is the relationship between the applied tension f and the mean chain end separation r, averaged over time or over a large number of chains at one instant in time.

Chains in isolation take up a wide variety of conformations,[1] governed by three factors: the statistics of random processes; a preference for certain sequences of bond arrangements because of steric and energetic restraints within the molecule; and the exclusion of some hypothetical conformations that would require parts of the chain to occupy the same volume in space. In addition, cooperative conformations are preferred for space-filling reasons in concentrated solutions or in the bulk state.

Flory (1969) has argued that the occupied-volume exclusion (repulsion) for an isolated chain is exactly balanced in the bulk state by the external (repulsive) environment of similar chains, and that the exclusion factor can therefore be ignored in the solid state. Direct observation of single-chain dimensions in the bulk state by inelastic neutron scattering gives values fully consistent with unperturbed chain dimensions obtained for dilute solutions in theta solvents[2] (Cotton et al., 1972), although intramolecular effects may distort the local randomness of chain conformation.

---

[1] Although the terms *configuration* and *conformation* are sometimes used interchangeably, the former has acquired a special meaning in organic stereochemistry and designates specific steric structures. *Conformation* is used here to denote a configuration of the molecule, which is arrived at by rotation of single-valence bonds in the polymer backbone.

[2] These are (poor) solvents in which repulsion between different segments of the polymer molecule is balanced by repulsion between polymer segments and solvent molecules.

Flory has again given compelling reasons for concluding that the chain end-to-end distance r in the bulk state will be distributed in accordance with Gaussian statistics for sufficiently long chains, even if the chains are relatively stiff and inflexible over short lengths (Flory, 1969). With this restriction to long chains it follows that the tension-displacement relation becomes a simple linear one,

$$f = Ar, \tag{1.1}$$

where f is the tensile force, r is the average distance between the ends of the chain, and A is inversely related to the mean square end-to-end distance r_0^2 for unstressed chains,

$$A = 3kT/r_0^2, \tag{1.2}$$

where k is Boltzmann's constant and T is the absolute temperature.

If the real molecule is replaced by a hypothetical chain consisting of a large number n of rigid, freely jointed links, each of length l (Figure 1.3), then

$$r_0^2 = nl^2. \tag{1.3}$$

In this case r_0^2 is independent of temperature because completely random link arrangements are assumed. The tension f in Eq. (1.1) then arises solely from an entropic mechanism; that is, from the tendency of the chain to adopt conformations of maximum randomness, and not from any energetic preference for one conformation over another. The tension f is then directly proportional to the absolute temperature T.

For real chains, consisting of a large number n of primary valence bonds along the chain backbone, each of length l,

$$r_0^2 = C_\infty nl^2, \tag{1.4}$$

where the coefficient C_∞ represents the degree to which this real molecule departs from the freely jointed model. C_∞ is found to vary from 4 to 10, depending on the chemical structure of the molecule and on temperature, because the

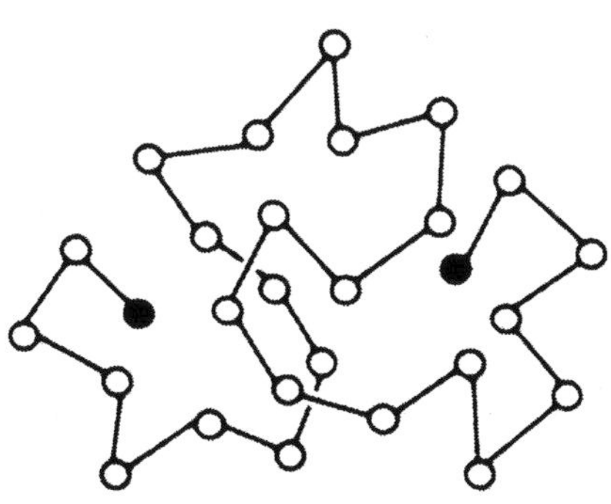

FIGURE 1.3 Model chain of freely jointed links.

energetic barriers to random bond arrangements are more easily overcome at higher temperatures (Flory, 1969). $C_\infty^{1/2} l$ may thus be regarded as the effective bond length of the real chain, a measure of the "stiffness" of the molecule.

Equation (1.1) is reasonably accurate only for relatively short distances r, less than about one-third of the fully stretched chain length (Cotton et al., 1972). Unfortunately, no good treatment exists for the tension in real chains at larger end separations. We must therefore revert to the model chain of freely jointed links, for which

$$f = (kT/l)L^{-1}(r/nl), \tag{1.5}$$

where L^{-1} denotes the inverse Langevin function. An expansion of this relation in terms of r/nl,

$$f = (3kTr/nl^2)[1 + (3/5)(r/nl)^2 + (99/175)(r/nl)^4 + (513/875)(r/nl)^6 + \cdots] \tag{1.6}$$

gives a useful indication of where significant departures from Eq. (1.1) may be expected.

Equation (1.5) gives a steeply rising relation between tension and chain end separation when the chain becomes nearly taut (Figure 1.4), in contrast to the Gaussian solution, Eq. (1.1), which becomes inappropriate for $r > \frac{1}{3}nl$. Rubber shows a similar steeply rising relation between tensile stress and elongation at high elongations. Indeed, experimental stress-strain relations closely resemble those calculated using Eq. (1.5) in place of Eq. (1.1) in the network theory of rubber elasticity (outlined in the following section). The deformation at which a small but significant departure is first found between the observed stress and that

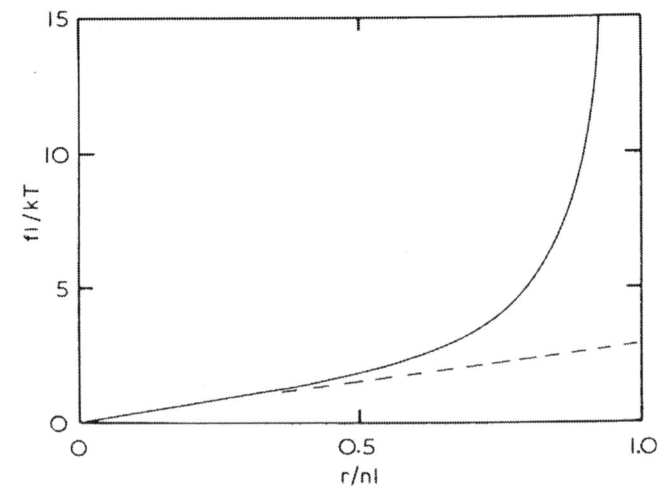

FIGURE 1.4 Tension-displacement relation for a freely jointed chain (Eq. (1.5)), ---, Gaussian solution (Eq. (1.1)) (from Gent, 1958).

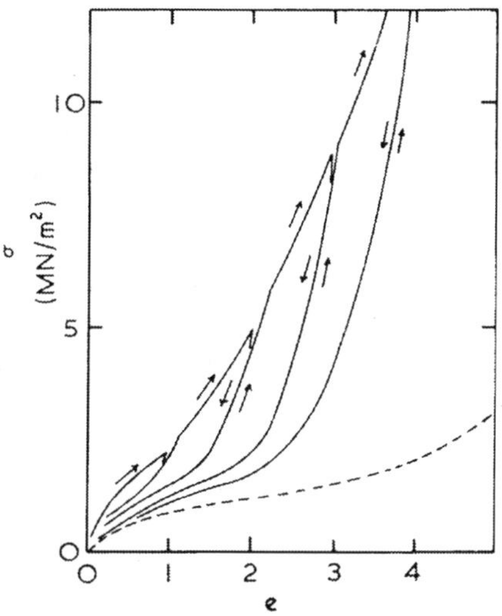

FIGURE 1.5 Stress-induced softening of a carbon black-filled vulcanizate of a copolymer of styrene and butadiene (25:75); ---, stress-strain curve of a corresponding unfilled vulcanizate (from Tobolsky and Mark, 1971).

predicted by small-strain theory, using Eq. (1.1), yields a value for the effective length l of a freely jointed link for the real molecular chain. This provides a direct experimental measure of molecular stiffness. The values obtained are relatively large, of the order of 5–15 main-chain bonds, for the only polymer that has been examined by this method so far, *cis*-1,4-polyisoprene (Morris, 1964).

Equation (1.5) has also been used to estimate the force at which a rubber molecule will become detached from a particle of a reinforcing filler (e.g., carbon black) when a filled rubber is deformed (Bueche, 1960, 1961). In this way, a general semiquantitative treatment has been achieved for stress-induced softening (Mullins effect) of filled rubbers (shown in Figure 1.5).

1.3 ELASTICITY OF A THREE-DIMENSIONAL NETWORK OF POLYMER MOLECULES

Some type of permanent structure is necessary to form a coherent solid and to prevent liquidlike flow of elastomer molecules. This requirement is met by incorporating a small number of intermolecular chemical bonds (crosslinks) to make a loose three-dimensional molecular network. Such crosslinks are generally assumed to form in the most probable positions, so that the long sections of molecules between them have the same spectrum of end-to-end lengths as a similar set of uncrosslinked molecules would have. Under Brownian motion each molecular section takes up a wide variety of conformations, as before, but now subject to the condition that its ends lie at the crosslink sites.

The elastic properties of such a molecular network are treated later. We consider first another type of interaction between molecules.

High-molecular-weight polymers form entanglements by molecular intertwining, with a spacing (in the bulk state) characteristic of the particular molecular structure (Fetters et al., 1999). Some representative values of the molecular weight M_e between entanglement sites are given in Table 1.1. Thus, a high-molecular-weight polymeric melt will show transient rubberlike behavior even in the absence of any permanent intermolecular bonds.

In a crosslinked rubber, many of these entanglements are permanently locked in (Figure 1.6), the more so the higher the degree of crosslinking. If they are regarded as fully equivalent to crosslinks, the effective number N of network chains per unit volume may be taken to be the sum of two terms N_e and N_c, arising from entanglements and chemical crosslinks, respectively, where

$$N_e = \rho N_A / M_e, \quad N_c = \rho N_A / M_c$$

and ρ is the density of the polymer, N_A is Avogadro's number, and M_e and M_c denote the average molecular weights between entanglements and between

TABLE 1.1 Representative Values of the Average Molecular Weight M_e Between Entanglements for Polymeric Melts[a]

Polymer	M_e	Polymer	M_e
Polyethylene	4000	Poly(isobutylene)	17,000
cis-1,4-Polybutadiene	7000	Poly(dimethylsiloxane)	29,000
cis-1,4-Polyisoprene	14,000	Polystyrene	35,000

[a]*Obtained from flow viscosity measurements.*

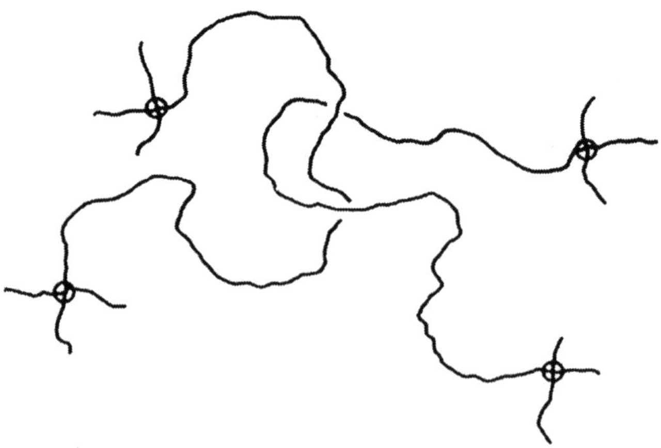

FIGURE 1.6 Sketch of a permanent entanglement (from Gent, 1958).

crosslinks, respectively. The efficiency of entanglements in constraining the participating chains is, however, somewhat uncertain, particularly when the number of chemical crosslinks is relatively small (Langley, 1968; Vilgis, 1992; Graessley, 2004). Moreover, the force-extension relation for an entangled chain will differ from that for a crosslinked chain (Prager and Frisch, 1967), being stiffer initially and nonlinear in form. The effective number N of molecular chains that lie between fixed points (i.e., crosslinks or equivalent sites of molecular entanglement) is therefore a somewhat ill-defined quantity, even when the chemical structure of the network is completely specified.

It is convenient to express the elastic behavior of the network in terms of the strain energy density W per unit of unstrained volume. The strain energy w for a single chain is obtained from Eq. (1.1) as

$$W = Ar^2/2. \tag{1.7}$$

For a random network of N such chains under a general deformation characterized by extension ratios $\lambda_1, \lambda_2, \lambda_3$ (deformed dimension/undeformed dimension) in the three principal directions (Figure 1.7), W is given by (Treloar, 1975)

$$W = NAr_f^2\left(\lambda_1^2 + \lambda_2^2 + \lambda_3^2 - 3\right)\Big/6, \tag{1.8}$$

where r_f^2 denotes the mean square end-to-end distance between chain ends (crosslink points or equivalent junctions) in the undeformed state. The close similarity of Eqs. (1.7) and (1.8) is evident, especially since $r^2 = (r_f^2/3)(\lambda_1^2 + \lambda_2^2 + \lambda_3^2)$.

For random crosslinking r_f^2 may be assumed to be equal to r_0^2, the corresponding mean square end-to-end distance for unconnected chains of the same molecular length. Because A is inversely proportional to r_0^2 (Eq. (1.2)), the only molecular parameter that remains in Eq. (1.8) is the number N of elastically effective chains per unit volume. Thus, the elastic behavior of a molecular network under moderate deformations is predicted to depend only on the number of molecular chains and not on their flexibility, provided that they are long enough to obey Gaussian statistics.

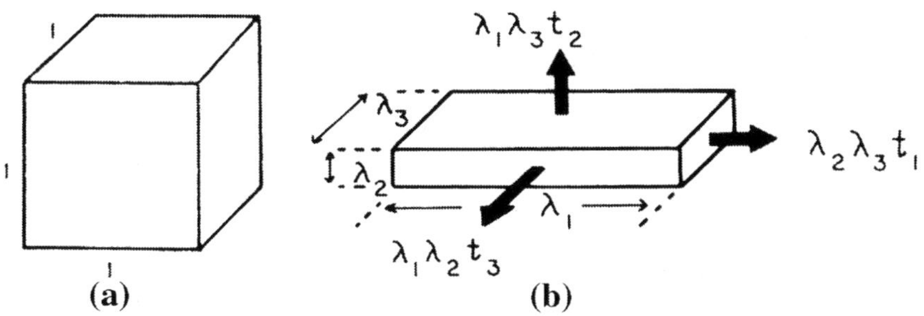

FIGURE 1.7 (a) Undeformed and (b) deformed states.

Although r_f^2 and r_0^2 are generally assumed to be equal at the temperature of network formation, they may well differ at other temperatures because of the temperature dependence of r_0^2 for real chains (Eq. (1.4)). Indeed, the temperature dependence of elastic stresses in rubbery networks has been widely employed to study the temperature dependence of r_0^2, as discussed elsewhere (Flory, 1969; Graessley, 2004).

Another way in which r_f^2 and r_0^2 may differ is when the network is altered after formation. For example, when the network imbibes a swelling liquid, r^2 for the swollen network will be increased by a factor λ_s^2 in comparison to its original value, where λ_s is the linear swelling ratio. At the same time the number of chains per unit volume will be decreased by a factor λ_s^{-3}. Thus, the strain energy density under a given deformation will be smaller for a swollen network by a factor λ_s^{-1}.

From the general relation for strain energy, Eq. (1.8), the elastic stresses required to maintain any given deformation can be obtained by means of virtual work considerations (Figure 1.7),

$$\lambda_2\lambda_3 t_1 = \partial W/\partial\lambda_1$$

with similar relations for t_2 and t_3. Because of the practical incompressibility of rubbery materials in comparison to their easy deformation in other ways, the original volume is approximately conserved under deformation. The extension ratios then obey the simple relationship

$$\lambda_1\lambda_2\lambda_3 = 1. \tag{1.9}$$

As a result, the stress-strain relations become

$$t_1 = \lambda_1(\partial W/\partial\lambda_1) - p, \quad \text{etc.}$$

where p denotes a possible hydrostatic pressure (which has no effect on an incompressible solid). Thus, only stress differences can be written explicitly

$$t_1 - t_2 = (NAr^2/3)\left(\lambda_1^2 - \lambda_2^2\right). \tag{1.10}$$

For a simple extension, say in the one-direction, we set $\lambda_1 = \lambda$, and $\lambda_2 = \lambda_3 = \lambda^{-1/2}$ (from Eq. (1.9)), and $t_2 = t_3 = 0$. Hence,

$$t\ (=t_1) = (NAr^2/3)(\lambda^2 - \lambda^{-1}). \tag{1.11}$$

It is customary to express this result in terms of the tensile force f acting on a test piece of cross-sectional area A_0 in the unstrained state, where

$$f/A_0 = t/\lambda.$$

The corresponding relation is shown in Figure 1.8. It illustrates a general feature of the elastic behavior of rubbery solids: although the constituent chains obey a linear force-extension relationship (Eq. (1.1)), the network does not. This feature arises from the geometry of deformation of randomly oriented chains. Indeed, the degree of nonlinearity depends on the type of deformation imposed. In simple shear, the relationship is predicted to be a linear one with a

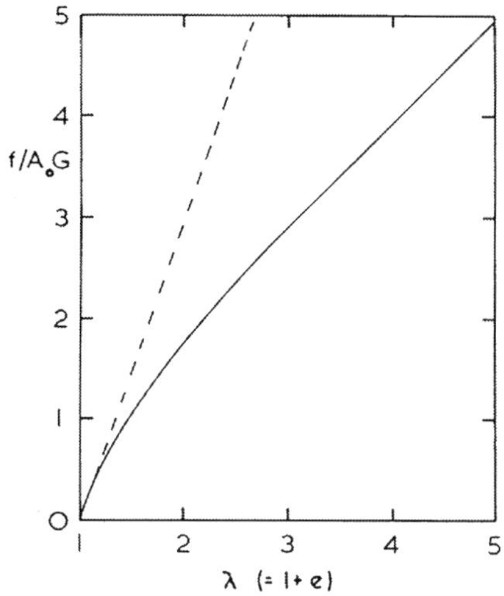

FIGURE 1.8 Force-extension relation for simple extension. ---, Linear relation obtaining at infinitesimal strains (from Gent, 1958).

slope (shear modulus G) given by

$$t_{12} = G\gamma, \quad G = NAr^2/3, \tag{1.12}$$

where γ is the amount of shear; for example, dx/dy.

Because rubbery materials are virtually incompressible in bulk, the value of Poisson's ratio is close to 0.5. Young's modulus E is therefore given by $3G$ to good approximation; however, the predicted relation between stress and tensile strain (extension), e ($=\lambda - 1$), is linear only for quite small extensions (Figure 1.8), so that Young's modulus is applicable only for extensions or compressions of a few percent.

All of these stress relations are derived from Eq. (1.8). They are valid therefore only for moderate deformations of the network; that is, for deformations sufficiently small for the chain tensions to be linearly related to their end-to-end distances r (Eq. (1.1)). Unfortunately, no correspondingly simple expression can be formulated for W using Eq. (1.5), the relationship for large strains of the constituent chains, in which the molecular stiffness parameter reappears. Instead, a variety of series approximations must be used, as in Eq. (1.6), to give close approximations to the behavior of rubber networks under large strains (Arruda and Boyce, 1993).

1.4 COMPARISON WITH EXPERIMENT

Although the treatment of rubber elasticity given in the preceding section is generally rather successful, certain discrepancies are found to occur. The first consists of observed stresses higher than predicted—for example, by Eq. (1.11)—and is often expressed by an additional contribution referred to as the C_2 term. This contribution is relatively large at small strains (although it is always the smaller part of the observed stress) and decreases in importance as the strain increases. It also decreases as the network is dilated by swelling with an inert liquid becoming zero at a swelling ratio of about 5. Thus, the "C_2 stress" appears to reflect a non-Gaussian characteristic of network chains, which is important only at small values of the chain end-to-end distance r. Indeed, Thomas (1955) has shown that the magnitude of the C_2 stress and its complex dependence on type and degree of strain, and on degree of swelling, can all be accurately described by a simple additional term in the relation for the strain energy w for a single network chain, Eq. (1.7), which becomes

$$2W = Ar^2 + Br^{-2}. \quad (1.13)$$

The second term clearly becomes insignificant at large values of r.

Further evidence bearing on the physical nature of the discrepancy is provided by two other observations: C_2 does not appear to be strongly dependent on temperature and therefore does not appear to be associated with the energetics of chain conformations; and it is closely correlated with the tendency of the polymer chains to form molecular entanglements. For example, those polymers with a high density of entanglements in the bulk state (Table 1.1) yield rubbery networks with a relatively high C_2 stress component (Graessley, 2004).

Finally, there is no evidence that isolated chains in theta solvents fail to conform to Gaussian statistics, so that the C_2 discrepancy appears to arise only when the molecular chains are tied into a network.

These varied aspects of the C_2 stress suggest that it is associated with entangled chains in networks (Figure 1.6) and specifically that it arises from restrictions on the conformations available to entangled chains, different from those operating at crosslink sites (Langley, 1968; Vilgis, 1992; Graessley, 2004). Prager and Frisch (1967) have pointed out that chains involved in model entanglements are governed by different statistics; their conclusions are quite consistent with what is known of the C_2 stress.

A second discrepancy between theory and experiment is found when the Gaussian part of the measured stresses is compared with the theoretical result for an ideal network. Numerical differences of up to 50% are obtained between the density of effective chains calculated from the observed stresses and that calculated from the chemistry of crosslinking. This discrepancy may be due to an error in the theoretical treatment as given here. James and Guth (1943, 1949) arrived at stresses only half as large as those given in Eq. (1.10), from a somewhat different theoretical standpoint.

A third and major discrepancy, already referred to, is found at large deformations when the network chains fail to obey Gaussian statistics, even approximately. Considerable success is achieved in this case by using Eq. (1.5) in place of Eq. (1.1) for chain tensions in the network.

Notwithstanding these discrepancies, the simple treatment of rubber elasticity outlined in this chapter has proved to be remarkably successful in accounting for the elastic properties of rubbers under moderate strains, up to about 300% of the unstrained length (depending on the length and flexibility, and hence the extensibility, of the constituent chains). It predicts the general form of the stress-strain relationships correctly under a variety of strains, the approximate numerical magnitudes of the stresses for various chemical structures, and the effects of temperature and of swelling the rubber with an inert mobile liquid on the elastic behavior. It also predicts novel second-order stresses, discussed later, which have no counterpart in classical elasticity theory. In summary, it constitutes a major advance in our understanding of the properties of polymeric materials.

1.5 CONTINUUM THEORY OF RUBBER ELASTICITY

A general treatment of the stress-strain relations of rubberlike solids was developed by Rivlin (1948, 1956), assuming only that the material is isotropic in elastic behavior in the unstrained state and incompressible in bulk. It is quite surprising to note what far-reaching conclusions follow from these elementary propositions, which make no reference to molecular structure.

Symmetry considerations suggest that appropriate measures of strain are given by three strain invariants, defined as

$$\begin{aligned} J_1 &= \lambda_1^2 + \lambda_2^2 + \lambda_3^2 - 3, \\ J_2 &= \lambda_1^2\lambda_2^2 + \lambda_2^2\lambda_3^2 + \lambda_3^2\lambda_1^2 - 3, \\ J_3 &= \lambda_1^2\lambda_2^2\lambda_3^2 - 1, \end{aligned}$$

where λ_1, λ_2, λ_3 are the principal stretch ratios (the ratios of stretched to unstretched lengths; Figure 1.7). Moreover, for an incompressible material, J_3 is identically zero, and hence only two independent measures of strain, J_1 and J_2, remain. It follows that the strain energy density W is a function of these two variables only:

$$W = f(J_1, J_2). \tag{1.14}$$

Furthermore, to yield linear stress-strain relations at small strains, W must be initially of second order in the strains e_1, e_2, e_3. Therefore, the simplest possible form for the strain energy function is

$$W = C_1 J_1 + C_2 J_2, \tag{1.15}$$

where C_1 and C_2 are elastic coefficients with a sum $2(C_1 + C_2)$ equal to the small-strain shear modulus G. Equation (1.15) was originally proposed by Mooney (1940), and is often called the Mooney-Rivlin equation. It is noteworthy that the first term corresponds to the relation obtained from the molecular theory of rubber elasticity, Eq. (1.8), if the coefficient C_1 is identified with $Nar^2/6 = \frac{1}{2}NkT(r/r_0)^2$.

On expanding Eq. (1.15) as a power series in strains e, where $e = \lambda - 1$, it is found to include all terms in e^2 and e^3. Thus it necessarily gives good agreement with experiment at small strains, say for values of e up to 10–20%, where higher powers of e are negligibly small. However, considerable confusion has arisen from its application at larger strains, for values of e of 100% or more, when it no longer holds. It is rather unfortunate that experimental stress-strain relations in simple extension appear to be in accord with Eq. (1.15) up to moderately large strains. This fortuitous agreement arises because the particular strain energy function obeyed by rubber, discussed later, depends on strain in such a way that the two stress-strain relations in tension are similar in form. Relations for other types of strain are quite different, even at modest strains (Rivlin and Saunders, 1951).

1.5.1 Stress-Strain Relations

Stresses can be obtained from the derivatives of the strain energy function W:

$$t_1 = \lambda_1(\partial W/\partial\lambda_1) - p. \tag{1.16}$$

Rewriting Eq. (1.16) in terms of the generic derivatives $\partial W/\partial J_1$ and $\partial W/\partial J_2$ yields

$$t_1 = 2\left[\lambda_1^2 \partial W/\partial J_1 - \left(1/\lambda_1^2\right)\partial W/\partial J_2\right] - p, \quad \text{etc.} \tag{1.17}$$

The functions $\partial W/\partial J_1$ and $\partial W/\partial J_2$ are denoted W_1 and W_2 hereafter.

Experimental measurements indicate that W_1 is approximately constant. However, the second term is far from constant even at moderate strains. Good agreement is obtained when it is expressed as a logarithmic function of J_2 (Gent and Thomas, 1958):

$$W = C_1 J_1 + C_2' Ln[(J_2 + 3)/3], \tag{1.18}$$

where C_2' is a constant. This form of the second term is in reasonably good numerical agreement with the predictions of Thomas's additional term in the strain energy function for a single chain, Eq. (1.13), and simpler in form.

Values of C_1 and C_2' are similar in magnitude, 0.25–0.5 MPa, for typical soft rubber vulcanizates. However, whereas C_1 is approximately proportional to the number of network strands per unit volume, C_2' appears to be rather constant, independent of the degree of crosslinking, and thus it is relatively more important for lightly crosslinked materials. As mentioned earlier, it appears to reflect physical restraints on molecular strands like those represented in the "tube" model of restricted configurations in the condensed state (Graessley, 2004)—restraints that diminish in importance as the deformation increases or the strands become more widely separated.

(i) Strain-Hardening at Large Strains

Rubber becomes harder to deform at large strains, probably because the long flexible molecular strands that comprise the material cannot be stretched indefinitely. The strain energy functions considered up to now do not possess this feature and therefore fail to describe behavior at large strains. Strain-hardening can be introduced by a simple modification to the first term in Eq. (1.18), incorporating a maximum possible value for the strain measure J_1, denoted J_m (Gent, 1996):

$$W = -C_1 J_m Ln(1 - J_1/J_m) + C_2' Ln[(J_2 + 3)/3]. \tag{1.19}$$

Equation (1.19) reduces to Eq. (1.18) when the strains are relatively small; that is, when the ratio J_1/J_m is small. Thus Eq. (1.19) is probably the simplest possible strain energy function that accounts for the elastic behavior to good approximation over the entire range of strains (Pucci and Saccomandi, 2002). It requires three fitting parameters, two of which are related to the small-strain shear modulus G:

$$G = 2[C_1 + (C_2'/3)]. \tag{1.20}$$

The molecular theory of rubberlike elasticity predicts that the first coefficient, C_1, is proportional to the number N of molecular strands that make up the three-dimensional network. The second coefficient, C_2', appears to reflect physical restraints on molecular strands like those represented in the "tube" model (Graessley, 2004) and is in principle amenable to calculation. The third parameter, J_m, is not really independent. When the strands are long and flexible, it will be given approximately by $3\lambda_m^2$, where λ_m is the maximum stretch ratio of an average strand. But λ_m^2 is inversely proportional to N for strands that are randomly arranged in the unstretched state (Treloar, 1975). J_m is therefore expected to be inversely proportional to C_1. Thus the entire range of elastic behavior arises from only two fundamental molecular parameters.

Considerable success has also been achieved in fitting the observed elastic behavior of rubbers by strain energy functions that are formulated directly in terms of the extension ratios $\lambda_1, \lambda_2, \lambda_3$ instead of in terms of the strain invariants I_1, I_2 (Ogden, 1984). Although experimental results can be described economically and accurately in this way, the functions employed are empirical and the numerical parameters used as fitting constants do not appear to have any direct physical significance in terms of the molecular structure of the material. On the other hand, the molecular elasticity theory, supplemented by a simple non-Gaussian term whose molecular origin is in principle within reach, seems able to account for the observed behavior at small and moderate strains with comparable success.

At moderate strains, the value of J_2 is often large enough for terms involving W_2 to be neglected. Some stress-strain relations are now derived using this approximation to illustrate how such calculations are carried out and to deduce under what conditions the deformations become unstable. Instabilities are interesting from a theoretical point of view because they occur suddenly, at a well-defined deformation, and they are often unexpected on the basis of classical elasticity theory. Moreover, a comparison of the observed onset of instability with the predictions of various strain energy functions W provides, at least in principle, a critical test for the validity of a proposed form for W. From a practical standpoint, unstable states are quite undesirable because the deformation becomes highly nonuniform, leading to premature failure.

(ii) Inflation of a Thin-Walled Tube

Inflation of a tube is described by extension ratios of λ_1 in the circumferential direction and λ_2 in the axial direction, with the wall thickness h becoming $h/\lambda_1\lambda_2$ because the rubber volume remains constant. The inflation pressure P gives rise to stresses in the circumferential and axial directions:

$$t_1 = \lambda_1^2\lambda_2 r\ P/h,$$
$$t_2 = \lambda_1^2\lambda_2 r\ P/2h, \tag{1.21}$$

where r is the tube radius in the unstrained state.

From Eq. (1.21), on putting the stress $t_3 = 0$, the undefined pressure p is obtained as:

$$p = -2W_1/\lambda_1^2\lambda_2^2. \tag{1.22}$$

(In a thin-walled tube of large radius the inflating pressure P is much smaller than the stresses t_1 and t_2 that it generates, and thus P can be neglected in comparison with the stress t_3 in determining p.) Inserting this result for p in Eqs. (1.21) and (1.22) yields a relation between the extension ratio λ_2 and the expansion ratio ν ($=\lambda_1^2\lambda_2$) of the internal volume of the tube:

$$2\lambda_2^3 = (\nu^2 + 1)/2\nu^2. \tag{1.23}$$

The relation between inflating pressure P and internal volume of the tube is then obtained as:

$$Pr/hC_1 = 2(\nu^2 - 1)[2\nu/(\nu^2 + 1)]^{1/3}/\nu^2[1 - (J_1/J_m)]. \tag{1.24}$$

This relation is plotted in Figure 1.9 for various values of the limiting strain measure J_m. The inflating pressure is seen to pass through a maximum at a volume expansion ratio between about 58% and 66%, depending on the value assumed for J_m. This feature suggests that larger expansions will be unstable. Indeed, thin-walled tubes undergo a strikingly nonuniform deformation at a critical inflation pressure, shown schematically in Figure 1.10. One portion of the tube becomes highly distended as a bubble or aneurysm while the rest is lightly inflated. The two stable deformations that can coexist at the same inflation pressure after the critical state is reached are shown schematically by the horizontal broken line in Figure 1.9. However, when J_m is infinitely large, the aneurysm is unbounded and failure would then occur immediately on reaching the critical pressure.

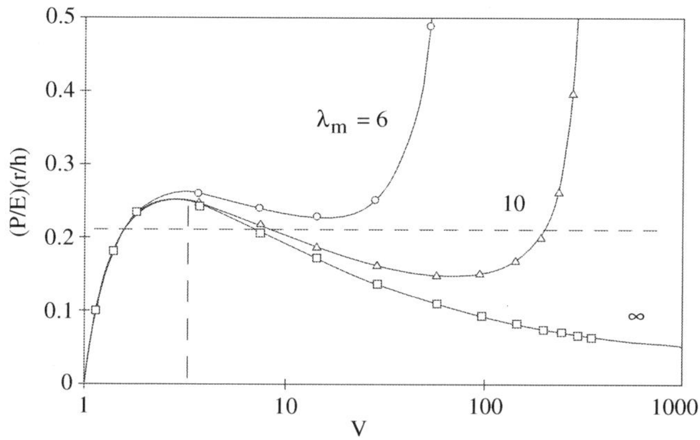

FIGURE 1.9 Pressure-volume relations for a thin-walled tube from Eq. (1.24) using various values for the maximum possible extension ratio λ_m. The vertical broken line denotes the onset of instability ($E = 6C_1$).

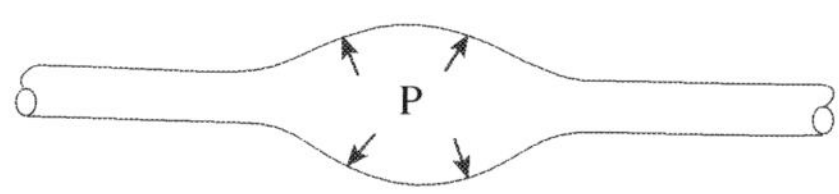

FIGURE 1.10 Sketch of an aneurysm in an inflated tube.

(iii) Inflation of a Thin-Walled Spherical Balloon

In this case, if the balloon radius expands by a factor λ, equibiaxial extensions of ratio λ will be set up in the balloon, with a shrinkage ratio $1/\lambda^2$ of the wall thickness to maintain the rubber volume constant. The circumferential stresses t_1 and t_2 are equal and given by

$$t_1 = t_2 = 2W_1(\lambda^2 - \lambda^{-4})/[1 - (J_1/J_m)] \qquad (1.25)$$

from Eq. (1.17), when the stress $t_3 = 0$ and $W_2 = 0$. The inflation pressure P is then given by:

$$Pr/hC_1 = 4(\lambda^{-1} - \lambda^{-7})/[1 - (J_1/J_m)], \qquad (1.26)$$

where r and h are the unstrained radius and wall thickness of the balloon. In this case the potential instability occurs even earlier, at a radial expansion ratio between 38% and 50% depending on the value chosen for J_m. In practice, the deformation becomes quite complex (Figure 1.11). The balloon remains roughly spherical in shape but one part is lightly stretched while the remainder is highly stretched. The two states of strain resemble the two deformations that are predicted at a given pressure after the critical point is reached.

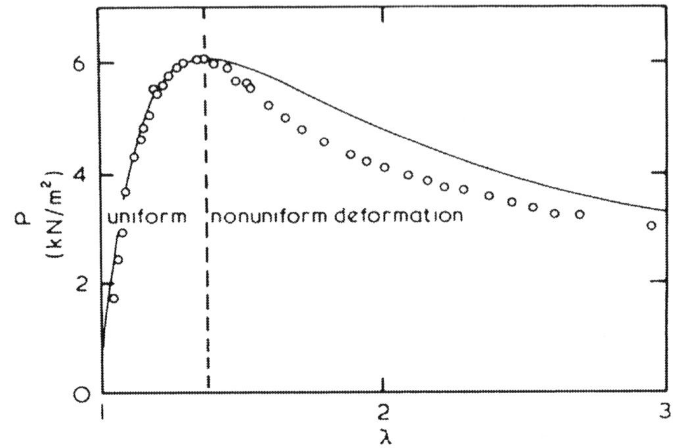

FIGURE 1.11 Inflation of a thin-walled spherical rubber balloon. Solid curve: Eq. (1.26) with $J_m = \infty$ (from Gent, 1958).

(iv) Inflation of a Thick-Walled Spherical Shell

The internal pressure P required to inflate a small spherical cavity in the center of a thick block can be obtained by integrating the contributions from concentric shells (thin-walled balloons) given in the preceding section. The result is (Gent and Lindley, 1958)

$$P/C_1 = 4\lambda_0^{-1} + \lambda_0^{-4} - 5 \tag{1.27}$$

for an infinitely extensible rubber ($J_m = \infty$), where λ_0 is the biaxial stretch ratio at the surface of the cavity. This relation does not exhibit a maximum and thus does not indicate that the deformation is unstable. However, at high values of λ_0 the pressure P asymptotes to a constant value of about $5C_1$; that is, about $5G/2$, where G is the small-strain shear modulus. For typical rubbery materials where G is about 0.5 MPa, the maximum pressure is thus about 1.2 MPa, or about 12 bar. Any small cavity will expand greatly at this rather modest inflation pressure. Internal fracture is therefore likely to occur in soft rubbery solids at inflation pressures or equivalently, triaxial tensions, of this amount. In practice, all rubbery solids are found to develop internal fractures when supersaturated with gases or liquids at pressures or triaxial tensions about equal to $5G/2$ (Gent and Lindley, 1958; Gent and Tompkins, 1969).

Note that the initial radius of the spherical cavity does not appear in Eq. (1.27). Thus, cavities of all sizes are predicted to inflate equally. However, we have neglected surface energy contributions that will tend to stabilize small cavities. When they are taken into account, it appears that only cavities having radii greater than about 100 nm will expand dramatically at the low pressures predicted by Eq. (1.27). Internal fractures suggest that vulcanized rubber must contain many precursor cavities of this effective size or larger.

(v) Surface Instability of Compressed or Bent Blocks

Biot (1965) showed that the surface of an elastic half-space will become unstable at critical values of strain ratios λ_1, λ_3 set up in two perpendicular directions in the surface. The critical condition is

$$\lambda_1^2 \lambda_3 = 0.2956. \tag{1.28}$$

When the block is subjected to unidirectional compression parallel to the surface, with free expansion permitted in the other two directions, then $\lambda_3 = 1/\lambda_1^{1/2}$ and Eq. (1.28) yields a critical value for λ_1 of 0.444. A large block of rubber in simple compression is therefore predicted to show a surface instability at a compression of 55.5%. (Beatty (1977) noted that various buckling and bulging modes of deformation are generally encountered before this, depending on the slenderness of the block.) If the block is subjected to equibiaxial compression parallel to the surface, then $\lambda_3 = \lambda_1$ and the critical compression becomes 33.3%.

When a thick elastic block (cuboid) is bent, the inner surface becomes compressed while the extension ratio λ_3 along the width is largely unchanged (at unity). Thus, from Eq. (1.28) an instability would be expected on the inner surface when λ_1 is 0.544; that is, when the surface is compressed by about 46%. Experimentally, sharp folds or creases appear suddenly in the inner surface of a bent block at a critical degree of bending (Gent and Cho, 1999); see Figure 1.12. However, the critical compression of the inner surface was considerably smaller than predicted by Biot's theory, 35% instead of 46%. It is not known why the instability occurred so much sooner than expected. Although rubber follows a more complex strain energy function than the simple form assumed here, it is unlikely that the difference would have such a large effect.

Rubber articles are often subjected to rather severe bending deformations, for example, in tires. Folds and creases in the interior may pass undetected. Nevertheless, they represent lines of high stress concentration and sites of possible failure. Folds ("Schallamach waves") also appear when soft rubber

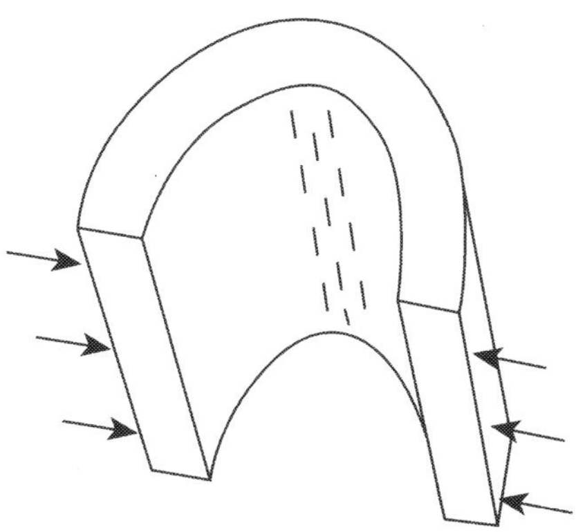

FIGURE 1.12 Sketch of a bent block showing creases that appear on the inner surface where the compressive strain is greatest.

slides over a rigid countersurface (Schallamach, 1971). They appear to be Biot creases caused by frictional compression of the surface.

(vi) Resistance of a Compressed Block to Indentation

When a block is subjected to a sufficiently large equibiaxial compression in the surface plane, it becomes unstable to small indentations. Green and Zerna (1975) expressed the relation between indentation force N and amount of indentation d as:

$$N/G = (8/3)R^{1/2}d^{3/2}f(\lambda), \tag{1.29}$$

where G is the shear modulus of the half-space material, R is the radius of the indentor, and $f(\lambda)$ is a function of the equibiaxial compression ratio λ, given by

$$f(\lambda) = (\lambda^9 + \lambda^6 + 3\lambda^3 - 1)/\lambda^4(\lambda^3 + 1). \tag{1.30}$$

Values of indentation force N for a given small indentation, from Eq. (1.27), are plotted in Figure 1.13 against the equibiaxial strain e parallel to the surface, where $e = \lambda - 1$. (N_0 denotes the value for an initially unstrained block, when $f(\lambda) = 2$.) The resistance to indentation is seen to decrease sharply as the compressive strain is increased, becoming zero at a compressive strain of 0.333, in agreement with Biot's result.

(vii) Torsional Instability of Stretched Rubber Rods (Gent and Hua 2004)

Another unstable state is encountered when a stretched rubber rod is subjected to large torsions. A kink suddenly appears at one point along the rod, Figure 1.14, and more kinks form on twisting the rod further.

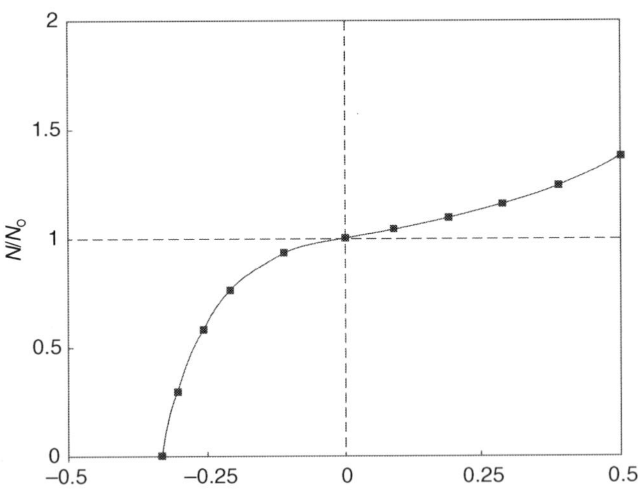

FIGURE 1.13 Force N for a small indentation vs. equibiaxial strain e parallel to the surface of a half-space. N_o denotes the force when $e = 0$ (from Green and Zerna, 1975).

(a) (b) (c)

FIGURE 1.14 Sketch of a "kink" that appears on twisting a stretched rubber rod (from Gent and Hua, 2004).

Minimization of the total elastic strain energy suggests that the rod will become unstable at a critical amount of torsion: part of the rod will unwind and form a tight ring while the remainder of the rod will become slightly more stretched. A simple criterion can be derived on this basis for the onset of "kinks." For a neo-Hookean material, Eq. (1.8), the condition for forming a kink becomes:

$$4(1 - 1/\lambda^3) = -(a^2\phi^2/\lambda) + 2\pi(a^2\phi^2/\lambda)/[\pi - (a\phi/\lambda^{1/2})], \qquad (1.31)$$

where ϕ is the critical amount of torsion at which uniform torsion becomes unstable, in terms of the imposed extension ratio λ and the rod radius a. Measured values for rods of different radius, stretched to extensions of up to 250%, were found to be in reasonably good agreement with Eq. (1.31), indicating that the sudden formation of kinks in twisted rubber rods is, indeed, a consequence of an elastic instability.

1.6 SECOND-ORDER STRESSES

Because the strain energy function for rubber is valid at large strains, and yields stress-strain relations which are nonlinear in character, the stresses depend on the square and higher powers of strain, rather than the simple proportionality expected at small strains. A striking example of this feature of large elastic deformations is afforded by the normal stresses t_{11},t_{22},t_{33} that are necessary to maintain a simple shear deformation of amount γ (in addition, of course, to simple shear stresses) (Rivlin, 1948, 1956). These stresses are predicted to increase in proportion to γ^2.

They are represented schematically in Figures 1.15 and 1.16 for two different choices of the arbitrary hydrostatic pressure p, chosen so as to give the

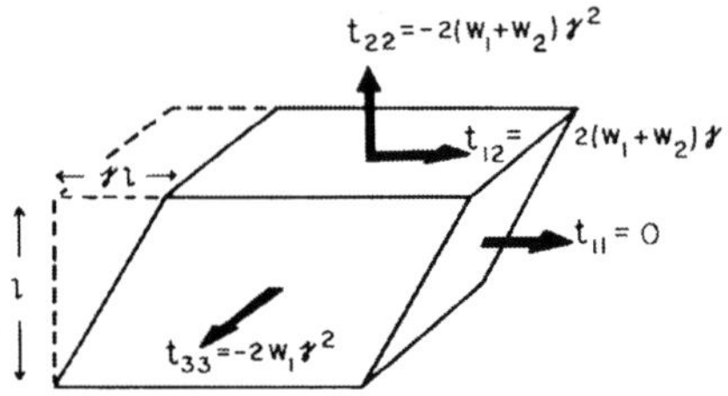

FIGURE 1.15 Stresses required to maintain a simple shear deformation of amount γ. The normal stress t_{11} is set equal to zero (from Gent, 1958).

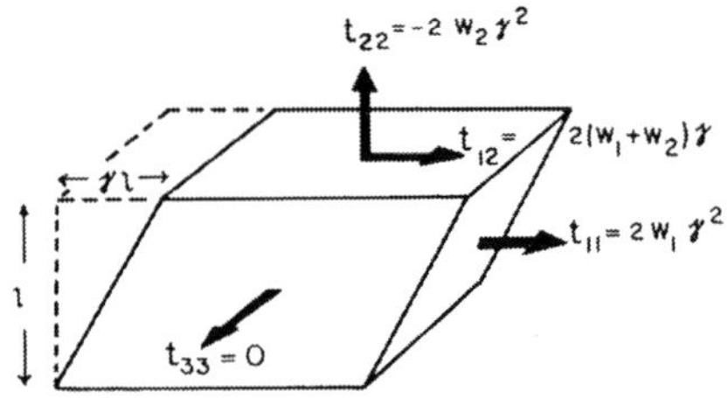

FIGURE 1.16 Stresses required to maintain a simple shear deformation of amount γ. The normal stress t_{33} is set equal to zero (from Gent, 1958).

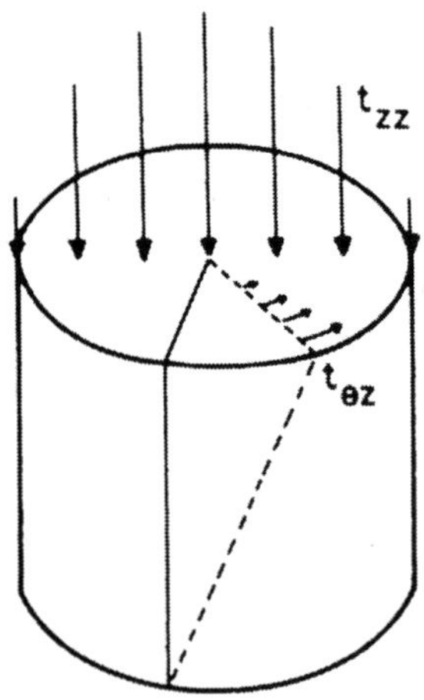

FIGURE 1.17 Sketch of a cylindrical rod under torsion showing the distribution of normal stress t_{zz} (corresponding to $-t_{22}$ in Figures 1.10 and 1.11) over the cross-section of the rod (from Treloar, 1975).

appropriate reference (zero) stress. In Figure 1.15, for example, the normal stress t_{11} in the shear direction is set equal to zero; this condition would arise near the front and rear surfaces of a sheared block. In Figure 1.16, the normal stress t_{33} is set equal to zero; this condition would arise near the side surfaces of a sheared block. In each case a *compressive* stress t_{22} is found to be necessary to maintain the simple shear deformation. In its absence the block would tend to increase in thickness on shearing.

When the imposed deformation consists of an inhomogeneous shear, as in torsion, the normal forces generated (corresponding to the stresses t_{22} in Figures 1.15 and 1.16) vary from point to point over the cross-section (Figure 1.17). The exact way in which they are distributed depends on the particular form of strain energy function obeyed by the rubber; that is, on the values of W_1 and W_2, which obtain under the imposed deformation state (Rivlin, 1947).

1.7 ELASTIC BEHAVIOR UNDER SMALL DEFORMATIONS

Under small deformations rubbers are linearly elastic solids. Because of high modulus of bulk compression (about 2000 MN/m^2) compared with the shear modulus G (about 0.2–5 MN/m^2), they may be regarded as relatively incompressible. The elastic behavior under small strains can thus be described by a single elastic constant G. Poisson's ratio is effectively 1/2, and Young's modulus E is given by $3G$, to good approximation.

A wide range of values for G can be obtained by varying the composition of the elastomer; that is, by changing the chemistry of crosslinking, oil dilution, and filler content. However, soft materials with shear moduli of less than about 0.2 MN/m^2 prove to be extremely weak and are seldom used. Also, particularly hard materials made by crosslinking to high degrees prove to be brittle and inextensible. The practical range of shear modulus, from changes in degree of crosslinking and oil dilution, is thus about 0.2–1 MN/m^2. Stiffening by fillers increases the upper limit to about 5 MN/m^2, but those fillers, which have a particularly pronounced stiffening action, also give rise to stress-softening effects like those shown in Figure 1.5, so that the modulus becomes a somewhat uncertain quantity.

It is customary to characterize the modulus, stiffness, or hardness of rubbers by measuring their elastic indentation by a rigid die of prescribed size and shape under specified loading conditions. Various nonlinear scales are employed to derive a value of hardness from such measurements (Soden, 1952). Corresponding values of shear modulus G for two common hardness scales are given in Figure 1.18.

Many rubber products are normally subjected to fairly small deformations, rarely exceeding 25% in extension or compression or 75% in simple shear. A good approximation for the corresponding stresses can then be obtained by conventional elastic analysis assuming linear relationships. One particularly

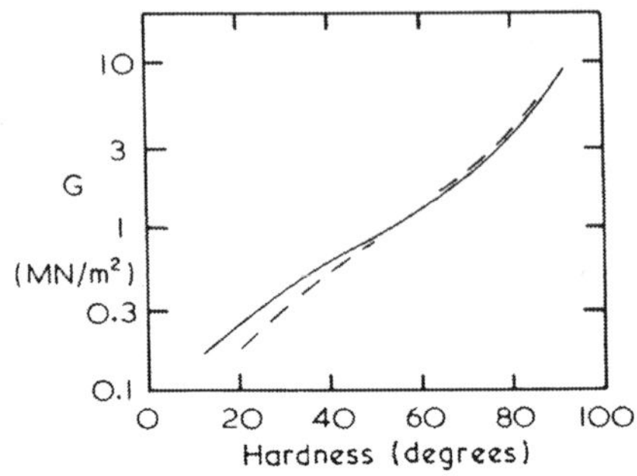

FIGURE 1.18 Relations between shear modulus G and indentation hardness: —, Shore A Scale; - - -, International Rubber Hardness Scale (from Tobolsky and Mark, 1971).

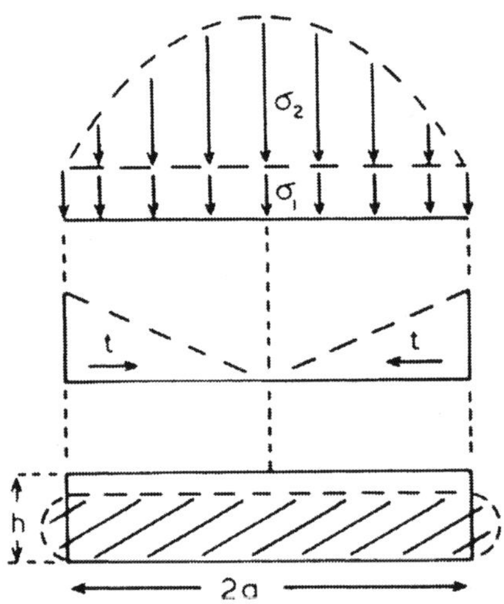

FIGURE 1.19 Sketch of a bonded rubber block under a small compression. The distributions of normal stress σ and shear stress t acting at the bonded surfaces are represented by the upper portions of the diagram (from Tobolsky and Mark, 1971).

important deformation is treated here: the compression or extension of a thin rubber block, bonded on its major surfaces to rigid plates (Figure 1.19). A general treatment of such deformations has been reviewed (Horton, 2002).

It is convenient to assume that the deformation takes place in two stages: a pure homogeneous compression or extension of amount e, requiring a uniform compressive or tensile stress $\sigma_1 = Ee$, and a shear deformation restoring points in the planes of the bonded surfaces to their original positions in these planes. For a cylindrical block of radius a and thickness h, the corresponding shear stress t acting at the bonded surfaces at a radial distance r from the cylinder axis is given by

$$t = Eer/h.$$

This shear stress is associated with a corresponding normal stress or pressure σ_2, given by

$$\sigma_2 = Ee(a^2/h^2)[1 - (r^2/a^2)]. \quad (1.32)$$

These stress distributions are shown schematically in Figure 1.19. Although they must be incorrect right at the edges of the block, because the assumption of a simple shear deformation cannot be valid at these points of singularity, they appear to provide satisfactory approximations over the major part of the bonded surfaces.

By integrating the sum of the normal stresses $\sigma_1 + \sigma_2$ over the bonded surface, the total compressive force F is obtained in the form (Gent, 1994)

$$F/\pi a^2 e = E[1 + (a^2/2h^2)] \equiv E'. \quad (1.33)$$

Clearly, for thin blocks of large radius the effective value of Young's modulus E' (given by the right-hand side of Eq. (1.33)) is much larger than the real value E because of the restraints imposed by the bonded surfaces. Indeed, for values of the ratio a/h greater than about 10, a significant contribution to the observed displacement comes from volume compression or dilation because E' is now so large that it becomes comparable to the modulus of bulk compression (Gent, 1994) (Figure 1.20).

A more accurate treatment of the compression of bonded blocks has been given by Horton et al. (2002) without invoking the assumption that a simple shear deformation holds right up to the bonded edges. They obtained a result of the same form as Eq. (1.33) but with the bracketed term on the right-hand side replaced by $[1.2 + (a^2/2h^2)]$. However, this term does not yield the correct value of unity for tall blocks; that is, when a/h is small, and it is equivalent to

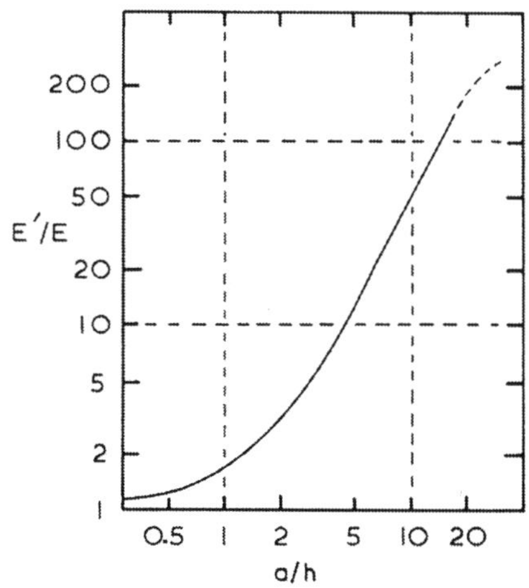

FIGURE 1.20 Effective value of Young's modulus E' for bonded blocks vs. ratio of radius to thickness a/h (from Tobolsky and Mark, 1971).

Eq. (1.33) for thin blocks of large radius, when *a*/*h* is large. It should therefore be regarded as a better approximation for blocks of intermediate size.

When a thin bonded block is subjected to tensile loading, a state of approximately equal triaxial tension is set up in the central region of the block. The magnitude of the stress in each direction is given by the tensile stress, or negative pressure, σ_2 at $r = 0$; that is, Eea^2/h^2, from Eq. (1.32). Under this outwardly directed tension a small cavity in the central region of the block will expand indefinitely at a critical value of the tension, of about 5*E*/6. Thus, if cavities are present in the interior of a bonded block, they are predicted to expand indefinitely (i.e., rupture) at a critical tensile strain e_c, given approximately by

$$e_c = 5h^2/6a^2$$

and at a corresponding critical value of the applied tensile load, obtained by substituting this value of *e* in Eq. (1.33). To avoid internal fractures of this kind, it is thus necessary to restrict the mean tensile stress applied to thin bonded blocks to less than about *E*/3.

In compression, on the other hand, quite large stresses can be supported. A stress limit can be calculated by assuming that the maximum shear stress, developed at the bonded edges, should not exceed *G*; that is, the maximum shear deformation should not exceed about 100%. This yields a value for the allowable overall compressive strain of *h*/3*a*, corresponding to a mean compressive stress of the order of *E* for disks with *a*/*h* between about 3 and 10. This calculation assumes that the approximate stress analysis outlined earlier is valid right at the edges of the block, and this is certainly incorrect. Indeed, the local stresses in these regions depend strongly on the detailed shape of the free surface in the neighborhood of the edge.

1.8 SOME UNSOLVED PROBLEMS IN RUBBER ELASTICITY

We turn now to some features of the elastic response of rubbery materials that are still not fully understood.

As normally prepared, molecular networks comprise chains of a wide distribution of molecular lengths. Numerically, small chain lengths tend to predominate. The effect of this diversity on the elastic behavior of networks, particularly under large deformations, is not known. A related problem concerns the elasticity of short chains. They are inevitably non-Gaussian in character and the analysis of their conformational statistics is likely to be difficult. Nevertheless, it seems necessary to carry out this analysis to be able to treat real networks in an appropriate way.

It is also desirable to treat network topology in greater detail; that is, to incorporate the functionality of crosslinks, their distribution in space, and loop formation. The effect of mutual interaction between chains in the condensed state appears to be accounted for satisfactorily by the "tube" model for uncrosslinked polymers, but its application to networks seems incomplete. But the problem in greatest need of attention is the response of highly filled elastomers to stress. Filled elastomers are not really elastic; their stress-strain relations are irreversible (see Figure 1.5), and it is therefore inappropriate to describe their response to stress by a strain energy function. Moreover, they appear to become anisotropic on stretching and to some degree after release. At present, the molecular processes that occur on deformation and the mathematical framework suitable for describing them are both unclear.

ACKNOWLEDGMENTS

We refer the reader to the classic survey of rubber elasticity by Treloar (1975) and to three recent reviews that give fuller accounts of the molecular theory (Graessley, 2004; Mark and Erman, 1988, 1992). The author thanks Mr. R.A. Paden for drawing several of the figures.

REFERENCES

Arruda, E.M., Boyce, M.C., 1993. J. Mech. Phys. Solids 41, 389.
Beatty, M.F., 1977. In: Rivlin, R.S. (Ed.), AMD Finite Elasticity, vol. 27. American Society of Mechanical Engineers, New York, p. 125.
Biot, M., 1965. Mechanics of Incremental Deformations. Wiley, New York.
Bueche, F., 1960. J. Appl. Polym. Sci. 4, 107.
Bueche, F., 1961. J. Appl. Polym. Sci. 5, 271.
Cotton, J.P., Farnoux, B., Jannink, G.J., 1972. J. Chem. Phys. 57, 290.
Fetters, L.J., Lohse, D.J., Graessley, W.W., 1999. J. Polym. Sci. Part B: Polym. Phys. 37, 1023.
Flory, P.J., 1969. Statistical Mechanics of Chain Molecules. Wiley-Interscience, New York.
Gent, A.N., 1958. J. Polym. Sci. Poly. Symp. 28, 625.
Gent, A.N., 1994. Rubber Chem. Technol. 67, 549.
Gent, A.N., 1996. Rubber Chem. Technol. 69, 59.
Gent, A.N., Cho, I.S., 1999. Rubber Chem. Technol. 72, 253.
Gent, A.N., Hua, K.C., 2004. Int. J. Non-Linear Mech. 39, 483.
Gent, A.N., Lindley, P.B., 1958. Proc. Roy. Soc. Lond. A 249, 195.
Gent, A.N., Thomas, A.G., 1958. J. Polym. Sci. 28, 625.
Gent, A.N., Tompkins, D.A., 1969. J. Appl. Phys. 40, 2520.
Graessley, W.W., 2004. Polymeric Liquids and Networks: Structure and Properties. Taylor and Francis Books, New York.
Green, A.E., Zerna, W., 1975. Theoretical Elasticity, 2nd ed. Clarendon Press, Oxford, p. 135 (Section 4.6).
Horton, J.M., Tupholme, G.E., Gover, M.J.C., 2002. ASME. J. Appl. Mech. 69, 836.
James, H.M., Guth, E., 1943. J. Chem. Phys. 11, 455.
James, H.M., Guth, E., 1949. J. Polym. Sci.4, 153.
Langley, N.R., 1968. Macromolecules 1, 348.
Mark, J.E., Erman, B., 1988. Rubberlike Elasticity: A Molecular Primer. John & Wiley Sons, New York.
Mark, J.E., Erman, B. (Eds.), 1992. Elastomeric Polymer Networks. Prentice-Hall, Englewood Cliffs, NJ.
Mooney, M., 1940. J. Appl. Phys. 11, 582.
Morris, M. C., 1964. J. Appl. Polm. Sci. 8, 545.
Ogden, R.W., 1984. Non-Linear Elastic Deformations. Ellis Harwood, Chichester, UK (Dover Publications, Mineola, NY, 1997, Chapter 7).
Prager, S., Frisch, H.L., 1967. J. Chem. Phys. 46, 1475.

Pucci, E., Saccomandi, G., 2002. Rubber Chem. Technol. 75, 839.
Rivlin, R.S., 1947. J. Appl. Phys. 18, 444.
Rivlin, R.S., 1948. Philos. Trans. Roy. Soc. Lond. Ser. A 241, 379.
Rivlin, R.S., 1956. In: Eirich, F.R. (Ed.), Rheology, Theory and Application, vol. 1. Academic Press, New York (Chapter 10).
Rivlin, R.S., Saunders, D.W., 1951. Philos. Trans. Roy. Soc. Lond. Ser. A 243, 251.
Schallamach, A., 1971. Wear 17, 301.
Soden, A.L., 1952. A Practical Manual of Rubber Hardness Testing. Maclaren, London.
Thomas, A.G., 1955. Trans. Faraday Soc. 51, 569.
Tobolsky, A.V., Mark, H.F. (Eds.), 1971. Polymer Science and Materials. Wiley, New York (Chapter 13).
Treloar, L.R.G., 1975. The Physics of Rubberlike Elasticity, 3rd ed. Clarendon Press, Oxford.
Vilgis, T.A., 1992. In: Mark, J.E., Erman, B. (Eds.), Elastomeric Polymer Networks. Prentice-Hall, Englewood Cliffs, NJ (Chapter 5).

Chapter 2

Polymerization: Elastomer Synthesis

Roderic P. Quirk[*] and Deanna L. Pickel[†]

[*] *The Maurice Morton Institute of Polymer Science, The University of Akron, Akron, OH, USA*
[†] *Center for Nanophase Materials Sciences, Oak Ridge National Laboratory Oak Ridge, TN, USA*

2.1 INTRODUCTION

The development of synthetic rubber played a special role in the history of polymerization chemistry. This was due primarily to the fact that attempts to synthesize rubber were made long before there was even the faintest idea of the nature of polymerization reactions. Such attempts began very soon after the elegant analytical work of Williams (1859) in 1860, which clearly demonstrated that *Hevea* rubber was "composed" of isoprene. Thus, Bouchardat (1879) was actually able to prepare a rubberlike substance from isoprene (which he obtained from rubber pyrolysis), using heat and hydrogen chloride. Tilden (1884) repeated this process in 1884 but used isoprene obtained from pyrolysis of turpentine to demonstrate that it was not necessary to use the "mother substance" of rubber itself. These explorations were soon followed by the work of Kondakow (1900) with 2,3-dimethylbutadiene that of Thiele (1901) with piperylene, and finally that of Lebedev (1910) on butadiene itself. Mention should also be made of the almost simultaneous, and apparently independent, discoveries in 1910 by Harries (1911) in Germany and Matthews and Strange (1910) in England of the efficient polymerization in isoprene by sodium.

Although all of these attempts had a noble purpose indeed, the means used could hardly be considered a contribution to science, as the transformation of the simple molecules of a diene into the "colloidal" substance known as rubber was then far beyond the comprehension of chemical science. As a matter of fact, the commercial production of synthetic rubber was already well established, at least in Germany and Russia, *before* Staudinger laid the basis for his macromolecular hypothesis during the 1920s (Staudinger, 1920). Even such relatively modern synthetic elastomers as polychloroprene and the poly(alkylene sulfides) were

The Science and Technology of Rubber. http://dx.doi.org/10.1016/B978-0-12-394584-6.00002-9

already in commercial production by 1930–1931. This was, of course, also before Carothers and coworkers' pioneering studies on the polymerization of chloroprene (Carothers et al., 1931)!

Hence, it is apparent that it was not the development of an understanding of polymerization that led to the invention of synthetic rubber, but perhaps the reverse. In contrast, it was the new science of organic macromolecules, whose foundations were established by Staudinger, which expanded rapidly during the 1930s and 1940s, and pointed the way to the synthesis of a vast array of new polymeric materials, including synthetic fibers and plastics and even new elastomers. This new science included the classical studies of polycondensation by Carothers and Flory and the establishment of the principles governing free radical chain addition reactions by Schulz, Flory, Mayo, and others (Flory, 1953a; Morawetz, 1985).

Thus it was that the paths of synthetic rubber and macromolecular science finally crossed and became one broad avenue (Morton, 1961). Hence today the design of a new elastomer or the modification of an old one requires the same kind of molecular architecture which applies to any other polymer and is based on an understanding of the principles of polymerization reactions.

2.2 CLASSIFICATION OF POLYMERIZATION REACTIONS AND KINETIC CONSIDERATIONS

Historically polymers have been divided into two broad classes: *condensation* polymers and *addition* polymers (Flory, 1953a; Carothers, 1929,1931). Flory (1953b) has defined these as follows:

Condensation polymers, in which the molecular formula of the structural unit (or units) lacks certain atoms present in the monomer from which it is formed, or to which it may be degraded by chemical means, and addition polymers, in which the molecular formula of the structural unit (or units) is identical with that of the monomer from which the polymer is derived.

Thus, an example of a condensation polymer would be a polyester, formed by the condensation reaction between a glycol and a dicarboxylic acid (with the evolution of water), whereas an addition polymer is exemplified by polystyrene, formed by the self-addition of styrene monomers.

Although these earlier definitions were based on the chain structure of the polymers, they were closely related, as just described, to the *mode of formation* as well. It soon became apparent that such a classification has serious shortcomings, as so-called polycondensates could result from "addition" polymerization reactions. For example, although Nylon 6 can be prepared by the polycondensation reaction of ε-aminocaproic acid (Braun et al., 1984), it is now synthesized by the ring-opening addition polymerization of ε-caprolactam (Sandler and Karo, 1992), and this process has a profound effect on the properties of the resulting polymer. This is, of course, due to basically the magnitude of the molecular weight of the final polymer.

Because it is the extraordinarily large size of the macromolecules which leads to their unusual properties, it would be most sensible to classify polymerization reactions in accordance with *the way in which they affect the molecular size and size distribution of the final product*, i.e., in terms of the mechanism of polymerization. On this basis, there appear to be only two basic processes whereby macromolecules are synthesized (Zhang et al., 2012; Penczek and Pretula, 2012; Moore, 1978; Saunders and Dobinson, 1976; Odian, 2004b; Penczek, 2002; Jenkins et al., 1996): (1) step-growth polymerization (polycondensation and polyaddition) and (2) chain-growth (chain) polymerization.

2.2.1 Polyaddition/Polycondensation

The distinguishing mechanistic feature of step-growth polymerization (Zhang et al., 2012; Penczek and Pretula, 2012; Moore, 1978; Saunders and Dobinson, 1976; Odian, 2004b; Penczek, 2002; Jenkins et al., 1996) is that all molecular species in the system can react with each other to form higher molecular weight species as shown in Eq. (2.1), where P_i is a species with a number-average number of monomer units per chain equal to i, P_j is a species with a number-average number of monomer units per chain equal to j, and P_{i+j} is a species with a number-average number of monomer units per chain equal to $i + j$. The kinetic consequence of this mechanism of polymer growth is that chain length increases monotonically with extent of reaction, i.e., with time of reaction, as shown in Figure 2.1a.

$$P_i + P_j \rightarrow P_{i+j}. \tag{2.1}$$

These step-growth polymerization reactions fall into two classes (Penczek, 2002; Jenkins et al., 1996):

Polycondensation: Growth of polymer chains proceeds by condensation reactions between molecules of all degrees of polymerization. A low-molar-mass by-product (AB) is also formed.

$$\text{A-R-A} + \text{B-R-B} \rightarrow \text{A-R-R-B} + \text{AB}. \tag{2.2}$$

Polyaddition: Growth of polymer chains proceeds by addition reactions between molecules of all degrees of polymerization.

$$\text{A-R-A} + \text{B-R-B} \rightarrow \text{A-R-AB-R-B}. \tag{2.3}$$

Here A and B are the functional end groups which react with each other. Examples of polycondensation can be seen in the formation of (1) polyesters and

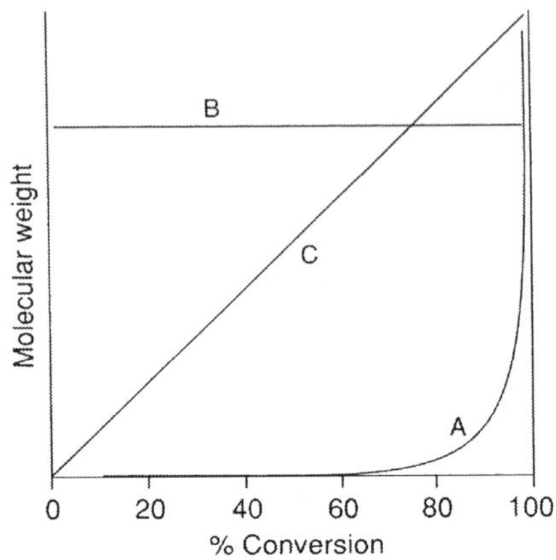

FIGURE 2.1 Variation of molecular weight with % conversion for (a) step-growth polymerization, (b) chain-growth polymerization, and (c) living chain-growth polymerization with no chain transfer and no chain termination.

(2) polyamides, where the A and B groups would be (1) hydroxyl and carboxyl and (2) amine and carboxyl, respectively, which would combine and split off a molecule of water (Braun et al., 1984; Sandler and Karo, 1992; Zhang et al., 2012; Moore, 1978). On the other hand, a polyaddition reaction (Eq. (2.2)) would be exemplified by the reaction of diisocyanates with glycols to form polyurethanes. In that case, of course, no by-products are formed.

The polymerizations shown in Eqs. (2.2) and (2.3) actually represent well-known reactions of small molecules, the only distinction being the minimum requirements of *difunctionality* of each molecule for polymer formation, which makes it possible for the product of each reaction to participate in further reactions. As a rule, the functional groups retain their reactivity regardless of the chain length (Flory, 1953c), so that these reactions follow the same kinetic rules as for simple molecules; however, in contrast to polyaddition reactions, polycondensations suffer from the serious problem of reversibility (e.g., hydrolysis, or "depolymerization") as a result of the possible accumulation of the by-product (e.g., water), and this must be taken into account. In general, because of the unfavorable equilibrium constant for polycondensation reactions (Saunders and Dobinson, 1976), the formation of high polymer requires removal of the small molecule by-products (Zhang et al., 2012).

In both of the foregoing types of reactions, two factors which govern the molecular weight of the polymer are the stoichiometry and the extent of reaction (Odian, 2004b). Thus, it is obvious that an excess of one type of end group will control the maximum chain length attainable, and this can be predicted if the initial molar ratio of functional groups is known. On the other hand, with equivalent amounts of the two types of end groups, the final chain length is theoretically limitless, i.e., infinite in size.

2.2.2 Chain Polymerization

The distinguishing mechanistic feature of chain-growth or chain polymerization (Zhang et al., 2012; Penczek and Pretula, 2012; Odian, 2004b; Penczek, 2002; Jenkins et al., 1996) is that chain growth (propagation) occurs only by addition of monomer to reactive sites present on the growing polymer molecules as shown in Eq. (2.4), where P_n^* is a polymer chain with a reactive site (*) and degree of polymerization of n, M is a monomer unit, and P_{n+1}^* is a polymer chain with a reactive site (*) and a degree of polymerization of $n + 1$.

$$P_n^* + M \rightarrow P_{n+1}^*. \quad (2.4)$$

This type of polymerization involves the successive addition of monomers to a growing chain, which is initiated by some reactive species (initiation). Such addition reactions may involve either multiple bonds or rings. The reactive species which initiate such chain reactions must be capable of opening one of the bonds in the monomer and may be either a radical, an electrophile, a nucleophile, or an organometallic species. Hence these polymerizations may proceed by a variety of possible mechanisms, depending on the electronic nature of the chain-carrying species, viz., free radical, cationic, anionic, and coordination, as illustrated by the following equations for reactions of double bonds with various types of initiating species:

Free radical:

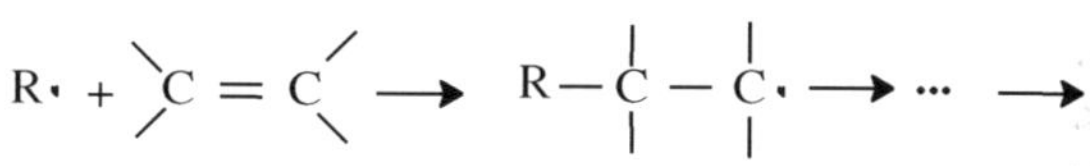

Cationic:

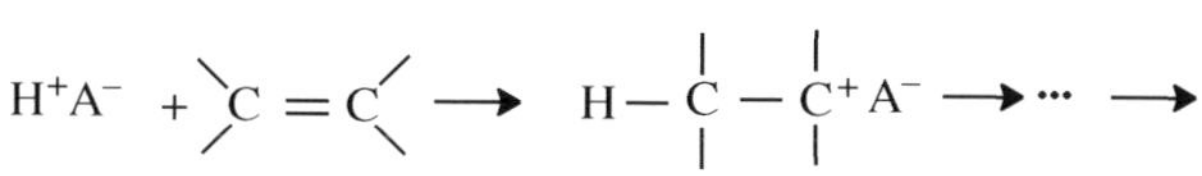

Anionic:

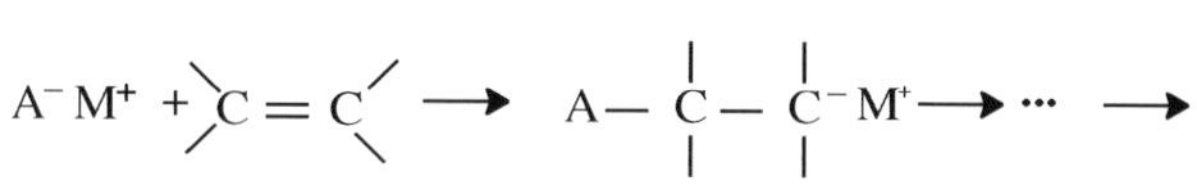

Coordination:

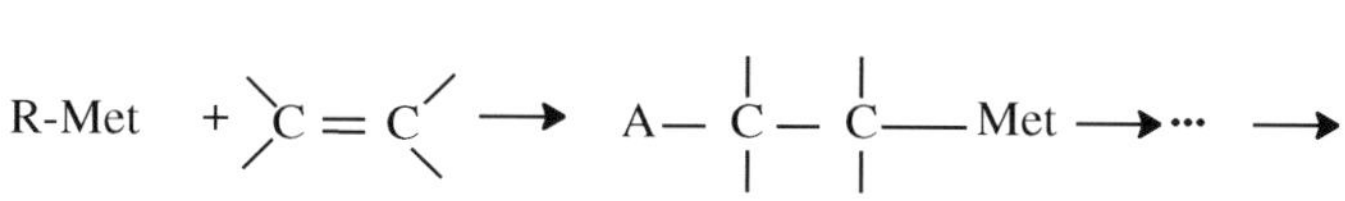

In these equations, the exact nature of the initiating and chain-carrying species can vary from essentially covalent for transition-metal organometallic species in coordination polymerization to ion pairs or free ions in ionic polymerizations, depending on the structure of the chain-carrying species, the counterion, the solvent, and the temperature.

A significant distinction between step polymerization and chain polymerization is that, in the latter, each macromolecule is formed by a "chain reaction" which is initiated by some activation step. Thus, at any given time during the polymerization, the reacting species present consist only of *growing chains and monomer molecules*, in addition to the "dead" polymer chains formed earlier by chain termination reactions. These growing chains may be very short-lived (e.g., free radicals or free ions) but may attain very long chain lengths during their brief lifetimes as illustrated in Figure 2.1b. On the other hand, they may have very long lifetimes (e.g., living polymers (Odian, 2004b; Penczek, 2002; Jenkins et al., 1996)) in which case the chain lengths may increase as a direct function of time of reaction as shown in Figure 2.1c. Hence, unlike the case of step polymerizations, the molecular weights in chain addition polymerization systems may or may not be directly related to time or extent of reaction (see Figure 2.1a).

2.3 POLYADDITION/POLYCONDENSATION

Although, as indicated earlier, polyaddition and polycondensation did not figure prominently in the early explorations of rubber synthesis, it was one of the earliest general methods used for polymerization, because of its relative simplicity. It is thus not surprising that the earliest truly synthetic resins and plastics were of the polycondensate type, such as phenol formaldehyde and polyester. The concept of linking together reactive end groups to build large molecules is fairly simple to comprehend and lends itself to a relatively simple mathematical analysis.

As stated previously, the kinetics of polyadditions and polycondensations follow the same rules as the simple monofunctional reactions (Zhang et al., 2012; Saunders and Dobinson, 1976; Odian, 2004b; Solomon, 1972), as the reactivity of the functional groups is maintained (Flory, 1953c) regardless of chain length. The only new feature is, of course, the growth in molecular size, and this has been amenable to a mathematical analysis (Flory, 1953a, p. 91). Considering the type of reactions defined in Eqs. (2.2) and (2.3), in the normal case, where the number of A and B groups is equal, the chain lengths are easily predictable as a function of the extent of reaction. Thus, if p represents the *fraction of end groups consumed* at any given time, then the number-average number of units per chain (X_n) is given by $1/(1-p)$. Thus,

$$M_n = M_o/(1-p), \qquad (2.5)$$

where M_n is the number-average molecular weight of the polymer and M_o is the molecular weight of a chain-repeating unit. The consequences of this simple relationship are profound. For example, when 50% of the functional groups have reacted, the number-average degree of polymerization is only 2. To prepare polymers with useful properties, molecular weights of at least 10,000 are required; this means that the degree of conversion of the functional groups must be greater than 99% for a repeating unit with a molar mass of 100 g ($X_n = 100$). It is obvious that relatively few reactions will qualify in terms of this rigorous requirement because of side reactions.

Because this type of polymerization is a completely random process, with all molecules having equal probability of reacting, the *distribution of molecular weights* corresponds to the most *probable, or binomial, distribution*, which is related to the extent of polymerization as follows (Flory, 1953d):

$$W_x = xp^{x-1}(1-p)^2, \tag{2.6}$$

$$N_x = p^{x-1}(1-p), \tag{2.7}$$

where W_x is the weight fraction of x-mers (chains having x units) and N_x is the mole fraction of x-mers. This distribution function can be used to calculate M_w, the weight-average molecular weight, as $M_w = M_o \Sigma x W_x$. It can be shown that the foregoing summation leads to the relation

$$M_w = \frac{(1+p)}{(1-p)}(M_o), \tag{2.8}$$

which then means that

$$M_w/M_n = 1 + p. \tag{2.9}$$

Hence the weight/number ratio of chain lengths in these systems undergoes a *steady increase with extent of reaction*, approaching an ultimate value of 2.

Thus, we see that polyaddition and polycondensation are characterized by the following features:

1. All molecules have equal probability of reacting.
2. The polymerization rates are essentially described by the concentrations and reactivity of the functional groups.
3. The chain lengths are monotonic functions of the extent of reaction and hence of the time of reaction.
4. The attainment of high molecular weights requires a high degree of conversion ($p \to 1$).

In those cases where at least one of the monomers has *more than two* functional groups, the added feature of branching chains is introduced, eventually leading to the formation of molecular networks (Flory, 1953e), i.e., gelation. This, of course, complicates the molecular size distribution but does not affect the kinetics of the polymerization.

The foregoing relationships of chain length to extent of reaction would then be expected to apply to such step polymerizations as are involved in the synthesis of poly(alkylene sulfides) from a dihalide and sodium polysulfide (polycondensation) or in the formation of the urethane polymers from glycols and diisocyanates (polyaddition). The polysulfide reaction is actually carried out in a suspension of the dihalide in an aqueous solution of the polysulfide, using a surfactant to stabilize the resulting polymer suspension.

The urethane polymers offer an interesting illustration of the characteristic molecular weights to be expected in this type of polymerization, which can be written as

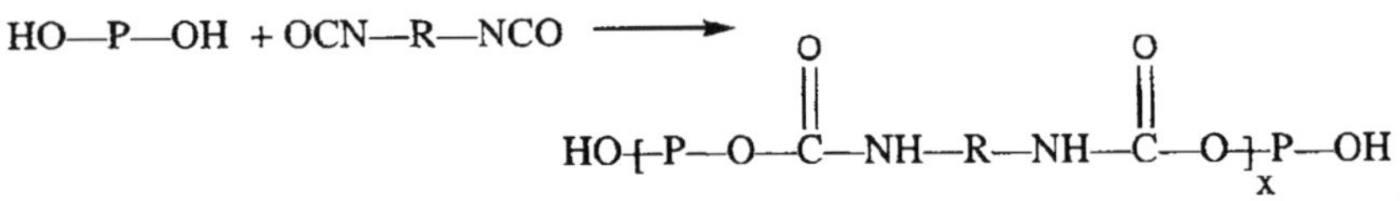

(2.10)

It should be noted that the *P* in Eq. (2.10) represents a low-molecular-weight polymer of a polyester or polyether-type (MW 2000), so that this is really a "chain extension" reaction. It turns out that the reaction between an isocyanate group and a hydroxyl goes to a high conversion, i.e., to approximately 98% ($p = 0.98$). Hence the value of x in Eq. (2.10) is about 50, and the final molecular weight of the urethane polymer is about 100,000. Such high molecular weights are, of course, due solely to the fact that this reaction goes so far toward completion, i.e., where the reactive functional groups can be reduced to concentrations of the order of 10^{-2} M.

2.4 CHAIN POLYMERIZATION BY FREE RADICAL MECHANISM

2.4.1 General Kinetics

The general kinetics for this mechanism (Eastmond, 1976b) involve the usual three primary steps of any chain reaction, i.e., initiation, propagation, and termination, as shown below. Initiation generally occurs by the formation of free radicals through the homolytic dissociation of weak bonds (e.g., in peroxides or azo compounds) or by irradiation. Termination reactions for vinyl polymers can occur either by combination (coupling), by disproportionation, or by a combination of both reactions (discussed next).

Initiation:

$$I \xrightarrow{k_i} 2R\bullet,$$

$$R\bullet + M \longrightarrow R\text{-}M\bullet.$$

Propagation:

$$M_j\bullet + M \xrightarrow{k_p} M_{j+1}\bullet.$$

Termination:

$$M_j\cdot + M_k\cdot \xrightarrow[(k_{\rm tc}+k_{\rm td})]{k_{\rm t}} \text{dead chains},$$

where I = initiator, M = monomer, R = initial free radical, and $M_j\cdot$ = propagating free radical.

Combination:

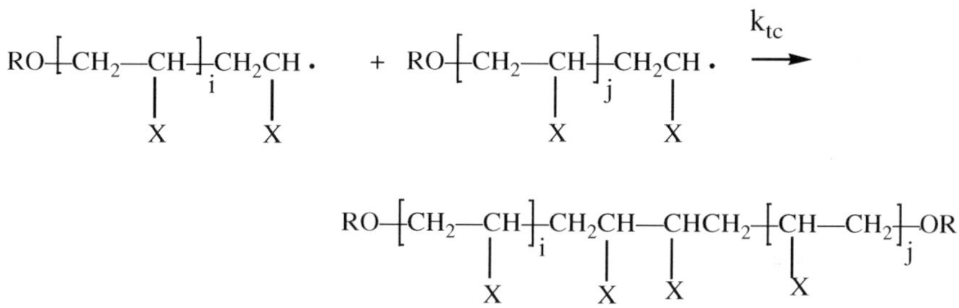

Disproportionation:

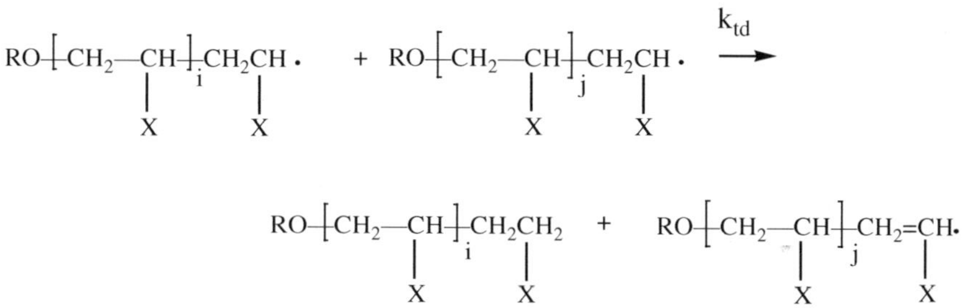

This sequence of steps then leads to the following simple kinetic treatment (Eastmond, 1976b):

$$\text{Rate of initiation} \quad R_{\rm t} = 2k_{\rm i}[I], \tag{2.11}$$

$$\text{Rate of propagation} \quad R_{\rm p} = k_{\rm p}[M_J\cdot][M], \tag{2.12}$$

$$\text{Rate of termination} \quad R_{\rm t} = 2k_{\rm t}[M_J\cdot]^2. \tag{2.13}$$

Assuming a *steady-state condition* where the rate of formation of radicals is equal to their rate of disappearance, i.e., $R_{\rm i} = R_{\rm t}$,

$$[M_J\cdot] = k_{\rm i}^{1/2}k_{\rm t}^{-1/2}[I]^{1/2} \tag{2.14}$$

and

$$R_{\rm p} = k_{\rm p}k_{\rm i}^{1/2}k_{\rm t}^{-1/2}[M][I]^{1/2}. \tag{2.15}$$

Equation (2.15) thus illustrates the dependency of the overall rate of polymerization on the concentrations of initiator and monomer. The half-power dependence of the rate on the initiator concentration appears to be a universal feature of the free radical mechanism and has been used as a diagnostic test for the operation of this mechanism.

Another important aspect of free radical polymerization is the dependency of the number-average degree of polymerization on initiator and monomer concentrations as shown in Eq. (2.16). Comparison with Eq. (2.15) shows that increasing the rate of initiation, by increasing the initiator concentration, increases the rate of polymerization but decreases the degree of polymerization, X_n, which corresponds to the number-average number of units per chain.

$$X_n = k_p k_i^{-1/2} k_t^{-1/2} [M][I]^{-1/2}. \tag{2.16}$$

The general nature of free radical chain polymerization deserves some special attention. Because of the high reactivity of the propagating chain radical, it can only attain a very short lifetime, several seconds at best. This results in a very low stationary concentration of propagating chain radicals (about 10^{-8} M in a homogeneous medium). During this short lifetime, however, each growing radical may still have the opportunity to add thousands of monomer units. Hence the chain length of the macromolecules formed in these systems has no direct relation to the extent of reaction, i.e., to the degree of conversion of monomer to polymer (see Eq. (2.16) and Figure 2.1). At all times during the polymerization, the reaction mixture contains only monomer, a very small concentration of propagating chains, and dead (nonpropagating) polymer, the latter usually of high molecular weight.

To illustrate more clearly the nature of free radical polymerization, it is instructive to examine the values of the individual rate constants for the propagation and termination steps. A number of these rate constants have been deduced, generally using nonstationary-state measurements such as rotating sector techniques and emulsion polymerization (Eastmond, 1976b). Recently, the IUPAC Working Party on "Modeling of kinetics and processes of polymerization" has recommended the analysis of molecular weight distributions of polymers produced in pulsed-laser-initiated polymerization (PLP) to determine values of propagation rate constants (Gilbert, 1996,1992). Illustrative values of propagation and termination rate constants are listed in Table 2.1 (Buback et al., 1995; Matheson et al., 1949,1951; Morton et al., 1952; Morton and Gadkary, 1956; Gadkary, 1956; Beuermann et al., 1997; Weerts et al., 1991; Hutchinson et al., 1993). Thus, although the chain-growth step can be seen to be a very fast reaction (several orders of magnitude faster than the rates of the step reactions of the end groups discussed in Section 2.3), it is still several orders of magnitude slower than the termination step, i.e., the reaction of two radicals. It is this high ratio of k_t/k_p which leads to the very low stationary concentration of growing radicals ($\sim 10^{-8}$ M) in these systems.

Although the three individual steps which combine to make up the chain reaction act as the primary control of the chain lengths (see Eq. (2.16)), "chain transfer" reactions can occur whereby one chain is terminated and a new one is

TABLE 2.1 Propagation and Termination Rate Constants in Radical Polymerization[a]

Monomer	k_p at 60°C (liters mole^{-1} sec^{-1})	k_t at 60°C (×10^{-7}) (liters mole^{-1} sec^{-1})
Styrene	176 (340)[b]	3.6
Methyl methacrylate	367 (830)[c]	1.0
Methyl acrylate	2100	0.5
Vinyl acetate	3700	7.4
Butadiene	100 (320)[d]	~100 (700)[d]
Isoprene	50	–
Chloroprene	(1270)[e]	–

[a]*Data taken from Matheson et al. (1949,1951), Morton et al. (1952), Morton and Gadkary (1956), and Gadkary (1956); data in parentheses determined by pulsed-laser-initiated polymerization*
[b]*Buback et al. (1995)*
[c]*Beuermann et al. (1997)*
[d]*Weerts et al. (1991)*
[e]*Hutchinson et al. (1993).*

initiated, without affecting the polymerization rate: such reactions will also, of course, affect the chain length. Chain transfer usually involves the homolytic cleavage of the most susceptible bond in molecules of solvent, monomer, impurity, etc., by the propagating radical, e.g.,

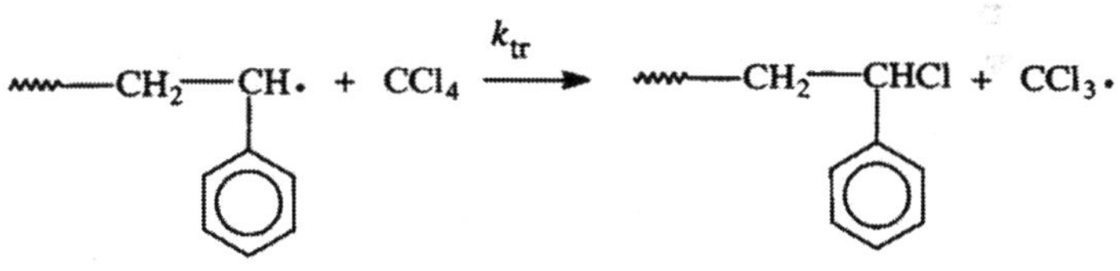

and can be designated as follows:

$$\text{Monomer transfer} \qquad M_J\cdot + M \xrightarrow{k_{trM}} M_J + M\cdot, \tag{2.17}$$

$$\text{Solvent transfer} \qquad M_j\cdot + S \xrightarrow{k_{trS}} M_j + S\cdot, \tag{2.18}$$

$$S\cdot + M \xrightarrow{k'_p} SM\cdot. \tag{2.19}$$

Hence the chain length of the polymer being formed at *any given instant* can be expressed as the ratio of the propagation rate to the sum of all the reactions leading to termination of the chain as

$$X_n = \frac{k_p[M_j\cdot][M]}{(k_{tc} + 2k_{td})[M_j\cdot]^2 + k_{trM}[M_j\cdot][M] + k_{trS}[M_j\cdot][S]}$$

or

$$\frac{1}{X_n} = \frac{(k_{tc} + 2k_{td})R_p}{k_p^2[M]^2} + \frac{k_{trM}}{K_p} + \frac{k_{trS}[S]}{k_p[M]}, \tag{2.20}$$

where X_n is the number-average number of units per chain, k_{tc} is the rate constant for termination by combination, and k_{td} is the rate constant for termination by disproportionation.

2.4.2 Molecular Weight Distribution

The chain length distribution of free radical addition polymerization can also be derived from simple statistics. Thus, for polymer formed at any given instant, the distribution will be the "most probable" and will be governed by the ratio of the rates of chain growth to chain termination,

$$W_x = xp^{x-1}(1-p)^2, \tag{2.21}$$

where p is the probability of propagation and $1 - p$ is the probability of termination (by disproportionation or transfer). This expression is, of course, identical to Eq. (2.6), except for the different significance of the term p. Unlike Eq. (2.6), however, it expresses only the *instantaneous* chain length for an increment of polymer, not the cumulative value for the total polymer obtained.

From Eq. (2.21) it follows that the number- and weight-average chain lengths X_n and X_w are expressed by

$$X_n = \frac{1}{1-p} \quad \text{and} \quad X_w = \frac{1+p}{1-p} \sim \frac{2}{1-p} \tag{2.22}$$

as p must always be close to unity for high polymers. Hence it follows again that

$$X_w/X_n = 2. \tag{2.23}$$

The value of X_w/X_n for the *cumulative polymer* may, of course, be much higher, depending on the changes in the value of p with increasing conversion. It should be noted, however, that this is valid only where the growing chains terminate by disproportionation or transfer, *not* by combination. It can be shown in the latter case (Flory, 1953f) that the increment distribution is much narrower, i.e.,

$$X_w/X_n = 1.5. \tag{2.24}$$

Thus, in summary, the kinetics of free radical polymerization are characterized by the following features:

1. Rate is directly proportional to the half-power of the initiator concentration.
2. Molecular weight is inversely proportional to the half-power of the initiator concentration.

3. The lifetime of the growing chain is short (several seconds) but a high molecular weight is obtained, leading to formation of high polymer at the outset of reaction.
4. No direct relation exists between extent of conversion and chain length.
5. Instantaneous chain length is statistical, but the cumulative value can be considerably broader because of changes in relative rates of propagation and termination.

2.4.3 Special Case of Diene Polymerization

As polydienes still constitute the backbone of the synthetic rubber industry, it is important to consider the special features which dienes exhibit in free radical polymerization. Despite the fact that this type of polymerization has played and is still playing the major role in industrial production of various polymers, it has never been successful in bulk or solution polymerization of dienes. This is an outcome of the kinetic features of the free radical polymerization of dienes, as indicated in Table 2.1. Thus, the relatively high k_t/k_p ratio (as compared with the other monomers shown) leads to very low molecular weights and very slow rates for polydienes prepared in homogeneous systems, as illustrated in Table 2.2 (Whitby and Crozier, 1932). It can be seen from these data that even in the case of these thermal uncatalyzed polymerizations, where the molecular weight would be at a maximum compared with catalyzed systems, it is still too low by at least an order of magnitude. These systems are also complicated by a competitive Diels-Alder reaction, leading to low molecular weight compounds, i.e., "oils."

It is therefore not surprising that the early investigators saw no promise in this mechanism of polymerization of butadiene, isoprene, etc., either by pure thermal initiation or by the use of free radical initiators, such as the peroxides. Instead they turned to sodium polymerization, which, although also rather slow and difficult to reproduce, at least yielded high-molecular-weight rubbery polymers from the dienes. Later, in the 1930s, when emulsion polymerization was introduced, it was found that this system, even though it involves the free radical

TABLE 2.2 Thermal Polymerization of Dienes[a]

Temperature (°C)	Time (h)	Isoprene			2,3-Dimethylbutadiene		
		Yield (%)		MW,	Yield (%)		MW,
		Oil	Rubber	Rubber	Oil	Rubber	Rubber
85	100	7.9	16.3	4600	–	–	–
85	250	–	–	–	2.7	19.6	3500
85	900	–	35.3	5700	–	49.7	3500
145	12.5	54.7	15.6	4000	11.1	15.6	2100

[a]*Data taken from Whitby and Crozier (1932).*

mechanism, leads to both fast rates and high molecular weights, conducive to the production of synthetic rubber. The special features of emulsion polymerization which lead to such surprising results are discussed in Section 2.5.

2.4.4 Controlled Radical Polymerization

There has been a revolution in free radical polymerization chemistry that began in the 1980s with the seminal patent of Solomon et al. (1986). These scientists found that it was possible to obtain controlled radical polymerization of monomers such a styrene and alkyl (meth) acrylates by effecting free radical polymerization in the presence of stable nitroxyl radicals as shown below. It has been found that these controlled polymerizations carried out

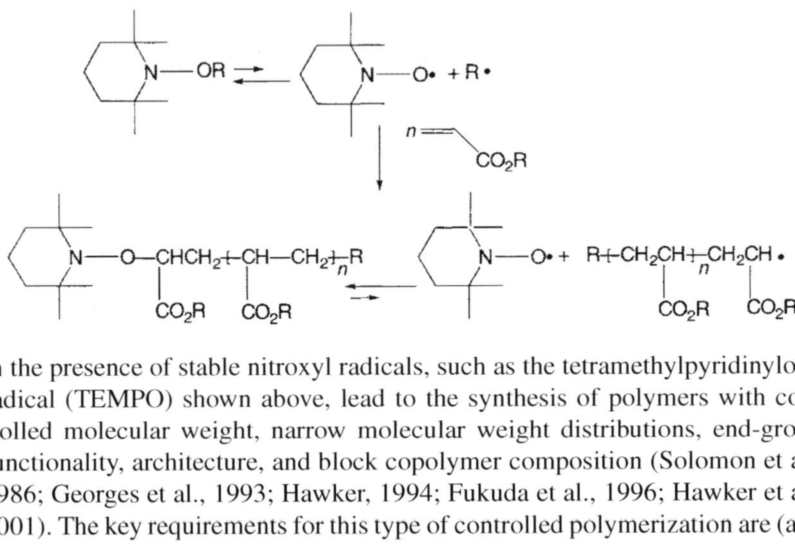

in the presence of stable nitroxyl radicals, such as the tetramethylpyridinyloxy radical (TEMPO) shown above, lead to the synthesis of polymers with controlled molecular weight, narrow molecular weight distributions, end-group functionality, architecture, and block copolymer composition (Solomon et al., 1986; Georges et al., 1993; Hawker, 1994; Fukuda et al., 1996; Hawker et al., 2001). The key requirements for this type of controlled polymerization are (a) a thermally labile bond that undergoes homolysis reversibly to form reactive radicals capable of initiating or propagating polymerization of vinyl monomers, (b) simultaneous formation of a stable radical that rapidly and reversibly combines with propagating radicals but which does not add to vinyl monomers, and (c) an equilibrium constant between radicals and covalent, dormant species that favors the dormant species. In order for a system of this type to be useful, the ratio of the concentration of active radical species to dormant species must be less than 10^{-5} (Fukuda and Goto, 2000). This implies that the majority of the lifetime of the chain is spent in the dormant stage. Successful systems must maintain an optimum amount of nitroxide such that polymerization can occur at an appreciable rate (Keoshkerian et al., 1998). It should be noted that radical-radical coupling can still occur, but it is minimized by the low concentration of propagating radicals (e.g., 10^{-8} M). Because termination still occurs, it is obviously inappropriate to call these polymerizations living, although these types of controlled radical polymerizations are often referred to as living in the literature.

The kinetics of the stable free radical polymerization are controlled by the persistent radical effect which has been clearly elucidated by Fischer (1997,1999).

Careful and extensive investigations of these nitroxide-mediated polymerizations (also referred to as stable free radical polymerization) have established optimum conditions for controlled radical polymerization of a variety of vinyl monomers (Matyjaszewski, 1998,2000). Variables examined include the structure of the nitroxide and the presence of other additives to control spontaneous polymerization of monomers such as styrene. It is noteworthy that in place of alkoxyamine initiators, a mixture of a normal free radical initiator such as an azo compound or a peroxide can also be used.

The application of these procedures to 1,3-dienes has presented problems. The rates of polymerization were observed to decrease and then stop due to a buildup of excess free nitroxide (Keoshkerian et al., 1998). An effective procedure for the controlled polymerization of isoprene at 145°C involved the addition of a reducing sugar such as glucose in the presence of sodium bicarbonate to react with the excess nitroxide (Keoshkerian et al., 1998). After 4 h, polyisoprene with $M_n = 21{,}000$ and $M_w/M_n = 1.33$ was obtained in 25% yield. The reaction of TEMPO-terminated polystyrene with either butadiene or isoprene resulted in the formation of the corresponding diblock copolymers that were characterized by ^{1}H NMR and SEC (Georges et al., 1998). No evidence for either polystyrene or polydiene homopolymers was reported.

An alternative procedure to reduce the concentration of excess nitroxide radicals has been reported by Hawker and coworkers (Benoit et al., 2000). They used the initiator shown below to successfully effect the controlled polymerization of isoprene. It was reported that the corresponding nitroxide has α-hydrogens that can decompose via disproportionation, thereby preventing buildup of excess nitroxide. Using this initiating/nitroxide system, it was possible to prepare a variety of polyisoprenes with controlled molecular weight, high 1,4-microstructure, and polydispersities that ranged from 1.07 for low molecular weights (e.g., $M_n = 5000$) to 1.20 for number-average molecular weights of 100,000. However, the required reaction conditions were 130°C and reaction times up to 48 h. Well-defined copolymers of isoprene and styrene or (meth) acrylates were also prepared at 120°C ($M_n \approx 17{,}000$; $M_W/M_n = 1.1$–1.2). This initiator system has also been used to copolymerize isoprene or chloroprene with dimethyl trans-1,3-butadiene-1-phosphonate, a P-containing monomer with potential applications as a flame retardant, anticorrosive, and/or antioxidant (Ajellal et al., 2008). Polymers with molecular weights up to 40,000 g/mol were prepared with various compositions. Broader molecular weight distributions were observed when chloroprene was used as the comonomer as compared to isoprene. Isoprene has also been polymerized using the nitroxide, N-tert-butyl-N-[1-diethylphosphono-(2,2-dimethylpropyl)]nitroxide (SG1), which has a higher dissociation rate constant (Bertin et al., 2005) and has been shown to exhibit excellent control over the polymerization of styrene (Lutz et al., 2001) and acrylates (Vinas et al., 2008).

Initial studies utilized a poly(tert-butyl acrylate)-SG1 macroinitiator for the chain extension of isoprene, but no kinetic data was reported (Matsuoka et al., 2009). More recently, Nicholas and coworkers reported the polymerization of isoprene with a variety of SG1-based alkoxyamine initiators, and investigated the effect of temperature, alkoxyamine concentration, and structure on the polymerization kinetics (Harrisson et al., 2011). Optimal control over the polymerization was observed at 115°C where 40% conversion was attained after 16 h, but only modest molecular weights were observed ($M_n \sim 8350$). The best control over the polymerization was observed with SG1-based alkoxyamines which contain secondary and tertiary nonacid groups.

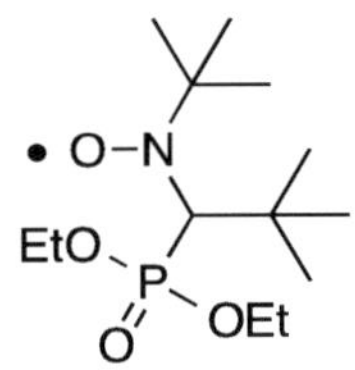

Several other methods for controlled radical polymerization have been developed and should be applicable to elastomer synthesis (Matyjaszewski, 1998, 2000). One of the other most important systems for controlled radical polymerization is atom transfer radical polymerization (ATRP) (Matyjaszewski and Xia, 2001). A transition metal (Mt) catalyst participates in an oxidation-reduction equilibrium by reversibly transferring an atom, often a halogen, from a dormant species (initiator or polymer chain) as shown below.

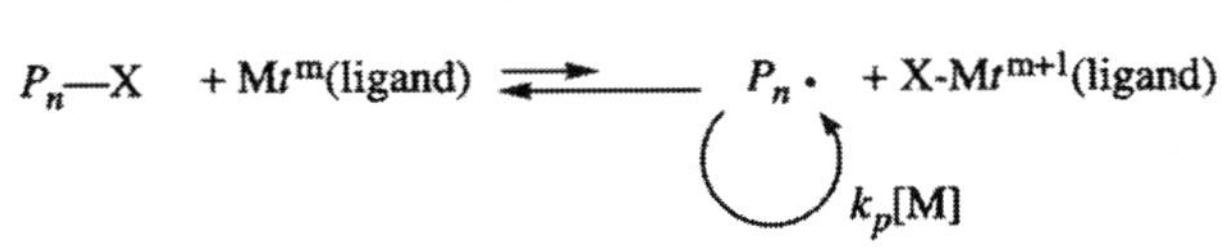

Although a variety of transition metal salts are effective, copper salts have been most extensively investigated. Unfortunately, polymerization of 1,3-dienes has not been successful by ATRP due to chelation of the copper catalyst by isoprene (Wootthikanokkhan et al., 2001).

Reversible addition-fragmentation chain transfer (RAFT) polymerization has been shown to be applicable to the polymerization of isoprene. The RAFT process is proposed to proceed by a degenerative chain transfer process that involves an initiator or radical source, monomer, and a chain transfer agent (CTA), and is controlled by the choice of the CTA (Moad et al., 2008; Favier and Charreyre, 2006). Each CTA consists of a stabilizing or destabilizing Z group and an R group that can efficiently reinitiate polymerization of the monomer. The choice of the R group, Z group, and monomer dictates the level of control over the polymerization. Perrier reported the bulk polymerization of isoprene by RAFT utilizing 2-(2-cyanopropyl) dithiobenzoate (CPDB) and 2-ethylsulfanylthiocarbonylsulfanylpropionic acid ethyl ester (ETSPE) as CTAs (Jitchum and Perrier, 2007). Controlled polymerizations were observed with the use of ETSPE at 115°C and a CTA:initiator ratio of 1:0.2, but molecular weights were relatively low (M_n 27,400) with M_w/M_n ranging from 1.30 to 1.85. Block copolymers with styrene and tert-butyl acrylate were also successfully prepared. Germack and Wooley also reported the RAFT polymerization of isoprene using S-1-dodecyl-S′-(r,r′-dimethyl-r″-acetic acid)trithiocarbonate as the CTA with various initiators (Germack and Wooley, 2007). Polyisoprenes with relatively narrow molecular weight distributions ($M_w/M_n \sim 1.2$–1.3) were obtained, but conversion (30%) and molecular weights ($M_n \sim 4700$) were low, and high radical concentrations were necessary. Block copolymers with styrene were prepared to demonstrate the retention of the active chain end (Germack and Wooley, 2007; Germack et al., 2006). High 1,4-microstructure was reported for all of the polyisoprenes prepared by RAFT. Attempts to prepare block copolymers containing isoprene or butadiene by RAFT in an emulsion polymerization proved unsuccessful (Bar-Nes et al., 2009). The polymerizations lacked molecular weight control, and significant crosslinking was observed as well as slow rates of polymerization.

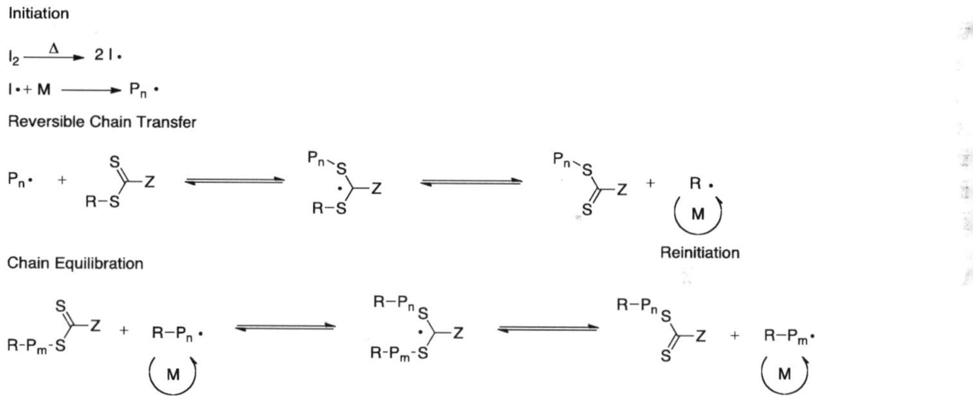

2.5 EMULSION POLYMERIZATION

2.5.1 Mechanism and Kinetics

Polymerization in aqueous emulsions, which has been widely developed technologically, represents a special case of free radical chain polymerization in a heterogeneous system (Poehlein, 1986; Piirma, 1982; Blackley, 1975; Napper and Gilbert, 1986; Gilbert, 1995; Tauer, 2003; Lovell and El-Aasser, 1997). Most emulsion polymerization systems comprise a water-insoluble monomer in water with a surfactant and a free radical initiator. Although it might be thought that polymerization of water-insoluble monomers in an emulsified state simply involves the direct transformation of a dispersion of monomer into a dispersion of polymer, this is not really the case, as evidenced by the following features of a true emulsion polymerization:

1. The polymer emulsion (or latex) has a much smaller particle size than the emulsified monomer, by several orders of magnitude.
2. The polymerization rate is much faster than that of the undiluted monomer, by one or two orders of magnitude.
3. The molecular weight of the emulsion polymer is much greater than that obtained from bulk polymerization, by one or two orders of magnitude.

It is obvious from the foregoing facts that the mechanism of emulsion polymerization involves far more than the mere bulk polymerization of monomer in a finely divided state. In fact, the very small particle size of the latex, relative to that of the original monomer emulsion, indicates the presence of a special mechanism for the formation of such polymer particles.

The mechanism of emulsion polymerization, as originally proposed by Harkins (1947), can best be understood by examining the components of this system, as depicted in Figure 2.2, for a typical "water-insoluble" monomer such as styrene (solubility = 0.07 g/L (Gardon, 1977)). The figure shows the various loci in which monomer is found, and which compete with each other for the available free radicals. Thus, in the initial stages, the monomer is found in three loci: dissolved in aqueous solution, as emulsified droplets, and within the soap micelles. Both the dissolved monomer and the relatively large monomer droplets represent minor loci for reaction with the initiator radicals (except, of course, in the case of highly water-soluble monomers). The large number of soap micelles containing imbibed monomer, however, represents a statistically important

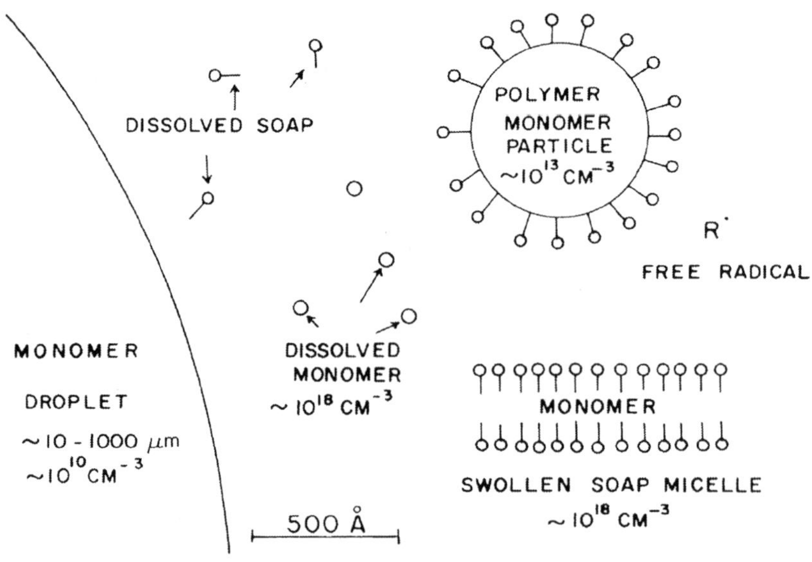

FIGURE 2.2 Loci in mechanism of emulsion polymerization.

locus for initiation of polymerization. It is thus not surprising that most of the polymer chains are generated within the monomer-swollen soap micelles. The very large number ($\sim 10^{15}$/ml) of very small polymer particles thus formed which are stabilized by adsorbing monolayers of soap depletes the available molecularly dissolved soap, thus destroying the soap micelles at an early stage of the polymerization ($\sim$10% conversion in the usual recipe). As all the available soap is distributed, and redistributed, over the surface of the growing particles, the amount of soap is the main factor controlling latex particle size.

During the second stage of the emulsion polymerization, therefore, the loci for available monomer consist of the dissolved monomer, the free monomer droplets, and the monomer imbibed by the numerous polymer particles. As before, the first two of these loci make a minor contribution, whereas the polymer-monomer particles provide a major locus for reaction with the initiator radicals diffusing from the aqueous phase. The major portion of the polymerization reaction apparently occurs within this large number of latex particles which are isolated from each other by electrostatic repulsion and kept saturated (Morton et al., 1954) with monomer diffusing from the monomer droplets. It is this aspect which leads to the unique characteristics of this system (Smith and Ewart, 1948). Thus, once an initiator radical enters a polymer monomer particle and initiates a chain, the latter must continue to propagate with the available monomer until *another* radical enters the same particle. In this way, the rate of chain termination is actually controlled by the *rate of entry* of radicals into the particles, and this generally increases the lifetime of the growing chains, and hence the chain length. Furthermore, because the growing chains are all located in *different* particles, they are unable to terminate each other, leading to a higher *concentration* of growing chains and a hence faster rate.

In this way, emulsion polymerization systems can simultaneously achieve a much faster rate and a much higher molecular weight than homogeneous systems. A comparison of the kinetic features of bulk and emulsion polymerization of styrene is given in Table 2.3. It is obvious at once that the main difference lies in the fact that the emulsion system is capable of raising the steady-state concentration of growing chain radicals by two to three orders of magnitude but *not* at the expense of increasing the termination rate which occurs in homogeneous solution (see Eq. (2.16))!

The situation described earlier, i.e., where radicals entering individual latex particles successively initiate and terminate growing chains, is referred to as *ideal emulsion polymerization*, as defined by the Smith-Ewart theory (Smith and Ewart, 1948). Under these conditions, the concentration of growing chains per unit volume of latex is easily predictable, because at any given time, *half of the particles will contain a growing chain.* In other words, the number of growing chains will be one-half the number of particles. As the latter is of the order of 10^{18} per liter, the concentration of growing chains is of the order of 10^{-6} M compared with 10^{-8} M for homogeneous polymerization systems. Because such growing chains are in an environment which is rich in monomer

TABLE 2.3 Comparison of Free Radical Polymerization Methods of Styrene

	Homogeneous	Emulsion
Monomer concentration (*M*)	5	5^a
Radical concentration (*M*)	10^{-8}	10^{-6}
Rate of polymerization at 60°C (%/h)	~2	100
Molecular weight (M_n)	10^5	10^7

[a]*Within latex particles.*

(within the monomer-polymer particles), it is not surprising that emulsion polymerization rates are one or two orders of magnitude higher than those of bulk polymerization, as shown in Table 2.3. Furthermore, this high radical concentration does not affect the radical lifetime, i.e., the chain size, which is governed solely by the availability of another free radical for termination and, thus, by the period between successive entries of radicals into particles. For a given rate of initiation, the time between radical entry depends on the number of latex particles; i.e., the radical lifetime (and molecular weight) increases with increasing numbers of particles. According to the theory of Smith and Ewart (1948), the number of polymer particles depends on both the initiator concentration and the surfactant concentration,

$$N \propto [I]^{2/5}[S]^{3/5}, \tag{2.25}$$

where $[I]$ is the concentration of initiator and $[S]$ is the concentration of surfactant.

The foregoing situation, of course, holds only for the ideal case, as defined earlier. If the growing chain within the latex particle undergoes some side reaction which transfers the radical activity out of the particle before the next radical enters, or if termination is not rapid when two radicals occupy the same particle, then the number of growing chains at any given time will be, respectively, either smaller or larger than one-half the number of particles. The latter case (more than one radical per particle) can occur, for example, if the particle size is sufficiently large and the termination rate too slow. The rate and molecular weight will then also be governed by other considerations than the interval between entry of successive radicals. Diagnostically, these situations can be distinguished from the ideal case by the effect of added initiator on the rate of polymerization after the formation of particles is complete. Thus, in either case, an increase in initiator concentration will lead to a faster rate of entry of radicals into particles and hence an increase in the number of radicals per particle, leading to an increase in polymerization rate. In contrast, in the ideal case, an increase in frequency of radical entries into particles should not affect the rate, as the particles will still contain a radical only half the time, even though the periods of chain growth will be shorter, leading to a lower molecular weight.

The ideal case of the Smith-Ewart treatment actually proposes a rather elegant method for obtaining the absolute value of the propagation rate constant k_p from emulsion polymerization systems, as shown in Eq. (2.26), where *N* is the number of particles per unit volume:

$$R_p = k_p[M]N/2. \tag{2.26}$$

Equation (2.26) leads to a solution for k_p from available knowledge of the rate R_p, the concentration of monomer in the monomer-polymer particles [*M*], and the number of particles, *N*. This method has been applied to several monomers and has been especially useful in the case of the dienes, where the classical method of photoinitiation poses difficulties. Some of these results are shown in Table 2.4 in the form of the usual kinetic parameters. The results obtained for styrene by photoinitiation techniques are included for comparison. It can be seen that the agreement is remarkably good, considering the widely different experimental methods used. Recent studies of the emulsion polymerization of butadiene have shown that the rate constant for propagation is even higher than previously estimated (see Table 2.1) (Weerts et al., 1991).

The data in Table 2.4 provide evidence that the slow rates and low molecular weights obtained in homogeneous free radical polymerization of these dienes are *not* due to a low rate constant for propagation but rather must be caused by a high rate constant for termination (as indicated in Table 2.1) (Matheson et al., 1949,1951; Morton and Gibbs, 1963). Hence, under the special conditions of emulsion polymerizations, where the termination rate is controlled by the rate of entry of radicals into particles, it becomes possible to attain both faster rates and higher molecular weights. It is this phenomenon which led to the rise of the emulsion polymerization system for the production of diene-based synthetic rubbers.

2.5.2 Styrene-Butadiene Rubber

(i) Kinetics and Molecular Weights

The most successful method developed for the production of a general-purpose synthetic rubber was the emulsion copolymerization of butadiene and styrene (SBR), which still represents the main process in use today (Blackley, 1975; Hofmann, 1989; Blow, 1971; Brydson, 1981; Bauer, 1979; Sun and Wusters, 2004; Demirors, 2003). The general principles of copolymerization will be discussed in a later section, but it is instructive at this point to examine the other main features of this system. The types of recipes used are seen in Table 2.5 (Bauer, 1979). The recipes shown are to be considered only as typical, as they are subject to many variations. It should be noted that the initiator in the 50°C recipe (hot rubber) is the persulfate, whereas in the 5°C recipe (cold rubber) the initiator consists of a redox system comprising the hydroperoxide-iron(II)-sulfoxylate-EDTA. In the latter case, the initiating radicals are formed by the reaction of the hydroperoxide with the ferrous iron, whose concentration is

TABLE 2.4 Propagation Rate Constants from Emulsion Polymerization

Monomer	k_p (liters $mole^{-1}$ sec^{-1}) at 60°C	E_p (kcal $mole^{-1}$)	$A_p(\times 10^{-7})$ (liters $mole^{-1}$ sec^{-1})	Reference
Butadiene	100	9.3	12	Morton et al. (1952)
Isoprene	50	9.8	12	Morton et al. (1952)
2,3-Dimethylbutadiene	120	9.0	9	Morton and Gibbs (1963)
Styrene	280	7.9	4	Morton et al. (1952)
Styrene	176	7.8	2.2	Matheson et al. (1949,1951)

TABLE 2.5 Typical SBR Emulsion Polymerization Recipes[a]

	SBR-1000[b]	SBR-1500[b]
Polymerization temperature (°C)	50	5
Time (h)	12	12
Conversion (%)	72	60–65
Ingredients		
Butadiene	71	71
Styrene	29	29
Water	190	190
Soap (fatty or rosin acid)	5	4.5–5
Potassium persulfate	0.3	–
n-Dodecanethiol	0.5	–
t-Dodecanethiol	–	0.2
p-Menthane hydroperoxide[c]	–	0.08
Trisodium phosphate ($Na_3PO_4 \cdot 10H_2O$)	–	0.5
Ferrous sulfate ($FeSO_4 \cdot 7H_2O$)	–	0.4
Sodium formaldehyde sulfoxylate	–	0.10
Tetrasodium salt of ethylenediamine tetraacetic acid (EDTA)	–	0.06

[a]*Parts by weight. Data taken from Bauer (1979).*
[b]*Commercial grade numbers assigned by the International Institute of Synthetic Rubber Producers to "hot" and "cold" SBR, respectively.*
[c]*Or pinane hydroperoxide.*

TABLE 2.6 Typical Properties of Emulsion-Polymerized SBR[a]

Property	Hot SBR	Cold SBR
Styrene content	24	24
Molecular weight		
Viscosity average	$1.5–4.0 \times 10^5$	2.8×10^5
Weight average	–	5×10^5
Number average	$0.3–1.0 \times 10^5$	$1.1–2.6 \times 10^5$

[a]*Data taken from Bauer (1979).*

controlled by the EDTA complexing agent; the sulfoxylate is needed to convert the oxidized ferric (III) back to ferrous iron. The phosphate salt serves as a stabilizing electrolyte for the latex.

In both recipes, the thiol acts as a chain transfer agent to prevent the molecular weight from attaining the excessively high values possible in emulsion polymerization systems (see Table 2.6). It acts in an analogous fashion to the solvent in Eqs. (2.18) and (2.19), except that the sulfur-hydrogen bond is extremely susceptible to be attacked by the growing chain radical, which is thus terminated by a hydrogen atom, forming the $RS\bullet$ radical which initiates growth of a new chain:

$$P\bullet + RS\text{-}H \xrightarrow{k_{tr}} P\text{-}H + RS\bullet, \quad (2.27)$$

$$RS\bullet + M \xrightarrow{k'_j} RS\text{-}M\bullet. \quad (2.28)$$

These thiols, which are known as "regulators," have transfer constants greater than 1, e.g., k_{tr}/k_p may be 3–4, so that only a small proportion is needed to reduce the molecular weight from several million to several hundred thousand. Diene-based polymers can undergo crosslinking reactions during the polymerization, which leads to the formation of insoluble "gel" rubber when the molecular weight becomes too high. Hence, thiol is used as "modifier" to prevent gel formation and keep the rubber processible. It is also necessary to stop the reaction at intermediate levels of conversion to minimize undesirable gel formation (see Table 2.7).

Shortly after Word War II, the American synthetic rubber industry began production of "cold" SBR, from which, it was found, superior tire rubber, especially as regards tread wear, could be prepared. Subsequent studies showed that the reduction in temperature from 50 to 5°C had little or no effect on the microstructure of the polydiene units (*cis*-1,4 versus *trans*-1,4 versus 1,2), or on the comonomer composition, but did exert a marked influence on the molecular weight distribution (Table 2.6). It was also shown (Morton and Salatiello, 1951) that the crosslinking reaction, i.e., addition of growing chains to polymer double bonds (mainly with 1,2 side chain units), was substantially reduced at these lower temperatures, thus reducing the tendency for gel formation at any given molecular weight.

Table 2.7 shows the maximum molecular weights of polybutadiene attainable at different polymerization temperatures, prior to gelation, expressed as the critical weight-average chain length, x_w, of the primary chains at the gel point.

TABLE 2.7 Crosslinking Parameters for Polybutadiene[a]

Temperature (°C)	Relative Crosslinking Rate, $r_x(\times 10^4)$[b]	χ_w of Primary Chains at Gel Point $(\times 10^4)$
60	1.98	2.15
50	1.36	3.13
40	1.02	4.18
0 (cal.)	0.16	26.3

[a]*Data taken from Saunders and Dobinson (1976).*
[b]$r_x = k_x/k_p$, *where* k_x *is the crosslinking rate constant and* k_p *is the propagation rate constant.*

Thus it can be seen that it is possible to increase the chain length by a factor of 9, without forming gel, by decreasing the polymerization temperature from 50 to 0°C. Hence, the amount of thiol chain transfer agent can also be reduced, and this improves the overall chain length distribution by avoiding the formation of the very low-molecular-weight fraction which results from the rapid reaction of the thiol in the early stages of the polymerization. Furthermore, this possibility of producing gel-free higher molecular weight SBR at reduced polymerization temperature enabled the preparation of a high Mooney viscosity (~100) polymer which could be plasticized by low-cost petroleum oils ("oil-extended" rubber), and still retain its advantageous mechanical properties. As a result, the cold SBR process accounts for more than 85% of the emulsion SBR produced (Sun and Wusters, 2004).

(ii) Chain Microstructure

As might be expected, the emulsion polymerization system does not alter the basic mechanism of free radical polymerization as regards the chain unit structure. The latter is, of course, independent of the type of free radical initiator used, in view of the "free" nature of the growing chain end radical. The temperature of polymerization does exert some influence, as shown by the data in Table 2.8, but not to a very great extent (Binder, 1954). It can be seen that the 1,2 side-chain vinyl content is rather insensitive to the temperature, whereas the *trans*-1,4 content increases with decreasing temperature, at the expense of the *cis*-1,4 content. The latter almost vanishes, in fact, at low temperatures and the polymer then attains its highest *trans*-1,4 content of about 80%. Hence this type of polybutadiene is sufficiently stereoregular to undergo a substantial amount of crystallization on cooling (Beu et al., 1948; Meyer, 1949). However, the introduction of the styrene comonomer is sufficient to destroy the chain regularity necessary for crystallization. Furthermore, it is the high-*cis*-1,4 polybutadiene which is desirable and not the *trans*-1,4 form, since the latter has a crystalline melting point of about 150°C and is not an elastomer at ambient temperature. As can be seen from Table 2.8, the possibility of attaining a high *cis*-1,4 content at a reasonably high polymerization temperature is quite remote.

Hence, it appears that these minor effects of temperature on the microstructure of the butadiene units cannot be expected to have any real influence on the properties of SBR.

2.5.3 Emulsion Polymerization of Chloroprene

(i) Kinetics

The only other diene that has been used extensively for commercial emulsion polymerization is chloroprene (2-chloro-1,3-butadiene) (Hofmann, 1989; Johnson, 1976; Stewart et al., 1985; Blackley, 1983; Musch and Magg, 1996). The chlorine substituent apparently imparts a marked reactivity to this monomer, since it polymerizes much more rapidly than butadiene, isoprene, or any other

TABLE 2.8 Chain Structure of Emulsion Polybutadiene and SBR[a]

Polymerization temperatures (°C)	Isomer (wt%) *cis*-1,4	*trans*-1,4	1,2
		Polybutadiene	
−33	5.4	78.9	15.6
5	13.0	69.9	16.5
50	19.0	62.7	18.8
70	20.8	59.4	19.8
		SBR	
−33	5.4	80.4	12.7
5	12.3	71.8	15.8
50	18.3	65.3	16.3
70	20.0	63.0	17.3
100	22.5	60.1	17.3

[a]*Data taken from Binder (1954).*

dienes (see Tables 2.1 and 2.4) $k_p(35°C) = 595\ L\ mol^{-1}\ sec^{-1}$ (Matheson et al., 1949). In fact, chloroprene is even more susceptible to spontaneous free radical polymerization than styrene and requires a powerful inhibitor for stabilization (Christie et al., 2001). It polymerizes extremely rapidly in emulsion systems, so that its rate must be carefully controlled.

Various recipes (Morton et al., 1956; Morton and Piirma, 1956) can be used for emulsion polymerization of chloroprene, with potassium persulfate as a popular initiator. A basic recipe (Neal and Mayo, 1954) which illustrates several interesting features about this monomer is shown in Table 2.9. Two

TABLE 2.9 Basic Recipe for Neoprene GN[a]

Ingredient	Parts by Weight
Chloroprene	100
Water	150
N wood rosin	4
Sulfur	0.6
Sodium hydroxide	0.8
Potassium persulfate	0.2–0.1
Latex stabilizer[b]	0.7

[a]*Temperature, 40° C; time, several hours; conversion, 90%. Data taken from Neal and Mayo (1954).*
[b]*A sodium salt of naphthalenesulfonic acid-formaldehyde condensation product.*

aspects of this recipe are especially noteworthy: the use of a rosin soap, and the presence of elemental sulfur. Rosin soaps are notorious as retarders in emulsion polymerization, as are most *polyunsaturated* fatty acids. Yet complete conversion can be attained within a few hours. With *saturated* fatty acid soaps, the reaction is almost completed (Morton et al., 1956) within 1 h at 40°C!

Sulfur copolymerizes (Neal and Mayo, 1954; Mochel and Peterson, 1949) with the chloroprene, forming di- and polysulfide linkages in the chain (as illustrated in Eq. (2.29)). The latex is

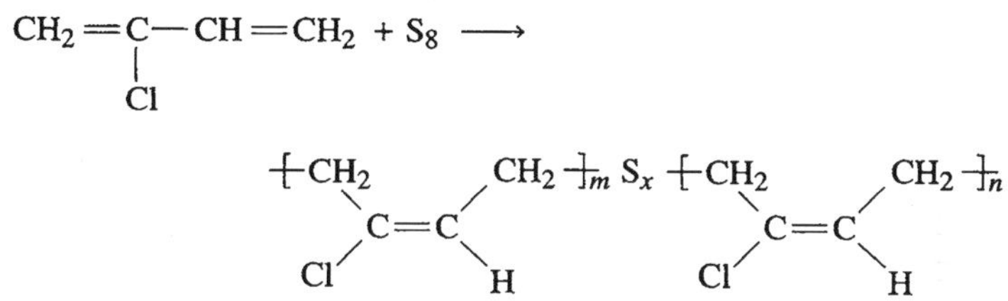

(2.29)

then treated with the well-known vulcanization accelerator, tetraethyl thiuram disulfide, which, by sulfur-sulfur bond interchange, degrades the crosslinked polychloroprene "gel" and renders it soluble and processible as schematically shown in Eq. (2.30). In this way it serves a purpose analogous to the thiol chain transfer agents in SBR

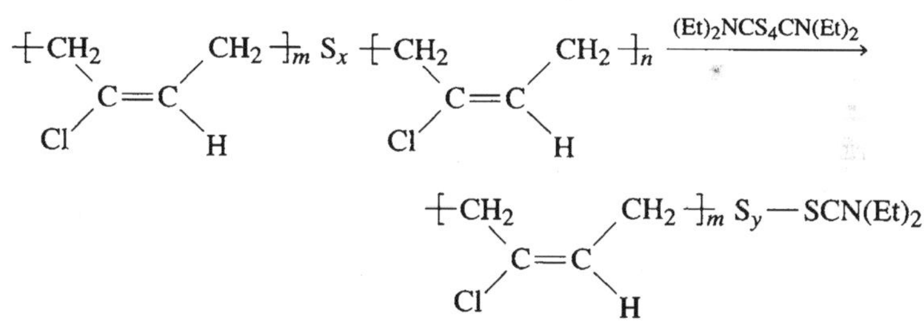

(2.30)

polymerization. As a matter of fact, the newer grades of polychloroprene are prepared (Stewart et al., 1985; Musch and Magg, 1996; Hargreaves, 1968) with the use of thiols and other chain transfer agents. The thiols have been found (Morton and Piirma, 1956) to yield narrower molecular weight distributions for chloroprene than for butadiene or isoprene, due to their much slower rate of disappearance in the presence of chloroprene. These mercaptan grades represent the most common, standard grades (Musch and Magg, 1996).

(ii) Chain Structure

Another feature of the emulsion polymerization of chloroprene that distinguishes it from that of the other dienes is the fact that it leads to a predominantly *trans*-1,4 chain microstructure. Thus, even at ambient polymerization temperature, the polychloroprene contains over 90% *trans*-1,4 units, as shown in

TABLE 2.10 Effect of Polymerization Temperature on Polychloroprene Chain Microstructure[a]

Temperature (°C)	Isomeric Chain Microstructure (%)					
	trans-1,4		Isomerized			
	Total	Inverted[b]	1,2	1,2[c]	3,4	*cis*-1,4
−150	~100	2.0	–	–	–	–
−40	97.4	4.2	0.8	0.6	0.5	0.8
−20	97.1	4.3	0.9	0.6	0.5	0.8
0	95.9	5.5	1.2	1.0	1.1	1.8
20	92.7	8.0	1.5	0.9	1.4	3.3
40	90.8	9.2	1.7	0.8	1.4	5.2
90	85.4	10.3	2.3	0.6	4.1	7.8

[a]*Data taken from Stewart et al. (1985), Coleman et al. (1977), Coleman and Brame (1978), and Ebdon (1978)*
[b]*4,1 enchainment.*
[c]$-CH_2-C=CHCH_2Cl$.

Table 2.10, which illustrates the effect of polymerization temperature on stereoregularity of the chain (Stewart et al., 1985; Petiaud and Pham, 1985; Coleman et al., 1977; Coleman and Brame, 1978; Ebdon, 1978). As expected, lower polymerization temperatures lead to a more stereoregular *trans*-1,4-polychloroprene. Because of the higher crystal melting point of the *trans*-1,4-polychloroprene ($T_m = 105°C$ (Garrett et al., 1970)), as compared with that of the *cis*-1,4-polyisoprene in Hevea rubber (~20°C), even the polymer containing as little as 80% *trans* units crystallizes readily on cooling, or on stretching. The melting point of emulsion polychloroprene is generally in the range of 40–50°C (Stewart et al., 1985). Hence emulsion polychloroprene is the only latex polymer which resembles natural rubber, in that it is sufficiently stereoregular to exhibit strain-induced crystallization (Johnson, 1976; Stewart et al., 1985). This, then, results in high tensile strength in gum vulcanizates, without the need of reinforcing fillers, just as in the case of Hevea rubber. This makes possible the use of polychloroprene in a variety of gum rubber products, endowing them with superior oil and solvent resistance (because of its polarity), as well as high strength.

2.6 COPOLYMERIZATION

2.6.1 Kinetics

Copolymerization involves the simultaneous chain polymerization of a mixture of two or more monomers (Hillmyer, 2012; Ham and Alfrey, 1964; Odian, 2004a; Tirrell, 1986). Aside from the general kinetic considerations which govern these chain reactions, as described earlier, there is imposed an additional feature, i.e., the relative participation of the different monomers during the growth of the chain. This new parameter is most important, since it controls the composition of the copolymer. Systems involving more than two monomers are difficult to resolve in this respect, but it has been found possible to treat the case of a pair of monomers with relative ease (Ham and Alfrey, 1964; Tirrell, 1989; Mayo and Walling, 1950; Mayo and Lewis, 1944).

In the chain addition polymerization of *two* monomers, *regardless of the mechanism involved*, the growing chain always must make a choice of reacting with one of the two monomers. Furthermore, there are *two* kinds of growing chains, depending on which type of monomer unit occupies the growing end. Thus, *four* types of propagation steps can be written as follows for any chain copolymerization of two monomers assuming that the reactivity of the chain end depends only on the chain end monomer unit (terminal model (Hillmyer, 2012; Klumperman, 2012)):

$$M_1^* + M_1 \xrightarrow{k_{11}} M_1^*, \tag{2.31}$$

$$M_1^* + M_2 \xrightarrow{k_{12}} M_2^*, \tag{2.32}$$

$$M_2^* + M_2 \xrightarrow{k_{22}} M_2^*, \tag{2.33}$$

$$M_2^* + M_1 \xrightarrow{k_{21}} M_1^*, \tag{2.34}$$

where M_1^* and M_1 refer to the growing chain and the monomer, respectively, as before, while the *subscripts* refer to the two kinds of monomers in the mixture. It can be seen that these four propagation reactions lead to *four* propagation rate constants, as shown. Hence the rate of consumption of each monomer may be expressed by the following equations:

$$d[M_1]/dt = k_{11}[M_1^*][M_1] + k_{21}[M_2^*][M_1], \tag{2.35}$$

$$d[M_2]/dt = k_{12}[M_1^*][M_2] + k_{22}[M_2^*][M_2]. \tag{2.36}$$

Since it is the *relative rate* of consumption of the two monomers which will decide the composition of the chain, it can be expressed by dividing Eq. (2.35) by Eq. (2.36) leading to Eq. (2.37).

$$\frac{d[M_1]}{d[M_2]} = \frac{k_{11}[M_1^*][M_1] + k_{21}[M_2^*][\mathrm{M}_1]}{k_{12}[M_1^*][M_2] + k_{22}[M_2^*][M_2]}. \tag{2.37}$$

It is obvious at once that Eq. (2.37) is quite intractable for direct use. However, it is possible to simplify it considerably by utilizing the "steady-state" treatment, analogous to the one previously described. This is done by assuming the rate of Eq. (2.32) to be equal to that of Eq. (2.34), and this leads to the equivalence

$$[M_2^*] = \left(\frac{k_{12}}{k_{21}}\right)[M_1^*]\frac{[M_2]}{[M_1]},$$

which, when inserted into Eq. (2.37), yields Eq. (2.38), after appropriate rearrangements are made,

$$\frac{d[M_1]}{d[M_2]} = \frac{[M_1]}{[M_2]}\left[\frac{r_1[M_1]+[M_2]}{r_2[M_2]+[M_1]}\right], \tag{2.38}$$

where $r_1 = k_{11}/k_{12}$ and $r_2 = k_{22}/k_{21}$. The parameters r_1 and r_2 are known as the monomer reactivity ratios, since they express the *relative* reactivity of each of the two kinds of growing chain ends with their "own" monomer as compared with the "other" monomer. They may in fact be considered as expressing the "homopolymerization" tendency of each type of monomer relative to crossover with the comonomer.

Equation (2.38), which relates the instantaneous composition of the copolymer ($d[M_1]/d[M_2]$) to the prevailing monomer concentrations, can be used to determine the values of r_1 and r_2. Many such values have been recorded (Ham and Alfrey, 1964; Mayo and Walling, 1950; Greenley, 1999; Eastmond, 1976a). Typical values of these parameters for styrene copolymerizations are shown in Table 2.11, which illustrates the wide variations that prevail. The *relative reactivity* actually expresses the relative reactivity of each of the monomers shown toward the styrene radical compared to the reaction with styrene monomer. Thus, the r_1 and r_2 values permit some conclusions about the expected composition of the copolymer obtained at any given monomer ratio.

For example, it can be deduced from Table 2.11 that a copolymer of styrene and maleic anhydride ($r_1 r_2 \Rightarrow 0$) would be strongly "alternating" (Odian, 2004b), since it would be *improbable* to have a sequence of two styrene unit, and highly improbable to have a sequence of two maleic anhydride units. Also, it would obviously be extremely difficult to prepare a copolymer of styrene and vinyl acetate, since the latter monomer would be virtually excluded from the styrene polymerization.

TABLE 2.11 Monomer Reactivity Ratios for Free Radical Copolymerizations with Styrene (M_1)[a]

Monomer (M_2)	r_1	r_2	Relative Reactivity ($1/r_1$)
Maleic anhydride	0.097	0.001	10.3
2,5-Dichlorostyrene	0.268	0.810	3.73
Methyl methacrylate	0.585	0.478	1.71
Methyl acrylate	0.871	0.148	1.15
Vinylidene chloride	1.839	0.087	0.54
Diethyl maleate	6.07	0.01	0.16
Vinyl acetate	18.8	0.02	0.05

[a]*Data taken from Greenley (1999).*

It is also obvious, from Eq. (2.38), that the copolymer composition would not necessarily correspond to the comonomer charge, depending on the values of r_1 and r_2. A desirable system would, of course, be one in which this were the case, i.e., where the comonomers enter into the copolymer in the ratio of their concentrations; i.e., where

$$\frac{d[M_1]}{d[M_2]} = \frac{[M_1]}{[M_2]}. \tag{2.39}$$

This is defined as an "azeotropic" copolymerization (Odian, 2004b), by analogy to the distillation of two miscible liquids. Equation (2.37) would apply under the conditions where

$$\frac{r_1[M_1] + [M_2]}{r_2[M_2] + [M_1]} = 1 \tag{2.40}$$

and this would be valid, for example, where $r_1 = r_2 = 1$. In that case, Eq. (2.39) could apply for *all charge ratios*, i.e., the two types of growing chains show no particular preference for either of the two monomers; this is described as a *random copolymerization* (Odian, 2004b). Also Eq. (2.40) would be valid when $r_1 = r_2$ and $[M_1] = [M_2]$, i.e., an azeotropic copolymerization would result *only at equimolar* charge ratios. *In general*, Eq. (2.40) is valid when

$$\frac{[M_1]}{[M_2]} = \frac{(r_2 - 1)}{(r_1 - 1)}. \tag{2.41}$$

This means that any copolymerization will be of an azeotropic type at the particular comonomer charge ratio indicated by Eq. (2.41). However, it also means that an azeotropic copolymerization is only possible when *both* r_1 and r_2 are either greater than 1 or less than 1.

It is important to emphasize that this kinetic treatment is valid for any chain polymerization mechanisms, i.e., free radical, cationic, anionic, and coordination. However, in the case of the ionic mechanisms, the type of initiator used and the nature of the solvent medium may influence the r_1 and r_2 values. This is due to the fact that the growing chain end in ionic systems is generally associated with a counterion, so that the structure and reactivity of such chain ends can be expected to be affected by initiator and the solvent. This will be discussed in Section 2.8.3.

2.6.2 Emulsion Copolymerization of Dienes

The three cases which involve copolymerizations leading to commercial synthetic rubbers are styrene-butadiene (SBR), butadiene-acrylonitrile (NBR), and chloroprene with various comonomers.

(i) Styrene-Butadiene (SBR)

A large number of studies have been made of the reactivity ratios in this copolymerization, both in homogeneous and emulsion systems, and the average

TABLE 2.12 Comonomer Composition of SBR[a]

Conversion (%)	Styrene in Copolymer (wt%)	
	Differential	Integral
0	17.2	–
20	18.8	17.9
40	20.6	18.7
60	23.3	19.7
80	29.5	21.2
90	36.4	22.5
95	45.0	–
100	(100)	(25.0)

[a]*Charge weight ratio, 75/25 butadiene/styrene; 50° C. Data taken from Fryling (1954).*

r values have been computed (Demirors, 2003; Uraneck, 1968; Fryling, 1954) for butadiene and styrene, respectively, as

$$r_B = 1.6, \quad r_S = 0.5$$

these values apply to solution and emulsion polymerization, presumably because neither of the monomers is particularly soluble in water, and both are quite insensitive to temperature. It appears, therefore, that the butadiene must enter the chain substantially faster than its charging ratio, and that each increment of polymer formed contains progressively more styrene. This is confirmed by the change in composition of the copolymer with conversion, as shown in Table 2.12. It can be seen that, at high conversion, the increment, or differential, composition becomes quite high in styrene content with concomitant loss of rubbery properties, even though the cumulative, or integral, composition still shows a low styrene content. This indicates the advisability of stopping the reaction at conversions not much higher than 60% (Hofmann, 1989).

It should be noted, too, that the r values for this system do not permit an azeotropic polymerization, as predicted by Eq. (2.39). With respect to the distribution of styrene monomer units in the copolymer, the monomer reactivity ratio product, $r_B r_S = 0.8$, is close to a value of 1.0, which would correspond to an "ideal" copolymerization (Odian, 2004b) which would correspond to a random distribution of styrene units along the chain. For an "ideal" copolymerization, the relative rates of incorporation of the two monomers are independent of the chain end unit as predicted by Eq. (2.42).

$$r_B = \frac{1}{r_S} \quad \text{therefore} \quad \frac{k_{BB}}{k_{BS}} = \frac{k_{SB}}{k_{SS}}. \tag{2.42}$$

It is reported that the number-average number of styrene units in a sequence is 1.2 as determined by high resolution gel permeation chromatography of ozonolysis products (Tanaka et al., 1986). The observed sequence distribution of monomer units was in accord with calculated values based on the monomer reactivity ratios (Tanaka et al., 1983).

(ii) Butadiene-Acrylonitrile (Nitrile Rubber)

According to Hofmann (1964), Mackey and Jorgensen (2000), and Hofmann (1984), the reactivity ratios of this pair of monomers at 50°C in *emulsion* polymerization are

$$r_B = 0.4, \quad r_{AN} = 0.04$$

and they decrease somewhat at lower temperatures, but not to a great extent. These ratios are no doubt influenced by the marked water solubility of the acrylonitrile compared to that of butadiene.

The foregoing r values lead to the following situation. In accordance with Eq. (2.39), an azeotropic copolymer is formed when the acrylonitrile charge is 35–40% by weight (or by mole), so that a constant composition is maintained throughout the polymerization. If the acrylonitrile charge is *below* this value, the initial copolymer is relatively rich in acrylonitrile, which progressively *decreases* with increasing conversion. However, if the acrylonitrile charge is higher than the "azeotrope," the initial copolymer contains less acrylonitrile than charged, but the acrylonitrile content *increases* with conversion. Since the commercial nitrile rubbers have nitrile contents from 10 to 40%, these considerations have a very practical significance. With respect to the distribution of comonomer units in the copolymer, the monomer reactivity ratio product, $r_B r_{AN} = 0.016$, is close to zero, which would correspond to an alternating distribution of comonomer units (Odian, 2004b).

(iii) Chloroprene

Polychloroprene is generally prepared commercially as a homopolymer, although small amounts of a comonomer are included in several grades of these elastomers. There are two good reasons for the paucity of chloroprene-based copolymers. In the first place, the homopolymer, as stated previously, has a high *trans*-1,4 chain structure and is therefore susceptible to strain-induced crystallization, much like natural rubber, leading to excellent tensile strength. It also has other favorable mechanical properties. Furthermore, chloroprene is *not* very susceptible to copolymerization by the free radical mechanism, as indicated by the r values in Table 2.13 (Greenley, 1980,1999). Thus, except for 2,3-dichloro-1,3-butadiene, chloroprene does not efficiently undergo copolymerization with other monomers. Hence, it is not surprising that the few copolymers of chloroprene available commercially contain only minor amounts of comonomers, which are included for their moderate effects in modifying the properties of the elastomers.

TABLE 2.13 Monomer Reactivity Ratios in Copolymerization of Chloroprene[a]

Monomer M_2	r_1	r_2
Styrene	5.98	0.025
Isoprene	2.82	0.06
Acrylonitrile	5.38	0.056
Methyl methacrylate	6.33	0.080
2,3-Dichloro-1,3-butadiene	0.31	1.98

[a]*Data taken from Greenley (1980,1999).*

2.7 CHAIN POLYMERIZATION BY CATIONIC MECHANISM

2.7.1 Mechanism and Kinetics

In these chain addition reactions, the active species is cationic in nature, initiated by strong acids, either of the protic or Lewis variety (Penczek and Pretula, 2012; Odian, 2004a; Kennedy and Iván, 1992; Kennedy and Maréchal, 1982; Kennedy, 1975; Sawamoto and Higashimura, 1989; Gandini and Cheradame, 1985; Matyjaszewski, 1996; Faust, 2012). Since most of these ionic polymerizations are carried out in nonaqueous solvents with low dielectric constants (Gandini and Cheradame, 1985), it is unlikely that the active species is a "free" ion, analogous to a free radical. A multiplicity of active species may be involved as propagating species as shown below by the spectrum of cationic species, one or more of which may be involved as active propagating species, especially in more polar solvents (Gandini and Cheradame, 1985; Puskas et al., 1976). Unfortunately, very little information

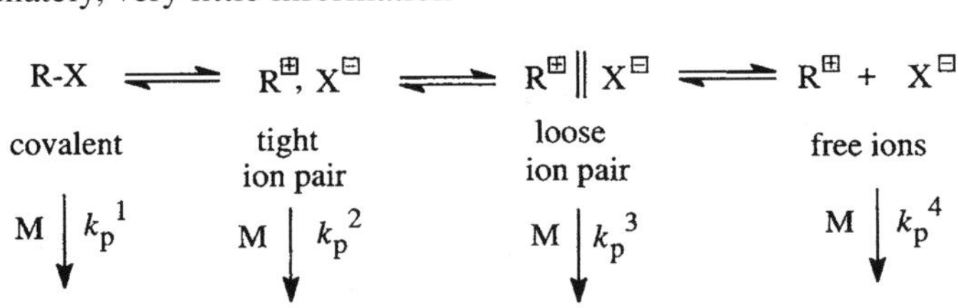

is available about the exact nature of the propagating species in cationic systems. This is mainly due to the inherent experimental difficulties, caused by high reactivity and sensitivity to impurities, especially to traces of water.

The most common initiators of cationic polymerization are Lewis acids, such as $AlCl_3$,BF_3,$SnCl_4$, and $TiCl_4$, although strong protic acids such as H_2SO_4 may also be used. Cationic polymerization is restricted to vinyl monomers with electron-donating or electron-delocalizing substituents, e.g., isobutylene, vinyl alkyl ethers, vinyl amines, styrene, and other conjugated hydrocarbons. These polymerizations are characterized by rapid rates at very low temperatures, e.g., isobutylene is polymerized almost instantaneously at $-100°C$ by $AlCl_3$. The presence of a hydrogen donor, such as water or a protic acid, as a cocatalyst, is usually a prerequisite, as has been shown (Norrish and Russell, 1952) in the case of isobutylene. On this basis, the Evans-Polanyi mechanism (Evans and Polanyi, 1947) proposed the following reaction sequence for the polymerization of isobutylene by BF_3 monohydrate:

Initiation:

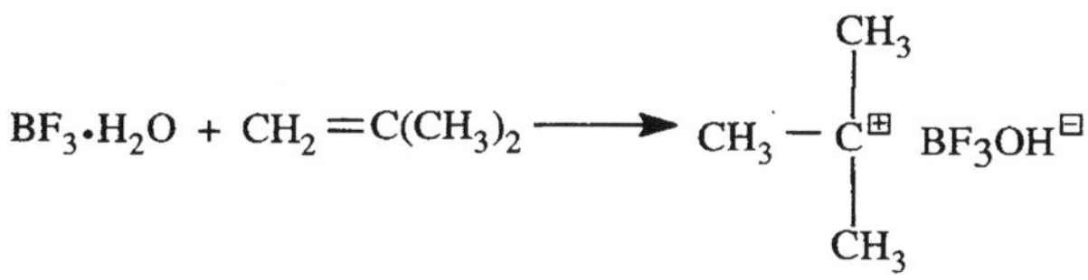

(2.43)

Propagation:

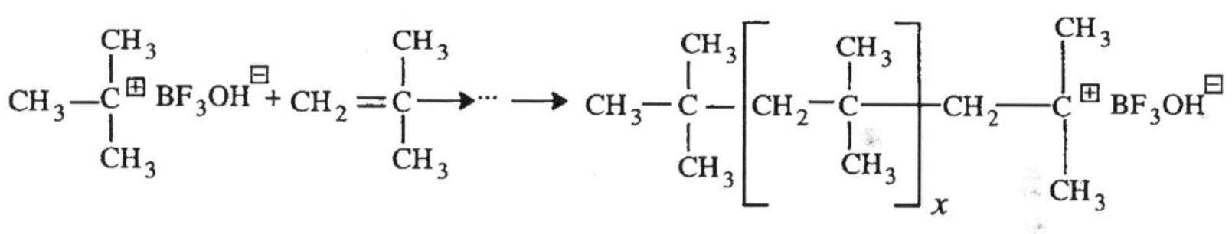

(2.44)

Termination:

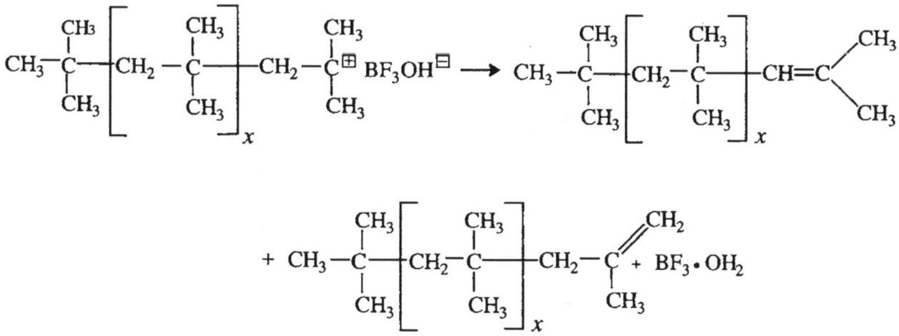

(2.45)

Chain transfer to monomer is also an integral step in the cationic polymerization of isobutylene, and it is this reaction which controls the molecular weight (Faust, 2012):

Chain Transfer:

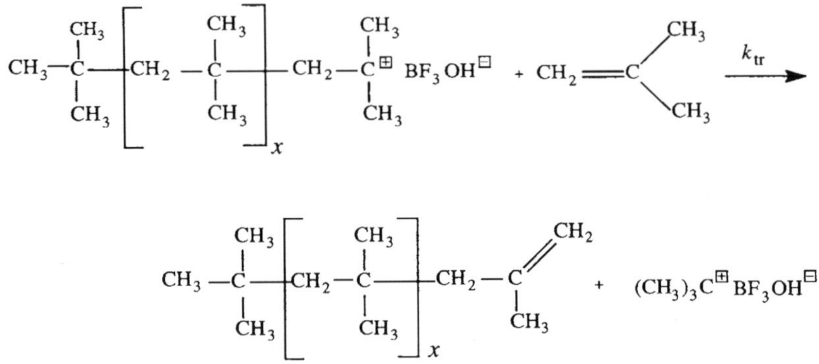

(2.46)

This mechanism has been supported by infrared and ^{1}H NMR spectroscopic data which provided evidence for the presence of the polymer end groups proposed by this mechanism (Kennedy and Maréchal, 1982; Kennedy, 1975; Pepper, 1950; Dainton and Sutherland, 1949; Puskas et al., 1976) i.e.,

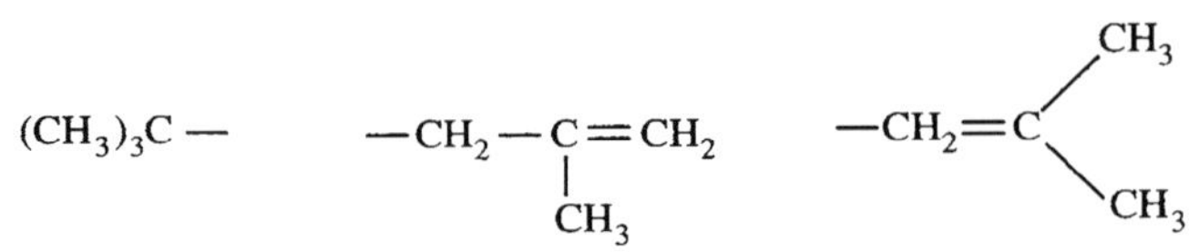

Furthermore, the use of the tracer complex $BF_3 \cdot D_2O$ showed (Dainton and Sutherland, 1949) that the polymer contained deuterium while the initiator became converted to $BF_3 \cdot H_2O$.

This mechanism actually involves initiation by addition of a proton (Eq. (2.43)), from the Bronsted acid formed by the Lewis acid and a coinitiator, to the monomer and subsequent termination of chain growth by loss of a proton to the initiator anion (Eq. (2.45)) or by chain transfer to monomer (Eq. (2.46)). The chain growth, therefore, occurs during the brief lifetime of the carbenium ion, and the initiator or a new carbenium ion is constantly regenerated. The chain transfer step would be expected to have no effect on rate, since the trimethylcarbenium ion should rapidly reinitiate chain growth, but the chain length will decrease. It should be pointed out again, at this point, that it is improbable, in view of the low dielectric constants of the solvents employed, that the active propagating species in these systems are free ions; therefore, the counterion has been depicted as being associated with the carbenium ion in each mechanistic step. This simplified mechanism does not consider the actual nature of the propagating cationic species (Winstein et al., 1954).

The foregoing simplified mechanism is amenable to kinetic analysis, using the steady-state method, as in the case of the free radical mechanism. Using the notation HA to designate the acid initiator, we can write:

Initiation:

$$\mathrm{HA} + M \xrightarrow{k_{\mathrm{j}}} \mathrm{H}M^{+}\mathrm{A}^{-}. \tag{2.47}$$

Propagation:

$$\mathrm{H}M^{+}\mathrm{A}^{-} + M \xrightarrow{k_{\mathrm{p}}} \cdots \rightarrow \mathrm{H}[M]_{x}M^{+}\mathrm{A}^{-}. \tag{2.48}$$

Termination:

$$\mathrm{H}[M]_{x}M^{+}\mathrm{A}^{-} \xrightarrow{k_{\mathrm{t}}} M_{x+1} + \mathrm{HA}. \tag{2.49}$$

Transfer:

$$\mathrm{H}[M]_{x}M^{+}\mathrm{A}^{-} + M \xrightarrow{k_{\mathrm{tr}}} M_{x+1} + \mathrm{H}M^{+}\mathrm{A}^{-}. \tag{2.50}$$

Rate of initiation:

$$R_{\mathrm{i}} = k_{\mathrm{i}}[\mathrm{HA}][M]. \tag{2.51}$$

Rate of termination:

$$R_{\mathrm{t}} = k_{\mathrm{t}}[\mathrm{H}M_{x}M^{+}\mathrm{A}^{-}]. \tag{2.52}$$

Rate of propagation:

$$R_{\mathrm{p}} = k_{\mathrm{p}}[\mathrm{H}M_{x}M^{+}\mathrm{A}^{-}][M]. \tag{2.53}$$

Rate of transfer to monomer:

$$R_{\mathrm{tr}} = k_{\mathrm{tr}}[\mathrm{H}M_{x}M^{+}\mathrm{A}^{-}][M]. \tag{2.54}$$

Here R_{t} has been assumed to be a first-order reaction, since the counterion A^{-} is considered to be specifically associated with the carbenium ion as shown in Eq. (2.45) and not as a separate species. Using the steady-state assumption, we equate R_{i} and R_{t} and thus obtain Eq. (2.55) for the steady-state concentration of growing chains.

$$[\mathrm{H}M_{x}M^{+}\mathrm{A}^{-}] = (k_{\mathrm{i}}/k_{\mathrm{t}})[\mathrm{HA}][M]. \tag{2.55}$$

Hence,

$$R_{\mathrm{p}} = -d[M]dt = (k_{\mathrm{p}}k_{\mathrm{t}}/k_{\mathrm{t}})[\mathrm{HA}][M]^{2}. \tag{2.56}$$

Here again, the propagation rate is virtually the polymerization rate, since the consumption of monomer by the initiation step is negligible. Unlike the free radical case, the rate here is *first order in initiator concentration*, obviously due to the first-order termination step.

Experimental verification of the foregoing kinetic scheme has been obtained in the case of the cationic polymerization of styrene (Pepper, 1949) and vinyl alkyl ethers (Eley and Richards, 1949), where the polymerization rate was indeed found to be dependent on the first power of the initiator and on the square of the monomer concentration. However, it should be noted that this simple kinetic scheme is not general for cationic polymerizations; even the steady-state assumption is not valid in many cationic polymerizations (Odian, 2004b).

The molecular weight of the polymer can again be expressed in terms of X_n, the number-average number of units per chain, which can be defined here as the ratio of the propagation rate to the sum of the rates of all processes leading to chain termination (including transfer). Hence from Eqs. (2.52–2.54),

$$X_n = \frac{R_p}{R_t + R_{tr}} = \frac{k_p[HM_xM^+A^-][M]}{k_t[HM_xM^+A^-] + k_{tr}[HM_xM^+A^-][M]}$$

or

$$1/X_n = k_{tr}/k_p + k_t/k_p[M]. \qquad (2.57)$$

Equation (2.57) provides a means of determining the relative contribution of the termination and transfer steps. Thus, if k_{tr} is largely relative to k_t, the molecular weight will be virtually independent of monomer concentration, but if the reverse is true, X_n will be directly proportional to $[M]$. Hence this relation lends itself to a simple experimental test, i.e., a plot of $1/X_n^o$ (the reciprocal of the *initial* X_n value) against $1/[M]^o$ (the reciprocal of the initial monomer concentration). It has actually been found that the polymerization of styrene by $SnCl_4$ in ethylene dichloride (Pepper, 1949), and of vinyl alkyl ethers by $SnCl_4$ in *m*-cresol (Eley and Richards, 1949), showed a dominance of termination over transfer, i.e., $X_n \alpha [M]$; however, for isobutylene polymerization catalyzed by $TiCl_4$ in *n*-hexane (Plesch, 1950), the observed polymer molecular weights were independent of monomer concentration, i.e., transfer appeared to predominate.

It is interesting to compare the nature of the individual steps in cationic polymerization with those of the free radical mechanism. Thus, unlike the situation in the latter case, the termination step in cationic polymerization may be expected to require a greater energy than that of propagation, since it involves σ-bond rupture compared to the low-energy attack of the growing carbenium ion on the π bond of the monomer. If indeed the termination step has a higher activation energy than that of propagation, then a rise in temperature should lead to an increase in termination relative to propagation and thus to a lower steady-state concentration of growing chains. The net result would thus be a *decrease* in polymerization rate and molecular weight, i.e., an apparent "negative" overall activation energy for polymerization. This might, of course, be partially or wholly offset if the activation energy of the initiation step were sufficiently high. However, in the majority of cases it appears that this is not the case, so that faster rates (and higher molecular weights) are indeed obtained at reduced temperatures (about −100°C) (Kresge et al., 1987). Kinetic studies have shown that the ion pair propagation rate constant is (5–6) × 10^8 L mol^{-1} sec^{-1} for the polymerization of isobutylene with $EtAlCl_2$ in hexanes/methyl chloride (60/40, vol/vol) at −80°C (Sipos et al., 2003).

2.7.2 Butyl Rubber

The only important commercial elastomer prepared by a cationic polymerization is butyl rubber, i.e., a copolymer of isobutene and isoprene. The latter monomer is incorporated in relatively small proportions (~1.5 mol% (Blackley, 1983)) in order to introduce sufficient unsaturation for sulfur vulcanization. The slurry process with aluminum chloride at −98 to −90°C in methyl chloride diluent can be described by the accompanying "flow sheet" (Faust, 2012; Kresge et al., 1987; Kennedy, 1968). In this process the polymerization is almost instantaneous and extensive cooling by liquid ethylene is required to control the reaction.

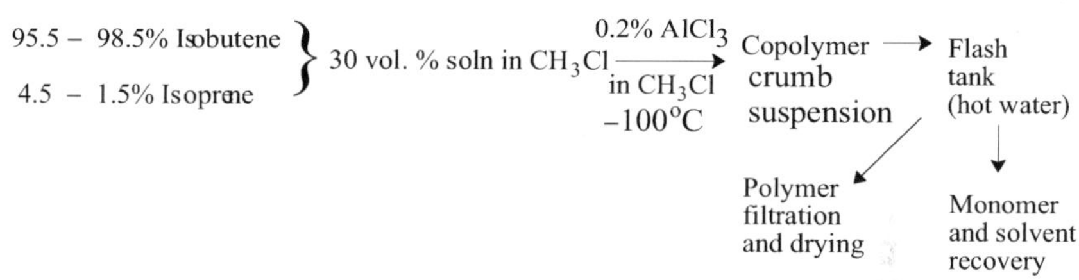

The molecular weights of butyl rubber grades are in the range of 300,000–500,000, and they are very sensitive (Kennedy, 1968) to polymerization temperature above −100°C. For example, a rise in polymerization temperature of 25°C can result in a 5-fold or 10-fold *decrease* in molecular weight, presumably due to the kinetic factors discussed previously (Kresge et al., 1987). The molecular weight distribution of butyl rubbers can be as high as $M_w/M_n = 3\text{–}5$ (Kresge et al., 1987), presumably because of the heterogeneous nature of the polymerization process.

Isoprene is used as the comonomer in butyl rubber (0.5–2.5 mol%) (Kresge et al., 1987) because the isobutene-isoprene reactivity ratios are more favorable for inclusion of the diene than those of the isobutene-butadiene pair (Kennedy, 1968). Thus, for the former pair, the r(isobutene) = 2.5, and r(isoprene) = 0.4 (Kennedy, 1968). It should be noted, however, that as discussed previously, such r values can be markedly influenced by the nature of the initiator and solvent used in the polymerization. The values just quoted are applicable to the commercial butyl rubber process, as described earlier. It has been shown that the isoprene unit enters the chain predominantly in a 1,4-configuration (Chen and Field, 1967).

2.7.3 Living Cationic Polymerizations

The controlled/living polymerization of alkyl vinyl ethers was reported in 1984 using the HI/I_2 initiating system in nonpolar solvents (Miyamoto et al., 1984). These polymerizations produced polymers with controlled molecular weights, narrow molecular weight distributions, and number average molecular weights that increased linearly with conversion. Shortly thereafter, Faust and Kennedy (1986) reported the discovery of the living cationic polymerization of isobutylene by initiation with cumyl acetate/boron trichloride in mixtures of chlorinated solvents plus n-hexane to obtain a homogeneous polymerization. Subsequent investigations have discovered a variety of living cationic polymerization systems (Matyjaszewski, 1996; Sipos et al., 2003; Hadjikyriacou et al., 2004). For controlled/living polymerization of isobutylene, tertiary halides are generally used in conjunction with strong Lewis acid co-initiators (BF_3,$SnCl_4$, $TiCl_4$,$EtAlCl_2$, and Me_2AlCl) (Matyjaszewski, 1996; Sipos et al., 2003; Hadjikyriacou et al., 2004). A key ingredient in many of these systems is a proton trap, such as 2,6-di-*tert*-butylpyridine, to suppress initiation by protons. The general features of living polymerization systems are analogous to those of controlled radical polymerization, i.e., a predominant, unreactive dormant species (covalent species, R-X or P-X) in equilibrium with a small concentration of reactive, propagating species (cationic ion pairs, P^+X^-) as shown below.

$$\text{R-X} + \text{A (Lewis acid)} \rightleftharpoons \text{R}^{\oplus}, \text{XA}^{\ominus}$$

$$\text{R}^{\oplus}, \text{XA}^{\ominus} + \text{monomer} \xrightarrow{k_p} \text{P}^{\oplus}, \text{XA}^{\ominus}$$

$$\underset{\text{active form}}{\text{P}^{\oplus}, \text{XA}^{\ominus}} \rightleftharpoons \underset{\text{dormant form}}{\text{P-X} + \text{A}}$$

2.7.4 Other Cationic Polymerizations: Heterocyclic Monomers

Although butyl rubber is by far the most important commercial elastomer to be synthesized by cationic polymerization, several heterocyclic monomers provide useful elastomeric materials via this mechanism also. Epichlorohydrin can be polymerized to high molecular weight using a complex catalyst formed from a trialkylaluminum compound and water as shown in Eq. (2.58) (Hofmann, 1989; Body and Kyllingstad, 1986; Vandenberg, 1969,1983). For copolymerizations with ethylene oxide, a catalyst formed from a trialkylaluminum compound, water, and acetylacetone is useful (Napper and Gilbert, 1986; Body and Kyllingstad, 1986). The mechanism proposed for these polymerizations is

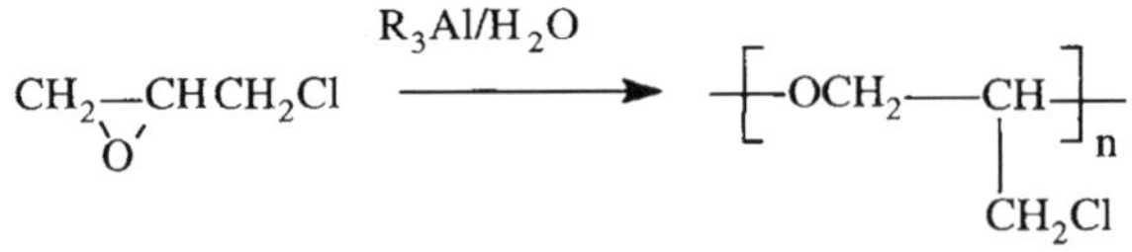

(2.58)

illustrated in Eq. (2.59), where coordination to two aluminum sites has been invoked to explain the stereochemical course of these polymerizations (Vandenberg, 1983). The average molecular weight of the homopolymer is 500,000, while the equimolar copolymer with ethylene oxide has molecular weights approaching 1,000,000 (Napper and Gilbert, 1986).

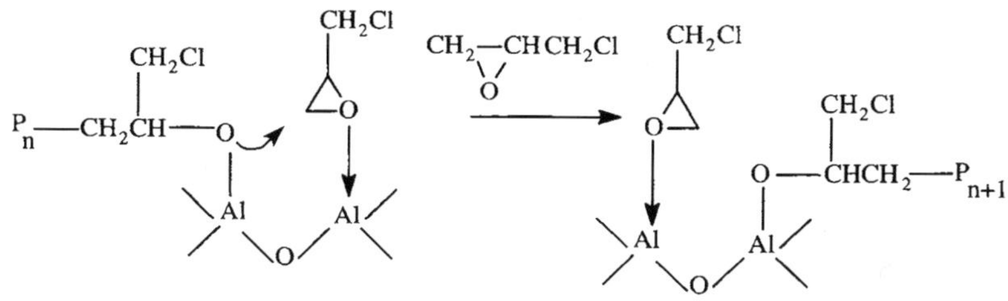

(2.59)

The cationic polymerization of tetrahydrofuran is used commercially to produce α,ω-dihydroxypoly(tetramethylene oxide) (PTMO glycol). Although this polymer is not used by itself as an elastomer, it is used as one of the elastomeric block components for preparation of segmented thermoplastic polyurethane (Meckel et al., 2004) and thermoplastic polyester (Adams et al., 2004) elastomers. The cationic polymerization of tetrahydrofuran (THF) is a living polymerization under proper experimental conditions (Inoue and Aida, 1984; Dreyfuss, 1982; Penczek et al., 1985; Kanaoka and Aoshima, 2012; Dreyfuss et al., 1989), i.e., it does not exhibit any termination step, very much like the analogous anionic polymerizations which are discussed in Section 2.8. However, these polymerizations are complicated by the fact that the ceiling temperature, where the free energy of polymerization is equal to zero, is estimated to be approximately 83 ± 2°C in bulk monomer solution (Dreyfuss and Dreyfuss, 1966); therefore, the polymerization is reversible and incomplete conversion is often observed, especially in the presence of added solvent. For example, at equilibrium in the bulk at 30°C, conversion is 72%; in a mixture of 37.5 vol.% CH_2Cl_2, conversion is only 27% (Dreyfuss, 1982). These factors limit the ability to prepare polytetrahydrofurans with controlled molecular weight and narrow molecular weight distributions which are often associated with living polymerizations. To the extent that equilibrium is approached, the polymer molecular weight distribution would broaden toward a statistical value of 2.

The most commonly used catalyst for the commercial polymerization of tetrahydrofuran is fluorosulfuric acid as shown in Eq. (2.60) (Dreyfuss et al., 1989).

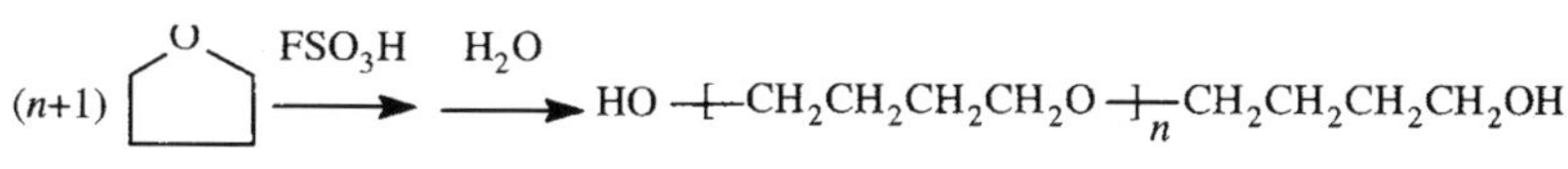

(2.60)

The mechanism of this cationic polymerization is quite different from the polymerization of isobutene (Eqs. (2.43)–(2.46)) in that the growing chain end is an oxonium ion intermediate in which the positive charge is located on oxygen atom rather than on carbon as shown in the following (Dreyfuss, 1982; Dreyfuss et al., 1989; Pruckmayr and Wu, 1978):

Initiation:

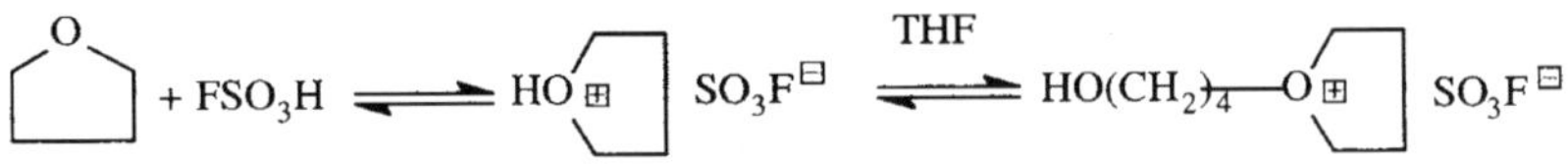

(2.61)

Propagation:

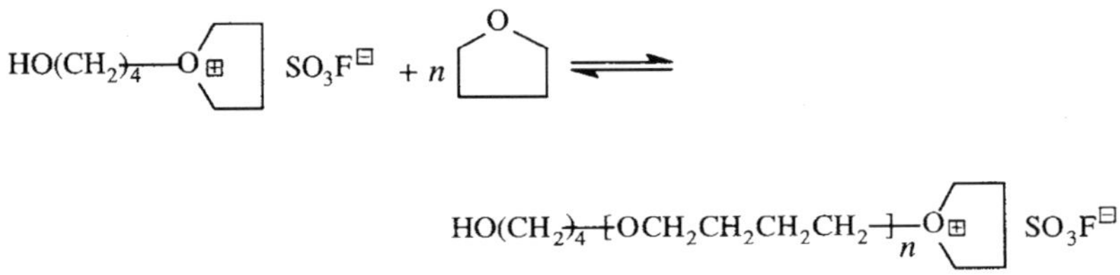

(2.62)

Another interesting aspect of this polymerization is the observation that the covalent ester is in equilibrium with the oxonium ion (Eq. (2.63)) and that both of these species can participate in propagation by reaction with monomer (Matyjaszewski et al., 1975).

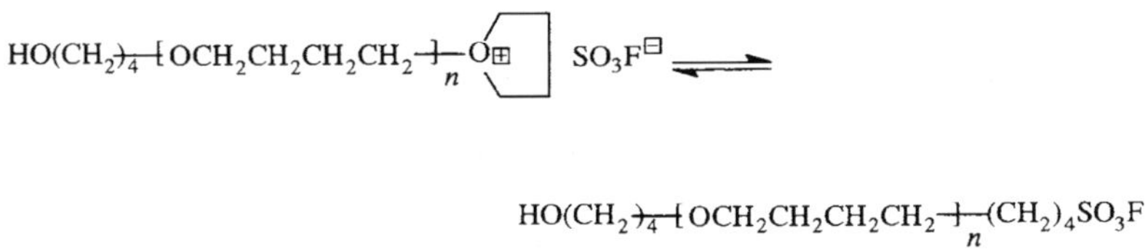

(2.63)

The commercial α,ω-dihydroxypoly(tetramethylene oxide) (PTMO glycol) polymers have molecular weights in the range of 600–3000 with molecular weight distributions in the range of 1.2–1.6 which is consistent with the equilibrium nature of these polymerizations (Dreyfuss et al., 1989).

2.8 CHAIN POLYMERIZATION BY ANIONIC MECHANISM

2.8.1 Mechanism and Kinetics

An anionic mechanism is proposed for those polymerizations initiated by alkali metal organometallic species, where there is good reason to assume that the metal is strongly electropositive relative to the carbon (or other) atom at the tip of the growing chain (Penczek and Pretula, 2012; Hsieh and Quirk, 1996; Young et al., 1984; Quirk, 2012; Morton, 1983; Szwarc, 1968; Rempp et al., 1988; Vanbeylen et al., 1988; Szwarc, 1983,1996). However, analogous to the discussion of the active species in cationic polymerization, a multiplicity of active species may be involved as propagating species in anionic polymerization as shown below (Hsieh and Quirk, 1996). In contrast to cationic polymerization, however, there is experimental evidence for the involvement of many of these species under certain experimental conditions (Szwarc, 1968,1983; Young et al., 1984).

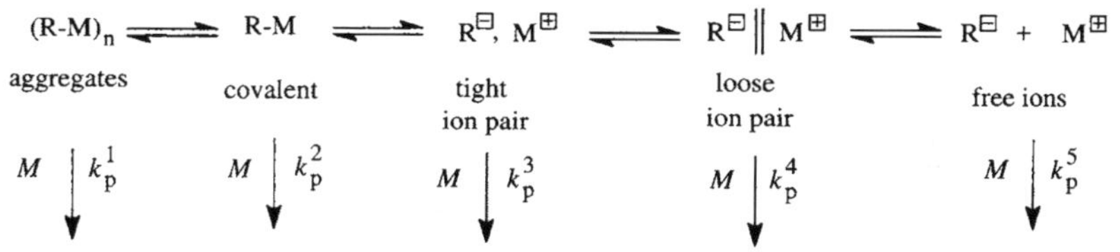

Although the ability of alkali metals, such as sodium, to initiate polymerization of unsaturated organic molecules has long been known (the earliest record dating back to the work of Matthews and Strange (1910) and Harries (1911), around 1910, on polymerization of dienes), the mechanism had remained largely obscure due to the heterogeneous character of this type of catalysis. The pioneering work of Higginson and Wooding (1952) on the homogeneous polymerization of styrene by potassium amide in liquid ammonia, and that of Robertson and Marion (1948) on butadiene polymerization by sodium in toluene, merely showed the important role of the solvent in participating in chain transfer reactions.

The true nature of homogeneous anionic polymerization only became apparent through studies of the soluble aromatic complexes of alkali metals, such as sodium naphthalene. These species are known to be radical anions (Paul et al., 1956; McClelland, 1964; Jones, 1968; Holy, 1974; Szwarc and Jagur-Grodzinski, 1972), with one unpaired electron stabilized by resonance and a high solvation energy, and are therefore chemically equivalent to a "soluble sodium." They initiate polymerization by an "electron transfer" process (Szwarc, 1968,1983; Smid et al., 2006), just as in the case of the metal itself, except that the reaction is homogeneous and therefore involves a much higher concentration of initiator. The mechanism of polymerization initiated by alkali metals (or their soluble complexes) can therefore be written as follows, using styrene as an example (Szwarc, 1968,1983):

Initiation:

$$\mathrm{Na} + \mathrm{CH_2{=}\underset{\displaystyle C_6H_5}{\underset{|}{CH}}} \rightleftharpoons \left[\mathrm{CH_2{=}\underset{\displaystyle C_6H_5}{\underset{|}{CH}}}\right]^{\dot{-}}\mathrm{Na}^{\oplus} \qquad (2.64)$$

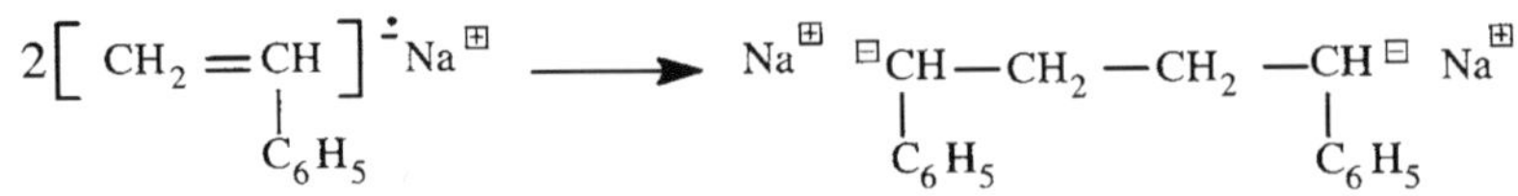

(2.65)

Propagation:

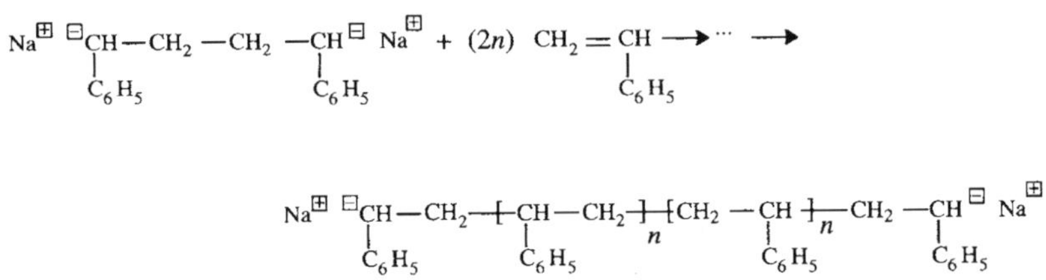

(2.66)

Thus the first step in the initiation reaction (Eq. (2.64)) involves a reversible electron transfer reaction from the alkali metal to the styrene monomer to form the styryl radical anion; in a rapid subsequent reaction, two radical anions couple to form a di-anion which can grow a polymer chain at both ends. In the case of the soluble alkali metal aromatic complexes, the overall initiation reaction is extremely fast, due to the high concentrations of radical anion ($\sim 10^{-3}$ M) and monomer (~ 1 M), and so is the subsequent propagation reaction. However, in the case of the alkali metal initiators, the electron transfer step (Eq. (2.64)) is very much slower, due to the heterogeneous nature of the reaction, so that the buildup of radical anions is much slower. In fact, there is evidence (Robertson and Marion, 1948) that, in such cases, a second electron transfer step can occur between the metal and the radical anion to form a di-anion, rather than coupling of the radical anions. In either case, the final result is a di-anion, i.e., a difunctional growing chain.

However, it was investigations of the homogeneous systems initiated by sodium naphthalene in polar solvents which demonstrated the special nature of anionic polymerization, i.e., the fact that a termination step may be avoided under certain circumstances, leading to the concept of "living" polymers (Szwarc et al., 1956). Since these are homogeneous systems, the stoichiometry of the reaction becomes apparent, i.e., two molecules of sodium naphthalene generate one chain. Furthermore, since all the chains are initiated rapidly and presumably have an equal opportunity to grow, their molecular weight distribution becomes very narrow, approximating the Poisson distribution (Flory, 1940,1953a). These aspects are obscured in the metal-initiated polymerizations owing to the continued slow initiation over a long period of time, leading to a great difference in the "age" of the growing chains and hence in their size distribution.

Polymerization initiated by electron transfer from a metal, or by an aromatic radical anion, represents only one of the anionic mechanisms. It is, of course, possible to consider separately those polymerizations initiated directly by organometallic compounds. Of the latter, the organolithium compounds are probably the best examples, since they are soluble in a wider variety of solvents and are relatively stable (Wakefield, 1974). Furthermore, it is these organometallic compounds which are used commercially for the preparation of synthetic elastomers (Hsieh et al., 1981; Hargis et al., 1987). The mechanism of these polymerizations is somewhat simpler than in the case of sodium naphthalene, since there is no electron transfer step; thus.

Initiation:

$$R^{\ominus}Li^{\oplus} + CH_2{=}CH(C_6H_5) \longrightarrow R{-}CH_2{-}CH(C_6H_5)^{\ominus}Li^{\oplus} \quad (2.67)$$

Propagation:

$$R{-}CH_2{-}CH(C_6H_5)^{\ominus}Li^{\oplus} + nCH_2{=}CH(C_6H_5) \longrightarrow R{+}CH_2{-}CH(C_6H_5){+}_n CH_2{-}CH(C_6H_5)^{\ominus}Li^{\oplus} \quad (2.68)$$

Termination by impurity or deliberate termination:

$$R{+}CH_2{-}CH(C_6H_5){+}_n CH_2{-}CH(C_6H_5)^{\ominus}Li^{\oplus} + H_2O \longrightarrow R{+}CH_2{-}CH(C_6H_5){+}_n CH_2{-}CH_2(C_6H_5) + LiOH \quad (2.69)$$

Hence each organolithium molecule generates one chain, and there is no termination of the growing chains or chain transfer reactions in the absence of adventitious impurities, such as water and acids, and if higher temperatures are avoided to prevent side reactions.

Unlike sodium naphthalene, which requires the presence of highly solvating solvents, such as tetrahydrofuran (THF), the organolithium systems can operate in various polar and nonpolar solvents such as ethers or hydrocarbons. However, the rates are much slower in the latter than in the former solvents. Hence, if the initiation reaction (Eq. (2.67)) is very much slower than the propagation reaction, the molecular weight distribution may be considerably broadened (Hsieh and McKinney, 1966). This does not, of course, vitiate the "living" polymer aspect of the polymerization, which has been shown (Morton et al., 1963) to operate in these systems, regardless of type of solvent, if side reactions do not intervene.

The absence of chain termination and chain transfer reactions in homogeneous anionic polymerization can lead to many novel synthetic routes. Thus, since each chain continues to grow when additional monomer is added, it is possible to synthesize block polymers by sequential addition of several monomers (Kahveci et al., 2012; Quirk and Kinning, 1989; Hadjichristidis et al., 2003a). Another possibility is the synthesis of linear chains with various functional end groups, by allowing the anionic polymer chain end to react with various electrophilic agents, e.g., with CO_2 to form –COOH groups (Quirk and Pickel, 2012; Quirk, 1992). In addition, linking reactions of polymer chains with multifunctional electrophilic reagents leads to the formation of "star-branched" polymers (Hsieh, 1976; Bauer and Fetters, 1978; Bywater, 1979; Hadjichristidis et al., 2001,2006). These possibilities are, of course, of considerable industrial interest.

In view of the unusual mechanism of anionic polymerization, especially the absence of termination and chain transfer reactions, the kinetics of these systems can be treated quite differently than for the other mechanisms. Thus it is possible, by suitable experimental techniques, to examine separately the rates of the initiation and propagation reactions (Bywater, 1976; Müller, 1989), since the stable organometallic chain ends are present in concentrations [10^{-3}–10^{-5} M] which are easily measured by ultraviolet-visible spectroscopy (Bywater et al., 1964). The propagation reaction is, of course, of considerable main interest and can be studied by making sure that initiation is complete. In this way, the kinetics of homogeneous anionic polymerization have been extensively elucidated with special reference to the nature of counterion and role of the solvent.

It has been found universally that, in accordance with Eqs. (2.66) and (2.68), the propagation rate is always first order with respect to monomer concentration, regardless of solvent system or counterion. However, in contradiction to the foregoing equations, the propagation rate dependency has generally been found to be *lower* than first order with respect to the concentration of growing chains, and the order was found to be strongly dependent on the nature of the solvent and counterion (Szwarc, 1983; Bywater, 1976; Müller, 1989; Bywater et al., 1964; Morton, 1969). Strongly solvating solvents, such as ethers and amines, lead to much faster rates than nonpolar solvents and affect the kinetics of these polymerizations quite differently than the hydrocarbon media, because more dissociated ionic species such as loose ion pairs and free ions are involved as propagating species (Szwarc, 1983). However, since the anionic synthesis of elastomers requires the use of lithium as counterion in hydrocarbon media, the following discussion will focus on the kinetics of these processes.

It would be expected that the kinetics of organolithium-initiated polymerization in hydrocarbon solvents would be simplified because of the expected correspondence between the initiator concentration and the concentration of propagating anionic species, resulting from the lack of termination and chain transfer reactions. However, in spite of intensive study, there is no general agreement on many kinetic aspects of these polymerizations. The complicating feature is that organolithium compounds are associated into aggregates in hydrocarbon solution, and the degree of aggregation depends on the structure of the organolithium compound, the concentration of organolithium compound, the solvent, and the temperature (Young et al., 1984; Wakefield, 1974; Brown, 1968; Wardell, 1982; Quirk and Monroy, 1995). In general, simple alkyllithium compounds are associated into hexamers or tetramers in hydrocarbon solution.

The kinetics of initiation for styrene and diene polymerization by alkyllithium compounds generally exhibit a fractional kinetic order dependence (e.g., 1/4 or 1/6) on the concentration of alkyllithium initiator. This can be rationalized in terms of the following steps:

Initiation:

$$(\mathrm{RLi})_n \underset{}{\overset{K_n}{\rightleftharpoons}} n\,\mathrm{RLi} \tag{2.70}$$

$$\mathrm{RLi} + \mathrm{CH_2{=}CH(C_6H_5)} \xrightarrow{k_i} \mathrm{R{-}CH_2{-}CHLi(C_6H_5)} \tag{2.71}$$

Thus, it is assumed that only the unassociated alkyllithium compound (formed by dissociation of the aggregate, Eq. (2.70)) reacts with monomer in the initiation step (Eq. (2.71)) so the rate of initiation can be expressed by Eq. (2.72).

$$R_{\mathrm{i}} = k_{\mathrm{i}}[\mathrm{RLi}][M]. \tag{2.72}$$

The equilibrium concentration of unassociated alkyllithium can be expressed in terms of Eq. (2.73).

$$[\mathrm{RLi}] = K^{1/n}[(\mathrm{RLi})_n]^{1/n}. \tag{2.73}$$

When this expression for [RLi] is substituted into Eq. (2.72), Eq. (2.74) is obtained.

$$R_{\mathrm{i}} = k_{\mathrm{i}}K^{1/n}[(\mathrm{RLi})_n]^{1/n}[M]. \tag{2.74}$$

A good example of this kinetic behavior was found in the study of the *n*-butyllithium-styrene system in benzene, in which a kinetic order dependency on *n*-butyllithium concentration was observed, consistent with the predominantly hexameric degree of association of *n*-butyllithium (Worsfold and Bywater, 1960). However, this expected correspondence between the degree of association of the alkyllithium compound and the fractional kinetic order dependence of the initiation reaction on alkyllithium concentration was not always observed (Young et al., 1984). One source of this discrepancy is the assumption that only the unassociated alkyllithium molecule can initiate polymerization. With certain reactive initiators, such as *sec*-butyllithium in hexane solution, the initial rate of initiation exhibits approximately a first-order dependence on alkyllithium concentration, suggesting that the aggregate can react directly with monomer to initiate polymerization (Bywater and Worsfold, 1967a). A further source of complexity is the cross-association of the initiator with the initiated polymer chain; in general, the cross-associated species exhibits a different degree of association and reactivity from the alkyllithium initiator (Schue and Bywater, 1969; Morton et al., 1970a). As a result of cross-association, only the initial rates of initiation can be used to determine the kinetic order dependence on initiator concentration. Unfortunately, these considerations have not always been recognized. It is interesting to note that the general reactivity of alkyllithiums as initiators is inversely related to the degree of aggregation (Selman and Hsieh, 1971), i.e., *sec*-butyllithium (tetramer) >*n*-butyllithium (hexamer) (Bywater and Worsfold, 1967a).

The kinetics of the propagation reaction in organolithium polymerization of styrenes and dienes in nonpolar solvents (i.e., hydrocarbons) have also been subjected to intensive study. For styrene polymerizations, a kinetic order dependence on chain end concentration is observed (Eq. (2.75)). Since it has been

$$R_p = k_p K_n^{1/2}[(\text{PsLi})_2]^{1/2}[M] \tag{2.75}$$

determined that poly(styryl)lithium is associated predominantly into dimers in hydrocarbon solution (Young et al., 1984). However, it is noteworthy that recent neutron scattering results have provided evidence for the presence of small amounts of higher-order aggregates (N > 100) in equilibrium with predominantly dimeric and some tetrameric species (Stellbrink et al., 1998). Thus, the observed kinetic order can be explained in terms of Eqs. (2.76 and 2.77), using the same reasoning as delineated for the initiation kinetics (Eqs. (2.70)–(2.74)). This explanation is based on the assumption that *only* the dissociated chain ends are active. However, evidence for the contribution of higher-order aggregates (>2) to the propagation kinetics has been reported from a 1H NMR study of PSLi propagation in cyclohexane (Mishima et al., 2008)

$$(\text{PsLi})_2 \underset{}{\overset{K_n}{\rightleftharpoons}} 2\ \text{PsLi} \tag{2.76}$$

$$\text{PsLi} + \text{CH}_2{=}\underset{\substack{|\\ \text{C}_6\text{H}_5}}{\text{CH}} \xrightarrow{k_p} \text{Ps—CH}_2\text{—}\underset{\substack{|\\ \text{C}_6\text{H}_5}}{\text{CHLi}} \tag{2.77}$$

The propagation kinetic order dependence on poly(dienyl)lithium chain end concentration for alkyllithium-initiated polymerization of dienes varies from 1/4 to 1/6 for butadiene and from 1/2 to 1/4 for isoprene (Young et al., 1984; Quirk, 2012; Bywater, 1976; Müller, 1989; Bywater and Worsfold, 1987). However, attempts to relate these kinetic orders to proposed higher states of association of poly(dienyl)lithium chain ends have proven to be complicated by the lack of agreement regarding the predominant degree of association of these species in hydrocarbon solution (Hsieh and Quirk, 1996). However, recent evaluation of the association states of poly(dienyl)lithium chain ends in benzene by small-angle neutron scattering, as well as both dynamic and static light scattering, indicates that predominantly tetrameric aggregates are in equilibrium with small amounts of higher-order aggregates ($n > 100$) (Niu et al., 2004; Stellbrink et al., 2002; Miyamoto et al., 2006; Matsuda et al., 2005,2007; Oishi et al., 2006,2007). Thus, the generally observed 0.25 kinetic order dependence on poly(dienyl)lithium chain-end concentration can be interpreted in terms of a predominantly tetrameric degree of aggregation.

In conclusion, it should be noted that the molecular weights and their distribution follow the rules originally discussed under living polymers (Quirk and Lee, 1992). This means that, regardless of the solvents and counterions used, if no termination, chain transfer, or side reactions occur, and if the initiation reaction is fast relative to the propagation reaction, then the stoichiometric molecular weight will be obtained and the molecular weight distribution will approach the Poisson distribution, i.e.,

$$X_w/X_n = 1 + 1/X_n, \tag{2.78}$$

where X_n is the number-average number of monomer units and X_w is the weight-average number of monomer units. This means that, in principle, a polymer chain of 100 units should have an X_w/X_n ratio of 1.01. This is, of course, impossible to prove experimentally, and it can be assumed that the real distribution would be somewhat broader, due for one thing to imperfect mixing in the reaction mixture. However, values of 1.05 for X_w/X_n are commonly found in these systems (Morton and Fetters, 1967; Fetters and Morton, 1974).

2.8.2 Chain Microstructure of Polydienes

Although the alkali metals, unlike the Ziegler-Natta systems, do not generally polymerize unconjugated olefins and are not known to lead to any tacticity, they do affect the chain microstructure of polydienes. Thus, the proportion of *cis*-1,4 and *trans*-1,4 addition versus the 1,2 (and 3,4 for polyisoprene) mode can be markedly affected by the nature of the counterion as well as the solvent. Ever since the discovery that lithium polymerization of isoprene can lead to a high *cis*-1,4 structure (Stavely et al., 1956), close to that of natural rubber, there have been many studies of these effects (Sipos et al., 2003; Bywater, 1976,1979; Hadjichristidis et al., 2001,2006). Table 2.14 shows some of these results for anionic polymerization of isoprene and butadiene. It is obvious from these data that the stereospecific high *cis*-1,4 polyisoprene is obtained only in the case of lithium in hydrocarbon solvents; the highest *cis* microstructure is also favored by high ratios of monomer to chain end (Worsfold and Bywater, 1978; Bywater, 1989; Morton and Rupert, 1983; Tobolsky and Rogers, 1959; Rembaum et al., 1962; Bywater and Worsfold, 1967b). Other solvents and/or counterions exert a dramatic effect in altering the chain microstructure to form 1,2 and 3,4 enchainments. Similar effects are observed with butadiene and other dienes (Hsieh and Quirk, 1996; Young et al., 1984; Foreman, 1968). However, in the

TABLE 2.14 Microstructure of Polydienes Prepared by Anionic Polymerization

Solvent	Chain microstructure (mole %)					
	Cation	*cis*-1,4	*trans*-1,4	1,2	3,4	Reference
Butadiene						
Hexane[a]	Li^+	30	62	8	–	Morton and Rupert (1983)
Cyclohexane[b]	Li^+	68	28	4	–	Morton and Rupert (1983)
None	Li^+	86	9	5	–	Morton and Rupert (1983)
Tetrahydrofuran (THF)	Li^+	6	6	88	–	Tobolsky and Rogers (1959)
Pentane	Na^+	10	25	65	–	Tobolsky and Rogers (1959)
THF	Na^+	0	9.2	90.8	–	Rembaum et al. (1962)
Pentane	K^+	15	40	45	–	Tobolsky and Rogers (1959)
Pentane	Rb^+	7	31	62	–	Tobolsky and Rogers (1959)
Pentane	Cs^+	6	35	59	–	Tobolsky and Rogers (1959)
Isoprene						
Cyclohexane[b]	Li^+	94	1	–	5	Worsfold and Bywater (1978)
Cyclohexane[a]	Li^+	76	19	–	5	Worsfold and Bywater (1978)
None	Li^+	96	0	–	4	Morton and Rupert (1983)
THF	Li^+	12[c]		29	59	Bywater and Worsfold (1967b)
Cyclohexane	Na^+	44[c]		6	50	Bywater (1989)
THF	Na^+	11[c]		19	70	Bywater (1989)
Cyclohexane	K^+	59[c]		5	36	Bywater (1989)
Cyclohexane	Cs^+	69[c]		4	27	Bywater (1989)

[a] *At monomer/initiator ratio of* ~17.
[b] *At monomer/initiator ratio of* 5×10^4.
[c] *Total of cis and trans forms.*

case of butadiene, the maximum *cis*-1,4 content attainable is much less than for isoprene; typical commercial polybutadienes prepared in hydrocarbon solution with butyllithium initiators have microstructures in the range of 36–44% *cis*-1,4, 48–50% *trans*-1,4, and 8–10% 1,2 microstructure (Duck and Locke, 1977). The effect of polar solvents, or of the more electropositive alkali metals, is to produce a high-1,2 polybutadiene.

This marked sensitivity of the stereochemistry of anionic polymerization to the nature of the counterion and solvent can be traced to the structure of the propagating chain end. The latter involves a carbon-metal bond which can have variable characteristics, ranging all the way from highly associated species with covalent character to a variety of ionic species (Hsieh and Quirk, 1996). The presence of a more electropositive metal and/or a cation-solvating solvents, such as ethers, can effect a variety of changes in the nature of the carbanionic chain end: (a) the degree of association of the chain ends can decrease or be eliminated; (b) the interaction of the cation with the anion can be decreased by cation solvation; (c) a more ionic carbon-metal bond will increase delocalization of the π electrons; and (d) polar solvents will promote ionization to form ion pairs and free ions. Direct evidence for these effects has been obtained from concentrated solution measurements (Morton and Fetters, 1964; Morton et al., 1970b), ^{1}H and ^{13}C NMR spectroscopy (Young et al., 1984; Santee et al., 1973), ultraviolet-visible spectroscopy (Young et al., 1984; Bywater and Worsfold, 1987; Hogen-Esch, 1978), and electrolytic conductance (Szwarc, 1968) measurements.

The control of chain structure and molecular weight afforded by the organolithium polymerization of dienes has, of course, been of great technological interest (Hsieh et al., 1981; Hargis et al., 1987; Halasa, 1981). Such product developments have been mainly in the form of (1) polybutadiene elastomers of various chain structures (Hargis et al., 1987; Duck and Locke, 1977; Halasa, 1981), functional end groups (Nagata et al., 1987), (2) liquid polybutadienes (Luxton, 1981), (3) butadiene-styrene copolymers (solution SBR (Demirors, 2003; Hsieh et al., 1981; Hargis et al., 1987; Halasa, 1981)), and (4) styrene-diene triblock copolymers (thermoplastic elastomers) (Holden and Hansen, 2004).

2.8.3 Copolymers of Butadiene

The possibilities inherent in the anionic copolymerization of butadiene and styrene by means of organolithium initiators, as might have been expected, have led to many new developments. The first of these would naturally be the synthesis of a butadiene-styrene copolymer to match (or improve upon) emulsion-prepared SBR, in view of the superior molecular weight control possible in anionic polymerization. The copolymerization behavior of butadiene (or isoprene) and styrene is shown in Table 2.15 (Ohlinger and Bandermann, 1980; Morton and Huang, 1979; Ells, 1963; Hill et al., 1983; Spirin et al., 1962). As indicated earlier, unlike the free radical type of polymerization, these anionic systems show a marked sensitivity of the reactivity ratios to solvent type (a similar effect is noted for different alkali metal counterions). Thus, in nonpolar solvents, butadiene (or isoprene) is preferentially polymerized initially, to the virtual exclusion of the styrene, while the reverse is true in polar solvents. This has been ascribed (Morton, 1983) to the profound effect of solvation on the structure of the carbon-lithium bond, which becomes much more ionic in such media, as discussed previously. The resulting polymer formed by copolymerization in hydrocarbon media is described as a tapered block copolymer; it consists of a block of polybutadiene with little incorporated styrene comonomer followed by a segment with both butadiene and styrene and then a block of polystyrene. The structure is schematically represented below:

[butadiene]-[B/S]-[styrene]

The data in Table 2.15 illustrate the problems encountered in such copolymerizations, since the use of polar solvents to assure a random styrene-diene

TABLE 2.15 Monomer Reactivity Ratios for Organolithium Copolymerization of Styrene and Dienes

Monomer 1	Monomer 2	Solvent	r_1	r_2	Reference
Styrene	Butadiene	Toluene	0.004	12.9	Ohlinger and Bandermann (1980)
Styrene	Butadiene	Benzene	0.04	10.8	Morton and Huang (1979)
Styrene	Butadiene	Triethylamine	0.5	3.5	Morton and Huang (1979)
Styrene	Butadiene	Tetrahydrofuran	4.0	0.3	Morton and Huang (1979)
Styrene	Isoprene	Benzene	0.26	10.6	Ells (1963)
Styrene	Isoprene	Tetrahydrofuran	9.0	0.1	Spirin et al. (1962)
Butadiene	Isoprene	Hexane	1.72	0.36	Hill et al. (1983)

copolymer of desired composition will, at the same time, lead to an increase in side vinyl groups (1,2 or 3,4) in the diene units (see Table 2.14). This is of course quite undesirable, since such chain structures result in an increase in the glass transition temperature (T_g) and therefore to a loss of good rubbery properties. Hence, two methods are actually used to circumvent this problem: (1) the use of limited amounts of polar additives such as tetrahydrofuran to accomplish a reasonable compromise between diene structure and monomer sequence distribution (Antkowiak et al., 1972); and (2) the addition of small amounts of potassium *t*-alkoxides (Wofford and Hsieh, 1969).

As mentioned earlier, the "living" nature of the growing chain in anionic polymerization makes this mechanism especially suitable for the synthesis of *block* copolymers, by sequential addition of *different* monomers. Since such copolymers have markedly different properties than simple copolymers, they will be discussed separately (in Section 2.10).

2.8.4 Terminally Functional Polydienes

Another characteristic of these homogeneous anionic polymerizations, as mentioned earlier, is their potential for the synthesis of polymer chains having reactive end groups. It was recently reported that chain end functionalization of high-molecular-weight polybutadiene and solution styrene-*co*-butadiene elastomers (SBR) with a derivative of Michler's ketone, 4,4′-bis(diethylamino) benzophenone, leads to tire tread formulations which have lower rolling resistance and good wet-skid resistance (Nagata et al., 1987). These effects were observed in spite of the low concentration of chain ends in these polymers (molecular weights >100,000) (Hargis et al., 1987).

The production of liquid short-chain *difunctional* polymers by anionic polymerization is of considerable technological interest and importance, and has attracted much attention in recent years, since it offers an analogous technology to that of the polyethers and polyesters used in urethane polymers. Such liquid "telechelic" polydienes could thus lead, by means of chain extension and crosslinking reactions, directly to "castable" polydiene networks (Hsieh et al., 1981,1987).

The most direct method of preparing telechelic polydienes utilizes a dilithium initiator which is soluble in hydrocarbon solution (Penczek and Pretula, 2012; Bandermann et al., 1985). The most expedient method of preparing such a dilithium initiator is to react 2 moles of an alkyllithium compound with a divinyl compound which will not homopolymerize. Unfortunately, because of the association behavior of organolithium compounds in hydrocarbon media (Wakefield, 1974; Brown, 1968; Wardell, 1982), many potential systems fail because they associate to form an insoluble network-like structure (Bandermann et al., 1985). Expediencies such as addition of Lewis bases can overcome solubility problems of dilithium initiators, however, such additives tend to produce high amounts of 1,2- and 3,4-microstructures (see Table 2.4). One exception is the adduct formed from the addition of two equivalents of *sec*-butyllithium to 1,3-*bis*(1-phenylethenyl)benzene as shown in Eq. (2.79) (Long et al., 1989; Tung and Lo, 1986). Although this is a hydrocarbon-soluble, dilithium initiator, it was found that bimodal molecular weight distributions are obtained; monomodal distributions can be obtained in the presence of lithium alkoxides or by addition of Lewis base additives (Quirk and Ma, 1991; Quirk et al., 2000). This initiator has also been used to prepare telechelic polymers in high yields (Quirk and Xu, unpublished data).

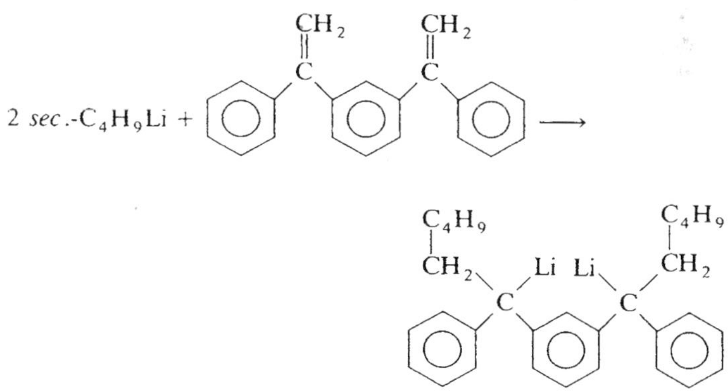

(2.79)

2.9 STEREOSPECIFIC CHAIN POLYMERIZATION AND COPOLYMERIZATION BY COORDINATION CATALYSTS

2.9.1 Mechanism and Kinetics

The term "Ziegler-Natta catalysts" refers to a wide variety of polymerization initiators generally formed from mixtures of transition metal salts of Group

IV to VIII metals and base metal alkyls of Group II or III metals (Penczek and Pretula, 2012; Boor, 1979; Ciardelli, 1992). It arose from the spectacular discovery of Ziegler et al. (1955) that mixtures of titanium tetrachloride and aluminum alkyls polymerize ethylene at low pressures and temperatures; and from the equally spectacular discovery by Natta (1955) that the Ziegler catalysts can stereospecifically polymerize monoolefins to produce tactic, crystalline polymers. As can be imagined, these systems can involve many combinations of catalyst components, not all of which are catalytically active or stereospecific. However, we shall be concerned here only with polymerizations involving the commercial elastomers, principally polyisoprene, polybutadiene (Duck and Locke, 1977; Zohuri et al., 2012; Teyssie et al., 1988), and the ethylene-propylene copolymers (Schöbel et al., 2012; Ver Strate, 1986; Davis et al., 1996; Noordermeer, 2003; Baldwin and Strate, 1972).

The mechanism of polymerization of alkenes using Ziegler-Natta-type catalysts is described as a coordination (Vandenberg, 1986) or insertion (Pino et al., 1987) polymerization process. The coordination terminology assumes that the growing polymer chain is bonded to a transition metal atom and that insertion of the monomer into the carbon-metal bond is preceded by, and presumably activated by, the coordination of the monomer with the transition metal center. Since coordination of the monomer may or may not be a specific feature of these polymerizations, the insertion terminology focuses on the proposal that these reactions involve a stepwise insertion of the monomer into the bond between the transition metal atom and the last carbon atom of the growing chain. It is important to note that the bonding of carbon atoms and transition metals is described as substantially covalent (Pino et al., 1987), in contrast to anionic organometallic species, such as organoalkali metal species, which are highly ionic.

Typical soluble catalysts for copolymerization of ethylene and propylene are formed from mixtures of vanadium salts with alkylaluminum chlorides, e.g., VCl_4 with either AlR_2Cl or $AlRCl_2$ where R = alkyl group (Ver Strate, 1986). A possible hexacoordinated metal structure for the resulting active catalyst is shown below.

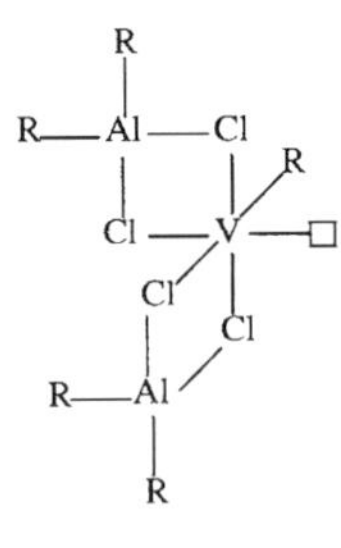

The important features of the active center in accord with the general model of Arlman and Cossee (1964) are: (1) an alkylated vanadium center, i.e., an R–V bond; and (2) an empty orbital on vanadium, represented by—h in the structure, which can be used to bond to the incoming monomer; and an oxidation state of +3 for vanadium (Zambelli and Tosi, 1974; Zambelli and Allegra, 1980; Corradini et al., 1985).

The formation of the active catalytic center from the reaction of the transition metal compound and an organoaluminum derivative is shown schematically in Eq. (2.80). Reduction to a lower valence state may accompany this alkylation reaction since it is generally considered that the active catalytic center has an oxidation state of +3 (Ver Strate, 1986).

Active center formation:

(2.80)

The steps involved in the chain polymerization of alkenes using this type of catalyst are shown in Eqs. (2.81)–(2.85) (Ver Strate, 1986).

Initiation:

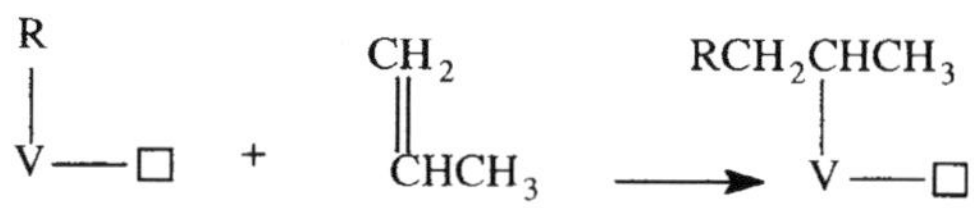

(2.81)

Propagation:

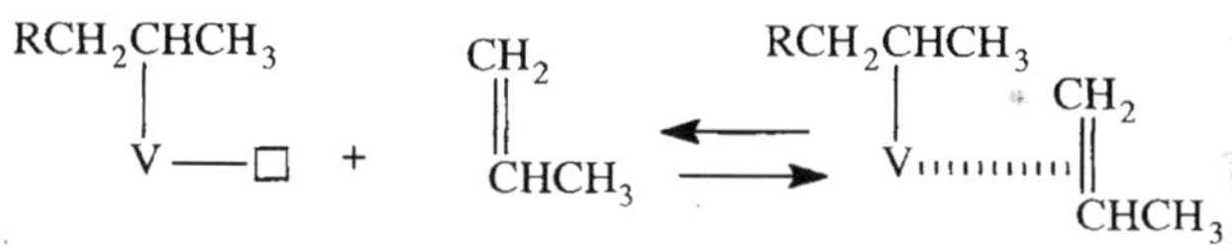

(2.82)

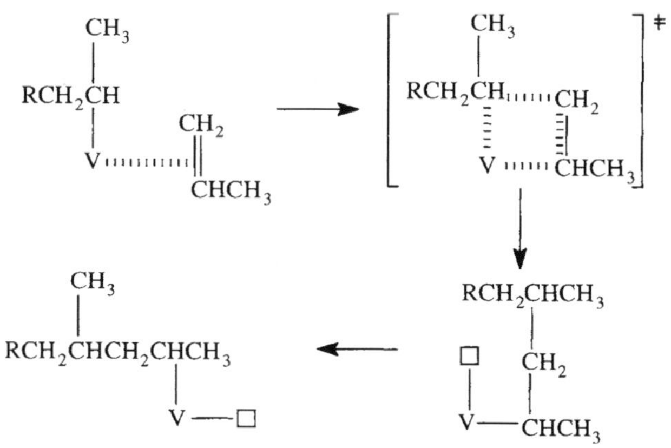

(2.83)

Termination:

$$\text{P-V} \longrightarrow \underset{\text{inactive polymer chain}}{\text{P}} + \underset{\text{inactive catalytic center}}{\text{V}_i} \quad (2.84)$$

Spontaneous transfer:

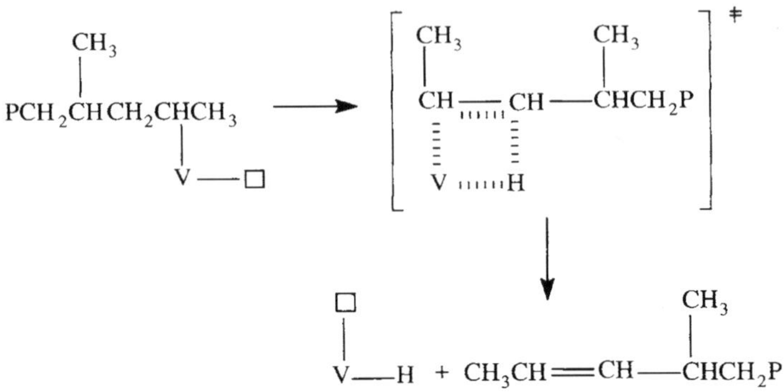

(2.85)

It should be noted that the monomer coordination step shown in Eq. (2.82) may not be a distinct step as discussed previously. An important feature of this mechanism which affects the stereospecificity of olefin polymerizations using these types of soluble catalysts is the fact that the insertion of the monomer into the transition metal-carbon bond involves a secondary insertion reaction, i.e., the more substituted carbon of the double bond in the monomer becomes bonded to the transition metal (Corradini et al., 1985). In contrast, a primary insertion mechanism to form a transition metal bond to the less substituted carbon on the double bond of the monomer Ti-CH_2CHR-P is involved in polymerizations using typical heterogeneous catalysts, e.g., from titanium halides and alkylaluminum compounds (Boor, 1979).

One of the models proposed to explain the stereospecificity for soluble vanadium-based catalysts postulates that it is the minimization of steric effects in the four-center transition state for monomer insertion (see Eq. (2.83)) which is responsible for the stereospecificity of the polymerization (Zambelli and Tosi, 1974; Corradini et al., 1989). Thus, it is considered that the *trans*-configuration minimizes steric effects in the transition state and this leads to a syndiotactic configuration of the polymer chain as shown below. In general, the kinetics

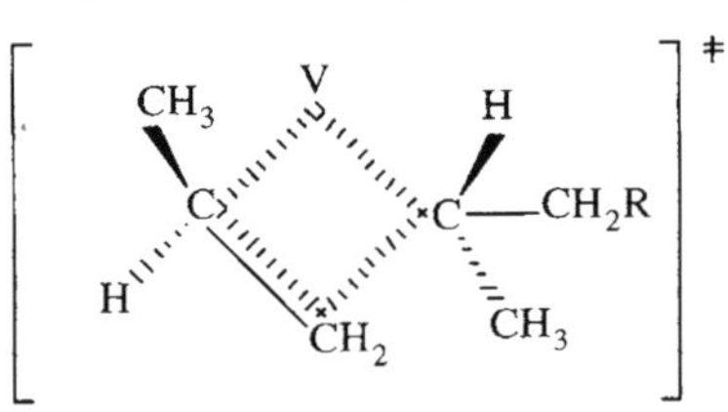

favored *trans* configuration

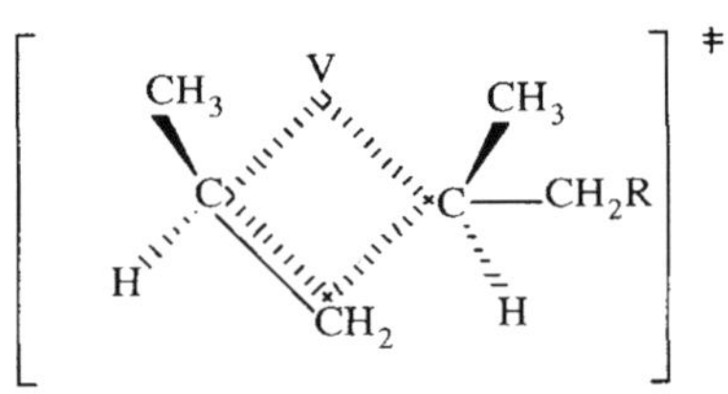

more hindered *cis* configuration

of alkene polymerizations using Ziegler-Natta-type catalysts are complicated by the multiplicity of active species, catalyst aging and deactivation effects, multiplicity of chain transfer processes, and often by the relatively rapid rates of polymerization (Boor, 1979; Schöbel et al., 2012).

2.9.2 Ethylene-Propylene Rubbers

The copolymerization of propylene with ethylene is complicated by the very unfavorable monomer reactivity ratios for propylene and other monomers with ethylene as shown in Table 2.16. In general, the less hindered ethylene monomer is favored in Ziegler-Natta copolymerizations by as much as two orders of magnitude for certain catalyst combinations. To obtain homogeneous copolymers, continuous processes are required utilizing incomplete conversions of the propylene comonomer (Ver Strate, 1986). A further aspect of the commercial preparation of ethylene-propylene rubbers is the inclusion of a third diene comonomer which introduces unsaturation into the final polymer to facilitate peroxide crosslinking reactions and to permit sulfur vulcanization; these terpolymers are called EPDM in contrast to the binary copolymers, which are designated as EPM. The following nonconjugated diene monomers are used commercially because they generate side-chain unsaturation rather than in-chain unsaturation which could lead to oxidative chain scission:

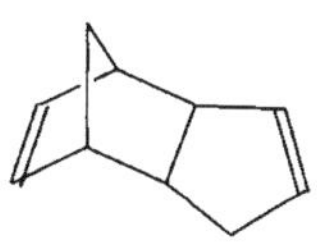

Dicyclopentadiene

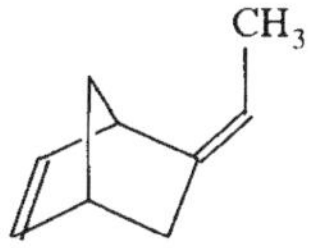

Ethylidene norbornene

$CH_2{=}CH{-}CH_2{-}CH{=}CH{-}CH_3$

1,4-Hexadiene

The compositions for more than 150 grades of EPDM clastomers are in the ranges of 40–90 mol% ethylene and 0–4 mol% diene (Ver Strate, 1986)). Thus, the structure of a typical EPDM elastomer with ethylidene norbornene as

TABLE 2.16 Monomer Reactivity Ratios for Copolymerization of Ethylene (M_1) and Propylene (M_2) with Ziegler-Natta Catalysts

Catalyst	Cocatalyst	Temperature (°C)	r_1	r_2
VCl_4	$Al(C_2H_5)_2Cl$	−10	13.7	0.021
		21	3.0	0.073
$VOCl_3$	$Al(i-C_4H_9)_2Cl$	30	16.8	0.052
$V(acac)_3$[a]	$Al(i-C_4H_9)_2Cl$	30	16	0.04
$\gamma-TiCl_3$	$Al(C_2H_5)_2Cl$	60	~8	0.05

[a]*Vanadium acetylacetonate.*

termonomer can be represented by the following structure:

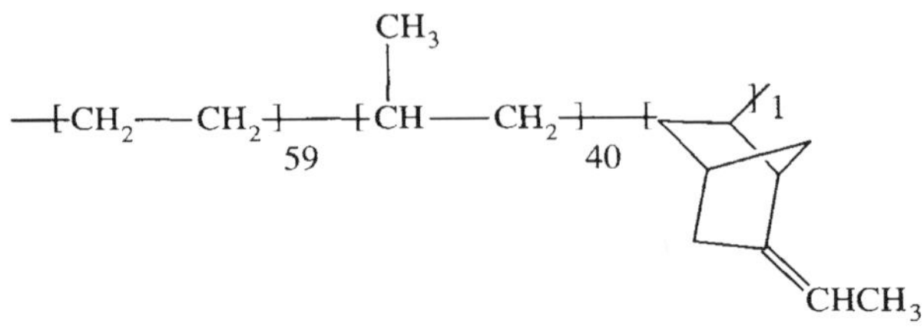

The compositions of EPDM elastomers are controlled by using the appropriate monomer feed ratio (see Eq. (2.38)) to obtain the desired composition in a continuous polymerization process. In general the excess propylene required is recycled. The molecular weights of EPDM polymers are controlled primarily by chain transfer reactions with added molecular hydrogen (Eqs. (2.86) and (2.87)), as is common with other Ziegler-Natta polymerizations (Boor, 1979).

$$\text{P-V} + \text{H}_2 \xrightarrow{k_{\text{tr}}} \text{P-H} + \text{V-H} \tag{2.86}$$

$$\text{V-H} + \text{CH}_2{=}\text{CHCH}_3 \xrightarrow{k_{\text{i}}} \text{V-}\underset{\substack{|\\ \text{CH}_3}}{\text{CH}}\text{—CH}_3 \tag{2.87}$$

In the past 20 years, there has been a revolution in the field of Ziegler-Natta and related catalysts for olefin polymerization. This revolution has resulted from the discovery of single-site, homogeneous metallocene catalysts that exhibit higher activity and the ability to readily incorporate more hindered comonomers with ethylene more uniformly along the polymer chain (Wasserman, 2003; Brintzinger et al., 1995). Metallocene catalysts contain one or two cyclopentadienyl rings coordinated to a transition metal such as titanium, zirconium, or hafnium. The higher activity of metallocene catalysts means that processes can be designed without the need for a catalyst removal step. The structure of a high activity, single-site metallocene catalyst is shown below. It is noteworthy that a metallocene cation is the proposed catalytically active species. For this type of catalyst generated with a different counterion, the monomer reactivity ratios for copolymerization of ethylene and propylene are r(ethylene) $= 1.35$ and r(propylene) $= 0.82$, indicating an almost random copolymerization behavior (Galimberti et al., 1999). These results can be compared with the copolymerization parameters for standard Ziegler-Natta catalysts in Table 2.16.

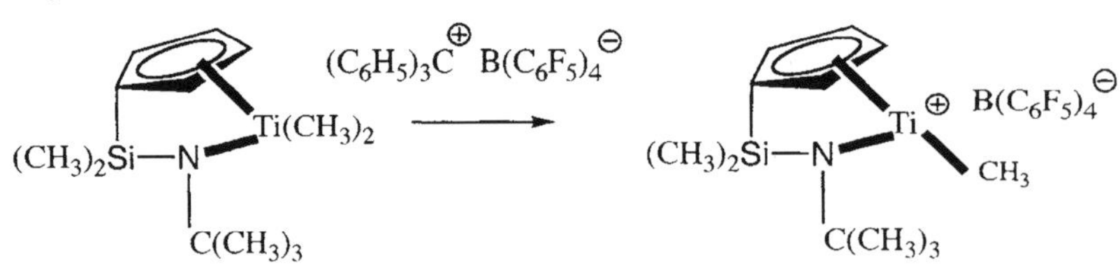

2.9.3 Polydienes

Shortly after the discovery of the Ziegler-Natta catalysts, it was found that analogous transition metal catalysts could also effect the stereospecific polymerization of dienes (Horne et al., 1956). The wide range of stereoregular polybutadienes which can be prepared with these catalysts is indicated in Table 2.17 (Porri and Giarrusso, 1989; Horne, 1983). The stereochemistry of polymerization is dependent upon the transition metal salt, the metal alkyl, temperature, additives, and the stoichiometry of the components. Commercial polybutadienes with high *cis*-1,4-microstructure are prepared using a wide range of transition metal catalysts, of which the most important are those derived from cobalt, nickel, neodymium, and titanium, analogous to those listed in Table 2.17.

The mechanism of stereospecific polymerization of 1,3-dienes is also categorized as an insertion polymerization and simplified representations of the stereoselectivity for *cis* (Eq. (2.88)) and *trans* (Eq. (2.89)) enchainments are shown below (Zohuri et al., 2012).

cis-stereospecificity

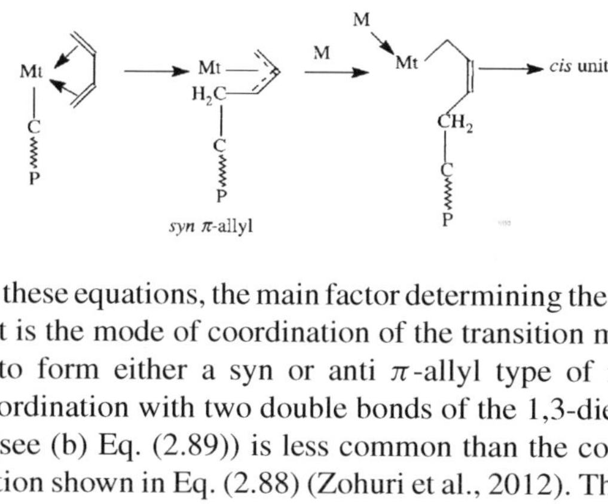

(2.88)

As indicated in these equations, the main factor determining the stereochemistry of enchainment is the mode of coordination of the transition metal center with the monomer to form either a syn or anti π-allyl type of intermediate. In general, the coordination with two double bonds of the 1,3-diene in an *s-trans* configuration (see (b) Eq. (2.89)) is less common than the coordination in an *s-cis* configuration shown in Eq. (2.88) (Zohuri et al., 2012). This interpretation is complicated by the fact that the syn and anti π-allyl complexes are in equilibrium. These simple

trans-stereospecificity

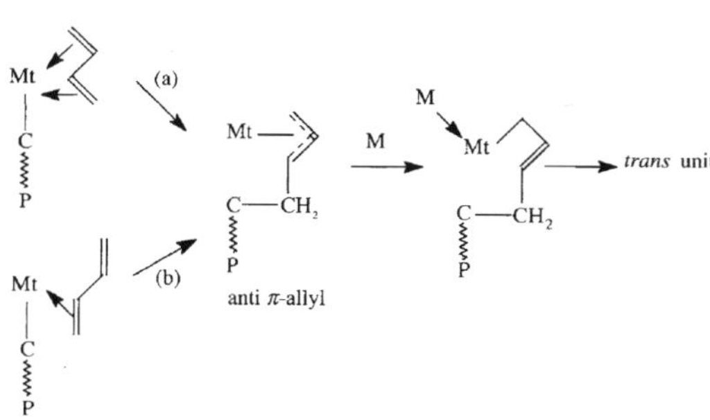

(2.89)

TABLE 2.17 Microstructure of Polydienes from Transition Metal-Initiated Polymerization[a]

Transition Metal salt/metal alkyl	Chain Microstructure (mole %)			
	cis-1,4	*trans*-1,4	1,2	3,4
		Isoprene		
$TiCl_4/AlR_3$ (1/1)	97			3[b]
α-$TiCl_3/Al(C_2H_5)_3$		98–100		
$Ti(OR)_4/AlR_3$ (1/7–8)	36			
		Butadiene		
$TiI_4/Al(i\text{-}C_4H_9)_3$ (1/4–5)	92–93	2–3	4–6	
Ni(octanoate)/$Al(C_2H_5)_3/BF_3$ (1/17/15)	96–97	2–3		
$CoCl_2/Al(C_2H_5)_2Cl$/pyridine·H_2O(1/1000/100)	98	1	1	
$NdCl_3/Al(i\text{-}C_4H_9)_3 \cdot nL$[c]	97	2.7	0.3	
$VCl_3/Al(C_2H_5)_3$		99–100		
$Co(acac)_3/AlR_3/CS_2$			99–100[d]	

[a]*Data taken from Porri and Giarrusso (1989).*
[b]*Horne (1983).*
[c]*L = Electron donor such as tetrahydrofuran or pyridine.*
[d]*Syndiotactic.*

mechanistic representations are reinforced by the observations that the stereochemistry of diene polymerizations can be altered by the addition of electron donors such as $N(C_2H_5)_3$,$P(OC_6H_5)_3$, or C_2H_5OH. Thus, addition of these electron donors changes the stereochemistry from highly *cis*-stereospecific to highly *trans*-stereospecific for butadiene with catalysts such as $Co(acac)_2/Al(C_2H_5)_2Cl$ (Zohuri et al., 2012). This is explained by assuming that the electron donor occupies one of the two coordination sites required for *cis* enchainment (see Eq. (2.88)) which forces the monomer to only coordinate with one site ((b) in Eq. (2.89)).

The most recent developments in catalysts for stereospecific polymerization of dienes have been in the area of the rare earth or Lanthanide catalysts, specifically the neodymium complexes (Hsieh and Yeh, 1985; Kerns et al., 2003). The advantages of these systems are high stereospecificity, high activity, control of molecular weight, and no gel formation (Hsieh and Yeh, 1985).

2.9.4 Polyalkenamers

Cyclic olefins undergo a very unusual type of ring-opening polymerization in the presence of certain transition metal catalysts (Olfstead, 1988; Kaminsky et al., 2012; Novak et al., 1992; Ivin, 1983; Bhowmick et al., 2001). This is illustrated in Eq. (2.90) for the ring-opening metathesis polymerization (ROMP)

(Novak et al., 1992) of cyclooctene to form polyoctenamer. Quite surprisingly, the double bond is maintained in the polymer, i.e., it is not a normal addition polymer. The generally accepted mechanism for these

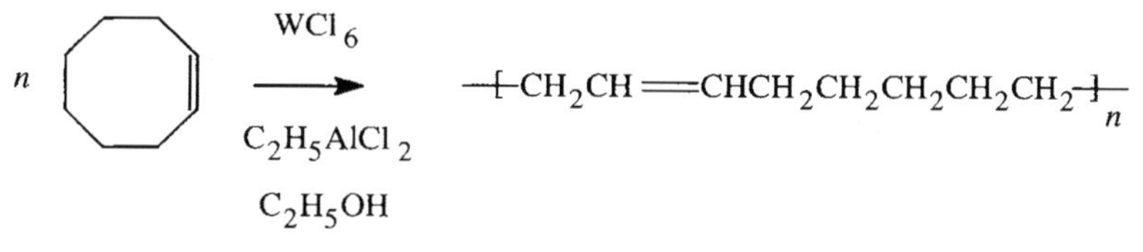

(2.90)

polymerizations proposes that the active propagating species is a metal carbene intermediate which undergoes a cycloaddition reaction with the cycloalkene to form a four-membered ring intermediate, i.e., a metallocyclobutane (Ivin, 1983). The metallocyclobutane then undergoes ring-opening to form a new metallocarbene propagating species as shown in the scheme below for polymerization of cyclopentene, where *Mt* represents the transition metal center,—h represents an empty orbital which is available for coordination with the double bond of the monomer, and P_n is the growing polymer chain with number-average number of monomer units equal to *n*. As indicated, these are reversible polymerizations, and an equilibrium distribution of monomer, cyclic oligomers, and high-molecular-weight polymer is produced.

Scheme for Metathesis Ring-Opening Polymerization:

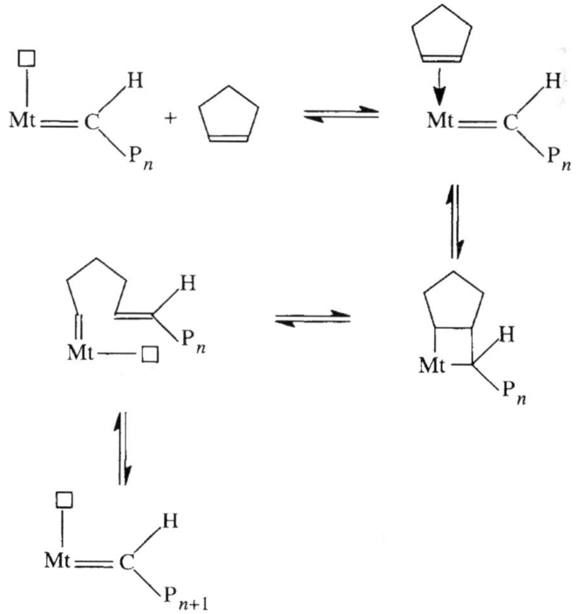

A possible reaction sequence for formation of the metal carbene is shown in Eq. (2.91), where [W] represents a tungsten catalyst center with its attendant ligands not specifically shown (Ivin, 1983).

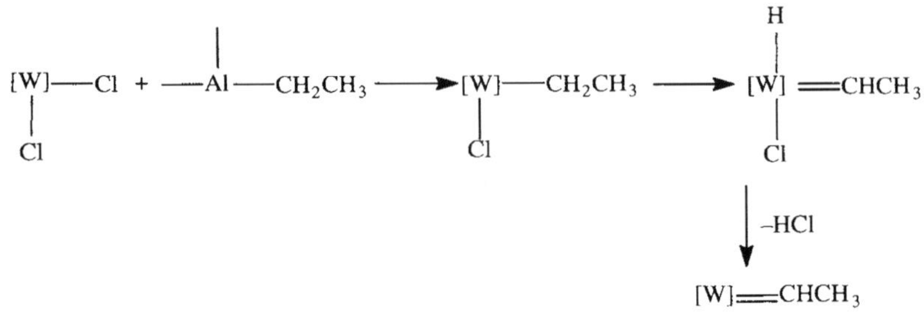

(2.91)

Although a variety of transition metal compounds can catalyze these ring-opening polymerizations, the most active catalysts are based on molybdenum, tungsten, and rhenium derivatives. These compounds are often used with organometallic cocatalysts, analogous to other transition metal catalysts for olefin and diene polymerization described in previous sections. The WCl_6/$EtAlCl_2$/EtOH catalyst system has been described as a commercially useful type of catalyst (Bhowmick et al., 2001). In general, the stereochemistry of the polymerization varies with the catalyst and reaction time. Polymerizable monomers of importance for elastomer synthesis include cyclopentene, cyclooctene, and 1,5-cyclooctadiene; it is noteworthy that cyclohexene is not polymerizable by this method, presumably because there is no ring strain to drive the polymerization (Schöbel et al., 2012). Another monomer of commercial significance is norbornene, which is very reactive; however, the resulting polymer has a relatively high glass transition temperature ($T_g = 35°C$ for 80% *trans* polymer) (Olfstead, 1988), but the glass transition temperature can be lowered to $-60°C$ with plasticizers (Bhowmick et al., 2001).

Since the ring-opening metathesis polymerization is a reversible polymerization, an equilibrium molecular weight distribution of cyclic oligomers and high-molecular-weight polymer is ultimately obtained (Ivin, 1983); for example, polyoctenamer generally consists of 10–15% cyclic oligomer and 85–90% polymer (Bhowmick et al., 2001). At short reaction times and high monomer concentrations, relatively high-molecular-weight polymer is formed as a result of kinetic control; the molecular weight decreases with increasing reaction time. The equilibration process also equilibrates the configuration of the double bonds in the polymer such that eventually an equilibrium distribution of configuration results also. Molecular weight control in ring-opening metathesis polymerization is achieved by addition of acyclic alkenes, which react with the growing chain to terminate chain growth and generate a new metal carbene initiator as shown in Eq. (2.92). Commercially available polyoctenamers (Vestenamers, Hüls, AG) have weight-average molecular weights of approximately 10^5 g/mole, variable *trans*-double bond contents (62–80%), and glass transition temperatures of -75 to $-80°C$ (Bhowmick et al., 2001).

An interesting aspect of the physical properties of polyoctenamers is that they undergo stress-induced crystallization (Olfstead, 1988). The commercial polymers described above have approximately 8 and 30% crystallinity for samples with 62 and 80% *trans*-double bond contents, respectively (Bhowmick et al., 2001).

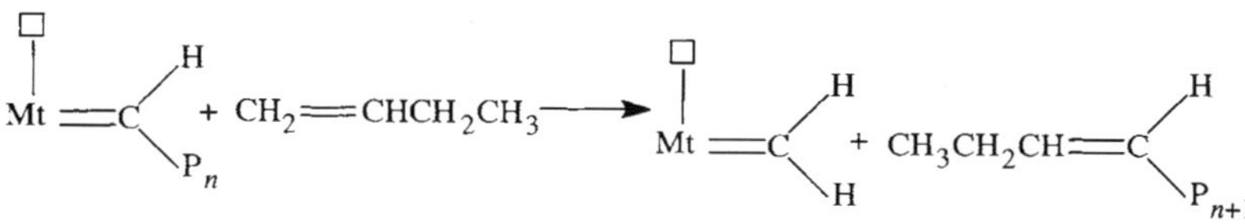

(2.92)

2.10 GRAFT AND BLOCK COPOLYMERIZATION*

The idea of graft (Kahveci et al., 2012; Quirk, 1984; Lutz and Peruch, 2012) or block (Riess and Hurtrez, 1985) copolymerization probably first arose as a means of modifying naturally occurring polymers, such as cellulose (cotton), rubber, or wool. Graft copolymerization, by analogy to the botanical term, refers to the growth of a "branch" of different chemical composition on the "backbone" of a linear macromolecule. In contrast, the related term, block copolymerization, refers to the specific case of growth of a polymer chain from the *end* of a linear macromolecule, thus leading to a composite linear macromolecule consisting of two or more "blocks" of different chemical structure. The importance of these types of polymer structures is basically due to the fact that polymer chains of different chemical structure, which are normally incompatible and form separate phases (because the small entropy of mixing is insufficient to overcome the mostly positive enthalpy of mixing), are chemically bonded to each other. This leads to the formation of microheterogeneities, which can have a profound effect on the mechanical properties of these heterophase systems when compared with the two homopolymers or with a physical mixture of the two polymers (Holden et al., 2004a).

As one might expect graft and block copolymerization can be accomplished by means of each of the three known mechanisms, i.e., radical, cationic, and anionic, each of which shows its own special characteristics (Odian, 2004b). Hence these mechanisms have been used wherever appropriate for the polymer and monomer involved. The examples quoted in the following discussions will deal primarily with elastomers.

2.10.1 Graft Copolymerization by Conventional Free Radical Reactions

This has been the most widely applied system for the formation of graft copolymers, since it provides the simplest method and can be used with a

* See also Chapters 12 and 13.

wide variety of polymers and monomers. It has not been very useful in the synthesis of block copolymers, as will become obvious from an examination of the methods used. These can be listed as follows.

(i) Chemical Initiation

This is still the most popular method for graft copolymerization of elastomers via free radicals. Free radicals (I) are generated from the same types of initiators which are used for free radical polymerization and copolymerization (see Section 2.4). In general, these radicals are formed in the presence of a polydiene elastomer and a monomer; therefore, there are several possible reactions of these initiator-derived radicals which can occur as shown in Eqs. (2.93)–(2.96). The competition between initiation of monomer polymerization (Eq. (2.93)) and reactions to form polymer-derived radicals (Eqs. (2.94)–(2.96)) is dependent on the reactivity of the initiating radical. Thus, no graft copolymer formation with styrene monomer is observed for either polybutadiene or polyisoprene when azobisisobutryonitrile is used to generate radicals, although good grafting efficiency was observed for benzoyl peroxide-generated radicals (Brydon et al., 1973; Cameron and Qureshi, 1980).

$$\mathrm{I} \longrightarrow \mathrm{I}\cdot$$

$$\mathrm{I}\cdot + \mathrm{M} \xrightarrow{k_1} \mathrm{I{-}M}\cdot \tag{2.93}$$

$$\mathrm{I}\cdot + -[\mathrm{CH_2CH{=}CHCH_2}]- \xrightarrow{k_2} \mathrm{I{-}H} + -[\dot{\mathrm{C}}\mathrm{HCH{=}CHCH_2}]- \tag{2.94}$$

$$\mathrm{I}\cdot + -[\mathrm{CH_2CH{=}CHCH_2}]- \xrightarrow{k_3} -[\mathrm{CH_2CH(I){-}}\dot{\mathrm{C}}\mathrm{HCH_2}]- \tag{2.95}$$

$$\mathrm{I}\cdot + -[\mathrm{CH_2CH(CH{=}CH_2)}]- \xrightarrow{k_3} -[\mathrm{CH_2CH}(\cdot\mathrm{CH{-}CH_2{-}I})]- \tag{2.96}$$

This result also indicates that growing polystyryl radicals do not abstract hydrogen from these polydienes to generate polymer-derived radicals. The competition between addition of initiator radicals to the double bonds in the polydiene (Eqs. (2.95) and (2.96)) and hydrogen abstraction (Eq. (2.94)) is also dependent on the initiator (Walling, 1967). Thus, *t*-butoxy radicals [$(CH_3)_3CO$] exhibit an unusual preference for hydrogen abstraction compared to alkyl radicals as shown in Table 2.18 (Walling, 1967). For radicals derived from benzoyl peroxide, the competition between the rate of hydrogen abstraction from polydiene (Eq. (2.94)) compared to addition of the initiator radical

TABLE 2.18 Radical Reactivity Toward Hydrogen Abstraction versus Addition to Double Bonds[a]

Radical	$k_{abstraction}/k_{addition}$ [b]
t-$(CH_3)_3CO\cdot$	30
$ROO\cdot$	1.0
$H_3C\cdot$	0.25
$RS\cdot$	Exclusive addition

[a]Data taken from Walling (1967).
[b]Ratio of rate constant for hydrogen abstraction (see Eq. (2.94)) versus addition to a double bond (see Eqs. (2.95) and (2.96))

to styrene monomer (Eq. (2.93)), e.g., k_{abstr}/k_{ad}^{S}, was found to be 1.2 for polyisoprene and 0.63 for polybutadiene (Cameron and Qureshi, 1980). With respect to the addition of initiating radicals to the double bond of the polydiene, it is reported that grafting is favored by higher 1,2-microstructure in the polydiene (Stein et al., 1974; Schmitt, 1979); the rate of addition to a vinyl side chain (Eq. (2.96)) is faster than addition to an in-chain double bond (Eq. (2.95)).

The formation of graft copolymer from the polymer-derived radicals generated in Eqs. (2.93)–(2.96) is shown in Eqs. (2.97) and (2.98), where *P* represents the polymeric backbone radicals. Finally, in order to control the molecular weight of the graft chains, chain transfer agents such as long-chain alkyl thiols can be added (see Eq. (2.27)) (Kulich and Kelley, 1985).

$$P\cdot + nM \longrightarrow \underset{\text{graft copolymer}}{P - [M]_n - M\cdot}, \tag{2.97}$$

$$P\cdot + I - [M]_n - M\cdot \longrightarrow \text{graft copolymer.} \tag{2.98}$$

Since all of these reactions are occurring simultaneously during the graft copolymerization, there is always the possibility of formation of *homopolymer* during the grafting reaction, via reaction of the initiator radical with monomer (Eq. (2.93)) and by chain transfer of the growing chain with species *other* than the polymer backbone (e.g., monomer, solvent, initiator). Therefore, in general, the graft copolymer will be contaminated with both the original backbone homopolymer as well as the monomer-derived homopolymer (Kahveci et al., 2012).

This type of graft copolymerization has been applied to the grafting of monomers like styrene and methyl methacrylate to natural rubber (Bhowmick et al., 2001), directly in the latex (Allen, 1963; Campbell, 1988). Similar methods have been developed for grafting the foregoing monomers, and many other vinyl monomers, to synthetic rubbers-like SBR, leading to a variety of plastic-reinforced elastomers and rubber-reinforced high-impact plastics (Kulich and Kelley, 1985; Battaerd and Tregear, 1967). In this case, grafting can also occur by the "copolymerization" of the monomer with the unsaturated bonds (mainly vinyl) in the polymer as described previously (see Eq. (2.96)); thus

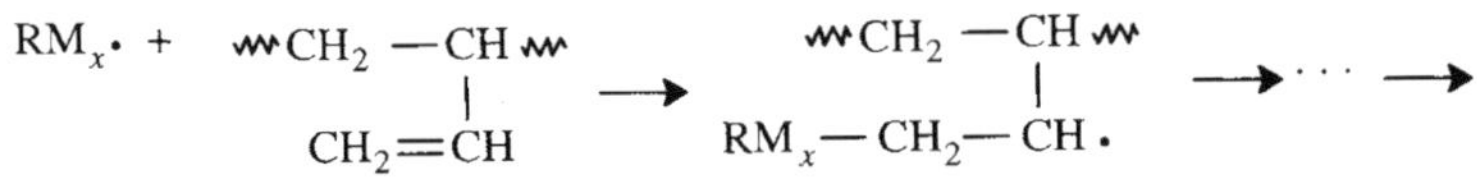

This reaction can, of course, also lead to crosslinking of the polymer chains, and this must be controlled.

(ii) Other Methods

Other methods of generating free radicals can also be used to initiate graft polymerization with elastomers, both natural and synthetic. These include irradiation of polymer-monomer mixtures by ultraviolet light (Cooper et al., 1959), high-energy radiation (Cockbain et al., 1959; Gupta and Anjum, 2003), and mechanical shear. The latter is of particular interest because of its unique mechanism, and has been extensively investigated.*

Thus it has been convincingly demonstrated that elastomers, when subjected to several mechanical shearing forces, undergo homolytic bond scission to form free chain radicals. The latter, when in the presence of oxygen, may then undergo various reactions, either becoming stabilized ruptured chains or reacting with other chains to form branched or crosslinked species (Pike and Watson, 1952; Ceresa, 1962). When blends of different elastomers are masticated, "interpolymers" are formed by the interaction of the radicals formed from the copolymers (Angier and Watson, 1955,1957). A further extension of such mechanochemical processes occurs when elastomers are masticated in the presence of polymerizable monomers, the chain radicals initiating polymerization and leading to formation of block and graft structures. This was clearly demonstrated in the case of natural rubber (Battaerd and Tregear, 1967) and of other elastomers (Angier et al., 1959). It should be noted that living anionic polymerizations (Hadjichristidis et al., 2003b) and controlled/living radical polymerizations such as atom transfer radical polymerization (ATRP) (Davis and Matyjaszewski, 2002) have been used to make well-defined graft copolymers. A very useful method is to copolymerize a well-defined macromonomer bearing a polymerizable chain end group with another monomer to generate a graft copolymer with a random distribution of well-defined graft branches (Quirk et al., 2004).

2.10.2 Block Copolymers by Controlled Radical Mechanisms

The advent of controlled radical polymerization methods, such as ATRP, RAFT, and NMP, has allowed for a wide variety of new block and graft polymers to be prepared due to the chemical variety of monomers which can be polymerized by these methods (Matyjaszewski and Tsarevsky, 2009; Matyjaszewski and Braunecker, 2007). Block copolymers can be polymerized by sequential monomer addition, but more often the first block is synthesized, isolated, and

* See also Chapter 11.

characterized, prior to chain extension with a second monomer. The order of monomer addition is of particular importance when preparing block copolymers via controlled radical polymerization methods. The order is often dependent on the polymerization mechanism utilized in the synthesis of the materials. For example, in ATRP the order of reactivity is as follows: acrylonitrile > methacrylates > styrene ≈ acrylates > acrylamides (Matyjaszewski and Braunecker, 2007). Steric effects dominate in the RAFT polymerization resulting in methacrylates being the most reactive (Moad et al., 2008; Chong et al., 1999,2003). As discussed in Section 2.4.2 there are very few reports on the polymerization of 1,3-dienes by controlled radical polymerization methods, leaving anionic polymerization as the major polymerization method to prepare copolymers composed of styrene and 1,3-dienes.

A great deal of work has also been reported on the ability to combine different polymerization mechanisms to prepare a diverse set of block copolymers by transformation of the active site or the use of dual initiators (Holden et al., 2004b). This generally involves either end-functionalization of the first block followed by the appropriate post-polymerization chemistry to prepare the desired macroinitiator. For example, anionic polymerization of either a difunctional isoprene of butadiene mid-block followed by NMP to form the second block of styrene has been reported (Acar and Matyjaszewski, 1999; Tong et al., 2000; Wang et al., 2000). There are also numerous reports on converting polymers made by cationic, ROMP, and step-growth polymerization to controlled radical polymerization macroinitiators (Matyjaszewski and Braunecker, 2007; Holden et al., 2004b). The controlled radical polymerization techniques have also been widely used to prepare multiblock copolymers (ABC, ABCA, etc.) which are emerging as materials of choice as researchers desire to explore new morphological space (Lodge, 2003; Bates et al., 2012). The synthesis of these types of materials will most definitely require the combination of multiple polymerization methods including anionic, ring-opening, cationic, and controlled radical polymerization.

2.10.3 Block Copolymers by Anionic Mechanism

One of the unique and useful features of living polymerizations is the ability to prepare block copolymers by sequential monomer addition (Webster, 1991). Although there are now living polymerization methodologies for cationic (Canale and Faust, 1999; Hadjikyriacou et al., 1995; Goethals and Du Prez, 2007; Sawamoto, 1991), Ziegler-Natta (Koo and Marks, 1999)/polyolefin (Domski et al., 2007), ring-opening metathesis (Bielawski and Grubbs, 2007), as well as development of reversible-deactivation radical polymerization methods (Jenkins et al., 2010) (e.g., atom-transfer radical polymerization (ATRP), nitroxide-mediated radical polymerization (NMP), and reversible addition-fragmentation chain-transfer (RAFT) polymerization) (Rizzardo et al., 1999; Sawamoto and Kamigaito, 2000; Coessens et al., 2001; Ouchi et al., 2009; Moad et al., 2009), living anionic polymerization, especially alkyllithium-initiated polymerization in hydrocarbon media, is the most important synthetic method for the synthesis of both well-defined polymers for research as well as for commercial processes (Hsieh and Quirk, 1996; Quirk, 2012; Morton, 1983; Kahveci et al., 2012; Quirk and Lee, 1992; Riess and Hurtrez, 1985; Holden et al., 2004a). By sequential addition of monomers, alkyllithium-initiated polymerization proceeds in homogeneous solution to form blocks wherein each block has a prescribed molecular weight, based on monomer-initiator stoichiometry, as well as a narrow molecular weight distribution ($M_w/M_n \leqslant 1.1$). As would be expected, such block copolymers are compositional homogeneous, due to the absence of any side reactions during the polymerization (e.g., termination, monomer transfer, branching).

Organolithium initiators have been particularly useful in this regard, since they are soluble in a variety of solvents (Wakefield, 1974) and since they can initiate the polymerization of a variety of monomers, such as styrene and its homologs, the 1,3-dienes, alkyl methacrylates, vinylpyridines, cyclic oxides and sulfides, and cyclic siloxanes (Hsieh and Quirk, 1996). Utilizing these monomers, various block copolymers have been synthesized (Riess and Hurtrez, 1985), some commercially, but the outstanding development in this area has been in the case of the ABA type of triblock copolymers, particularly polystyrene-*b*-polydiene-*b*-polystyrene (S-D-S), and their hydrogenated analogs (Holden et al., 2004a).

The ABA triblocks which have been most exploited commercially are of the styrene-diene-styrene type, prepared by sequential polymerization initiated by alkyllithium compounds as shown in Eqs. (2.99)–(2.101) (Holden et al., 2004a; Legge, 1987; Handlin et al., 2007). The behavior of these block copolymers illustrates the special characteristics of block (and graft) copolymers, which are based on the general *incompatibility* of the different blocks (Bates and Fredrickson, 1990). Thus for a typical "thermoplastic elastomer" (Holden et al., 2004a), the polystyrene end blocks (~15,000–20,000 MW) aggregate into a separate phase, which forms a microdispersion within the matrix composed of the polydiene chains (50,000–70,000 MW) (Meier, 1969; Helfand, 1975; Leibler, 1980). A schematic representation of this morphology is shown in Figure 2.3. This phase separation, which occurs in the melt (or swollen) state, results, at ambient temperatures, in a network of elastic polydiene chains held together by glassy polystyrene microdomains. Hence these materials behave as virtually crosslinked elastomers at ambient temperatures, but are completely thermoplastic and fluid at elevated temperatures.

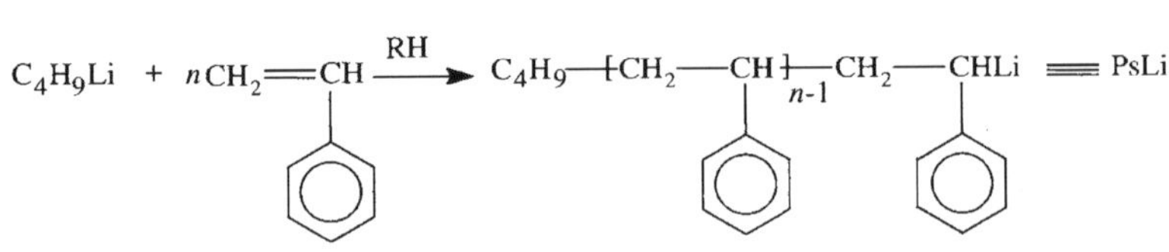

(2.99)

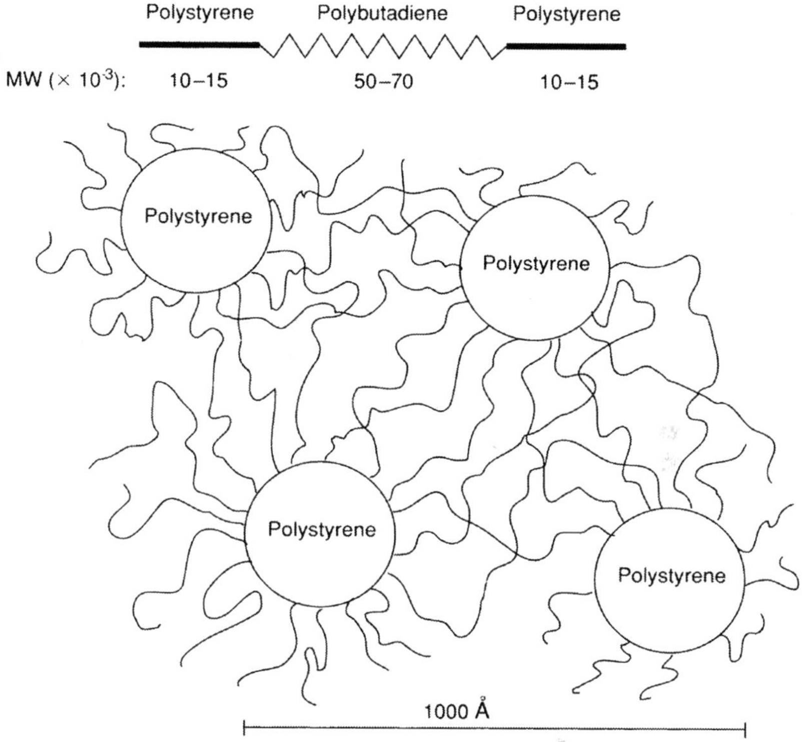

FIGURE 2.3 Structure of thermoplastic elastomers from ABA triblock copolymers. (From Morton, 1971.)

$$\text{PsLi} + m\,\text{CH}_2{=\!=}\text{CH}{-\!-}\text{CH}{=\!=}\text{CH}_2 \longrightarrow \text{Ps}{-\!-}[\text{CH}_2\text{CH}{=\!=}\text{CHCH}_2]_{m-1}{-\!-}\text{CH}_2\text{CH}{=\!=}\text{CHCH}_2\text{Li} \equiv \text{Ps-PBDLi} \quad (2.100)$$

(2.101)

It is important to note that the morphology of ABA block copolymers is dependent primarily on the relative composition of the block components

(Kahveci et al., 2012; Bates and Fredrickson, 1990; Meier, 1969; Helfand, 1975; Leibler, 1980; Abetz and Simon, 2005). For example, as the styrene content increases, the morphology changes from spherical polystyrene domains to cylindrical; increasing styrene content over a narrow composition range results in the formation of a unique, gyroid morphology with co-continuous styrene and diene domains; further increases form lamellar arrays of both phases and eventually phase inversion to form a continuous polystyrene phase. The analogous morphologies, i.e., gyroid to cylindrical to spherical

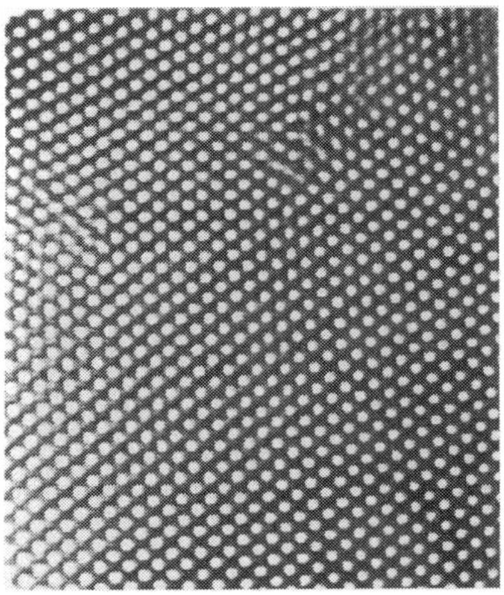

FIGURE 2.4 Transmission electron photomicrograph of an ultrathin film of a styrene-isoprene-styrene triblock copolymer (MW 16,200–75,600–16,2000) × 100,000.

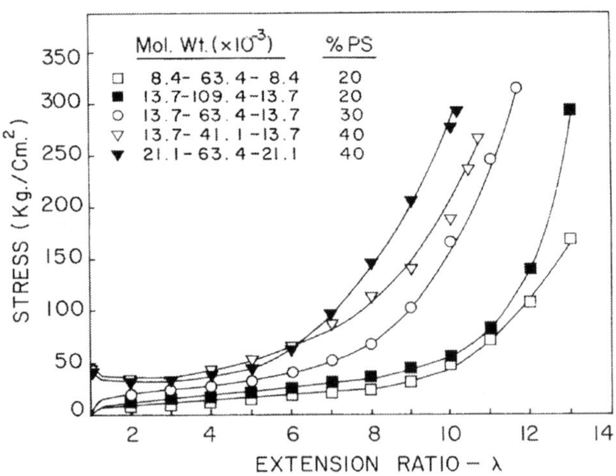

FIGURE 2.5 Tensile properties of styrene-isoprene-styrene triblock copolymers. (From Morton, 1971.)

polydiene domains, are formed as the diene content decreases. The properties of the ABA triblock copolymers are dependent and vary with the composition (morphology).

Such "thermoplastic elastomers" are very attractive technologically, since they can be heat-molded like thermoplastics, yet exhibit the behavior of rubber vulcanizates (Holden et al., 2004a). As would be expected, their structure, morphology, and mechanical properties have been studied extensively (Morton, 1971; Holden et al., 1969; Smith and Dickie, 1969; Hamley, 2004). An electron photomicrograph of a typical styrene–isoprene-styrene (SIS) triblock film is shown in Figure 2.4, while the tensile properties of a series of such triblock copolymers are shown in Figure 2.5. The unusually high tensile strength of these elastomers, better than that of conventional vulcanizates, is ascribed both to the remarkable regularity of the network, as illustrated in Figure 2.5, and to the energy-absorbing characteristics of the polystyrene domains, which yield and distort under high stress (Morton, 1971).

This interesting behavior of the ABA triblock copolymers is not a unique feature of the styrene-diene structure, but can be found in the case of other analogous chemical structures. Thus thermoplastic elastomers have been obtained from other triblock copolymers, where the dienes have been replaced by cyclic sulfides (Morton et al., 1971), cyclic siloxanes (Morton et al., 1974), or alkyl acrylates (Jerome, 2004); poly(alkyl methacrylate) end blocks have also been investigated (Jerome, 2004).

2.10.4 Block Copolymers by Cationic Mechanism

The discovery of living cationic polymerization has provided methods and technology for the synthesis of useful block copolymers, especially those based on elastomeric polyisobutylene (Kennedy and Puskas, 2004). It is noteworthy that isobutylene can only be polymerized by a cationic mechanism. One of the most useful thermoplastic elastomers prepared by cationic polymerization is the polystyrene-*b*-polyisobutylene-*b*-polystyrene (SIBS) triblock copolymer. This polymer imbibed with anti-inflammatory drugs was one of the first polymers used to coat metal stents as a treatment for blocked arteries (Sipos et al., 2005). The SIBS polymers possess an oxidatively stable, elastomeric polyisobutylene center block and exhibit the critical enabling properties for this application including processing, vascular compatibility, and biostability (Faust, 2012). As illustrated below, SIBS polymers can be prepared by sequential monomer addition using a difunctional initiator with titanium tetrachloride in a mixed solvent (methylene chloride/methylcyclohexane) at low temperature (−70 to −90°C) in the presence of a proton trap (2,6-*di*-*t*-butylpyridine). To prevent formation of coupled products formed by intermolecular alkylation, the polymerization is terminated prior to complete consumption of styrene. These SIBS polymers exhibit tensile properties essentially the same as those of vulcanized butyl rubber.

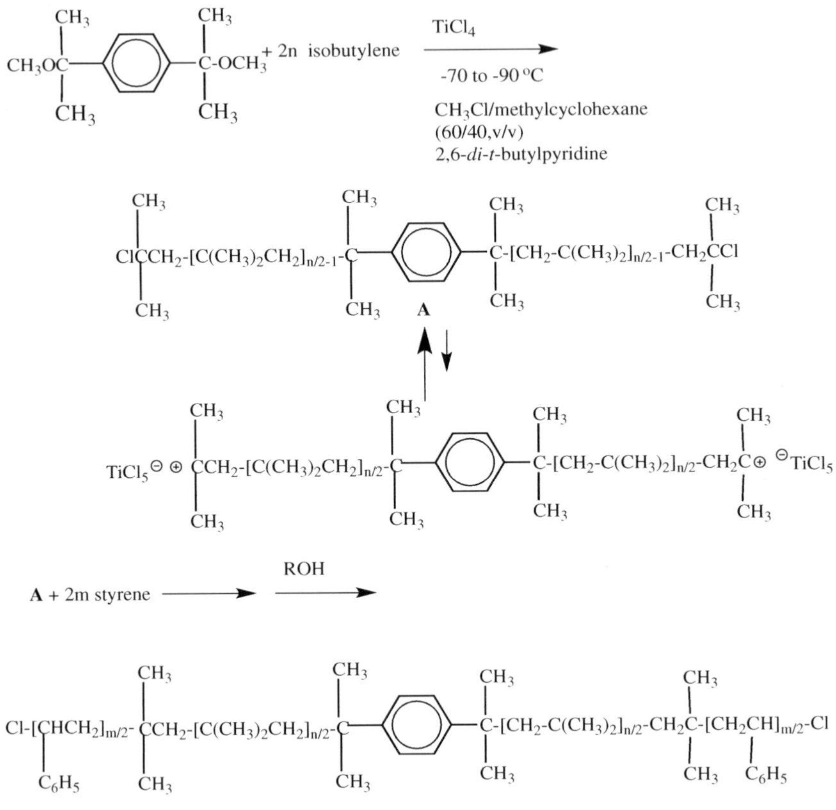

polystyrene-*b*-polyisobutylene-*b*-polystyrene

2.10.5 Block Copolymers by Ziegler-Natta (Insertion) Mechanism

The development of new living polymerization chemistries involving transition metal propagating centers has provided new methods for synthesis of polyolefin block copolymers (Holden et al., 2004a; Coates et al., 2002). Using Ziegler-Natta-type chemistry, the living insertion polymerization of unactivated olefins, such as ethylene and propylene, coupled with the ability to control stereochemistry, has provided a powerful tool for the synthesis of new polyolefin architectures of academic and commercial interest. For example, Sita and coworkers (Keaton et al., 2001) utilized zirconium acetamidinate initiators (e.g., **49**) with a borate cocatalyst, **2**, $[PhNMe_2H][B(C_6F_5)_4]$, to form poly(vinylcyclohexene-*block*-hexene-*block*-vinylcyclohexene) in chlorobenzene at $-10°C$. The resulting polymer exhibited a narrow molecular weight distribution (1.08). This catalyst produced highly isotactic polyvinylcyclohexene (mmmm > 95%) by a living polymerization mechanism as determined by kinetic analysis.

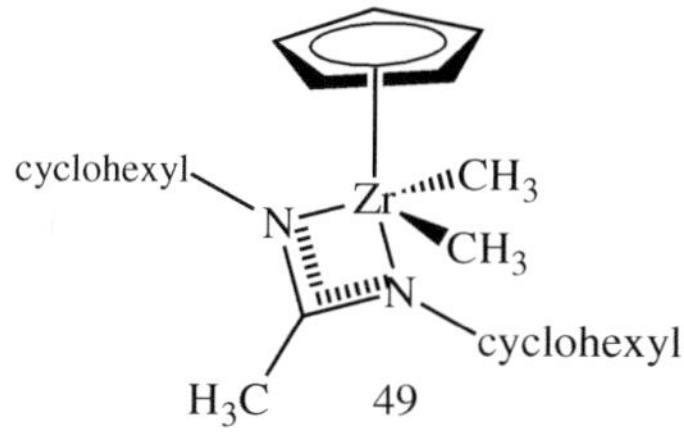

49

Sita and coworkers (Harney et al., 2006; Giller et al., 2011) have described the use of an analogous initiator, **50**, activated with the borate cocatalyst, **2**, to produce either living, stereoblock polypropylene (iPP) with an mmmm pentad level of 71% (94% stereoselectivity) when the ratio of [**50**]/[**2**] = 1, or to produce atactic polypropylene (aPP) when [**50**]/[**2**] = 0.5. In order to modulate the stereochemistry between these two blocks, compound **51** was added in sequence. Compound **51** effects methyl group transfer either with the active center from **50** or with **2** and the cationic product of demethylation of 51 is inactive as a polymerization initiator. When 0.5 equivalents of **51** are added to the iPP block system ([**50**]/[**2**] = 1), the catalyst system is converted to the aPP system ([**50**]/[**2**] = 0.5) and an iPP-*block*-aPP diblock sequence is generated. When [**50**]/[**2**] = 0.5, degenerative-transfer (DT) living Ziegler-Natta polymerization process has been proposed to occur. This proceeds via a rapid and reversible transfer of methyl groups between active (cationic Zr) and dormant (neutral Zr) centers. Because the dormant state is configurationally unstable with respect to a metal-centered epimerization process ($k_{epimerization} \gg k_{propagation}$), an atactic PP is produced. Addition of 0.5 equivalents of **2** then can be used to demethylate the dormant (neutral Zr) centers and generate the third iPP block forming the triblock copolymer, iPP-*block*-aPP-*block*-iPP.

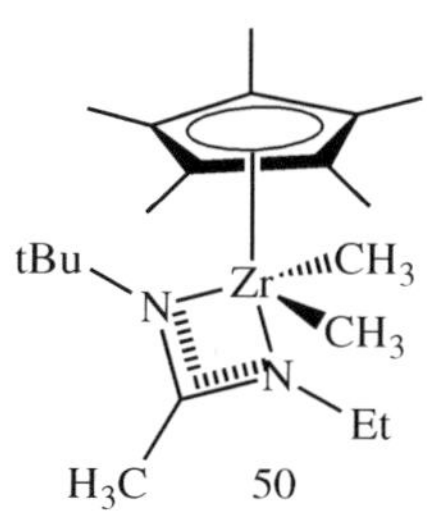

50

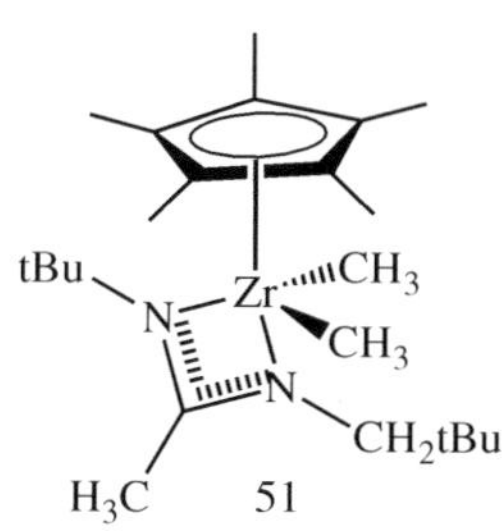

51

Coates and coworkers Tian et al., 2001 have developed the fluorinated phenoxy-imine complexes (e.g., **50**) with MAO (methylalumoxane) cocatalyst. At 0°C this system produced highly syndiotactic polypropylene ([rrrr] = 0.96) with a

narrow molecular weight

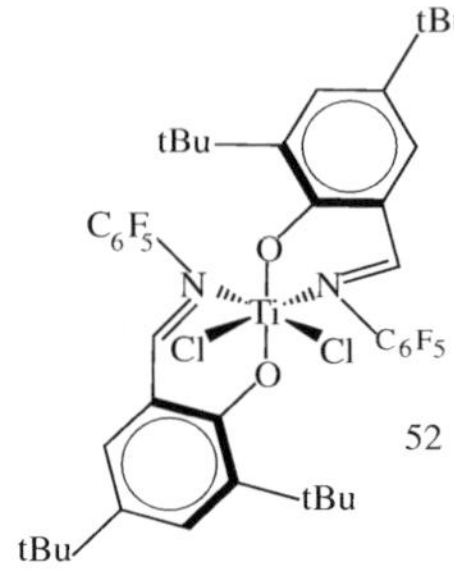

52

distribution (ca. 1.1); the absence of chain transfer was documented by the absence of olefin end groups and a linear relationship between molecular weight and polymer yield. Using this catalyst system, well-defined block copolymers ($M_w/M_n = 1.12$) with crystalline, syndiotactic polypropylene domains ($T_m = 131°C$), and amorphous ethylene-*co*-propylene (EPR) segments ($T_g = -45°C$) were produced.

It is important to note that any molecular architecture that provides a thermoplastic block chemically bonded to an elastomeric block, which is in turn bonded to another thermoplastic segment, should exhibit the properties of a thermoplastic elastomer. For example, grafting thermoplastic branches onto an elastomeric backbone produces thermoplastic elastomer behavior (Bhowmick et al., 2001; Quirk et al., 2004). Other examples are the segmented-type polymers—$[AB]_n$—with alternating hard and soft segments; thus, a variety of segmented polyesters and polyurethanes with polyether or polyester soft segments exhibit properties of thermoplastic elastomers (Bhowmick et al., 2001; Holden et al., 2004a; Abouzahr, 1985).

REFERENCES

Abetz, V., Simon, P.F.W., 2005. Phase behaviour and morphologies of block copolymers. Adv. Polym. Sci. 189, 125–212.

Abouzahr, S., 1985. In: Folkes, M.J. (Ed.), Processing, Structure, and Properties of Block Copolymers. Elsevier, London, p. 165.

Acar, M.H., Matyjaszewski, K., 1999. Block copolymers by transformation of living anionic polymerization into controlled/"living" atom transfer radical polymerization. Macromol. Chem. Phys. 200 (5), 1094–1100.

Adams, R.K., Hoeschele, G.K., Witsiepe, W.K., 2004. In: Holden, G., Kricheldorf, H.R., Quirk, R.P. (Eds.), Thermoplastic Elastomers, 3rd ed. Hanser Publishers, Munich, Germany, 2004, p. 183.

Ajellal, N., Thomas, C.M., Carpentier, J.F., 2008. Controlled radical polymerization of conjugated 1,3-dienes with methyl 1,3-butadiene-1-phosphonate. Polymer 49 (20), 4344–4349.

Allen, P.W., 1963. In: Bateman, L. (Ed.), The Chemistry and Physics of Rubberlike Substances. Applied Science Publication Ltd., London, p. 97.

Angier, D.J., Watson, W.F., 1955. Mastication of rubber 2. Interpolymerization of natural rubber and neoprene on cold milling. J. Polym. Sci. 18 (87), 129–140.

Angier, D.J., Watson, W.F., 1957. Trans. Inst. Rubber Ind. 33, 22.

Angier, D.J., Ceresa, R.J., Watson, W.F., 1959. Mastication of rubber 8. Preparation of block polymers by mechanical shear of polymer-monomer systems. J. Polym. Sci. 34 (127), 699–708.
Antkowiak, A.E., Tate, D.P., Oberster, A.E., et al., 1972. Temperature and concentration effects on polar-modified alkyllithium polymerizations and copolymerizations. J. Polym. Sci. Part A: Polym. Chem. 10 (5), 1319–1334.
Arlman, E.J., Cossee, P., 1964. Ziegler-Natta catalysis 3. Stereospecific polymerization of propene with the catalyst system TiCl3-AlEt3. J. Catal. 3 (1), 99–104.
Baldwin, F.P., Strate, G.V., 1972. Polyolefin elastomers based on ethylene and propylene. Rubber Chem. Technol. 45 (3), 709–881.
Bandermann, F., Speikamp, H.D., Weigel, L., 1985. Bifunctional anionic initiators: a critical study and overview. Makromol. Chem. 186 (10), 2017–2024.
Bar-Nes, G., Hall, R., Sharma, V., et al., 2009. controlled/living radical polymerization of isoprene and butadiene in emulsion. Eur. Polym. J. 45 (11), 3149–3163.
Bates, F.S., Fredrickson, G.H., 1990. Block copolymer thermodynamics—theory and experiment. Annu. Rev. Phys. Chem. 41, 525–557.
Bates, F.S., Hillmyer, M.A., Lodge, T.P., et al., 2012. Multiblock polymers: panacea or pandora's box? Science 336 (6080), 434–440.
Battaerd, H.A.J., Tregear, G.W., 1967. Graft Copolymers. Interscience Publishers, New York.
Bauer, R.G., 1979. Kirk-Othmer Encyclopedia of Chemical Technology, 3rd ed., vol. 8. Wiley, New York, p. 611.
Bauer, B.J., Fetters, L.J., 1978. Synthesis and dilute-solution behavior of model star-branched polymers. Rubber Chem. Technol. 51 (3), 406–436.
Benoit, D., Harth, E., Fox, P., et al., 2000. Accurate structural control and block formation in the living polymerization of 1,3-dienes by nitroxide-mediated procedures. Macromolecules 33 (2), 363–370.
Bertin, D., Gigmes, D., Marque, S.R.A., et al., 2005. Polar, steric, and stabilization effects in alkoxyamines C-ON bond homolysis: a multiparameter analysis. Macromolecules 38 (7), 2638–2650.
Beuermann, S., Buback, M., Davis, T.P., et al., 1997. Critically evaluated rate coefficients for free-radical polymerization 2. Propagation rate coefficients for methyl methacrylate. Macromol. Chem. Phys. 198 (5), 1545–1560.
Beu, K.E., Reynolds, W.B., Fryling, C.F., et al., 1948. X-ray studies of low-temperature polybutadiene and butadiene-styrene copolymers. J. Polym. Sci. 3 (4), 465–480.
Bhowmick, A.K., Stein, C., Stephens, H.L., 2001. In: Bhowmick, A.K., Stephens, H.L. (Eds.), Handbook of Elastomers. M. Dekker, New York, p. 775.
Bielawski, C.W., Grubbs, R.H., 2007. Living ring-opening metathesis polymerization. Prog. Polym. Sci. 32 (1), 1–29.
Binder, J.L., 1954. Microstructures of polybutadienes and butadiene-styrene copolymers. Ind. Eng. Chem. 46 (8), 1727–1730.
Blackley, D.C., 1975. Emulsion Polymerization. Applied Science Publishers, London.
Blackley, D.C., 1983. Synthetic Rubbers: Their Chemistry and Technology. Applied Science Publishers, Essex.
Blow, C.M., 1971. Rubber Technology and Manufacture. Hewnes-Butterworths, London.
Body, R.W., Kyllingstad, V.L., 1986. In: Kroschwitz, J.I. (Ed.), Encyclopedia of Polymer Science and Engineering, vol. 6. Wiley-Interscience, New York, p. 307.
Boor, J., 1979. Ziegler-Natta Catalysts and Polymerizations. Academic Press, New York.
Bouchardat, G., 1879. C. R. H. Acad. Sci. 89, 1117.
Braun, D., Cherdron, H., Kern, W., 1984. Practical Macromolecular Organic Chemistry. Harwood Academic Publishers, New York, p. 255.
Brintzinger, H.H., Fischer, D., Mulhaupt, R., et al., 1995. Stereospecific olefin polymerization with chiral metallocene catalysts. Angew. Chem. Int. Ed. Engl. 34 (11), 1143–1170.
Brown, T.L., 1968. Nuclear magnetic resonance studies of organometallic exchange processes. Acc. Chem. Res. 1 (1), 23–32.
Brydon, A., Burnett, G.M., Cameron, G.G., 1973. Free-radical grafting of monomers to polydienes 1. Effect of reaction conditions on grafting of styrene to polybutadiene. J. Polym. Sci. Part A: Polym. Chem. 11 (12), 3255–3269.

Brydson, J.A., 1981. In: Whelan, A., Lee, K.S. (Eds.), Developments in Rubber Technology, vol. 2. Applied Science Publishers, London, p. 21.
Buback, M., Gilbert, R.G., Hutchinson, R.A., et al., 1995. Critically evaluated rate coefficients for free-radical polymerization 1. Propagation rate coefficient for styrene. Macromol. Chem. Phys. 196 (10), 3267–3280.
Bywater, S., 1976. In: Bamford, C.H., Tipper, E. (Eds.), Comprehensive Chemical Kinetics, vol. 15. Elsevier, New York, p. 1.
Bywater, S., 1979. Preparation and properties of star-branched polymers. Adv. Polym. Sci. 30, 89–116.
Bywater, S., 1989. In: Eastmond, G.C., Ledwith, A., Russo, S., Sigwalt, P. (Eds.), Comprehensive Polymer Science, vol. 3. Pergamon Press, Oxford, UK, p. 433.
Bywater, S., Worsfold, D.J., 1967a. Alkyllithium anionic polymerization initiators in hydrocarbon solvents. J. Organomet. Chem. 10 (1), 1–6.
Bywater, S., Worsfold, D.J., 1967b. Anionic polymerization of isoprene: ion and ion-pair contributions to polymerization in tetrahydrofuran. Can. J. Chem. 45 (16), 1821–1824.
Bywater, S., Worsfold, D.J., 1987. In: Hogen-Esch, T.E., Smid, J. (Eds.), Recent Advances in Anionic Polymerization. Elsevier, New York, p. 109.
Bywater, S., Worsfold, D.J., Johnson, A.F., 1964. Electronic spectra of some anionic polymerization systems. Can. J. Chem. 42 (6), 1255–1260.
Cameron, G.G., Qureshi, M.Y., 1980. Free-radical grafting of monomers to polydienes 4. Kinetics and mechanism of methyl-methacrylate grafting to polybutadiene. J. Polym. Sci. Part A: Polym. Chem. 18 (11), 3149–3161.
Campbell, D.S., 1988. In: Roberts, A.D. (Ed.), Natural Rubber Science and Technology. Oxford University Press, Oxford, p. 679.
Canale, P.L., Faust, R., 1999. Capping of living polystyryl cations with ditolylethylene: model reactions with dimeric cations. Macromolecules 32 (9), 2883–2888.
Carothers, W.H., 1929. Studies on polymerization and ring formation I. An introduction to the general theory of condensation polymers. J. Am. Chem. Soc. 51 (8), 2548–2559.
Carothers, W.H., 1931. Polymerization. Chem. Rev. 8 (3), 353–426.
Carothers, W.H., Williams, I., Collins, A.M., et al., 1931. Acetylene polymers and their derivatives. II. A new synthetic rubber: chloroprene and Its polymers. J. Am. Chem. Soc. 53 (11), 4203–4225.
Ceresa, R.J., 1962. Block and Graft Copolymers. Butterworths, London, p. 65.
Chen, H.Y., Field, J.E., 1967. Configuration of isoprene structure in butyl rubber by time-averaging high resolution nuclear magnetic resonance. J. Polym. Sci. Polym. Lett. Ed. 5 (6), 501–503.
Chong, Y.K., Le, T.P.T., Moad, G., et al., 1999. A more versatile route to block copolymers and other polymers of complex architecture by living radical polymerization: the RAFT process. Macromolecules 32 (6), 2071–2074.
Chong, Y.K., Krstina, J., Le, T.P.T., et al., 2003. Thiocarbonylthio compounds [Sc(Ph)S-R] in free radical polymerization with reversible addition-fragmentation chain transfer (RAFT polymerization). Role of the free-radical leaving group (R). Macromolecules 36 (7), 2256–2272.
Christie, D.I., Gilbert, R.G., Congalidis, J.P., et al., 2001. Spontaneous polymerization in the emulsion polymerization of styrene and chlorobutadiene. Macromolecules 34 (15), 5158–5168.
Ciardelli, F., 1992. In: Aggarwal, S.L., Russo, S. (Eds.), Comprehensive Polymer Science, First Supplement. Pergamon Press, Oxford, p. 67.
Coates, G.W., Hustad, P.D., Reinartz, S., 2002. Catalysts for the living insertion polymerization of alkenes: access to new polyolefin architectures using Ziegler-Natta chemistry. Angew. Chem. 41 (13), 2236–2257.
Cockbain, E.G., Pendle, T.D., Turner, D.T., 1959. Formation of graft polymers by gamma-irradiation of natural rubber latex and methyl methacrylate. J. Polym. Sci. 39 (135), 419–426.
Coessens, V., Pintauer, T., Matyjaszewski, K., 2001. Functional polymers by atom transfer radical polymerization. Prog. Polym. Sci. 26 (3), 337–377.
Coleman, M.M., Brame, E.G., 1978. C-13 NMR characterization of polychloroprene microstructure 2. Relaxation, noe, and quantitative measurements. Rubber Chem. Technol. 51 (4), 668–676.

Coleman, M.M., Tabb, D.L., Brame, E.G., 1977. 13C NMR characterization of polychloroprene microstructure I. Spectral assignments. Rubber Chem. Technol. 50 (1), 49–62.
Cooper, W., Sewell, P.R., Vaughan, G., 1959. Radiation graft copolymerization in aqueous dispersion. J. Polym. Sci. 41 (138), 167–176.
Corradini, P., Guerra, G., Pucciariello, R., 1985. New model of the origin of the stereospecificity in the synthesis of syndiotactic polypropylene. Macromolecules 18 (10), 2030–2034.
Corradini, P., Busico, V., Guerra, G., 1989. In: Eastmond, G.C., Ledwith, A., Russo, S., Sigwalt, P. (Eds.), Comprehensive Polymer Science, vol. 4. Pergamon Press, Oxford, p. 29.
Dainton, F.S., Sutherland, G., 1949. Application of infrared analysis to elucidate the mechanism of the boron trifluoride catalyzed vapor phase polymerization of isobutene at room temperature. J. Polym. Sci. 4 (1), 37–43.
Davis, K.A., Matyjaszewski, K., 2002. Statistical, gradient, block, and graft copolymers by controlled/living radical polymerizations. Adv. Polym. Sci. 159, 1–169.
Davis, S.C., Hellens, W.V., Zahalka, H.A., et al., 1996. In: Salamone, J.C. (Ed.), Polymeric Materials Encyclopedia. CRC Press, Boca Raton, p. 2264.
Demirors, M., 2003. In: Kroschwitz, J.I. (Ed.), Encyclopedia of Polymer Science and Technology, vol. 4. Wiley-Interscience, New York, p. 229.
Domski, G.J., Rose, J.M., Coates, G.W., et al., 2007. Living alkene polymerization: new methods for the precision synthesis of polyolefins. Prog. Polym. Sci. 32 (1), 30–92.
Dreyfuss, P., 1982. Poly(Tetrahydrofuran). Gordon and Breach, New York.
Dreyfuss, M.P., Dreyfuss, P., 1966. p-Chlorophenyldiazonium hexafluorophosphate as a catalyst in polymerization of tetrahydrofuran and other cyclic ethers. J. Polym. Sci. Part A: Polym. Chem. 4 (9), 2179–2200.
Dreyfuss, P., Dreyfuss, M.P., Pruckmayr, G., 1989. In: Kroschwitz, J.I. (Ed.), Encyclopedia of Polymer Science and Engineering, vol. 16. Wiley-Interscience, New York, p. 649.
Duck, E.W., Locke, J.M., 1977. In: Saltman, W.M. (Ed.), The Stereo Rubbers. Wiley, New York, p. 139.
Eastmond, G.C., 1976a. Free radical polymerization. In: Bamford, C.H., Tipper, C.F.H. (Eds.), Comprehensive Chemical Kinetics, vol. 14A. Elsevier, New York, p. 333.
Eastmond, G.C., 1976b. Free radical polymerization. In: Bamford, C.H., Tipper, C.F.H. (Eds.), Comprehensive Chemical Kinetics, vol. 14A. Elsevier, New York, p. 1.
Ebdon, J.R., 1978. Microstructure of polychloroprene determined by C-13 nuclear magnetic resonance. Polymer 19 (10), 1232–1233.
Eley, D.D., Richards, A.W., 1949. The kinetics of ionic polymerizations 1. The polymerization of vinyl octyl ether catalyzed by iodine. Trans. Faraday Soc. 45 (5), 425–436.
Ells, F.R., 1963. Ph.D. Dissertation, The University of Akron, Akron, OH.
Evans, A.G., Polanyi, M., 1947. Discussion on some aspects of the chemistry of macromolecules. J. Chem. Soc. 252–257.
Faust, R., 2012. Cationic polymerization of nonpolar vinyl monomers. In: Matyjaszewski, K., Moller, M. (Eds.), Polymer Science: A Comprehensive Reference, vol. 3. Elsevier BV, Amsterdam, pp. 501–526.
Faust, R., Kennedy, J.P., 1986. Living carbocationic polymerization 3. Demonstration of the living polymerization of isobutylene. Polym. Bull. 15 (4), 317–323.
Favier, A., Charreyre, M., 2006. Experimental requirements for an efficient control of free-radical polymerizations via the reversible addition-fragmentation chain transfer (RAFT) process. Macromol. Rapid Commun. 27 (9), 653–692.
Fetters, L.J., Morton, M., 1974. Resolution of state of association of poly(dienyl)lithium chain ends in hydrocarbon solvents. Macromolecules 7 (5), 552–559.
Fischer, H., 1997. The persistent radical effect in living radical polymerization. Macromolecules 30 (19), 5666–5672.
Fischer, H., 1999. The persistent radical effect in controlled radical polymerizations. J. Polym. Sci. Part A: Polym. Chem. 37 (13), 1885–1901.
Flory, P.J., 1940. Molecular size distribution in ethylene oxide polymers. J. Am. Chem. Soc. 62 (6), 1561–1565.
Flory, P.J., 1953a. Principles of Polymer Chemistry. Cornell University Press, Ithaca, NY (Chapter 1).
Flory, P.J., 1953b. Principles of Polymer Chemistry. Cornell University Press, Ithaca, NY, p. 37.
Flory, P.J., 1953c. Principles of Polymer Chemistry. Cornell University Press, Ithaca, NY, p. 75.

Flory, P.J., 1953d. Principles of Polymer Chemistry. Cornell Univeristy Press, Ithaca, NY, p. 318.
Flory, P.J., 1953e. Principles of Polymer Chemistry. Cornell University Press, Ithaca, NY, p. 347.
Flory, P.J., 1953f. Principles of Polymer Chemistry. Cornell University Press, Ithaca, NY, pp. 335–336.
Foreman, L.E., 1968. In: Kennedy, J.P., Törnqvist, E.G.M., (Eds.), Polymer Chemistry of Synthetic Elastomers. Interscience Publishers, New York, p. 491.
Fryling, C.F., 1954. In: Whitby, G.S. (Ed.), Synthetic Rubber. John Wiley, New York, p. 257.
Fukuda, T., Goto, A., 2000. Controlled/living radical polymerization. In: Matyjaszewski, K. (Ed.), ACS Symposium Series 768, vol. 768. American Chemical Society, Washington, DC.
Fukuda, T., Terauchi, T., Goto, A., et al., 1996. Well-defined block copolymers comprising styrene acrylonitrile random copolymer sequences synthesized by living radical polymerization. Macromolecules 29 (8), 3050–3052.
Gadkary, S.D., 1956. M.S. Thesis, University of Akron, Akron, Ohio.
Galimberti, M., Mascellani, N., Piemontesi, F., et al., 1999. Random ethene/propene copolymerization from catalyst system based on a constrained geometry half-sandwich complex. Macromol. Rapid Commun. 20 (4), 214–218.
Gandini, A., Cheradame, H., 1985. In: Kroschwitz, J.I. (Ed.) Encyclopedia of Polymer Science and Engineering, vol. 2. Wiley-Interscience, New York, p. 729.
Gardon, J.L., 1977. In: Schildknecht, C.E. (Ed.), Polymerization Processes. Wiley-Interscience, New York (Chapter 6).
Garrett, R.R., Hargreaves, C.A., Robinson, D.N., 1970. Structurally regular trans-1,4-poly(chloroprene). J. Macromol. Sci. Chem. A 4 (8), 1679–1703.
Georges, M.K., Veregin, R.P.N., Kazmaier, P.M., et al., 1993. Narrow molecular weight resins by a free-radical polymerization process. Macromolecules 26 (11), 2987–2988.
Georges, M.K., Hamer, G.K., Listigovers, N.A., 1998. Block copolymer synthesis by a nitroxide-mediated living free radical polymerization process. Macromolecules 31 (25), 9087–9089.
Germack, D.S., Wooley, K.L., 2007. Isoprene polymerization via reversible addition fragmentation chain transfer polymerization. J. Polym. Sci. Part A: Polym. Chem. 45 (17), 4100–4108.
Germack, D.S., Harrisson, S., Brown, G.O., et al., 2006. Influence of the structure of nanoscopic building blocks on the assembly of micropatterned surfaces. J. Polym. Sci. Part A: Polym. Chem. 44 (17), 5218–5228.
Gilbert, R.G., 1992. Consistent values of rate parameters in free-radical polymerization systems. Pure Appl. Chem. 64 (10), 1563–1567.
Gilbert, R.G., 1995. Emulsion Polymerization: A Mechanistic Approach. Academic Press, San Diego.
Gilbert, R.G., 1996. Critically-evaluated propagation rate coefficients in free radical polymerizations I. Styrene and methyl methacrylate. Pure Appl. Chem. 68 (7), 1491–1494.
Giller, C., Gururajan, G., Wei, J., et al., 2011. Synthesis, characterization, and electrospinning of architecturally-discrete isotactic-atactic-isotactic triblock stereoblock polypropene elastomers. Macromolecules 44 (3), 471–482.
Goethals, E.J., Du Prez, F., 2007. Carbocationic polymerizations. Prog. Polym. Sci. 32 (2), 220–246.
Greenley, R.Z., 1980. Recalculation of some reactivity ratios. J. Macromol. Sci. Chem. A14 (4), 445–515.
Greenley, R.Z., 1999. In: Brandrup, J., Immergut, E.H. (Eds.), Polymer Handbook, 4th ed. Wiley-Interscience, New York, pp. II-181.
Gupta, B., Anjum, N., 2003. Plasma and radiation-induced graft modification of polymers for biomedical applications. Adv. Polym. Sci. 162, 35–61.
Hadjichristidis, N., Pitsikalis, M., Pispas, S., et al., 2001. Polymers with complex architecture by living anionic polymerization. Chem. Rev. 101 (12), 3747–3792.
Hadjichristidis, N., Pispas, S., Floudas, G., 2003a. Block Copolymers Synthetic Strategies, Physical Properties, and Applications. Wiley-Interscience, Hoboken, NJ.
Hadjichristidis, N., Pispas, S., Pitsikalis, M., et al., 2003b. In: Kroschwitz, J.I. (Ed.), Encyclopedia of Polymer Science and Technology, vol. 6. Wiley-Interscience, New York, p. 348.
Hadjichristidis, N., Iatrou, H., Pitsikalis, M., et al., 2006. Macromolecular architectures by living and controlled/living polymerizations. Prog. Polym. Sci. 31 (12), 1068–1132.
Hadjikyriacou, S., Fodor, Z., Faust, R., 1995. Synthetic applications of nonpolymerizable monomers in living cationic polymerization: functional polyisobutylenes by end-quenching. J. Macromol. Sci. Pure Appl. Chem. A32 (6), 1137–1153.

Hadjikyriacou, S., Acar, M.H., Faust, R., 2004. Living and controlled polymerization of isobutylene with alkylaluminum halides as coinitiators. Macromolecules 37 (20), 7543–7547.
Halasa, A.F., 1981. Recent advances in anionic-polymerization. Rubber Chem. Technol. 54 (3), 627–640.
Ham, G.E., Alfrey, T., 1964. Copolymerization. Interscience Publishers, New York.
Hamley, I.W., 2004. Developments in Block Copolymer Science and Technology. Wiley, Hoboken, NJ.
Handlin, D.L., Trenor, S., Wright, K., 2007. Applications of thermoplastic elastomers based on styrenic block copolymers. In: Matyjaszewski, K., Gnanou, Y., Leibler, L. (Eds.), Macromolecular Engineering, vol. 4. Wiley, Weinheim, pp. 2001–2032.
Hargis, I.G., Livigni, R.A., Aggarwal, S.L., 1987. In: Whelan, A., Lee, K.S. (Eds.), Developments in Rubber Technology, vol. 4. Elsevier Applied Science, London, p. 1.
Hargreaves, C.A., 1968. In: Kennedy, J.P., Tö rnqvist, E.G.M. (Eds.), Polymer Chemistry of Synthetic Elastomers. Interscience Publishers, New York, p. 233.
Harkins, W.D., 1947. A general theory of the mechanism of emulsion polymerization. J. Am. Chem. Soc. 69 (6), 1428–1444.
Harney, M.B., Zhang, Y.H., Sita, L.R., 2006. Discrete, multiblock isotactic-atactic stereoblock polypropene microstructures of differing block architectures through programmable stereomodulated living Ziegler-Natta polymerization. Angew. Chem. 45 (15), 2400–2404.
Harries, C.D., 1911. Über Kohlenwasserstoffe Der Butadienreihe Und Über Einige Aus Ihnen Darstellbare Künstliche Kautschukarten: I. Über Synthesen Von Kohlenwasserstoffen Der Butadienreihe. Über Isopren Oder?-Methylbutadien. Justus Liebigs Ann. Chem. 383 (2–3), 157–227.
Harrisson, S., Couvreur, P., Nicolas, J., 2011. Sg1 nitroxide-mediated polymerization of isoprene: alkoxyamine structure/control relationship and alpha. Omega-chain-end functionalization. Macromolecules 44 (23), 9230–9238.
Hawker, C.J., 1994. Molecular-weight control by a living free-radical polymerization process. J. Am. Chem. Soc. 116 (24), 11185–11186.
Hawker, C.J., Bosman, A.W., Harth, E., 2001. New polymer synthesis by nitroxide mediated living radical polymerizations. Chem. Rev. 101 (12), 3661–3688.
Helfand, E., 1975. Block copolymers, polymer-polymer interfaces, and theory of inhomogeneous polymers. Acc. Chem. Res. 8 (9), 295–299.
Higginson, W.C.E., Wooding, N.S., 1952. Anionic polymerisation 1. The polymerisation of styrene in liquid ammonia solution catalysed by potassium amide. J. Chem. Soc. 760–774.
Hill, D.J.T., O'Donnell, J.H., O'Sullivan, P.W., et al., 1983. Anionic co-polymerization of butadiene and isoprene: applicability of the terminal model to high conversion. Polym. Bull. 9 (6–7), 292–298.
Hillmyer, M.A., 2012. Polymer synthesis. In: Matyjaszewski, K., Möller, M. (Eds.), Polymer Science: A Comprehensive Reference, vol. 1. Elsevier, Amsterdam, pp. 31–45.
Hofmann, W., 1964. Nitrile rubber. Rubber Chem. Tech. 37 (2), 1–262.
Hofmann, W., 1984. Recent developments in nitrile rubber technology. Prog. Rubber Technol. 46, 43–84.
Hofmann, W., 1989. Rubber Technology Handbook. Hanser Publishers, Munich, West Germany.
Hogen-Esch, T.E., 1978. Ion-pairing effects in carbanion reactions. Adv. Phys. Org. Chem. 15, 153–266.
Holden, G., Hansen, D.R., 2004. In: Holden, G., Kricheldorf, H.R., Quirk, R.P. (Eds.), Thermoplastic Elastomers, 3rd ed., Hanser Publishers, Munich, Germany, p. 45.
Holden, G., Bishop, E.T., Legge, N.R., 1969. Thermoplastic Elastomers. J. Polym. Sci. Polym. Symp. 26 (1), 37–57.
Holden, G., Kricheldorf, H.R., Quirk, R.P., 2004a. Thermoplastic Elastomers, 3rd ed. Hanser Publishers, Munich, Germany.
Holden, G., Kricheldorf, H.R., Quirk, R.P., 2004b. Thermoplastic Elastomers, 3rd ed. Hanser Publishers, Munich, Germany (Chapter 14).
Holy, N.L., 1974. Reactions of radical-anions and dianions of aromatic-hydrocarbons. Chem. Rev. 74 (2), 243–277.
Horne, S., 1983. In: Quirk, R.P. (Ed.), Transition Metal Catalyzed Polymerizations: Alkenes and Dienes. Harwood, New York, p. 527.

Horne, S.E., Kiehl, J.P., Shipman, J.J., et al., 1956. Ameripol Sn: a cis-1,4-polyisoprene. Ind. Eng. Chem. 48 (4), 784–791.
Hsieh, H.L., 1976. Synthesis of radial thermoplastic elastomers. Rubber Chem. Technol. 49 (5), 1305–1310.
Hsieh, H.L., McKinney, O.F., 1966. Relationship between heterogeneity index and kinetic ratio of anionically polymerized polymers. J. Polym. Sci. Polym. Lett. Ed. 4 (11), 843–847.
Hsieh, H.L., Quirk, R.P., 1996. Anionic Polymerization Principles and Practical Applications. Marcel Dekker, New York.
Hsieh, H.L., Yeh, H.C., 1985. Polymerization of butadiene and isoprene with lanthanide catalysts: characterization and properties of homopolymers and copolymers. Rubber Chem. Technol. 58 (1), 117–145.
Hsieh, H.L., Farrar, R.C., Udipi, K., 1981. Anionic polymerization: some commercial applications. Chemtech 11 (10), 626–633.
Hutchinson, R.A., Aronson, M.T., Richards, J.R., 1993. Analysis of pulsed-laser-generated molecular-weight distributions for the determination of propagation rate coefficients. Macromolecules 26 (24), 6410–6415.
Inoue, S., Aida, T., 1984. In: Ivin, K.J., Saegusa, T. (Eds.), Ring-Opening Polymerization, vol. 1. Elsevier, New York, p. 185.
Ivin, K.J., 1983. Olefin Metathesis. Academic Press, London.
Jenkins, A.D., Kratochvil, P., Stepto, R.F.T., et al., 1996. Glossary of basic terms in polymer science. Pure Appl. Chem. 68 (12), 2287–2311.
Jenkins, A.D., Jones, R.G., Moad, G., 2010. Terminology for reversible-deactivation radical polymerization previously called controlled radical or living radical polymerization. Pure Appl. Chem. 82 (2), 483–491.
Jerome, R., 2004. In: Holden, G., Kricheldorf, H.R., Quirk, R.P. (Eds.), Thermoplastic Elastomers, 3rd ed. Hanser Publishers, Munich, Germany, p. 444.
Jitchum, V., Perrier, S., 2007. Living radical polymerization of isoprene via the RAFT process. Macromolecules 40 (5), 1408–1412.
Johnson, P.R., 1976. Polychloroprene Rubber. Rubber Chem. Technol. 49 (3), 650–702.
Jones, M.T., 1968. In: Kaiser, E.T., Kevan, L. (Eds.), Radical Ions. Interscience, New York.
Kahveci, M.U., Yagci, Y., Avgeropoulos, A., et al., 2012. Well-defined block copolymers. In: Matyjaszewski, K., Möller, M. (Eds.), Polymer Science: A Comprehensive Reference, vol. 6. Elsevier, Amsterdam, pp. 455–509.
Kaminsky, W., Boggioni, L., Tritto, I., 2012. Cycloolefin polymerization. In: Matyjaszewski, K., Möller, M. (Eds.), Polymer Science: A Comprehensive Reference, vol. 3. Elsevier, Amsterdam, pp. 843–873.
Kanaoka, S., Aoshima, S., 2012. Cationic polymerization of polar monomers. In: Matyjaszewski, K., Möller, M. (Eds.), Polymer Science: A Comprehensive Reference, vol. 3. Elsevier, Amsterdam, pp. 527–558.
Keaton, R.J., Jayaratne, K.C., Henningsen, D.A., et al., 2001. Dramatic enhancement of activities for living Ziegler-Natta polymerizations mediated by exposed zirconium acetamidinate initiators: the isospecific living polymerization of vinylcyclohexane. J. Am. Chem. Soc. 123 (25), 6197–6198.
Kennedy, J.P., 1968. In: Kennedy, J.P., Tornqvist, E. (Eds.), Polymer Chemistry of Synthetic Elastomers, vol. I. Interscience Publishers, New York, p. 291.
Kennedy, J.P., 1975. Cationic Polymerization of Olefins: A Critical Inventory. Wiley, New York.
Kennedy, J.P., Iván, B., 1992. Designed Polymers by Carbocationic Macromolecular Engineering: Theory and Practice. Hanser, Munich.
Kennedy, J.P., Maréchal, E., 1982. Carbocationic Polymerization. Wiley, New York.
Kennedy, J.P., Puskas, J.E., 2004. Thermoplastic elastomers by carbocationic polymerization. In: Holden, G., Kricheldorf, H.R., Quirk, R.P. (Eds.), Thermoplastic Elastomers. Hanser Publishers, Munich (Chapter 12).
Keoshkerian, B., Georges, M., Quinlan, M., et al., 1998. Polyacrylates and polydienes to high conversion by a stable free radical polymerization process: use of reducing agents. Macromolecules 31 (21), 7559–7561.
Kerns, M., Henning, S., Rachita, M., 2003. In: Kroschwitz, J.I. (Ed.), Encyclopedia of Polymer Science and Technology, vol. 5. Wiley-Interscience, New York, p. 317.

Klumperman, B., 2012. Controlled composition: statistical, gradient, and alternating copolymers. In: Matyjaszewski. K., Möller, M. (Eds.), Polymer Science: A Comprehensive Reference, vol. 6. Elsevier, Amsterdam, pp. 433–453.

Kondakow, I., 1900. Ueber Das Anormale Verhalten Der Poly-Haloïdverbindungen Zu Alkoholischer Kalilauge. J. Prakt. Chem. 62 (1), 166–188.

Koo, K., Marks, T.J., 1999. Efficient new routes to functionalized polyolefins. Chemtech 29 (10), 13–19.

Kresge, E.N., Schatz, R.H., Wang, H.-C., 1987. In: Kroschwitz, J.I. (Ed.), Encyclopedia of Polymer Science and Engineering, vol. 8. Wiley-Interscience, New York, p. 423.

Kulich, D.M., Kelley, P.D., 1985. In: Kroschwitz, J.I. (Ed.), Encyclopedia of Polymer Science and Engineering, vol. 1. Wiley-Interscience, New York, p. 388.

Lebedev, S.V., 1910. Zh. Russ. Fiz.-Kim. Ova. Chast. 42, 949.

Legge, N.R., 1987. Thermoplastic elastomers. Rubber Chem. Technol. 60 (3), G83–G117.

Leibler, L., 1980. Theory of microphase separation in block co-polymers. Macromolecules 13 (6), 1602–1617.

Lodge, T.P., 2003. Block copolymers: past successes and future challenges. Macromol. Chem. Phys. 204 (2), 265–273.

Long, T.E., Broske, A.D., Bradley, D.J., et al., 1989. Synthesis and characterization of poly(tert-butyl methacrylate-b-isoprene-b-tert-butyl methacrylate) block copolymers by anionic techniques. J. Polym. Sci. Part A: Polym. Chem. 27 (12), 4001–4012.

Lovell, P.A., El-Aasser, M.S., 1997. Emulsion Polymerization and Emulsion Polymers. Wiley, New York.

Lutz, P.J., Peruch, F., 2012. Graft copolymers and comb-shaped homopolymers. In: Matyjaszewski, K., Möller, M. (Eds.), Polymer Science: A Comprehensive Reference, vol. 6. Elsevier, Amsterdam, pp. 511–542.

Lutz, J.F., Lacroix-Desmazes, P., Boutevin, B., 2001. The persistent radical effect in nitroxide mediated polymerization: experimental validity. Macromol. Rapid Commun. 22 (3), 189–193.

Luxton, A.R., 1981. The preparation, modification and applications of nonfunctional liquid polybutadienes. Rubber Chem. Technol. 54 (3), 596–626.

Mackey, D., Jorgensen, A.H., 2000. Nitrile rubber. In: Kirk-Othmer (Ed.), Encyclopedia of Chemical Technology. Wiley, New York, p. 1–15.

Matheson, M.S., Auer, E.E., Bevilacqua, E.B., et al., 1949. Rate constants in free radical polymerizations 1. Methyl methacrylate. J. Am. Chem. Soc. 71 (2), 497–504.

Matheson, M.S., Auer, E.E., Bevilacqua, E.B., et al., 1951. Rate constants in free radical polymerization 3. Styrene. J. Am. Chem. Soc. 73 (4), 1700–1706.

Matsuda, Y., Sato, T., Oishi, Y., et al., 2005. Association of polybutadiene living anions in cyclohexane. J. Polym. Sci. Pol. Phys. 43 (12), 1401–1407.

Matsuda, Y., Nojima, R., Sato, T., et al., 2007. Reversed micelle of polybutadiene living anions in cyclohexane. Macromolecules 40 (5), 1631–1637.

Matsuoka, H., Suetomi, Y., Kaewsaiha, P., et al., 2009. Nanostructure of a poly(acrylic acid) brush and its transition in the amphiphilic diblock copolymer monolayer on the water surface. Langmuir 25 (24), 13752–13762.

Matthews, F.E., Strange, E.H., 1910. Improvements in the manufacture of synthetic Caoutchouc and the like. British Patent 24,790.

Matyjaszewski, K., 1996. Cationic Polymerizations: Mechanisms, Synthesis, and Applications. Marcel Dekker, New York.

Matyjaszewski, K. (Ed.), 1998. Controlled/Living Radical Polymerization. ACS Symposium Series, vol. 685. American Chemical Society, Washington, DC.

Matyjaszewski, K. (Ed.), 2000. Controlled/Living Radical Polymerization. ACS Symposium Series, vol. 768. American Chemical Society, Washington, DC.

Matyjaszewski, K., Braunecker, W.A., 2007. In: Matyjaszewski, K., Gnanou, Y., Leibler, L. (Eds.), Macromolecular Engineering: Precise Synthesis, Materials Properties, Applications: 1. Synthetic Techniques. Wiley, Weinheim (Chapter 5).

Matyjaszewski, K., Tsarevsky, N.V., 2009. Nanostructured functional materials prepared by atom transfer radical polymerization. Nat. Chem. 1 (4), 276–288.

Matyjaszewski, K., Xia, J.H., 2001. Atom transfer radical polymerization. Chem. Rev. 101 (9), 2921–2990.

Matyjaszewski, K., Kubisa, P., Penczek, S., 1975. Kinetics and mechanism of cationic polymerization of tetrahydrofuran in solution 1. THF-Ccl4 system. J. Polym. Sci. Part A: Polym. Chem. 13 (4), 763–784.
Mayo, F.R., Lewis, F.M., 1944. Copolymerization: I. A. basis for comparing the behavior of monomers in copolymerization; the copolymerization of styrene and methyl methacrylate. J. Am. Chem. Soc. 66 (9), 1594–1601.
Mayo, F.R., Walling, C., 1950. Copolymerization. Chem. Rev. 46 (2), 191–287.
McClelland, B.J., 1964. Anionic free radicals. Chem. Rev. 64 (3), 301–315.
Meckel, W., Goyert, W., Wieder, W., et al., 2004. In: Holden, G., Kricheldorf, H.R., Quirk, R.P. (Eds.), Thermoplastic Elastomers, 3rd ed. Hanser Publishers, Munich, Germany, p. 15.
Meier, D.J., 1969. Theory of block copolymers I. Domain formation in A-B block copolymers. J. Polym. Sci. Polym. Symp. 26 (1), 81–98.
Meyer, A.W., 1949. Effects of polymerization temperature on structure. Ind. Eng. Chem. 41 (8), 1570–1577.
Mishima, E., Matsumiya, Y., Yamago, S., et al., 2008. Kinetics of living anionic polymerization of polystyrenyl lithium in cyclohexane. Polym. J. 40 (8), 749–762.
Miyamoto, M., Sawamoto, M., Higashimura, T., 1984. Living polymerization of isobutyl vinyl ether with the hydrogen iodide iodine initiating system. Macromolecules 17 (3), 265–268.
Miyamoto, N., Yamauchi, K., Hasegawa, H., et al., 2006. Aggregation behavior of polyisoprene chain ends during living anionic polymerization as investigated by time-resolved small-angle neutron scattering. Physica B 385, 752–755.
Moad, G., Rizzardo, E., Thang, S.H., 2008. Radical addition-fragmentation chemistry in polymer synthesis. Polymer 49 (5), 1079–1131.
Moad, G., Rizzardo, E., Thang, S.H., 2009. Living radical polymerization by the RAFT process: a second update. Aust. J. Chem. 62 (11), 1402–1472.
Mochel, W.E., Peterson, J.H., 1949. The structure of neoprene 2. Determination of end-groups by means of radiosulfur. J. Am. Chem. Soc. 71 (4), 1426–1432.
Moore, J.A., 1978. Macromolecular Synthesis, Collective Volume 1. Wiley, New York.
Morawetz, H., 1985. Polymers. The Origins and Growth of a Science. Wiley, New York.
Morton, M., 1961. Rubber Plast. Age. 42, 397.
Morton, M., 1969. In: Ham, G.E. (Ed.), Vinyl Polymerization Part II. M. Dekker, New York, p. 211.
Morton, M., 1971. In: Bikales, N. (Ed.). Encyclopedia of Polymer Science and Technology, vol. 15. Wiley, New York, p. 508.
Morton, M., 1983. Anionic Polymerization: Principles and Practice. Academic Press, New York.
Morton, M., Fetters, L.J., 1964. Homogeneous anionic polymerization V. Association phenomena in organolithium polymerization. J. Polym. Sci. A: Gen. Pap. 2 (7), 3311–3326.
Morton, M., Fetters, L.J., 1967. Homogeneous anionic polymerization of unsaturated monomers. J. Polym. Sci. Macromol. Rev. 2 (1), 71–113.
Morton, M., Gadkary, S.D., 1956. In: 130th Meet. Am. Chem. Soc., Atlantic City, NJ.
Morton, M., Gibbs, W.E., 1963. Emulsion polymerization of 2,3-dimethylbutadiene-1,3. J. Polym. Sci. A: Gen. Pap. 1 (8), 2679–2695.
Morton, M., Huang, L.-K., 1979. Unpublished data, Huang, L.-K. Ph.D. Dissertation, The University of Akron, Akron, OH.
Morton, M., Piirma, I., 1956. Emulsion polymerization of chloroprene 2. Molecular weights. J. Polym. Sci. 19 (93), 563–577.
Morton, M., Rupert, J.R., 1983. In: Bailey Jr., F.E. (Ed.), Initiation of Polymerization. ACS Symposium Series, vol. 212. American Chemical Society, Washington, DC, p. 283.
Morton, M., Salatiello, P.P., 1951. Cross-linking reaction in butadiene polymerization. J. Polym. Sci. 6 (2), 225–237.
Morton, M., Salatiello, P.P., Landfield, H., 1952. Absolute propagation rates in emulsion polymerization 2. Butadiene in hydroperoxide-polyamine systems. J. Polym. Sci. A 8 (2), 215–224.
Morton, M., Kaizerman, S., Altier, M.W., 1954. Swelling of latex particles. J. Colloid Sci. 9 (4), 300–312.
Morton, M., Cala, J.A., Altier, M.W., 1956. Emulsion polymerization of chloroprene 1. Mechanism. J. Polym. Sci. 19 (93), 547–562.
Morton, M., Hall, J.L., Rembaum, A.A., 1963. Homogeneous anionic polymerization 2. Molecular weight of polystyrene initiated by lithium alkyls. J. Polym. Sci. A: Gen. Pap. 1 (1), 461–474.

Morton, M., Pett, R.A., Fetters, L.J., 1970a. Cross-association of polyisoprenyllithium with ethyllithium. Macromolecules 3 (3), 333–337.
Morton, M., Fetters, L.J., Pett, R.A., et al., 1970b. Association behavior of polystyryllithium, polyisoprenyllithium, and polybutadienyllithium in hydrocarbon solvents. Macromolecules 3 (3), 327–332.
Morton, M., Kammereck, R.F., Fetters, L.J., 1971. Polymerisation and block copolymerisation of cyclic sulphides. Br. Polym. J. 3 (3), 120–128.
Morton, M., Kesten, Y., Fetters, L.F., 1974. Structure and properties of poly(alpha-methylstyrene-b-dimethylsiloxane-b-alpha-methylstyrene). Polym. Prepr. Am. Chem. Soc. Div. Polym. Chem. 15, 175.
Müller, A.H., 1989. In: Eastmond, G.C., Ledwith, A., Russo, S., Sigwalt, P. (Eds.), Comprehensive Polymer Science, vol. 3. Pergamon Press, Oxford, UK, p. 387.
Musch, R., Magg, H., 1996. In: Salamone, J.C. (Ed.), Polymeric Materials Encyclopedia, vol. 2. CRC Press, Boca Raton, Florida, p. 1238.
Nagata, N., Kobatake, T., Watanabe, H., et al., 1987. Effect of chemical modification of solution-polymerized rubber on dynamic mechanical-properties in carbon-black-filled vulcanizates. Rubber Chem. Technol. 60 (5), 837–855.
Napper, D.H., Gilbert, R.G., 1989. In: Eastmond, G.C., Ledwith, A., Russo, S., Sigwalt, P. (Eds.), Comprehensive Polymer Science, vol. 4. Pergamon, Oxford, p. 171.
Natta, G., 1955. Une nouvelle classe de polymeres d alpha-olefines ayant une regularite de structure exceptionnelle. J. Polym. Sci. 16 (82), 143–154.
Neal, A.M., Mayo, L.R., 1954. In: Whitby, G.S. (Ed.), Synthetic Rubber. John Wiley, New York, p. 770.
Niu, A.Z., Stellbrink, J., Allgaier, J., et al., 2004. A new view of the anionic diene polymerization mechanism. Macromol. Symp. 215, 1–15.
Noordermeer, J.W.M., 2003. In: Kroschwitz, J.I. (Ed.), Encyclopedia of Polymer Science and Technology, vol. 6. Wiley-Interscience, New York, p. 178.
Norrish, R.G.W., Russell, K.E., 1952. Polymerization of isobutene by stannic chloride. Trans. Faraday Soc. 48 (1), 91–98.
Novak, B.M., Risse, W., Grubbs, R.H., 1992. The development of well-defined catalysts for ring-opening olefin metathesis polymerizations (ROMP). Adv. Polym. Sci. 102, 47–72.
Odian, G., 2004a. Principles of Polymerization, 4th ed. Wiley-Interscience, Hoboken, New Jersey (Chapter 5).
Odian, G., 2004b. Principles of Polymerization, 4th ed. Wiley-Interscience, Hoboken, New Jersey.
Ohlinger, R., Bandermann, F., 1980. Kinetics of the propagation reaction of butadiene-styrene copolymerization with lithium-organic compounds. Macromol. Chem. 181 (9), 1935–1947.
Oishi, Y., Watanabe, H., Kanaya, T., et al., 2006. Dynamics of monofunctional polybutadienyl lithium chains aggregated in benzene. Polym. J. 38 (3), 277–288.
Oishi, Y., Matsumiya, Y., Watanabe, H., 2007. Kinetics of anionic polymerization of polybutadienyl lithium in benzene: an osmotic effect on propagation process. Polym. J. 39 (4), 304–317.
Olfstead, E.A., 1988. In: Kroschwitz, J.I. (Ed.), Encyclopedia of Polymer Science and Engineering, vol. 11. Wiley-Interscience, New York, p. 287.
Ouchi, M., Terashima, T., Sawamoto, M., 2009. Transition metal-catalyzed living radical polymerization: toward perfection in catalysis and precision polymer synthesis. Chem. Rev. 109 (11), 4963–5050.
Paul, D.E., Lipkin, D., Weissman, S.I., 1956. Reactions of sodium metal with aromatic hydrocarbons. J. Am. Chem. Soc. 78 (1), 116–120.
Penczek, S., 2002. Terminology of kinetics, thermodynamics, and mechanisms of polymerization. J. Polym. Sci. Part A: Polym. Chem. 40 (11), 1665–1676.
Penczek, S., Pretula, J.B., 2012. Fundamental aspects of chain polymerization. In: Matyjaszewski, K., Möller, M. (Eds.), Polymer Science: A Comprehensive Reference, vol. 3. Elsevier, Amsterdam, pp. 3–38.
Penczek, S., Kubisa, P., Matyjaszewski, K., 1985. Cationic ring-opening polymerization 2. Synthetic applications. Adv. Polym. Sci. 68 (69), 1–298.
Pepper, D.C., 1949. Friedel-crafts polymerizations 2. The kinetics of polymerization of styrene by stannic chloride. Trans. Faraday Soc. 45 (4), 404–411.
Pepper, D.C., 1950. Sci. Proc. R. Dublin Soc. 25, 131.

Petiaud, R., Pham, Q.T., 1985. Polychloroprenes 1. Study of the microstructure using H-1 nuclear magnetic resonance. J. Polym. Sci. Part A: Polym. Chem. 23 (5), 1333–1342.

Piirma, I., 1982. Emulsion Polymerization. Academic Press, New York.

Pike, M., Watson, W.F., 1952. Mastication of rubber 1. Mechanism of plasticizing by cold mastication. J. Polym. Sci. 9 (3), 229–251.

Pino, P., Giannini, U., Porri, L., 1987. In: Kroschwitz, J.I. (Ed.), Encyclopedia of Polymer Science and Engineering, vol. 8. Wiley-Interscience, New York, p. 147.

Plesch, P.H., 1950. The low-temperature polymerisation of isobutene. J. Chem. Soc. 543–556.

Poehlein, G.W., 1986. In: Kroschwitz, J.I. (Ed.), Encyclopedia of Polymer Science and Engineering, vol. 6. Wiley-Interscience, New York, p. 1.

Porri, L., Giarrusso, A., 1989. In: Eastmond, G.C., Ledwith, A., Russo, S., Sigwalt, P. (Eds.), Comprehensive Polymer Science, vol. 4. Pergamon Press, Oxford, UK, p. 53.

Pruckmayr, G., Wu, T.K., 1978. Polymerization of tetrahydrofuran by proton acids. Macromolecules 11 (4), 662–668.

Puskas, I., Banas, E.M., Nerheim, A.G., 1976. Nature of double-bond in low-molecular weight polyisobutylenes and polybutene copolymers. J. Polym. Sci. Polym. Symp. 56, 191–202.

Quirk, R.P., 1984. Recent advances in controlled grafting of elastomers. Rubber Chem. Technol. 57 (3), 557–582.

Quirk, R.P., 1992. In: Aggarwal, S.L., Russo, S. (Eds.), Comprehensive Polymer Science, First Supplement. Pergamon Press, Oxford, UK, p. 83.

Quirk, R.P., 2012. Anionic polymerization of nonpolar monomers. In: Matyjaszewski, K., Möller, M. (Eds.), Polymer Science: A Comprehensive Reference, vol. 3. Elsevier BV, Amsterdam, pp. 559–590.

Quirk, R.P., Kinning, D.J., 1989. In: Allen, G., Bevington, J.C. (Eds.), Comprehensive Polymer Science, vol. 7. Pergamon Press, Oxford, UK, p. 1.

Quirk, R.P., Lee, B., 1992. Experimental criteria for living polymerizations. Polym. Int. 27 (4), 359–367.

Quirk, R.P., Ma, J.J., 1991. Dilithium initiators based on 1,3-bis(1-phenylethenyl)benzene—tetrahydrofuran and lithium SEC-butoxide effects. Polym. Int. 24 (4), 197–206.

Quirk, R.P., Monroy, V.M., 1995. Anionic initiators. In: Kroschwitz, J.I. (Ed.), Kirk-Othmer Encyclopedia of Chemical Technology, 4th ed., vol. 14. Wiley, New York, pp. 461–476.

Quirk, R.P., Pickel, D.L., 2012. Controlled end-group functionalization (including telechelics). In: Matyjaszewski, K., Möller, M. (Eds.), Polymer Science: A Comprehensive Reference, vol. 6. Elsevier, Amsterdam, pp. 351–412.

Quirk, R.P., Xu, D. Unpublished Work, University of Akron.

Quirk, R., Yoo, T., Lee, Y., et al., 2000. Applications of 1,1-diphenylethylene chemistry in anionic synthesis of polymers with controlled structures. Adv. Polym. Sci. 153, 67–162.

Quirk, R.P., Pickel, D.L.G., Schulz, G.O., 2004. In: Holden, G., Kricheldorf, H.R., Quirk, R.P. (Eds.), Thermoplastic Elastomers, 3rd ed. Hanser Publishers, Munich, Germany, p. 323.

Rembaum, A., Tobolsky, A.V., Morrow, R.C., et al., 1962. Some results of cesium-initiated diene polymerization and copolymerization. J. Polym. Sci. 61 (171), 155–165.

Rempp, P., Franta, E., Herz, J.E., 1988. Macromolecular engineering by anionic methods. Adv. Polym. Sci. 86, 145–173.

Riess, G., Hurtrez, G., 1985. In: Kroschwitz, J.I. (Ed.), Encyclopedia of Polymer Science and Engineering, vol. 2. Wiley-Interscience, New York, p. 324.

Rizzardo, E., Chiefari, J., Chong, B.Y.K., et al., 1999. Tailored polymers by free radical processes. Macromol. Symp. 143, 291–307.

Robertson, R.E., Marion, L., 1948. The mechanism of sodium-initiated polymerization of diolefins. Can. J. Res. Sect. B 26 (9), 657–667.

Sandler, S.R., Karo, W., 1992. Polymer Syntheses, 2nd ed. Academic Press Inc., San Diego, CA, p. 127.

Santee, E.R., Malotky, L.O., Morton, M., 1973. Analysis of chain unit structure of poly(butadiene) by means of the Hr-300 NMR Spectrometer. Rubber Chem. Technol. 46 (5), 1156–1165.

Saunders, J.H., Dobinson, F., 1976. In: Bamford, C.H., Tipper, C.F.H., (Eds.), Comprehensive Chemical Kinetics, vol. 15. Elsevier, New York (Chapter 7).

Sawamoto, M., 1991. Modern cationic vinyl polymerization. Prog. Polym. Sci. 16 (1), 111–172.

Sawamoto, M., Higashimura, T., 1989. In: Kroschwitz, J.I. (Ed.), Encyclopedia of Polymer Science and Engineering, vol. Supplemental. Wiley-Interscience, New York, p. 399.

Sawamoto, M., Kamigaito, M., 2000. Controlled synthesis of functionalized polymers by transition-metal-mediated living radical polymerization. Macromol. Symp. 161, 11–18.
Schmitt, B.J., 1979. Polymer alloys: their structure, morphology, and properties. Angew. Chem. Int. Ed. Engl. 18 (4), 273–295.
Schöbel, A., Winkenstette, M., Anselment, T.M.J., et al., 2012. Copolymerization of alkenes and polar monomers by early and late transition metal catalysts. In: Matyjaszewski, K., Möller, M. (Eds.), Polymer Science: A Comprehensive Reference, vol. 3. Elsevier, Amsterdam, pp. 779–823.
Schue, F., Bywater, S., 1969. Exchange reactions between lithium alkyl and alkenyl aggregates in hydrocarbon solution. Macromolecules 2 (5), 458–461.
Selman, C.M., Hsieh, H.L., 1971. Effect of aggregate size on alkyllithium initiated polymerizations. J. Polym. Sci. Polym. Lett. Ed. 9 (3), 219–224.
Sipos, L., De, P., Faust, R., 2003. Effect of temperature, solvent polarity, and nature of lewis acid on the rate constants in the carbocationic polymerization of isobutylene. Macromolecules 36 (22), 8282–8290.
Sipos, L., Som, A., Faust, R., 2005. Controlled delivery of paclitaxel from stent coatings using poly(hydroxystyrene-b-isobutylene-b-hydroxystyrene) and its acetylated derivative. Biomacromolecules 6 (5), 2570–2582.
Smid, J., Van Beylen, M., Hogen-Esch, T.E., 2006. Perspectives on the contributions of michael szwarc to living polymerization. Prog. Polym. Sci. 31 (12), 1041–1067.
Smith, T.L., Dickie, R.A., 1969. Viscoelastic and ultimate tensile properties of styrene-butadiene-styrene block copolymers. J. Polym. Sci. Polym. Symp. 26 (1), 163–187.
Smith, W.V., Ewart, R.H., 1948. Kinetics of emulsion polymerization. J. Chem. Phys. 16 (6), 592–599.
Solomon, D.H., 1972. Step-Growth Polymerizations. Dekker, New York.
Solomon, D.H., Rizzardo, E., Cacioli, P., 1986. Polymerization Process and Polymers Produced Thereby. US 4,581,429.
Spirin, Y.L., Medvedev, S.S., Gantmakher, A.R., et al., 1962. Polymerization catalyzed by lithium and lithium alkyl. J. Polym. Sci. 58 (166), 1181–1189.
Staudinger, H., 1920. Über polymerisation. Chem. Ber. 53 (6), 1073–1085.
Stavely, F.W., Dunbrook, R.F., Forman, L.E., et al., 1956. Coral rubber: a cis-1,4-polyisoprene. Ind. Eng. Chem. 48 (4), 778–783.
Stein, D.J., Fahrbach, G., Adler, H., 1974. Crosslinking reactions in rubber particles of high-impact polystyrene. Angew. Makromol. Chem. 38, 67–79.
Stellbrink, J., Willner, L., Jucknischke, O., et al., 1998. Self-assembling behavior of living polymers. Macromolecules 31 (13), 4189–4197.
Stellbrink, J., Allgaier, J., Willner, L., et al., 2002. Real time SANS study on head group self-assembly for lithium based anionic polymerizations. Polymer 43 (25), 7101–7109.
Stewart Jr., C.A., Takeshita, T., Coleman, M.L., 1985. In: Kroschwitz, J.I. (Ed.), Encyclopedia of Polymer Science and Engineering, vol. 3. Wiley, New York, p. 441.
Sun, H.N., Wusters, J.P., 2004. In: Kroschwitz, J.I. (Ed.), Kirk-Othmer Encyclopedia of Chemical Technology, 4th ed., vol. 4. Wiley-Interscience, New York, p. 365.
Szwarc, M., 1968. Carbanions, Living Polymers, and Electron Transfer Processes. Interscience Publishers, New York.
Szwarc, M., 1983. Living polymers and mechanisms of anionic polymerization. Adv. Polym. Sci. 49, 1–177.
Szwarc, M., 1996. Ionic Polymerization Fundamentals. Hanser Publishers, Munich.
Szwarc, M., Jagur-Grodzinski, J., 1972. In: Szwarc, M. (Ed.), Ions and Ion Pairs in Organic Reactions. Wiley-Interscience, New York, p. 1.
Szwarc, M., Levy, M., Milkovich, R., 1956. Polymerization initiated by electron transfer to monomer: a new method of formation of block polymers. J. Am. Chem. Soc. 78 (11), 2656–2657.
Tanaka, Y., Sato, H., Nakafutami, Y., et al., 1983. Sequence distribution of styrene butadiene copolymer initiated by N-butyllithium. Macromolecules 16 (12), 1925–1928.
Tanaka, Y., Sato, H., Adachi, J., 1986. Sequence distribution of commercial SBR by ozonolysis-GPC measurement. Rubber Chem. Technol. 59 (1), 16–26.
Tauer, K., 2003. In: Kroschwitz, J.I. (Ed.), Encyclopedia of Polymer Science and Technology, vol. 6. Wiley-Interscience, New York, p. 410.

Teyssie, P., Hadjiandreou, P., Julemont, M., et al., 1988. In: Quirk, R.P. (Ed.), Transition Metal Catalyzed Polymerizations. Cambridge University Press, Cambridge, p. 639.
Thiele, J., 1901. Zur Kenntniss Des Piperylens Und Tropilidens. Justus Liebigs Ann. Chem. 319 (2), 226–230.
Tian, J., Hustad, P.D., Coates, G.W., 2001. A new catalyst for highly syndiospecific living olefin polymerization: homopolymers and block copolymers from ethylene and propylene. J. Am. Chem. Soc. 123 (21), 5134–5135.
Tilden, W.A., 1884. On the decomposition of Terpenes by heat. J. Chem. Soc. 45, 410–420.
Tirrell, D.A., 1986. In: Kroschwitz, J.I. (Ed.), Encyclopedia of Polymer Science and Engineering, 2nd ed., vol. 4. Wiley, New York, p. 192.
Tirrell, D.A., 1989. In: Eastmond, G.C., Ledwith, A., Russo, S., Sigwalt, P. (Eds.), Comprehensive Polymer Science, vol. 3. Pergamon, Oxford, p. 195.
Tobolsky, A.V., Rogers, C.E., 1959. Isoprene polymerization by organometallic compounds. J. Polym. Sci. 40 (136), 73–89.
Tong, J.D., Ni, S.R., Winnik, M.A., 2000. Synthesis of polyisoprene-b-polystyrene block copolymers bearing a fluorescent dye at the junction by the combination of living anionic polymerization and atom transfer radical polymerization. Macromolecules 33 (5), 1482–1486.
Tung, L.H., Lo, G.Y., 1986. In: Lal, J., Mark, J.E., (Eds.), Advances in Elastomers and Rubber Elasticity. Plenum Press, New York, p. 129.
Uraneck, C.A., 1968. In: Kennedy, J.P., Törnqvist, E.G.M. (Eds.), Polymer Chemistry of Synthetic Elastomers. Interscience Publishers, New York, p. 158.
Vanbeylen, M., Bywater, S., Smets, G., et al., 1988. Developments in anionic polymerization: a critical review. Adv. Polym. Sci. 86, 87–143.
Vandenberg, E.J., 1969. Epoxide polymers: synthesis stereochemistry structure and mechanism. J. Polym. Sci. Part A: Polym. Chem. 7 (2), 525–567.
Vandenberg, E.J., 1983. In: Price, C.C., Vandenberg, E.J. (Eds.), Coordination Polymerization. Plenum Press, New York, p. 11.
Vandenberg, E.J., 1986. In: Kroschwitz, J.I. (Ed.), Encyclopedia of Polymer Science and Engineering, vol. 4. Wiley-Interscience, New York, p. 174.
Ver Strate, G., 1986. In: Kroschwitz, J.I. (Ed.), Encyclopedia of Polymer Science and Engineering, vol. 6. Wiley-Interscience, New York, p. 522.
Vinas, J., Chagneux, N., Gigmes, D., et al., 2008. Sg1-based alkoxyamine bearing a N-succinimidyl ester: a versatile tool for advanced polymer synthesis. Polymer 49 (17), 3639–3647.
Wakefield, B.J., 1974. The Chemistry of Organolithium Compounds. Pergamon Press, Oxford.
Walling, C., 1967. Some aspects of the chemistry of alkoxy radicals. Pure Appl. Chem. 15 (1), 69–80.
Wang, X.-S., Luo, N., Ying, S.-K., et al., 2000. The synthesis of Aba block copolymers by means of 'living'/controlled radical polymerization using hydroxyl-terminated oligomers as precursor. Eur. Polym. J. 36 (1), 149–156.
Wardell, J.L., 1982. In: Wilkinson, G., Stone, F.G.A., Abel, E.W. (Eds.), Comprehensive Organometallic Chemistry: The Synthesis, Reactions, and Structures of Organometallic Compounds, vol. 1. Pergamon Press, Oxford, p. 43.
Wasserman, E.P., 2003. In: Kroschwitz, J.I. (Ed.), Encyclopedia of Polymer Science and Technology, vol. 7. Wiley-Interscience, New York, p. 35.
Webster, O.W., 1991. Living polymerization methods. Science 251 (4996), 887–893.
Weerts, P.A., German, A.L., Gilbert, R.G., 1991. Kinetic aspects of the emulsion polymerization of butadiene. Macromolecules 24 (7), 1622–1628.
Whitby, G.S., Crozier, R.N., 1932. Studies of polymers and polymerization: IV. Observations on the polymerization of isoprene and 2,3-dimethylbutadiene-1,3. Can. J. Res. 6 (2), 203–225.
Williams, C.G., 1859. On Isoprene and Caoutchine. Proc. Roy. Soc. London 10, 516–519.
Winstein, S., Clippinger, E., Fainberg, A.H., et al., 1954. Salt effects and ion-pairs in solvolysis. J. Am. Chem. Soc. 76 (9), 2597–2598.
Wofford, C.F., Hsieh, H.L., 1969. Copolymerization of butadiene and styrene by initiation with alkyllithium and alkali metal tert-butoxides. J. Polym. Sci. Part A: Polym. Chem. 7 (2), 461–469.
Wootthikanokkhan, J., Peesan, M., Phinyocheep, P., 2001. Atom transfer radical polymerizations of (meth)acrylic monomers and isoprene. Eur. Polym. J. 37 (10), 2063–2071.

Worsfold, D.J., Bywater, S., 1960. Anionic polymerization of styrene. Can. J. Chem. 38 (10), 1891–1900.

Worsfold, D.J., Bywater, S., 1978. Lithium alkyl initiated polymerization of isoprene: effect of cis-trans isomerization of organolithium compounds on polymer microstructure. Macromolecules 11 (3), 582–586.

Young, R.N., Quirk, R.P., Fetters, L.J., 1984. Anionic polymerizations of non-polar monomers involving lithium. Adv. Polym. Sci. 56, 1–90.

Zambelli, A., Allegra, G., 1980. Reaction-mechanism for syndiotactic specific polymerization of propene. Macromolecules 13 (1), 42–49.

Zambelli, A., Tosi, C., 1974. Stereochemistry of propylene polymerization. Adv. Polym. Sci. 15, 31–60.

Zhang, M., June, S.M., Long, T.E., 2012. Principles of step-growth polymerization (polycondensation and polyaddition). In: Matyjaszewski, K., Möller, M. (Eds.), Polymer Science: A Comprehensive Reference, vol. 5. Elsevier, Amsterdam, pp. 7–47.

Ziegler, K., Holzkamp, E., Breil, H., et al., 1955. The Mulheim normal pressure polyethylene process. Angew. Chem. 67 (19–2), 541–547.

Zohuri, G.H., Albahily, K., Schwerdtfeger, E.D., et al., 2012. Metallocene alkene polymerization catalysts. In: Matyjaszewski, K., Möller, M. (Eds.), Polymer Science: A Comprehensive Reference, vol. 3. Elsevier, Amsterdam, pp. 673–697.

Chapter 3

Structure Characterization in the Science and Technology of Elastomers

C.M. Roland
Naval Research Laboratory, Chemistry Division, Code 6120, Washington, DC, USA

3.1 INTRODUCTION

Early structural characterization of polymers focused on solution properties and their relationship to molecular weight (Dawkins, 1986; Booth and Price, 1989; Yamakawa, 1971; Flory, 1969). Subsequently spectroscopic and chromatographic techniques were developed, and reviews are widely available (Tanaka, 1991; Campbell and White, 1989; Baldwin and Ver Strate, 1972; Hsu, 2004; Stuart, 2002; Koenig, 1999). This chapter describes various characterization techniques, including discussion of the classical methods of analysis, as well as NMR, SANS, and so on. The main modifications for the fourth edition include the addition of newer methods and results and an update of the references.

Knowledge of chemical structure (backbone composition, compositional and sequence distribution) and the macrostructure of a rubber (molecular weight and its distribution, the degree and type of branching) allows inferences to be drawn concerning the mechanical properties (Chapters 4, 5, and 10), the rheology (Chapters 6 and 14), curing behavior (Chapter 7), filler reinforcement (Chapter 8), and chemical reactivity (Chapter 11). However, relationships between fundamental molecular variables and the properties of condensed matter are semiquantitative at best, and correlation never guarantees causation. Following the earlier editions, methods for molecular weight determination are discussed at greater length than spectroscopic methods. This bias is based on the idea that areas in which common problems might arise, with individuals making their own data interpretation, should receive greater attention.

The Science and Technology of Rubber. http://dx.doi.org/10.1016/B978-0-12-394584-6.00003-0

3.2 CHEMICAL COMPOSITION

Elemental analysis of polymers, which refers to determination of the elements and their respective quantities comprising the material, can often be carried out by the methods used for low molecular weight organic compounds (Dawkins, 1986; Booth and Price, 1989; Stuart, 2002; Braun, 1996; Mitchell, 1992; Collins et al., 1973; Tyler, 1967). This is particularly true for methods involving combustion of the sample. Thus, C, H, and N can be determined on milligram samples by complete combustion followed by gas chromatographic analysis of the evolved gases. Sulfur and halogens are also easily determined after combustion, by titration of sulfate or SO_2 for S, and by potentiometric titration with $AgNO_3$ for halogens after treatment of the gases with NaOH and hydrazine sulfate, for example. Interference by nitrogen on sulfur tests can be a problem. Quantities as low as ppm metals can be determined quickly by X-ray techniques.

Elemental analysis reveals only which atoms are present. Determination of the chemical structure usually requires spectroscopic methods. The moieties in the polymer absorb and emit radiation at characteristic frequencies. Skeletal bond transitions can be detected in the infrared and Raman spectra, electronic transitions typical of unsaturated bonds correspond to ultraviolet and visible wavelengths, atomic nuclei with magnetic moments are probed using magnetic resonance experiments.

For basic information, infrared spectroscopy (Hsu, 2004; Stuart, 2002; Koenig, 1999; Ishida, 1987; Messerschmidt and Harthcock, 1988) (invariably Fourier transform infrared, FTIR) is a straightforward technique. Thin films of elastomers, obtained by casting or molding 20–30 mg of sample, can be measured directly. Most analyses make use of the mid-infrared region (4000–400 cm^{-1}), with sample identified by comparisons using widely available spectral libraries (Hsu, 2004; http://www.spectroscopynow.com). If the polymer is crosslinked, it must be microtomed, or spectra obtained in reflection, most commonly using attenuated total reflection (ATR) (Fuhrmann and Karger-Kocsis, 2003; Shield and Ghebremeskel, 2003; McCarley and Bunge, 2003; Rodrigues et al., 2001; Sen et al., 2001) or photoacoustic FTIR (Peckj et al., 1991; Waddell and Parker, 1992). The latter refers to the measurement of sound arising from the selective absorption of modulated light. The sample surface is heated by the absorption, which in turns heats the adjacent air, producing an alternating pressure wave. Sub-milligram-sized samples can be analyzed using infrared microscopy (Messerschmidt and Harthcock, 1988; Stebounova et al., 2003; Keilmann, 2003, 2002; Dumas, 2003; Wetzel and LeVine, 1999; Palanker et al., 2000; Kempfert et al., 2001; Froud et al., 2003; Szep et al., 2003; Maeda et al., 2003; Berger et al., 2003).

Raman spectroscopy yields analogous information and is complementary to infrared spectroscopy. At least for molecules with a center of symmetry, vibrations that are infrared inactive are generally Raman active, and vice versa.

Since the detected light is scattered rather than transmitted by the sample, Raman spectra are readily obtained on crosslinked specimens, and the method has been applied to the study of rubber vulcanization (Tanaka, 1991; Koenig, 1999; Zhao et al., 2006). Interferences due to fluorescence can be avoided by using a longer wavelength source (Rabolt and Chase, 2000). Spatial resolutions of a few hundred nanometers can be achieved using Raman microscopy (Ghanbari-Siahkali et al., 2003; Zerda et al., 2003; Dieing et al., 2011; Smitthipong et al., 2008; Wang et al., 2011).

Whereas Raman and infrared are useful for "fingerprinting" (i.e., identifying chemical structure), extension of the spectroscopic measurements into the visible and ultraviolet (UV) regions is done primarily for quantitative analyses (Grum, 1972; Raubner et al., 2002; Braun et al., 1993). Extinction coefficients for conjugated unsaturated structures are very large.

The foregoing spectral absorption methods can yield quantitative results, although calibration is required. With nuclear magnetic resonance spectroscopy (NMR) (Koenig, 1999; Stuart, 2002; Cheng, 1991; Kinsey, 1990; Wang et al., 1993), the absorption intensity is directly proportional to the amount of the particular isotope present; consequently, ratios of absorption intensities in proton NMR, for example, can be used to determine the number of chemically distinct protons in a sample. The characteristic NMR resonance frequency (e.g., "chemical shift") depends on chemical environment, and therefore the specific chemical nature of the material can be identified.

Chemical information obtained using ^{1}H and ^{13}C NMR is usually obtained on samples in solution (liquid-state NMR) in order to improve resolution. However, ^{13}C spectra can also be obtained on neat specimens, such as rubber. This is possible as long as there is sufficient molecular motion to average the orientation-dependent variation in chemical shift of chemically identical atoms (chemical shift anisotropy, CSA). Chemical shifts in ^{13}C NMR spectra span a much wider range than in proton NMR, and therefore the former provides better spectral resolution. However, the Nuclear Overhauser effect (NOE) and other nuclear relaxation processes cause the ^{13}C absorption intensities to deviate from direct proportionality to the number of carbon atoms. Thus, unless specific techniques are utilized, ^{13}C NMR spectral intensities using standard liquid-state NMR acquisition methods are not quantitative.

This is also true when applying solid-state NMR techniques (Kinsey, 1990) to obtain high-resolution ^{13}C spectra of more rigid samples. Using a combination of magic angle spinning (MAS) and proton decoupling (via high power radiofrequency irradiation of proton resonances), the line-broadening effects of CSA and ^{13}C-proton dipole-dipole interactions can be removed. Relatively high-resolution ^{13}C NMR spectra can be obtained, and these methods can be used for various investigations of cured elastomers (Madan et al., 2012; Litvinov et al., 2011; Wu et al., 2011; Goga et al., 2008; Kotani et al., 2007).

NMR not only yields chemical information, but can also be used to analyze polymer tacticity, sequence lengths, short-chain branching, and crystallinity.

From elemental analyses, infrared, and NMR methods, the chemical repeat units, including blockiness, can be determined, although quantification is more difficult for low concentrations of a monomer unit or functional group, or for atoms other than C and H. An alternative approach is based on the chemical reactivity of the particular group (Smith and Patterson, 1986), although the reaction may be slow due to the low concentration of the group. Examples include acid groups titrated with base; quantitative addition reactions of many olefins with ozone or halogens; esterification of hydroxyl groups with anhydrides, with subsequent titration to evaluate hydroxyl content. In many cases, the procedure has been established for small molecule materials.

Pyrolysis (Hammond and Lehrie, 1989; Vitalini and Scamporrino, 1992; Ghebremeskel et al., 1996; Yamada et al., 1991; Schulten et al., 1989) of samples can produce characteristic fragments, which may be analyzed by gas chromatography (GC) (Stuart, 2002) or mass spectrometry (MS) (Lattimer, 1990). Since the relationship between fragments and the original polymer is often complex, this technique is a last recourse, applied, for example, to insoluble polymers. Combined GC/MS has been used to analyze the volatile components in natural rubber (Hoven et al., 2003). In ozonolysis (Stuart, 2002), an unsaturated sample is reacted to form an unstable intermediate, ozonide, which is then further reacted for chemical identification. Ozonolysis of rubber is usually combined with GC analysis (Kawahara et al., 1998; Tanaka et al., 1987; Tanaka and Sato, 1986; Miller and Tobias, 1978). In secondary ion mass spectrometry (SIMS), the sample surface is irradiated with an ion beam, followed by mass spectrometry of the emitted secondary ions (Leeson et al., 1997; Ruch et al., 2003; van Gennip et al., 2004). SIMS has found various applications in rubber (Ruch et al., 2003; Winesett and Tsou, 2008) including surface analysis (Fulton, 2006; Van Ooij and Rangarajan, 1988; Briggs, 1989) and studies of carbon black interaction (Mathew et al., 2008; Betrand et al., 2002; Betrand and Weng, 1999). Pyrolysis can also serve as a fingerprinting technique for routine analyses. In thermogravimetric analysis (TGA) (Sircar, 1992, 1997), the polymer degradation by-products volatize, whereby the residue provides a measure of the carbon black or other filler content.

Electron paramagnetic resonance (EPR), or electron spin resonance (ESR), can be used to detect types and quantities of free radicals. Such information is of value in studying the chemistry occurring during degradation and fracture of polymeric materials (Capancioni et al., 2003; Hauck et al., 1997; Hinojosa et al., 1972; Devries, 1971; Partridge et al., 1993; Pace and Roland, 1991). EPR can also be applied to study carbon black and other fillers in polymers (Brosseau et al., 2001; Tang et al., 1994; Hommel et al., 1993).

3.3 SEQUENCE DISTRIBUTION OF REPEAT UNITS

Elastomers are usually comprised of various additives (a typical formulation includes more than a dozen ingredients), as well as possible impurities. Additionally, the polymer itself may be a copolymer of different monomer units, or the rubber may be a blend of two or more polymers. In order to determine whether the elastomer contains additives or impurities, or is itself heterogeneous, the components of the mixture must be separated (i.e., fractionated). Many methods are available to separate components, with the latter then identified by techniques such as those shown in Table 3.1. Precipitation and/or dissolution methods simply rely on differences in solubility of the components (Ver Strate et al., 1988). If the molecular weight distribution is broad, a fractionation scheme is combined with one of the more sophisticated analyses described later. If there is intramolecular heterogeneity, "cross" or "orthogonal" fractionation, which requires at least two fractionation mechanisms, may be needed.

The technique of temperature rising elution fractionation (TREF) (Garcia et al., 2011; Feng and Hay, 1998; Mirabella, 1993; Wild, 1991) has been

TABLE 3.1 Characterization of Chemical Composition

Technique	Principle of Operation
Elemental analysis	Analysis of products of decomposition
Infrared absorption $1 < \lambda(\mu m) < 16$	Characteristic vibrational frequencies
Raman scattering	Characteristic vibrational frequencies
Nuclear magnetic resonance (NMR)	Characteristic transition energies of any nucleus with a magnetic moment (^1H, ^2H, ^{13}C, ^{129}Xe, etc.)
Ultraviolet (UV) and visible light absorption $0.2 < \lambda(\mu m) < 0.8$	Characteristic energies of electronic transitions
Functional group analysis	Analyze for known reactions of chemical moiety
Pyrolysis/ozonolysis	Pyrolysis with chromatographic or mass spectrographic identification of fragments
Electron paramagnetic resonance (EPR)	Energy of unpaired electron spin transitions, which depends on chemical environment
Thermogravimetric analysis (TGA)	Weight loss due to temperature-dependent decomposition and evaporation
Mass spectrometry (MS)	Mass of fragments reflects chemical composition
Secondary ion mass spec (SIMS)	Characteristic ions emitted from surface

developed to measure the compositional distribution of semicrystalline polymers. Polymer is dissolved off a substrate as temperature is raised through the melting region, so that discrimination is based on differences in crystallizability of the fractions. A similar method uses supercritical fluids (Dekmezian et al., 1990). TREF can provide information about the sequence distribution, since longer sequences of a stereoregular repeat unit are more crystallizable.

Size exclusion chromatography (SEC) (Stuart, 2002; Ruckebusch et al., 2008; Zimina et al., 1992; Dekmezian et al., 1990; Grubsic-Gallot et al., 1990), also referred to as gel permeation chromatography, utilizes differences in hydrodynamic volume. It is discussed in more detail later. A very basic technique is adsorption chromatography (Glockner, 1989; Xu et al., 2000), in which separation arises from variation in the retention of the chain units or functional groups in a mobile phase, due to their interaction with a stationary surface. Other techniques rely on rates of sedimentation (Yamada et al., 1997; Stanley and Strey, 2003) and diffusion-adsorption phenomena (thin layer chromatography, TLC) (Tacx et al., 1988; Mori and Mouri, 1989).

Different arrangements of the monomer units give rise to different chemical shifts and scalar couplings (splittings) in the NMR spectra. Using selection rules and empirical knowledge of chemical shifts, chemical structures can be assigned. Since chemical shifts in ^{13}C NMR spectra are larger than in proton spectra, subtle structural differences can be seen for carbon atoms separated by up to five bonds from the point of difference. Thus, ^{13}C NMR can be very useful for determining the distribution of chain units in the polymerz backbone. An example of this is seen in Figure 3.1, which shows ^{13}C NMR

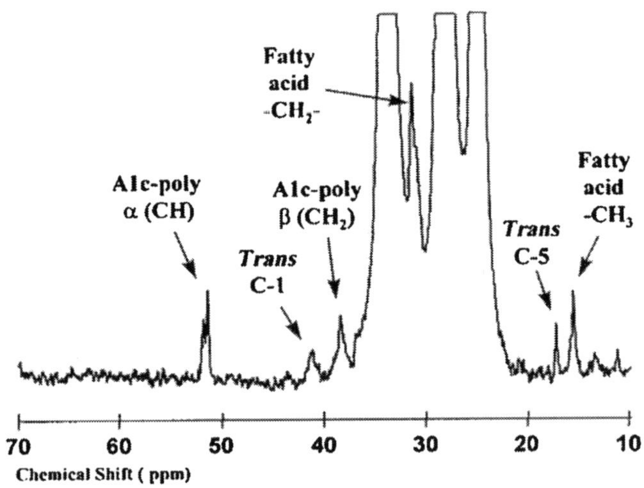

FIGURE 3.1 Magnified portion of ^{13}C NMR spectrum of natural rubber vulcanized to half its maximum torque. The peak at 16 ppm arises due to cis-to-trans isomerization (Mori and Koenig, 1998).

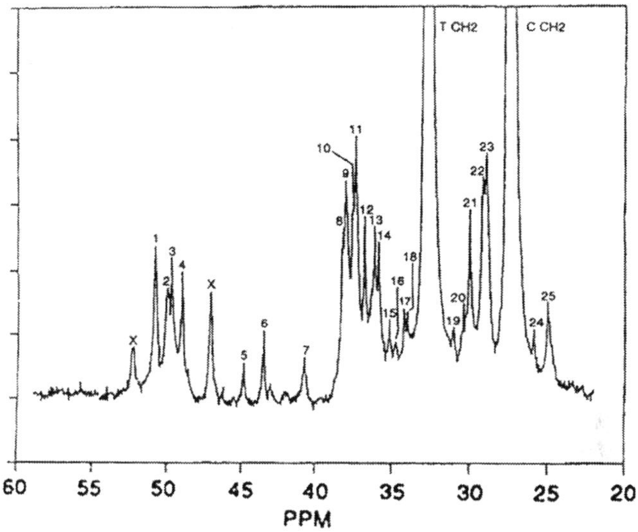

FIGURE 3.2 Magnified ^{13}C NMR spectrum of *cis*-1,4-polybutadiene after curing. The arrows designate new peaks appearing due to the vulcanization. The first seven represent methine carbons and the other 18 are due to methylene carbons (Clough and Koenig, 1989).

spectra of sulfur-cured natural rubber (NR) reinforced with carbon black (Mori and Koenig, 1998). There is a small peak reflecting the *trans* isomer content, which grows in intensity as the vulcanization proceeds. Thus, the NMR measurements yield quantitative information about the extent of *cis-trans* isomerization accompanying the curing of NR. Similar results are found for *cis*-1,4-polybutadiene rubber (Zaper and Koenig, 1988; Clough and Koenig, 1989; Koenig, 2000). Figure 3.2 illustrates the plethora of changes in the NMR spectrum accompanying curing.

Differences in the arrangement of monomer units along the chain backbone are another form of heterogeneity. Chemically different monomer units may occur in sequences of varying length, while identical monomer units can have different geometrical arrangement (stereoisomers), yielding different properties. The stereo-regularity of the polymer can often be determined using the same techniques employed for chemically distinct units, with NMR and infrared the most useful.

When the sequences in the copolymer are longer than 6–8 carbons, techniques other than NMR are needed to directly determine their length. The use of pyrolysis followed by GC-MS analysis has been proposed to find the long sequences as fragments in the pyrolyzate, but the data produced are complicated and difficult to interpret (Tosi, 1968; Yamada et al., 1990; Tulisalo et al., 1985; Hu, 1981).

3.4 CHAIN ARCHITECTURE

The viscoelastic response of amorphous polymers at elevated temperatures is governed to a significant extent by the weight average molecular weight, M_w, the presence of any long chain branching, and the MWD (Tuminello et al., 1993; Graessley and Ver Strate, 1980; Doi, 1993; Ngai and Plazek, 1995; Wasserman, 1995; Majeste et al., 2003). Even the properties of cured elastomers may reflect the length of the original chains, since chain ends, whose concentration is inversely proportional to the number-average molecular weight M_n, represent defects.

3.4.1 Molecular Weight and Its Distribution

The chain length distribution is usually presented as a plot of the mole fraction or weight average of molecules versus molecular weight. The various average molecular weights represent the moments of the chain length distribution

$$\overline{M}_j = \sum_{i=1}^{\infty} N_i M_i^j \Big/ \sum_{i=1}^{\infty} N_i M_i^{j-1}, \tag{3.1}$$

where $i = 1,2$, and 3 yield the number average M_n, weight average M_w, and z average M_z, respectively. For a uniform distribution of chain lengths, $M_n = M_w = M_z$. The common random MWD gives $M_w/M_n = 2$ and $M_z/M_w = 1.5$, while for the Schulz-Zimm distribution, $M_n + M_z = 2M_w$ and $1 < M_z/M_w < 2$. Figure 3.3 shows a typical molecular weight distribution.

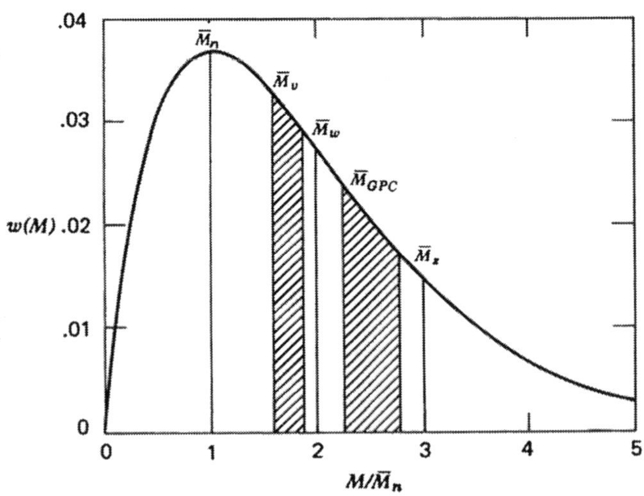

FIGURE 3.3 Representative molecular weight distribution. $M_{GPC} = \sum w_i M_i^{1+a} / \sum w_i M_i$, where a is the Mark-Houwink exponent (Collins et al., 1973).

The number-average molecular weight determines the colligative properties (i.e., those that depend only on the number of dissolved molecules) of polymer solutions (Table 3.2). Measurements of freezing point depression (cryoscopy) or boiling point elevation (ebulliometry), in principle, will yield the same information for macromolecules as for small molecules (Bonnar et al., 1958; Barth and Mays, 1991; Cooper, 1989). These are absolute techniques, which do not require calibration. The change in (melting or boiling) temperature, ΔT, follows a relationship of the form

$$\Delta T/kc = M_n^{-1} + A_1 c + A_2 c^2 + \cdots, \tag{3.2}$$

where c is the concentration, A_i are the virial coefficients, and k is a constant that depends on the solvent, temperature, and the instrument. The number of moles of macromolecules present at mass concentrations sufficiently low to obtain ideal solutions is very small. Therefore, the anticipated change, ΔT, in boiling or freezing points is correspondingly small; for $M_n = 10^5$ g/mol, $\Delta T \sim 10^{-5}$–10^{-4} °C. Thus, these classic techniques are not used above $M_n \sim 10^4$ g/mol. Moreover, cryoscopy and ebulliometry are time consuming and have poor accuracy.

For low molecular weight samples, the preferred method is by vapor pressure osmometry (VPO) (Barth and Mays, 1991; Cooper, 1989). This technique is based on the decrease of vapor pressure of a solvent due to the presence of dissolved polymer. The different equilibrium vapor pressures cause a difference in condensation rate on two matched thermistors, contained in a chamber saturated with solvent vapor. One thermistor is coated with solvent and the other with a solution of the polymer. More solvent condenses on the solution, raising its temperature. The consequent temperature difference is measured, and by calibration, M_n can be determined. VPO is fast and can yield M_n as high as about 5×10^4 g/mol. Commercial vapor pressure osmometers are available.

Membrane osmometry (Barth and Mays, 1991; Cooper, 1989; Brown and Verdier, 1972) relies on the lowering of the activity (free energy) of the solvent by dissolution of a solute, to yield a direct determination of M_n. When a polymer solution is brought in contact with pure solvent, the concentration gradient induces mixing by diffusion. If a semipermeable membrane is placed between the pure solvent and solution, the polymer is trapped but solvent can pass. Equilibrium cannot be attained, but if the polymer solution is in a closed cavity, a pressure develops on the solution side, which eventually stops the solvent flow. The magnitude of this "osmotic" pressure, Π, will depend only on the number of polymer molecules present, at least in the absence of polymer-polymer interactions (i.e., at low concentrations). The equation relating these quantities is analogous to the relation above for the change in melting and boiling points. Measurements at several concentrations with extrapolation of Π/c to $c = 0$ yields M_n. Even for high molecular weight materials, the osmotic pressure is significant. For a 1% solution of a material of $M_n = 10^5$ g/mol,

TABLE 3.2 Characterization of Molecular Weight and Its Distribution

Technique	Variables Measured	Principle of Operation	Range (g/mol)
Membrane osmometry (MO)	Number average molecular weight; virial coefficient	Osmotic pressure due to diffusion of solvent through membrane impermeable to polymer	
Vapor pressure osmometry	Number average molecular weight	Vapor pressure lowering	$<5 \times 10^4$
Cryoscopy	Number average molecular weight	Freezing point depression	$<10^4$
Ebulliometry	Number average molecular weight	Boiling point elevation	$<10^4$
Light scattering (LS)	Weight average molecular weight Virial coefficient Radius of gyration	Light scattering intensity from solution is proportional to the molecular weight of the solute Angular dependence related to particle size	$>10^3$
Neutron scattering	Weight average molecular weight Virial coefficient Radius of gyration	Scattering intensity is proportional to nuclear cross section	
Intrinsic viscosity	Intrinsic viscosity Viscosity average molecular weight	Solution viscosity depends on polymer molecular weight	$>10^3$
Size exclusion (SEC) or gel permeation (GPC) chromatography	Molecular weights and distribution	Permeation of polymer from a flowing solution into a porous stationary phase	$<10^7$
Field flow fractionation (FFF)	Molecular weights and distribution	External field effects separation of flowing solute	$<10^7$
Ultracentrifugation	Molecular weight averages	Concentration gradient in a large gravitational field is related to molecular weights	$>5 \times 10^2$
Sedimentation	Molecular weights Sedimentation coefficient	Rates of diffusion in a gravitational field related to molecular weights	$>5 \times 10^2$
Melt viscosity	Dynamic or steady-state viscosity	Empirical correlation of viscosity and M_W	–

the pressure is about 250 Pa at 25 °C. Generally, the technique is useful for molecular weights in the range from 10^4 to almost 10^6 g/mol.

A frequent difficulty with the method is that Π/c may become nonlinear in c for higher concentrations. If this is due to agglomeration, solvent or temperature changes can sometimes eliminate the problem. To get measurable osmotic pressures for larger M_n, the range of c must be increased. A_2 decreases with M_n in good solvents ($A_2 \sim M^{-0.2}$), but not as fast as c is increased. Therefore, nonlinear plots are often encountered when studying high M_n samples. In order to linearize the data, $(\Pi/c)^{1/2}$ can be plotted versus c. The greatest difficulties with this method relate to imperfections in the membrane. Sometimes the polymer can diffuse through the membrane, with steady-state values indicating permeation of molecules having $M < 5 \times 10^3$ g/mol. This can lead to significant errors, ca. 10%, and even larger for distributions skewed to low molecular weights.

Light scattering is another absolute technique for the determination of molecular weights ($>10^3$ g/mol) (Stuart, 2002; Barth and Mays, 1991; Cooper, 1989; Tanaka, 1999; Berne and Pecora, 1976; Boothroyd and Fetters, 1991; Schulz et al., 1957; Casassa, 1972; Shultz and Stockmayer, 1969; Suzuki et al., 1973). (We are only concerned with static light scattering for structural information; however, inelastic light scattering is a powerful technique for studying polymer dynamics. From calibration, it can also yield molecular weight determinations (Berne and Pecora, 1976).) Light passing through a medium is scattered due to regions of different refractive index, n; thus, polymer molecules dissolved in a solvent having a refractive index different from the polymer will scatter light. Ignoring the contribution from the solvent (which is small compared to the macromolecular scattering), isolated molecules (no interparticle interference) that are much smaller than the wavelength of the light (no intramolecular interferences) will scatter light with an intensity given by

$$I(q) = I_0 \left[\frac{4\pi^2 n_o^2}{\lambda^4 N_A} \left(\frac{\partial n}{\partial c} \right)_T^2 \right] Mc, \tag{3.3}$$

where q $\left(\equiv \frac{4\pi}{\lambda} \sin\left(\frac{\theta}{2}\right)\right)$ is the momentum transfer and θ is the scattering angle, I_0 is the intensity scattered through $q = 0$, λ is the wavelength, N_a is Avagadro's number, and n_0 and n are the respective solvent and solution refractive indices. The change in the refractive index with concentration is approximately equal to the ratio of the relative refractive index and the polymer mass density, $(\partial n/\partial c)_T = (n_{\text{poly}} - n_0)/\rho$. Generally, $\partial n/\partial c|_T < 0.2$ mL/g for polymer solutions.

In a given experiment, the quantity in brackets in Eq. (3.3) is a constant, K, so

$$R(q) \equiv I(q)/I_0 = KMc, \tag{3.4}$$

where $R(q)$ is the Rayleigh ratio. Polymer chains are large enough ($>\lambda/10$) that light scattered from different parts of the molecule have different phase shifts; thus, Eq. (3.4) is modified by the molecular structure factor, $P(q)$

$$R(q) = KMcP(q). \tag{3.5}$$

This structure factor can be approximated at small scattering angles as

$$P(q) \approx 1 - \frac{1}{3}q^2 S_g^2 + \cdots, \tag{3.6}$$

which is valid for particles of any shape. S_g is the root-mean-square radius of gyration of the polymer, and small angle is defined by the condition $q \ll S_g^{-1}$. A plot of $R(q)$ versus q^2 at small q is a straight line whose intercept yields the molecular weight, and the ratio of slope to intercept yields the radius of gyration. Since the reciprocal of the Rayleigh ratio is more sensitive to small variations than R itself, Eq. (3.5) is usually rewritten (using the fact that $1+x = [1-x]^{-1}$ for small x)

$$\frac{Kc}{R(q)} = M^{-1}\left(1 + \frac{1}{3}q^2 S_g^2\right). \tag{3.7}$$

Thus, $R(q)^{-1}$ is plotted versus q^2, with the molecular weight and S_g obtained via extrapolation to $q = 0$.

The foregoing assumes independent scattering from each particle. At the usual polymer concentrations, even though the visual turbidity remains negligible, waves scattered from different chains interfere. This interparticle interference causes the scattering intensity to deviate from a linear dependence on c. To correct for the concentration effect, a power series expansion is assumed

$$\frac{Kc}{R(q)} = M_w^{-1}\left(1 + \frac{1}{3}q^2 \left\langle S_g \right\rangle^2\right) + 2A_2 c + \cdots, \tag{3.8}$$

where A_2 is called the second virial coefficient. Note that in consideration of a distribution of molecular masses, M is now the weight average and the obtained radius of gyration is the mean-square z average. Application of Eq. (3.8) requires extrapolation to $c = 0$, in addition to the extrapolation to zero angle. These two extrapolations are carried out simultaneously via a Zimm plot (Figure 3.4). A proper Zimm plot yields both the radius of gyration and the virial coefficient, consistent with the M_w obtained for the polymer. Deviations from the linearity at lower angles indicate artifacts, such aggregates, gel, dust, and so on (Vavra et al., 1967). With rubbers, clarification problems can be especially serious due to the possibility of removing some of the polymer along with the contaminants. Using a laser beam having a small cross-sectional area, it is sometimes possible to see "between" gel or dust particles, which are then detected only intermittently as "spikes" of large scattering intensity. This may alleviate to some extent the need for solution clarification. Supplementing the data with dynamic light scattering measurements has been proposed as a means to correct for the contribution from large impurities (Lindner and Glatter, 2000).

Multiple detectors, from 3 to as many as 18, facilitate the extrapolation to zero angle. In principle, a scattering detector can be placed at a sufficiently

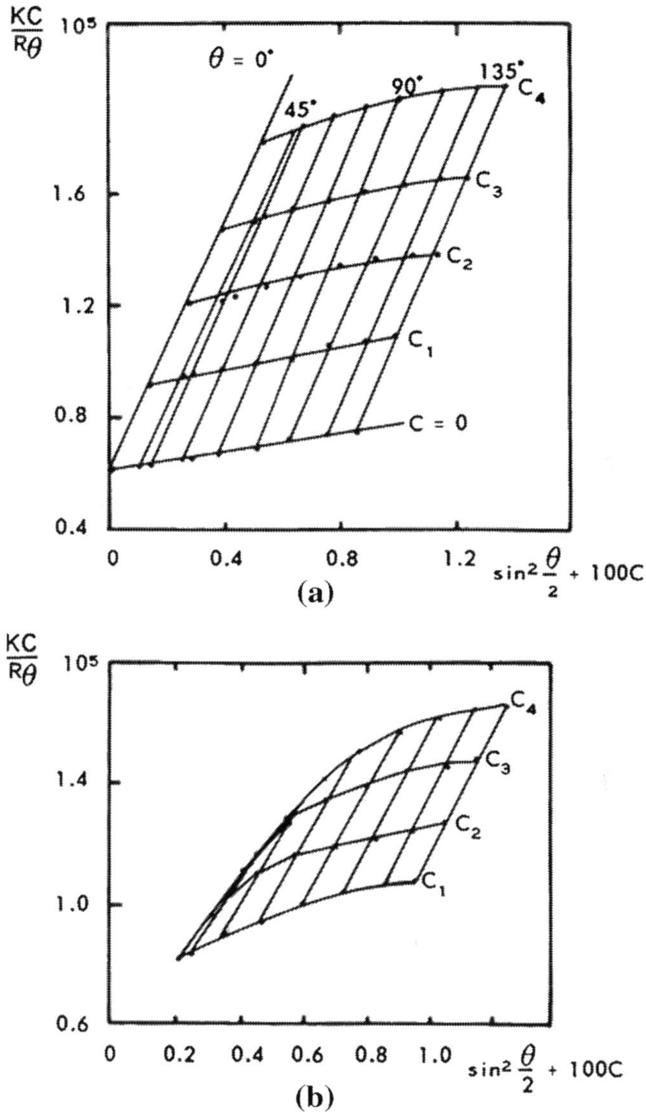

FIGURE 3.4 Zimm plots of commercial polyvinylchloride in THF: (a) purified solution with expected linear variation of $R(q)^{-1}$, (b) as received material, showing curvature due to aggregation (Vavra et al., 1967).

low angle, that such extrapolation of the scattering intensity to $q = 0$ becomes unnecessary, at least if the molecular weights are not too large. When a light scattering detector is attached to a chromatograph, the concentration of polymer is low, to some extent minifying the need for extrapolation to zero concentration. Of course, the radius of gyration and virial coefficient are not obtained. However, in this manner, a molecular weight determination by light scattering can be completed expeditiously.

Structural information can also be obtained from measurements of small-angle neutron scattering (SANS) (Higgins and Benoit, 1997; Wignall, 1996; Lohse, 1994; Hammouda, 2010). Scattering of neutrons is due to their interaction with nuclei. It differs from light scattering, in that contrast arises from differences in neutron scattering length, rather than refractive index differences. Neutron cross-sections are commonly expressed in terms of the relevant correlation function. Elastic, coherent scattering is proportional to the spatial Fourier transform of the pair-correlation function. The angular distribution of coherent scattering is reinforced by constructive interference, and thus yields structural and conformational information. Conventional SANS probes dimensions in the range from 1 to 100 nm length scales, while so-called ultra-small angle neutron scattering (USANS) probes length scales up to tens of μm (Sharp et al., 2009). The advent of the Bonse-Hart camera, which utilizes multiple reflection crystals before and after the sample (Bonse and Hart, 1965), has overcome the historical problem of poor signal-to-noise ratios in measurements at very high resolution.

Inelastic coherent and incoherent scattering are proportional to the space and time Fourier transforms of, respectively, the pair-correlation and the self-correlation functions. Analogous to dynamic light scattering, these are used to probe polymer dynamics (Arbe et al., 2003; Richter, 2003; Stepanek et al., 2002; Annis et al., 2001).

The neutron scattering cross-sections of carbon and oxygen are completely coherent. ^{1}H has a very large incoherent scattering cross-section and a small coherent cross-section, while the opposite is true for deuterium. One advantage of SANS is that isotopic labeling (usually the replacement of hydrogen by deuterium) can be used to provide selective contrast. Such labeling has minimal chemical effect, enabling $\langle S_g \rangle$ and M_w to be determined in situations when light scattering cannot be used. Dust is also not a problem with SANS. An obvious disadvantage of the method is the greater cost of preparing labeled materials.

One important area in which SANS has yielded information not otherwise available is determining chain dimensions in bulk polymers. Samples can be a couple of millimeters in thickness, with up to 50% of the incident neutrons scattered. The mean-square radius of gyration and M_w can be determined for a deuterated version of a polymer mixed with its unlabeled analog. No Zimm analysis is required, despite overlap and interpenetration of the chain molecules. The only requirement is that both isotopic species have the same M_w and MWD. A model-independent analysis of the scattering used the Gunier approximation

$$I(q) = I_0 \exp\left(-\frac{1}{3} q^2 S_g^2\right) \tag{3.9}$$

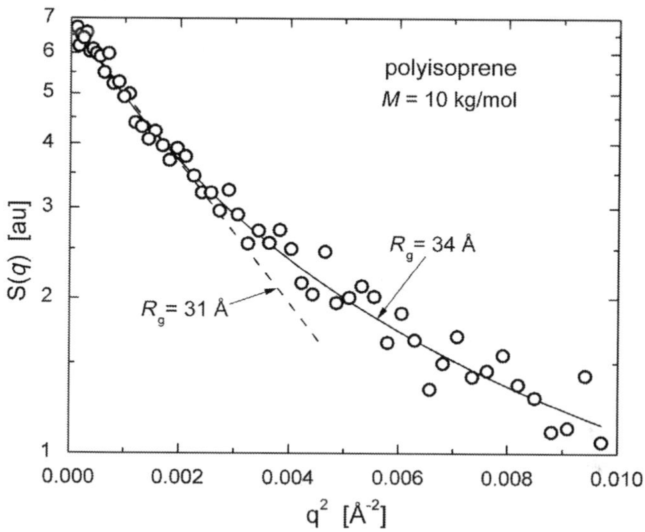

FIGURE 3.5 SANS from blend of deuterated and conventional polyisoprene ($M_w = 10^4$ g/mol), plotted in the Gunier form (Eq. (3.9)) yielding a straight line at small angles. The solid line is the fit to the Debye form for Gaussian coils (Eq. (3.10)). Data from Akcasu et al. (1980).

valid at small angles. Rubber is usually comprised of flexible chains, so that the single chain structure factor is expected to have the Debye form for a Gaussian coil (Berne and Pecora, 1976):

$$P(q) = \frac{2}{q^4 R_g^4}\left[\exp\left(-q^2 R_g^2\right) - \left(1 - q^2 R_g^2\right)\right]. \tag{3.10}$$

This result is shown for polyisoprene in Figure 3.5 (Akcasu et al., 1980). Direct SANS measurements of the size of chains in the melt validated Flory's original hypothesis that polymers behave ideally in the bulk. Other applications of SANS to elastomers include investigations of microscopic aspects of network deformation (Gronski et al., 1990; Boue et al., 1991; Westermann et al., 2001) and the effect of fillers, such as carbon black and silica, on network deformation (Botti et al., 2003; Westermann et al., 1999; Zhang et al., 2001).

The two sources of neutrons are nuclear reactors and particle accelerators (spallation sources). Neutron scattering facilities in North America include the National Institute of Standards and Technology, Oak Ridge National Laboratory, the Los Alamos Neutron Science Center, Argonne National Laboratory, Chalk River (Canada) Neutron Beam Laboratory, and the University of Missouri (Columbia). Generally, access involves a proposal system for nonproprietary work, with proprietary studies on a fee basis. There are also SANS facilities in Europe, Asia, and Australia.

TABLE 3.3 Solution Viscosity Definitions ($\eta \equiv$ Solution Viscosity; $\eta_0 \equiv$ Solvent Viscosity)

Viscosity	Relative	Specific	Reduced	Inherent	Intrinsic
Definition	$\eta_{rel} = \eta/\eta_0$	$\eta_{sp} = \eta_{ret} - 1$	$\eta_{red} = \eta_{sp}/C$	$\eta_{inh} = \ln(\eta_{rel})/C$	$[\eta] = \lim_{C \to 0} \eta_{red}$

In addition to the scattering methods, techniques such as cryoscopy, ebulliometry, and membrane osmometry give absolute molecular weight determinations, with the results independent of branching (Table 3.2). Methods that depend on the size of the molecule in solution, such as the intrinsic viscosity, size exclusion chromatography (SEC), sedimentation, and so on, require calibration, and the results are a function of polymer geometry, in particular long chain branching (LCB) (Table 3.2).

Another parameter that can be related to molecular weight is the relative viscosity, defined as the ratio of the viscosity of a polymer solution η to the viscosity of the solvent η_0 (see Table 3.3). Reliable for molecular weights $>10^3$ g/mol, the viscosities can be determined by measuring flow times through capillary tubes (diameters $\sim$1 mm), usually with gravity as the driving force for the flow. Automated instrumentation is widely available. Rotational and oscillatory type viscometers are used where a uniform, well-defined, or low shear rate is required. The ratio of the two viscosities, η/η_0, is called the relative viscosity, and the difference between that number and unity is the "specific" viscosity. Extrapolation to zero concentration of the specific viscosity divided by the concentration yields the intrinsic viscosity $[\eta]$. The intrinsic viscosity is related to molecular weight M as (Flory, 1969).

$$[\eta] = \Phi \langle r^2 \rangle^{3/2} / M, \tag{3.11}$$

in which $\langle r^2 \rangle$ is the mean-square end-to-end distance and Φ is a constant ($= 2.6 \times 10^{21}$ for r in cm). Equation (3.11) can be written as

$$[\eta]_\Phi = K M^{1/2}, \tag{3.12}$$

where Θ refers to Theta conditions, and $K(= \Phi(\langle r^2 \rangle / M^{3/2}))$ is constant for sufficiently high molecular weight (ca. 5000 g/mol). Working at Θ conditions is not always feasible, since the temperature control is exacting and there is a tendency for precipitation due to the limited solubility. A convenient alternative is to use the Mark-Houwink relation (Barth and Mays, 1991; Cooper, 1989; Zeng et al., 2006; Kurata and Tsunashima, 1999),

$$[\eta] = K M_v^a, \tag{3.13}$$

where $M_n < M_v < M_w$ (Figure 3.3), and a ranges from 0.5 to about 0.8 for rubbery polymers. Both K and a depend on polymer, solvent, and temperature, and must be determined empirically using polymers of known molecular weight. For good solvents, $[\eta]$ for a polymer varies only $\pm 30\%$, and it can be estimated empirically with reasonable accuracy (Van Krevelen, 1990; Kuo et al., 1990).

A convenient and very common method of molecular weight determination ($<10^7$ g/mol) is Size Exclusion Chromatography (SEC), or the more specific term, Gel Permeation Chromatography (GPC) (Stuart, 2002; Ruckebusch et al., 2008; Dekmezian et al., 1990; Grubsic-Gallot et al., 1990; Kuo and Provder, 1987). A dilute solution of the polymer is passed through a column packed with small (e.g., 5 μm) porous articles. The pore diameters are about the size of the dissolved region polymer chains. As the polymer solution passes through the column, polymer molecules diffuse into and out of pores. The accessibility of a given pore is determined by the hydrodynamic volume of the dissolved molecules. Since small molecules fit into more pores, their elution time from the column is longer. The flow can actually be stopped for minutes with little distortion of the chromatogram. The measured parameter is the elution volume V_e, which is a function of the molecular sizes in the solution. Based on calibration under identical conditions of the column, using narrow MWD standards, the molecular weight is directly related to V_e.

Empirically, it is known that if different polymers have the same elution volume from a given GPC column, the product $[\eta]M$ will be constant, where $[\eta]$ is determined under the same conditions (T and solvent). This follows from the Einstein-Simha relation, whereby the quantity $[\eta]M$ is proportional to the hydrodynamic volume. The hydrodynamic volume depends on temperature and solvent, and is roughly half the volume calculated from S_g. Using the "universal calibration method," molecular weight determinations can be made for a polymer using known standards of a different polymer (Grubisic et al., 1967; Puskas and Hutchinson, 1993; Kuo et al., 1993). These molecular weight standards are commercially available for various polymers. From Eq. (3.13),

$$M_U = \left[\frac{K_S}{K_U}\right]^{1/(a_U+1)} M_S^{\frac{a_S+1}{a_U+1}}, \tag{3.14}$$

where the subscripts U and S refer to the unknown and standard samples, respectively.

The resolution of a calibration curve becomes poorer at very high and low molecular weights; that is, large differences in molecular weight are reflected in small changes in elution volume. Note that additives, such as antioxidants, will also elute, so that these may be confused with a low molecular weight fraction of the polymer. Various aspects of the SEC experiment can be adjusted to improve resolution for a given determination (Barth and Mays, 1991; Cooper, 1989; Rooney and Ver Strate, 1981). The resolution is improved, *inter alia*, by smaller packing particles in the column, although this increases the potential for shear degradation of the sample. The presence of sample heterogeneity, such as blends or branching, exacerbates the resolution problem. SEC detectors include differential refractometers and UV/visible detectors,

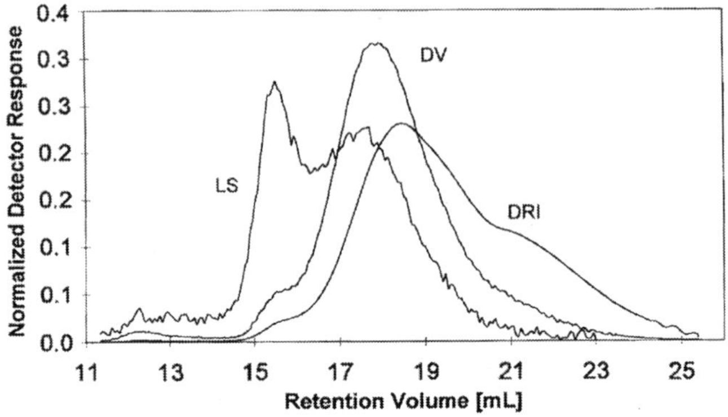

FIGURE 3.6 Chromatograms of a blend of linear and branched polyesters for light scattering (LS), viscometric (DV), and differential refractive index (DRI) detectors. The response of the LS is superior for larger molecular weights, while the DRI extends the sensitivity to larger retention volumes (Balke and Mourey, 2001).

which give a signal proportional to the polymer concentration, as well as viscometric and light scattering detectors, whose signal depends on both the concentration and the molar mass (see earlier). Multiple detectors are employed to facilitate deconvolution of overlapping retention volumes (see Figure 3.6) (Gaborieau et al., 2007; Medrano et al., 2003; Provder et al., 1995; Brun and Liq, 1998; Nagy, 2003; Wang et al., 2000; Balke and Mourey, 2001). For example, the effluent from the column passes through a light scattering cell or a viscometer, and then through a differential refractometer. The latter serves as a concentration detector, and is usually last in line, since it is most easily disrupted by downstream back pressure.

A potential error in the SEC analysis is peak broadening due to nonuniformities in the pore structure of the GPC or inherent to the flow through the column. The resolution of a column can be characterized by the ratio of the spread of the elution volume for a monodisperse sample to the peak of the retention volume. At the extremes of the range for a broad MWD sample, a given elution volume can shift by as much as 50% in molecular weight for repeat measurements on the same sample. This type of peak broadening is due to diffusion, mixing, and concentration effects, and perhaps in part to a dependence of the hydrodynamic volume on concentration. Another source of error for very high molecular weight samples (e.g., $M \sim 10^6$ g/mol or higher) is mechanical degradation (chain scission) during transport through the column (Rooney and Ver Strate, 1981).

The classical method of solvent-nonsolvent fractionation according to MWD and compositional distribution relies on solubility differences among the various species. The method is empirical and tedious, involving characterization of phase-separated "cuts." However, fractionations can be carried out with minimal equipment, and for some polymers are the only source of narrow compositional and MWD standards for techniques such as SEC or $[\eta]$. TREF, described earlier, is an automated implementation of the fractionation procedure.

The term Field Flow Fractionation (FFF) (Barth and Mays, 1991; Cooper, 1989; Messaud et al., 2009; Williams and Lee, 2006; Lee and Molnar, 1995; Schimpf et al., 2000; Lee et al., 2000) refers to a family of one-phase chromatographic techniques, carried out in thin flow channels. In principle, FFF yields absolute molecular weights, although in practice calibrations similar to SEC are used. An external field is applied perpendicular to the laminar flow of solvent in a channel. The parabolic velocity profile of the flow causes separation according to the position of molecules with respect to the wall. Molecules having weaker interaction with the field remain further from the walls of the channel, thereby eluting sooner. The external field may be a flow field, an electric or magnetic field, or a thermal field. In Thermal Field Flow Fractionation (TFFF) (Lee and Williams, 2010; Kim et al., 2006; Stegeman et al., 1994; Lee and Molnar, 1995; Lewandowski et al., 2000) a temperature gradient of tens of degrees is imposed, causing migration and accumulation of the solute polymer at the cold wall, in competition with Brownian motion. The transport coefficient governing the thermal diffusion depends on chemical composition, as well as molecular weight. For a given solute, smaller molecules, having the larger diffusion coefficients, migrate more toward the center, and are thus eluted first by the flow. In addition to discrimination according to molecular size, TFFF can be used to measure thermal diffusion coefficients of chain molecules (vanAsten et al., 1996; Nguyen et al., 1998).

FFF is fast and provides good mass resolution. Attractive features include the absence of a substrate and the channel being too large to be plugged by gel (useful since polymer samples often contain gel). The method can be applied to molecular weights up to ca. 10^7 g/mol, and commercial instruments are available.

An absolute method for molecular weight determination is matrix-assisted laser desorption ionization time-of-flight mass spectrometry (MALDI-TOF) (Kona et al., 2005; Creel, 1993; Nielsen, 1999; Cho et al., 2001). The sample is dispersed in a UV-absorbing matrix (e.g., *trans*-cinnamic acid or 2,5-dihydroxybennzoic acid). Irradiation with a UV laser induces evaporation of ionized polymer chains, which are then detected using TOF. The technique requires relatively narrow MWD samples. Alternative ionization methods have been employed, such as electrospray ionization mass spectrometry (ESI-MS), which may have advantages for certain polymer end groups (Vana et al., 2002). TFFF and MALDI-TOF can be coupled to analyze polydisperse samples and polymer mixtures (Kassalainen and Williams, 2003).

Sedimentation relies on gravity to drive the separation of macromolecules according to their size, with high rotation speeds (up to 70,000 rpm) used to accelerate the process (Stuart, 2002). Ultracentrifugation (Scott et al., 2005; Laue, 1990) emerged as a versatile, powerful method with the development of new detection methods, based on Schlieren optics, UV/VIS absorption,

Rayleigh interference, fluorescence, light scattering, turbidity, and multiwavelength UV/VIS optics (Karabudak et al., 2010). It is widely applied to the analysis of proteins, providing the ability to examine the quaternary structural state, as well as conformational changes (Chou et al., 2011; Schuck, 2000). Nonequilibrium, sedimentation velocity measurements are employed, which are quicker but less rigorous than equilibrium experiments. Ultracentrifugation is one of few techniques yielding the z-average molecular weight, and it can also provide information concerning sample homogeneity.

The methods discussed heretofore involve polymer solutions. Neat samples are analyzed directly via melt viscosity measurements. At shear rates lower compared to the rate of Brownian motion of the chains, the melt viscosity is independent of shear rate (Newtonian behavior), $\eta = \eta_0$. Thus, the viscosity-M_w relationship for a particular temperature can provide an accurate M_w determination (see Figure 3.7; Colby et al., 1987; Abdel-Goad et al., 2004). Given the high molecular weight of elastomers, the main difficulty is attaining zero-shear-rate conditions. At ambient temperatures, $\eta_0 > 10^7$ Pa s. Note that "cold flow" of unvulcanized rubber usually requires weeks or longer. The obvious way to reduce the viscosity is to make measurements at elevated temperatures. However, especially for unsaturated rubbers, degradation may occur before steady-state conditions can be attained. The situation is worse if the polymer has long chain branching.

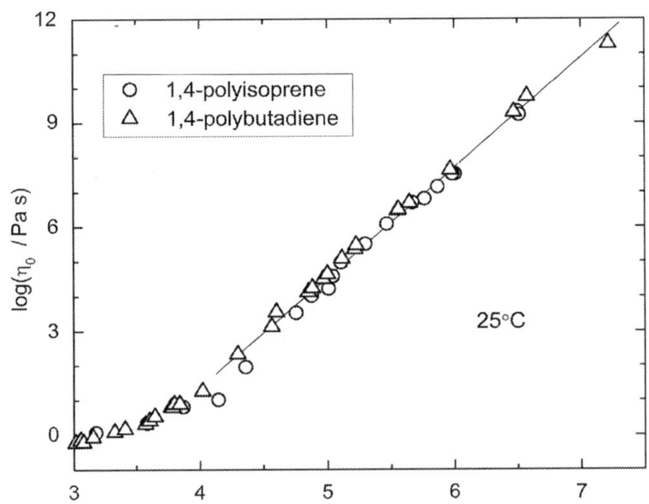

FIGURE 3.7 Zero-shear-rate viscosities of 1,4-polybutadiene (Colby et al., 1987) and 1,4-polyisoprene (Abdel-Goad et al., 2004). The line is the best-fit to molecular weights at least three times the entanglement molecules (= 1850 and 6190 for PB and PI, respectively), which gives a power-law slope equal to 3.2.

3.4.2 Branching

Side-chains on a polymer backbone are described as either short-chain branching (SCB) or long-chain branching (LCB), the latter defined as branches having molecular weights at least a few times the entanglement molecular weight. Branching affects various properties, for example the geometry and size of the chain molecule (Figure 3.8), affording means to characterize the degree of branching (Grest et al., 1996).

The intrinsic viscosity is reduced by LCB (viz. Eq. (3.10)), with a smaller ratio of the respective mean-square radii of gyration of branched and unbranched polymers having the same M_w:

$$g = \frac{\langle S^2 \rangle_{\rm br}}{\langle S^2 \rangle_{\rm lin}}, \quad g' = \frac{[\eta]_{\rm br}}{[\eta]_{\rm lin}}. \tag{3.15}$$

The parameter g is less than unity for polymers with LCB, and sometimes defined as the ratio of the root-mean-square values, $S_{g,\rm br}/S_{g,\rm lin}$. The calculation of g for a given branch structure is straightforward, but radii of gyration are hard to measure experimentally. Intrinsic viscosities are readily measurable, but relating g' to g, and thus to branch structure is difficult. Calculations of g' involve dilute solution theory, as discussed following Eq. (3.10), with the

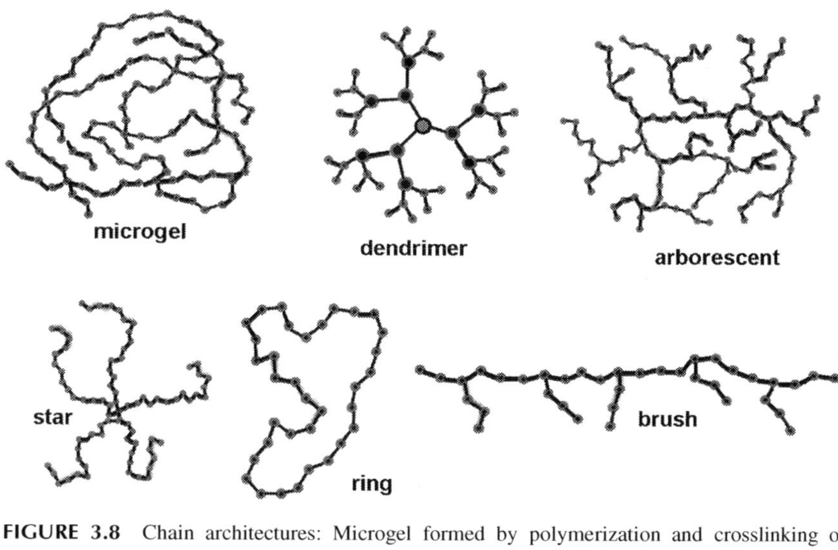

FIGURE 3.8 Chain architectures: Microgel formed by polymerization and crosslinking of monomers dispersed as an emulsion; dendrimer formed by cascade synthesis to yield (in this example) a third-generation tiered polymer; arborescent branched polymer, which lacks the regularity of a dendrimer; star polymer, in which each arm originates from the same unit and terminates in a dangling end; ring or cyclic polymer; brush or comb polymer, in which dangling chains are anchored to a linear backbone (Roland, 2011).

type of branching explicitly considered. Alternatively, g may be evaluated using model polymers. Determination of absolute levels of branching from intrinsic viscosity measurements is not possible, since it depends on the nature of the branching, including the junction functionality and branch architecture (random, star-branched, dendrimeric, etc.). However, relative values of g' can be used to assess relative degree of branching.

With light scattering measurement of the radius of gyration (see Eq. (3.6)), direct comparisons of g values can be made. In Θ solvents, the experimental g values generally agree with calculations, with some exceptions (Small, 1975).

SEC can be used to detect branching (Zheng et al., 2011, 2012a,b; Zhang et al., 2012; Ponomarenko et al., 2012; Puskas et al., 2012; Suarez and Coto, 2011; Samperi et al., 2011; Castignolles and Gaborieau, 2010; Minari et al., 2010). Since the product $M[\eta]$ is constant for a fixed value of the SEC elution volume, SEC data can be obtained for a homogeneous fraction of a sample of unknown branching, and the ratio of its intrinsic viscosity to $[\eta]$ of the linear polymer having the same elution volume is calculated. This M calculated from its intrinsic viscosity is then compared to the molecular weight of the unknown determined by an absolute technique.

An empirical relationship for broad MWD samples gives an average value

$$g'' = \frac{M_{v,\text{lin}}}{M_{w,LS}} \times \left(\frac{M_w}{M_v}\right)_{\text{GPC}}, \tag{3.16}$$

in which $M_{v,\text{lin}}$ is the molecular weight calculated from the intrinsic viscosity assuming the polymer is linear, $M_{w,LS}$ is from light scattering measurements, and $\left(\frac{M_w}{M_v}\right)$ is taken from the elution volumes assuming no branching. As an example, for NBS 1476 (a polyethylene standard with LCB), $g'' = 0.4$, using $[\eta] = 1.04$ dl/g in decalin, $(M_w/M_v)_{\text{SEC}} = 1.16$, $M_{w,LS} = 125{,}000$ g/mol. The number of branches yielding this value of the branching parameter depends on the nature of branching. For random tetrafunctional branches, $g = 0.4$ corresponds to 20 chain ends per weight average chain.

Advanced SEC analytical techniques take advantage of online light scattering and viscometry [III B 4-10]. With the SEC calibrated for the product $[\eta]M$ versus elution time for a particular polymer species, together with $[\eta]_{\text{Br}}$ or M_w^{Br} measurements, g factors can be calculated as the ratio of the measured intrinsic viscosity to the value calculated for the corresponding linear polymer (i.e., the linear polymer that would have eluted at the same time).

If well-characterized samples are available, a calibration of viscoelastic properties for the neat material, such as the melt viscosity, can be used to assess LCB. Below a certain molecular weight (which is a few times the entanglement molecular weight), branching reduces the melt viscosity, since the molecular size is smaller for fixed molecular weight. Since LCB inhibits motion along the chain contour (reptation), it can effect an enormous increase in the viscosity and terminal relaxation times. A polymer with LCB is always more viscous than its linear counterpart of equal M_w (Grest et al., 1996; Kraus and Gruver, 1965; Ngai and Roland, 1997; Robertson et al., 2001), and for this reason LCB is introduced by design in some commercial rubbers to reduce cold flow (i.e., the room temperature creep of rubber during storage). Substantial LCB may require use of a polymer with a low number-average molecular weight, in order to retain a viscosity low enough for processing; however, the consequent high concentration of chain ends may be deleterious to cured properties, for example heat buildup or strength.

For rubbers such as NR (or synthetic 1,4-polyisoprene) (Santangelo and Roland, 1998) and polybutadiene (both 1,2- and 1,4-isomers) (Carella et al., 1986), LCB increases the temperature sensitivity of the viscosity and terminal relaxation time. Thus, by comparing apparent activation energies, or, in the more usual case where the behavior is non-Arrhenius, the temperature dependence of the ratio of relaxation times for an unknown and a linear sample of the same polymer, inferences can be drawn concerning LCB (Figure 3.9). For polymers such as 1,4-polyisoprene, which have a dipole moment parallel to the chain, dielectric measurements of the normal mode can be used to measure the temperature dependence and thus assess the presence of LCB.

Note that LCB does not enhance the T dependence of all rubbers. For example, the temperature dependence is the same for linear and branched polyisobutylene (Robertson et al., 2001; Santangelo et al., 1999), polydimethysiloxane (Roland and Santangelo, 2002), and hydrogenated 1,2-polybutadiene (Carella et al., 1986).

Although short branches do not suppress reptation and thus have a weak effect on the rheology, SCB can alter many physical properties, indirectly through suppression of crystallization and directly by changing the shape and flexibility of the chain molecules. The degree of SCB can be measured by NMR (Feast et al., 1997) and FTIR (Blitz and McFaddin, 1994). A high concentration of SCB increases the spatial extent of the chain, to reduce steric congestion of the side-groups with themselves and with the repeat units of the backbone. Thus, polymers with substantial SCB are less flexible. A measure of this property is the persistence length, l_p, defined as the distance over which bond orientations become uncorrelated. For a freely jointed chain l_p equals one bond length, while for a completely rigid rod polymer, the persistence length just equals the chain length. An empirical equation relating l_p to the number branches is (Ramachandran et al., 2008)

$$l_p = l_p^{\infty} - a_{\mathrm{SCB}} \exp\left(-n/b_{\mathrm{SCB}}\right) \tag{3.17}$$

in which l_p^{∞} is the persistence length for complete branching, and a_{SCB} and b_{SCB} are constants, the former specific to the polymer chain and the latter depending on the type of branches.

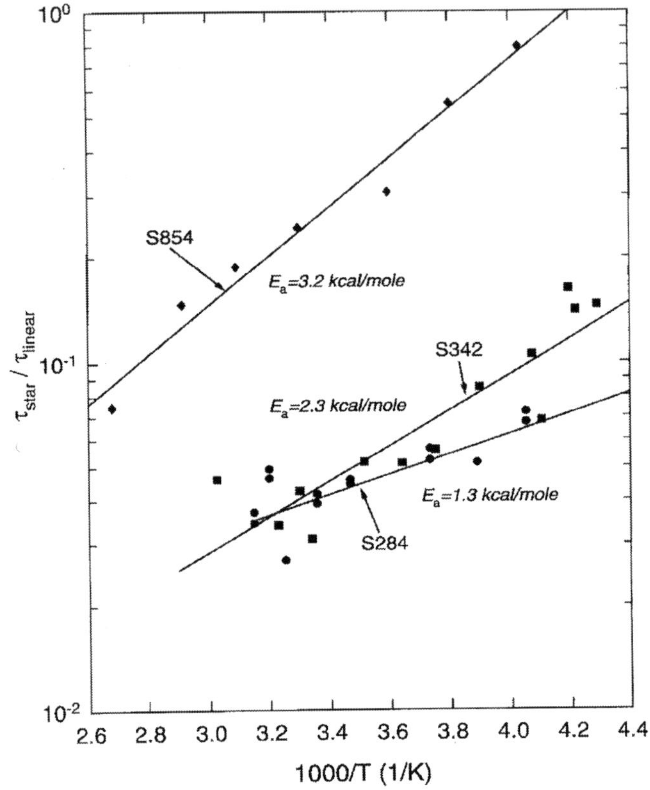

FIGURE 3.9 Terminal relaxation times of 3-arm star 1,4-polyisoprenes having M_w(g/mol) = 284,000 (S284), 342,000 (S342), and 854,000 (S854), normalized by the relaxation time for linear 1,4-polyisoprene. The excess activation energy due to the LCB is ~400 J per mole of branch entanglements. (Reprinted from Santangelo and Roland, 1998).

Ramachandran et al. (2011) used Eq. (3.9) to obtain average coil size and a power-law to describe the rod-like character at high q

$$I(q) = B_f q^{-d_f}. \tag{3.18}$$

B_f is a constant prefactor and the exponent d_f represents the dimension of the fractal object; typically it is assumed $d_f = 1$. From fits to SANS scattering data the persistence length can be extracted. Figure 3.10 (Ramachandran et al., 2011, 2008) shows the results for the stiffening of the polyethylene due to the presence of short-chain branches.

3.4.3 Gel

The presence of gel suppresses flow of a rubber. In gel-sol studies crosslinks are introduced in a controlled fashion (e.g., using peroxide or ionizing radiation), and the initial molecular weight is determined from the gel dose. The analysis,

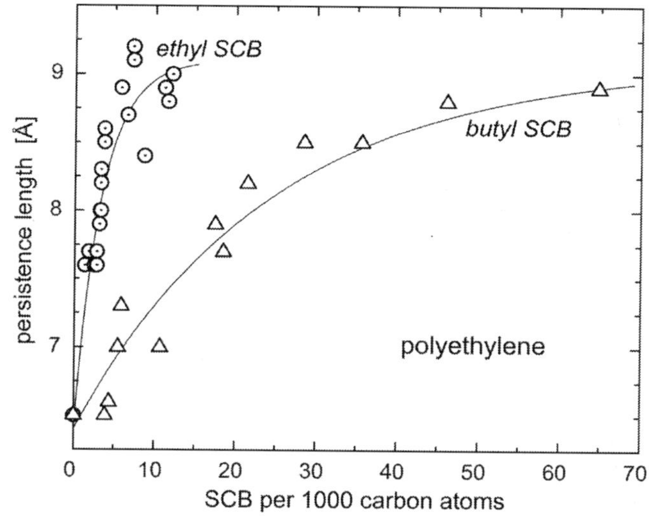

FIGURE 3.10 Persistence length of polyethylene from SANS versus the number of ethyl (circles) (Ramachandran et al., 2011) or butyl (triangles) (Ramachandran et al., 2008) short-chain branches per 1000 carbon atoms. The solid lines are fits to Eq. (3.17) with $l_p^{\infty} = 9.1$ Å, $a_{SCB} = 2.7$, and $b_{SCB} = 3.4$ and 24.6, respectively, for ethyl and butyl SCB.

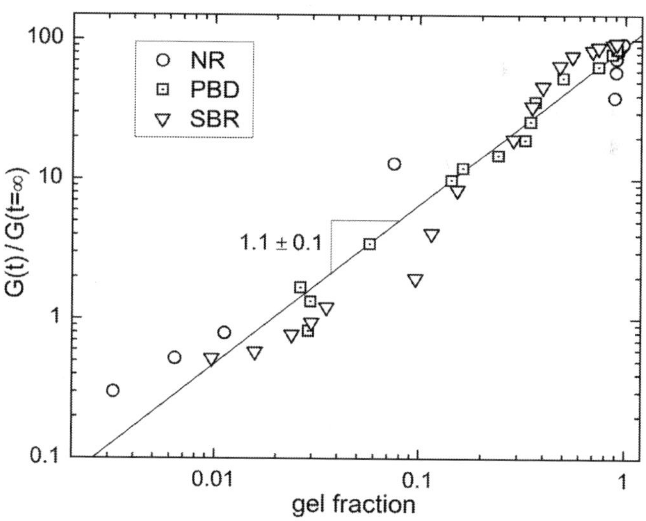

FIGURE 3.11 Rheometer torque normalized by the fully cured plateau value versus the gel content during sulfur vulcanization of three rubbers (Zhang et al., 2010).

however, requires knowledge of the rates of both crosslinking and chain scission. As crosslinking proceeds beyond the gel point, the partition between gel and sol is governed by the MWD of the sample, which in principle can be used to measure MWD (Bohm and Tveekrem, 1982).

Networks with crosslink densities only slightly beyond the gel point have minimal mechanical strength. Thus, care must be taken in dissolution of the soluble fraction to avoid damaging the network. If the gel fraction data extrapolate to a negative amount of crosslinking for gelation, the implication is that the original material contained gel. Although a gelled polymer cannot flow, large stresses imposed during milling or mixing induce nonuniform flow of a lightly crosslinked rubber; the latter also have measurable Mooney viscosities. During curing the gel content continually increases, but values approaching 100% are attained only after the rheometer torque attains a plateau value (Figure 3.11 (Zhang et al., 2010)). Polymers dissolved in a solvent can associate to yield aggregated networks (Okabe et al., 1992; Horsky and Bohdanecky, 1990). This phenomenon is common with polymers that crystallize (e.g., EPDM, polyvinylchloride, polyurethanes). Such reversible gelation depends on the solvent and temperature.

3.5 GLASS TRANSITION AND SECONDARY RELAXATION PROCESSES

When a measurement of polymer dynamics is made using one of a number of probes of relaxation (for example, mechanical or dielectric spectroscopy, light or neutron scattering), a maximum in the susceptibility, corresponding to maximum absorption, is observed at a frequency equivalent to that for the relevant molecular or segmental motions. Thus, in the limit of small perturbations, for which the material response is linear, the relaxation directly reflects the equilibrium Brownian motion. This follows from fluctuation-dissipation theory, originally proposed to explain Johnson noise in electrical conductors (Roland, 2011). The gigantic size of polymer molecules provides for an enormous number of degrees of freedom, and thereby motion encompassing many decades of time over a broad range of length scales (see Figure 3.12). The longest relaxation processes involve transport of the chain center of mass, and these are termed the terminal dynamics. At shorter times, relaxation involves motion of progressively smaller portions of the chain molecule. The rubbery plateau reflects the transient entanglement network, whose dissolution gives rise to a peak in the loss modulus (and in the dielectric loss for polymers having a dipole moment parallel to the chain (Runt and Fitzgerald, 1997; Kremer and Schonhals, 2003)). At frequencies beyond the rubbery plateau, chain motions are unaffected by entanglements. In the softening zone, the modulus is a strong function of frequency, with the behavior referred to as Rouse dynamics. (The Rouse theory was originally developed for dilute polymer solutions, but with small modification can be applied to unentangled polymer melts (Roland, 2011).) Near the end of the transition zone, the portion of a chain involved in the relaxation is too short for Gaussian behavior (that is, its end-to-end distance does not have a Gaussian distribution). These "sub-Rouse modes" (Ngai and

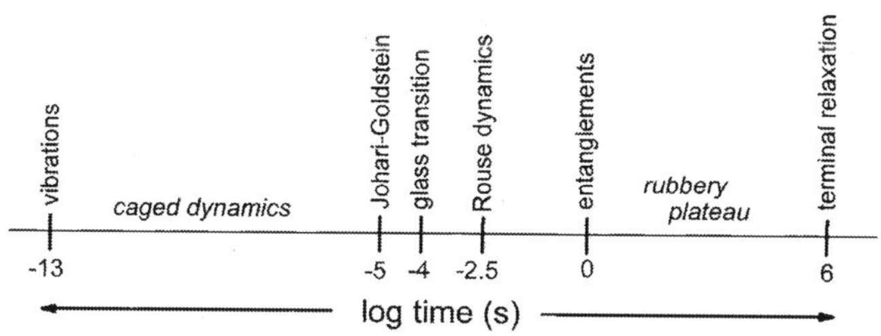

FIGURE 3.12 Schematic indicating range of motions and corresponding frequencies for 1,4-polyisoprene ($M_w = 500{,}00$ g/mol) at -40 °C.

Plazek, 1995) become the local segmental dynamics at the highest frequencies prior to the onset of glassy behavior.

The local segmental dynamics are associated with the rubber-to-glass transition that occurs close to T_g. Only in the glass transition zone of the viscoelastic spectrum are the chain modes and local segmental modes close enough in frequency to be measured simultaneously with conventional spectroscopies. Since the chain modes and the local segmental dynamics have different temperature dependences (Santangelo and Roland, 1998; Plazek et al., 1995; Roland et al., 2001), their mutual contribution to the viscoelastic response causes a breakdown of the time-temperature superposition principle in the glass transition zone. Thus, although master curves for the chain dynamics can be constructed that extend from the end of the softening zone through the terminal relaxation, and master curves of the local dynamics are possible, within the softening zone, the shape of the viscoelastic spectrum changes with temperature (see Figure 3.13).

While all relaxation times depend on temperature and pressure, only the global motions (viscosity, terminal relaxation time, steady-state recoverable compliance) are functions of M_w (and to a lesser extent MWD). The glass transition temperature of rubbers is independent of molecular weight because chain ends for high polymers are too sparse to affect this bulk property (Figure 3.14; Bogoslovov et al., 2010). The behavior can be described by the empirical Fox-Flory equation (Fox and Flory, 1954):

$$T_g = T_{g,\infty} - k_{FF}/M_n. \tag{3.19}$$

Note the use of the number-average molecular weight in Eq. (3.19), since it is a measure of the concentration of chain ends.

At frequencies faster than for segmental relaxation, or at temperatures below T_g, secondary relaxation process can be observed, especially in dielectric spectra. In polymers, many of these secondary processes involve motion of pendant groups. However, the slowest secondary relaxation, referred to as the Johari-Goldstein process (Ngai and Paluch, 2004), involves all atoms in the repeat unit (or the entire molecule for low M_w materials). This process is referred to as the Johari-Goldstein relaxation, and it serves as the precursor to the prominent glass transition.

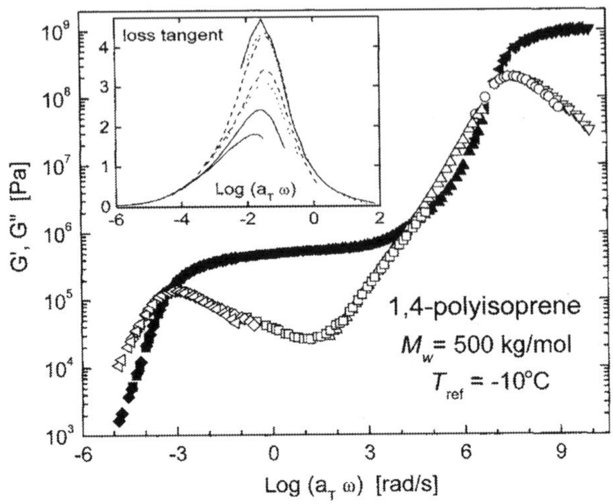

FIGURE 3.13 Apparent master curves (reference temperature = −10 °C) for the storage (solid symbols) and loss (hollow symbols) moduli of *cis*-1,4-polyisoprene (M_w = 500,000 g/mol). The breakdown of time-temperature superpositioning, barely evident in the softening zone, is seen clearly in the loss tangent peak, which is shown in the inset for temperatures from −66 °C to −48 °C (the peak height decreases with increasing temperature) (Santangelo and Roland, 1998).

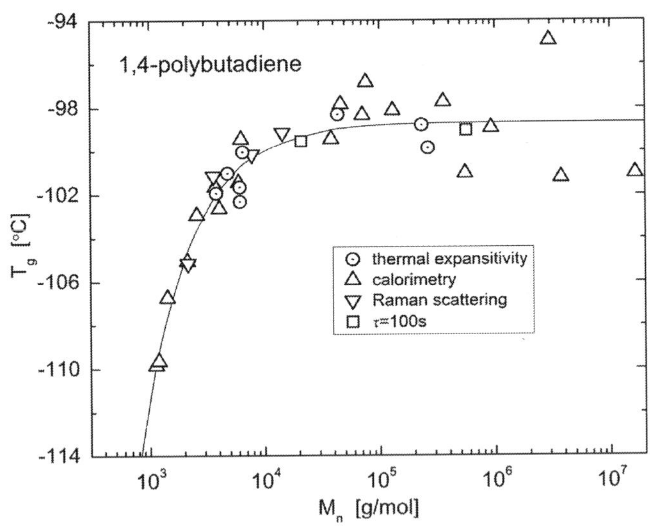

FIGURE 3.14 Glass transition temperatures measured by the change in thermal expansion coefficient (circles), from calorimetry (triangles), by Raman scattering (inverted triangles), and by relaxation spectroscopy for $\tau_\alpha = 100$ (square). The solid line is the fit to the Fox-Flory equation with $k_{FF} = 12$ kg/mol and $T_{g,\infty} = 174.4$K (Bogoslovov et al., 2010).

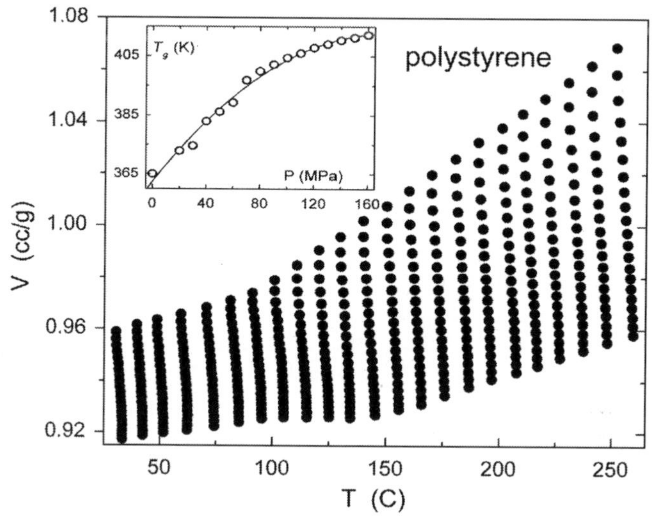

FIGURE 3.15 Specific volume of polystyrene (M_w=110,000 g/mol) as a function of temperature at various pressures. The inset shows the variation of T_g with pressure (Roland and Casalini, 2003).

The T_g of an elastomer is below room temperature, and for conventional applications it is usually lower than −30 °C; otherwise, at ambient temperature, the material would lack the flexibility associated with rubbery behavior and be overly sensitive to temperature fluctuations. The glass transition is a second-order transition because, unlike melting or boiling, there is no discontinuity in the volume or heat capacity at T_g, but only a discontinuous change in the rate of change (i.e., in the thermal expansion coefficient and specific heat (Figure 3.15) (Roland and Casalini, 2003). Experimentally, T_g simply reflects the temperature at which local segmental motions become slow relative to the experimental timescale. This means that values of T_g are a function of both the measurement frequency and, to a lesser extent, the particular relaxation process under observation. There exists a host of properties that can serve as indicators of T_g (Figure 3.14). A common one is the softening point; however, changes in stiffness of a material can reflect melting of crystalline domains, rather than a transition from the glassy state. T_g is affected by the presence of co-monomers in the chain backbone, similar to the behavior of miscible blends, and by nonpolymeric ingredients in the formulation. Both plasticization (lowering of T_g) and anti-plasticization (increase of T_g) can be brought about by addition of lower and higher T_g components, respectively.

In experiments in which temperature is scanned to determine T_g, the measurement frequency is approximately the ratio of the rate at which T is changed to the transition temperature. In such nonisothermal measurements, it is preferable to approach T_g from the equilibrium state (i.e., by cooling), since the properties in the glassy state change due to physical aging. Physical aging refers to structural relaxation below T_g, associated with the slow decrease of the density to its equilibrium value. This causes changes in physical properties, including a shift of the glass transition temperature. A metric of the degree of nonequilibrium is the fictive temperature, T_f, also referred to as the structural temperature. T_f is the temperature at which the enthalpy of the material would be the equilibrium value.

The most common means to assess T_g is via heat capacity (C_p) measurements, usually carried out in differential scanning calorimeters (DSC). The temperature of the sample is changed at a fixed rate, with the heat flow monitored. An older and largely obsolete variation is differential thermal analysis (DTA), in which heat flow is programmed, and the consequent rate of temperature change is measured. With either method, a discontinuity in the response evidences a thermal transition. In a DSC experiment, T_g is usually taken to be the midpoint of the change in heat capacity; alternatively, T_g can be defined as the temperature of the intersection of the extrapolated baseline with the tangent of the maximum slope (Sircar, 1997). If the polymer chain is envisioned as a string of flexible beads, the head capacity change at T_g, which is almost entirely configurational, is about 11 J/K per mole of beads (Wunderlich, 2007). When DSC measurements are made during heating, the sample begins in a nonequilibrium state; thus, physical aging during the measurement can cause an endothermic peak to appear at the onset of the glass transition. In this situation, the fictive temperature is more informative than the apparent transition temperature. A graphical method for determining T_f is illustrated in Figure 3.16.

In modulated (or alternating) DSC (MDSC) (Hutchinson, 2003; Hohne et al., 2002; Simon, 2001; Pielichowski and Flejtuch, 2002), an oscillating heating (or cooling) rate is superimposed on the usual linear temperature ramp. There are two principal ways to interpret the data: obtain the complex heat capacity and decompose into the real (stored) and imaginary (dissipated) heat capacity, or separate the measured heat flow into reversing and nonreversing components (Schawe, 1995). In either method the objective is to distinguish time-independent effects from kinetic processes. MDSC is useful for studying transitions too weak to be detected by conventional DSC, analysis of multiple endotherms, and separating preexisting crystallinity in the sample from that occurring during the DSC measurement. A development in MDSC (or differential calorimetry) is the quasi-isothermal method, in which a temperature oscillation is applied without any temperature ramp, and the procedure is repeated for a range of fixed temperatures. (This is the usual method of performing isothermal dynamic mechanical and dielectric relaxation spectroscopies.) The reversing heat capacity is obtained from the measurements, with the total heat capacity determined by a conventional DSC experiment. Quasi-isothermal MDSC is used to study crystallization in polymers (Cebe, 2005) and can also provide accurate absolute heat capacities (Krishnan and Nagarajan, 2010).

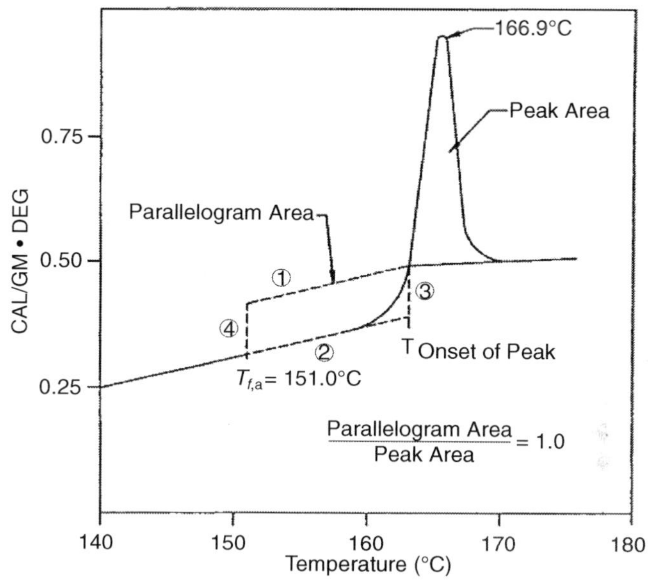

FIGURE 3.16 DSC of a cured epoxy resin. To determine the fictive temperature, a parallelogram is constructed to the low temperature side of the enthalpy peak, with horizontal sides parallel to the measured heat capacity. The onset of the peak defines the high temperature boundary, and the low temperature boundary (i.e., T_f) is chosen such that the area of the parallelogram equals the peak area.

A recent development in thermal analysis that has been applied to polymers is nanocalorimetry, in which a chip with one or two sensors is used to make heat capacity measurements on thin films (Mileva et al., 2009; Mathot et al., 2011). Sample masses of a few tens of nanograms can be analyzed, with sensitivities of picoJoule per K. Because of the small thermal mass, temperature ramps as high as 10^4 K/s can be achieved.

3.6 MORPHOLOGY

Analysis of the morphology of an elastomer includes not only characterizing the molecular structure of the polymer and of the compounding ingredients, but may also extend to the supermolecular scale: the filler network structure (see Chapter 8), the phase morphology in heterogeneous blends and thermoplastic elastomers, and any aggregates (clusters) in rubbers, for example, containing ionic groups (e.g., carboxylated rubber). A survey of experimental techniques is given in Figure 3.17. Selected aspects of rubber morphology are discussed in the following sections.

3.6.1 Orientation

Rubber elasticity arises from the orientation of chain segments, and the degree of this orientation underlies the mechanical properties. The most facile way to

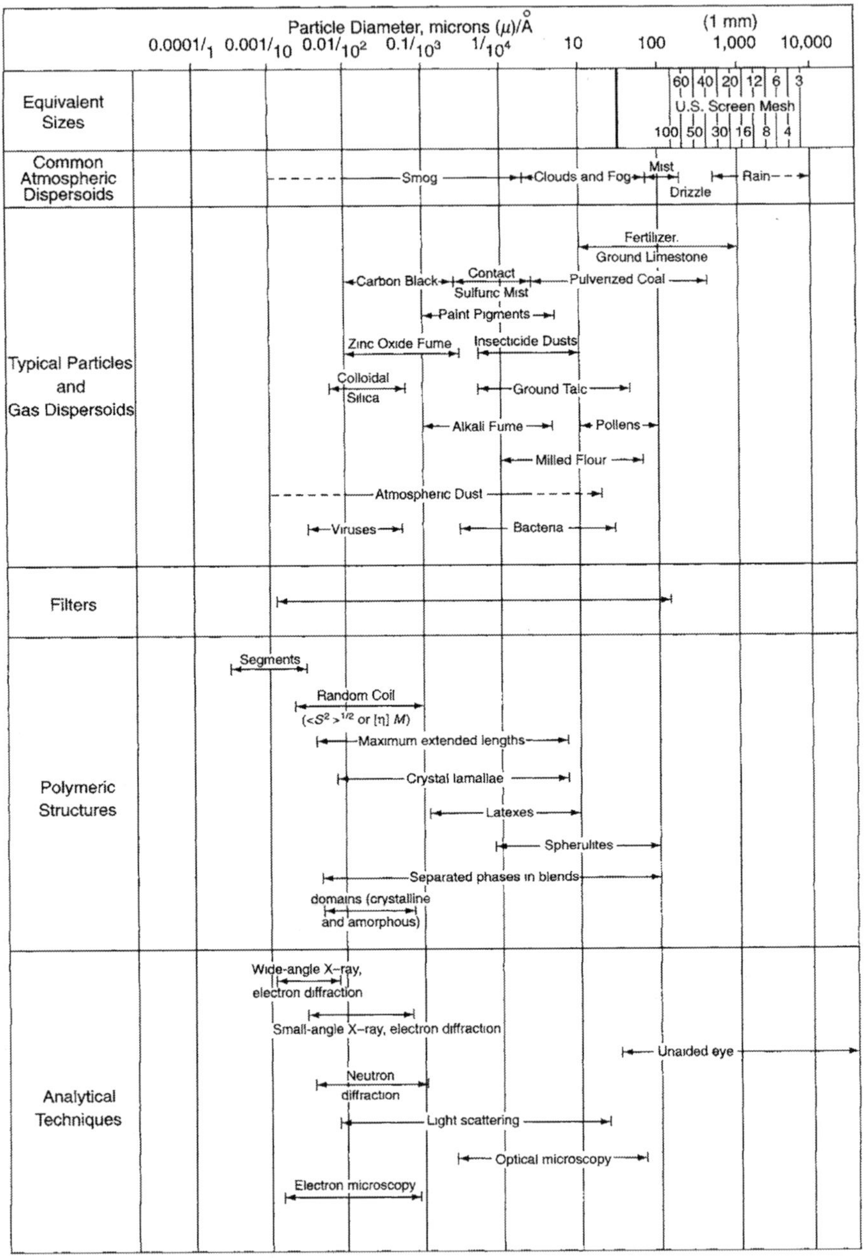

FIGURE 3.17 Characterization of particles.

quantify the orientation is from the (optical) birefringence, defined as the difference in refractive indices for two perpendicular directions. This birefringence depends on the optical anisotropy of the chain units and their degree of orientation. The mechanical stress is proportional to this same orientation factor; namely, the stress optical rule in Chapter 6.

In a semicrystalline material the observed birefringence is the sum of the Δn from the oriented amorphous phase and from the crystalline regions, which provides the measurement as a means to quantify the crystallinity in polymers. However, the contribution from form birefringence, due to the distortion of light waves traversing the boundary between the amorphous and crystalline phases (or any phases with different refractive indices), can cause the degree of crystallinity to be overestimated.

Another method of measuring polymer orientation is infrared dichroism, which refers to the ratio of the IR absorption for light having orthogonal polarizations. IR dichroism yields the same information as birefringence measurements (i.e., the second Legendre polynomial, $f_2(\phi)$), but with the advantage of chemical selectivity. The orientation of specific groups can be determined, as well as the orientation of the components in mixture (Zemel and Roland, 1992a; Myers and Cooper, 1994; Everall and Bibby, 1997; Bruell et al., 2010). Analysis requires knowledge of the orientation of the vibrational transition moment with respect to the chain axis. A few percent of deuterated polymer can be added to the sample, so that the infrared measurements can be carried out on thick samples, while still conforming to Beer's law (absorbance < ca. 0.7). Thicker samples also enable simultaneous measurement of the stress (Kannan and Kornfield, 1994). The complementary technique, polarized Raman spectroscopy, is also used to characterize orientation (Jiang et al., 2008; Park et al., 2011; Bower et al., 1995).

In polarized fluorescence measurements, the sample is illuminated by polarized light, with the emitted intensity measured for polarization parallel and perpendicular to the excitation radiation (Bur et al., 2000; Hennecke et al., 1992). A probe dye can be added to enhance sensitivity. The experiment yields both $f_2(\phi)$ and the fourth Legendre polynomial,

$$f_4(\phi) = \langle 35\cos^4\phi - 30\cos^2\phi + 3\rangle/8.$$

The latter is a more strongly decreasing function of ϕ than is $f_2(\phi)$, so that $f_4(\phi)$ reflects the more oriented chains in the distribution of orientations. For uniaxial strain, these two functions alone are adequate to describe the orientation (Bower, 1981).

Another method to characterize orientation is deuterium NMR (Dupres et al., 2009; Ekanayake et al., 2000; Valic et al., 2003). Deuterons have a spin quantum number equal to 1, and therefore a nuclear quadrupole moment. Coupling between the quadrupole moment and the two possible NMR transitions (resulting from the three Zeeman levels of the spin = 1 nucleus) yields differences in the energy levels of the two transitions, giving rise to a doublet. The separation in this doublet is proportional to $f_2(\phi)$ (which depends on the orientation of the chemical bond axis of the deuteron to the applied external magnetic field); thus, the magnitude of the observed splitting yields directly the orientation.

Orientation in crosslinked elastomers primarily reflects the configurational entropy and intramolecular conformational energy of the chains. However, as first shown by deuterium NMR experiments on silicone rubber (Deloche and Samulski, 1981; Sotta et al., 1987), unattached probe molecules and chains become oriented by virtue of their presence in a deformed network. This "nematic coupling" effect is brought about intermolecular interactions (excluded volume interactions and anisotropic forces) which can cause "nematic coupling" (Zemel and Roland, 1992a; Tassin et al., 1990). The orientation is only locally effective, so it makes a negligible contribution to the stress (Doi and Watanabe, 1991), and the chains retain their isotropic dimensions (Sotta et al., 1987).

Wide-angle X-ray scattering can be used to evaluate orientation, but it is more suited to semicrystalline polymers. In principle, scattering methods can yield all orientation functions, although experimental limitations restrict determinations to $f_4(\phi)$. Using high flux synchrotron X-ray sources, transient orientations can be measured. This has been used to study the behavior of the amorphous phase during stretching of NR (Figure 3.18; Toki et al., 2003, 2011).

3.6.2 Blends

Although many rubbers can be mixed to form a blend that is homogeneous, the overwhelming majority of such mixtures are phase-separated on the microscopic scale. This does not mean the latter are not useful, since the only requirements are that a satisfactory dispersion is obtained and there is sufficient compatibility to avoid macroscopic demixing of the components. Thermodynamic miscibility, which implies segmental mixing of the components, is rare, although a number of thermodynamically miscible elastomer blends are known. The phase size in a heterogeneous blend will depend on the compatibility of the components, the relative viscosities of the components, and the mixing conditions (Roland and Bohm, 1984; Avgeropoulos et al., 1976; Paul and Barlow, 1980; Peng et al., 2011; Sirisinha et al., 2003; Chen et al., 2006). The particle size distribution can be determined from electron micrographs, using image analysis software (Cheng and Nauman, 2004; Qu et al., 2004). If the image contrast is poor, some underestimation of domain sizes can result, due to poorly discerned peripheries. In microtomed samples, errors can arise if the section thickness is less than the particle size (Weibel, 1973).

In principle, the dispersed phase size distribution can be determined from small-angle scattering experiments, using either X-rays or neutrons. Since a comparatively large volume is sampled by the incident beam, scattering

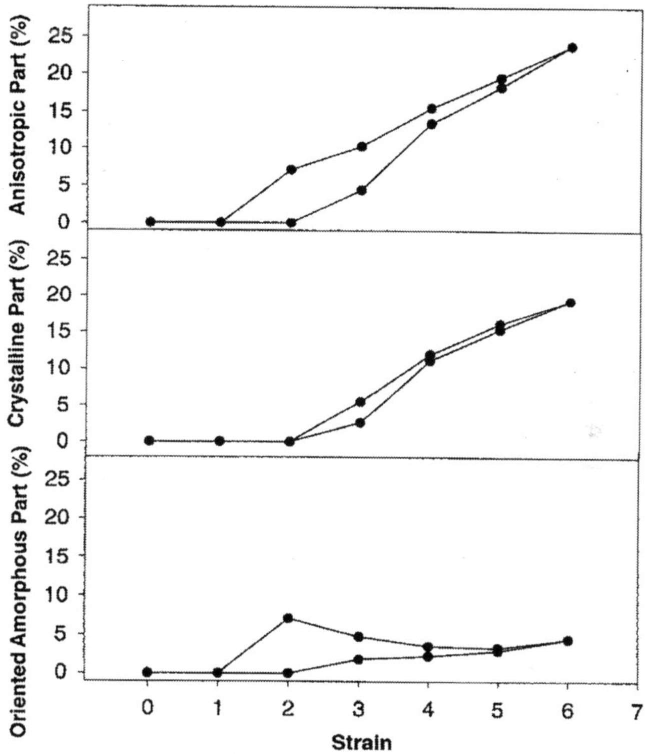

FIGURE 3.18 Total anisotropic fraction, crystal fraction, and oriented amorphous fraction of sulfur vulcanized NR stretched slowly at ambient temperature. The upper points correspond to extension, and the lower to retraction. For strains >200%, crystallization is induced, resulting in a decrease in the amorphous phase orientation.

methods are less subject to bias than micrographs. However, an inherent problem in obtaining structural information from scattering experiments is that only a limited range of reciprocal space is probed, although this limitation has been largely overcome with the advent of USANS (Sharp et al., 2009). The size distribution cannot be uniquely determined from a scattering curve. Only "scattering equivalent" distributions are obtained, based on assumptions about the particle shape and the functional form of the distribution. In Figure 3.19 the dispersed particle size of 1,4-polybutadiene in polychloroprene is shown, as determined by transmission electron microscopy (TEM) and small-angle scattering measurements (Roland and Bohm, 1984). Scanning transmission X-ray microscopy, based on the chemical specificity of near edge X-ray absorption, has potential for investigating heterogeneous morphologies, although it requires a synchrotron source (Winesett et al., 2003). Figure 3.20 demonstrates its utility for resolving blend morphologies without the requirement for staining.

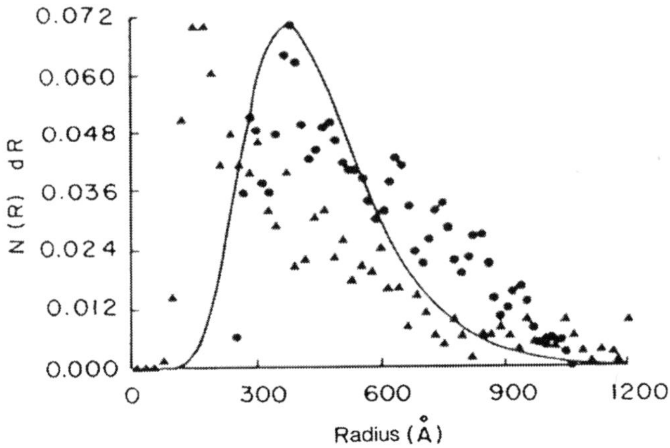

FIGURE 3.19 Size of dispersed phase in blend of 5% 1,4-polybutadiene in polychloroprene, as determined by analysis of TEM images (triangles), numerical transformation of combined SANS and SAXS data (circles), and best-fit to scattering curves assuming log normal distribution for particle sizes (solid line) (Roland and Bohm, 1984).

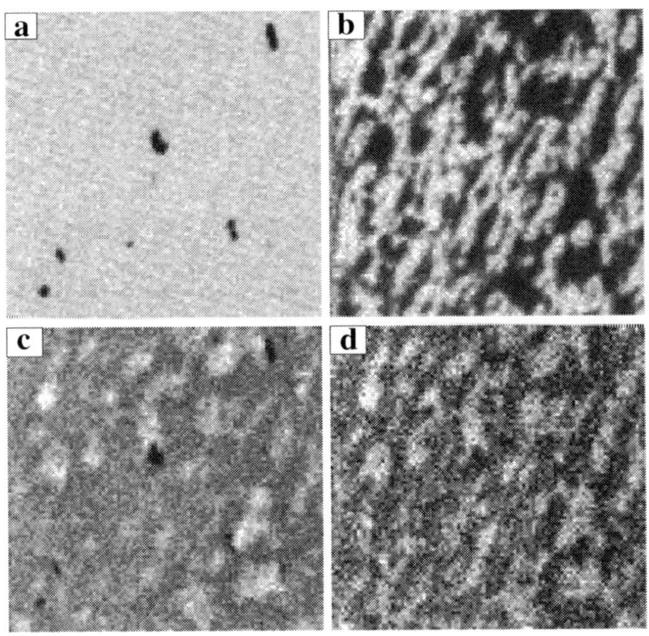

FIGURE 3.20 Scanning transmission X-ray micrographs ($10 \mu m^2$) of polybutadiene/poly(isobutylene-co-4-methylstyrene) (30/70) blend containing 20 phr carbon black, at increasing photon energies: (a) lowest energy—only carbon black structure evident; (b) higher energy—unsaturated polymer also evident; (c) and (d) highest photon energies—polyisobutylene now apparent with reversed contrast. Darker regions in the images depict locations of higher absorbing material and lighter regions are more transparent (Winesett et al., 2003).

A simple method of determining whether the phase morphology is homogeneous is by calorimetry. Observation of two transitions, corresponding to the respective T_g of each component, indicates a phase-separated morphology. However, a single transition does not guarantee thermodynamic miscibility, especially if the component T_gs are close. On the other hand, spectroscopic measurements can reveal two distinct relaxation peaks, even for a thermodynamically miscible blend (Miller et al., 1990; Roland and Ngai, 1991; Alegria et al., 1994; Ngai and Roland, 1995). Referred to as "dynamic heterogeneity," this arises when the components have very different intrinsic mobilities. The motion of each component is determined both by its chemical composition and by its local environment. The latter is averaged out when the phase morphology is homogeneous, which usually results in a single blend T_g. However, if the intrinsic mobilities are very different, two transitions can be observed, as shown in Figure 3.21. The transition, measured either by calorimetry or relaxation spectroscopy, of dynamically heterogeneous, miscible blends is usually quite broad (Roland, 1987).

When the respective component glass transition temperatures are close, the blend T_g is not a useful measure of blend homogeneity. In fact, excess mixing volumes and specific interactions can cause anomalous behavior. The T_g of such a blend can be lower (as seen in polychloroprene/epoxidized polyisoprene blends (McGrath and Roland, 1994)) or higher (as seen in polylepichlorohydrin/polyvinylmethylether blends (Alegria et al., 1995)), than T_g of either neat component. In blends of polymers having nearly equivalent

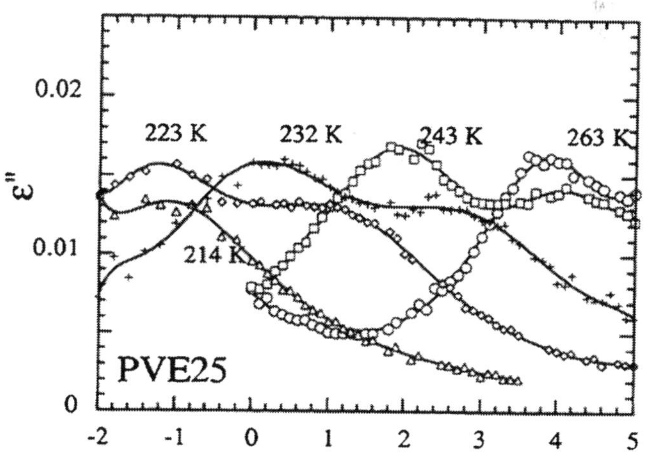

FIGURE 3.21 Isothermal dielectric loss curves for a thermodynamically-miscible blend of 25% 1,2-polybutadiene with natural rubber. Notwithstanding the homogenous morphology, the respective mobilities of the components differ, whereby two peaks are observed in the spectrum. Reprinted from Alegria et al. (1994).

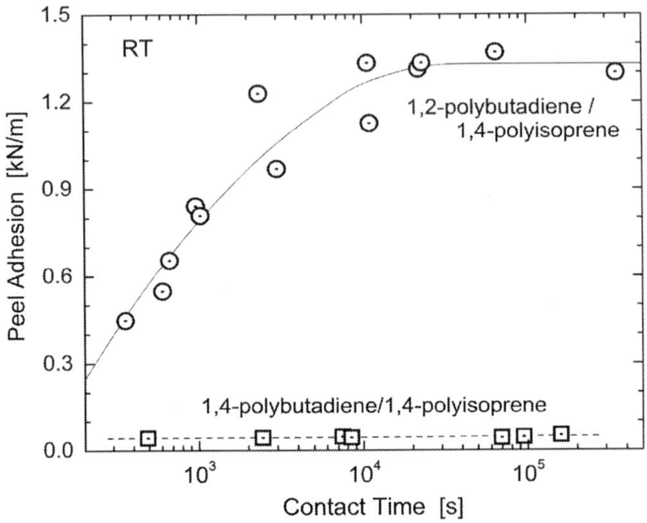

FIGURE 3.22 Resistance to separation of strips of two miscible rubbers (circles) and two immiscible rubbers (squares). Whereas adhesion of the latter remains low, the former is time dependent as the chains interdiffuse. At long times the interface is gone and the peel adhesion becomes constant (Roland, 1987).

T_g, spontaneous interdiffusion of the two materials can serve as an indication of thermodynamic miscibility (Figure 3.22; Roland, 1987).

^{129}Xe NMR has proven useful as a probe for blends (Miller, 1993). Xenon is highly polarizable, so that even van der Waals interactions produce large changes in its NMR chemical shift. When dissolved in a heterogeneous polymer blend, two ^{129}Xe NMR lines are observed, if the domains are large (relative to the diffusion time of the xenon). On the other hand, a single resonance is consistent with miscibility and yields an upper bound on the domain size. The technique is most useful for rubbery materials, so that the spectral lines are sharp. Various groups have used ^{129}Xe NMR to investigate the phase morphology of blends (Wachowicz et al., 2008; Wolak et al., 2003; Miyoshi et al., 2002; da Silva et al., 2000; Walton et al., 1992, 1993; McGrath and Roland, 1994). Figure 3.23 shows the changes in the ^{129}Xe NMR spectrum during redissolution of a blend of 1,4-polyisoprene and 1,4-polybutadiene after heating above the lower critical solution temperature (LCST).

A problem with phase-separated blends is obtaining a uniform dispersion of the compounding ingredients. For example, the distribution of curatives can be skewed by diffusion from one phase into the other. This problem is more significant when the component polymers have substantially different solubility parameters. Various methods have been proposed to assess the crosslink distribution in rubber blends (Tinker, 1995). Since the glass transition behavior is affected by vulcanization, especially for high degrees of crosslinking,

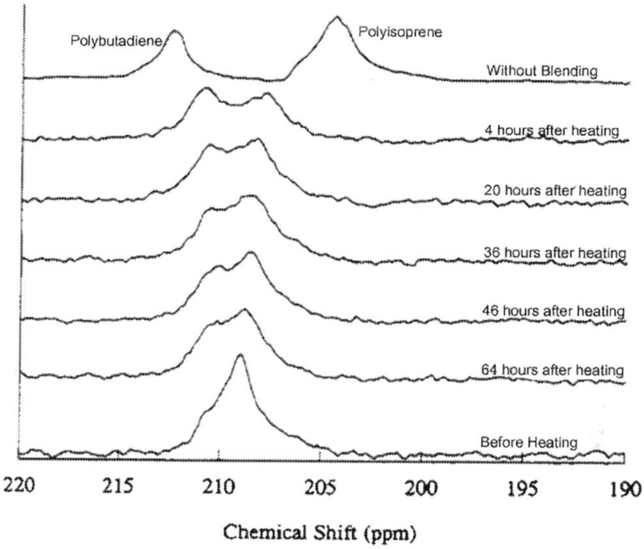

FIGURE 3.23 NMR spectra of ^{129}Xe dissolved in a homogeneous blend (bottom), and at various times after heating above the LCST, whereby phase separation transpires. The uppermost curve was measured on the pure polymers, simultaneously in the NMR tube but physically separated (Walton et al., 1993).

measurement of T_g (Shutilin, 1985; Tinker, 1990) or the local segmental relaxation (Roland, 1994) of the components can yield information about crosslink distributions. However, the method is insensitive for low degrees of crosslinking and requires that the components have significantly different T_gs. If the crosslinking is low enough that a substantial fraction of the polymer remains soluble, analysis of the respective sol and gel fractions can potentially enable the relative crosslinking of the phases to be assessed (Shvydkaya et al., 1980). There have been attempts to use TEM of swollen rubber blends to investigate crosslink distributions (Tinker, 1995). NMR imaging can detect spatially varying crosslink densities. Although the technique has been applied to study oxidation of rubber (Blumler and Blumich, 1997), the spatial resolution is too coarse (>10 μm) for blend studies.

NMR of swollen rubber has been used to determine crosslink distributions in blends (Loadman and Tinker, 1989; Brown et al., 1992; Cook, 1999). Swelling enhances chain mobility, and the isotropic motion of nuclei averages local fields, thereby narrowing the spectral lines. This allows individual resonances to be characterized. Crosslinking constrains motion and makes it more anisotropic, and thereby the incoherent averaging is less complete, broadening line widths (Figure 3.24). Empirical correlations with crosslink density are required for quantitative results. A variation on this technique is the use of ^{129}Xe, taking advantage of the shift dependence on the crosslink density (Parker et al., 2007).

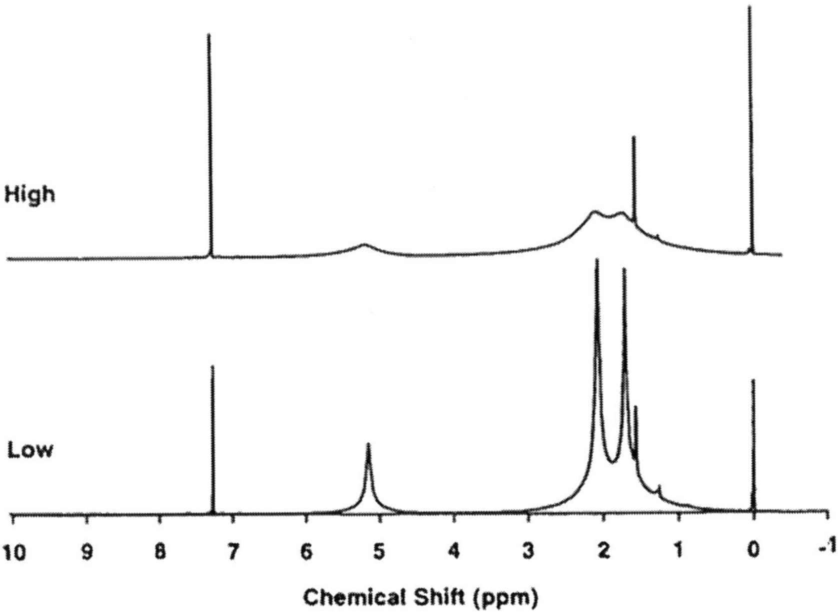

FIGURE 3.24 ^{1}H NMR spectra of swollen NR elastomers. The spectrum in the upper panel is for a sample having a crosslink density 4.5 times larger than that for the lower panel. The resonance at 5.2 ppm, due to the olefinic protons, can be utilized to determine relative crosslink densities (Tinker, 1995).

Crystallization in miscible blends can occur with rejection of the noncrystallizing component, so that its concentration in the amorphous phase increases. Alternatively, if it can be accommodated in the unit cell, it may be entrapped, with consequent alteration in the mean unit cell volume (Tomlin and Roland, 1993). In NR, there is also a shift to formation from α-lamellae to the β-lamellar form (Zemel and Roland, 1992b) (Figure 3.25). These crystal structures have the same unit cell, but the latter has a greater fold-surface free energy. Thus, the noncrystallizing blend component is more readily accommodated into the fold plane at the crystal surface.

3.6.3 Crystallinity

Rubbery behavior—large, reversible extensibility—implies an absence of crystallinity, and this is usually the case for undeformed elastomers. However, small extents of crystallization may be present at ambient temperature in some elastomers, including EPDM with high ethylene content, epichlorohydrin rubber, and polypropylene oxide. The crystallites in these materials can act as reinforcing agents. Many thermoplastic elastomers have crystalline domains that function as reversible crosslinks (Rzymski and Radusch, 2005; Bhowmick and Stephens, 2001).

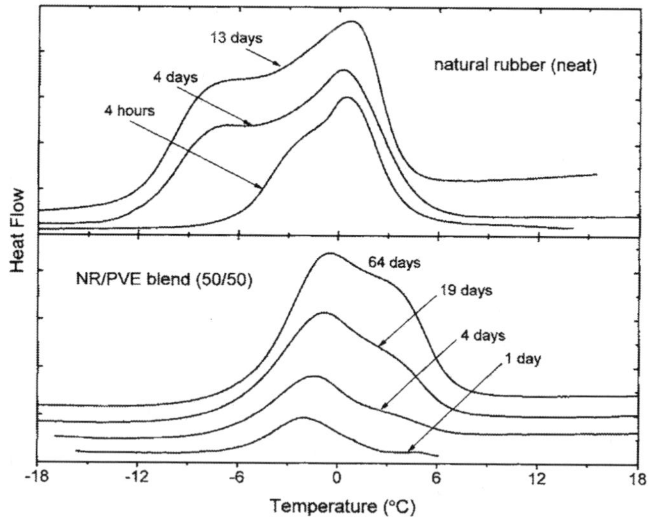

FIGURE 3.25 DSC melting endotherms in NR, pure (upper panel) and blended 50/50 with 1,2-polybutadiene (lower panel), after crystallization at −25 °C for the indicated time periods. The neat rubber initially crystallizes into the more stable α-lamellae, with substantial β-lamellae only forming at higher extents of crystallization. Miscible blending causes preferential crystallization of the lower melting β-lamellae (Zemel and Roland, 1992b).

If there is sufficient regularity of their backbone structure, amorphous rubbers crystallize at lower temperatures; this mainly affects storage behavior. Unoriented NR crystallizes through lamellar growth into radial spherulites, having two morphologically differing forms, the more stable α-lamella and the slower forming β-lamella (Andrews et al., 1971). Both lamellar types have the same crystal unit cell, but differ with respect to growth rates, lamellar thicknesses, and morphologies. As mentioned, the β-lamella have a higher fold-surface free energy. In order to accommodate the presence of *trans* units, synthetic *cis*-1,4-polyisoprene tends to crystallize more in the β-form, since the fold surface better tolerates noncrystallizing units (Edwards, 1975). A similar alteration in the crystal morphology of 1,4-polyisoprene is caused by miscible blending (see Figure 3.23) (Zemel and Roland, 1992b).

Although elastomers are usually amorphous, strain-induced crystallization occurs in rubbers such as *cis*-1,4-polybutadiene, butyl rubber, and NR, the most important commercial elastomer. Crystallization under stress, discovered 200 years ago (Gough, 1805), increases the modulus and most failure properties of rubber, and is essential to performance in many applications (Magill, 1995; Gent and Zhang, 2002; Hamed and Rattanasom, 2002; Santangelo and Roland, 2001). For NR, the propensity for strain-crystallization correlates directly with the failure properties (Choi and Roland, 1996). Crystallization of natural rubber under strain transpires through row nucleation of lamellae, whose growth proceeds perpendicular to the strain direction (Edwards,

1975; Andrews et al., 1971). The latter, secondary crystallization, can be quite slow (Mitchell and Meier, 1968), and its rate is unaffected by strain (Andrews et al., 1971). However, the row nucleation rate is greatly enhanced by orientation, effecting rapid initial crystallization (Krigbaum and Roe, 1964; Ziabicki and Jarecki, 1978; Elyashevich, 1982; Roland and Sonnenschein, 1991). The time for strain-induced crystallization of NR is less than 60 ms at RT (Mitchell and Meier, 1968), so that synchrotron X-ray sources (Toki et al., 2003, 2011) are required for on-the-fly measurements.

A prerequisite for high levels of strain-crystallization is steric purity of the polymer backbone, and this accounts for the better performance of natural in comparison to synthetic *cis*-1,4-polyisoprene (Brock and Hackathorn, 1972; Cooper and Smith, 1963). Reinforcing filler influences the crystallization behavior (Trabelsi et al., 2003a; Mark, 2003; Sharaf et al., 1995; Hamed and Park, 1999), as do the other compounding ingredients. The denatured proteins and other hydrocarbon-insoluble contaminants in NR primarily affect the rate of crystallization. For example, it has been shown that acetone extraction (Gent, 1955) and deproteinization of natural rubber (Burfield, 1984) both reduce the isotropic crystallization rate, presumably by reducing nucleation sites. In contrast, the degree of crystallinity attained in the absence of orientation is governed by the backbone microstructure, specifically the length of *cis*-1,4 sequences, not by the nonrubber constituents (Burfield and Tanaka, 1987).

Since the presence of impurities primarily affects crystal nucleation rather than growth (Gent, 1954, 1955), the degree of crystallinity and its dependence on

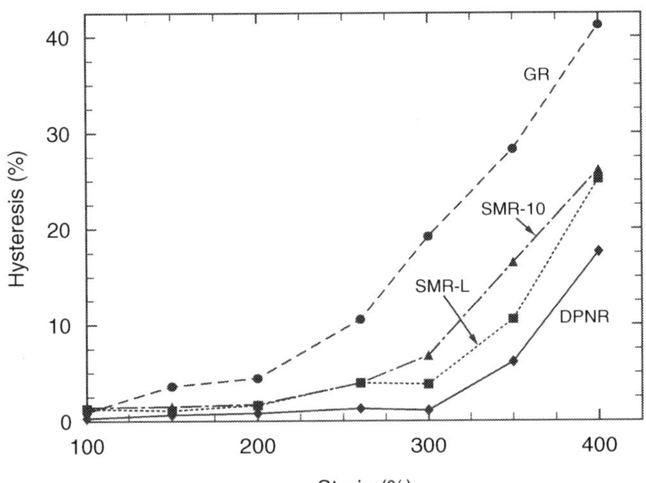

FIGURE 3.26 The mechanical energy dissipated relative to the total input strain energy for four grades of NR, stretched at RT to various extent. Strain crystallization at ca. 250% or higher results in more marked hysteresis (Choi and Roland, 1997).

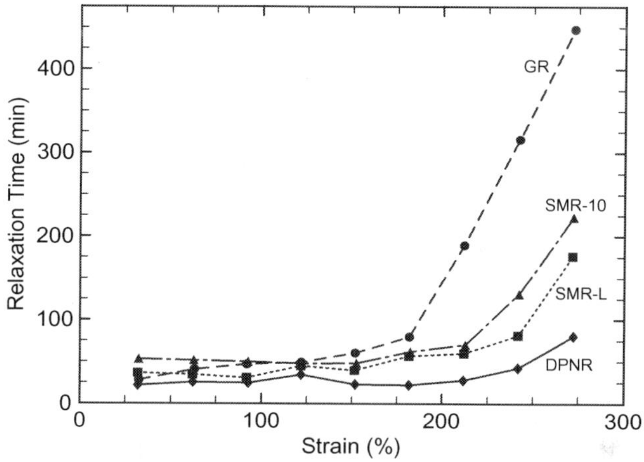

FIGURE 3.27 The time over which stress decay was observed for four NR elastomers. Initially viscoelasticity governs the relaxation time; at higher strains crystallization commences (Choi and Roland, 1997).

strain are of less interest in accounting for the failure properties of rubber. However, the minimum strain required for crystallization determines the stress concentration necessary to induce crystallization in the vicinity of a crack, as well as its spatial extent. Thus, this minimum strain plays a governing role in mechanical performance. Among the various grades of NR, the greater the purity, the higher the strain required to induce crystallization (Choi and Roland, 1997).

Methods to study crystallization of deformed elastomers include X-ray diffraction (Toki et al., 2003, 2011; Mitchell, 1984; Oono et al., 1973; Trabelsi et al., 2003b), optical birefringence (Treloar, 1947; Taylor and Darin, 1955), infrared or Raman spectroscopy, electron microscopy (Shimizu et al., 1998), dilatometry (Gent and Zhang, 2001, 2002), NMR (Kameda and Asakura, 2003), and mechanical measurements (Choi and Roland, 1997; Gent, 1954). Strain-induced crystallization is manifested in the latter by both greater hysteresis (Figure 3.26) and a longer time for stress decay (Figure 3.27). However, the shape of the stress-strain curve during extension does not obviously reveal the onset of crystallization (Toki et al., 2003, 2011; Choi and Roland, 1997).

3.6.4 Defects

It is well known that elastomers, like virtually all solid materials, have preexisting, "naturally occurring" flaws (Roland and Smith, 1985). By intensifying local stresses, such flaws exert an influence on the failure properties of elastomers. More recently, interest in these flaws has increased, due to concerns about their potential for reducing the barrier performance of rubber films. This performance is crucial in the use of latex rubber products, such as surgical gloves and condoms, to block transmission of the submicron-sized particles responsible for AIDS, hepatitis, and other viral diseases (Roland, 1993).

The passage of viral-sized particles through ostensibly intact NR latex films has been directly observed in the laboratory (Carey et al., 1992; Kiernan, 1996), and indirectly observed in the field (The New York Times, 1994). Evidence for such flaws in NR comes from microscopic observations (Arnold et al., 1988), as well as water absorption measurements, wherein the initial rapid uptake suggests the existence of capillary channels (Gazeley et al., 1988).

Since failure properties are affected by the intrinsic defects in rubber, they offer a means to characterize these flaws. The strength of rubber is influenced by the presence of defects having an apparent size in the range between 5 and 70 μm (Bueche, 1959; Braden and Gent, 1960; Gent, 1978; Roland and Sobieski, 1989). Thus, a means to estimate the size of the largest inherent flaw is by measuring the strength of samples after the introduction of cuts of varying size. The data are extrapolated to the strength of the uncut material, to yield the inherent crack size. If the rubber strain-crystallizes, the precuts must be small, so that such crystallization of the bulk material transpires prior to failure (Thomas and Whittle, 1970). For a linearly elastic material, the strength will vary as the square-root of the flaw size (Gent, 1978), while more generally, the strain energy to break is proportional to the flaw size (Bell et al., 1982). In Table 3.4 the flaw size deduced from the strain energy to break is listed for various grades of NR (Choi and Roland, 1996).

Since fatigue failure of rubber is envisioned as growth of intrinsic flaws, measurement of fatigue lifetimes (e.g., deformation cycles to failure) can provide a measure of the intrinsic flaw size (Gent et al., 1964; Lake and Lindley, 1965; Lake, 1983). Table 3.4 includes the flaw sizes determined from the fatigue life of these sample NR compounds (Choi and Roland, 1996).

The values determined for the inherent flaws only represent effective sizes, corresponding to a given degree of stress concentration (Greensmith, 1964). This stress concentration also depends on the shape of a crack (e.g., bluntness), as well as the dissipative capacity of the material itself (Hamed et al., 1996). Although compounding variables such as crosslink density strongly affect the

TABLE 3.4 Intrinsic Flaw Size of Natural Rubbers (Choi and Roland, 1996)

NR Grade	Intrinsic Flaw Size (μm)	
	Strain Energy	Fatigue Life
Guayule rubber	29	26
SMR-10	29	31
SMR-L	26	17
Deproteinized NR	16	10

failure properties of rubber, their influence on the intrinsic defect size is more modest (Lake and Lindley, 1965; Lake, 1983; Hamed, 1983). The measured flaw size is unaffected by temperature (Braden and Gent, 1960), although it can vary with carbon black type (Dizon et al., 1973). The degree of dispersion of compounding ingredients can influence the strength of rubber, apparently by a flaw-initiation mechanism (May, 1964).

ACKNOWLEDGMENTS

I am grateful to D.J. Lohse and G. ver Strate for making available in electronic form the second edition of this chapter, which has served as the basis for some of the text in the ensuing two versions. This work was supported by the Office of Naval Research.

REFERENCES

Abdel-Goad, M., Pyckhout-Hintzen, W., Kahle, S., Allgaier, J., Richter, D., Fetters, L.J., 2004. Macromolecules 37, 8135.
Akcasu, A.Z., Summerfield, G.C., Jahsan, S.N., Han, C.C., Kim, C.Y., Yu, H., 1980. J. Polym. Sci. Polym. Phys. 18, 863.
Alegria, A., Colmenero, J., Ngai, K.L., Roland, C.M., 1994. Macromolecules 27, 4486.
Alegria, A., Elizetxea, C., Cendoya, I., Colmenero, J., 1995. Macromolecules 28, 8819.
Andrews, E.H., Owen, P.G., Singh, A., 1971. Proc. Royal Soc. London A 324, 79.
Annis, B.K., Kim, M.H., Alamo, R., Pyda, M., 2001. J. Polym. Sci. Polym. Phys. 39, 2852.
Arbe, A., Colmenero, J., Farago, B., Monkenbusch, M., Buchenau, U., Richter, D., 2003. Chem. Phys. 292, 295.
Arnold, S.G., Whitman, J.E., Fox, C.H., Cottler-Fox, M.H., 1988. Nature 335, 19.
Avgeropoulos, G.N., Weissert, F.C., Biddison, P.H., Bohm, G.G.A., 1976. Rubber Chem. Technol. 49, 93.
Baldwin, F.P., Ver Strate, G., 1972. Rubber Chem. Technol. 44, 709.
Balke, S.T., Mourey, T.H., 2001. J. Appl. Polym. Sci. 81, 370.
Barth, H., Mays, J., 1991. Modern Methods of Polymer Characterization. Wiley, New York.
Bell, C.L.M., Stinson, D., Thomas, A.G., 1982. Rubber Chem. Technol. 55, 66.
Berger, L., Kausch, H.H., Plummer, C.J.G., 2003. Polymer 44, 5877.
Berne, B.J., Pecora, R., 1976. Dynamic Light Scattering. Wiley, New York (Chapter 8).
Betrand, P., Weng, L.T., 1999. Rubber Chem. Technol. 72, 384.
Betrand, P., Weng, L.T., Lauer, W., Zimmer, R., 2002. Rubber Chem. Technol. 75, 627.
Bhowmick, A.K., Stephens, H.L. (Eds.), 2001. Handbook of Elastomers, 2nd ed. Marcel Dekker, New York.
Blitz, J.P., McFaddin, D.C., 1994. J. Appl. Polym. Sci. 51, 13.
Blumler, P., Blumich, B., 1997. Rubber Chem. Technol. 70, 468.
Bogoslovov, R.B., Hogan, T.E., Roland, C.M., 2010. Macromolecules 43, 2904.
Bohm, G.G.A., Tveekrem, J.O., 1982. Rubber Chem. Technol. 55, 575.
Bonnar, R.U., Dimbat, M., Stross, F.H., 1958. Number-Average Molecular Weights—Fundamentals and Determination. Interscience, New York, NY.
Bonse, U., Hart, M., 1965. Appl. Phys. Lett. 7, 238.
Booth, C., Price, C. (Eds.), 1989. Comprehensive Polymer Science, vol. 1. Polymer Characterization, Pergammon, NY.
Boothroyd, A., Fetters, L.J., 1991. Macromolecules 24, 5215.
Botti, A., Pyckhout-Hintzen, W., Richter, D., Urban, V., Straube, E., Kohlbrecher, J., 2003. Polymer 44, 7505.
Boue, G., Bastide, J., Buzier, M., Lapp, A., Herz, J., Vilgis, T.A., 1991. Coll. Polym. Sci. 269, 195.
Bower, D.I., 1981. J. Polym. Sci. Polym. Phys. 19, 93.
Bower, D.I., Lewis, E.L.V., Ward, I.M., 1995. Polymer 36, 3473.
Braden, M., Gent, A.N., 1960. J. Appl. Poly. Sci. 3, 100.

Braun, D., 1996. Simple Methods for Identification of Plastics, 3rd ed. Hanser, NY.
Braun, D., Rettig, R., Rogler, W., 1993. Angew. Makromol. Chem. 211, 165
Briggs, D., 1989. Brit. Polym. J. 21, 3.
Brock, M.J., Hackathorn, M.J., 1972. Rubber Chem. Technol. 45, 1301.
Brosseau, C., Molinie, P., Boulic, F., Carmona, F., 2001. J. Appl. Phys. 89, 8297.
Brown, J., Verdier, P., 1972. J. Res. Natl. Bur. Stand. A76, 161.
Brown, P.S., Loadman, M.J.R., Tinker, A.J., 1992. Rubber Chem. Technol. 65, 744.
Bruell, R., Raquel, M., Rode, K., 2010. Macro. Chem. Phys. 211, 2233.
Brun, Y., 1998. J. Liq. Chromatography & Related Technol. 21, 1979.
Bueche, F., 1959. Rubber Chem. Technol. 32, 1269.
Bur, A.J., Roth, S.C., Thomas, C.L., 2000. Rev. Sci. Instr. 71, 1516.
Burfield, D.R., 1984. Polymer 25, 1823.
Burfield, D.R., Tanaka, Y., 1987. Polymer 28, 907.
Campbell, D., White, J., 1989. Polymer Characterization. Chapman and Hall, NY.
Capancioni, S., Schwach-Abdellaoui, K., Kloeti, W., Herrmann, W., Brosig, H., Borchert, H.H., Heller, J., Gurny, R., 2003. Macromolecules 36, 6135.
Carella, J.M., Gotro, J.T., Graessley, W.W., 1986. Macromolecules 19, 659.
Carey, R.F., et al., 1992. Sexually Transmitted Diseases 19, 230.
Casassa, E., 1972. Polym. J. 3, 517.
Castignolles, P., Gaborieau, M., 2010. J. Separation Sci. 33, 3564.
Cebe, P., 2005. J. Polym. Sci. Polym. Phys. Ed. 43, 629.
Chen, X., Xi, J., Guo, B.H., 2006. J. Appl. Polym. Sci. 102, 3201.
Cheng, H., 1991. In: Barth, H., Mays, J. (Eds.), Modern Methods of Polymer Characterization, Chemical Analysis, vol. 113. Wiley, N.Y., p. 409.
Cheng, M.H.Y., Nauman, E.B., 2004. J. Polym. Sci. Polym. Phys. 42, 603.
Cho, D., Park, S., Kwon, K., Chang, T., Roovers, J., 2001. Macromolecules 34, 7570.
Choi, I.S., Roland, C.M., 1996. Rubber Chem. Technol. 69, 591.
Choi, I.S., Roland, C.M., 1997. Rubber Chem. Technol. 70, 202.
Chou, C.Y., Hsieh, Y.H., Chang, G.G., 2011. Methods 54, 76.
Clough, R.S., Koenig, J.L., 1989. Rubber Chem. Technol. 62, 908.
Colby, R.H., Fetters, L.J., Graessley, W.W., 1987. Macromolecules 20, 2237.
Collins, E., Bares, J., Billmeyer, F., 1973. Experiments in Polymer Science. Wiley, NY.
Cook, S., 1999. Kaut. Gummi Kunst. 52, 350.
Cooper, A. (Ed.), 1989. Determination of Molecular Weight, Chemical Analysis Series, vol. 103. Wiley, New York.
Cooper, W., Smith, R.K., 1963. J. Polym. Sci. A 1, 159.
Creel, H., 1993. Trends Polym. Sci. 1, 336.
da Silva, N.M., Tavares, M.I.B., Stejskal, E.O., 2000. Macromolecules 33, 115.
Dawkins, J.V. (Ed.), 1986. Developments in Polymer Characterization, vol. 1–5. Elsevier, NY.
Dekmezian, A., Morioka, T., Campo, C., 1990. J. Pol. Sci. Phys. 28, 1908.
Dekmezian, A., Morioka, T., Campo, C., 1990. J. Polym. Sci. Polym. Phys. 28, 1908.
Deloche, B., Samulski, E.T., 1981. Macromolecules 14, 575.
Devries, K.L., 1971. J. Polym. Sci. C Polym. Symp. 32, 325.
Dieing, T., Hollricher, O., Toporski, J. (Eds.), 2011. Confocal Raman Microscopy, Springer.
Dizon, E.S., Hicks, A.E., Chirico, V.E., 1973. Rubber Chem. Technol. 46, 231.
Doi, M., 1993. In: Thomas, E.L. (Ed.), Structure and Properties of Polymers. Wiley-VCH (Weinheim).
Doi, M., Watanabe, H., 1991. Macromolecules 24, 740.
Dumas, P.J., 2003. de Phys. IV 104, 359.
Dupres, S., Long, D.R., Albouy, P.A., Sotta, P., 2009. Paul. Macromolecules 42, 2634.
Edwards, B.C., 1975. J. Polym. Sci., Polym. Phys. 13, 1387.
Ekanayake, P., Menge, H., Schneider, H., Ries, M.E., Brereton, M.G., Klein, P.G., 2000. Macromolecules 33, 1807.
Elyashevich, G.K., 1982. Adv. Poly. Sci. 43, 205.
Everall, N.J., Bibby, A., 1997. Appl. Spect. 51, 1083.
Feast, W.J., Keeney, A.J., Kenwright, A.M., Parker, D., 1997. Chem. Comm. 18, 1749.
Feng, Y., Hay, J.N., 1998. Polymer 39, 6723.
Flory, P.J., 1969. Statistical Mechanics of Chain Molecules. Oxford Univ. Press, NY.

Fox, T.G., Flory, P.J., 1954. J. Polym. Sci. 14, 315.
Froud, C.A., Hayward, I.P., Laven, J., 2003. Appl. Spect. 57, 1468.
Fuhrmann, I., Karger-Kocsis, J., 2003. J. Appl. Polym. Sci. 89, 1622.
Fulton, S.W., 2006. Rubber Chem. Technol. 79, 790.
Gaborieau, M., Gilbert, R.G., Gray-Weale, A., Hernandez, J.M., Castignolles, P., 2007. Macro. Theor. Simul. 16, 13.
Garcia, R.A., Coto, B., Exposito, M.T., Suarez, I., Fernandez-Fernandez, A., Caveda, S., 2011. Macromol. Res. 19, 778.
Gazeley, K.F., Gorton, A.D.T., Pendle, T.D., 1988. In: Roberts, A.D. (Ed.), Natural Rubber Science and Technology. Oxford Univ. Press Oxford (Chapter 3).
Gent, A.N., 1954. Inst. Rubb. Ind. 30, 139.
Gent, A.N., 1954. Trans. Faraday Soc. 50, 521.
Gent, A.N., 1955. J. Polym. Sci. 18, 321.
Gent, A.N., 1978. In: Eirich, F.R. (Ed.), Science and Technology of Rubber. Academic Press, New York (Chapter 10).
Gent, A.N., Zhang, L.Q., 2001. J. Polym. Sci. Polym. 39, 811.
Gent, A.N., Zhang, L.Q., 2002. Rubber Chem. Technol. 75, 923.
Gent, A.N., Lindley, P.B., Thomas, A.G., 1964. J. Appl. Poly. Sci. 8, 455.
Ghanbari-Siahkali, A., Almdal, K., Kingshott, P., 2003. Appl. Spect. 57, 1482.
Ghebremeskel, G.N., Sekinger, J.K., Hoffpauir, J.L., Hendrix, C., 1996. Rubber Chem. Technol. 69, 874.
Glockner, G., 1989. In: Booth, C., Price, C. (Eds.), Comprehensive Polymer Science, vol. 1. Polymer Characterization. Pergammon, New York, 1989. p. 313.
Goga, N.O., Demco, D.E., Kolz, J., Ferencz, R., Haber, A., Casanova, F., Bluemich, B., 2008. J. Mag. Res. 192, 1.
Gough, J., 1805. Proc. Lit. and Phil. Soc. Manchester, 2nd Ser., 1, 288.
Graessley, W., Ver Strate, G., 1980. Rubber Chem. Technol. 53, 842.
Greensmith, H.W., 1964. J. Appl. Polym. Sci. 8, 1113.
Grest, G.S., Fetters, L.J., Huang, J.S., Richter, D., 1996. In: Prigogine, I., Rice, S.A. (Eds.), Advances in Chemical Physics XCIV. Wiley, NY 1996.
Gronski, W., Forster, F., Pyckhout-Hintzen, W., Springer, T., 1990. Makromolekulare Chem. Macromol. Symp. 40, 121.
Grubisic, Z., Rempp, P., Benoit, H., 1967. J. Polym. Sci. 21, 753.
Grubsic-Gallot, Z., Lingelser, J.P., Gallot, Y., 1990. Polym. Bull. 23, 389.
Grum, F., 1972. In: Weissberger, A., Rossiter, B. (Eds.). Visible and Ultraviolet Spectrophotometry. Physical Methods of Chemistry, Part III B. vol. 1. Wiley Interscience, New York, p. 207.
Hamed, G.R., 1983. Rubber Chem. Technol. 56, 244.
Hamed, G.R., Park, B.H., 1999. Rubber Chem. Technol.72, 946.
Hamed, G.R., Rattanasom, N., 2002. Rubber Chem. Technol. 75, 935.
Hamed, G.R., Kim, H.J., Gent, A.N., 1996. Rubber Chem. Technol. 69, 807.
Hammond, T., Lehrie, R., 1989. Brit. Polym. J. 21, 23.
Hammouda, B., 2010. J. Macromol. Sci. C 50, 14.
Hauck, D., Fink, G., Chwatinski, C., Blumler, P., Blumich, B., Unseld, K., 1997. Kauch. Gummi Kunstat. 50, 392.
Hennecke, M., Kurz, K., Fuhrmann, J., 1992. Macromolecules 25, 6190.
Higgins, J.S., Benoit, H.C., 1997. Polymers and Neutron Scattering. Clarendon Press, Oxford.
Hinojosa, O., Arthur, J.C., Nakamura, Y., 1972. J Polym. Sci. C Polym. Symp. 37, 27.
Hohne, G., Merzlyakov, M., Schick, C., 2002. Thermochim. Acta 391, 51.
Hommel, H., Touhami, A., Legrand, A.P., 1993. Makromol. Chem. Macro. Chem. Phys. 194, 879.
Horsky, J., Bohdanecky, M., 1990. Eur. Pol. J. 26, 1190.
Hoven, V.P., Rattanakaran, K., Tanaka, Y., 2003. Rubber Chem. Technol. 76, 1128.
Hsu, S.L., 2004. Handbook of Vibrational Spectroscopy: A Companion for Polymer Scientists. Wiley, NY
Hu, J.C., 1981. Anal. Chem. 53, 942.
Hutchinson, J.M., 2003. J. Therm. Anal. Calorimetry 72, 619.
Ishida, H., 1987. Rubber Chem. Technol. 60, 497.
Janca, J., Ananieva, I.A., Menshikova, A.Y., Evseeva, T.G., 2004. J. Chromatography B800, 33.
Jiang, S., Bae, S.C., Granick, S., 2008. Langmuir 24, 1489.

Kameda, T., Asakura, T., 2003. Polymer 44, 7539.
Kannan, R.M., Kornfield, J.A., 1994. J. Rheol. 38, 1127.
Karabudak, E., Wohlleben, W., Colfen, H., 2010. Eur. Biophys. J. 39, 397.
Kassalainen, G.E., Williams, S.K.R., 2003. Anal. Chem. 75, 1887.
Kawahara, S., Inomata, Y., Kakubo, T., Tanaka, Y., Hatada, K., Ute, K., Miyatake, N., 1998. Rubber Chem. Technol. 71, 277.
Keilman, F., 2002. Vibr. Spect. 29, 109.
Keilmann, F., 2003. J. Biol. Phys. 29, 195.
Kempfert, K.D., Coel, B., Troost, P., Lavery, D.S., 2001. Amer. Lab. 33, 22.
Kiernan, V., 1996. New Scientist, March 23, p. 7.
Kim, W.S., Eum, C.H., Molnar, A., Yu, J.S., Lee, S., 2006. Analyst 131, 429.
Kinsey, R., 1990. Rubber Chem. Technol. 63, 407.
Koenig, J.L., 1999. Spectroscopy of Polymers, 2nd ed. Elsevier, NY.
Koenig, J.L., 2000. Rubber Chem. Technol. 73, 385.
Kona, B., Weidner, S.M., Friedrich, J.F., 2005. Int. J. Polym. Anal. Char. 10, 85.
Kotani, M., Dohi, H., Kimura, H., Muraoka, K., Kaji, H., 2007. Macromolecules 40, 9451.
Kraus, G., Gruver, J.T., 1965. J. Polym. Sci. A 3, 105.
Kremer, F., Schonhals, A. (Eds.), 2003. Broadband Dielectric Spectroscopy. Springer, Berlin.
Krigbaum, W.R., Roe, R.J., 1964. J. Polym. Sci. A2, 4391.
Krishnan, R., Nagarajan, K., 2010. J. Therm. Anal. Cal. 102, 1135.
Kuo, C.Y., Provder, T., 1987. ACS Symp. Series 352, 2.
Kuo, C.Y., Provder, T., Koehler, M.E., 1990. J. Liq. Chromatography 13, 3177.
Kuo, C.Y., Provder, T., Koehler, M.E., 1993. ACS Symp. Series 521, 231.
Kurata, M., Tsunashima, Y., 1999. In: Brandrup, J., Immergut, E.H., Grulke, E.A. (Eds.), Polymer Handbook, 4th ed. Wiley, New York (Chapter 7).
Lake, G.J., 1983. Prog. Rubber Technol. 45, 89.
Lake, G.J., Lindley, P.B., 1965. J. Appl. Poly. Sci. 9, 1233.
Lattimer, R., 1990. Rubber Chem. Technol. 63, 298.
Laue, T.M., 1990. In: Kroschwitz, J.I. (Ed.), Concise Encyclopedia of Polymer Science and Engineering. Wiley, New York, 1990, p. 1225.
Lee, S.H., Molnar, A., 1995. Macromolecules 28, 6354.
Lee, D., Williams, S.K.R., 2010. J. Chromat. 1217, 1667.
Lee, S., Eum, C.H., Plepys, A.R., 2000. Bull. Korean Chem. Soc. 21, 69.
Leeson, A.M., Alexander, M.R., Short, R.D., Briggs, D., Hearn, M.J., 1997. Surf. Interfac. Anal. 25, 261.
Lewandowski, L., Sibbald, M.S., Johnson, E., Mallamaci, M.P., 2000. Rubber Chem. Technol. 73, 731.
Lindner, H., Glatter, O., 2000. Macromol. Symp. 162, 81.
Litvinov, V.M., Orza, R.A., Klueppel, M., van Duin, M., Magusin, P.C.M.M., 2011. Macromolecules 44, 4887.
Loadman, M.J.R., Tinker, A.J., 1989. Rubber Chem. Technol. 62, 234.
Lohse, D.J., 1994. Rubber Chem. Technol. 67, 367.
Madan, B., Naskar, N.K., Perera, K., Moreland, C., Hodge, T., Wallace, K., Beckham, H.W., Smith, D.W., 2012. Rubber Chem. Technol. 85, 68.
Maeda, Y., Yamamoto, H., Ikeda, I., 2003. Macromolecules 36, 5055.
Magill, J.H., 1995. Rubber Chem. Technol. 68, 507.
Majeste, J.C., Carrot, C., Stanescu, P., 2003. Rheo. Acta 42, 432.
Mark, J.E., 2003. Prog. Polym. Sci. 28, 1205.
Mathew, T., Datta, R.N., Dierkes, W.K., Noordermeer, J.W.A., van Ooij, W.J., 2008. Rubber Chem. Technol. 81, 209.
Mathot, V., Pyda, M., Pijpers, T., Poel, G.V., van de Kerkhof, E., van Herwaarden, S., van Herwaarden, F., Leenaers, A., 2011. Thermochim. Acta 522, 36.
May, W., 1964. Trans. Inst. Rubber Ind. 40, T109.
McCarley, K.D., Bunge, A.L., 2003. Int. J. Pharm. 250, 169.
McGrath, K.J., Roland, C.M., 1994. Rubber Chem. Technol. 67, 629.
Medrano, R., Laguna, M.T.R., Saiz, E., Tarazona, M.P., 2003. Phys. Chem. Chem. Phys. 5, 151.
Messaud, F.A., Sanderson, R.D., Runyon, J.R., Ray, J., Otte, T., Pasch, H., Williams, S.K.R., 2009. Prog. Polym. Sci. 34, 351.

Messerschmidt, R., Harthcock, M. (Eds.), 1988. Infrared Microspectroscopy, Practical Spectroscopy Series, vol. 6. Marcel Dekker, NY.
Mileva, D., Androsch, R., Zhuravlev, E., Schick, C., 2009. Macromolecules 42, 7275.
Miller, J.B., 1993. Rubber Chem. Technol. 66, 455.
Miller, G.H., Tobias, R.H., 1978. Rubber Chem. Technol. 51, 977.
Miller, J.B., McGrath, K.J., Roland, C.M., Trask, C.A., Garroway, A.N., 1990. Macromolecules 23, 4543.
Minari, R.J., Rodriguez, V.I., Estenoz, D.A., Vega, J.R., Meira, G.R., Gugliotta, L.M., 2010. J. Appl. Polym. Sci. 116, 590.
Mirabella, F.M., 1993. Polymer 34, 1729.
Mitchell, G.R., 1984. Polymer 25, 1562.
Mitchell, J. (Ed.), 1992. Applied Polymer Analysis and Characterization, vol. 2. Hanser, NY.
Mitchell, J.C., Meier, D.J., 1968. J. Polym. Sci. Polym. Phys. 6, 1689.
Miyoshi, T., Takegoshi, K., Terao, T., 2002. Macromolecules 35, 141.
Mori, M., Koenig, J.L., 1998. J. Appl. Polym. Sci. 70, 1391.
Mori, S., Mouri, M., 1989. Anal. Chem. 61, 2171.
Myers, C.W., Cooper, S.L., 1994. Appl. Spect. 48, 72.
Nagy, D.J., 2003. Amer. Lab. 35, 38.
Ngai, K.L., Paluch, M., 2004. J. Chem. Phys. 120, 857.
Ngai, K.L., Plazek, D.J., 1995. Rubber Chem. Technol. 68, 376.
Ngai, K.L., Roland, C.M., 1995. Macromolecules 28, 44033.
Ngai, K.L., Roland, C.M., 1997. J. Polym. Sci. Polym. Phys. 35, 2503.
Nguyen, M., Beckett, R., Pille, L., Solomon, D.H., 1998. Macromolecules 31, 7003.
Nielsen, M.W.F., 1999. Mass. Spectrom. Rev. 18, 309.
Okabe, M., Mitsui, K., Oranaka, H., Jahahashi, M., Masuda, H., 1992. Polymer J. 24, 653.
Oono, R., Miyasaka, K., Ishikawa, K., 1973. J. Polym. Sci. 11, 1477.
Pace, M.D., Roland, C.M., 1991. Polymer 32, 1027.
Palanker, D.V., Simanovskii, D.M., Huie, P., Smith, T.I., 2000. J. Appl. Phys. 88, 6808.
Parker, W.O., Ferrando, A., Ferri, D., Canepari, V., 2007. Macromolecules 40, 5787.
Park, M.S., Wong, Y.S., Park, J.O., Venkatraman, S.S., Srinivasarao, M., 2011. Macromolecules 44, 2120.
Partridge, R., Symons, M.C.R., Wyatt, J.L., 1993. J. Chem. Soc. Faraday Trans. 89, 1285.
Paul, D.R., Barlow, J.W., 1980. J. Macromol. Sci. Rev. Macromol. Chem. C18, 108.
Peckj, M.C.P., Samus, M.A., Killgoar, R.C., Carter, R.O., 1991. Rubber Chem. Technol. 64, 610.
Peng, X., Huang, Y., Xia, T., Kong, M., Li, G., 2011. Eur. Polym. J. 47, 1956.
Pielichowski, K., Flejtuch, K., 2002. Polimery 47, 784.
Plazek, D.J., Chay, I.-C., Ngai, K.L., Roland, C.M., 1995. Macromolecules 28, 6432.
Ponomarenko, S.A., Rasulova, N.N., Luponosov, Y.N., Surin, N.M., Buzin, M.I., Leshchiner, I., Peregudova, S.M., Muzafarov, A.M., 2012. Macromolecules 45, 2014.
Provder, T., Urban, M.W., Barth, H.G. (Eds.), 1995. Chromatographic Characterization of Polymers: Hyphenated and Multidimensional Techniques, Adv. Chem. Series 247, Amer. Chem. Soc. Washington, DC.
Puskas, J.E., Hutchinson, R., 1993. Rubber Chem. Technol. 66, 742.
Puskas, J.E., Burchard, W., Heidenreich, A.J., Dos Santos, L., 2012. J. Polym. Sci. Polym. Chem. Ed. 50, 70.
Qu, X.W., Shang, S.R., Liu, G.D., Zhang, S.W., Zhang, Y., Zhang, L.C., 2004. J. Appl. Polym. Sci. 91, 1685.
Rabolt, J.F., Chase, D.B. (Eds.), 2000. Fourier Transform Raman Spectroscopy: From Concept to Experiment. Academic Press, NY.
Ramachandran, R., Beaucage, G., Kulkarni, A.S., McFaddin, D., Merrick-Mack, J., Galiatsator, V., 2008. Macromolecules 41, 9802.
Ramachandran, R., Beaucage, G., McFaddin, D., Merrick-Mack, J., Galiatsatos, V., Mirabella, F., 2011. Polymer 52, 2661.
Raubner, V.M., Jordan, R., Nuyken, O., Lippert, T., Hauer, M., Schnyder, B., Wokaun, A., 2002. Appl. Surf. Sci. 197, 786.
Richter, D., 2003. J. Appl. Cryst.36, 389.
Robertson, C.G., Roland, C.M., Paulo, C., Puskas, J.E., 2001. J. Rheol. 45, 759.

Rodrigues, F.H.A., Santos, E.F., Feitosa, J.P.A., Ricardo, N.M.P.S., de Paula, R.C.M., 2001. Rubber Chem. Technol.74, 57.
Roland, C.M., 1987. Macromolecules 20, 2557.
Roland, C.M., 1993. Rubber World 208, 15.
Roland, C.M., 1994. Macromolecules 27, 4242.
Roland, C.M., 2011. Viscoelastic Behavior of Rubbery Materials. Oxford Univ. Press.
Roland, C.M., Bohm, G.G.A., 1984. J. Polym. Sci. Pt. B-Polym. Phys. 22, 79.
Roland, C.M., Casalini, R., 2003. J. Chem. Phys. 119, 1838.
Roland, C.M., Ngai, K.L., 1991. Macromolecules 24, 2261.
Roland, C.M., Santangelo, P.G., 2002. J. Non-Cryst. Solids 307, 835–841.
Roland, C.M., Smith, C.R., 1985. Rubber Chem. Technol.58, 806.
Roland, C.M., Sobieski, J.W., 1989. Rubber Chem. Technol. 62, 683.
Roland, C.M., Sonnenschein, M.F., 1991. Polym. Eng. Sci. 31, 1434.
Roland, C.M., Ngai, K.L., Santangelo, P.G., Qiu, X.H., Ediger, M.D., Plazek, D.J., 2001. Macromolecules 34, 6159.
Rooney, J.G., Ver Strate, G., 1981. In: Cazes, J. (Ed.), Liquid Chromatography of Polymers and Related Materials III, Marcel Dekker.
Ruch, D., Muller, J.F., Migeon, H.N., Boes, C., Zimmer, R., 2003. J. Mass. Spect. 38, 50.
Ruckebusch, C., Vilmin, F., Coste, N., Huvenne, J.P., 2008. Appl. Spect. 62, 791.
Runt, J.P., Fitzgerald, J.J. (Eds.), 1997. Dielectric Spectroscopy of Polymeric Materials. American Chemical Society, Washington, DC.
Rzymski, W.M., Radusch, H.J., 2005. Polimery 50, 249.
Samperi, F., Battiato, S., Puglisi, C., Scamporrino, A., Ambrogi, V., Ascione, L., Carfagna, C., 2011. J. Polym. Sci. Polym. Chem. Ed. 49, 3615.
Santangelo, P.G., Roland, C.M., 1998. J. Non-Cryst. Solids 235, 709.
Santangelo, P.G., Roland, C.M., 1998. Macromolecules 31, 3715.
Santangelo, P.G., Roland, C.M., 2001. Rubber Chem. Technol. 74, 69.
Santangelo, P.G., Roland, C.M., Puskas, J.E., 1999. Macromolecules 32, 1972.
Schawe, J., 1995. Thermochim. Acta 260, 16.
Schimpf, M.E., Caldwell, K., Giddings, J.C., 2000. Field-Flow Fractionation Handbook. Wiley, New York.
Schuck, P., 2000. Biophys. J. 78, 1606.
Schulten, H.-R., Plage, B., Lattimer, R.P., 1989. Rubber Chem. Technol. 62, 698.
Schulz, G.V., Altgelt, K., Cantow, H.J., 1957. Rubber Chem. Technol. 30, 805.
Scott, D.J., Harding, S.E., Rowe, A.J. (Eds.), 2005. Modern Analytical Ultracentrifugation: Techniques and Methods. Royal Society of Chemistry, UK.
Sen, S., Mabuni, C., Walsh, D., 2001. J. Appl. Polym. Sci. 82, 672.
Sharaf, M.A., Kloczkowski, A., Mark, J.E., 1995. Rubber Chem. Technol. 68, 601.
Sharp, M.A., Pranzas, P.K., Schreyer, A., 2009. Adv. Eng. Matl. 11, 441.
Shield, S.R., Ghebremeskel, G.N., 2003. J. Appl. Polym. Sci. 88, 1653.
Shimizu, T., Tsuji, M., Kohjiya, S., 1998. Mat. Sci. Res. Internat.4, 117.
Shultz, A.R., Stockmayer, W.H., 1969. Macromolecules 2, 178.
Shutilin, Y.F., 1985. Polym. Sci. USSR 27, 2386.
Shvydkaya, N.P., Shumanov, L.A., Semenova, L.P., Lykin, A.S., 1980. Kauch. Rezina 11, 19.
Simon, S.L., 2001. Thermochim. Acta 374, 55.
Sircar, A.K., 1992. Rubber Chem. Technol. 65, 503.
Sircar, A.K., In: Turi, E.A. (Ed.), Thermal Characterization of Polymeric Materials, 2nd ed., Academic Press, 1997.
Sirisinha, C., Saeoui, P., Guaysomboon, J., 2003. J. Appl. Polym. Sci. 90, 4038.
Small, P.A., 1975. Adv. Polym. Sci. 18, 1.
Smith, W.T., Patterson, J., 1986. Anal. Chem. 58, 102R.
Smitthipong, W., Gadiou, R., Vidal, L., Wagner, P., Nardin, M., 2008. Vibr. Spect. 46, 8.
Sotta, P., Deloche, B., Herz, J., Lapp, A., Durand, D., Rabadeux, J.-C., 1987. Macromolecules 20, 2769.
http://www.spectroscopynow.com
Stanley, C.B., Strey, H.H., 2003. Macromolecules 36, 6888.
Stebounova, L, Akhremitchev, B.B., Walker, G.C., 2003. Rev. Sci. Instr. 74, 3670.
Stegeman, G., Vanasten, A.C., Kraak, J.C., Poppe, H., Tijssen, R., 1994. Anal. Chem. 66, 1147.

Stepanek, P., Morkved, T.L., Krishnan, K., Lodge, T.P., Bates, F.S., 2002. Physica A 314, 41.
Stuart, B.H., 2002. Polymer Analysis. Wiley, NY.
Suarez, I., Coto, B., 2011. Eur. Polym. J. 47, 2331.
Suzuki, H., Leonis, C., Gordon, M., 1973. Makromol. Chem.172, 227.
Szep, A., Marosfoi, B., Bertalan, G., Anna, P., Marosi, G., 2003. Macromol. Symp. 202, 269.
Tacx, J., Ammerdorffer, J., German, A., 1988. Polymer 29, 2087.
Tanaka, Y., 1991. Rubber Chem. Technol. 64, 325.
Tanaka, T. (Ed.), 1999. Experimental Methods in Polymer Science. Academic Press, San Diego, CA.
Tanaka, Y., Sato, H., Adachi, J., 1987. Rubber Chem. Technol. 60, 25.
Tang, H., Liu, Z.Y., Piao, J.H., Chen, X.F., Lou, Y.X., Li, S.H., 1994. J. Appl. Polym. Sci. 51, 1159.
Tassin, J.-F., Baschwitz, A., Moise, J.-Y., Monnerie, L., 1990. Macromolecules 23, 1879.
Taylor, G.R., Darin, S.R., 1955. J. Appl. Phys. 20, 1075.
The New York Times, 1994. March 22, p. C3.
Thomas, A.G., Whittle, J.M., 1970. Rubber Chem. Technol. 43, 222.
Tinker, A.J., 1990. Rubber Chem. Technol. 63, 503.
Tinker, A.J., 1995. Rubber Chem. Technol. 68, 461.
Toki, S., Sics, I., Ran, S.F., Liu, L.Z., Hsiao, B.S., 2003. Polymer 44, 6003.
Toki, S., Takagi, R., Ito, M., Hsiao, B.S., 2011. Polymer 52, 2453.
Tomlin, D.W., Roland, C.M., 1993. Polymer 34, 2665.
Tosi, C., 1968. Adv. Polym. Sci. 5, 451.
Trabelsi, S., Albouy, P.A., Rault, J., 2003a. Macromolecules 36, 9093.
Trabelsi, S., Albouy, P.A., Rault, J., 2003b. Macromolecules 36, 7624.
Treloar, L.R.G., 1947. Trans. Faraday Soc. 43, 284.
Tulisalo, J., Seppala, J., Hastbacka, K., 1985. Macromolecules 18, 1144.
Tuminello, W., Buck, W., Kerbow, D., 1993. Macromolecules 26, 499.
Tyler, W., 1967. Rubber Chem. Technol. 40, 238.
Valic, S., Judeinstein, P., Deloche, B., 2003. Polymer 44, 5263.
Vana, P., Albertin, L., Barner, L., Davis, T.P., Barner-Kowollik, C., 2002. J. Polym. Sci. Polym. Polym. Chem. 40, 4032.
vanAsten, A.C., Kok, W.T., Tijssen, R., Poppe, H., 1996. J. Polym. Sci. Polym. Phys. 34, 297.
van Gennip, W.J.H., Thune, P.C., Dijkstra, J.B., Niemantsverdriet, J.W., 2004. Appl. Phys. Lett. 84, 1789.
Van Krevelen, D., 1990. Properties of Polymers, 3rd ed. Elsevier, Amsterdam.
Van Ooij, W.J., Rangarajan, V., 1988. Rubber Chem. Technol. 61, 594.
Vavra, J., Lapcik, J., Sabados, J., 1967. J. Polym. Sci. A-2 5, 1305.
Ver Strate, G., Cozewith, C., Ju, S., 1988. Macromolecules 21, 3360.
Vitalini, D., Scamporrino, E., 1992. Polymer 33, 4597.
Wachowicz, M., Gill, L., Wolak, J., White, J.L., 2008. Macromolecules 41, 2832.
Waddell, W.H., Parker, J.R., 1992. Rubber Chem. Technol. 65, 836.
Walton, J.H., Miller, J.B., Roland, C.M., 1992. J. Polym. Sci. Polym. Phys. 30, 527.
Walton, J.H., Miller, J.B., Roland, C.M., Nagode, J.B., 1993. Macromolecules 26, 4052.
Wang, H.T., Bethea, T., Harwood, H., 1993. Macromolecules 26, 715.
Wang, B.L., Mukataka, S., Kokufuta, E., Ogiso, M., Kodama, M., 2000. J. Polym. Sci. Polym. Phys. 38, 214.
Wang, J., Chen, B., Yan, F., Xue, Q., Zhao, F., 2011. Wear 272, 176.
Wasserman, S.H., 1995. J. Rheology 39, 601.
Weibel, E.R., 1973. In: Hoyat, M.A. (Ed.), Principles and Techniques of Electron Microscopy, vol. 3. Van Nostrand, New York (Chapter 6).
Westermann, S., Pyckhout-Hintzen, W., Richter, D., Straube, E., Egelhaaf, S., May, R., 2001. Macromolecules 34, 2186.
Westermann, S., Kreitschmann, M., Pyckhout-Hintzen, W., Richter, D., Straube, E., Farago, B., Goerigk, G., 1999. Macromolecules 32, 5793.
Wetzel, D.L., LeVine, S.M., 1999. Science 285, 1224.
Wignall, G.D., 1996. In: Mark, J.E. (Ed.), Physical Properties of Polymers Handbook, Amer. Inst. Phys. Woodbury, NY (Chapter 22).
Wild, L., 1991. Adv. Polym. Sci. 98, 1.
Williams, S.K.R., Lee, D., 2006. J. Separation Sci. 29, 1720.

Winesett, D.A., Tsou, A.H., 2008. Rubber Chem. Technol. 81, 265.
Winesett, D.A., Ade, H., Smith, A.P., Urquhart, S.G., Dias, A.J., Stevens, P., 2003. Rubber Chem. Technol. 76, 803.
Wolak, J., Jia, X., Gracz, H., Stejskal, E.O., White, J.L., Wachowicz, M., Jurga, S., 2003. Macromolecules 36, 4844.
Wu, G.L., Zeng, S.J., Ou, E.C., Yu, P.R., Xiong, Y.Q., Xu, W.J., 2011. J. Appl. Polym. Sci. 120, 1162.
Wunderlich, B., 2007. J. Appl. Polym. Sci. 105, 49.
Xu, S.H., Wells, P.S., Tao, Y.M., Yun, K.S., Parcher, J.F., 2000. ACS Symp. Series 748, 96.
Yamada, T., Okumoto, T., Ohtani, H., Tsuge, S., 1990. Rubber Chem. Technol. 63, 191.
Yamada, T., Okumoto, T., Ohtani, H., Tsuge, S., 1991. Rubber Chem. Technol. 64, 708.
Yamada, H., Manas-Zloczower, I., Feke, D.L., 1997. Powder Technol. 92, 163.
Yamakawa, H., 1971. Modern Theory of Polymer Solutions. Harper, NY.
Zaper, A.M., Koenig, J.L., 1988. Macromol. Chem. 189, 1239.
Zemel, I.S., Roland, C.M., 1992a. Polymer 33, 4522.
Zemel, I.S., Roland, C.M., 1992b. Polymer 33, 3427.
Zeng, W., Du, Y., Xue, Y., Frisch, H.L., 2006. In: Mark, J.E. (Ed.), Physical Properties of Polymers Handbook, Amer. Inst. Phys. Woodbury, NY (Chapter 17).
Zerda, T.W., Song, G., Waddell, W.H., 2003. Rubber Chem. Technol. 76, 769.
Zhang, Y.M., Ge, S., Tang, B., Koga, T., Rafailovich, M.H., Sokolov, J.C., Peiffer, D.G., Li, Z., Dias, A.J., McElrath, K.O., Lin, M.Y., Satija, S.K., Urquhart, S.G., Ade, H., Nguyen, D., 2001. Macromolecules 34, 7056.
Zhang, P., 2010. F. Zhao, Y. Yuan, X. Shi, and S. Zhao. Polymer 51, 257.
Zhang, C.H., Niu, H., Dong, J.Y., 2012. Polym. Bull. 68, 949.
Zhao, Q., Tannenbaum, R., Jacob, K.I., 2006. Carbon 44, 1740.
Zheng, Y., Turner, W., Zong, M.M., Irvine, D.J., Howdle, S.M., Thurecht, K.J., 2011. Macromolecules 44, 1347.
Zheng, Y., Thurecht, K.J., Wang, W.X., 2012. J. Polym. Sci. Polym. Chem. Ed. 50, 629.
Zheng, J., Liu, F., Lin, Y.C., Zhang, Z.J., Zhang, G.C., Wang, L., Liu, Y., Tang, T., 2012. Macromolecules 45, 1190.
Ziabicki, A., Jarecki, L., 1978. Coll. Polym. Sci. 256, 332.
Zimina, T., Fell, A., Castledine, J., 1992. Polymer 33, 4129.

Chapter 4

The Molecular Basis of Rubberlike Elasticity

B. Erman* **and J.E. Mark†**

*Department of Chemical and Biological Engineering, Koc University, Rumeli Feneri Yolu 34450, Istanbul, Turkey

†Department of Chemistry, University of Cincinnati, Cincinnati, OH 45221-0172, USA

4.1 INTRODUCTION

Rubberlike materials consist of relatively long polymeric chains having a high degree of flexibility and mobility, which are joined into a network structure. The requirement of flexibility and mobility is associated with very high deformability. As a result of an externally imposed stress, the long chains may alter their configurations, an adjustment that takes place relatively rapidly because of the high chain mobility. The requirement of having the chains linked into a network structure is associated with solidlike features, where the chains are prevented from flowing relative to each other under external stresses. As a result, a typical rubber may be stretched up to about 10 times its original length. On removal of the external force, it rapidly recovers its original dimensions, with essentially no residual or nonrecoverable strain. As a result of these unique mechanical properties, rubbers find important usage ranging from automobile tires to heart valves, and gaskets in jet planes and space vehicles.

In ordinary solids, such as crystalline or amorphous glassy materials, an externally applied force changes the distance between neighboring atoms, resulting in interatomic or intermolecular forces. In these materials, the distance between two atoms may be altered by no more than a fraction of an angstrom if the deformation is to be recoverable. At higher deformations, the atoms slide past each other, and either flow takes place or the material fractures. The response of rubbers on the other hand is almost entirely intramolecular.

Externally applied forces transmitted to the long chains through the linkages let their extremities change the conformations of the chains, and each chain acts like a spring in response to the external stress. The molecular mechanisms relating to rubberlike elasticity were recognized in the early 1930s. Rigorous

The Science and Technology of Rubber. http://dx.doi.org/10.1016/B978-0-12-394584-6.00004-2

statistical mechanical theories describing the mechanical behavior of rubbers were given by Guth and James (1941), Wall and Earlier (1942), Flory (1942), and Flory and Rehner (1943a,1943b). The present understanding of the molecular basis of rubber elasticity owes much to these early theories. They give an idealized picture of rubber elasticity, but at the same time form the basis of more advanced molecular theories of such elasticity, which describe the effects of intermolecular entanglements observed in real networks. Developments in the field from the beginning up to the present have been reviewed in several monographs (see for example Treloar (1975) and Erman and Mark (1989), and the more recent monograph by Erman and Mark (1997)).

In this review we first discuss the structural features of networks that contribute to the stress upon deformation. We discuss the simple classical models of elasticity and the departures from these simple models. Specifically, we differentiate between two classes of models, (1) the constraint models, which assume that the total elastic energy of the network equals the sum of the individual network chain energies, and (2) the trapped entanglement models, which assume that entanglements that are trapped during the crosslinking stage contribute additionally to the network elastic energy. We also give the molecular interpretation of coefficients obtained from the phenomenological theories. The area of swollen gels and responsive gels is among the most widely studied systems over the past several years. Recent work in this area is discussed. We then discuss the thermoelastic (force-temperature) behavior of networks. The final topics involve neutron scattering experiments (that allow the direct determination of chain dimensions in undeformed and deformed networks), and some recent experimental studies on the elasticity of single-polymer chains.

4.2 STRUCTURE OF A TYPICAL NETWORK

A network is obtained by linking polymer chains together, and this linkage may be either physical or chemical. Physical linking can be obtained by (1) absorption of chains onto the surface of finely divided particulate fillers, (2) formation of small crystallites, (3) coalescence of ionic groups, or (4) coalescence of glassy blocks. These physical crosslinks are, in general, not permanent and may disappear on swelling or increase in temperature. The corresponding networks are referred to as "physical" or "thermoreversible" and are not considered in the present review. Refer to Burchard and Ross-Murphy (1990) for further reading on such materials.

Chemical crosslinks may be obtained by randomly joining segments in already formed chains, by random copolymerization, or by end-linking functionally-terminated chains. Sulfur cures, peroxide cures, and high-energy irradiations are familiar methods of random crosslinking. Copolymerization monomers where at least one type has three or more reactive sites also lead to randomly crosslinked networks. Formation of networks by end-linking individual chains by ϕ-functional linkages is the most appropriate method of forming well-defined structures, where the functionality ϕ of a linkage is defined as the number of chains meeting at the junction. A network is called perfect if its junctions have a functionality of at least 3, it has no dangling chains (chains attached to the network only at one end), and it has no loops (chains with two ends meeting at the same junction). Properties of perfect networks are discussed in this chapter. Refer to Erman and Mark (1997) and to the original work by Flory (1953) and Flory (1982) for the structure and properties of imperfect networks.

The structure of a perfect network may be defined by two variables, the cycle rank ξ and the average junction functionality ϕ. Cycle rank is defined as the number of chains that must be cut to reduce the network to a tree. The three other parameters used often in defining a network are (1) the number of chains ν, (2) the number of junctions μ, and (3) the molecular weight M_c of chains between two junctions. They may be obtained from ξ and ϕ using the relations

$$\nu = \frac{\xi}{\left(1 - \frac{2}{\phi}\right)}, \tag{4.1}$$

$$\mu = \frac{2\nu}{\phi}, \tag{4.2}$$

$$M_c = \frac{\left(1 - \frac{2}{\phi}\right)\rho N_A}{\frac{\xi}{V_0}}, \tag{4.3}$$

where ρ is the density, V_0 is the reference volume of the network, and N_A is Avogadro's number.

The cycle rank completely defines the connectivity of a network and is the only parameter that contributes to the elasticity of a network as will be discussed further in the following section on elementary molecular theories. In several other studies, contributions from entanglements that are trapped during crosslinking are considered in addition to the chemical crosslinks (Graessley, 1982). The trapped entanglement model is also discussed in the following.

In a typical elastomer, the number of skeletal bonds in a network chain range from about 100 to 700 (Flory, 1976). Networks with chains shorter than 100 bonds have low extensibilities. Those having chains much larger than 700 bonds may have very high extensibilities but are too weak to serve as load-carrying materials. It is possible to prepare bimodal networks, however, by end-linking very short and very long chains to form networks of significant toughness (Mark, 1985).

4.3 ELEMENTARY MOLECULAR THEORIES

The basic postulate of elementary molecular theories of rubber elasticity states that the elastic free energy of a network is equal to the sum of the elastic free energies of the individual chains. In this section, the elasticity of the single chain is discussed first, followed by the elementary theory of elasticity of a network. Corrections to the theory coming from intermolecular correlations, which are not accounted for in the elementary theory, are discussed in Section 4.4.

4.3.1 Elasticity of the Single Chain

The chemical structure of a polymer chain determines its statistical properties, such as its average dimensions in space and its flexibility. These parameters, in turn, affect various properties of a network consisting of these chains. A detailed understanding of the single chain is therefore important.

A short sequence of the poly(dimethylsiloxane) chain is shown as an example in Figure 4.1a.

The silicone and oxygen atoms are located in alternating order along the backbone, with the CH_3 constituting the side groups. The backbone structure between the $i-$ first and $i+$ second backbone bonds is shown in Figure 4.1b. Lengths of the bonds l_i and bond angles ϕ_i, shown in the figure, are approximately constant. Torsional rotations may take place relatively easily, however, about the skeletal bonds. The torsional rotation angle for the ith bond is shown by ϕ in the figure. Large rotations may take place about the skeletal bonds as a result of which the chain may take different spatial conformations, one of which is shown in Figure 4.1c. The vector $\mathbf{r}$ shows the instantaneous end-to-end vector of the chain.

In most molecular theories of rubberlike elasticity, the individual chains are approximated by the freely jointed or the freely rotating chain model. In reality, however, rotations about each bond are subject to potentials that arise from the intrinsic torsional potential of the backbone bond and from steric attractive and repulsive forces from neighboring atoms along the chain. The energy to which the bond torsional angle ϕ_i is subject is shown schematically in Figure 4.1d. The "energy map" shown in this figure exhibits three minima, referred to as the three "isomeric states" or "isomeric minima." Of course the numbers of isomeric states may be smaller or larger than 3, depending on the chain architecture. The three minima shown in Figure 4.1d, which are approximately spaced at 120° intervals, are referred to as the *trans* (t), *gauche*+ (g+), and *gauche*− (g−) states. The shape of a chain changes continuously and rapidly as each bond fluctuates about an isomeric minimum, with an amplitude of the order of ±60°, and occasionally goes over the energy peak to another isomeric minimum. The rate of transition from one isomeric minimum to another, which is on the order of one per nanosecond at sufficiently high temperatures, depends primarily on the temperature and on the height of the energy barriers shown in Figure 4.1d. These transitions determine the dynamics of the chain, and determine in part its glass transition temperature. Equilibrium properties of the chain, on the other hand, are influenced by the energy levels of the isomeric minima, as well as their locations. The *trans* state in a randomly configured chain with bond torsional

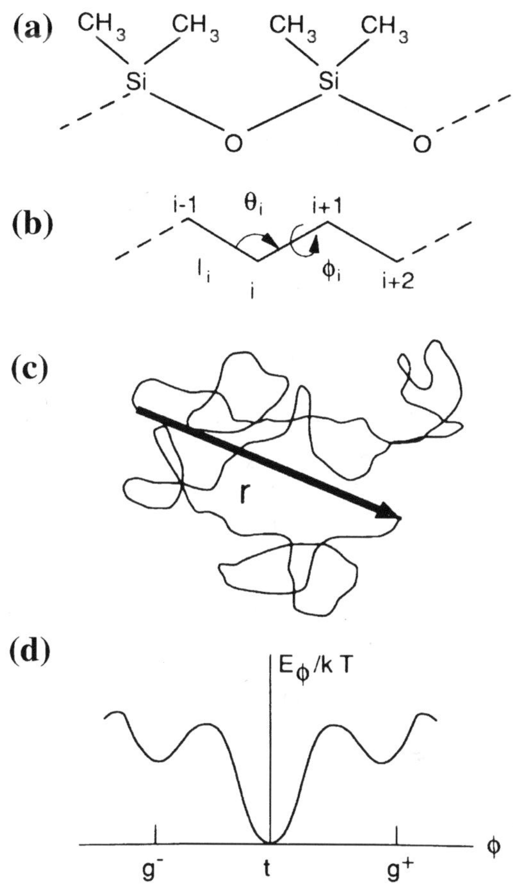

FIGURE 4.1 (a) A short sequence of the poly(dimethylsiloxane) (PDMS) chain; (b) Specification of the backbone structure; (c) Typical spatial conformation; (d) Typical conformational energy map for different rotational angles ϕ_i.

energies as shown in Figure 4.1d is more populated than the g+ and g− states. The average number of bonds in the t, g+, and g− states is determined according to the Boltzmann distribution of statistical mechanics (Flory, 1969; Mattice and Suter, 1994). According to statistical mechanical arguments, the number of each type of isomeric state in a chain remains essentially the same when the chain is stretched at its two ends. The change in the end-to-end vector takes place by the redistribution of the isomeric states along the chain. As the number of each type of isomeric state remains the same, the total internal energy of the chain remains constant during stretching. The elasticity of the chain resulting from redistribution of isomeric states is referred to as entropic elasticity, and a major part of the elasticity of a network is entropic. If part of the work done in deformation is used to change the relative populations of isomeric states, the bond angles, and the chain lengths, a change in internal energy takes place, which results in an "energetic" component of the elasticity. The relationship of the entropic and energetic components to molecular constitution in a network is discussed in the following sections.

The vector **r** joining the two ends of the chain takes different values resulting from rotations about the individual bonds. For chains with more than about 50 skeletal bonds, the probability $W(\mathbf{r})dx\,dy\,dz$ that one end of **r** is at the origin and the other end is in an infinitesimal volume $dV = dx\,dy\,dz$ is satisfactorily represented by the Gaussian function (Flory and Yoon, 1974; Yoon and Flory, 1974).

$$W(\mathbf{r})dx\,dy\,dz = \left(\frac{3}{2\pi\langle r^2\rangle_0}\right)^{3/2} \exp\left(-\frac{3r^2}{2\langle r^2\rangle_0}\right) dx\,dy\,dz. \tag{4.4a}$$

Here, $\langle r^2\rangle_0$ represents the average of the squared end-to-end vectors, and the subscript zero indicates that the chain is in the unperturbed or so-called theta state (Flory, 1953). It is now well established that chains in the bulk undiluted state are in the unperturbed state. Equation (4.4a) represents the probability distribution of the vectorial quantity **r**. A less detailed form of representation is the distribution $w(\mathbf{r})$ showing the probability that the magnitude r of **r** has a certain value irrespective of the direction. Thus, the probability that the chain end-to-end length is in the range r to $r + dr$ irrespective of its direction is

$$w(r)dr = \left(\frac{3}{2\pi\langle r^2\rangle_0}\right)^{3/2} \exp\left(-\frac{3r^2}{2\langle r^2\rangle_0}\right) 4\pi r^2\,dr. \tag{4.4b}$$

A schematic representation of the Gaussian function of Eq. (4.4b) is given by the dashed curve in Figure 4.2. The abscissa is normalized by dividing the end-to-end distance by the contour length of the chain, and the ordinate is made nondimensional (unitless) by multiplying $w(r)$ by the contour length.

For chains having fewer than 50 bonds, such as the short chains in a bimodal network for example, the distribution departs markedly from the Gaussian limit. Among various representations of $w(r)$ for short chains are the Hermite series (Flory and Yoon (1974, 1974)), the Fixman-Alben distribution, and Monte Carlo simulations (Erman and Mark, 1988). The Fixman-Alben distribution is given by

$$w(r)dr \propto \exp\left(-ar^2 + b^2r^4\right)4\pi r^2\,dr, \tag{4.4c}$$

where a and b are coefficients. The Fixman-Alben distribution and Monte Carlo simulations for a poly(dimethylsiloxane) chain of 20 skeletal bonds are compared with the Gaussian approximation in Figure 4.2.

The molecular theories of networks to be presented in the following paragraphs are based on the Gaussian picture of the individual network chains. With reference to the form of the distribution function, these theories are referred to as "Gaussian theories."

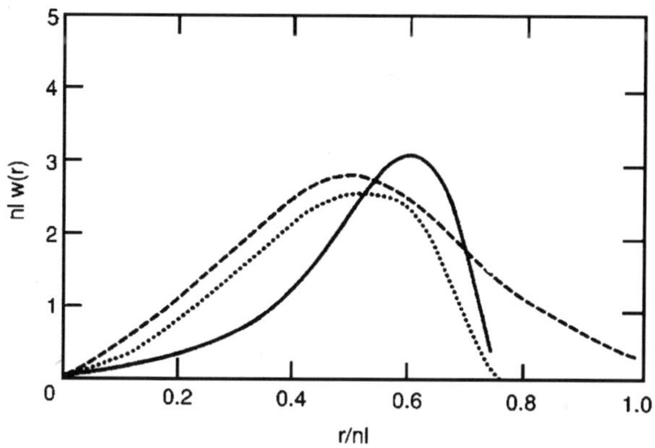

FIGURE 4.2 Distributions for the end-to-end distance of a PDMS chain having $n = 20$ skeletal bonds of length $l = 25$. The Fixman-Alben distribution (dotted curve) and that from a Monte Carlo simulation (solid curve) are compared with the Gaussian approximation (dashed curve).

The elastic free energy A_{el} of a Gaussian chain is related to the probability distribution $W(r)$ by the thermodynamic expression (Erman and Mark, 1997)

$$A_{el} = C(T) - kT \ln W(r), \tag{4.5}$$

where $C(T)$ is a function only of temperature T, and k is the Boltzmann constant. Substituting Eq. (4.4a) into Eq. (4.5) leads to

$$A_{el} = A * (T) + \left(\frac{3kT}{2\langle r^2 \rangle_0} \right) r^2. \tag{4.6}$$

Here, $A * (T)$ is a function of temperature alone. Equation (4.6) represents the elastic free energy of a Gaussian chain with ends fixed at a separation of r. The average force required to keep the two ends at this separation is obtained from the thermodynamic expression (Flory, 1969)

$$f = \left(\frac{\partial A_{el}}{dr} \right)_T, \tag{4.7}$$

$$= \left(\frac{3kT}{\langle r^2 \rangle_0} \right) r, \tag{4.8}$$

where Eq. (4.8) is obtained by substituting Eq. (4.6) into Eq. (4.7). The subscript T denotes differentiation at fixed temperature.

Equation (4.8) shows that the single chain behaves like a linear spring, with spring constant equal to $3kT/\langle r^2 \rangle_0$.

4.3.2 The Elastic Free Energy of the Network

The total elastic free energy ΔA_{el} of the network relative to the undeformed state is obtained by summing Eq. (4.6) over the ν chains of the network

(Mark and Erman, 2007):

$$\Delta A_{\mathrm{el}} = \frac{3kT}{2\langle r^2\rangle_0}\sum_{\nu}(r^2 - \langle r^2\rangle_0), \tag{4.9}$$

$$= \frac{3\nu kT}{2}\left(\frac{r^2}{\langle r^2\rangle_0} - 1\right), \tag{4.10}$$

where $\langle r^2\rangle = \sum r^2/\nu$ is the average square of the end-to-end vectors of chains in the deformed network. Substituting

$$\langle r^2\rangle = \langle x^2\rangle + \langle y^2\rangle + \langle z^2\rangle \tag{4.11}$$

in Eq. (4.10) and using the fact that chain dimensions are isotropic in the undeformed state

$$\langle x^2\rangle_0 = \langle y^2\rangle_0 = \langle z^2\rangle_0 = \frac{\langle r^2\rangle_0}{3}, \tag{4.12}$$

one obtains

$$\Delta A_{\mathrm{el}} = \frac{\nu kT}{2}\left[\frac{\langle x^2\rangle}{\langle x^2\rangle_0} + \frac{\langle y^2\rangle}{\langle y^2\rangle_0} + \frac{\langle z^2\rangle}{\langle z^2\rangle_0} - 3\right]. \tag{4.13}$$

The ratios of mean-squared dimensions appearing in Eq. (4.13) are microscopic quantities. To express the elastic free energy of a network in terms of the macroscopic state of deformation, an assumption has to be made relating microscopic chain dimensions to macroscopic deformation. Their relation to macroscopic deformations imposed on the network has been a main area of research in the area of rubberlike elasticity. Several models have been proposed for this purpose, which are discussed in the following sections. Before, we describe the macroscopic deformation, stress, and the modulus of a network.

4.3.3 The Reduced Stress and the Elastic Modulus

The state of macroscopic deformation may be characterized by considering the deformation of a rectangular prism, with extension ratios $\lambda_x,\lambda_y,\lambda_z$, along the x, y, and z directions, respectively, as

$$\lambda_x = \frac{L_x}{L_{x0}}, \quad \lambda_y = \frac{L_y}{L_{y0}}, \quad \lambda_z = \frac{L_z}{L_{z0}}, \tag{4.14}$$

where L_{x0},L_{y0},L_{z0} are the sides of the prism before deformation and L_x,L_y,L_z are the corresponding sides in the deformed state. For the sake of simplicity, we consider here dry networks; that is, networks in the absence of a diluent both in the state of formation and during deformation. However, we discuss effects of swelling in section 4.6. Also, we consider only the uniaxial case. The true stress τ (i.e., force per unit deformed area), resulting from a uniaxial force acting on a cross-section of the network sample, is obtained by the Treloar relations (Treloar, 1975)

$$\tau = 2V^{-1}\left[\lambda_i^2\left(\frac{\partial \Delta A_{\mathrm{el}}}{\partial \lambda_i^2}\right) - \lambda_j^2\left(\frac{\partial \Delta A_{\mathrm{el}}}{\partial \lambda_j^2}\right)\right]_{T,V}, \tag{4.15}$$

where the subscript i denotes the direction of the applied uniaxial force and j denotes one of the other two directions on which only a hydrostatic pressure is acting. The expression in the brackets is evaluated at constant temperature and volume as identified by the subscripts T and V. Refer to Treloar (1975) and Ogden (1997) for applications to other states of stress.

The engineering stress σ defined as the force per unit undeformed area follows from Eq. (4.15) as

$$\sigma_i = \frac{\tau}{\lambda}. \tag{4.16}$$

The reduced force $[f^*]$ is defined according to

$$[f^*] = \frac{\tau}{(\lambda^2 - \lambda)}. \tag{4.17}$$

In the limit of small deformations, the reduced force equates to the shear modulus of the sample; that is,

$$G = \lim_{\lambda \to 1}[f^*]. \tag{4.18}$$

The expressions given in this section, which are explained in more detail in Erman and Mark (1988), are general expressions. In the next section, we introduce two network models that have been used in the elementary theories of elasticity to relate the microscopic deformation to the macroscopic deformation: the affine and the phantom network models.

The Affine Network Model

One of the earlier assumptions regarding microscopic deformation in networks is that the junction points in the networks move affinely (linearly) with macroscopic deformation. It follows that chain end-to-end vectors deform affinely also, and

$$\langle x^2\rangle = \lambda_x^2\langle x^2\rangle_0, \quad \langle y^2\rangle = \lambda_y^2\langle y^2\rangle_0, \quad \langle z^2\rangle = \lambda_z^2\langle z^2\rangle_0. \tag{4.19}$$

Substituting Eq. (4.19) into Eq. (4.13) and using Eq. (4.1) lead to

$$\Delta A_{\mathrm{el,affine}} = \frac{1}{2}\left(\frac{\phi}{\phi - 2}\right)\xi kT\left(\lambda_x^2 + \lambda_y^2 + \lambda_z^2 - 3\right). \tag{4.20}$$

A more rigorous statistical analysis (Flory, 1953) gives an additional volume term, $-\mu kT(V/Vo)$, in Eq. (4.20). This term does not appear in the simplified derivation presented here.

The true stress for the uniaxial case is obtained by substituting Eq. (4.20) into Eq. (4.15) as

$$\tau = \left(\frac{\phi}{\phi - 2}\right)\frac{\xi kT}{V_0}(\lambda^2 - \lambda^{-1}). \tag{4.21}$$

The shear modulus G_{af} of an affine network is obtained from Eqs. (4.17) and (4.18) as

$$G_{\mathrm{af}} = \frac{\nu kT}{V_0}, \tag{4.22}$$

where V_0 is the volume during the formation of the network.

The Phantom Network Model

The instantaneous vector $\mathbf{r}$ joining two junctions at the extremities of a network chain may be written as the sum of a time-averaged mean $\bar{\mathbf{r}}$ and the instantaneous fluctuation $\Delta\mathbf{r}$ from this mean; that is,

$$\mathbf{r} = \bar{\mathbf{r}} + \Delta\mathbf{r}. \tag{4.23}$$

According to the phantom network model, the fluctuations $\Delta\mathbf{r}$ are independent of deformation and the mean $\bar{\mathbf{r}}$ deform affinely with macroscopic strain. Squaring both sides of Eq. (4.23) and averaging over all chains give

$$\langle \mathbf{r^2} \rangle = \langle \bar{\mathbf{r}}^\mathbf{2} \rangle + \langle (\Delta\mathbf{r})^2 \rangle. \tag{4.24}$$

The average of the quantity $\bar{\mathbf{r}} \cdot \Delta\mathbf{r}$ has been equated to zero in obtaining Eq. (4.24) inasmuch as this quantity is equally likely to be positive or negative because of the fluctuating term $\Delta\mathbf{r}$, and the average over all possible occurrences vanishes. Equation (4.24) is valid in both the deformed and undeformed states. It may be written in terms of λ_x, λ_y, and λ_z as

$$\begin{aligned}\langle r^2 \rangle &= \left(\frac{\lambda_x^2 + \lambda_y^2 + \lambda_z^2}{3}\right)\langle \bar{r}^2 \rangle_0 + \langle (\Delta r)^2 \rangle \\ &= \left[\left(\frac{\lambda_x^2 + \lambda_y^2 + \lambda_z^2}{3}\right)\left(1 - \frac{2}{\phi}\right) + \frac{2}{\phi}\right]\langle r^2 \rangle_0.\end{aligned} \tag{4.25}$$

Equation (4.25) is obtained by use of

$$\begin{aligned}\langle \bar{r}^2 \rangle_0 &= \left(1 - \frac{2}{\phi}\right)\langle r^2 \rangle_0, \\ \langle (\Delta r)^2 \rangle_0 &= \frac{2}{\phi}\langle r^2 \rangle_0.\end{aligned} \tag{4.26}$$

These two relations result from the phantom network model. Their derivations are given elsewhere (Flory, 1976; Mark and Erman, 2007).

Using Eq. (4.26) in Eq. (4.25) and substituting the resulting expression into Eq. (4.13) lead to

$$\Delta A_{\mathrm{el,phantom}} = \frac{1}{2}\xi kT\left(\lambda_x^2 + \lambda_y^2 + \lambda_z^2 - 3\right). \tag{4.27}$$

Comparison of the expressions for the elastic free energies for the affine and phantom network models shows that they differ only in the front factor. Expressions for the elastic free energy of more realistic models than the affine and phantom network models are given in the following section.

The true stress for the phantom network model is obtained by substituting Eq. (4.27) into Eq. (4.15) as

$$\tau = \frac{\xi kT}{V_0}(\lambda^2 - \lambda^{-1}) \tag{4.28}$$

and the shear modulus G_{ph} is obtained from Eqs. (4.17) and (4.18) as

$$G_{\mathrm{ph}} = \frac{\xi kT}{V_0}. \tag{4.29}$$

Equation (4.29) shows that the modulus is proportional to the cycle rank ξ, and that no other structural parameters contribute to the modulus. The number of entanglements trapped in the network structure does not change the cycle rank. Possible contributions of these trapped entanglements to the modulus cannot therefore originate from the topology of the phantom network.

4.4 MORE ADVANCED MOLECULAR THEORIES

4.4.1 The Constrained Junction Model

The constrained-junction model was formulated in order to explain the decrease of the elastic moduli of networks upon stretching. It was first introduced by Ronca and Allegra (1975) and Flory (1977). The model assumes that the fluctuations of junctions are diminished below those of the phantom network because of the presence of entanglements and that stretching increases the range of fluctuations back to those of the phantom network. As indicated by the second part of Eq. (4.26), the fluctuations in a phantom network are substantial. For a tetrafunctional network, the mean-square fluctuations of junctions amount to as much as half of the mean-square end-to-end vector of the network chains. The strength of the constraints on these fluctuations is measured by a parameter κ, defined as

$$\kappa = \frac{\langle(\Delta R)^2\rangle}{\langle(\Delta s)^2\rangle}, \tag{4.30}$$

where $\langle(\Delta R)^2\rangle$ and $\langle(\Delta s)^2\rangle$ denote, respectively, the mean-square dimension of junction fluctuations in the phantom network and in the entanglement domain. If the range of fluctuations decreases to zero because of entanglements, κ becomes infinitely large. If the effect of entanglements is nil, then $\kappa = 0$. The κ parameter is proportional to the number of junctions in the volume occupied by a given network chain. Thus,

$$\kappa = I\langle r^2\rangle_0^{3/2}(\mu/V_0), \tag{4.31}$$

where I is the constant of proportionality. For a network with tetrafunctional junctions, κ may be written as

$$\kappa = I(N_A d/2)^{3/2}(\langle r^2\rangle_0/M)^{3/2}(\xi/V_0)^{-1/2}, \tag{4.32}$$

where d is the density of the network and M is the molecular weight of chains between two crosslinks.

The elastic free energy of the constrained-junction model is given by the expression

$$\Delta A_{\text{el}} = \frac{1}{2}\xi kT\left\{\sum_{i=1}^{3}(\lambda_i^2 - 1) + \frac{\mu}{\xi}\sum_{i=1}^{3}\left[B_i + D_i - \ln(1+B_i) - \ln(1+D_i)\right]\right\}, \tag{4.33}$$

where

$$B_i = \kappa^2(\lambda_i^2 - 1)(\lambda_i^2 + \kappa) \quad D_i = \lambda_i^2\kappa^{-1}B_i. \tag{4.34}$$

The true stress for the uniaxial case is obtained from Eqs. (4.15) and (4.33) as

$$\tau = \frac{\xi kT}{V_0}\left\{(\lambda - \lambda^{-2}) + \frac{\mu}{\xi}\left[\lambda K(\lambda^2) - \lambda^{-2}K(\lambda^{-1})\right]\right\}, \tag{4.35}$$

where

$$K(\lambda^2) = B\left\lfloor \dot{B}(B+1)^{-1} + \kappa^{-1}(\lambda^2\dot{B} + B)(B + \kappa\lambda^{-2})^{-1}\right\rfloor, \tag{4.36}$$

$$\dot{B} = \frac{\partial B}{\partial \lambda^2} = B[(\lambda^2 - 1)^{-1} - 2(\lambda^2 + \kappa)^{-1}]. \tag{4.37}$$

The reduced force is given as

$$[f*] = \frac{\xi kT}{V_0}\left\{1 + \frac{\mu}{\xi}\left[\frac{\lambda K(\lambda^2) - \lambda^{-2}K(\lambda^{-1})}{\lambda - \lambda^{-2}}\right]\right\}. \tag{4.38}$$

The shear modulus of the constrained-junction model is obtained in the limit of small deformations as

$$G = \left[1 + \frac{\mu}{\xi}\left(\frac{\kappa^2 + 1}{(1+\kappa)^4}\kappa^2\right)\right]G_{\text{ph}}. \tag{4.39}$$

Equation (4.39) shows that for nonzero values of the parameter κ the shear modulus of the constrained-junction model is larger than the phantom network shear modulus. For the affine limit, $\kappa \to \infty$, the shear modulus is

$$G = \left(1 + \frac{\mu}{\xi}\right) G_{\text{ph}} = \frac{\phi}{\phi - 2} G_{\text{ph}}. \tag{4.40}$$

Equation (4.40) shows that the small deformation shear modulus of an affine network increases indefinitely over the phantom network modulus as junction functionality approaches 2.

Models of rubber elasticity have been reviewed for finite deformation and compared with experimental data by Boyce and Arruda (2000). A hybrid model of the Flory-Erman model for low stretch deformation and the Arruda-Boyce model for large stretch deformation is shown to give an accurate predictive description of Treloar's classic data over the entire stretch range for all deformation states.

4.4.2 Entanglement Models

These models assume that intermolecular contributions result from entanglements trapped into the network structure during crosslinking and are elastically effective upon deformation. The idea that the entanglements permanently trapped into the network structure contribute to the elasticity of the network originated in the work of Langley and Polmanteer (1974) (see for example Ferry, 1980; Graessley, 2007). The deformation dependence of this contribution based on molecular arguments is modeled by several authors as reviewed by Heinrich et al. (2003), Kaliske and Heinrich (1999), Heinrich and Kaliske (1997), Everaers and Kremer (1995) and Mergell and Everaers (2001). The basic idea of entanglement contribution has its roots in the fact that uncrosslinked systems can exhibit transient but significant elastic response resulting from the tube-like confinement of chains that deform upon the application of stress, as reviewed by McLeish (2002). Edwards introduced the tube concept and the first statistical mechanics description of trapped entanglements and their contribution to deformed elastomers. Among several treatments of entanglement contributions, the slip-link model of Edwards and Vilgis is the most transparent, which we outline below. For several other treatments, refer to the work of Heinrich et al. (1988).

The slip-link model simplifies the tube model by incorporating the effects of entanglements along the chain contour into the elastic free energy. According to the mechanism of the slip link, sketched in Figure 4.3, a link joins two different chains, which may slide a distance along the contour of the chains. The elastic free energy resulting from this model is

$$A_{\text{el}} = \frac{1}{2} N_c kT \left\{ \sum_{i=1}^{3} \lambda_i^2 + \frac{N_s}{N_c} \sum_{i=1}^{3} \left[\frac{(1+\eta)\lambda_i^2}{1+\eta\lambda_i^2} + \log\left(1 + \eta\lambda_i^2\right) \right] \right\}, \tag{4.41}$$

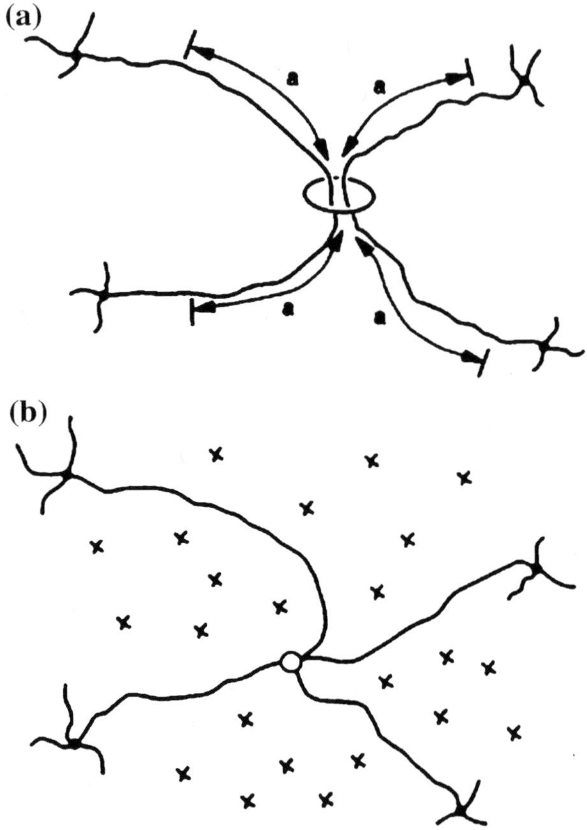

FIGURE 4.3 Schematic drawing of a slip link, with its possible motions along the network chains specified by the distances a, and its locking into position as a cross-link.

where N_c and N_s are the number of chemical crosslinks and slip links, respectively, and $\eta = 0.2343$. The first term on the right-hand side of Eq. (4.41) is the contribution to the elastic free energy from the phantom network. The effect of entanglements enters as a further contribution and is proportional to the number of slip links.

The elastic free energy of the constrained-junction model, similar to that of the slip-link model, is the sum of the phantom network free energy and that due to the constraints. Both the slip-link and the constrained-junction model free energies reduce to that of the phantom network model when the effect of entanglements diminishes to zero. One important difference between the two models, however, is that the constrained-junction model free energy equates to that of the affine network model in the limit of infinitely strong constraints, whereas the slip-link model free energy may exceed that for an affine deformation, as may be observed from Eq. (4.41).

4.4.3 Contribution of Trapped Entanglements to the Modulus

The cycle rank ξ of a network denotes the number of chains that have to be cut in order to reduce the network to a tree. The moduli of the phantom, affine, slip-link, and constrained-junction models are all proportional to the cycle rank. The cycle rank is independent of the number of trapped entanglements in the crosslinked system. A network in which chains are highly entangled has the same cycle rank as one with no entanglements. Therefore these models categorically reject contributions from trapped entanglements to the modulus. However, a large body of experiments has shown that a certain fraction of trapped entanglements contribute to the modulus (Graessley, 1982; Langley, 1968). The contributions to the modulus are given by the widely used Langley equation:

$$G = G_{\text{ch}} + T_e G_N^0. \tag{4.42}$$

Here, the modulus G is given as the sum of the modulus G_{ch} due to chemical crosslinks and the trapped entanglement term $T_e G_N^0$, where T_e (called the Langley trapping factor) is the fraction of trapped entanglements that contribute to the modulus, and G_N^0 is the plateau modulus related to the molecular weight M_e between entanglement by the expression

$$G_N^0 = \frac{\rho RT}{M_e}. \tag{4.43}$$

According to the arguments based on the constrained-junction model, the term G_{ch} should equate to the phantom network modulus onto which contributions from entanglements are added.

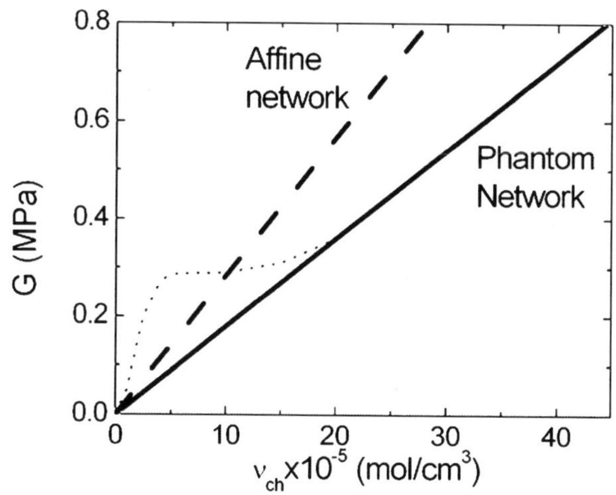

FIGURE 4.4 Experimental results of Oppermann and Rennar (1987) shown by the dotted line. The moduli converge to the phantom network result to bring about high degrees of cross-linking. The results of Erman and Wagner (1980) fall on the straight line and fully agree with the phantom network model.

Experimental determination of the contributions above those predicted by the reference phantom network model has been controversial. Experiments of Oppermann and Rennar (1987) on endlinked poly(dimethylsiloxane) networks, represented by the dotted points in Figure 4.4, indicate that contributions from trapped entanglements are significant for low degrees of end-linking but are not important when the network chains are shorter. Experimental results of Erman and Wagner (1980) on randomly crosslinked poly(ethyl acrylate) networks fall on the solid line and indicate that the observed high deformation limit moduli are within the predictions of the constrained-junction model.

4.5 PHENOMENOLOGICAL THEORIES AND MOLECULAR STRUCTURE

The elastic free energy given by the elementary and the more advanced theories is symmetric functions of the three extension ratios λ_x, λ_y, and λ_z. One may also express the dependence of the elastic free energy on strain in terms of three other variables, which are in turn functions of λ_x, λ_y, and λ_z. In phenomenological theories of continuum mechanics, where only the observed behavior of the material is of concern rather than the associated molecular mechanism, these three functions are chosen as

$$\begin{aligned} I_1 &= \lambda_x^2 + \lambda_y^2 + \lambda_z^2, \\ I_2 &= \lambda_x^2\lambda_y^2 + \lambda_x^2\lambda_z^2 + \lambda_y^2\lambda_z^2, \qquad (4.44) \\ I_3 &= \lambda_x^2\lambda_y^2\lambda_z^2, \\ \Delta A_{\text{el}} &= \Delta A_{\text{el}}(I_1, I_2, I_3). \qquad (4.45) \end{aligned}$$

The most general form of the elastic free energy may be written as a power series

$$\Delta A_{\text{el}} = \sum_{i,j,k=1}^{\infty} C_{ijk}(I_1 - 3)^i (I_2 - 3)^j (I_3 - 1)^k, \qquad (4.46)$$

where C_{ijk} are the phenomenological coefficients. The simple case of the phantom and affine networks is obtained as the first term of the series

$$\Delta A_{\text{el}} = C_{100}(I_1 - 3). \qquad (4.47)$$

The elastic free energy of the so-called Mooney-Rivlin solid is obtained from Eq. (4.46) as

$$\Delta A_{\text{el}} = C_{100}(I_1 - 3) + C_{010}(I_2 - 3). \qquad (4.48)$$

The reduced force follows from Eqs. (4.15) and (4.17) as

$$[f^*] = 2C_1 + \frac{2C_2}{\lambda}, \qquad (4.49)$$

where $C_1 = C_{100}/V_0$ and $C_2 = C_{010}/V_0$. For large deformations, the reduced force equates to $2C_1$, which may be identified with the phantom network model modulus. For small deformations, $2C_2$ may be obtained by equating Eq. (4.49) to Eq. (4.39) for the constrained-junction model. Thus,

$$2C_1 = G_{\text{ph}}, \tag{4.50}$$

$$2C_2 = \frac{\mu}{\xi}\left(\frac{\kappa^2+1}{(1+\kappa)^4}\kappa^2\right)G_{\text{ph}}. \tag{4.51}$$

Further references to the phenomenological treatment may be found in Treloar (1975), Ogden (1997), Erman and Mark (1997) and Mark (1973).

4.6 SWELLING OF NETWORKS AND RESPONSIVE GELS

Throughout the preceding discussion, the networks were assumed to be formed in the dry state and tested in the dry state. In recent years, much emphasis has been placed on swelling of networks and their phase transitions under different activities of the network-solvent system. Large-scale volume transitions triggered by small changes in environmental variables directed attention to possible uses of swollen gels in the field of responsive materials technologies. The transition involves the gel exuding solvent, for example upon decrease in temperature. The resulting shrinkage (syneresis) is widely known as "gel collapse," and is shown schematically in Figure 4.5. In the following discussion, charged systems will be considered in particular because the presence of charges facilitates the volume phase transitions in swollen gels.

The change in free energy of a network upon swelling is taken as the sum of the change in the elastic free energy, ΔA_{el}, and the change in free energy of

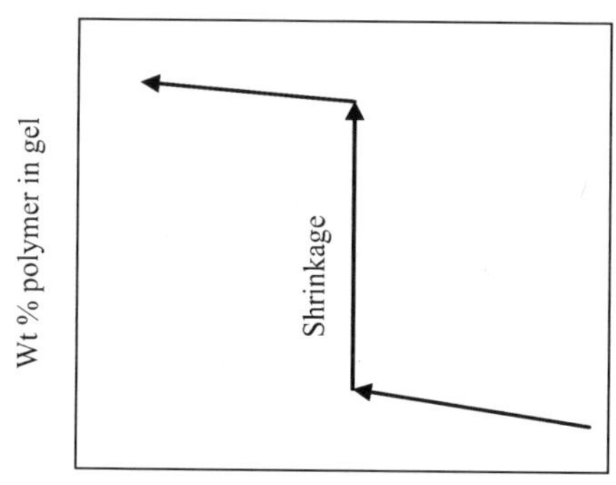

FIGURE 4.5 A gel exuding solvent upon decrease in temperature, with the shrinkage (syneresis) generally described as "gel collapse."

mixing, ΔA_{mix}, and the contributions from ionic groups ΔA_i:

$$\Delta A = \Delta A_{\text{el}} + \Delta A_{\text{mix}} + \Delta A_i, \tag{4.52}$$

where ΔA_{el} may be taken as any of the expressions resulting from a model, ΔA_{mix} is the free energy of mixing, and ΔA_i is the contribution of the ionic groups on the chains. The total chemical potential $\Delta\mu_1$ of solvent in the swollen network is obtained for the constrained-junction model as

$$\begin{aligned}\frac{\Delta\mu_1}{RT} &= \ln(1 - v_2) + v_2 + \chi v_2^2 \\ &\quad + \frac{1}{\lambda}\frac{\rho V_1}{M_c}\left[1 + \frac{\mu}{\xi}K(\lambda^2)\right] - i\nu\left(\frac{V_1}{V_0 N_A}\right)\left(\frac{v_2}{v_{20}}\right),\end{aligned} \tag{4.53}$$

where v_2 is the volume fraction of polymer, χ is the Flory interactions parameter, V_1 is the molar volume of solvent, M_c is the molecular weight of a network chain, and λ is the extension ratio, defined for the swelling case as:

$$\lambda = \left(\frac{V}{V_0}\right)^{1/3} = \left(\frac{n_1 V_1 + x V_1 n_2}{V_0}\right). \tag{4.54}$$

Here, x is the number of repeat units in one network chain, n_1 is the number of solvent molecules, n_2 is the total number of network chains in the system, i is the number of ionic groups on the chains, ν is the number of chains. v_{20} is the volume fraction of chains during the formation of the network.

Equating the chemical potential to zero gives a relationship between the equilibrium degree of swelling and the molecular weight M_c. The relation for $M_{c,\text{ph}}$ is obtained for a tetrafunctional phantom network model as:

$$M_{c,\text{ph}} = -\frac{\frac{1}{2}\rho V_1\left(\frac{v_2}{v_{20}}\right)^{1/3}}{\ln(1 - v_2) + v_2 + \chi v_2^2 - i\nu\left(\frac{V_1}{V_0 N_A}\right)\left(\frac{v_2}{v_{20}}\right)}, \tag{4.55}$$

where v_2 denotes the equilibrium degree of swelling.

For the affine network model, the molecular weight between crosslinks $M_{c,\text{af}}$ is obtained as

$$M_{c,\text{af}} = -\frac{\rho V_1\left(v_2^{1/3} - \frac{v_2}{2v_{20}}\right)}{\ln(1 - v_2) + v_2 + \chi v_2^2 - i\nu\left(\frac{V_1}{V_0 N_A}\right)\left(\frac{v_2}{v_{20}}\right)}. \tag{4.56}$$

Alternatively, the chemical potential expression may be solved for v_2, leading to a value for the degree of swelling of the network. The solution shows that the degree of swelling increases as the chain length between crosslinks increases. The dominant forces that operate in swollen uncharged gels are van der Waals forces, hydrogen bonds, hydrophobic forces, and forces resulting from chain entropy.

When the network chains contain ionic groups, there will be additional forces that affect their swelling properties. Translational entropy of counter-ions, Coulomb interactions, and ion pair multiplets are forces that lead to interesting phenomena in ion-containing gels. These phenomena were studied in detail by Khokhlov (1992), Khokhlov and Philippova (1996, 2002), Osada and Khokhlov (2002) and Philippova et al. (1995). The free energy of the networks used by Khokhlov and collaborators is

$$\Delta A = \Delta A_{\text{mix}} + \Delta A_{\text{el}} + \Delta A_{\text{trans}} + \Delta A_{\text{Coulomb}}, \tag{4.57}$$

where ΔA_{trans} and $\Delta A_{\text{Coulomb}}$ are the contributions to the elastic free energy of the networks from the translational entropy of the counter-ions and the free energy of Coulomb interactions. Several interesting features of gels are obtained through the use of Eq. (4.57). A network chain of a polyampholyte gel contains both positive and negative charges. The liquid phase in the swollen polyampholyte gel may contain additional counter-ions. The theoretical and experimental literature on polyampholyte gels was reviewed recently by Nisato and Candau (2002). In ion-containing gels, when ion-containing groups are fully dissociated the gel swells excessively, because of the tendency of the free counter-ions to occupy as much space as possible. In the other extreme case, called the ionomer regime, counter-ions are condensed on oppositely charged monomer units, forming ion pairs followed by formation of multiplets. This decreases the osmotic pressure of the gel and results in its collapse. The conditions for ion pair formation and physical and chemical factors leading to gel swelling and collapse have been discussed by Philippova (2000).

4.7 ENTHALPIC AND ENTROPIC CONTRIBUTIONS TO RUBBER ELASTICITY: THE FORCE-TEMPERATURE RELATIONS

The major component of elasticity of a network arises from the "entropic elasticity" of the individual chains. This was the basic assumption of the early molecular theories of rubber elasticity (Erman and Mark, 1997). A closer consideration of the statistics of the single chain shows that the rotational isomeric states allowable to each torsion angle of the chain are not of the same energy, and stretching a chain or changing the temperature may move it from one isomeric minimum to a more favorable one. This results in an energetic contribution to the elasticity of a chain. Thus the total force acting on a network may be written as the sum of an entropic contribution, f_s, and an energetic contribution, f_e

$$f = f_s + f_e. \tag{4.58}$$

Force-temperature relations lead to a quantitative assessment of the relative amounts of entropic and energetic components of the elasticity of the network.

In uniaxial deformation, the energetic contribution to the total elastic force (Treloar, 1975; Erman and Mark, 1997; Mark and Erman, 2007; Mark, 1973;

Flory, 1961; Flory and Ciferri, 1960) is given by the thermodynamically exact relation

$$\frac{f_e}{f} \equiv -T\left[\frac{\partial \ln (f/T)}{dT}\right]_{L,V}. \tag{4.59}$$

The subscripts L and V denote that differentiation is performed at constant length and volume. To carry out the differentiation indicated in Eq. (4.59), an expression for the total tensile force f is needed. One may use the expression given by Eq. (4.28) for the phantom network model. Applying the right-hand side of Eq. (4.59) to Eq. (4.28) leads to

$$\frac{f_e}{f} = \frac{T d \ln \langle r^2 \rangle_0}{dT}. \tag{4.60}$$

Equation (4.60) is important because the right-hand side relates to a microscopic quantity, $\langle r^2 \rangle_0$, and the left-hand side is the ratio of the energetic component of the force to the total force, both macroscopic quantities. It should be noted that Eq. (4.60) is obtained by using a molecular model. Experimentally, the determination of the force at constant volume is not easy. For this reason, expressions for the force measured at constant length and pressure p or constants α and p are used. These expressions are

$$\frac{f_e}{f} \equiv -T\left[\frac{\partial \ln (f/T)}{dT}\right]_{L,p} - \frac{\beta T}{(a^3 - 1)}, \tag{4.61}$$

$$\frac{f_e}{f} \equiv -T\left[\frac{\partial \ln (f/T)}{dT}\right]_{\alpha,p} - \frac{\beta T}{3}, \tag{4.62}$$

where β is the thermal expansion coefficient of the network. It should be noted, however, that both of these equations are derived on the basis of the equation of state for simple molecular models and therefore are not quantities based purely on experimental data. Values of the energetic contribution for some typical elastomers are given in Table 4.1.

TABLE 4.1 Some Typical Values for f_e/f

Elastomer	f_e/f
Natural rubber	0.18
cis-1,4-Polybutadiene	0.13
Poly(dimethylsiloxane)	0.20
Elastin	0.26

4.8 DIRECT DETERMINATION OF MOLECULAR DIMENSIONS

Until recently, knowledge of chain dimensions in elastomeric networks (in both undeformed and deformed states) relied on results obtained by measurements at macroscopic length scales and/or the use of a molecular model. The fact that chains take random configurations was based, for example, on the agreement of measured moduli with those predicted by the appropriate theory. Similarly, the fact that the deformations of the chains lie between those of the affine and phantom models followed from the agreement between macroscopically measured moduli and those predicted by theoretical models. Developments in spectroscopic techniques during the last decade, however, shifted the emphasis to a molecular picture of the network. With the help of these experiments, chain dimensions and their transformations under external strain are observed directly at molecular length scales, and inferences from macroscopically measured quantities are no longer necessary. Small-angle neutron scattering is presently the most powerful technique for the determination of chain dimensions and their transformations under strain.

The technique of neutron scattering and its application to polymers in the dilute and bulk states, to blends, and to networks are described in several review articles (Ullman, 1980; Higgins and Benoit, 1994; Maconnachie and Richards, 1978; Picot, 1982; Sperling, 1984; Lohse, 1986). The major general conclusion of these studies is that the dimensions of chains in a network are identical to their unperturbed dimensions in the bulk uncrosslinked state. These findings were followed by the demonstration (Beltzung et al., 1982) that the radii of gyration of poly(dimethylsiloxane) (PDMS) chains in networks are identical to those in the uncrosslinked bulk state. The effects of (1) swelling and (2) uniaxial extension of networks on chain dimensions have been studied extensively since 1980 (Beltzung et al., 1982; Hinckley, 1978; Bastide and Boue, 1986; Bastide et al., 1984; Boue et al., 1987; Boue et al., 1986; Clough et al., 1980; Yu et al., 1986), leading in general to the following conclusions:

- Transformations of chain dimensions in general fall between the predictions of the affine and phantom models.
- The effect of the state of dilution during crosslinking on transformations of chain dimensions is very pronounced.

Small-angle neutron scattering has also been applied to the analysis of networks that were relaxing after a suddenly applied constant uniaxial deformation (Boue et al., 1991). Results of dynamic neutron scattering measurements of Allen et al. (1972, 1973, 1971) indicate that segments of network chains diffuse around in a network, and the activation energies of these motions are smaller than those obtained for the center of mass motion of the whole chains. Measurements by Ewen and Richter (1987) and Oeser et al. (1988) on PDMS networks with labeled junctions show that the fluctuations of junctions are substantial and equate approximately to those of a phantom network model. Their results also indicated that the motions of the junctions are diffusive and are similar to those expected from the Rouse model, and that motions of the junctions are much slower than those of deuterated free chain ends.

4.9 SINGLE-MOLECULE ELASTICITY

A single molecule can be deformed by simply elongating it; that is, by increasing the chain's end-to-end separation. These types of experimental investigations typically focus on force-displacement measurements on single molecules. This is a very active area of research, with much of the progress being summarized in regularly appearing review articles (Chu, 1991; Janshoff et al., 2000; Granzier and Pollack, 2000; Hugel and Seitz, 2001; Strick et al., 2001; Zhang and Zhang, 2003; Slater, 2004). A large body of experiments on single molecule elasticity is on biological molecules, mostly on DNA (Bustamante et al., 2003) as well as large biological molecules such as titin (Tskhovrebova et al., 1997) and polysaccharides (Rief et al., 1997), which are reviewed by Zlatanova et al. (2000) and Zhang and Zhang (2003). Several conformational features of single chain synthetic polymers have also been investigated. Among these are the effect of side groups on chain elasticity (Zhang et al., 2000), effect of *cis* and *trans* content (Zhang and Zhang, 2003), and the behavior of suprastructure in synthetic polymers such as helices and micelles (Zhang and Zhang, 2003).

Developing techniques for grasping the single chains to be elongated are one of the main challenges in this area, and developing sufficiently sensitive methods for measuring the stresses and strains involved are another. With regard to the required attachments, it is obviously advantageous to have this occurance at the *ends* of the chains, and this is accomplished by either having the chains terminate with carefully chosen functional groups or with micrometer-sized beads. In the first case, the functional groups can be bonded onto complementary groups on a probe. In the second approach, the bead can be grasped using a micropipette or a laser beam (acting as an "optical tweezer"). Some less controlled experiments have been carried out by simply having one part of the chain physically or chemically adsorbed onto a surface with another part similarly adsorbed onto a probe. The probe in all these cases is typically the cantilever of what is essentially an atomic force microscope. The degree to which it is moved is a measure of the strain (in the range of nanometers), and its deflection a measure of the force of deformation (generally in the range of piconewtons, pN).

The utility of such measurements in the area of rubberlike elasticity is illustrated in the following section.

4.9.1 Gaussian Versus Non-Gaussian Effects

Such elongation experiments can also provide important information on the retraction of stretched single chains (Chu, 1991; Schroeder et al., 2003; Perkins et al., 1994a,b; Perkins et al, 1995). These studies were carried out on chains labeled so as to be directly observable in fluorescence microscopy. Also, the experiments are carried out in a solvent such as water, as was done in the case of the experiments on single λ-DNA molecules illustrated in Figure 4.6 (Chu, 1991). The chains were stretched to desired values of the elongation, and then permitted to relax. The circles represent some experimental results obtained, and the curve shown represents the expected behavior when the retractive force f is proportional to the end-to-end distance r remaining at that stage of the

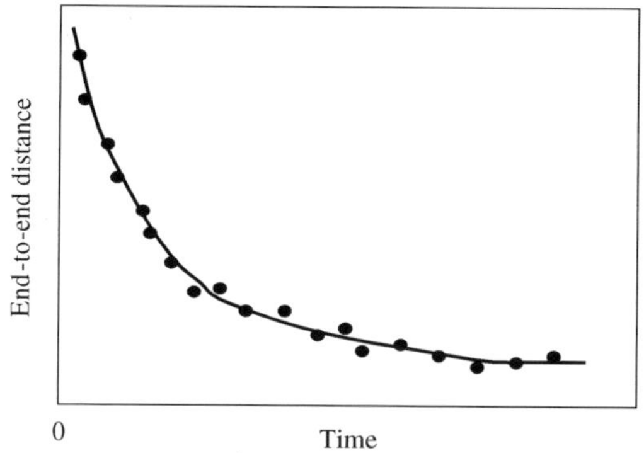

FIGURE 4.6 Relaxation of a λ-DNA molecule when the maximum extent of stretching is in the Gaussian region. Qualitative sketch based on experimental results presented elsewhere (Schroeder et al., 2003).

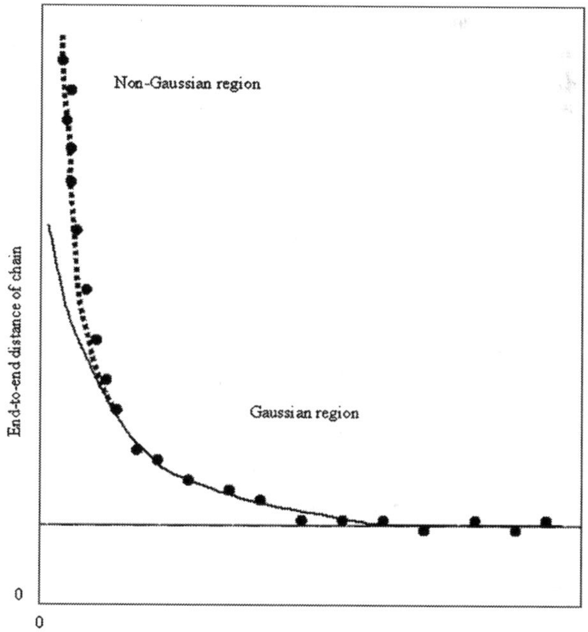

FIGURE 4.7 Relaxation of the same molecules as in Figure 4.6 when the maximum extent of stretching was beyond the Gaussian region. Qualitative sketch based on experimental results presented elsewhere (Perkins et al., 1994a,b, 1995).

retraction. The results indicate that for moderate extensions the stretched DNA chains are still in the Gaussian region, for which Eq. (4.8), $f = (3kT/\langle r^2 \rangle_o)r$, predicts the observed proportionality.

In other experiments on this same system, the DNA chains were in fact stretched close to the limits of their extensibility (Perkins et al., 1994a). The results for this more complicated case are shown schematically in Figure 4.7. At the higher elongations, the chains are clearly in the non-Gaussian region, as evidenced by a much more pronounced initial drop-off in the retractive force. This is then followed by a Gaussian decrease in f once the elongation is sufficiently small to be Gaussian (approximately two-thirds of full extension (Erman and Mark, 1997)).

Although such studies are inherently very interesting, they are not directly relevant to the many unresolved questions in the area of rubberlike elasticity that involve the interactions *among the chains* making up an elastomeric network.

REFERENCES

Allen, G. et al., 1971. Thermoelasticity. Trans. Faraday Soc. 67, 1278.
Allen, G. et al., 1972. Faraday Symp. Chem. Soc. 6, 169.
Allen, G., Higgins, J.S., Wright, C.J., 1973. Faraday Symp. Chem. Soc. 7, 348.
Bastide, J., Boue, F., 1986. Physica A 140, 251.
Bastide, J., Duplessix, R., Picot, C., 1984. Macromolecules 17, 83.
Beltzung, M. et al., 1982. PDMS orientation neutron scattering. Macromolecules 15, 1594.
Boue, F. et al., 1986. Europhys. Lett. 1, 637.
Boue, F. et al., 1987. Prog. Coll. Polym. Sci. 75, 152.
Boue, F. et al., 1991. Coll. Polym. Sci. 269, 195.
Boyce, M.C., Arruda, E.M., 2000. Constitutive models of rubber elasticity: a review. Rubber Chem. Technol. 73 (3), 504–523.
Burchard, W., Ross-Murphy, S.B. (Eds.), 1990. Physical Networks: Polymers and Gels. Elsevier, London.
Bustamante, C., Bryant, Z., Smith, S.B., 2003. Ten years of tension: single-molecule DNA mechanics. Nature 421 (6921), 423–427.
Chu, S., 1991. Laser manipulation of atoms and particles. Science 253, 861–866.
Clough, S.B., Maconnachie, A., Allen, G., 1980. Macromolecules 13, 774.
Erman, B., Mark, J.E., 1988. Use of the Fixman-Alben distribution function in the analysis of non-Gaussian rubberlike elasticity. J. Chem. Phys. 89, 3314–3316.
Erman, B., Mark, J.E., 1989. Rubber-like elasticity. Ann. Rev. Phys. Chem. 40, 351–74.
Erman, B., Mark, J.E., 1997. Structures and Properties of Rubberlike Networks. Oxford University Press, New York.
Erman, B., Wagner, W., Flory, P.J., 1980. Macromolecules 13, 1554–1558.
Everaers, R., Kremer, K., 1995. Entanglements, foundations of rubber elasticity. Macromolecules 7291, 28.
Ewen, B., Richter, D., 1987. Festkoerperprobleme 27, 1.
Ferry, J.D., 1980. Viscoelastic Properties of Polymers, third ed. Wiley, New York.
Flory, P.J., 1942. J. Chem. Phys. 10, 51.
Flory, P.J., 1953. Principles of Polymer Chemistry. Cornell University Press, Ithaca NY.
Flory, P.J., 1961. Thermoelasticity. Trans. Faraday Soc. 57, 829.
Flory, P.J., 1969. Statistical Mechanics of Chain Molecules. Interscience, New York.
Flory, P.J., 1976. Phantom networks. Proc. R. Soc. Lond. A 351, 351.
Flory, P.J., 1977. The effect of constraints (Pearce book). In: Pearce, E.M., Schaefgen, J.R. (Eds.), Contempory Topics in Polymer Chemistry. Plenum, New York.

Flory, P.J., 1982. Macromolecules 15, 99.
Flory, P.J., Ciferri, A., Hoeve, C.A.J., 1960. Thermoelasticity. J. Polym. Sci. 45, 235.
Flory, P.J., Rehner, J., 1943a. Swelling. J. Chem. Phys. 11, 512.
Flory, P.J., Rehner, J., 1943b. Swelling. J. Chem. Phys. 11, 521.
Flory, P.J., Yoon, D.Y., 1974. Moments of PE. J. Chem. Phys. 61, 5358.
Graessley,W.W., 1982. Adv. Polym. Sci. 47, 67.
Graessley, W.W., 2007. Polymeric Liquids and Networks: Structure and Properties. Taylor and Francis, New York.
Granzier, H.L., Pollack, G.H. (Eds.), 2000. Elastic Filaments of the Cell. Kluwer Academic, New York.
Guth, E., James, H.M., 1941. Phantom Theory. Ind. Eng. Chem. 33, 624.
Heinrich, G., Kaliske, M., 1997. Theoretical and numerical formulation of a molecular based constitutive tube-model of rubber elasticity. Comput. Theor. Polym. Sci. 7 (3-4), 227–241.
Heinrich, G., Straube, E., Helmis, G., 1988. Rubber Elasticity of Polymer Networks: Theories Polymer Physics. Springer, Berlin/Heidelberg, pp. 33–87.
Heinrich, G. et al., 2003. The thermoelasticity of rubberlike materials and related constitutive laws. J. Macromol. Sci., Pure Appl. Chem. A40 (1), 87–93.
Higgins, J.S., Benoit, H., 1994. Neutron Scattering From Polymers. Clarendon Press, Oxford.
Hinckley, J.A. et al., 1978. Macromolecules 11, 836.
Hugel, T., Seitz, M., 2001. The study of molecular interactions by AFM force spectroscopy. Makromol. Rapid Commun. 22, 989–1016.
Janshoff, A. et al., 2000. Force spectroscopy of molecular systems—single molecule spectroscopy of polymers and biomolecules. Angew. Chem. Int. Ed. 39, 3213–3237.
Kaliske, M., Heinrich, G., 1999. An extended tube-model for rubber elasticity: Statistical-mechanical theory and finite element implementation. Rubber Chem. Technol. 72 (4), 602–632.
Khokhlov, A.R., 1992. In: Dusek, K. (Ed.), Responsive Gels: Volume Transitions I. Springer, Verlag: Berlin, p. 125.
Khokhlov, A.R., Philippova, O.E., 1996. In: Webber, S.E. et al. (Ed.), Solvents and Self-Organization of Polymers. Kluwer Academic Publishers, Dordrecht, The Netherlands, p. 197.
Khokhlov, A.R., Philippova, O.E., 2002. In: Osada, Y., Khokhlov, A.R. (Eds.), Polymer Gels and Networks. Marcel Dekker, New York, p. 163.
Langley, N.R., 1968. Effective strand densities. Macromolecules 1, 348.
Langley, N.R., Polmanteer, K.E., 1974. J. Polym. Sci., Polym. Phys. Ed. 12, 1023–1034.
Lohse, D.J., 1986. The use of small angle neutron scattering in polymer research. Polym. News 112, 8–16.
Maconnachie, A., Richards, R.W., 1978. Polymer 19, 739.
Mark, J.E., 1973. Thermoelastic properties of rubberlike networks and their thermodynamic and molecular interpretation. Rubber Chem. Technol. 46, 593.
Mark, J.E., 1985. Bimodal networks and networks reinforced by the in-situ precipitation of silica. Brit. Polym. J. 17, 144.
Mark, J.E., Erman, B., 2007. Rubberlike Elasticity: A Molecular Primer. Cambridge University Press.
Mattice, W.L., Suter, U.W., 1994. Conformational Theory of Large Molecules: The Rotational Isomeric State Model in Macromolecular Systems. Wiley, New York.
McLeish, T.C.B., 2002. Tube theory of entangled polymer dynamics. Adv. Phys. 51 (6), 1379–1527.
Mergell, B., Everaers, R., 2001. Tube models for rubber-elastic systems. Macromolecules, 34 (16), 5675–5686.
Nisato, G., Candau, S.J., 2002. In: Osada, Y., Khokhlov, A.R. (Eds.), Polymer Gels and Networks. Marcel Dekker, New York, p. 131.
Oeser, R. et al., 1988. Dynamic fluctuations of crosslinks in a rubber: a neutron-spin-echo study. Phys. Rev. Lett. 60, 1041.
Ogden, R.W., 1997. Nonlinear Elastic Deformations, Dover Publications, New York, p. 499.
Oppermann, W., Rennar, N., 1987. Model PDMS elastomers. Prog. Coll. Polym. Sci. 75, 49.
Osada, Y., Khokhlov, A.R. (Eds.), 2002. Polymer Gels and Networks. Marcel Dekker, New York.
Perkins, T.T. et al., 1994a. Relaxation of a single DNA molecule observed by optical microscopy. Science 264, 822–826.
Perkins, T.T., Smith, D.E., Chu, S., 1994b. Direct Observation of Tube-Like Motion of a Single Polymer Chain. Science 264, 819–822.

Perkins, T.T. et al., 1995. Stretching of a single tethered polymer in a uniform flow. Science 268, 83–87.
Philippova, O.E., 2000. Polym. Sci. Ser. C. 42, 208.
Philippova, O.E. et al., 1995. Transitions in polyelectrolyte networks. Macromolecules 28, 3925.
Picot, C., 1982. In: Static and Dynamic Properties of the Polymeric Solid State. Reidel: Dordrecht, p. 127.
Rief, M. et al., 1997. Single molecule force spectroscopy on polysaccharides by atomic force microscopy. Science 275 (5304), 1295–1297.
Ronca, G., Allegra, G., 1975. Elasticity. J. Chem. Phys. 63, 4990.
Schroeder, C.M. et al., 2003. Observation of polymer conformational hysteresis in extensional flow. Science 301, 1515–1519.
Slater, G.W. et al., 2004. Deformation, stretching, and relaxation of single-polymer chains: fundamentals and examples. Soft. Mater. 2, 155–182.
Sperling, L.H., 1984. Characterization of polymer conformation and morphology through small-angle neutron scattering—a literature review. Polym. Eng. Sci. 24, 1–21.
Strick, T. et al., 2001. The manipulation of single biomolecules. Phys. Today 53 (10), 46–51.
Treloar, L.R.G., 1975. The Physics of Rubber Elasticity, third ed. Clarendon Press, Oxford.
Tskhovrebova, L. et al., 1997. Elasticity and unfolding of single molecules of the giant muscle protein titin. Nature 387 (6630), 308–312.
Ullman, R., 1980. Annu. Rev. Mater. Sci. 10, 261.
Wall, F.T., Earlier, J., 1942. Chem. Phys. J. Chem. Phys. 10, 132–134.
Yoon, D.Y., Flory, P.J., 1974. Moments of PE. J. Chem. Phys. 61, 5366.
Yu, H. et al., 1986. Advances in Elastomers and Rubber Elasticity. In: Lal, J.J., Mark, J.E. (Eds.), Plenum Press, New York, p. 407.
Zhang, W., Zhang, X., 2003. Single molecule mechanochemistry of macromolecules. Prog. Polym. Sci. 28, 1271–1295.
Zhang, W. et al., 2000. Single polymer chain elongation of Poly(N-Isopropylacrylamide) and Poly(Acrylamide) by atomic force microscopy. J. Phys. Chem. B. 104, 10258–10264.
Zlatanova, J., Lindsay, S.M., Leuba, S.H., 2000. Single molecule force spectroscopy in biology using the atomic force microscope. Prog. Biophys. Mol. Biol. 74 (1-2), 37–61.

Chapter 5

The Viscoelastic Behavior of Rubber and Dynamics of Blends

K.L. Ngai,[*] S. Capaccioli,[*] and D.J. Plazek[†]

[*]*Dipartimento di Fisica, Universitá di Pisa, Largo B. Pontecorvo 3, I-56127, Pisa, Italy*
[†]*University of Pittsburgh, Pittsburgh, PA 15261, USA*

5.1 INTRODUCTION

Most rubber is produced from cross-linkable high molecular weight linear polymers with low glass temperatures (Ferry, 1980; Flory, 1953; Treloar, 1975; Mark and Erman, 1988; Mark et al., 2004; Gent, 2004). The high molecular weight is necessary to obtain high extensibility in the ultimate elastomer and the low glass temperature is required to obtain resilience. These precursors are collections of entangled linear molecules that ultimately are free to flow past one another and hence are viscoelastic liquids (Ferry, 1980; Mark et al., 2004; Plazek and Berry, 1986). They are viscoelastic by virtue of their time-dependent mechanical response, which reflects the sluggish configurational changes of the molecules. Upon being crosslinked sufficiently, a chemical molecular network (rubber or elastomer) is formed that transforms the polymer into a viscoelastic solid, which does not flow. Like its precursor polymer, the viscoelastic properties are strongly dependent on time or frequency, temperature, pressure, and the presence of swelling solvent or filler. Among viscoelastic solids, rubber has the unique characteristic of preserving material integrity even when subjected to high stresses or strains, although the viscoelastic behavior is highly dependent on the large stresses or strains (Treloar, 1975; Mark et al., 2004; Payne and Whittaker, 1971; Mullins, 1959; Smith, 1977). However, for sufficiently small stresses and strains, the viscoelastic behavior becomes invariant and the linear viscoelastic regime prevails (Ferry, 1980). In this chapter, we describe the linear viscoelastic properties of rubber and its dependence on various parameters, including crosslink density, chemical structure, and molecular weight, with experimental data principally coming from measurements on a series of well-characterized and fully cured bisphenol-A-based epoxy resins (Choy and

The Science and Technology of Rubber. http://dx.doi.org/10.1016/B978-0-12-394584-6.00005-4

Plazek, 1986; Plazek and Choy, 1989; Plazek and Frund, 1990; Plazek and Chay, 1991); as well as some polybutadienes and fluorinated elastomers (Plazek et al., 1983; Plazek et al., 1988; Plazek and Rosner, 1998). Some nonlinear viscoelastic behavior is discussed.

NOMENCLATURE

A	empirical constant in VFTH equation
a_T	shift factor
B	bulk compliance
B^*	complex dynamic bulk compliance
B'	bulk storage compliance
B''	bulk loss compliance
C	empirical constant in VFTH equation, or rate of tearing
C_1, C_2	coefficient in WLF equation
C_p	specific heat
$C_{\text{ph}}(t)$	correlation function of junction for a perfect network of phantom chains
$C_J(t)$	correlation function of junction
c_1	constant
c_2	constant
c_3	constant
c_4	constant
c_5	constant
D	tensile compliance
D^*	complex dynamic tensile compliance
D'	tensile storage compliance
D''	tensile loss compliance
E	Young's (tensile) modulus
E^*	complex dynamic tensile modulus
E'	tensile storage modulus
E''	tensile loss modulus
E_a	primitive activation energy
E_a^*	measured or apparent activation energy
f	free volume fraction, or functionality
f_1	proportionality constant
f_c	elastic stress in the constrained junction model of Flory
f_{NM}	frequency of normal mode

f_{ph}	elastic stress in the phantom chains model
$F_s(Q,t)$	intermediate scattering function
G	shear modulus
$G(t)$	shear relaxation modulus
G^*	complex dynamic shear modulus
G'	shear storage modulus
G''	shear loss modulus
G_e	equilibrium modulus
G_g	glassy shear modulus
H	shear relaxation spectrum, or enthalpy
J	shear compliance
$J(t)$	shear creep compliance
J^*	complex dynamic shear compliance
J'	shear storage compliance
J''	shear loss compliance
J_d	delayed shear compliance
J_e	equilibrium shear compliance
J_g	glassy shear compliance
$J_p(t)$	reduced shear compliance
$J_r(t)$	recoverable shear compliance
J_s	steady state shear compliance
$\hat{J}(\omega)$	spectral density function
K	bulk modulus
K^*	complex dynamic bulk modulus
K'	bulk storage modulus
K''	bulk loss modulus
k	proportionality factor
L	shear retardation spectrum
$L_p(t)$	reduced shear retardation spectrum
M_x	average molecular weight per crosslinked unit
M_c	average molecular weight of a network strand
n	coupling parameter of segmental relaxation
n_i	coupling parameter of the *i*th mode of a polymer strand anchored on both ends to the network junctions
n_J	coupling parameter of network junction relaxation
P	pressure
Q	cooling or heating rate, or scattering vector
R	gas constant

S	entropy
S_c	configurational entropy
$S(Q,t)$	dynamic structure factor
T	temperature
T_f	fictive temperature
T_g	glass temperature
T_0	reference temperature
T_∞	VFTH temperature
T_{gf}	glass transition temperature of the fast component
T_{gs}	glass transition temperature of the slow component
$T_{g,\mathrm{TSDC}}$	glass transition temperature from thermal stimulated depolarized current
t	time
t_c	crossover time of the Coupling Model
t_{cure}	time of curing
V	specific volume
β	Andrade coefficient
β_R	Kohlrasuch exponent of constrained Rouse mode
γ_0	imposed shear strain
$\gamma(t)$	shear strain
$\Delta\varepsilon_\beta$	relaxation strength of the JG relaxation
δ	phase angle between stress and strain
ζ_0	monomeric friction coefficient
λ	retardation time
η	shear viscosity
κ	Flory's parameter in the constrained junction model
ρ	density
ϕ_{eff}	effective concentration
ϕ_{self}	self-concentration factor
σ_0	imposed shear stress
$\sigma(t)$	shear stress
τ	relaxation time
τ_J	network junction relaxation time
τ_α	segmental relaxation time
$\tau_{o\alpha}$	primitive relaxation time of the Coupling Model
τ_β	Johari-Goldstein secondary relaxation time
τ_{oJ}	primitive relaxation time of the network junction
$\tau_{\alpha f}$	segmental relaxation time of the fast component in binary polymer blend

$\tau_{\alpha s}$	segmental relaxation time of the slow component in binary polymer blend
τ_{JG}	Johari-Goldstein relaxation time
τ_{NMR}	segmental relaxation time from deuteron NMR measurement
$\hat{\tau}_R$	effective Rouse relaxation time
τ_R^0	Rouse relaxation time
τ_{Ri}	relaxation time of *i*th Rouse mode
τ_∞	prefactor
τ_∞^*	prefactor
ω	frequency in rad/sec

Since rubbers typically are based on amorphous polymers, there is a glass temperature below which the structure falls out of equilibrium. As in other amorphous polymers, the molecular segmental mobility determines the glass temperature of a rubber (Ferry, 1980; Mark et al., 2004; Plazek and Berry, 1986; Ngai et al., 1997a; Ngai and Plazek, 1995; Plazek and Ngai, 1996). The change from glassy to rubbery behavior occurs when the randomly oriented polymer chains between crosslinks take over to govern the viscoelastic response. There are common features as well as major differences in the viscoelastic behavior of amorphous polymers and rubbers. A better understanding of both systems is gained by comparing and interpreting them in terms of theoretical models.

Miscible blends of two polymers can generate beneficial viscoelastic properties for application not found in either one of the two component polymers. The segmental relaxation and the global chain dynamics of the two components in the blend can be significantly modified depending on the composition and the components, resulting in properties not seen before in polymers. The changes are not only interesting and challenging for explanation, but also they can be used as critical tests of theories or models of polymer chain dynamics and segmental dynamics proposed for polymers. The changes on both components are expected to be most significant if the two components have large difference in mobility at the same temperature, a situation realized in highly asymmetric polymer blends that have a large difference in the glass transition temperatures of the two components. The substantial changes of properties in component dynamics of these highly asymmetric polymer blends, particularly the anomalous ones, proffer excellent opportunity for discovering new physics of polymer dynamics that are not found by experiment on conventional polymers. The new physics discovered have impact on the current state-of-the-art understanding of polymer dynamics and viscoelasticity. The last section is devoted to a critical assessment of the current trend of research, followed by a review of the complete list of the extraordinary properties exhibited by the components of these polymer blends, and explanation of all of them by the Coupling Model.

5.2 DEFINITIONS OF MEASURED QUANTITIES, $J(T)$, $G(T)$, AND $G^*(\omega)$; AND SPECTRA $L(\log \lambda)$ AND $H(\log \tau)$

Viscoelastic behavior is a time-dependent mechanical response and usually is characterized with creep compliance, stress-relaxation, or dynamic mechanical measurements. Since time is an additional variable to deformation and force, to obtain unique characterizing functions in these measurements one of the usual variables is held constant.

5.2.1 Creep and Recovery

In a shear creep experiment a shearing stress σ_o is created in a previously relaxed material and held constant while the resulting shear strain $\gamma(t)$ increases monotonically with time t.

Given a sufficiently long time of creep, the velocity of creep will decelerate to zero and $\gamma(t)$ attains an equilibrium limit if a viscoelastic solid is being measured. On the other hand, if the material is a viscoelastic liquid, the velocity of creep will decelerate to a finite constant value. Viscoelastic steady state is achieved, and $\gamma(t)$ increases indefinitely. The creep experiment has a second part when the stress is set to zero after a period of creeping. A portion or all of the strain accumulated during creeping is then recovered as a function of time for a viscoelastic liquid or solid, respectively. For a viscoelastic liquid, the portion that is permanent deformation and irrecoverable reflects the contribution of viscous flow to the total deformation accumulated during creep. Since a viscoelastic solid does not flow, all of its creep deformation is recoverable.

When the strains or the strain rates are sufficiently small, the creep response is linear. In this case, when the time-dependent strain is divided by the fixed stress, a unique creep compliance curve results; that is, at each time there is only one value for this ratio, which is the compliance—$\gamma(t)/\sigma_o \equiv J(t)$. The unique shear creep compliance function $J(t)$ (Pa^{-1} or cm^2/dyne, $1\ \text{Pa}^{-1} = 0.1\ \text{cm}^2/\text{dyne}$) obtained for an amorphous polymer has the usual contributions

$$J(t) = \gamma(t)/\sigma_o = J_g + J_d\psi(t) + t/\eta \equiv J_r(t) + t/\eta, \tag{5.1}$$

where J_d is a delayed compliance; $\psi(t)$ is a normalized memory function, which is equal to zero when $t = 0$ and is one when $t = \infty$; and η (Pa s or poise) is the shear viscosity (10 poise = 1 Pa s). J_g is called the glassy compliance, which represents the long-time limit of strains that accrue so fast that their time dependence cannot be observed within the usually accessible experimental window, even at the glass temperature, where many molecular motions are very sluggish. The t/η term reflects the permanent viscous deformation. $J_r(t)$ is the recoverable shear compliance, which can be obtained from creep recovery measurements. For a viscoelastic solid η is operationally infinite, since normally a molecular network is present to preclude any permanent deformation. The counterpart of Eq. (5.1) for a viscoelastic solid is

$$J(t) = \gamma(t)/\sigma_o = J_g + J_d\psi(t). \tag{5.2}$$

The creep and recovery experiment is the only characterization of viscoelastic behavior that is readily comprehended since according to Eq. (5.1) contributions to the creep compliance are *additive* in the strain. The molecular processes involved are simply short- and long-range configurational orientations and viscous flow reflecting the permanent increasing separation of the centers of gravity of neighboring polymer molecules. It has also been shown that the solvent in polymer solutions contributes additively to the creep strain (Riande et al., 1995).

With all the viscoelastic functions it is important to note the limiting values or forms that are qualitatively independent of the molecular structure. For a viscoelastic liquid, $\lim_{t\to 0} J(t) = J_g$, $\lim_{t\to\infty} J(t) = t/\eta$, and $\lim_{t\to\infty} J_r(t) = J(t) - t/\eta = J_g + J_d \equiv J_s$. The last limiting value J_s is called the steady-state recoverable shear compliance. It is the maximum recoverable strain per unit stress, which reflects the maximum configurational orientation achievable at the present stress.

For a viscoelastic solid $J_r(t) = J(t)$, $\lim_{t\to 0} J(t) = J_g$, and $\lim_{t\to\infty} J(t) = J_g + J_d \equiv J_e$. A different notation J_e is to denote the equilibrium compliance of a solid.

5.2.2 Stress Relaxation

After a constant shear strain is created in a previously relaxed material, the resulting shear stress decays with ensuing time to zero for a viscoelastic liquid and to a finite equilibrium value for a viscoelastic solid. The shear stress relaxation modulus (Pa or dynes/cm^2)

$$G(t) = \sigma(t)/\gamma_0 = G_e + [G(0) - G_e]\varphi(t), \tag{5.3}$$

where γ_o is the imposed fixed strain; $\phi(t)$ is the relaxation function decreasing from $\phi(0) = 1$ at $t = 0$ to $\phi(\infty) = 0$ at $t = \infty$; and G_e is the equilibrium modulus, which is finite for a viscoelastic solid and zero for viscoelastic liquid. The time-dependent stresses arising from different molecular mechanisms are *not additive*, and hence it is difficult if not impossible to isolate and characterize each one of them individually. Nevertheless $G(t)$ is connected to $J(t)$ by the convolution integral equation, $\int_0^t G(s)J(t-s)ds = 1$, from which one function can be calculated from the other by a numerical procedure (Hopkins and Hamming, 1957).

5.2.3 Dynamic Mechanical Measurements

Sinusoidal stresses or strains of constant frequency are applied to a sample until a steady sinusoidal strain or stress results, with a fixed phase angle between the input and the output. For example for a sinusoidal shear strain,

$$\gamma(t) = \gamma_o \sin \omega, \tag{5.4}$$

where γ_o is the strain amplitude and ω the angular frequency, the stress σ will oscillate sinusoidally as

$$\sigma(t) = \sigma_o \sin(\omega t + \delta). \tag{5.5}$$

Since the stress always leads the strain, the phase angle δ is positive. Using the trigonometric formula for the sine of the sum of two angles, Eq. (5.5) can be rewritten in terms of the shear storage modulus $G'(\omega)$ and the loss modulus $G''(\omega)$ as

$$\sigma(t) = \gamma_o G'(\omega) \sin(\omega t) + G''(\omega) \cos(\omega t), \tag{5.6}$$

where $G'(\omega) = (\sigma_o/\gamma_o)\cos\delta$ and $G''(\omega) = (\sigma_o/\gamma_o)\sin\delta G'$ is a measure of the elastic energy stored and recovered and G'' is a measure of the energy dissipated as heat in the cyclic deformation. The ratio $G''(\omega)/G'\omega)$ is $\tan\delta$. Although usually not written out explicitly, δ is a function of ω. The complex dynamics shear modulus $G^*(\omega)$ is defined by

$$G^*(\omega) = G'(\omega) + iG''(\omega) = (\sigma_o/\gamma_o)\exp(i\delta). \tag{5.7}$$

It can be derived (Ferry, 1980) from the general expression of calculating the stress corresponding to any strain history that the G' and G'' are related to the stress relaxation modulus $G(t)$ in Eq. (5.3) as follows:

$$G'(\omega) = G_e + \omega \int_0^\infty [G(t) - G_e] \sin \omega t, \tag{5.8}$$

$$G''(\omega) = \omega \int_0^\infty [G(t) - G_e] \cos \omega t. \tag{5.9}$$

Again, G_e in Eqs. (5.8) and (5.9) is zero for a liquid but is the equilibrium modulus for a viscoelastic solid. For a liquid, it can be shown that $\lim_{\omega\to 0} G'(\omega) = \omega^2\eta^2 J_s$ and $\lim_{\omega\to 0} G''(\omega) = \omega\eta$. For the solid, $\lim_{\omega\to 0} G'(\omega) = G_e$ and $\lim_{\omega\to 0} G''(\omega) = \omega \int_0^\infty [G(t) - G_e]dt$.

In an analogous manner, the steady sinusoidal strain in response to the applied sinusoidal stress of constant frequency is expressed in terms of the (in phase) storage compliance $J'(\omega)$ and the (90° out of phase) loss compliance $J''(\omega)$ as

$$\gamma(t) = \sigma_o[J'(\omega) \sin(\omega t) - J''(\omega) \cos(\omega t)], \tag{5.10}$$

where $J'(\omega) = (\gamma_o/\sigma_o)\cos\delta$ and $J''(\omega) = (\gamma_o/\sigma_o)\sin\delta$. The complex dynamics shear compliance $J^*(\omega)$ is defined by

$$J^*(\omega) = J'(\omega) - iJ''(\omega) = (\gamma_o/\sigma_o)\exp(-i\delta). \tag{5.11}$$

The minus sign in Eqs. (5.10) and (5.11) is the consequence of the strain lagging behind the stress. It can be shown that, for a viscoelastic liquid, the dynamic creep compliances are related to the creep compliance $J(t)$ by the one-sided Fourier transforms:

$$J'(\omega) = J_s - \omega \int_0^{\infty} [J_s - J(t) + t/\eta] \sin \omega t \, dt, \tag{5.12}$$

$$J''(\omega) = 1/\omega\eta + \omega \int_0^{\infty} [J_s - J(t) + t/\eta] \cos \omega t \, dt. \tag{5.13}$$

For a viscoelastic solid, the relations are given by

$$J'(\omega) = J_e - \omega \int_0^{\infty} [J_e - J(t)] \sin \omega t \, dt \tag{5.14}$$

$$J''(\omega) = \int_0^{\infty} [J_e - J(t)] \cos \omega t \, dt. \tag{5.15}$$

It follows immediately from Eqs. (5.7) and (5.11) that

$$J^*(\omega) = 1/G^*(\omega). \tag{5.16}$$

For the liquid, the low frequency limits are: $\lim_{\omega \to 0} J'(\omega) = J_s$ and $\lim_{\omega \to 0} J''(\omega) = 1/\omega\eta$. For the solid, they are $\lim_{\omega \to 0} J'(\omega) = J_e$ and $\lim_{\omega \to 0} J''(\omega) = \omega \int_0^{\infty} [J_e - J(t)] dt$.

(i) Tensile and Bulk Moduli: Tensile and Bulk Compliance

Although the viscoelastic functions and their interrelations have been given earlier for shear deformation, they apply to tensile and bulk deformations as well. The analogs of $J(t)$, $J'(\omega)$, $J''(\omega)$, $G(t)$, $G'(\omega)$, and $G''(\omega)$, for tensile deformation are the tensile compliance $D(t)$, the tensile storage compliance $D'(\omega)$, the tensile loss compliance $D''(\omega)$, the tensile modulus $E(t)$, the tensile storage modulus $E'(\omega)$, and the tensile loss modulus $E''(\omega)$, respectively. For bulk deformation the corresponding quantities are the bulk compliance $B(t)$, the bulk storage compliance $B'(\omega)$, the bulk loss compliance $B''(\omega)$, the tensile bulk modulus $K(t)$, the bulk storage modulus $K'(\omega)$, and the bulk loss modulus $K''(\omega)$. All previous equations hold for tensile (bulk) deformation after all quantities have been changed in connotation from shear to tensile (bulk). It can be shown that the three compliances and moduli are related by the following equations:

$$D(t) = \frac{J(t)}{3} + \frac{B(t)}{9}, \tag{5.17}$$

$$E(t) = \left[\frac{1}{3G(t)} + \frac{1}{9K(t)}\right]^{-1}. \tag{5.18}$$

If the specimen is highly incompressible such that $K(t) \gg G(t)$, and $B(t) \ll J(t)$, we have the simpler relations:

$$E(t) = 3G(t), \tag{5.19}$$

$$D(t) = J(t)/3. \tag{5.20}$$

These relations enable one to relate the shear viscoelastic functions to their tensile counterparts. At high compliance levels, rubbers are highly incompressible and the proportional relation between the tensile and shear moduli and compliances holds. However at lower compliances approaching J_g, the Poisson ratio μ (which in an elongational deformation is $-d\ln w/d\ln l$, where w is the specimen's width and l is its length) is less than 1/2. Equations (5.19) and (5.20) are then no longer exact. For a glass $\mu \sim 1/4$. When $G(t) = K(t)$, $E(t) = 2.25G(t)$.

5.3 THE GLASS TEMPERATURE

The glass temperature T_g is a function of the rate of cooling and the pressure at which it was determined. At a specified rate of cooling and pressure, T_g is a corresponding state variable for the viscoelastic properties and applications of noncrystalline polymers (Mark et al., 2004; Plazek and Ngai, 1996). It is a material characterizing parameter. For example, if T_g is much higher than the temperature of application, the polymer is a hard glass and may be suitable for applications as engineering plastics. If T_g is sufficiently lower, the polymer is rubbery and may be used in the rubber industry. Although references are made to "the transition from the rubbery state to the glassy state," no such transition occurs because the rubber and the glass are in the same thermodynamic state; only the molecular mobility is different. The kinetic change from rubbery to glassy behavior is reflected in the change in enthalpy H or volume V. For example, as seen in Figure 5.1 at a fixed rate of cooling Q_1 a liquid's H and V decreases along an equilibrium line until the liquid's approach to its equilibrium configuration becomes so slow that it cannot keep up with the diminishing temperature. At all lower temperatures the specific volume is greater than its equilibrium value and glassy behavior is observed. At a slower rate of cooling Q_2 equilibrium can be approximated to a lower temperature and a lower T_g is observed. The intersection of the equilibrium liquid line with the nonequilibrium glass line is defined as T_g. Thus T_g is a decreasing function of the cooling rate Q.

Upon cessation of cooling below T_g the enthalpy and volume slowly but incessantly decrease toward their equilibrium values. During this decrease other kinetic properties, such as the rate of creep, slow down. This deceleration of kinetic processes is called physical aging and is reflected in Figure 5.1 by the

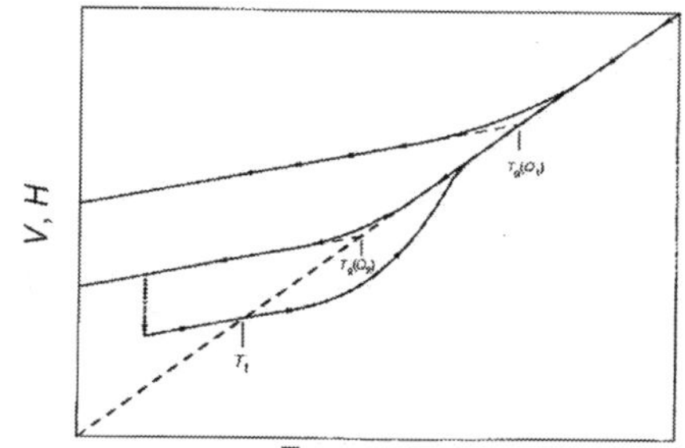

FIGURE 5.1 Schematic plots of the variation of Volume V and enthalpy H with temperature. The uppermost line represents cooling from equilibrium liquid at a more rapid rate Q_1. The line in the middle represents cooling at a slower rate Q_2. The thin lines are extrapolations of the glass lines to higher temperatures. Their intersections with the equilibrium liquid line (thicker dashed line) define the glass temperatures, $T_g(Q_1)$ and $T_g(Q_2)$. The downward pointing arrow indicates isothermal physical aging for a period of time. The low lying line represents heating at the rate Q_2 of the aged glass to restore thermal equilibrium at some higher temperature.

dotted line. These effects are reversible, in that these changes are erased when the material is taken to temperatures above T_g where equilibrium properties are readily achieved (Struik, 1978; Kovacs et al., 1979; McKenna, 1989; Hodge, 1983; Moynihan et al., 1976). The rate of volume contraction greatly decelerates as the temperature is decreased, and it becomes impossible to wait for the achievement of equilibrium at temperatures below about $T_g - 25°C$. If physical aging is terminated after a time t_{age} and the annealed glass is heated at a constant rate, V initially increases along a denser glass line, undershoots the equilibrium volume line, and finally returns to equilibrium at some temperature higher than $T_g(Q_2)$. Explanation of this behavior is beyond the scope of this chapter but can be found in other texts (Tool, 1946). The temperature at which the glass line intersects the equilibrium line is called the fictive temperature T_f of the annealed glass.

5.4 VISCOELASTIC BEHAVIOR ABOVE T_g

5.4.1 Isothermal Measurements of Time or Frequency Dependence

The viscoelastic response of equilibrium rubber networks can be obtained by measuring the shear and tensile moduli or compliances as a function of time, or the corresponding dynamic moduli and compliances as a function of frequency. As discussed in Section 5.2, the measurements of any viscoelastic function can be converted to another viscoelastic function.

The Epons 828, 1001, 1002, 1004, and 1007 fully cured with stoichiometric amounts of DDS are examples of well-characterized networks. Therefore mechanical measurements on them offer insight into the viscoelastic properties of rubber networks. The shear creep compliance $J(t)$ of these Epons were measured near above their glass temperatures (Choy and Plazek, 1986; Plazek and Choy, 1989; Plazek and Chay, 1991). From the statistical theory of rubber elasticity (Ferry, 1980; Flory, 1953; Treloar, 1975; Mark and Erman, 1988; Mark et al., 2004; James and Guth, 1947; James and Guth, 1953; Flory, 1976; Flory, 1979b; Flory and Erman, 1982) the equilibrium modulus G_e is proportional to the product $T\rho$, where ρ is the density at temperature T, and hence the equilibrium compliance J_e is proportional to $(T\rho)^{-1}$. Thus $J(t)$ is expected to be proportional to $(T\rho)^{-1}$, and $J(t)T\rho$ is the quantity that should be compared at different temperatures. Actually the reduced creep compliance

$$J_p(t) = J(t)T\rho/T_0\rho_0 \qquad (5.21)$$

is the quantity of choice in plotting against time, where T_0 is a reference temperature and ρ_0 is the density at T_0. This is used so that we can read off from such a plot the actual magnitudes of $J(t)$ measured at the reference temperature T_0. Illustrative results are shown in Figure 5.2, where $J_p(t)$ determined on Epon 1007/DDS at seven temperatures between 99.8°C and 127.3°C with $T_0 = 100.7$°C are presented. The measured $J_p(t)$ extends from the glassy level slightly above 10^{-10} cm^2/dyn up to a firm rubbery compliance close to 10^{-7} cm^2/dyn.

5.4.2 Temperature Dependence

Traditionally it is assumed that the temperature dependences of the retardation times of all viscoelastic modes or mechanisms of polymers are proportional to one and the same monomeric friction coefficient ζ_0 (Ferry, 1980; Mark et al., 2004; Plazek and Berry, 1986; Doi and Edwards, 1986). For rubber networks, the viscoelastic modes include those with shorter retardation times responsible for volume change and the glass temperature, and the longer retardation times of polymer strands between crosslinks contributing to rubbery deformation and J_e. Thus, the retardation times $\lambda(T)$ of all the viscoelastic modes contributing to $J_p(t)$ at any temperature T are related to that at a chosen reference temperature T_0 by the same multiplicative factor given by

$$a_T = \zeta_0(T)/\zeta_0(T_0). \qquad (5.22)$$

This means that the measured $J_p(t)$ at (T, t) is equivalent to that at $(T_0, t/a_T)$, as seen in Figure 5.2 by the measured compliance moving horizontally to shorter times along the logarithmic time scale with increasing temperature. This time-temperature equivalence of the viscoelastic function is referred to as thermorheological simplicity. In practice, the test of time-temperature equivalence of experimental data is carried out by successively shifting the data measured at T, first for T closest T_0, along the $\log(t)$ axis by a displacement $-\log a_T$ to overlap with each other to form a reduced curve. Measurements for

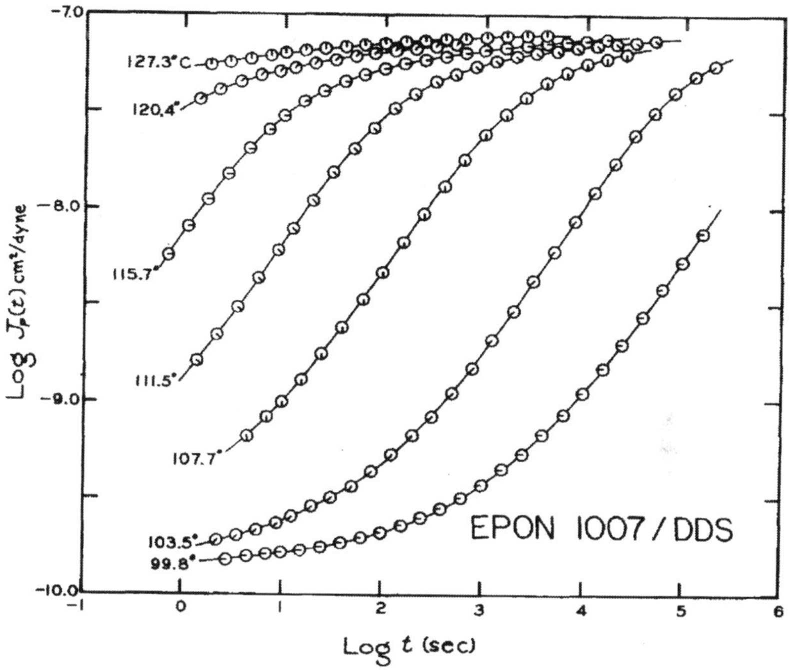

FIGURE 5.2 Reduced shear creep compliance curves $J_p(t)$ cm^2/dyn, determined on Epon 1007/DDS at seven temperatures, as indicated, presented logarithmically as a function of logarithmic time t.

$T > T_0$ are shifted to longer times, and measurements for $T < T_0$ are shifted to shorter times. A well-defined reduced curve means the viscoelastic response is thermorheologically simple (Schwarzl and Staverman, 1952). It represents $\log J_p(t)$ at T_0 over an extended time range. The time scale shift factors a_T that were used in the reduction of the creep compliance curves to obtain the reduced curve constitute the temperature dependence. a_T is fitted to an analytical form, which is often chosen to be the Williams-Landel-Ferry (WLF) equation (Ferry, 1980),

$$\log a_T = -C_1(T - T_0)/(C_2 + T - T_0), \tag{5.23}$$

or equivalently the Vogel-Fulcher-Tammann-Hesse (VFTH) equation (Ferry, 1980),

$$a_T = A\exp[C/(T - T_\infty)]. \tag{5.24}$$

For the chosen reference temperature T_0, $A = \exp[-C/(T_0 - T_\infty)]$. The parameter $C/2.303$ is equal to the product, C_1C_2, of the two constants in the WLF equation, and $T_\infty = T_0 - C_2$. From the WLF or the VFTH equation for the temperature dependence of a_T, the reduced data curve can be constructed for another choice of the reference temperature T_0.

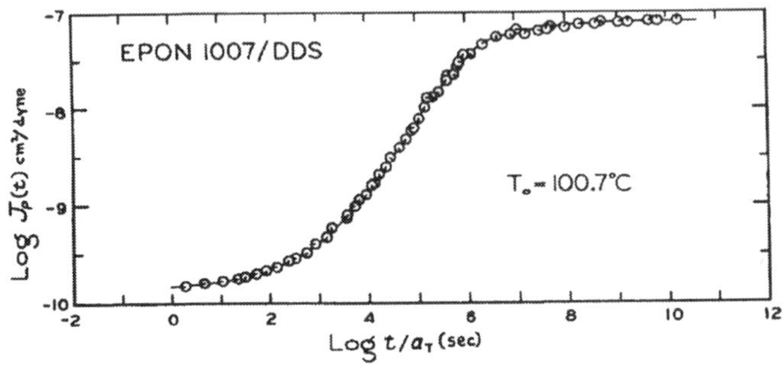

FIGURE 5.3 Reduced shear creep compliance curves $J_p(t)$ of Epon 1007/DDS shifted to superimpose with the curve at the reference temperature 100.7°C shown logarithmically versus the logarithm of the reduced time t/a_T.

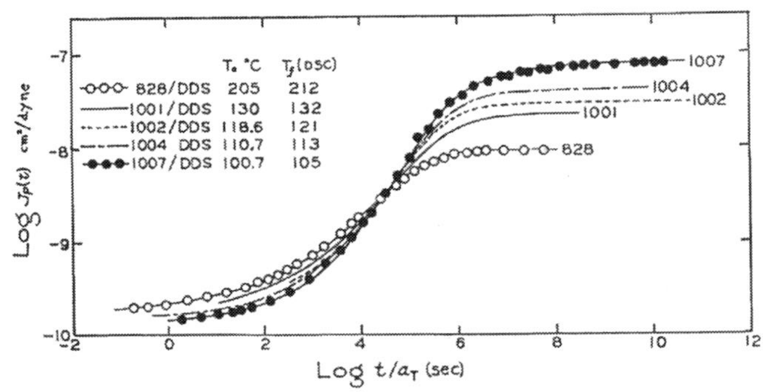

FIGURE 5.4 Comparison of reduced shear creep compliance curves of Epon 828, 1001, 1002, 1004, and 1007/DDS plotted logarithmically against time at the reference temperatures indicated that are close to the respective T_gs.

The procedure was carried out on the measured $J_p(t)$ of Epon 1007/DDS in Figure 5.2. The reduced curve extending over 10 decades of time with the choice T_0 of 100.7°C is shown in Figure 5.3. This test of time-temperature equivalence is eminently successful as found in other molecular network polymers (Plazek and Chay, 1991; Plazek et al., 1983; Plazek et al., 1988).

The measured $J_p(t)$ curves of the other epoxy resins studied, Epon 1004, 1002, 1001, and 828/DDS, were successfully reduced at chosen reference temperatures and altogether the results are shown in Figure 5.4 as functions of the reduced time t/a_T in a double logarithmic plot. This comparison plot was constructed by requiring all the reduced curves to cross at a compliance level of $\log J_p(t) = -8.5$. The choice of T_0 is 100.7, 110.7, 118.6, 130, and 205 for the DDS crosslinked 1007, 1004, 1002, 1001, and 828 Epons, respectively.

The time scale shift factors a_T that were determined in the reduction of the creep compliance curves to obtain the reduced curves shown in Figure 5.4 are presented in Figure 5.5. The logarithm of a_T is plotted as a function of the reciprocal absolute temperature. The temperature dependence data can be fitted

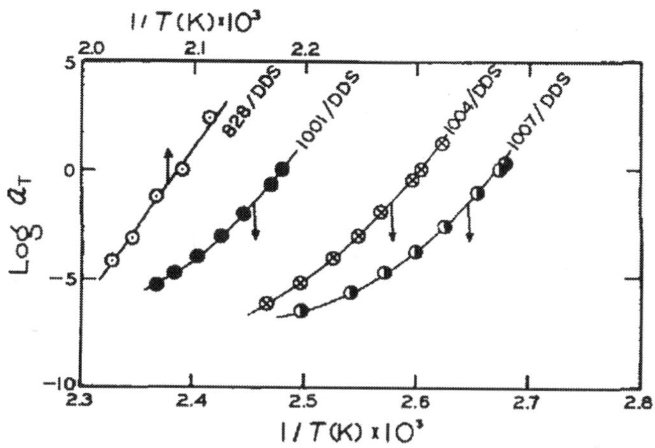

FIGURE 5.5 Logarithmic temperature shift factors log a_T plotted as functions of the reciprocal absolute temperature T/K for the four indicated epoxy resins which were obtained from the reduction process used in producing the curves in Figure 5.8.

to either the WLF Eq. (5.23) or the VFTH Eq. (5.24), except for the tightest network 828/DDS, which appears to be Arrhenius.

5.4.3 The Equilibrium Compliance J_e

It can seen from Figure 5.4 that the equilibrium compliance J_e decreases uniformly from the 1007/DDS to the 828/DDS and is expected on the basis of the kinetic theory of rubberlike elasticity, since the concentration of network chains increases and the molecular weight per crosslinked unit, M_x, decreases in the same order. The M_x values calculated as $\rho RT J_e$ are remarkably close to the molecular weight values of the starting epoxy resins.

5.5 VISCOELASTIC BEHAVIOR OF OTHER MODEL ELASTOMERS

Qualitatively crosslinked rubbers share common properties, although quantitatively they depend on molecular architecture and the chemical type of the polymer and the crosslinking agent. Here we show and compare the viscoelastic behavior of various model network systems.

5.5.1 Fluorinated Hydrocarbon Elastomers ("Viton")

The starting material is a copolymer of vinylidene fluoride and hexafluoropropylene with molecular weight in the neighborhood of 2.0×10^5. The vulcanization recipe is 100 parts copolymer, 3 parts high activity magnesia, 3 parts calcium hydroxide, and 1–4 parts curing agents. Three "Viton" fluoroelastomers with different degrees of crosslinking obtained from DuPont

TABLE 5.1 Physical Properties of Fluoroelastomers

Sample	Density 25°C (g/cm³)	log G_e(Pa)[a]	log G_e(Pa)[b]	M_c (g/mol)	M_c[c] (g/mol)
11-A	1.837	5.76	5.80	7790	7220
10-A	1.833	6.05	5.94	4060	5220
10-B	1.830	6.22	6.17	2720	3070
Sample	**C_1 (kPa)**	**C_2 (kPa)**	**log 2 (C_1 + C_2)[d]**	**SI, g/g[e]**	
11-A	150	280	5.93	3.01	
10-A	300	200	6.00	2.42	
10-B	620	0	6.09	2.13	

[a] *From compression measurements on swollen samples.*
[b] *Estimated from compliance curve, $T_0 = -20°C$.*
[c] *From estimate of J_e (creep).*
[d] *C_1 and C_2 in Pa. Mooney-Rivlin Equation constants. For the equation, see Ferry (1980).*
[e] *Weight swelling index measured in methyl ethyl ketone at 25° C.*

were studied. The average molecular weights of the network chains M_c were estimated to be 2700 for sample 10-B, 4100 for 10-A, and 7800 for 11-A. These parameters, together with some other physical properties of the three samples, are listed in Table 5.1.

(i) Creep Compliance Data

Torsional shear measurements yielded isothermal shear creep compliances, $J(t)$, that ranged from the glassy level 10^{-10} cm²/dyne (10^{-9} Pa^{-1}) to near equilibrium values in the neighborhood of the usual 10^{-7} cm²/dyne (10^{-6} Pa^{-1}). The reduced isothermal shear compliance, $J_p(t)$, calculated by Eq. (5.21), obeys time-temperature superposition, and the well-defined reduced curves are shown logarithmically in Figure 5.6 plotted against the reduced time scale, t/a_T, where a_T is the usual time scale shift factor. The most highly crosslinked sample has the highest T_g, −19.6°C, as expected. In addition to raising T_g, the increasing level of crosslinking depresses the equilibrium compliance, J_e. This depression is, of course, rationalized by the classical kinetic theory of rubberlike elasticity, which concludes that J_e is inversely proportional to the concentration of elastic elements; that is, the number of polymeric network chains per unit volume. The approach to the different J_e values is clearly seen in Figure 5.6 for the three fluoroelastomers studied.

(ii) Temperature Dependence of the Shift Factors

The horizontal logarithmic time scale shifts that are required to superpose the data obtained at different temperatures are the logarithms of the a_T shift factors. The a_T values thus reflect the principal temperature dependence of the viscoelastic process. It was possible to represent the time scale temperature

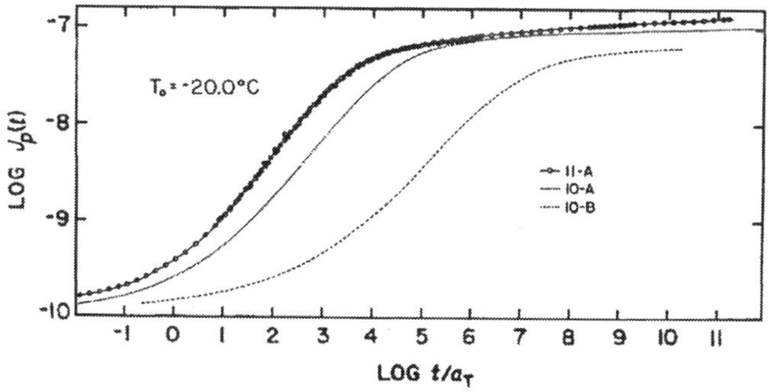

FIGURE 5.6 Logarithmic plot of the reduced shear creep compliance curves $J_p(t)$ (in $cm^2/dyn = 10\ Pa^{-1}$) against the reduced time t/a_T (in seconds). The reference temperature of reduction, T_0, is $-20.0°C$ for all curves. Data points are shown only for Sample 11-A.

dependences of the three samples with a single VFTH equation (5.24) in which only one parameter, T_∞, which reflects the change in T_g, varies with the level of crosslinking. The fit achieved is shown in Figure 5.7. The atmosphere in which the measurements were made is important since samples measured in air contain the moisture absorbed under ambient conditions, whereas those measured in rough vacuum (ca. 10^{-2} torr = 1.3 Pa) are at least partially dried. The constant C in the VFTH equation is 990°C for all the samples measured. The difference, Δ, between T_g and T_∞ is 32.8°C for 11-A, 31.3°C for 10-A, and 30.4°C for 10-B. Within experimental uncertainty, Δ is thus shown to be constant, as is usually observed for a homologous series of polymers or polymer solutions over sizeable concentration range (Plazek and O'Rourke, 1971; Berry and Fox, 1968).

(iii) Retardation Spectra

Retardation spectra L_p were determined from the $J_p(t)$ curves of Figure 5.10. The results reduced to $T_o = -20°C$ are shown in Figure 5.8, where the principal maximum concentration of retardation mechanisms is shown to be at correspondingly longer times as T_g is increased with additional crosslinking. The logarithmic curves at shorter reduced times showing a slope of 1/3 for all three fluoroelastomers are consequences of the assumed Andrade creep (Plazek, 1960; Plazek and O'Rourke, 1971), $J(t) \equiv J_g + \beta t^{1/3}$ at short times used to determine J_g.

A significant feature to be noted is the presence of the long-time or terminal peak, which grows in size with decreasing crosslink concentration. The sample 10-B, with the highest crosslink density, exhibits but a slight shoulder at long times, whereas Sample 11-A, with about one-third the crosslink density, exhibits a distinct long-time terminal peak that reflects greater viscoelastic losses at longer times presumably arising from adjustments of the entanglement network.

The effect is seen in proper perspective when the characteristics of the different networks are viewed at corresponding-state temperatures (usually temperatures equidistant from T_g). Such a comparison of the retardation spectra is made in Figure 5.14. When the reference temperatures shown in Figure 5.9

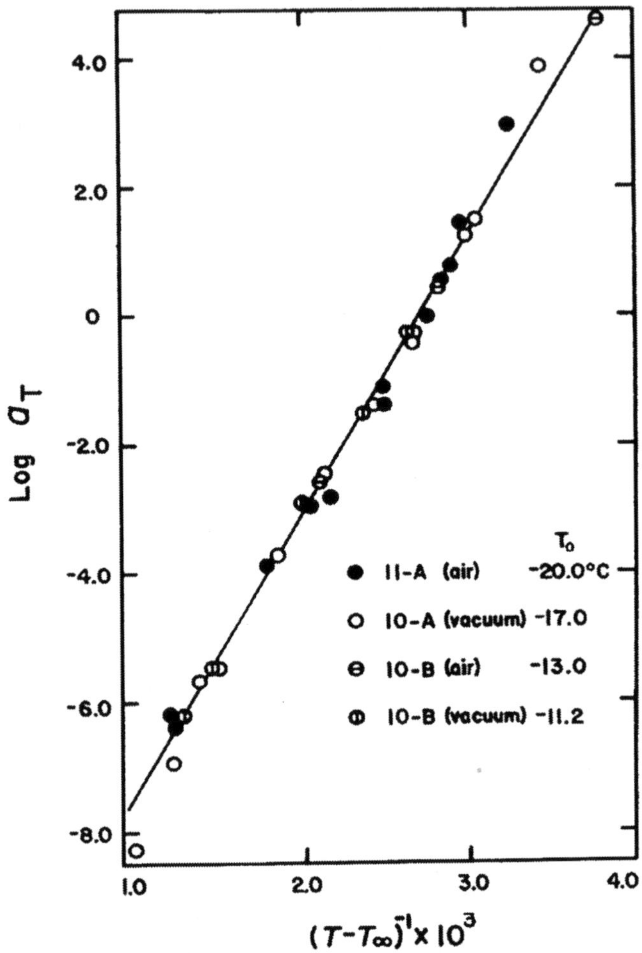

FIGURE 5.7 Logarithmic plot of the time scale shift factors against temperature differences. The atmosphere in which the measurements were made is either in air containing the moisture absorbed under ambient condition or in a rough vacuum that is partially dry. $T_\infty = -57, -54, -50, -48$°C, respectively, for Samples 11-A (air), 10-A (vacuum), 10-B (air), 10-B (vacuum).

are chosen, the difference, $\Delta' = T_o - T_g$, becomes 4.2, 5.3, and 6.1°C for 11-A, 10-A, and 10-B, respectively, and the short-time portions of the spectra coincide rather closely. From the difference in the ordinate of the curves in Figure 5.9, we can infer that at log $(\lambda/a_T) = 10$, Sample 11-A would creep approximately 100 times faster than Sample 10-B.

(iv) Derived Dynamic Mechanical Properties

Once the retardation spectra are known over an extended time scale, it is possible to calculate numerically the components of the complex dynamic shear

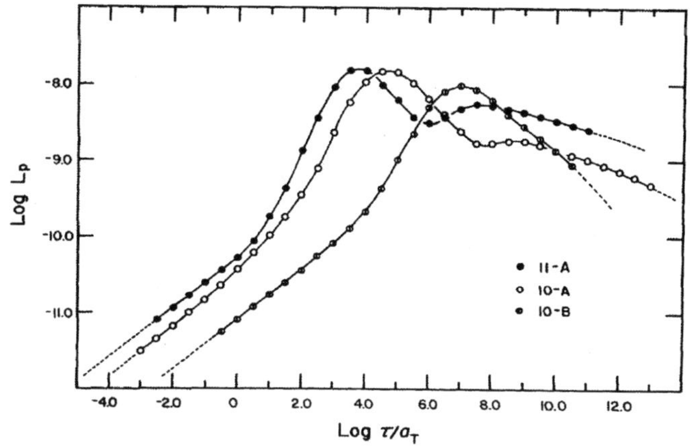

FIGURE 5.8 Logarithmic plot of the retardation spectra as functions of the logarithmic reduced retardation times, λ/a_T (in seconds). The reference temperature of reduction, T_0, is $-20.0°C$ for all curves.

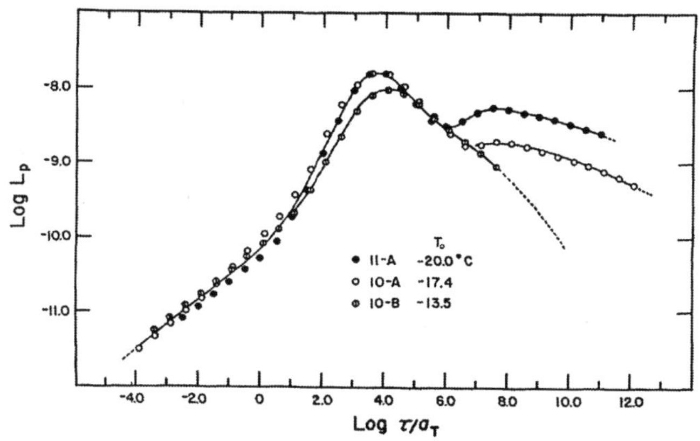

FIGURE 5.9 Logarithmic plot of the retardation spectra against log λ/a_T reduced to the reference temperature indicated for correspondence at short times in the primary softening dispersion.

compliance, $J'_p(\omega) - iJ''_p(\omega)$. With the $J'_p(\omega)$ and $J''_p(\omega)$ values, the storage, $G'_p(\omega)$, and loss, $G''_p(\omega)$, dynamic moduli can be calculated algebraically by Eq. (5.16). The results for $J'_p(\omega)$ and $G''_p(\omega)$ are shown in Figures 5.10 and 5.11, respectively. The reference temperatures chosen in these plots are the same as those for the presentation of the retardation spectra in Figure 5.9 except for a slight (one degree) change for Sample 10-B. The $J'_p(\omega)$ curves can be seen to be the semiquantitative mirror images of the $J_p(t)$ curves. The limiting low frequency value for $J'_p(\omega)$ is the equilibrium compliance J_e, which is the long time limit of the $J_p(t)$ curve for these viscoelastic solids. The small change in shape of the primary softening dispersion is partly a reflection of a significant

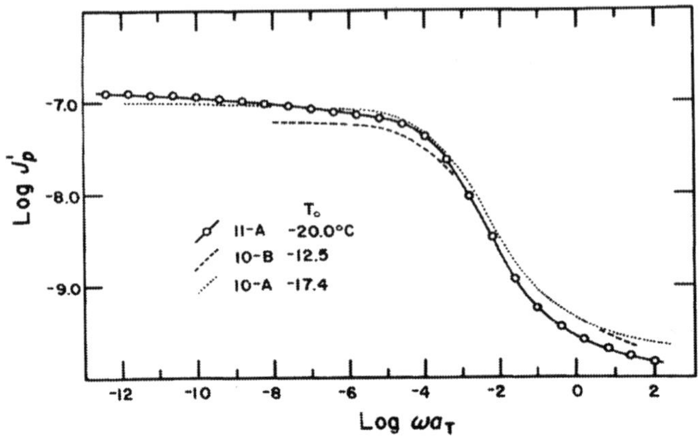

FIGURE 5.10 Logarithmic comparison plot of the reduced dynamic storage compliance, J_p' (in $\mathrm{cm}^2/\mathrm{dyne} = 10\ \mathrm{Pa}^{-1}$), against the logarithm of the reduced frequency $\omega a_T\,(\mathrm{s}^{-1})$. The reduced reference temperatures give correspondence in the softening dispersion and match the loss tangent primary maxima.

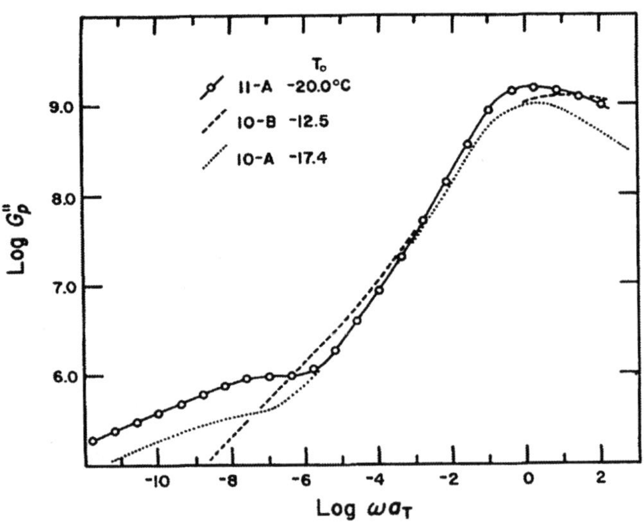

FIGURE 5.11 Logarithmic comparison plot of the reduced dynamic loss modulus, G_p'' (in $\mathrm{dyne/cm}^2 = 0.1\ \mathrm{Pa}$), against the logarithm of the reduced frequency, $\omega a_T\,(\mathrm{s}^{-1})$. The reduced reference temperatures give correspondence in the softening dispersion and match the positions of the loss tangent primary maxima.

effect of the level of crosslinking on the viscoelastic loss in this region of the frequency scale. Since the amount of energy dissipated per cycle of deformation is proportional to G'', the reduced loss modulus $G_p''(\omega)$ is of interest, as well as the loss tangent, which is a measure of the relative energy loss. Figure 5.11 shows that the low frequency losses in the more lightly crosslinked elastomers

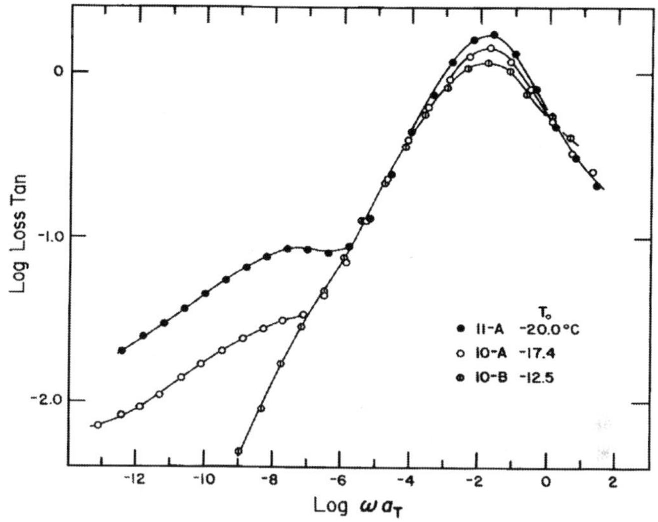

FIGURE 5.12 Logarithmic plot of the loss tangent against the logarithm of the reduced frequency, $\omega a_T\,(s^{-1})$, reduced to reference temperatures that match the positions of the primary maxima.

are significantly greater at corresponding-state temperatures. In the softening dispersion, the three curves superpose roughly, and the distinctions are not clear.

However, the loss tangent curves (Figure 5.12) show that Sample 11-A, with the lowest concentration of network chains, exhibits the largest relative energy dissipation in the primary softening dispersion. It is clear that the relative ranking of energy losses is the same in both the primary softening dispersion and in the low frequency region; that is, Sample 10-B, with the tightest molecular network, shows the least amount of relative energy dissipation.

5.5.2 Urethane-Crosslinked Polybutadiene Elastomers (Plazek et al., 1988)

Three urethane-crosslinked polybutadiene elastomers (TB-1, TB-2, and TB-3) of varying crosslinking levels, along with a similarly crosslinked styrene-butadiene copolymer (HTSBR) and two polybutadiene polymers randomly crosslinked with dicumyl peroxide (PB-1 and PB-2), have been investigated to determine their viscoelastic behavior. Elsewhere, TB-1, TB-2, and TB-3 have been designated as HTPB-1, HTPB-2, and HTPB-3, respectively.

Torsional creep measurements were made on the urethane-crosslinked polybutadiene elastomers at temperatures between −68°C and 25°C. The average molecular weight of a networks chain, M_c, is 3400, 5200, and 8300 for TB-1, TB-2, and TB-3, respectively. The reduced shear creep compliance $J_p(t/a_T)$ curves obtained for the three samples are shown in Figure 5.13. The reference temperatures are chosen to be 7.4°C, 0.0°C, and 17.0°C for TB-1, TB-2, and TB-3, respectively, so that superposition is achieved at shorter times

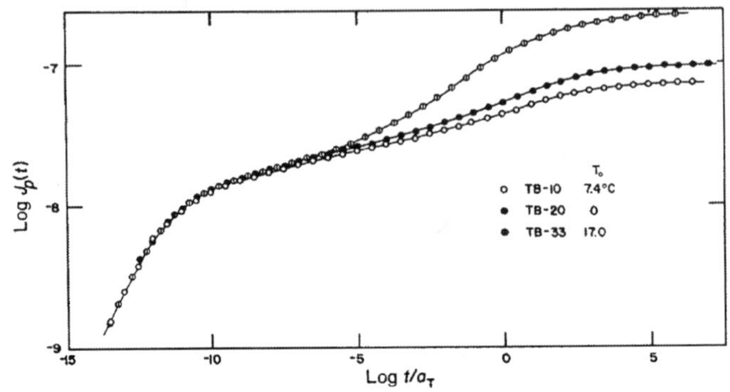

FIGURE 5.13 The logarithm of the reduced shear creep compliance curves, $J_p(t)$ (in Pa^{-1}), for the three urethane-end linked polybutadiene elastomers displayed as a function of the logarithm of the reduced time t/a_T (in seconds). The reference temperatures of reduction are chosen so that superposition is achieved at short times in the primary softening dispersion. (○) TB-1, 74°C, (•) TB-2, 0°C, (⊖) TB-3, 17°C.

in the primary softening dispersion. The most loosely urethane-crosslinked TB-3 has the largest $J_e = 2.5 \times 10^{-6}\ Pa^{-1}$. There is a plateau intermediate between the glassy compliance J_g (not reached in these measurements) and J_e and its level is about $2.0 \times 10^{-7}\ Pa^{-1}$ in all three samples. The network chain density does not affect the form of the time-dependent response up to and including the intermediate plateau in the $J_p(t)$ curves. Only the terminal dispersion (i.e., the approach to J_e) is influenced. The shift factors, a_T, that were used to obtain the reduced $J_p(t/a_T)$ curves surprisingly have the Arrhenius temperature dependence for all three samples, and not the WLF form found in the Epons and the Vitons. This departure to the Arrhenius form is attributed to the nature of the crosslinking units (Plazek et al., 1988).

Reduced retardation spectra, L_p, calculated from the creep compliance $J_p(t/a_T)$ curves shown in Figure 5.18, are presented logarithmically in Figure 5.14 as a function of the logarithm of the reduced retardation time, λ/a_T. The chosen reference temperatures bring the short-time spectral contributions together, showing the similarity between the networks of different chain density. The contributions to the terminal peak do indeed reflect the network topological differences. The more lightly crosslinked networks are markedly more dissipative at relatively longer times, that is, in the time scale region populated by the mechanisms contributing to the long-time or terminal peak in L_p. The objective measure of the width of the plateaus seen in the $J_p(t/a_T)$ curves is the separation of the peaks in $L_p(\lambda/a_T)$. These terminal peaks and their separation from the softening dispersion peak are remarkably large (12–13 decades of time). Usually spectral peaks dominate a half-dozen logarithmic units of the time scale, and the separation of the terminal peak from the softening peak (Berry and Fox, 1968) is a strong function of the network density (Ferry, 1980; Plazek, 1966). Nevertheless, the inverse relationship of viscoelastic

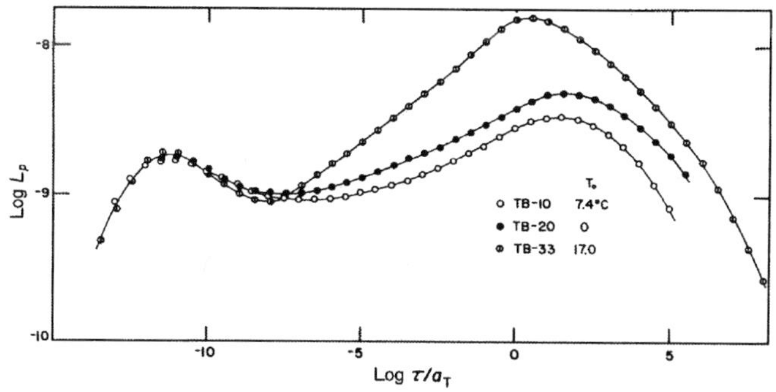

FIGURE 5.14 The logarithm of the reduced retardation spectrum, L_p, shown as function of the logarithm of the reduced retardation time, λ/a_T, for the three urethane-end linked polybutadiene elastomers. (○) TB-1, (•) TB-2, (⊖)TB-3. The response has been reduced to corresponding state temperatures of 74°C, 0°C, and 17°C, respectively, for the primary softening transition.

dissipation at long times to network density as first seen in the response of natural rubber networks (Ferry, 1980) is confirmed phenomenologically in the urethane-crosslinked polybutadiene elastomers, even if their viscoelastic spectra are not completely understood on a molecular level.

5.5.3 Comparisons Between Different Elastomers

The logarithm of the reduced shear creep compliance $J_p(t/a_T)$ of several elastomers is compared at 0°C in Figure 5.15 as a function of the logarithm reduced time scale log (t/a_T). The elastomers include (1) the urethane-crosslinked polybutadiene elastomer, TB-2, described in Section 5.2.2; (2) the randomly crosslinked fluorinated hydrocarbon elastomer Viton 10-A, described in Section 5.2.1; (3) two polybutadiene elastomers, PB-1 and PB-2, randomly crosslinked with 0.8 and 0.3 g of dicumyl peroxide for 100 g of polybutadiene (Plazek et al., 1988); and (4) a styrene-butadiene copolymer (Plazek et al., 1988) similar to the urethane-crosslinked polybutadiene elastomers. However, we shall not review the various viscoelastic functions measured on natural rubbers with varying levels of crosslinking (Plazek, 1966; Thirion and Chasset, 1965; Ferry et al., 1964). Stress relaxation, dynamic mechanical, and creep measurements were made on the same series of natural rubber samples, respectively, by Thirion and Chasset (1965), Ferry et al. (1964), and Plazek (1966).

The equilibrium shear compliances J_e indicated at long times are all about 1.0×10^{-6} Pa^{-1} except for that of the PB-1 sample, which is unexpectedly somewhat higher. The fact that the PB-1 softening dispersion is found at shorter times than that of the PB-2 elastomer has to be attributed to its higher *cis* to *trans* ratio of placements, 0.80 as opposed to 0.67, which reflects a lower T_g. The fact that the rate of creep of the Viton elastomer is approximately five orders of magnitude slower in the softening dispersion than that of the PB-2 is believed to be a reflection of a T_g, which is 70°C higher. What is most unexpected is that

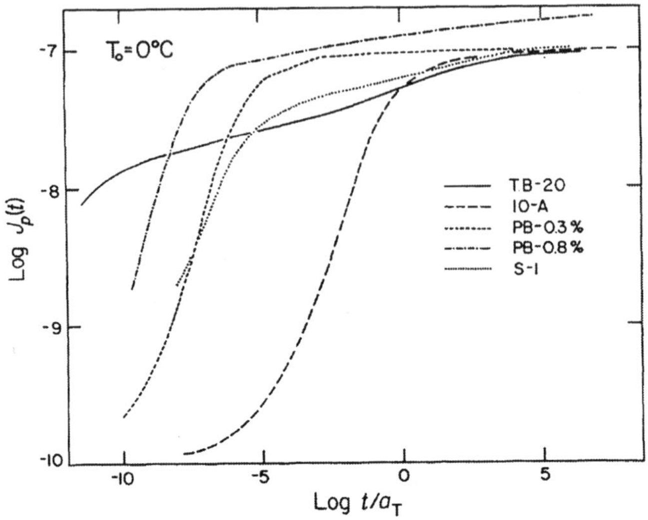

FIGURE 5.15 Double logarithmic plot of the reduce shear creep compliance curves, $J_p(t)$ (in Pa^{-1}), as a function of the logarithm of the reduced retardation time, t/a_T, for five elastomers at $T_0 = 0°C$. Solid line (HTPB-2), long dashed line (Viton), short dashed line (PB-2), dashed-dotted line (PB-1), dotted line (HTSBR).

the softening of dispersion of the urethane-crosslinked elastomers is found at surprising short times; for example, the intermediate plateau of TB-2 is reached near $\log(t/a_T) = -12$ in spite of the fact that its T_g is some 13°C higher than that of the PB-2 material. Yet the latter's intermediate plateau is approached at about $\log(t/a_T) = -6$; some seven orders of magnitude later. Hence the material that is further above its T_g creeps slower in the softening zone.

However, if the creep compliance curves are compared at their respective T_gs, we see in Figure 5.16 that the softening dispersions are, within experimental uncertainty, at the same place in the time scale of response. Specifically the positions of the four $J_p(t/a_T)$ curves at a compliance level of 1.0×10^{-8} Pa^{-1} appear to be spread on a time scale by not much more than one decade of time. Relative uncertainties of T_g values of $\pm 1.5°C$ can account for this spread in positions. Until more precise relative T_gs can be measured we can tentatively surmise that at T_g all polymers at the same rate are deep in the softening zone. This conclusion appears reasonable when we consider that short-range chain dynamics should determine both creep rates just above the glassy level as well as changes in the local liquid structure, the kinetics of which determine T_g.

5.5.4 Other Viscoelastic Measurements

Measurements of linear and nonlinear viscoelastic behavior of elastomers have a long history. Instead of reviewing the works done in the past by various workers

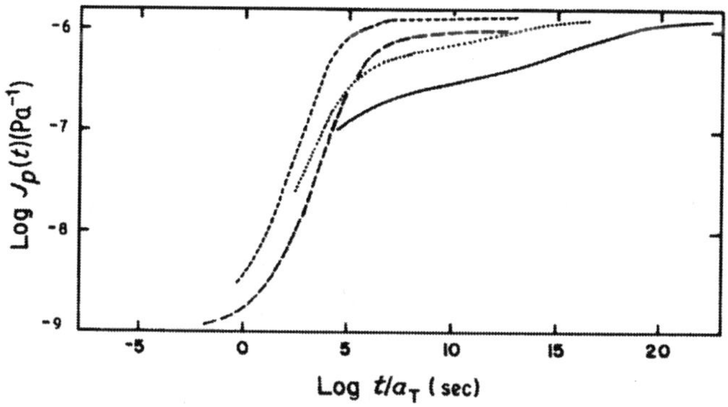

FIGURE 5.16 Double logarithmic plot of the reduced shear creep compliance, $J_p(t)$, as a function of the reduced retardation time, t/a_T, for four elastomers at $T_0 = T_g$. Solid line (HTPB-2), long dashed line (Viton), short dashed line (PB-2), dotted line (HTSBR).

using different techniques, we choose to present the viscoelastic properties of elastomers through what we believe to be accurate measurements on chemically and physically well-characterized elastomers. We hope these sets of data bring out the generic viscoelastic properties of elastomers and their dependence on variables such as the crosslink density and glass temperature. There are remaining differences from one family of elastomers to another, and they should reflect the influence of other variations in molecular structure of the polymer and the crosslinking agent. Nevertheless, the reader should be aware of the previous works on the same subject. Here we can cite only some of them (Plazek, 1966; Thirion and Chasset, 1965; Ferry et al., 1964; Leaderman, 1955; Payne, 1960; Leaderman, 1962; Dicke and Ferry, 1966; Valentine et al., 1968; Payne and Whittaker, 1971; Saunders and Ferry, 1974; Gent, 2004; Isono and Ferry, 1984; Arai and Ferry, 1986a; Arai and Ferry, 1986b).

5.6 THEORETICAL INTERPRETATION OF VISCOELASTIC MECHANISMS AND ANOMALIES

5.6.1 Breakdown of Thermorheological Simplicity of Low Molecular Weight Polymer

Rubber is a viscoelastic solid formed by crosslinking a polymer, which is initially a viscoelastic liquid. In spite of this difference there still are some common issues in understanding the physics of the glass temperature and the viscoelastic mechanisms in the softening dispersion (i.e., called the glass-rubber transition zone in Ferry (1980). A case in point can be taken by comparing the viscoelastic behavior of the neat epoxy resin Epon 1001F (Plazek and

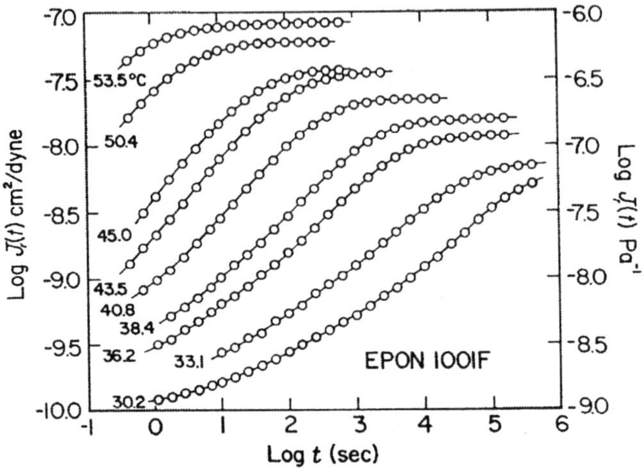

FIGURE 5.17 Logarithmic presentation of the recoverable shear compliance, $J_r(t)$, of Epon 1001F as a function of the logarithm of time t at nine temperatures as indicated. Dramatic loss of long-time viscoelastic mechanisms is evident when temperature is decreased toward T_g.

Chay, 1991) with that of the fully cured network Epon 1001F/DDS discussed in Section 5.5. The glass temperature, T_g, of Epon 1000F is 31°C, which is significantly lower than the value of 127°C of the fully cured Epon 1000F/DDS. The increase in the T_g in going from the oligomer to the rubber can be rationalized (at least in part) by the decrease in the specific volume. However, it is not so easy to understand the difference in the temperature dependence of the softening dispersion of the two materials. As shown previously, the reduced shear compliance curves, $J_p(t)$, measured at different temperatures of Epon 1001F/DDS were reduced to a common curve, $J_p(t/a_T)$, at a reference temperature, T_0. This thermorheological simplicity of the softening dispersion of Epon 1001F/DDS was not found in neat Epon 1001F. The shear viscosity, η, was determined in the temperature range of 30–77°C. The recoverable shear compliance, $J_r(t)$ (defined by Eq. 5.1), was obtained at nine temperatures from 30.2°C to 53.5°C and is presented in Figure 5.17.

The steady-state recoverable shear compliance, $J_s = \lim_{t\to\infty} J_r(t)$, shows a strong temperature dependence. A dramatic 10-fold decrease of J_s is seen in Figure 5.17 to occur with a 20°C decrease in temperature toward T_g. Consequently, the log $J_r(t)$ curves cannot be reduced to a common curve. The thermorheological complexity exhibited by neat Epon 1001F is similar to that found in many other low-molecular-weight polymers (Ngai et al., 1997a; Ngai and Plazek, 1995; Plazek and O'Rourke, 1971; Cochrane et al., 1980; Ngai et al., 1992a; Schönhals and Schlosser, 1993; Plazek, 1996; Plazek et al., 1992; Ngai et al., 1993b; Ngai, 1979; Ngai et al., 1988; Ngai and Rendell, 1990). A good example is the low-molecular-weight polystyrene of 12,300 molecular weight chosen for comparison because J_s at high temperatures is about the same

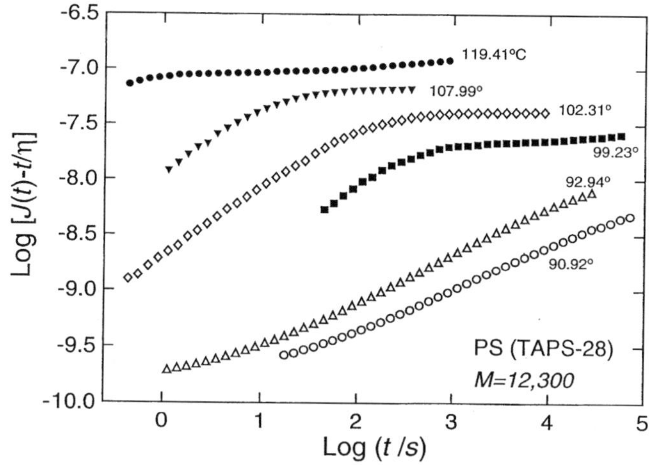

FIGURE 5.18 Logarithmic presentation of the recoverable shear compliance $J_r(t)$ of a nearly monodisperse polystyrene sample with molecular weight of 12,300 as a function of the logarithm of time t at six temperatures as indicated. Dramatic loss of long-time viscoelastic mechanisms is evident when temperature is decreased toward T_g.

as neat Epon 1001F. The $J_r(t)$ data of this nearly monodisperse polystyrene sample are shown in Figure 5.18. The decrease of J_s when temperature is lowered to approach T_g is quantitatively similar in the two cases. This anomalous but general viscoelastic behavior of low-molecular-weight polymers certainly contradicts the assumption that the relaxation times of all relaxation mechanisms or modes simply shift along the logarithmic time scale and are proportional to the monomeric friction factor ζ_o. This assumption, if correct, would mean that the polymer was thermorheologically simple, which is clearly not the case. It should be of fundamental interest in the study of the dynamics of polymers. Unfortunately, possibly because of the difficulty of finding an explanation, it seems to have been ignored in the development of polymer viscoelasticity in the past few decades.

(i) The Coupling Model

As far as we know, at present only one explanation has been given for the breakdown of thermorheological simplicity in low-molecular weight polymers (Ngai et al., 1997a; Ngai and Plazek, 1995; Plazek et al., 1994; Plazek et al., 1992; Ngai et al., 1993b), and that is based on the Coupling Model (Ngai, 1979; Ngai et al., 1988; Ngai and Rendell, 1990; Ngai et al., 1992b; Tsang and Ngai, 1997; Ngai and Tsang, 1999; Ngai and Rendell, 1997; Ngai, 2001). It is beyond the scope of this chapter to review in detail this model and the explanation it provides. A brief description of the essential points is given here. The segmental relaxation responsible for "structural relaxation" and the glass behavior moderated by molecular crowding and interaction is a cooperative process involving complex time-dependent motions of many molecules. Although all local segments attempt configurational rearrangements with primitive rate, $\tau_{0\alpha}^{-1}$, and primitive relaxation function,

$$\phi(t) = \exp(-t/\tau_{0\alpha}), \tag{5.25}$$

not all attempts can be successful due to constraints arising from intermolecular interaction or coupling between the local segments. The initially successful (unsuccessful) ones are appropriately called the fast (slow) local segmental motions. Since the many-molecule dynamics are stochastic, at some later time the fast (slow) motions become slow (fast). Immediate parallel consequences of such complex many-molecule dynamics are: (1) it takes more decades of time for the segmental relaxation to complete its course (time scale stretching) and the characteristic relaxation time of the relaxation, τ_α, is much longer than $\tau_{0\alpha}$; (2) the relaxation process is dynamically heterogeneous, meaning that at any time there are fast and slow moving molecules and they exchange roles over the time scale, τ_α, of the entire segmental (α) relaxation process (Ngai and Rendell, 1990; Schmidt-Rohr and Spiess, 1991). The Kohlrausch's (1847) correlation function,

$$\phi(t) = \exp[-(t/\tau_\alpha)^{1-n}], \tag{5.26}$$

where the exponent, $(1-n)$, is a fraction of unity, and τ_α is a characteristic segmental relaxation time, is perhaps one of the functions with the least number of parameters that stretch the course of the segmental relaxation over a number of decades. The time scale stretching increases with n. Slowing down implies that $\tau_\alpha \gg \tau_{0\alpha}$. A priori the relation between τ_α and $\tau_{0\alpha}$ is unknown. However the Coupling Model provides a relation between these two relaxation times. At sufficiently short times, primitive relaxation prevails and proceeds unimpeded. It is only after some rather well defined time, t_c, that the many-molecule dynamics slow down the relaxation. Hence the correlation function is Eq. (5.25) for $t < t_c$, and crossover to Eq. (5.38) for $t > t_c$, smoothly within a small neighborhood of t_c. The crossover time, t_c, is temperature independent and its value is determined by the intermolecular interaction potential. For polymers, $t_c \approx 1\text{–}2 \times 10^{-12}$ s (Ngai and Rendell, 1997; Ngai, 2001; Colmenero et al., 1992). This crossover is expected on theoretical grounds (Ngai, 1979; Ngai et al., 1992b; Ngai and Rendell, 1990), borne out by rigorous results of simple models (Tsang and Ngai, 1997; Ngai and Tsang, 1999), and supported by experimental data (Ngai, 2001; Colmenero et al., 1992; Ngai, 1994). Taking advantage of this general crossover property of the correlation function and requiring quasi-continuity of its two parts at t_c, we have the relation,

$$\tau_\alpha = [t_c^{-n}\tau_{0\alpha}]^{1/(1-n)}, \tag{5.27}$$

between the two relaxation times via the "coupling parameter" n, which appears in the fractional exponent of the Kohlrausch correlation function (Eq. (5.26)).

This relation has been used to explain the anomalous properties of τ_α, due to the complex many-molecule dynamics, from the corresponding normal and transparent properties $\tau_{0\alpha}$ of a simple independent relaxation (Ngai et al., 1997a; Ngai, 2001; Ngai, 1994).

In polymers having no side-group, the appearance of a secondary relaxation at higher frequencies than the segmental relaxation is intriguing. Examples include polybutadiene (PB) and even polyisoprene (PI). The existence of a secondary relaxation in PB has been well known (Kudlik et al., 1999; Casalini et al., 2001), but in PI it was found only recently (Roland et al., 2004). The dielectric relaxation data of PB and PI showing both the segmental relaxation and the secondary relaxation is presented in Figures 5.19 and 5.20. Equally intriguing is the appearance of secondary relaxation in rigid small molecular glass-formers such as toluene and chlorobenzene (Johari and Goldstein, 1970; Johari, 1976; Ngai and Paluch, 2004). Because there is no internal degree of freedom these secondary relaxations must originate from some local motion of the entire molecule. Such secondary relaxations are of the Johari-Goldstein (JG) kind (Johari and Goldstein, 1970; Johari, 1976; Ngai and Paluch, 2004). They are supposedly universal, existing in all glass-formers, and are considered to be the precursor of the primary structural relaxation. Some criteria for distinguishing JG relaxation from other garden variety of secondary relaxations have been established based on the close relationship of its properties with the primary relaxation (Ngai and Paluch, 2004). The primitive or independent relaxation of the Coupling Model (Eq. (5.37)) is the precursor of the cooperative (i.e., intermolecularly coupled) α-relaxation (Eq. (5.26)) and it entails the motion of all parts of the molecule, but is a local process. Thus, these attributes of the primitive relaxation are shared with the JG relaxation and it can be expected that the independent relaxation time, $\tau_{0\alpha}$, is approximately located near the most probable relaxation time, τ_β, of the JG relaxation at all temperatures, T, and pressures, P. The relation complementary to Eq. (5.27), $\tau_{0\alpha} = (t_c)^n(\tau_\alpha)^{1-n}$, enables $\tau_{0\alpha}$ to be calculated from τ_α and n in the Kohlrausch function (Eq. (5.38)), that fits the time dependence of the α-relaxation. For polymers and small molecular liquids, t_c has the approximate value of 2×10^{-12} s. Remarkably,

$$\tau_{0\alpha}(T,P) \approx \tau_\beta(T,P) \tag{5.28}$$

is found to hold quantitatively in many glass-formers (Ngai and Paluch, 2004; Ngai, 2003; Ngai and Paluch, 2003; Correzi et al., 2002; Ngai and Capaccioli, 2004). The correspondence between the two relaxation times of PB and PI brings the primitive relaxation of the Coupling Model to life.

(ii) Explanation of Thermorheological Complexity

From the properties of the Johari-Goldstein, which is connected to the primitive local segmental dynamics relaxation, we can deduce that the primitive local segmental relaxation time, $\tau_{0\alpha}$, or its friction coefficient depends on the specific

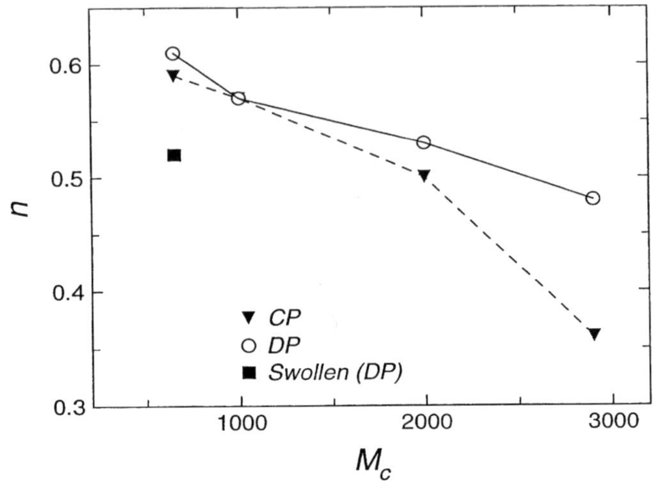

FIGURE 5.19 Plot of the coupling parameter of junction dynamics n_J determined from experimental data by Shi et al. (Shi et al., 1993) for four polymer networks with different molecular weights between crosslinks, M_c. Solid inverted triangles and open circles are from ^{31}P NMR data taken using cross (CP) and direct (DP) polarization, respectively. The solid square is for a swollen sample with $M_c = 650$. The lines are drawn to guide the eyes.

volume, V (or free volume fraction f), and/or entropy, S (or the configurational entropy S_c). Since the primitive relaxation as well as the JG relaxation is a local process involving the entire monomer, their friction coefficient can be identified as the monomeric friction coefficient, ζ_0; that is,

$$\tau_{0\alpha} \propto \zeta_0. \tag{5.29}$$

We hasten to point out that in the literature ζ_0 is commonly associated with the friction coefficient of τ_α and the relaxation times of all other longer time viscoelastic mechanisms (e.g., the entire relaxation or retardation spectrum). This common association of ζ_0 with τ_α is not used in the Coupling Model (CM), which instead identifies ζ_0 as the friction coefficient of $\tau_{0\alpha}$ (Ngai et al., 1997a; Ngai et al., 1987; Ngai et al., 1992a; Plazek et al., 1994; Ngai et al., 1992a; Plazek et al., 1992; Ngai et al., 1993b), although the dependences of ζ_0 on V (f) and/or S (S_c) are retained.

The whole purpose of the CM is to take intermolecular coupling of the local segmental relaxation into account, and in the process transform $\tau_{0\alpha}$ to τ_α. It follows from Eqs. (5.39) and (5.41) that

$$\tau_\alpha \propto [\zeta_0]^{1/(1-n)}. \tag{5.30}$$

The exponent, $1/(1-n)$, is larger than one. Hence τ_α has a stronger dependence on V (f) and/or S (S_c), and on temperature and pressure, than $\tau_{0\alpha}$ or τ_β.

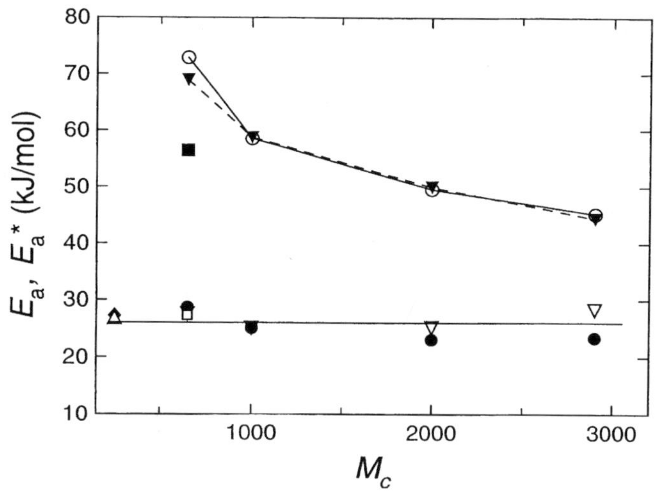

FIGURE 5.20 Solid inverted triangles and open circles are the apparent activation enthalpy, E_a^*, obtained from ^{31}P NMR data taken using cross (CP) and direct (DP) polarization, respectively (Shi et al., 1993). The solid square is E_a^* for a swollen sample with $M_c = 650$. The open inverted triangles and solid circles are E_a calculated by Eq. (5.49) from E_a^* and n_J for the cross (CP) and direct (DP) polarization experiments, respectively. The open square is E_a calculated from E_a^* and n_J for the swollen sample. The solid diamond and open triangle are E_a determined directly from experiment for sample with $M_c = 250$ using cross (CP) and direct (DP) polarization, respectively.

The softening dispersions of entangled low-molecular-weight polymers are often modeled by the Rouse modes modified for undiluted polymers. From their very definition only involving the coordinates of a single chain, the Rouse modes are not intermolecularly coupled, and their relaxation times, τ_{Ri}, are proportional to the monomeric friction coefficient, ζ_0; that is,

$$\tau_{Ri} \propto \zeta_0. \tag{5.31}$$

Thus the relaxation time τ_α of the segmental relaxation has stronger temperature dependence than τ_{Ri} of the Rouse modes in the softening dispersion, albeit the former is shorter than the latter. This difference in temperature dependence immediately explains the breakdown of thermorheological simplicity of low-molecular-weight polymers (Figures 5.17 and 5.18) (Ngai and Plazek, 1995; Plazek and O'Rourke, 1971; Plazek et al., 1994). It also leads to an explanation of the observed decrease of J_s when temperature is decreased toward T_g, which was given in Ngai et al. (1993b).

The disparity between the temperature dependences of τ_α and τ_{Ri} (Eqs. (5.30) and (5.31)) depends on the size of the coupling parameter, n. Polystyrene has a larger n (=0.63) than polyisobutylene (=0.45) (Ngai and Plazek, 1995; Ngai et al., 1996; Ngai et al., 1997b). Hence we expect the degree of breakdown of thermorheological simplicity is lesser in polyisobutylene (PIB) than polystyrene (PS). This expectation, as well as other predicted differences in viscoelastic properties of PIB and PS, was confirmed by experiments (Plazek et al., 1992; Ngai et al., 1993b). When a low-T_g diluent is added to PS, the severity of intermolecular coupling in neat PS is mitigated, and the coupling parameter n is decreased. Consequently, in solutions of PS there is lesser breakdown of thermorheological simplicity, and the decrease of the steady-state compliance, J_s, with falling temperature is suppressed. These changes in viscoelastic behavior from neat PS to solutions of PS were found experimentally (Ngai et al., 1997c), and the viscoelastic spectrum of PS in solution resembles that of PIB (Ngai and Plazek, 2002). Later we shall discuss the dynamics of junctions in crosslinked polymers and their changes upon addition of a diluent (swelling). The junction dynamics are also well described by the Coupling Model, and are characterized by the junction coupling parameter, n_J (Shi et al., 1993; Roland and Ngai, 1991). Decrease of n_J on dilution was found in the experimental data of junction dynamics (Shi et al., 1993; Roland and Ngai, 1991), in analogy to the change in segmental dynamics of PS upon addition of a diluent.

5.6.2 Thermorheological Simplicity of Elastomers

The polymer strands in a network are tied at both ends to crosslink junctions, and obviously their relaxation cannot be modeled as Rouse modes. Furthermore, the crosslink junctions of a network are highly constrained. These constraints give rise to intermolecular coupling of the junction motions. As shown in Section 5.5.4 to follow, the junction dynamics are well described by the Coupling Model with the Kohlrausch stretched exponential correlation function, $\exp[-(t/\tau_J)^{1-n_J}]$, with a sizeable coupling parameter, n_J (Shi et al., 1993; Roland and Ngai, 1991). τ_J is the slowed down junction relaxation time caused by intermolecularly coupling. In analogy to Eq. (5.27) for segmental relaxation, τ_J is related to its primitive junction relaxation time, τ_{0J}, by

$$\tau_J = [t_c^{-n_J} \tau_{0J}]^{1/(1-n_J)}. \tag{5.32}$$

In the Coupling Model all primitive relaxation times, including $\tau_{0\alpha}$ and τ_{0J}, are proportional to the monomeric friction coefficient, ζ_0. Hence from Eq. (5.32) the temperature dependence of τ_J depends on n_J and is given by

$$\tau_J \propto [\zeta_0]^{1/(1-n_J)}. \tag{5.33}$$

Since the junctions are intermolecularly coupled and have significantly large n_J, so is the longer length scale modes, i, of the polymer strands anchored on both ends to the junctions, which now have nonzero coupling parameters, n_i, and obviously they cannot be modeled by Rouse modes. The temperature dependences of their relaxation times, $\tau_i \propto [\zeta_0]^{1/(1-n_i)}$, are no longer that different from τ_α because n_i is comparable to n_J and n. Thus thermorheological complexity of the viscoelastic response of crosslinked networks is minimized and cannot be detected by time-temperature super-positioning.

Intuitively, junctions are expected to be more effective in constraining viscoelastic mechanisms having larger length scales (i.e., longer wavelengths and retardation times). The largest effect should occur for the mode whose length scale corresponds to the molecular weight between crosslinks, M_x, which is responsible for the equilibrium compliance of the network.

5.6.3 Changes of the Segmental Relaxation Time and the Johari-Goldstein Relaxation Time with Crosslink Density

We have pointed out at the end of Section 5.7 that the JG relaxation was found in the uncrosslinked Epon 828 and its relaxation time, τ_β, is in good agreement with $\tau_{0\alpha}$. We have also discussed that the coupling parameters, n, of the segmental relaxation of the repeat units near the junctions are comparable to the coupling parameter, n_J, of the junctions (see the following subsection). As a result there is an increase of the average n of the segmental relaxation with the increase of crosslink density. From Eq. (5.27), such increase of n gives rise to concomitant increase of the segmental relaxation time, τ_α, and hence the glass temperature, T_g, which experimentally is indeed the case. On the other hand, $\tau_{0\alpha}$ is unchanged if the crosslink density is not too high to have an effect on the specific volume and configurational entropy. We expect then from $\tau_{0\alpha}(T) \approx \tau_\beta(T)$ (Eq. (5.28)) that, on crosslinking Epon 828, the JG relaxation time should not differ much from that found in neat Epon 828, while τ_α and the glass temperature increase significantly. Isochronal (Parthun and Johari, 1995; Alig and Johari, 1993) and isothermal (Beiner and Ngai, 2005) dielectric relaxation data of Epon 828 have found that indeed τ_β changes little and has about the same Arrhenius temperature dependence, while T_g increases significantly with time of curing. At T_g, τ_α is the same for all samples by definition, but the separation distance, $\log[\tau_\alpha(T_g)] - \log[\tau_\beta(T_g)]$, increases with curing time (Parthun and Johari, 1995; Alig and Johari, 1993; Beiner and Ngai, 2005). This trend is predicted by the Coupling Model through Eq. (5.27), which shows the increase of $\log[\tau_\alpha(T_g)] - \log[\tau_{0\alpha}(T_g)]$ with n caused by increasing crosslink density, and $\tau_{0\alpha}(T_g) \approx \tau_\beta(T_g)$ (Eq. (5.28)).

5.6.4 Junction Dynamics

(i) Experimental Data

The relaxation dynamics of junctions in polymer networks have not been well known until the advent of solid-state ^{31}P NMR spin-lattice relaxation measurements in a series of poly(tetrahydrofuran) networks with tris(4-isocyanatophenyl)-thiophosphate junctions (Shi et al., 1993). The junction relaxation properties were studied in networks with molecular weights between crosslinks, M_c, ranging from 250 to 2900. The dominant mechanism for ^{31}P nuclear spin lattice relaxation times measured over a wide range of temperatures were fit satisfactorily by spectral density functions, $\hat{J}(\omega)$, derived from the Fourier transforms of the Kohlrausch stretched exponential correlation functions,

$$C_J(t) = \exp[-(t/\tau_J)^{1-n_J}], \tag{5.34}$$

where τ_J and n_J are the junction correlation time and coupling parameter, respectively. In the temperature range of measurements high above T_g, τ_J was appropriately assumed to have an Arrhenius temperature dependence of

$$\tau_J = \tau_\infty^* \exp(E_a^*/RT). \tag{5.35}$$

From these fits Shi et al. obtained the coupling parameter, n_J, the preexponential factor, τ_∞^*, and the activation enthalpy, E_a^*, for the series of networks of different crosslink densities as well as a swollen sample.

(ii) Coupling Model Explanation (Roland and Ngai 1991; Ngai et al. 1993a)

n_J was found to increase with crosslink density and decrease with the addition of a solvent at constant crosslink density (Shi et al., 1993). These trends of n_J are in accord with the interpretation of n_J as the coupling parameter because of the increase of intermolecular constraints with the density of crosslinks. There is a significant increase of the activation enthalpy, E_a^*, with higher crosslink density, which is compensated by a concomitant decrease of the preexponential factor, τ_∞^*. On the other hand, on swelling the polymer network with $M_c = 650$, E_a^* decreases while τ_∞^* increases. The experimental data on n_J, E_a^*, and τ_∞^* of the networks are summarily shown in Figures 5.19–5.21.

The Arrhenius temperature dependence of τ_J implies the same for the primitive relaxation time of the junction, τ_{0J}, which is written out explicitly as

$$\tau_{0J} = \tau_\infty \exp(E_a/RT). \tag{5.36}$$

The relation between τ_J and τ_{0J}, the counterpart of Eq. (5.27), has been given by Eq. (5.32). The latter spawns two separate relations,

$$E_a^* = E_a/(1-n_J) \tag{5.37}$$

and

$$\tau_\infty^* = [t_c^{-n_J}\tau_\infty]^{1/(1-n_J)} = \tau_\infty[\tau_\infty/t_c]^{n_J/(1-n_J)}. \tag{5.38}$$

These expressions qualitatively explain the experimentally observed increase of E_a^* and decrease of τ_∞^* with crosslink density through the increase of n_J. Quantitatively we can calculate E_a and τ_∞ by Eqs. (5.37) and (5.38), respectively, from the experimental values of n_J, E_a^*, and τ_∞^* for each network, and $t_c = 2 \times 10^{-12}$ s.[1] The calculated values of E_a and τ_∞, shown in

[1] At the time of writing the reference, the value of t_c was not determined directly by experiment, and values about 10 times greater were used.

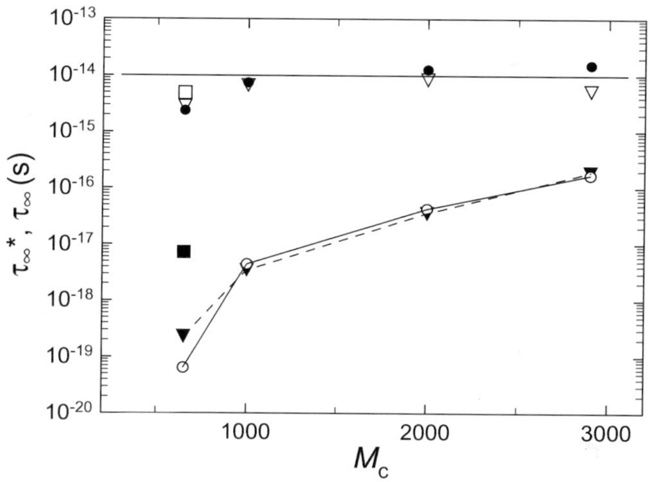

FIGURE 5.21 Solid inverted triangle and open circles are the apparent preexponential factor, τ_∞^*, determined by Shi et al. (1993) from their experimental data for four polymer networks with different molecular weights between crosslinks, M_c, using cross (CP) and direct (DP) polarization, respectively. The solid square is τ_∞^* for a swollen sample with $M_c = 650$. The open inverted triangles and solid circles are τ_∞ calculated by Eq. (5.50) from τ_∞^* and n_J for the cross (CP) and direct (DP) polarization experiments, respectively, with $t_c = 2 \times 10^{-12}$s. The open square is τ_∞ calculated from τ_∞^* and n_J for the swollen sample.

Figures 5.20 and 5.21, respectively, are roughly constant ($E_a \approx 27$ kJ/mol and $\tau_\infty \approx 10^{-14}$ s), indicating that all the crosslink junctions have about the same mobility had their motion not been slowed down by intermolecular coupling to various degrees depending on the crosslink density. The nearly constant $E_a \approx 27$ kJ/mol can be interpreted as some barrier enthalpy and $\tau_\infty \approx 10^{-14}$ s as the reciprocal of some reasonable attempt frequency.

(iii) Similarity of Flory's Constrained Junction Model for Elasticity to the Coupling Model for Junction Dynamics

The application of the Coupling Model to junction dynamics turns out to be similar in spirit as the application of the constrained junction model of Flory (1979a,1985) and Flory and Erman (1982) to the elasticity of networks. Calculation of the stress-strain relationship for a real network requires analysis of the response of a given chain to the imposition of a bulk deformation, described classically by two extremes, the affine (Treloar, 1973) and the phantom (Treloar, 1975; James and Guth, 1947; James and Guth, 1953) model. Real networks exhibit a strain dependence of their elastic stress that is at variance with the predictions of either the affine or the phantom network models. This is not surprising since the junctions in a real polymer network fluctuate away from positions corresponding to affine displacement, while interference from neighboring chains reduces the magnitude of such fluctuations from that available to a phantom network (Flory, 1976; Ronca and Allegra, 1975). In the constrained junction model of Flory and coworkers, the fluctuation of the junctions is limited to a domain of constraints imposed by steric hindrances from neighboring segments, with the range and position of these domains changing with deformation. Flory introduced the parameter κ, defined in terms of the number of junctions in the volume occupied by a network chain, as a measure of the severity of the local constraints relative to those imposed by the phantom network. The contribution to the force from the constraints is zero for $\kappa = 0$. Thus the same arguments given by Flory for the importance of considering local constraints on junctions (as caused by interactions with neighboring chains) in his treatment of elasticity of polymer networks justify the need of using models, such as the Coupling Model, which explicitly considers the effect on the dynamics of these constraints when describing junction relaxation.

Flory's constraint junction model and the Coupling Model are concerned with the effects of constraints on junctions in polymer networks. They address different manifestations of constraints on network junctions imposed by surrounding chains. Flory was concerned with the consequent restriction on the configurations available to the network, which affects its elastic energy. The Coupling Model focuses on the manner in which intermolecular constraints slow down the motion of the network junctions and cause the modification of the correlation function of the junctions for a perfect network of phantom chains, $C_{\text{ph}}(t)$. From the idea that phantom chains can pass one another, we expect that $C_{\text{ph}}(t)$ has an exponential time dependence, $\exp(-t/\tau_{J0})$. Following the general physical principle behind the Coupling Model, the constraints on junctions will modify $C_{\text{ph}}(t)$ to $C_J(t)$ as given by Eq. (5.34) at times after $t_c \approx 2 \times 10^{-12}$ s.

In the Coupling Model, the magnitude of n increases with the severity of the constraints relative to those imposed by the phantom network. Thus, n bears a similarity to the quantity κ in the constraint junction model. The elastic stress in the constrained junction model, f_c, differs from that for phantom chains, f_{ph}, in a perfect network (no dangling ends). The ratio f_c/f_{ph} is a measure of the contribution to the stress from local constraints on junction fluctuations to that for phantom chains. Similarly, in the Coupling Model we may use the logarithm of the ratio $\log(\tau_\infty/\tau_\infty^*)$ and the ratio E_a^*/E_a as gauges of modification of the junction relaxation by constraints. Typically τ_∞ is of the order of 10^{-14} s (see also Figure 5.21). This, together with $t_c \approx 2 \times 10^{-12}$ s, indicates that the ratio τ_∞/t_c is much less than unity. Hence from Eqs. (5.37) and (5.38), both $\log(\tau_\infty/\tau_\infty^*)$ and the difference $(E_a^*/E_a - 1)$ increase with n or the severity of constraints, and both quantities vanish at $n = 0$ (corresponding to a phantom network). These dependences on n are analogous to the dependence of the quantity f_c in Flory's model.

The proffered analogy is supported by an examination of the dependences on crosslink density, diluent concentration, crosslink functionality, and macroscopic strain on the junction constraints of the Flory model with n,

$\log(\tau_\infty/\tau_\infty^*)$, and $(E_a^*/E_a - 1)$ in the Coupling Model. The results of these comparisons of the Flory model and the Coupling Model are summarized in Table 5.2.

5.7 COMPONENT DYNAMICS OF HIGHLY ASYMMETRIC POLYMER BLENDS

5.7.1 Intermolecularly Coupled Segmental Relaxation and Interchain Coupled Chain Dynamics in Highly Asymmetric Polymer Blends

The study of both the chain dynamics and the segmental dynamics of a component in miscible blends of two polymers is worthwhile because there are changes in the properties not seen before due to the presence of the other component. The changes are not only interesting and challenging for explanation, but also they possibly can be used as critical tests of theories or models of polymer chain dynamics and segmental dynamics proposed for polymers. The changes on both components are expected to be very significant if the two components have a large difference in mobility at the same temperature, a situation realized when there is a large difference in their T_gs. The substantial changes of properties, and particularly the anomalous ones, offer excellent opportunity for discovering new physics of polymer dynamics that cannot be extracted from experimental data of neat polymers. This is the spirit in which we and collaborators have followed in the study of component dynamics in polymer blends, starting from the initial attempts (Roland and Ngai, 1991; Ngai and Roland, 2004a Ngai and Roland, 2004b), and in mixtures of small molecular glass-forming liquids (Capaccioli et al., 2012; Mierzwa et al., 2008; Kessairi et al., 2008; Ngai, 2011), and aqueous mixture (Ngai, 2011; Ngai et al., 2011). We emphasize the importance of these observed changes of dynamic properties, especially the anomalous ones. On the other hand, the focus of other workers in component dynamics of polymer blends is on the explanation and prediction of the changes of the glass transition temperature (Lodge and McLeish, 2000; Lodge et al., 2006; Zhao et al., 2009; Chen et al., 2008; Arrese-Igor et al., 2010), the segmental (Lutz et al., 2003; Lartigue et al., 1997; Zhao et al., 2008; Jin et al., 2004; Sakai et al., 2004; Maranas, 2007) and the terminal relaxation times (Niedzwiedz et al., 2007; Brodeck et al., 2010; Diddens et al., 2011; Ngai and Wang, 2011; Arrese-Igor et al., 2011; Moreno and Colmenero, 2008; Arrese-Igor et al., 2012), and the monomeric friction factor of chain diffusion (Chen et al., 2008; Haley and Lodge, 2005; Zeroni et al., 2008) of the components in polymer blends.

The essential difference of polymer blends from mixtures of van der Waals glass-forming liquids is the chain connectivity of each component. This was correctly pointed out by Lodge and McLeish (LM) (Lodge and McLeish, 2000).

TABLE 5.2 Similarity Between Flory's Constrained Junction Model for Elasticity and the Coupling Model for Junction Dynamics

Network Feature	Anticipated Effect	Flory Model	Coupling Model			^{31}P NMR Experiment		
		f_c	n	$E_a^*/E_a - 1$	$\log(\tau_\infty/\tau_\infty^*)$	n	$E_a^*/E_a - 1$	$\log(\tau_\infty/\tau_\infty^*)$
Higher crosslink density	More firmly embedded junctions	*Higher*	*Higher*	*Higher*	*Higher*	*Higher*	*Higher*	*Higher*
Diluent	Reduced severity of constraints	*Lower*	*Lower*	*Lower*	*Lower*	*Lower*	*Lower*	*Lower*
Higher crosslink functionality	More constrained junctions	*Higher*	*Higher*	*Higher*	*Higher*			
Extension	Alleviation of constraints	*Lower*	*Lower*	*Lower*	*Lower*			

They went on to suggest the local composition is altered from the average value, ϕ, to an effective concentration, $\phi_{\text{eff}} = \phi_{\text{self}} + (1 - \phi_{\text{self}})\phi$, where ϕ_{self} accounts for the excess of segments due to the chain connectivity. The modified Fox equation, $T_g(\phi) = [\phi_{\text{eff}}/T_{g1} + (1 - \phi_{\text{eff}})/T_{g2}]^{-1}$, is then used to obtain $T_g(\phi)$ of the component. The segmental relaxation time in the blend, $\tau_{seg}(\phi)$, is calculated by modifying the Vogel-Fulcher-Tammann-Hesse (VFTH) equation obtained for the component when it is in its neat state, $\tau_{\text{seg}} = \tau_\infty \exp[B/(T - T_0)]$, by only replacing T_0 by $T_0(\phi)$. Here $T_0(\phi) = T_0 + [T_g(\phi) - T_g]$. In other words, the difference between $T_0(\phi)$ of component in the blend and T_0 of the neat polymer is exactly the same as the difference between $T_g(\phi)$ and T_g of the neat polymer. The remaining VFTH parameters, B and τ_∞, are assumed not to have changed. Thus, in the LM model, the component segmental relaxation time is given by $\tau_{\text{seg}}(\phi) = \tau_\infty \exp[B/(T - T_0(\phi))]$.

Lodge and McLeish suggested that ϕ_{self} can be calculated from the Kuhn length of the component. However, the ϕ_{self} calculated does not work in many cases, and it has to be treated as an adjustable parameter to yield agreement with experimental results. In some polymer blends, the application of the LM model leads to a self-concentration value significantly larger (Zhao et al., 2009; Arrese-Igor et al., 2010; Lutz et al., 2003) or anomalously much smaller (Zhao et al., 2009) than anticipated by the model. In some cases, the LM model was declared unable to predict the observed behavior of either component (Zeroni et al., 2008). Bedrov and Smith (2005) performed molecular dynamics simulations of model miscible polymer blends consisting of chemically realistic 1,4-polybutadiene (CR-PBD) as the slow component (higher T_g) and PBD chains with reduced dihedral barriers as the fast component (LB-PBD) with lower T_g. They found that only $\phi_{\text{self}} = 0.0$ provides a better description of the α-relaxation times for the LB-PBD component in the 10% LB-PBD blend with CR-PBD, indicating within the LM formulation that the segmental relaxation of the fast component is determined entirely by the bulk composition of the blend. This is contrary to the spirit of the LM model, which assumes that the local, dynamically relevant environment for a segment differs in composition from the bulk blend due to chain connectivity. The situation of the 50% LB-PBD blend is worse; Bedrov and Smith found even for the best with $\phi_{\text{self}} = 0.0$, the Lodge-McLeish model prediction still underestimates the slowing down of the α-relaxation of the LB-PBD component. Roland et al. (2005) applied their data of the PVME/P2CS blends, and found that the modified Fox equation gives $\phi_{\text{self}} = 0.437$ for the P2CS *less than* the actual concentration. The result contradicts the enrichment due to the intramolecularly bonded neighboring segments. These problems of the LM model are found in many other studies not cited here. Despite the failures, it is not abandoned, and continues to be a central theme in considering the component dynamics of polymer blends.

We should not lose sight of the following facts on the LM model. First, the concept of self-concentration in the LM model does not apply to mixtures of small molecular glass-formers, and yet the component dynamics are similar to polymer blends. Second, ad hoc assumptions are made in arriving at the results of the model; for example, the VFTH parameters, B and τ_∞, of the polymer in the blend are the same as those of the neat polymer, and the arbitrary formula used for $T_0(\phi)$. Third, the use of the LM model is limited to fitting the mean segmental relaxation time of a component in the blend, and hence cannot address other properties of the component dynamics. Fourth, the only link of the LM model to the glass transition problem is the empirical Fox equation on which it is based, and therefore it offers no help to understand the more fundamental problem of glass transition in the simpler case of a pure polymer. These shortcomings of the LM model do not seem to deter other workers on polymer blends to apply the model. This is because some experimentalists would like to make comparison of their measurements of mean segmental relaxation time and T_g of the components with theoretical predictions, and the LM model provides the predictions from the known parameters of the pure polymer and one adjustable parameter. However, the physics and the factors determining the segmental relaxation and T_g quantitatively are not fully understood even for the pure polymer, otherwise the problem of glass transition in neat polymers would have been declared solved, which is not the case by popular consent (Chang, 2008). Thus, based on an empirical relaxation with added assumption and the frequent use of a fitting parameter, the LM model does not shed any light into the fundamental understanding of polymer blend dynamics. Segment self-concentration is certainly a relevant consideration for the effective average concentration of each polymer component. However, the majority of current papers published in the literature on polymer blend dynamics are making the LM the center of attention and applying it solely to explain the component segmental relaxation time and T_g as the principal goal. We think this is a dangerous trend in the research on polymer blend dynamics because several intriguing component dynamic properties, which may be the key for gaining a fundamental understanding and which have an impact on the fundamentals of polymer segmental and terminal chain dynamics, are left by the wayside.

The purpose of this chapter is to bring back into focus and for open discussion the other segmental and chain dynamic properties of components in polymer blends, particularly those found recently in polymer blends with large difference in T_g of the two components, which we consider important but not sufficiently emphasized in the current literature. The properties are new and not found in any homopolymer, suggesting novel physics are involved. An abbreviated review of these properties given in the next section is necessary because they have never been collected in one place and considered altogether by any published work so far. Either some of the properties reviewed have not been explained by anyone, or previous explanations by others are shown to be invalid. We accentuate this unfavorable situation in the current research field by formulating these unresolved problems as a number of outstanding properties. The task of explaining completely this collection of properties is challenging, but is absolutely necessary if the objective is to fully solve the problems. In Section 5.2.3, we offer an explanation for all the properties by the Coupling Model (CM). Strong support of the premise of the CM in application to component dynamics in highly asymmetric polymer blends for segmental α-relaxation is given by quasielastic neutron scattering experiments by Sakai et al. (2004), Garcıa Sakai et al. (2008), and Garcıa Sakai et al. (2005), and for unentangled chain dynamics by quasieleastic neutron scattering by Niedzwiedz et al. (2007), and molecular dynamics simulations by Brodeck et al. (2010) and Diddens et al. (2011) and elaborated in Ngai and Wang (2011). By no means do we claim that the Coupling Model is the only way to explain these properties. On the contrary, we encourage others to seek possibly better, alternative explanations. In our opinion, leaving all these outstanding properties unexplained is not an option if the goal is to truly solve the problem of component dynamics in highly asymmetric polymer blends. The conclusion is presented in the final section.

5.7.2 Anomalous Component Dynamics of Polymer Blends

As stated in the Introduction, the anomalous properties of component dynamics in polymer blends are our emphasis, and they are considered one by one in the following subsections. These anomalous properties mostly were found experimentally in polymer blends with a large difference in T_g of the two components, henceforth referred to as highly asymmetric polymer blends (HAPB). This is unsurprising because the larger changes of component dynamics in HAPB make the anomalous properties more evident and demanding for explanation. In the following subsections, the anomalous properties of component segmental and global chain dynamics in polymer blends are brought out. To make them stand out, their descriptions are put into separate paragraphs and are numbered for easy reference in Section 5.2.3, where they are explained theoretically.

(i) Segmental and Global Chain Dynamics of PEO in Blends with PMMA

(i).1 Background of Property (i)

The segmental dynamics of poly(ethylene oxide) (PEO) in blends with poly(methyl methacrylate) (PMMA) have received much attention because they are highly asymmetric polymer blends (HAPB) (Lutz et al., 2003; Lartigue et al., 1997; Lodge et al., 2006; Zeroni et al., 2008; Mpoukouvalas and Floudas, 2008; Goulart Silva et al., 2000). Puzzling properties exhibited by the PEO component were first found by Lutz et al. (2003) in deuteron NMR experiments at high frequencies from 31 MHz to 76 MHz corresponding to times of order of nanoseconds. The blends are composed of deuterated PEO (d_4PEO) with $M_w = 1.25 \times 10^5$ g/mol and PMMA with $M_w = 1.06 \times 10^5$ g/mol. We formulate these puzzling properties as Property (i) as follows.

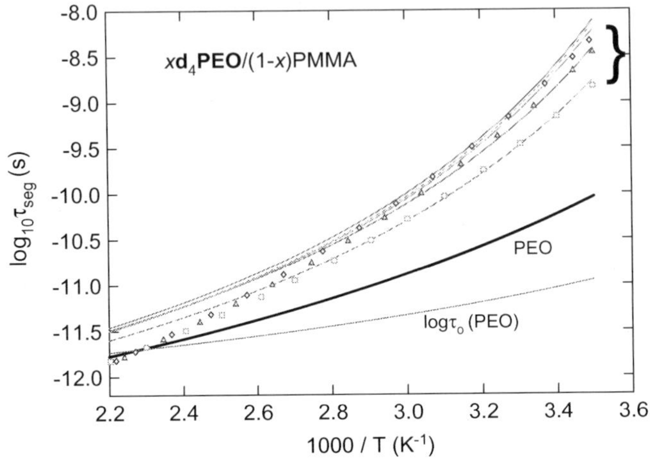

FIGURE 5.22 The segmental relaxation times for PEO neat (bold solid line) and in blends with PMMA (dashed lines) containing 3% to 30% PEO (from top to bottom) from deuteron NMR measurements. Fits to the data by the VFT law given by Lutz et al. (2003) are shown but not the data themselves. Also shown is the independent relaxation time for PEO (dotted line, using $n = 0.5$), which lies close to the characteristic time, $t_c = 2$ ps. The most probable relaxation times calculated for a number of temperatures by the CM equation for concentrations of PEO at 3% (diamonds, $\hat{n} = 0.76$), 10% (triangles, $\hat{n} = 0.75$), and 30% (squares, $\hat{n} = 0.715$), respectively.

(i).2 Property (i)

High frequency deuteron NMR experiments have found that segmental relaxation times, $\tau_{\alpha f}$, of the fast PEO component in the range, $-8.5 \geqslant \log(\tau_{\alpha f}/\mathrm{s}) \geqslant -11.5$, are nearly independent of composition of the blends from 3% to 30% d4PEO over the temperature regime studied, and thus surprisingly $\tau_{\alpha f}$ is hardly influenced by the presence of the PMMA (Lutz et al., 2003; Lartigue et al., 1997). For example, $\tau_{\alpha f}$ of 0.5% and 3% PEO dissolved at near tracer compositions in a PMMA matrix are retarded by less than one order of magnitude than pure PEO samples. The result holds for a wide range of temperatures extending to well below the glass transition of the PMMA matrix, where the segmental relaxation times of PMMA are about 12 orders of magnitude longer than $\tau_{\alpha f}$ of PEO. The deuteron NMR data of $\tau_{\alpha f}$ of the PEO component in blends from 3% to 30% d4PEO are shown in Figure 5.22, together with the neutron scattering data in the main figure and in the inset, which will be introduced and discussed later.

Another deuteron NMR study of $\tau_{\alpha f}$ of PEO in blends with PVAc shows the same behavior (Zhao et al., 2008). Throughout the range $-9 \geqslant \log(\tau_{\alpha f}/s) \geqslant -11.8$, the $\tau_{\alpha f}$ of 2%, and 50% PEO differ by about half a decade. The difference between $\tau_{\alpha f}$ of 2% and 100% PEO increases with the increase of $\tau_{\alpha f}$, but no more than about a decade in the same range. The effect seems general.

The following contrasting experimental facts also need to be explained. The anomalous behavior of $\tau_{\alpha f}$ of component PEO determined in the range, $-8.5 \leqslant \log \tau_{\alpha f} \leqslant -11.5$, by deuteron NMR is not found in $\tau_{\alpha f}$ of the fast component in other HAPB when probed by other techniques at much lower frequencies than deuteron NMR. Examples include dielectric relaxation measurements of $\tau_{\alpha f}$ of other HAPB at much longer times, say in the range $-6 \leqslant \log \tau_{\alpha f} \leqslant 2$ as shown in Ngai and Roland (2004a, 2004b). The T_g of the PEO component in blends with PMMA detected by DSC changes significantly with composition on decreasing concentration of PEO down to 20% PEO (Ngai et al., 2011; Mpoukouvalas and Floudas, 2008; Goulart Silva et al., 2000). In another HAPB of PVME with PS, by using thermally stimulated depolarization current technique (Leroy et al., 2002), the T_g of the PVME component shows pronounced increase on lowering PVME concentration down to 20%. Both dielectric relaxation at low frequencies and DSC have found high sensitivity of $\tau_{\alpha f}$ of PI to change in composition in blends with PtBS, another HAPB, down to 20% PI (Arrese-Igor et al., 2010). In these measurements involving long relaxation time of the fast component in the HAPB mentioned earlier, unlike deuteron NMR measurements at high frequencies, large increase of $\tau_{\alpha f}$ together with stronger temperature dependence compared with the pure fast component was observed, and the effects increase with increasing presence of the slow component (data shown later in Section 5.2.3). The very different results obtained at short time scales by deuteron NMR need explanation. Later we show the same anomalous behavior of PEO in blends with PMMA by quasielastic neutron scattering (QENS) measurements as deuteron NMR. The time scales of QENS are comparable to deuteron NMR.

(i).3 Background of Property (ii)

Studies have been made on the terminal relaxation, chain diffusion, and chain normal modes of the fast component in HAPB by simulations (Brodeck et al., 2010; Diddens et al., 2011; Moreno and Colmenero, 2008), neutron scattering (Niedzwiedz et al., 2007), and dielectric relaxation experiments (Arrese-Igor et al., 2012). Several remarkable properties of the chain dynamics of the fast component have been found by these studies.

An example from recent works is the study of the dynamics of unentangled PEO chains in 35% PEO/65% PMMA by quasielastic neutron scattering (Niedzwiedz et al., 2007), and in 25% PEO/75% PMMA by molecular dynamics simulations (Brodeck et al., 2010). The Rouse model has the mean-square displacement, $\langle r^2(t)\rangle$, of a chain segment, which increases proportionally to the square root of time according to

$$< r^2(t) >= 2(W(T)l^4 t/\pi)^{0.5}, \tag{5.39}$$

where $W(T) = 3k_B T/\zeta(T)l^2$ is the elementary Rouse relaxation rate, $\zeta(T)$ is the friction coefficient, and l the segment length (Niedzwiedz et al., 2007;

Brodeck et al., 2010). In the Gaussian approximation for displacements, the segmental self-correlation function relates directly to $\langle r^2(t)\rangle$, resulting in the intermediate self-dynamic structure factor having the Gaussian form (Niedzwiedz et al., 2007),

$$S(Q,t) = \exp\left[- Q^2 < r^2(t) > /6\right] = \exp\left[- (Q^2 W^{0.5} l^4 \pi^{-0.5}/3) t^{0.5}\right]. \tag{5.40}$$

The time dependence is the stretched exponential function of time, $\exp[-(t/\tau_R)^\beta]$, with $\beta = 0.5$. The experiments performed at the backscattering instruments by Niedzwiedz et al. (2007) are capable of measuring the self-motion of PEO chains up to 1 ns. The mean square displacements of PEO chains in the 35% PEO/65% PMMA sample were obtained by Fourier transformation of backscattering spectra at $Q = 2.4$ and 3.2 nm^{-1} and temperatures from 350 K to 400 K. The results show the PEO chain dynamics are in agreement with the Rouse model prediction for $t \leqslant 1$ ns (see Figure 1 in Niedzwiedz et al. (2007)).

Additional measurements of the PEO chain motion in the PMMA matrix up to 80 ns were made using the neutron spin echo spectrometers (Niedzwiedz et al., 2007). The measured PEO single chain dynamic structure factor, $S(Q,t)/S(Q)$, for all small momentum transfer Q from 1 nm to 3 nm^{-1}, shows its decay is much slower than that expected from the Rouse rate determined at times shorter than 1 ns by the backscattering instruments. The slowing down was quantified by fitting to an effective smaller Rouse rate. The fits show retardations by factors of 4–20, depending on PEO content and temperature. Moreover, the time dependence of $S(Q,t)/S(Q)$ is at variance with the Rouse model. The crossover time from Rouse behavior to retarded and non-Rousean behavior determined by the combination of neutron backscattering and neutron spin echo scattering experiments is about 1 ns, independent of temperature T and Q (Niedzwiedz et al., 2007).

MD simulations have obtained the incoherent scattering functions $F_s(Q,t)$ of PEO chains in the blend for $0.1 < Q < 0.4$ Å^{-1}, $300 \leqslant T \leqslant 500$ K up to almost 100 ns. In the same Q, T, and time ranges, Rouse behavior was found before for pure PEO (Niedzwiedz et al., 2008) as evidenced by the stretched exponential time dependence, $\exp[-(t/\tau)^\beta]$, of $F_s(Q,t)$ with $\beta = 0.5$. In order to compare with neutron scattering experimental results, $F_s(Q,t)$ of PEO chains in the blend with PMMA from simulations was considered in the restricted time range from 60 ps to 2 ns (Brodeck et al., 2010) corresponding roughly to the dynamic range of the backscattering instruments. It was found in this restricted time range that the stretched exponential function with $\beta = 0.5$ is a good description of $F_s(Q,t)$ (see Figure 2 in Brodeck et al. (2010)), and the averaged τ has the Q^{-4}-dependence (see Figure 4b in Brodeck et al. (2010)). All these features are characteristics of Rouse dynamics, and thus verifying that Rouse dynamics holds for $t < 2$ ns, again independent of T and Q. Notwithstanding, the $\beta = 0.5$ fit of the restricted time range does not describe the full time decay of $F_s(Q,t)$, which is more stretched and retarded in the longer time range, consistent with the finding of neutron scattering results (Niedzwiedz et al., 2007). At longer times past $t_c = 1$–2 ns, the decay of $F_s(Q,t)$ becomes slower and the relaxation time departs from the Rouse model in dependencies on time, momentum transfer, and temperature. To explain the anomalies, Niedzwiedz et al. (2007) and Brodeck et al. (2010) had proposed the random Rouse model (RRM) in which each monomer has different mobility taken from a broad log-normal distribution. Despite the success of the RRM, Diddens et al. (2011) extracted the distribution of friction coefficients from the MD simulations of a PEO/PMMA blend and found the distribution is much narrower than expected from the RRM used to explain the data.

The neutron scattering and MD simulation data of the unentangled PEO chains in 35% PEO/65% PMMA summarized earlier clearly demonstrate at times shorter than about 1 ns in neutron scattering and 2 ns in MD simulations that the relaxation is describable by the Rouse model, independent of T and Q chosen for the studies. However, beyond 1 or 2 ns, it is slowed down and departs from Rouse dynamics. Hence there is crossover in the dynamics of PEO chains at the crossover time $t_c \approx 1$–2 ns, which is expected by the Coupling Model (CM) for interacting PEO chains in the blends (Ngai and Wang, 2011; Ngai, 2011). It turns out that the data can be explained by utilizing alone the observed crossover of PEO chain dynamics at t_c in the framework of the CM (Ngai and Wang, 2011). The observation is a special case of a general property of relaxation in interacting systems, which is the crossover from independent relaxation to coupled many-body relaxation at some t_c determined by the interaction potential (Ngai, 2011). Due to its fundamental nature, this property of unentangled chain dynamics of PEO in blends with PMMA is formulated as Property (ii) as follows.

(i).4 Property (ii).

Neutron scattering and molecular dynamics simulations have shown that the global self-dynamics of unentangled PEO chains in blends with PMMA follow the Rouse model at times shorter than 1–2 ns, but at longer times it becomes slower and departs from the Rouse model in dependencies on time, momentum transfer, and temperature (Niedzwiedz et al., 2007; Brodeck et al., 2010). In contrast, unentangled pure PEO obeys the Rouse dynamics at all times of the neutron scattering experiments (Niedzwiedz et al., 2008). Thus, the crossover from Rouse dynamics to non-Rouse dynamics of unentangled PEO chains in blends is due to the presence of PMMA, and originates from interchain coupling/constraint by the much less mobile PMMA. This property is in accord with the crossover from one-body primitive relaxation (Rouse modes in this case) to many-body relaxation at time t_c (Ngai, 2011) determined by the interchain coupling/constraint (exerted by the components of the blend in the present case).

(i).5 Background of Property (iii)

Another remarkable property of the unentangled chain dynamics of the fast component in HAPB related to Property (ii) was found first by molecular dynamics simulations of bead-spring model by Moreno and Colmenero (2008) and later by Arrese-Igor et al. (2012) using dielectric relaxation of unentangled polyisoprene (PI) in blends with a much slower component of poly(*tert*-butylstyrene) (PtBS) also unentangled. The chain dynamics of unentangled PI in pure PI are well described by the Rouse model with Rouse normal modes indexed by $p = 1, \ldots N-1$ (Ferry, 1980). The model predicts each of its modes has exponential time dependence, and their relaxation times, τ_p, including the longest relaxation time, τ_1, is proportional to $(N/p)^2$, or $(M/p)^2$, where M is the molecular weight. Remarkably, the MD simulations of the bead-spring model of a HAPB found that the Rouse modes of the fast component has stretched exponential time dependence, and the dependence of τ_p on $(N/p)^2$ is modified to $(N/p)^{x(T)}$. The exponent $x(T)$ is temperature dependent. At high temperatures and short τ_1, $x(T)$ has the value of ≈ 2, as predicted by the Rouse model. Moreno et al. suggested at high temperatures the disparity in dynamic of the two components is not large and hence deviation from Rouse model predictions is minor. In view of Property (ii) and as an analog, a possible reason for this to happen is that τ_1 becomes shorter than or comparable to the crossover time $t_c \approx 1$ or 2 ns at high temperatures. Most interesting is the result that $x(T)$ increases with decreasing temperature and reaches the value of about 3.5 at the lowest T studied by simulations. Moreno suggested this has to do with the large disparity between dynamics of the two components. This is certainly correct, but a more fundamental reason is the modification of the single chain Rouse dynamics to coupled chain dynamics after crossing t_c or Property (ii). The dielectric relaxation data of the normal modes of PI in the 35% PI blends with PtBS (Arrese-Igor et al., 2012) had confirmed the MD results by the observation of the broadening of the loss peak and the enhanced $(N/p)^{x(T)}$-dependence of $\hat{\tau}_1 \equiv \hat{\tau}_R$ of PI in the blend.

The fractional exponent, $\beta_R(T)$, of the stretched correlation function of the $p = 1$ mode had been obtained by simulations at various temperatures. Arrese-Igor et al. showed they can obtain the $(N/p)^{x(T)}$-dependence of the Rouse time $\hat{\tau}_R$ in the blend from the $(N/p)^2$-dependence of τ_R^0 of the Rouse model by raising it to the power of $1/\beta_R(T)$. The operation is expressed explicitly by

$$\hat{\tau}_R \propto [\tau_R^0]^{1/\beta_R(T)} \propto (N/p)^{2/\beta_R(T)}. \tag{5.41}$$

This equation indeed gives good account of the observed $(N/p)^{x(T)}$-dependence of $\hat{\tau}_R$ by applying the $\beta_R(T)$ from simulations. They attempted to justify this equation by taking an equation from Schweizer's (1989) formally exact, nonlinear generalized Langevin equation (GLE) for a flexible probe polymer in a dense entangled polymer melt (see Eq. (3.50) in Schweizer). They rewrote this same equation in the form

$$\Phi_1(t) = \exp\left[-\frac{\xi_0}{\tau_R^0}\int_0^t \frac{dt'}{\xi_0 + \xi(t')}\right]. \tag{5.42}$$

They compared the stretched exponential time dependence of $\Phi_1(t)$ they obtained by MD simulations with the formal Eq. (5.42), and asserted that the integral in the equation must have the time dependence of t^β in order that they be the same. However, this argument is tantamount to a conjecture that remains to be verified.

It is noteworthy that when Schweizer worked out his equation for entangled polymer melts, he did not obtain a stretch exponential that has exponent $\beta = 0.57$ or 0.59 that would have let him obtain the experimental $M^{3.5}$ or the $M^{3.4}$ dependence of the terminal relaxation time in order to be consistent with the experiment. On the other hand, as hinted by Schweizer (1989), Property (ii) of the CM when applied to terminal chain relaxation time $\hat{\tau}_R$ of PI in blends with PtBS leads to the relation,

$$\hat{\tau}_R = \left[t_{cR}^{-(1-\beta_R(T))}\tau_R^0\right]^{1/\beta_R(T)}, \tag{5.43}$$

which has already been established from the observation of the crossover at $t_{cR} = 1$ or 2 ns by neutron scattering experiment (Niedzwiedz et al., 2007) and simulations (Brodeck et al., 2010; Ngai and Wang, 2011). From Eq. (5.43) and the Rouse relation, $\tau_R^0 \propto (N/p)^2$, we immediately have the explanation of the $(N/p)^{x(T)}$-dependence with $x(T) = 2/\beta_R(T)$ found by simulations. The lower the temperature, the larger the disparity in mobility of the two components, and the more severe are the constraints imposed on the normal modes of the more mobile component, and hence the larger the coupling parameter or the smaller is $\beta_R(T)$ in Eq. (5.43). As recognized by Schweizer, the same Eq. (5.43) has been successfully applied to entangled polymers to explain viscosity and self-diffusion molecular weight dependence and temperature dependence (McKenna et al., 1985; Ngai and Plazek, 1986a; Ngai et al., 1997a).[2]

In view of the remarkable changes of dispersion and molecular weight dependence of normal modes of unentangled chains of the fast component in HAPB found by MD simulations and dielectric relaxation experiment, such chain normal modes are highlighted as Property (iii), which is summarized as follows.

(i).6 Property (iii)

The chain normal modes of unentangled fast component polymer in HAPB have dynamics totally changed from the Rouse dynamics seen in the pure polymer, especially at low temperatures where there is large disparity in mobility of the two components. The relaxation time, $\hat{\tau}_1 \equiv \hat{\tau}_R$, becomes much longer

2 For a review, see Ngai (2011).

than τ_R^0 of the Rouse model, the correlation function is changed from exponential function to the stretched exponential function with exponent $\beta(T)$, and the dependence of $\hat{\tau}_R$ is changed from the Rouse model prediction of $\tau_R^0 \propto (N/p)^2$ to the $(N/p)^{x(T)}$-dependence of $\hat{\tau}_R$, where $x(T) = 2/\beta_R(T)$. The deviations from Rouse dynamics decrease on increasing temperature or decreasing $\hat{\tau}_R$ as $\beta_R(T) \rightarrow 1$.

Before we go to the next property, it is worthwhile to mention a very similar situation as PI in blends with PtBS, encountered before in dielectric normal mode relaxation measurements by Adachi and coworkers of nearly monodisperse PI trapped in crosslinked natural rubber networks (Poh et al., 1987), and in much less mobile PB networks (Adachi et al., 1989). The experimental data were reviewed by Adachi and Kotaka (1990) and Adachi and Kotaka (1993). The PB networks used were loosely crosslinked with the molecular weight between crosslinks, $M_x > M_e$, where M_e is the molecular weight between entanglements. The dielectric loss peak is significantly broader than the loss peak of entangled pure monodisperse PI melt, and the relaxation time of the former is much longer than the latter. These changes of PI when introduced into the PB network are similar to those experienced by the unentangled PI chains in blends. The dielectric relaxation loss curve, $\varepsilon''(\omega)$, is well fit by the Fourier transform of the stretch exponential function with the entanglement coupling parameter, $n \equiv (1-\beta)$, equal to 0.45 (Ngai et al., 1997a). This value of n immediately explains the observed $(M_{PI})^{3.6}$-dependence of the dielectric relaxation time of the PI chains, just like the use of $\beta_R(T)$ in Eq. (5.43) to explain the $(N/p)^{x(T)}$-dependence of the PI relaxation time $\hat{\tau}_R$ in the blend.

(i).7 Background of Property (iv)

The puzzling behaviors of $\tau_{\alpha f}$ of PEO in blends with PMMA measured at high frequencies by deuteron NMR (Lutz et al., 2003), summarized as Property (i), have drawn considerable interest from other workers on segmental and terminal chain dynamics of these blends (Lodge et al., 2006; Chen et al., 2008; Niedzwiedz et al., 2007; Arrese-Igor et al., 2011; Haley and Lodge, 2005; Zeroni et al., 2008; Genix et al., 2005; Ngai, 2007). In 2005, Haley and Lodge (2005) measured the tracer diffusion coefficient of 0.3% unentangled PEO (M = 1000 g/mol) in a matrix of PMMA (M = 10,000 g/mol) over a temperature range from 125°C to 220°C by forced Rayleigh scattering. Also measured is the dynamic viscosities of blends of two different high-molecular-weight PEO tracers (M = 440,000 and 900,000 g/mol) in the same PMMA matrix at temperatures ranging from 160°C to 220°C. The monomeric friction factors determined by diffusion and rheology were in good agreement. Surprisingly to them, the magnitude, as well as the temperature dependence of the monomeric friction factors of the PEO tracer diffusion in the PMMA matrix, was much greater than that of the segmental dynamics of PEO ($M_w = 1.25 \times 10^5$ g/mol) at a concentration of 3% in a matrix of PMMA ($M_w = 1.06 \times 10^5$ g/mol) measured by Niedzwiedz et al. (2007) using the deuteron NMR at the high

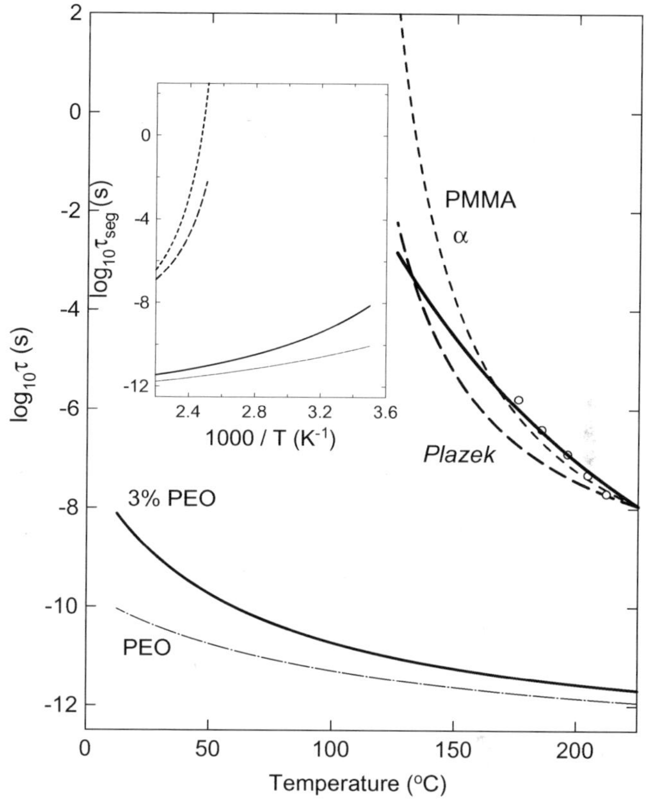

FIGURE 5.23 Comparison between the temperature dependences of the segmental relaxation time for (1) the PEO homopolymer (thin dashed-dotted line) (Lutz et al., 2003), and (2) 3% of PEO in PMMA (Lutz et al., 2003) (thick solid line) with that of (3) the dielectric segmental relaxation time of PMMA (Bergman et al., 1998) (thin dashed line), (4) the terminal relaxation time of PMMA Zawada et al. (1992) shifted downward by 9.1 decades (open circles), (3) the shift factor of the softening dispersion of PMMA given by Ngai and Plazek (1996) shifted vertically downward by 0.82 decade to coincide with the dielectric segmental relaxation time at the highest temperature (thick dashed line), and (4) the monomeric friction factor of the PEO tracers in PMMA matrix from Haley and Lodge (2005) (continuous line). The inset show the same data of (1), (2), (3), and (5) plotted against reciprocal temperature.

frequencies from 31 MHz to 76 MHz. As shown in Figure 5.23, the difference is about two orders of magnitude at 220°C and increases with decreasing temperature to more than six orders of magnitude at 125°C. Hence, in PMMA matrices, the PEO tracer diffusion relaxation time is a much stronger function of temperature than the corresponding PEO segmental relaxation time obtained by high frequency deuteron NMR.

In Figure 5.23 we reproduce for high molecular weight PMMA its dielectric segmental relaxation time (Bergman et al., 1998), the shift factor of the Rouse dynamics in the softening dispersion measured by Ngai and Plazek (1996) from 114°C to 189°C, and the shift factor of the entangled terminal dispersion obtained by Zawada et al. (1992) from 176°C to 212°C. All these PMMA shift factors have much stronger temperature dependence than the segmental relaxation time of 3% PEO in PMMA matrix measured by high frequency deuteron NMR. The same statement applies to the segmental relaxation time of 0.5% PEO in PMMA matrix measured by high frequency NMR.

Zhao et al. (2008) also found the high frequency segmental dynamics of 2% PEO in blends with PVAc determined by deuteron NMR show weaker temperature dependence than the terminal dynamics of 5% PEO in the same blend determined by Urakawa et al. (2006) from viscoelastic measurements. By the way, Urakawa et al. found problems with the use of LM models to fit their data, and they suggested the necessity of taking into account the effect of the intermolecular interaction in the LM model, in accord with the CM. The much stronger temperature dependence of the PEO terminal relaxation time than PEO segmental dynamics is not unique to PEO tracer in PMMA matrix, and a few percent of PEO in PVAc. Niedzwiedz et al. (2007) had found this effect earlier by comparing the terminal relaxation time of PEO in the 20% PEO/80% PMMA blend (Colby, 1989; Zawada et al., 1992) to the segmental dynamics obtained by high frequency deuteron NMR in a common temperature region. This means that the ratio of $\tau_{\mathrm{terminal}}/\tau_{\mathrm{seg}}$ is not constant in the blend, as assumed in some models of blend dynamics such as the LM model. Conventional theories of polymer viscoelasticity have the same monomeric friction factors and temperature dependence for segmental and global dynamics, with τ_{term} and τ_{α} having the same temperature dependence far above the glass transition temperature (Plazek, 1982; Roland et al., 2001) including pure PEO (He et al., 2004). Hence, the anomalous behavior of the PEO segmental and global dynamics in PMMA matrices defies explanation, which we formulate as Property (iv).

(i).8 Property (iv)

While the T-dependence of segmental relaxation and terminal relaxation at high temperatures is not very different for pure and unentangled PEO, the T-dependence of terminal relaxation time, τ_{term}, of PEO chains in the 0.3, 1.0, and 20% PEO blends with PMMA, found by diffusion and rheological measurement of Haley and Lodge (2005), Zeroni et al. (2008), and Ngai (2007), is much stronger than that of segmental relaxation time τ_{α} from high frequency deuteron NMR. This effect seen in the faster component is much reduced in several other polymer blends that do not have such large difference in T_g of the two components as PEO/PMMA. These include PEO in PVAc (Zhao et al., 2008), PB in PVE (He et al., 2004), and PI in PS (Zhao et al., 2008; He et al., 2005). The effect is absent for PI in PVE (Min et al., 2001; Haley et al., 2003; Ngai and Capaccioli, 2007). It is important to point out that the segmental relaxation of the faster component in *all* the blends discussed earlier was measured by high frequency deuteron NMR and $\tau_{\alpha f}$ obtained at very short times (Lutz et al., 2003). We hasten to point out that the relation between the temperature dependences of the terminal relaxation and the segmental relaxation is summarily different from the preceding if much longer $\tau_{\alpha f}$ were obtained by low frequency techniques including dielectric relaxation or calorimetry instead of high frequency deuteron NMR measurements. With the much longer $\tau_{\alpha f}$ measured by these two low frequency techniques, the results from experiments in another HAPB, 20% PI/80% PtBS show the opposite relation. That is the segmental relaxation time $\tau_{\alpha f}$ has stronger temperature dependence than the terminal chain relaxation time, τ_{term} (Arrese-Igor et al., 2010; Arrese-Igor et al., 2011).

(i).9 Background of Property (v)

The segmental α-relaxation data of 20% PEO/80% PMMA taken by Garcıa Sakai et al. (2008) and Garcıa Sakai et al. (2005) in the ps-ns time range reveal the crossover from a fast (F) relaxation of the PEO component to a slow (S) relaxation near $t_c = 1$ ps, close to $t_c = 2$ ps assumed in the 2004 paper (Ngai and Roland, 2004b) to explain the deuteron NMR data of Lutz et al. (2003). The relaxation times, $\tau_{\text{QENS},F}(T)$ and $\tau_{\text{QENS},S}(T)$, of the fast and slow relaxations of PEO, respectively, obtained from QENS by Sakai et al. for pure PEO and PEO in blends with 10, 20, and 30% PEO at momentum transfer $Q = 1.3$ Å^{-1}, are shown in Figure 5.24 (large-size symbols). For comparison, shown also are the $\tau_{\text{NMR},S}(T)$ from deuteron NMR (Lutz et al., 2003) (only the VF fits of the data of compositions ranging from 30% to 3% PEO are shown by lines), and the calculated $\tau(T)$ by the CM, exactly as shown before in Figure 1 of the 2004 paper (Ngai and Roland, 2004b). More detailed comparison of $\tau_{\text{QENS},S}(T)$ and $\tau_{\text{NMR},S}(T)$ will be given later in Section (i).2. Included in the inset of Figure 5.24 are two data points, $\tau_G(T)$, from QENS experiment on 25% PEO/75% PMMA by Genix et al. (2005), which was used by Colmenero and Arbe (CA) to demonstrate the onset of the Arrhenius T-dependence of $\tau_{\alpha f}(T)$ of confined PEO component (Colmenero and Arbe, 2007). Here we focus on the good agreement between $\tau_G(T)$ and $\tau_{\text{QENS},S}(T)$ in the inset as well as between $\tau_{\text{QENS},S}(T)$ and $\tau_{\text{NMR}}(T)$ in the main figure. From the agreement, it is clear that the $\tau_G(T)$ data from QENS experiment of Genix et al. are no different from $\tau_{\text{QENS},S}(T)$ and $\tau_{\text{NMR}}(T)$. Thus, the nature of $\tau_G(T)$ is automatically explained via Property (i) along with $\tau_{\text{NMR}}(T)$ and $\tau_{\text{QENS},S}(T)$. Since there is no indication of crossover from VFTH T-dependence to Arrhenius T-dependence at any temperature of $\tau_{\text{NMR}}(T)$ from NMR measurements, the assertion of the existence of such a crossover of $\tau_G(T)$ to Arrhenius T-dependence by CA (Colmenero and Arbe, 2007) based on two data points of $\tau_G(T)$ in their Figure 10 is not supported.

It is noteworthy that the proposed crossover by CA to confined motions of the 20% PEO component in blends with PMMA is at some $\tau_{\text{cross}}(T_{\text{cross}})$ shorter than 10^{-10} s with T_{cross} significantly higher than T_{gf}. This scenario for dynamics confinement is further illustrated by a "cartoon" given by CA as Figure 8 in Colmenero and Arbe (2007), with the crossover occurring at $\tau_{\text{cross}}(T_{\text{cross}}) \approx 10^{-10}$ s. While in another highly asymmetric polymer blend, 20% PVME/80% PS, the proposed $\tau_{\text{cross}}(T_{\text{cross}})$ is of the order of 10^{-5} s (see Figure 9 in Colmenero and Arbe (2007)). In 20% PI/80% PtBS (Arrese-Igor et al., 2010), which has larger difference in T_g of the two components than 20% PVME/80% PS, the crossover to Arrhenius was not observed in the segmental α-relaxation time, $\tau_{\alpha f}(T)$, of the fast PI component at least for all times shorter than 10^{-1} s and 10^{-2} s in the blends with PtBS having molecular weights $M_n = 1300$ g/mol and 2300 g/mol, respectively (Arrese-Igor et al., 2010). If the crossover of PI to local motion exists, τ_{cross} has to be longer than these very long times, and T_{cross} has to be close to either the DSC or dielectric T_{gf} of the PI component. Since 25% PEO/75% PMMA, 20% PVME/80% PS, and 20% PI/80% PtBS all are HAPB, if the crossover is real, it is hard to understand why their $\tau_{\text{cross}}(T_{\text{cross}})$ can differ by ten orders of magnitude, and the relation of T_{cross} to T_{gf} can be so different. This quandary further casts doubt on the validity of the dynamic confinement hypothesis.

In several HAPB, using calorimetry or the thermally stimulated depolarization current (TSDC) technique, separate glass transition temperatures, T_{gf} and T_{gs}, of the two components had been detected. These blends include 25% PEO/75% PMMA (Lodge et al., 2006), x% PVME/(1 − x)PS with $x = 0.30$ (Lorthioir et al., 2003a, Lorthioir et al., 2003b) and 0.20 (Leroy et al., 2002), x% PI/(1 − x) PtBS with $x = 0.25$, 0.30, and up to 50 (Zhao et al., 2009), and miscible blends of polystyrene/poly(vinyl methyl ether), poly(*o*-chlorostyrene)/poly(vinyl methyl ether), and poly(styrene-*block*-*n*-butylacrylate)/polystyrene (Miwa et al., 2005). Basically, the detection of glass transition of the faster component by calorimetry or TSDC means detection of its segmental α-relaxation with relaxation time $\tau_{\alpha f}$ typically of the order of 10^2 or 10^3 s depending on the scanning rate. In the "confinement" scenario, the transformation of the dynamics of the fast component to "localized motions with low degree of cooperativity" (Colmenero and Arbe, 2007) cannot be responsible for the glass transition temperature T_{gf} detected by DSC. This is because localized motions with low degree of cooperativity, like secondary relaxations in polymers, usually are not detectable by DSC. Besides, had it been possible to detect the localized motions by DSC, the temperature at which the Arrhenius T-dependence of the PEO component assumed by CA to reach 10^2 s will be much lower than that found by DSC in R (Lodge et al., 2006; Zeroni et al., 2008; Mpoukouvalas and Floudas, 2008; Goulart Silva et al., 2000).

The compositions of the blends are usually given in terms of weight percent. Because the monomeric molecular weight of the slower component is usually larger than that of the faster component, the mole% of the faster

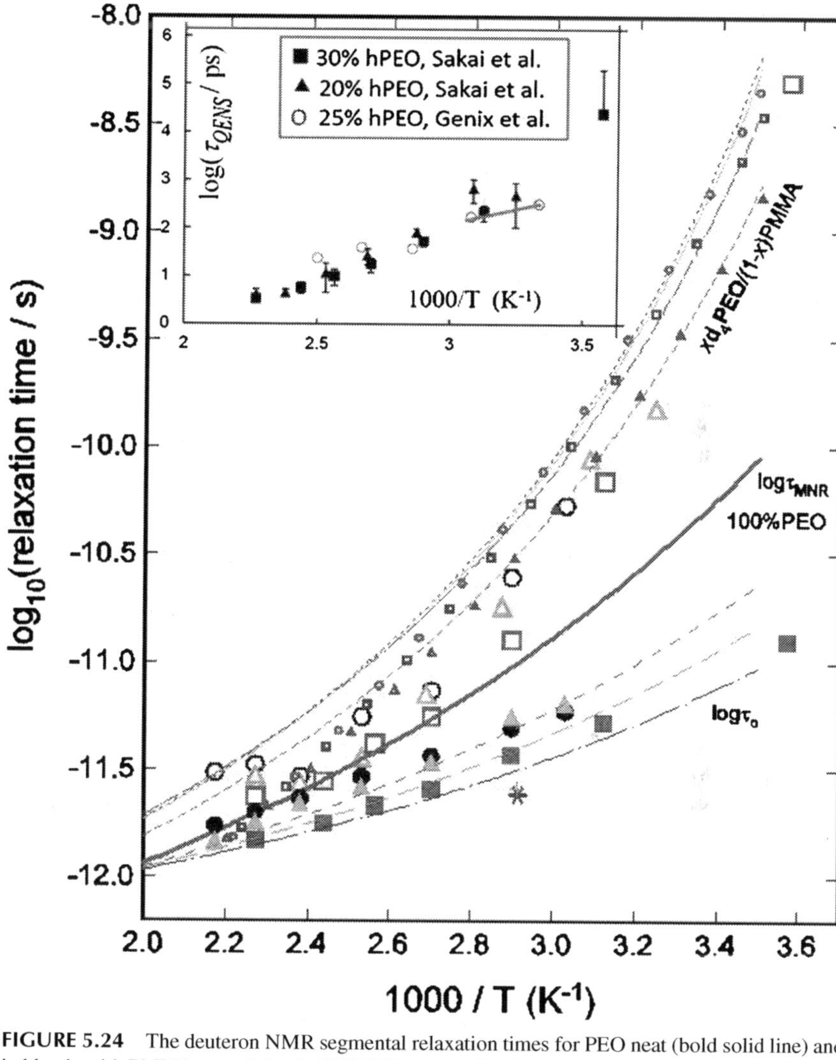

FIGURE 5.24 The deuteron NMR segmental relaxation times for PEO neat (bold solid line) and in blends with PMMA containing 3–30% PEO (upper thin lines are the fits to the data that are not shown), as reported by Lutz et al. Shown by the dashed-dotted line at the bottom is the independent relaxation time for neat PEO calculated using $n = 0.67$, the value reported by Lutz et al. (2003). The small-size symbols on top are the most probable relaxation times calculated for the lower (○, $\hat{n} = 0.84$), mid-range (□, $\hat{n} = 0.83$), and higher concentrations (△, $\hat{n} = 0.81$) of PEO in the blends, respectively. The large open and closed symbols are the relaxation times of the slow and fast processes, respectively, found by QENS experiments (Garcıa Sakai et al., 2008; Garcıa Sakai et al., 2005) at $Q = 1.3$ Å^{-1}: (circles) 10%, (triangles) 20%, (squares) 30%, and (+ and ∗) 100% PEO. The inset shows the $\tau_{QENS,S}$ of PEO in blends with 30, 25, and 20% PEO in PMMA. The thick line in the inset is drawn to duplicate the construction as done in Colmenero and Arbe (2007) of the Arrhenius temperature dependence of the hypothesized confined PEO relaxation from the two data points (O) obtained from the QENS data of Genix et al. (2005).

component is larger than the corresponding wt.%. For example, 20 wt.% of PEO in blend with PMMA, is equivalent to about 43 mol% of PEO, 20 wt.% of PI in blend with PtBS is equivalent to about 40 mol% of PI, and 20% of PVME in blend with PS is equivalent to about 28 mol% of PVME. The larger mol% of the faster component in the HAPB studied means the repeat units of the fast component responsible for the segmental motion are not overwhelmingly outnumbered by the repeat units of the slow component in these blends. There are still plenty of mobile repeat units of the faster component within the surrounding, and the segmental motion of the faster component is not confined to local motion as conceived by the confinement hypothesis, at least for the compositions of the HAPB studied with 20 wt.% of the fast component. Only in the limit of dilute concentration of the faster component will immobilization of its dynamics be realized, but this is not the case for all the relevant HAPB considered here.

It turns out that the data of Lorthioir et al. as well as the follow-up by Schwartz et al. (2007) have fundamental values. We start by pointing out that from experiments two relaxations originating from the PVME component were found in the PVME/PS blends. Let us consider first the case of the 30% PVME blend in the paper by Lorthioir et al. By inspection of their Figure 3 and also the upper inset of Figure 5.25, here, it is clear there are two resolved peaks in the isochronal dielectric measurements at 1 Hz. Another way to show the presence of two distinct processes is via the dielectric glass transition temperature, $T_{g,\mathrm{diel}} \approx 280$ K, at 1 Hz of the fast PVME component given in Figure 1b of Lorthioir et al. The responsible relaxation process has relaxation time $\tau_{\alpha f}(T_{g,\mathrm{diel}})$ equal to $10^{-0.8}$ s (corresponding to 1 Hz) at $T_{\mathrm{gf}} \approx 280$ K. On the other hand, deduced from isothermal dielectric loss data and as shown in their Figure 7 (see also the main part of Figure 5.25 here), there is another dielectric relaxation process (supposedly the confined α'-relaxation in the confinement scenario) of the PVME component with relaxation time of about $10^{-4.2}$ s at $T = T_{g,\mathrm{diel}} \approx 280$ K, which is 3.4 decades shorter than $\tau_{\alpha f}(T_{g,\mathrm{diel}}) = 10^{-0.8}$ s of the other process. Thus from the isochronal and isothermal dielectric data, we have two coexisting and yet distinctly different relaxation processes, all originating from the PVME component.

Another way to measure glass transition temperature of the PVME component in the blends with PS was demonstrated by Leroy et al. (2002) using thermally stimulated depolarization current (TSDC). This technique utilizes the dipoles in the PVME chains to sense their specific T_g and thus enables the glass transition temperature, $T_{g,\mathrm{TSDC}}$, of the dielectrically active component PVME in the blend to be measured. Leroy et al. obtained $T_{g,\mathrm{TSDC}} \approx 293$ K for PVME in 20% PVME/80% PS (see also Figure 6a of Colmenero and Arbe (2007)). Since the TSDC measurements were made at the same heating rate of 10 K/min as usually done in DSC experiments (10 K/min), we can expect the relaxation time, $\tau_{\alpha f}(T_{g,\mathrm{TSDC}} = 293$ K), is in the range from $10^{2.0}$ to $10^{3.0}$ s. Again, from the relaxation map of $\tau_{\alpha'}$ provided by Lorthioir et al. and Figure 5.25, it indicates $\tau_{\alpha'}(T_{g,\mathrm{TSDC}} = 293$ K) is $\approx 10^{-4.9}$ s. The TSDC relaxation time and

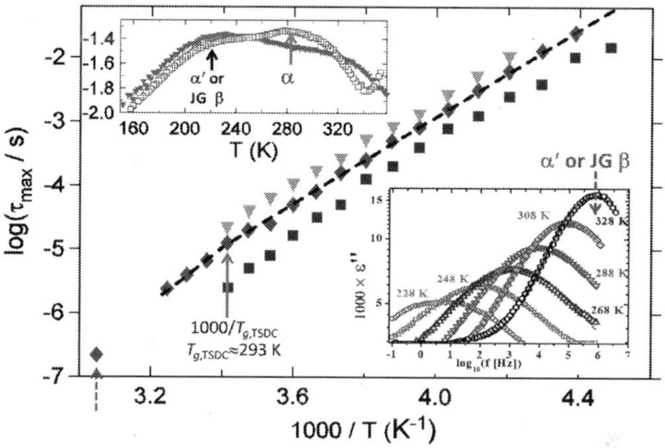

FIGURE 5.25 Plot of the characteristic relaxation time, τ_{max}, of the α'-relaxation for the PVME component in three blends x% PVME/(1 − x)% PS as functions of temperature obtained by Lorthioir et al. (2003a). Inverted triangles for $x = 30$, diamonds for $x = 20$, and squares for $x = 10$. The arrows indicate the reciprocal of the glass transition temperature, $T_{g,TSDC} = 293$ K, for the 20% PVME blend. The lower inset is obtained by digitizing Figure 5 in (Lorthioir et al., 2003a). It shows the isothermal spectra of the α'-relaxation in the 20% PVME blend. From the peak frequencies, f_{max}, some of the τ_{max} in the main figure are determined. The shortest τ_{max} in the figure corresponds to f_{max} at 328 K indicated by the arrow in the inset. The very broad α-relaxation is not resolved in the isothermal spectra (lower inset), but its possible location at 328 K is suggested to be in the vicinity of 1–0.1 Hz. The upper inset shows the isochronal loss data at 1 Hz of LAC for $x = 30$ (inverted triangles) and $x = 20$ (open squares). The arrow near 220 K indicates the α'- or JG β-relaxation in the Arrhenius regime of the main figure. The arrow near 280 K indicates the α-relaxation for $x = 30$.

the dielectric relaxation time of PVME component in this blend are separated by 7–8 decades at 293 K, and hence the relaxation detected by TSDC is a process distinctly different from the α'-relaxation albeit both originate from the PVME component in 20% PVME/80% PS. These two different processes coexist at the same temperature such as $T_{g,TSDC} = 293$ K.

This is all made clearer by putting the isochronal (at 1 Hz) and isothermal dielectric loss data of $\tau_{\alpha'}(T)$ for PVME in the blends x% PVME/(1 − x)% PS with $x = 30, 20$, and 10, together with $T_{g,TSDC} \approx 293$ K for the 20% PVME/80% PS blend in Figure 5.25, and described as follows. The isochronal data are shown in the upper inset. The presence of two peaks for $x = 30$ clearly indicate two processes from PVME. The PS component is not seen in this isochronal because of the small dielectric strength and its location at higher temperatures. For $x = 20$, the peak at higher temperature is not resolved and appears as a shoulder. Some of the $\tau_{\alpha'}(T)$ for the 20% PVME/80% PS blend in the main figure are obtained from the loss peak frequency determined from the isothermal loss spectra of the α'-relaxation published by Lorthioir et al. and reproduced by digitization in the lower inset of Figure 5.25. The $\tau_{\alpha'}(T = 328\ \text{K})$ corresponding to the peak frequency at the highest temperature 328 K is included and shown by the red closed diamond having the shortest relaxation time in the main part of the figure. This data point of $\tau_{\alpha'}(T = 328\text{ K})$ was not shown by Lorthioir et al. in their Figure 7. Together with few others at temperatures higher than $T_{g,\text{TSDC}} \approx 293\ K$ it demonstrates the change of temperature dependence of $\tau_{\alpha'}(T)$ on crossing $T_{g,\text{TSDC}} \approx 293$ K. This property is typical of JG β-relaxation of the fast component in mixtures of van der Waals glass-formers, which coexists with the α-relaxation also of the fast component (Capaccioli et al., 2012; Mierzwa et al., 2008; Kessairi et al., 2008; Ngai, 2011; Ngai et al., 2011), and also in neat glass-formers where of course it is accompanied by the omnipresent α-relaxation (Ngai, 2011).

It is worthwhile to point out the width of the frequency dispersion of the segmental α-relaxation of PVME in blends x% PVME/$(1 - x)$% PS with smaller $x = 0.25, 0.20$, and 0.10 are much broader than the blends with $x \geqslant 0.5$, particularly when the relaxation time $\tau_{\alpha f}(T)$ becomes long. The isothermal spectra of the α-relaxation for blends with $x \leqslant 0.30$ were seldom shown because they are so broad, and only isochronal spectra were presented, but even that cannot determine with precision the characteristic α-relaxation time of the PVME component. For blends with $x \geqslant 0.5$, the α-loss peak of the PVME is clearly resolved as shown by Cendoya et al. for $x = 0.65$ and 0.50 (see Figure 1 in Cendoya et al. (1999)). The α-relaxation times $\tau_{\alpha f}(T)$ obtained by dielectric, NMR, and QENS combined have VFT temperature dependence practically from 10^{-10} to about 10 s (Cendoya et al., 1999). No crossover to Arrhenius T-dependence is found anywhere, at least for all $\tau_{\alpha f}(T)$ shorter than about 1 s. Extrapolating this VFT dependence down to 10^2 s, the dielectric $T_{g,\text{diel}}$ determined by $\tau_{\alpha f}(T_{g,\text{diel}}) = 10^2$ s is in agreement with that ($\approx$265 K for $x = 0.50$) determined by Leroy et al. (2002) using TSDC. All these facts ensure the dielectric loss peaks observed in blends with $x \geqslant 0.50$ are the α-relaxation of PVME, and the Johari-Goldstein (JG) β-relaxation (Johari and Goldstein, 1970; Johari, 1976; Ngai and Paluch, 2004) are not resolved. We caution that the secondary relaxation at high frequencies found by Cendoya et al. in blends with $x \geqslant 0.5$ is not the JG β-relaxation of PVME in the blend. It is already seen in the dielectric spectra of pure PVME (Urakawa et al., 2001), and it originates from intramolecular motion of a moiety in the side group (labeled as γ_{PVME} in Figure 5.26).

The dielectric loss data of the 50% PVME/50% PS blend taken at ambient pressure by Schwartz et al. (2007) belong also to the α-relaxation, a conclusion we reached from the agreement of the relaxation times with those of Cendoya et al. (1999). The situation is different in blends x% PVME/$(1 - x)$% PS with smaller $x = 0.25$ and 0.20, where the α-loss peak of PVME is broadened immensely and becomes difficult or impossible to resolve by dielectric spectroscopy. In the isothermal dielectric loss spectra by Schwartz et al. (2007) for the 25% PVME blend and by Lorthioir et al. for the 20% PVME blend, observed instead is the resolved JG β-relaxation (called α' relaxation by Lorthioir et al. (2003a)). We have already shown in connection with the 20% PVME blend in Figure 5.25 that the observed JG β-relaxation is distinctly different from the α-relaxation because the latter was found by TSDC, and it has much longer

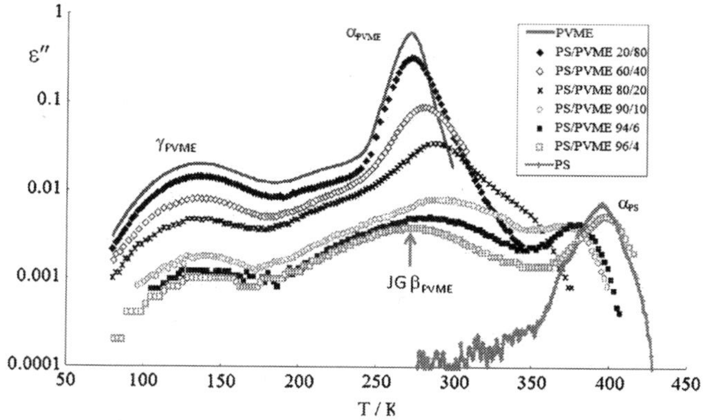

FIGURE 5.26 Temperature dependence of dielectric loss at 1 kHz for blends of PVME with PS at several compositions (Urakawa et al., 2002). Figure provided courtesy of K. Adachi.

relaxation time $\tau_{\alpha f}$ than the dielectrically observed JG β-relaxation time τ_{JGf} at the $T_{g,TSDC}$.

Schwartz et al. (2007) made the same mistake of identifying the dielectrically observed relaxation of the PVME component in the 25% PVME blend at ambient and elevated pressures as the α-relaxation with VFT-dependence at higher temperatures and crossing over to Arrhenius dependence at lower temperatures. At ambient pressure and $T = 300$ K, from Figure 5 of Schwartz et al., the logarithm of the dielectric α-relaxation time of the 50% PVME blend, $\log[\tau_{\alpha f}(T = 300 \text{ K})]$, is about -5.25. At the same $T = 300$ K and ambient pressure, the logarithm of the dielectrically observed relaxation time of the 25% PVME blend shown in Figure 6 of Schwartz et al. (or Figure 5.27, here) has a value nearly the same as -5.25. This comparison shows that the dielectrically observed relaxation time of the 25% PVME blend cannot be interpreted as the α-relaxation time $\tau_{\alpha f}$ because $\tau_{\alpha f}$ of the 25% PVME blend has to be much longer than that of the 50% PVME blend at the same temperature and pressure. Another way to show the same is to use once more the glass transition temperature, $T_{g,TSDC}$, of the PVME component determined by TSDC. This has been obtained for the 35% and 20% PVME blends, but not the 25% PVME blend. Nevertheless, $T_{g,TSDC} \approx 289$ K of the 25% PVME blend can be determined by the interpolation provided by Leroy et al. (2002). This estimate is supported by the extrapolation of the dielectric T_g of the PVME component to 25% PVME given by Lorthioir et al. (2003a), which is also $\approx$290 K. The relaxation at ambient pressure obtained by Schwartz et al. dielectrically for the 25% PVME blend has relaxation time $\approx 10^{-5}$ s at 289 K (see Figure 5.27), and it cannot be the α-relaxation for otherwise it contradicts the observation by TSDC that $\tau_{\alpha f}$ at $T_{g,TSDC} \approx 289$ K is much longer. There is indeed a change of temperature dependence of the dielectric relaxation time from VFT to Arrhenius in the neighborhood of $T_{g,TSDC} \approx 290$ K, but this is perfectly

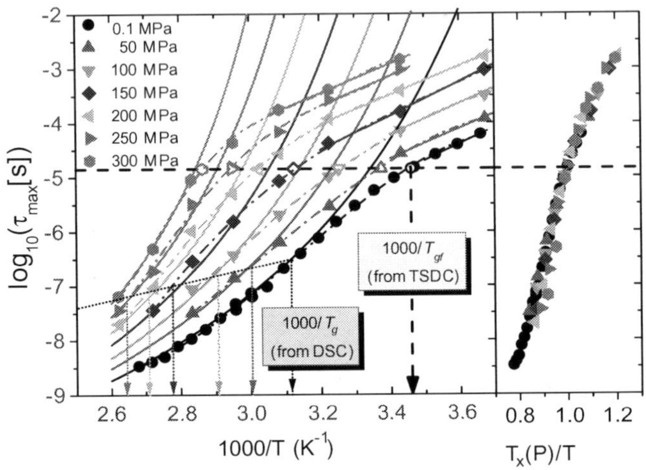

FIGURE 5.27 Logarithm of the relaxation time τ_{JG} vs inverse temperature at different pressures (from bottom to top: $P = 0.1$, 50, 100, 150, 200, 250, and 300 MPa) for PVME in PS (25/75). Data from Schwartz et al. (2007) in which they interpret the observed as α-relaxation with VFT dependence and crossover to Arrhenius dependence due to confinement by the PS matrix. Solid curved lines represent the VFT fits of the experimental data at higher temperatures. The vertical dashed black and the thinner black continuous arrows indicate at ambient pressure the locations of $1000/T_{g,TSDC}$, with $T_{g,TSDC} \approx 289$ K of the PVME component and $1000/T_{g,DSC}$, with $T_{g,DSC} \approx 321$ K of the blend as a whole obtained by DSC. Results obtained by interpolation of TSDC and DSC data from Leroy et al. (2002). The colored vertical arrows mark the positions where the system (likely the slow majority component PS) enters into the glassy state, as determined by PVT data. Note the approximate agreement between this location and $1000/T_{g,DSC}$ at ambient pressure. The horizontal black dashed line marks the $\tau_{JG}(P,T) \approx 10^{-4.8}$ s. The dotted-dashed color lines are polynomial fits drawn to suggest the crossover of temperature dependence of $\tau_{JG}(P,T)$ at $T_X(P)$ where $\tau_{JG}(P,T_X(P)) \approx 10^{-4.8}$ s independent of P. The open symbols mark the change of concavity at the crossover, which also occurs at the same values of $\tau_{JG}(P,T_X(P))$ for all $T_X(P)$. Inset shows the master-curve obtained by plotting $\tau_{JG}(P,T)$ versus $T_X(P)/T$.

consistent from the behavior of the JG β-relaxation of the PVME component, as found in the fast component of binary mixtures of van der Waals liquids.

The best proof of invalidity of the confinement scenario is provided by the DSC and dielectric relaxation data from Arrese-Igor et al. (2010) on the highly asymmetric blends of polyisoprene (PI) of molecular weight $M_n = 2700$ with poly(tert-butylstyrene) (PtBS) of two different molecular weights $M_n = 1300$ and $M_n = 2300$. Their DSC measurements confirm the presence of two separate glass transition temperatures of the PI and the PtBS components for blends with less than 50% of PI. The components T_{gf} and T_{gs} of PI and PtBS in 35% and 20% PI blends from DSC are indicated by arrows in Figures 5.28 and 5.29, where the dielectric relaxation times data are presented in an Arrhenius plot. Like that discussed before in the PEO/PMMA blends, the detection of T_{gf} of the fast component by DSC, PI in the present case has basically ruled out the confinement scenario. This is because the relaxation time of order of 100 s obtained by DSC has to be that of the α-relaxation of the PI component.

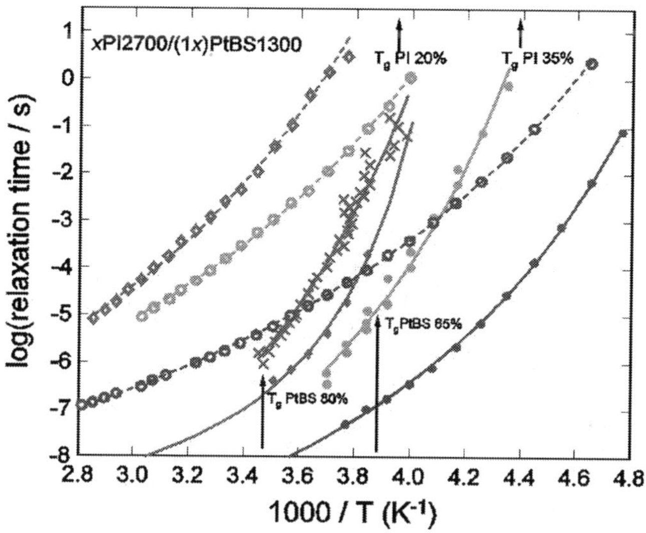

FIGURE 5.28 The NM relaxation time $\hat{\tau}_R$(open symbols) and the segmental α-relaxation time $\tau_{\alpha f}$ (closed symbols) of PI27 in pure PI27 (squares), 35% PI27 blend (circles), and 20% PI27 blend (diamonds) with PtBS1300. Data from Arrese-Igor et al. (2010) and Arrese-Igor et al. (2011), and replotted as a new figure. The x symbols represent the characteristic times obtained from isochronal representation of the data for 20% PI blends. The location of the $1000/T_g$ of the two components for the two blends are indicated by the arrows accompanied by value of T_g.

The localized relaxation of PI in the confinement scenario analogous to secondary relaxation cannot be detected by DSC even in neat PI. The dielectric data of 35% and 20% PI in blends with PtBS give direct proof that there is absolutely no crossover of T-dependence of the α-relaxation time $\tau_{\alpha f}$ of PI as predicted by the confinement scenario. This can be seen in Figure 5.29, where $\tau_{\alpha f}$ of PI in 35% and 20% PI blends follows the VFT dependence on decreasing temperature, all the way down to long times of the order 1 s and 10^{-1} s, respectively. The dielectric T_{gf} can be determined by extrapolating the VFT dependence of $\tau_{\alpha f}$ down to temperature at which it is equal to 100 s. The dielectric T_{gf} is lower than the DSC T_{gf} but this is caused by the fact that $\tau_{\alpha f}$ is determined by the loss peak frequency rather than the average $\langle\tau_{\alpha f}\rangle$. The average $\langle\tau_{\alpha f}\rangle$ is longer than $\tau_{\alpha f}$ because the loss peak is broadened much more at the lower frequency flank, and a higher dielectric T_{gf} comparable to the DSC T_{gf} would be obtained had $\langle\tau_{\alpha f}\rangle$ been used to determine it.

Thus, from both the DSC and the dielectric relaxation data cited earlier, the crossover of $\tau_{\alpha f}$ of PI in the HAPB of 35% and 20% PI with PtBS from VFT to Arrhenius dependences is not found at any temperature. This is the most direct proof that the confinement scenario is unreal. Arrese-Igor et al. (2010) admitted that the crossover predicted by the confinement scenario is not observed on $\tau_{\alpha f}$ of PI in the HAPB, but still maintained a vestige of confined dynamics by invoking the marked decrease of the intensity and increase of width as temperature decreases of the α-loss peak of PI in the 20% PI blend,

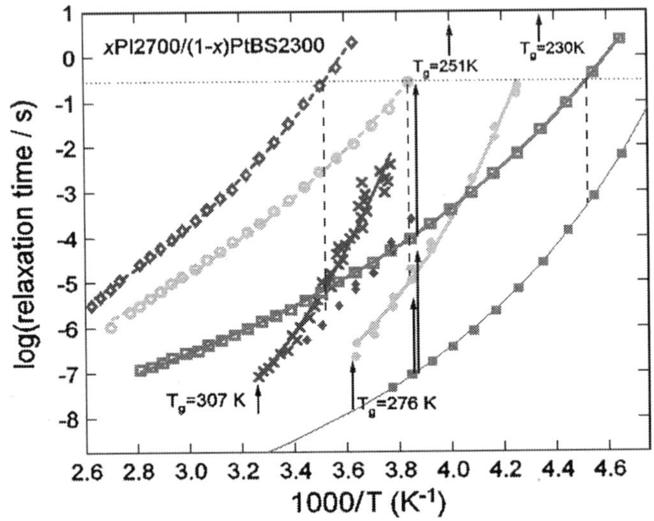

FIGURE 5.29 The NM relaxation time, $\hat{\tau}_R$ (open symbols), and the segmental α-relaxation time, $\tau_{\alpha f}$ (closed symbols), of PI27 in pure PI27 (squares), 35% PI27 blend (circles), and 20% PI27 blend (diamonds) with PtBS2300. Data from Arrese-Igor et al. (2010) and Arrese-Igor et al. (2011), and replotted as a new figure. The x symbols represent the characteristic times obtained from isochronal representation of the data for 20% PI blends. The horizontal dotted line indicates $\log(\hat{\tau}_R/s) \approx -0.5$, corresponding to $\log(f_{NM}/\text{Hz}) = 0.7$. The vertical thin dashed lines indicate increasing separation between $(\log \hat{\tau}_R/\text{s})$ and $\log(\tau_{\alpha f}/\text{s})$, at constant $\log(\hat{\tau}_R/\text{s}) \approx -0.5$, on decreasing PI27 content. The location of the $1000/T_g$ of the two components for the two blends are indicated by the arrows accompanied by the value of T_g. The tips of the arrows located slightly to the right of $1000/260\ \text{K}^{-1} = 3.85\ \text{K}^{-1}$ are the calculated values of $\log(\hat{\tau}_R/\text{s})$ and $\log(\tau_{\alpha f}/\text{s})$ by the CM (see text).

shown by Figure 7 in their paper. The experimental findings of Lorthioir et al. and Schwartz et al. on the x% PVME/$(1-x)$% PS blends are valuable for fundamental understanding of the dynamics of the fast component in the highly asymmetric polymer blends in general. In pure PVME, the JG β-relaxation has not been resolved by isothermal or isochronal dielectric spectroscopy. This is probably due to the weak intermolecular coupling of pure PVME as suggested by the chemical structure of the repeat unit, which is effectively like replacing one hydrogen atom of the repeat unit of polyethylene by $-$O-CH_3 unit. The flexibility of the oxygen linkage leads to weak intermolecular constraints and coupling parameter, n. From either the empirically established correlation between the ratio τ_α/τ_{JG} and n or between τ_α/τ_0 and n from the CM (Capaccioli et al., 2012; Mierzwa et al., 2008; Kessairi et al., 2008; Ngai, 2011; Ngai, 1998), the α- and the JG β-relaxations in pure PVME are too close together, and they merge to show up as a single broad loss peak. On blending PVME with PS, the presence of the much slower PS component imposes stronger intermolecular constraints on the PVME component, and hence increase of n and the ratio τ_α/τ_{JG}, resulting in the JG β-relaxation being resolved in blends rich in PS. This phenomenon had been seen in mixtures of small molecular van der Waals liquids (Arrese-Igor et al., 2011; Moreno and Colmenero, 2008) and in the family of poly(n-butyl methacrylate-*stat*-styrene) random copolymers on increasing the styrene content (Ngai and Plazek, 1995; Ngai and Plazek, 1986b), but never in polymer blends until the work of Lorthioir et al. and Schwartz et al. By blending PVME with higher concentration of PS, the JG β-relaxation is separated further away from the coexisting α-relaxation of the PVME component, and it is resolved for the first time by them. Because this is an important finding of a fundamental property of polymer blends, it is highlighted as another property that needs to be addressed as follows.

(i).10 Property (v).

The α- and the JG β-relaxation are two separate and distinct processes originating from the faster PVME component and coexisting in blends of 20, 25, and 30% PVME with PS. Albeit the α-relaxation is too broad to be resolved by dielectric spectroscopy, the α-relaxation time $\tau_{\alpha f}$ follows the VFT dependence on decreasing temperature all the way down to long times of the order 10^2 s, where glass transition at T_{gf} had been detected by TSDC. The JG β-relaxation time τ_{JGf} changes T-dependence at T_{gf} from a super-Arrhenius dependence above T_{gf} to Arrhenius dependence below T_{gf}. At T_{gf}, τ_{JGf} is of the order of 10^{-5} s. These properties apply to other HAPB such as PEO with PMMA or PI with PtBS, although other factors and experimental difficulty hamper detection of the JG β-relaxation of PEO and PI in these HAPB.

(i).11 Background of Property (vi)

Presented also in Figures 5.30 and 5.31 of the last section are the normal mode relaxation times $\hat{\tau}_R$ of PI in the 35% and 20% PI blends with PtBS. The data on the normal modes are taken from another paper by Arrese-Igor et al. (2011). From the frequency, $f_{\text{max,NM}}$, at the maxima of the normal mode relaxation of PI27 ($M_n = 2700$) in these blends with PtBS13 ($M_n = 1300$) and PtBS23 ($M_n = 2300$), the $\hat{\tau}_R$ are obtained from the identity $\hat{\tau}_R = 1/(2\pi f_{\text{max,NM}})$. These data considered here are just one of the molecular weights of unentangled PI in blends with PtBS13 and PtBS23, discussed before in Section (i).5 in connection with Property (iii). There we have already mentioned that the correlation function of the normal modes in the blend is a stretched exponential with a temperature-dependent exponent $\beta_R(T)$, and the dependence of $\hat{\tau}_R$ is changed from the Rouse model prediction of $\tau_R^0 \propto (M_n)^2$ to the $(M_n)^{x(T)}$-dependence, where $x(T) = 2/\beta_R(T)$.

In this section, we compare the normal mode relaxation with the segmental relaxation, and discuss another remarkable property that emerges from the comparison. The property is best described by the figure published by Arrese-Igor et al. (2011), which is reproduced here as Figure 5.30. In this figure, $f_{\text{max,NM}} \approx 0.5$ Hz is the same for all blends ranging from pure PI to 20%

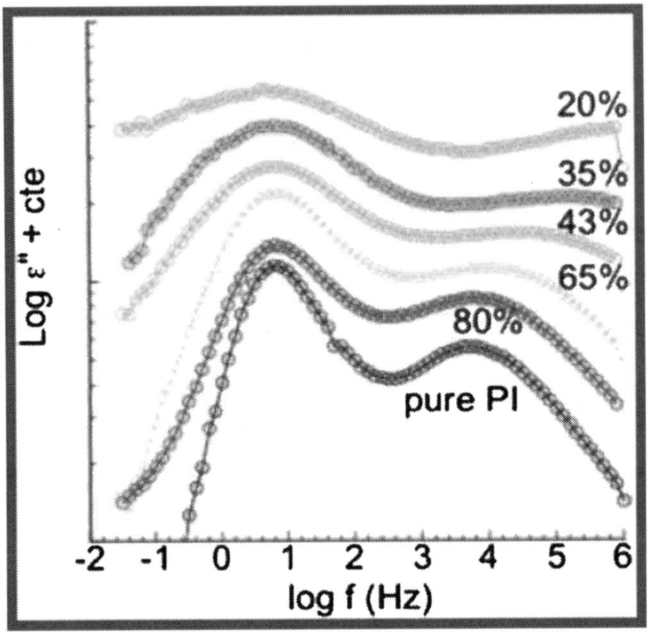

FIGURE 5.30 Imaginary part of the relative dielectric permittivity for PI27 and PI27 in PI27/PtBS23 blends at different concentrations and temperatures. The $f_{max,NM}$ is the same for all cases. The data indicate increase separation between the NM and segmental relaxation at constant $f_{max,NM}$ on decreasing content of PI27 in the blends. Different curves have been shifted in the y-axis for clarity. Reproduced from Arrese-Igor et al. (2011) by permission of the American Chemical Society.

of PI27 in blends with PtBS. The data indicate increased separation between the NM time, $\hat{\tau}_R$, and segmental relaxation time, $\tau_{\alpha f}$, of PI27 on decreasing concentration of PI27 in the blends at constant τ_{NM}. The separation between the two relaxations increases from about three decades for pure PI to almost six decades when PI27 content is reduced to 20%. It is intuitively obvious as well as shown by experiments mentioned before that both the segmental α- and the NM relaxations of PI27 slow down on increasing the content of PtBS with higher T_g in the blend. However, challenging is precisely how the observed behavior can be explained. The impression one may get from the figure is like that described by Arrese et al.:

"For low PI content blends, however, the NM-relaxation is much slower relative to the α-relaxation than the behavior observed for the homopolymer." This impression is fortified in their Abstract by the statement: "Finally, the terminal dynamics of PI component shows a stronger T-dependence than its segmental dynamics, and the effect is more pronounced the higher the PtBS content. As a result, the separation between the maxima of both relaxations ranges from 3 decades in pure PI and PI $\geqslant$50% content blends to ~5 decades for 20% PI blends."

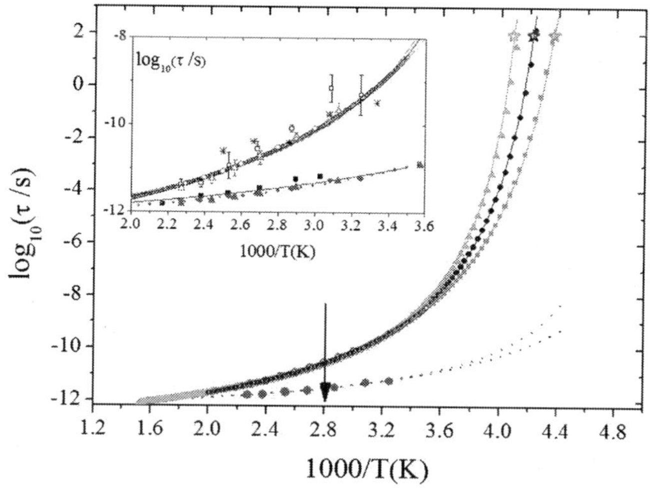

FIGURE 5.31 In the main figure, open circles are deuteron NMR data and fit of segmental α-relaxation time of the PEO component from Lutz et al. in the 20% PEO–80% PMMA blend (see Figure 4 in Lutz et al. (2003). The vertical arrow locates the position of $T_g = 356$ K for the PMMA component in the 20% PEO–80% PMMA blend (see Figure 4 in Lutz et al. (2003). The three lines are VFT fits obtained by fitting NMR data of the PEO component in the 20% PEO–80% PMMA blend and for it to reach T_g with the three possible values of, respectively, $T_g = 245$ K (Mpoukouvalas and Floudas, 2008), $T_g = 237$ K (Goulart Silva et al., 2000), and 229 K (Goulart Silva et al., 2000) (see text). The closed circles are the relaxation times of the fast process obtained from QENS by Garcıa Sakai et al. (2008), and identified as τ_0. The two dotted lines are fits to τ_0 by two possible VFT temperature dependences, VFT1 and VFT2 (see text). The three kinds of symbols are obtained from the CM Eq. (5.38) by substituting VFT1 for τ_0 and applying the n parameter represented by the symbols in Figure 41b. The inset is an enlargement of data and fits over the short timescale. In addition to the same symbols and lines of the main figure, it shows QENS data for PEO component in the 25% PEO–75% PMMA blend from a backscattering spectrometer at $Q = 1.02$ Å^{-1} (Genix et al., 2005) (asterisks). Also from QENS data at $Q = 1.02$ Å^{-1} (Garcıa Sakai et al., 2008) are open squares indicating the slow relaxation, and the closed squares the fast relaxation for the PEO component in the 20% PEO–80% PMMA blend (squares). The closed and open triangles are the same for the 30% PEO–70% PMMA blend. The line is the primitive relaxation time of PEO obtained from NMR data of pure PEO.

The core of the statement, "*the terminal dynamics of PI component shows a stronger T-dependence than its segmental dynamics*" is not consistent with the data of $\hat{\tau}_R$ and $\tau_{\alpha f}$ shown in Figures 5.28 and 5.29, except if one considers the extrapolated values of $\tau_{\alpha f}$ to high temperatures by the VFT fit to data at lower temperatures as performed by the red solid line in Figure 5.28. Past viscoelastic and dielectric measurements on pure polymers (Ngai and Plazek, 1995) including PS (Ngai and Plazek, 1986b), PVAc (Plazek, 1980), PI (Schönhals, 1993), and atactic polypropylene (aPP) (Roland et al., 2001; Plazek and Plazek, 1983; Fytas and Ngai, 1988) all have found that the terminal relaxation has weaker temperature dependence than segmental relaxation at temperatures not far above T_g, and approaches to become about the same at high temperatures. This is the case also for the 20% and 35% PI blends with PtBS as shown in Figures 5.28 and 5.29, especially for the 35% PI blend where the terminal relaxation has weaker temperature dependence than segmental relaxation for all measurement temperatures. This makes more intriguing the observation of the increase of separation between the NM time $\hat{\tau}_R$ and segmental relaxation time $\tau_{\alpha f}$ of PI27 on decreasing concentration of PI27 in the blends at constant $\hat{\tau}_R$. For this reason the observation is highlighted as Property (vi) that needs to be explained for anyone interested in a full understanding of the segmental and terminal relaxations of components in HAPB.

(i).12 Property (vi).

On decreasing the content of the fast PI component in blends with PtBS, both the segmental relaxation and the NM have an increase in nonexponentiality (more stretched) and relaxation time. The effects increase with decrease of temperature. Observed at constant NM relaxation time equal to $\hat{\tau}_R \approx 3.2 \times 10^{-2}$ s is an increase in separation between the NM time $\hat{\tau}_R$ and segmental α-relaxation time $\tau_{\alpha f}$ of PI on decreasing the concentration of PI in the blends. This nontrivial relation between the two processes deserves an explanation.

(i).13 Background of Property (vii)

More fundamental properties of the PVME component in the PVME blends with PS than summarized as Property (v) had been found by Schwartz et al. (2007) on elevating pressure. The dielectric loss spectra of the 75% PVME and 50% PVME blends they published come from the α-relaxation of the PVME component, because the dielectric relaxation time at ambient pressure reaches 1 s at temperatures in the neighborhood of 254 K and 264 K, which are the glass transition temperatures $T_{g,\text{TSDC}}$ determined by TSDC (Leroy et al., 2002) of the two blends, respectively. The α-loss spectra of the PVME component in these two blends obey temperature-pressure superposition; that is, spectra measured at different combinations of pressure and temperature while maintaining the loss peak frequency $f_{\alpha f}$ constant have the same frequency dispersion and the same stretch exponent, $(1-n)$, in the Kohlrasuch function fitting the α-loss peak. This finding of Schwartz et al. is not new. For a component of polymer blend, it was shown before in a 65% PVME blend with PS by Alegrıa et al. (2002), in a 50/50 blend of PVME and poly(2-chlorosytrene) (P2CS) by Roland et al. (2005), and in a 50% poly(cyclohexyl methacrylate) (PCHMA) blend with poly(α-methylstyrene) (PαMS) (Roland and Casalini, 2007). This is a general property of dynamics in polymer blends, which is highlighted as follows.

(i).14 Property (vii).

The shape of the segmental α-relaxation peak of the faster component in polymer blends depends only on the segmental α-relaxation time, $\tau_{\alpha f}$, and is otherwise independent of thermodynamic conditions; that is, various combinations of pressure and temperature that keep $\tau_{\alpha f}$ constant.

(i).15 Background of Property (viii)

Schwartz et al. (2007) also have made dielectric relaxation measurements on the 25% PVME/75% PS blend at ambient and elevated pressure. Already shown in the previous section, the loss spectra they observed in the 25% PVME blend are not from the α-relaxation (not to be confused with the case of the 75% and 50% PVME blends), but are from the JG β-relaxation of the PVME component. The temperature-pressure superposition of the loss spectra of the JG β-relaxation also was found by Schwartz et al. A well-known property of JG β-relaxation is that its relaxation time, $\tau_{\mathrm{JG}}(T)$, changes its temperature dependence from Arrhenius below glass transition temperature to a stronger T-dependence above it. This occurs in neat glass-formers (Ngai, 2011), as well as in mixtures of van der Waals glass-formers where the glass transition temperature is that of the fast component, T_{gf} (Capaccioli et al., 2012; Mierzwa et al., 2008; Kessairi et al., 2008). We have seen this also in the ambient pressure data of $\tau_{\mathrm{JG}}(T)$ from Lorthioir et al. (see Figure 5.25), and also in the ambient and elevated pressure data of Schwartz et al. (2007) shown in Figure 6, and reproduced here as Figure 5.27 with illustrations added to make the following points clear. In this figure, the vertical broken red[3] and the blue arrows indicate at ambient pressure the location of $1000/T_{g,\mathrm{TSDC}}$, with $T_{g,\mathrm{TSDC}} \approx$ 289 K of the PVME component and $1000/T_{g,\mathrm{DSC}}$, with $T_{g,\mathrm{DSC}} \approx 321$ K of the blend obtained by interpolation of TSDC and DSC data, respectively, from Leroy et al. (2002). While there is a clear change in T-dependence of $\tau_{\alpha'}(P,T)$ or τ_{JG} (P,T) at $T_{g,\mathrm{TSDC}} \approx 289$ K, there no evidence of such at $T_{g,\mathrm{DSC}} \approx 321$ K.

The data taken by Schwartz et al. at elevated pressures P were not taken at sufficiently closely spaced temperatures to exactly locate the crossover in T-dependence of $\tau_{\mathrm{JG}}(P,T)$, but the presence of a change of its curvature on decreasing temperature ensures there is a crossover. In Figure 5.27, lines are drawn to connect points at lower temperatures where the curvature has changed. Aided by these lines, the crossover seems to occur uniformly at the same value of the relaxation time, $\tau_{\mathrm{JG}}(P,T) \approx 10^{-4.8}$ s, independent of P. Moreover, the relaxation time at the crossover is close to that of the 20% PVME blend at $T_{g,\mathrm{TSDC}} \approx 293$ K shown previously in Figure 5.25. This property of $\tau_{\alpha'}(P,T)$ or τ_{JG} (P,T) found by Schwartz et al. is another proof that the observed process is the JG β-relaxation of the PVME component.

Since upon applying pressure the relation between the α-relaxation times and the glass transition temperatures, T_{gf} and T_{gs}, of PVME and PS changes,

---

[3] For interpretation of color in Figures 5.24, 5.27, and 5.32 the reader is referred to the web version of this book.

the constancy of $\tau_{JG}(P,T)$ at the crossover is remarkable. For this reason, this result is formulated as another notable property of the dynamics of the fast component in HAPB as follows.

(i).16 Property (viii).

The temperature dependence of the JG β-relaxation time, τ_{JG}, of the PVME component in a 25% PVME/75% PS blend at ambient pressure, $P_a = 0.1$ MPa, changes from VFT-dependence to Arrhenius T-dependence at the glass transition temperature, T_{gf}, of the PVME component. At the crossover, the relaxation time, $\tau_{JG}(P_a,T_{gf})$, is approximately $10^{-4.8}$ s. At any of the elevated pressures, P, up to 300 MPa, the same change was observed, and it occurs at temperature, T_P, where $\tau_{JG}(P,T_P)$ is about the same as $\tau_{JG}(P_a,T_{gf}) \approx 10^{-4.8}$ s.

Unfortunately, the glass transition temperature, $T_{gf}(P)$, of the PVME component at any of the elevated pressures was not measured, and it cannot be verified that T_P is approximately the same as $T_{gf}(P)$, as in the case of ambient pressure. The colored vertical arrows in Figure 5.27 mark the positions where the system (likely the slow majority component PS) enters into the glassy state, as determined by PVT data. Nevertheless, it is reasonable to assume that the crossover at elevated pressure is also at $T_{gf}(P)$ like at ambient pressure. If this assumption is correct, and the fact that the α-relaxation time of the PVME component, $\tau_\alpha(P,T_{gf}(P))$, is the same, say, 10^2 s at $T_{gf}(P)$ for all P including P_a, then Property (iv) can be restated by the following alternative.

(i).17 Property (viii′).

The ratio of the α-relaxation and the JG β-relaxation times, $\tau_\alpha(P,T_{gf}(P))/\tau_{JG}(P,T_{gf}(P))$, at the glass transition temperature, $T_{gf}(P)$, of the PVME component is independent of pressure, P.

Restated this way, Property (viii′) of the PVME component in the blend is shared by some neat glass-formers (Capaccioli et al., 2012; Ngai, 2011) and mixtures of van der Waals liquids (Mierzwa et al., 2008; Kessairi et al., 2008; Ngai, 2011), and the explanation of it readily follows from the coinvariance of τ_α, τ_{JG}, and n to changes of pressure and temperature at either constant τ_α or constant τ_{JG}, found by experiments and computer simulations (Capaccioli et al., 2012; Mierzwa et al., 2008; Kessairi et al., 2008; Ngai, 2011; Ngai et al., 2005; Bedrov and Smith, 2011; Roland et al., 2010), and also from the CM (Eq. (5.27)) when combined with $\tau_{JG} \approx \tau_0$.

Schwartz et al. have provided data showing the change in relation between T_{gf} and T_{gs} on elevating pressure, which are reproduced here in Figure 5.27. The intersection of the dotted diagonal line in the figure with each curve is the temperature, T_{PS}, below which the component PS is out of equilibrium, and therefore it is the temperature at which confinement effect starts according to the confinement scenario of Lorthioir et al. (Lorthioir et al., 2003a; Colmenero and Arbe, 2007). It can be seen by inspection that the value of $\tau_{\alpha'}(P,T_{PS})$ of the confinement scenario ($\tau_{JG}(P, T_{PS})$ in our interpretation) changes significantly with pressure and bears no relation with the relaxation time $\approx 10^{-4.8}$ s at the crossover for all pressures, providing yet another doubt on the confinement hypothesis.

(i).18 Background of Property (ix)

Concentration fluctuations are prevalent at some compositions of polymer blends, resulting in spatial heterogeneity and broadening of the frequency dispersion of the segmental α-relaxation in polymer blends (Shears and Williams, 1973; Zetsche et al., 1990). This factor was the center of attention in models by blend dynamics by Zetsche and Fischer (1994) and Katana et al. (1995), the CM (Roland and Ngai, 1991; Roland and Ngai, 1992), and by Kumar et al. (1996) and Kamath et al. (2003). Observed experimentally, the frequency dispersion of the segmental α-relaxation of the fast component exhibits marked dependence on temperature and composition, particularly in the case of HAPB. This was first shown by Zetsche et al. (1990) and Zetsche and Fischer (1994) for a 0.6PVME/0.4PS blend, and also by Cendoya et al. for 0.5PVME/0.5PS blends (see Figure 1c in Cendoya et al. (1999)). One particularly interesting feature is that the dielectric loss spectrum of the PVME component is broadened more on the low frequency side of the peak. The effect becomes more pronounced as temperature is lowered toward T_{gf} of the PVME component, and time-temperature superposition fails badly.

Similar behavior and breakdown of time-temperature superpositioning was found in the global chain relaxation of PI in a 50% PI/50% PtBS blend by Chen et al. (2008) and Watanabe et al. (2007), and in a 43% PI/57% PtBS blend by Arrese-Igor et al. (2011). Thus, another outstanding problem in the spectral dispersion of both the segmental and global chain relaxations of the fast component in HAPB to address is formulated as Property (ix) as follows.

(i).19 Property (ix)

Caused by concentration fluctuations, the α-relaxation of the fast component is spatially heterogeneous. Slower α-relaxation naturally occurs in location with environment richer in the slow component, and its friction factor has stronger temperature dependence, consistent with the observed asymmetric broadening of the dispersion as observed experimentally. Consequently, time-temperature superposition fails. The same effect was found in the global chain dynamics of the fast component. This property of segmental α-relaxation caused by concentration fluctuations was a major concern of theory of component dynamics in polymer blends at earlier times, but it seems to be forgotten or not emphasized in current activities. Instead the emphasis was put on fitting the T_g of the components to the LM model or considering the possibility of the "confinement" scenario. However, these two models in their present forms cannot address this property.

5.7.3 Explanation of Properties (i)–(ix)

Molecular glass-formers including amorphous polymers are constituted of basic structural units interacting with each other with nontrivial potential. Naturally, motion of units cannot be independent and certain many-body consideration is required for a fundamental understanding of the segmental α-relaxation and chain relaxation in polymers. Usually the potential is anharmonic, and the phase space exhibits chaos. Various versions of the CM (Ngai, 1979; Ngai et al., 1988; Ngai and Rendell, 1990; Ngai et al., 1992b; Tsang and Ngai, 1997; Ngai and Tsang, 1999) based on nonlinear Hamiltonian mechanics have shown that chaos in the phase space is responsible for slowing down the relaxation, changing the time dependence of the correlation function from $\exp(-t/\tau_0)$ for one-body and independent relaxation to the Kohlrausch stretched exponential, $\exp[-(t/\tau)^{1-n}]$, for many-body relaxation. However, the onset of chaos is at some time, t_c, before which one-body relaxation prevails. Thus the overall correlation function, $\phi(t)$, exhibits a crossover from $\exp(-t/\tau_0)$ to $\exp[-(t/\tau)^{1-n}]$. The crossover takes place over a neighborhood centered at t_c within which $\phi(t)$ and its derivatives are continuous. The intermolecular potential determines t_c for segmental α-relaxation, while interchain coupling/constraint determine t_c for chain relaxation and diffusion. The crossover has been found by quasielastic neutron scattering and molecular dynamics simulations for segmental α-relaxation in pure polymers (Colmenero et al., 1992; Ngai, 2011) to occur at t_c = 1–2 ps, and at longer t_c = 1–2 ns for chain relaxation of entangled homopolymers (Ngai, 2011; Richter et al., 1989; Richter et al., 1990) and of the component PEO in blends with PMMA (Niedzwiedz et al., 2007; Brodeck et al., 2010). From the crossover, relation between the many-body relaxation time, τ, and one-body and independent (primitive) relaxation time, τ_0, given by

$$\tau = [(t_c)^{-n}\tau_0]^{1/(1-n)} = \tau_0(\tau_0/t_c)^{n/(1-n)}, \tag{5.44}$$

if $\tau \gg t_c$, and

$$\tau = [(1-n)(t_c)^{-n}\tau_0]^{1/(1-n)} = (1-n)^{1/(1-n)}\tau_0(\tau_0/t_c)^{n/(1-n)}, \tag{5.45}$$

if otherwise. For consideration of segmental relaxation in polymers, because of polymer chain connectivity, the correlation function of the primitive relaxation is not the exponential function but rather the Hall-Helfand function (Hall and Helfand, 1982). Notwithstanding, the slow many-body relaxation is still well described by the stretched exponential (Ngai and Rendell, 1991). Despite not being an exponential function of time, the Hall-Helfand function still has a primitive relaxation rate, $(\tau_0)^{-1}$, and Eqs. (5.44) and (5.45) remain unchanged.

(i) Explanation of Property (i)

(i).1 The Previous Attempt

When applied to the fast component of polymer blends and binary mixtures (Ngai and Roland, 2004a; Capaccioli and Ngai, 2005), the key input of the CM is the increase of the most probable coupling parameter, n (taken from a distribution due to concentration fluctuation), of the fast component on increasing the concentration of the slow component due to enhanced intermolecular coupling. This is supported experimentally by the broadening of the α-relaxation observed even at low concentration of the fast component where concentration fluctuation is not important (Kessairi et al., 2008; Lutz et al., 2003; Capaccioli and Ngai, 2005). In the study of the fast component by low frequency dielectric and mechanical relaxation or solid state ^{13}C MAS NMR (Ngai and Roland, 2004b), the primitive relaxation time, τ_0, is much longer than $t_c = 1$–2 ps because the measured $\tau_{\alpha f}$ is exceedingly long compared to $t_c = 1$ ps. Large change of $\tau_{\alpha f}$ is expected from Eq. (5.56) by the change of n on accompanying the change in the composition of the blend. This is normally the case and observed in low frequency measurements. The situation and consequences are very different for measurements made at high frequencies by deuteron NMR (Lutz et al., 2003) and quasielastic neutron scattering (Garcıa Sakai et al., 2008; Garcıa Sakai et al., 2005) experiments at temperatures where τ_0 is not much longer than $t_c = 1$ ps in the study of PEO in blends with PMMA. In this case, since the ratio τ_0/t_c is not large, Eq. (5.45) predicts small change of $\tau_{\alpha f}$ for the same change of n or the same change in the composition of the blend, which show exceedingly large change of $\tau_{\alpha f}$ in low frequency dielectric relaxation measurements. This is essentially the explanation of the insensitivity of $\tau_{\alpha f}$ of PEO in blends with PMMA to change of composition observed by high frequency deuteron NMR (e.g., Property (i)) given in the 2004 publication (Ngai and Roland, 2004b). At that time, experimental data of τ_0 was not available for either pure PEO or PEO as a component in blends. To proceed, use was made of the CM equation to calculate $\tau_0(T)$ of pure PEO from its $\tau(T)$ obtained by deuteron NMR (Lutz et al., 2003) and taking $n = 0.5$. Assumption was made that $\tau_0(T)$ of component PEO in blends with PMMA is the same as that of pure PEO. $\tau_{\alpha f}(T)$ for the PEO component in the blends was calculated by Eq. (5.56), and the value of n larger than 0.5 was needed to fit the deuteron NMR data of PEO component in the blends. For blend compositions ranging from 3% to 30% PEO, the deuteron NMR data of $\tau_{\alpha f}(T)$, denoted before by $\tau_{\mathrm{NMR}}(T)$, which change slightly with composition and temperature, had been fitted well by n ranging from 0.715 (for 30% PEO) to 0.760 (for 3% PEO). The data of $\tau_{\alpha f}(T)$ (lines) and the fits (small-size symbols at discrete temperatures) together with the assumed $\tau_0(T)$ of component PEO in blends (line) are reproduced in Figures 5.24 and 5.26.

The situation is very different in the case where $\tau_0(T)$ is much longer than that of the deuteron NMR experiment, the same $n = 0.50$ for pure PEO, 0.715 for 30% PEO, and 0.760 for 3% PEO will have their calculated $\tau_{\alpha f}(T)$ differ by orders of magnitude. This case would be realized experimentally had it been possible to measure $\tau_{\alpha f}(T)$ of component PEO in the same blends by low frequency dielectric relaxation. Unfortunately, this is not possible even by the detailed dielectric measurements carried out by Jin et al. (2004). However it is possible by dielectric measurements in the HAPB, x% PI/(1 − x)% PtBS (Arrese-Igor et al., 2010), where the great sensitive of $\tau_{\alpha f}(T)$ to concentration of PI was found. In summary, the 2004 paper (Ngai and Roland, 2004b) has demonstrated with assumptions that the anomalous PEO component dynamics from deuteron NMR data (e.g., Property (i)) is due to $\tau_0(T)$ not much longer than t_c in the experimental temperature range.

(i).2 Support from Quasielastic Neutron Scattering Experiments

After the 2004 paper had been published, quasielastic neutron scattering (QENS) experiments on xPEO/(1 − x)PMMA blends with x = 10, 20, 30, and 100% were performed, and results reported by Garcıa Sakai et al. (2008) and Garcıa Sakai et al. (2005). By using deuterium labeling, the dynamics of the PEO component and its characteristic relaxation times were determined over spatial scales from 3 to 10 Å. It is only by now that we come to realize that the QENS data not only support the explanation of the deuteron NMR data made in the 2004 paper, but also provide solid evidence for the existence of the primitive relaxation time, $\tau_0(T)$, of the PEO component. In fact the data taken in the ps-ns time range reveal the crossover from a fast (F) relaxation of the PEO component to a slow (S) relaxation near $t_c = 1$ ps, close to $t_c = 2$ ps assumed in the 2004 paper (Ngai and Roland, 2004b). The fast F-process of PEO in the blends as well as in pure PEO observed by QENS (Garcıa Sakai et al., 2008) is exactly the primitive process of the CM. The relaxation times, $\tau_{\mathrm{QENS},F}(T)$ and $\tau_{\mathrm{QENS},S}(T)$, of the F- and S-relaxations of PEO, respectively, obtained from QENS by Sakai et al. for pure PEO and PEO in blends with 10, 20, and 30% PEO at momentum transfer $Q = 1.3$ Å^{-1} are shown in Figure 5.24 (large-size symbols). For comparison, shown also are the $\tau_{\mathrm{NMR},S}(T)$ from NMR (only the VF fits of the data of compositions ranging from 30 to 3% PEO are shown by lines), and the calculated $\tau(T)$ by the CM, exactly as shown before in Figure 1 of the 2004 paper. As can be seen by inspection, $\tau_{\mathrm{QENS},F}(T)$ is insensitive to blending with PMMA, exactly the same as assumed for $\tau_0(T)$ in the 2004 paper (Ngai and Roland, 2004b). The temperature dependence of $\tau_{\mathrm{QENS},F}$ is Arrhenius with activation energy of 10–15 kJ/mol, independent of composition and scattering vector Q, consistent with conformational transition in PEO (Garcıa Sakai et al., 2008). The Q-dependence of $\tau_{\mathrm{QENS},F}$ is a power law $Q^{-\lambda}$ with λ varying between 2.0 and 2.5 (Garcıa Sakai et al., 2008). All these properties of $\tau_{\mathrm{QENS},F}$ are indications that the fast process is the one-body or primitive α-relaxation of the CM.

The slow QENS relaxation time, $\tau_{\mathrm{QENS},S}(T)$, is slightly shorter than $\tau_{\mathrm{NMR},S}(T)$ from NMR, but the temperature dependence is similar except at short times. Actually the QENS data of $\tau_{QENS,S}(T)$ should be more accurate than NMR data in the short time range because the highest frequency in NMR measurement is 76 MHz corresponding to $10^{-8.7}$ s. Remarkably, the value of $n \approx 0.5$ equivalent to the experimental value of the stretch exponent, $(1-n) \approx 0.5$, for pure PEO, and $n \approx 0.75$ from $(1-n) \approx 0.25$ for 20% PEO from QENS experiments are close to the values used in the 2004 paper to calculate $\tau_{\alpha f}(T)$ by the CM equation. Notwithstanding, it is possible that the actual coupling parameter n of pure PEO in the ps-ns range is not as large as 0.50 as obtained directly from the stretch exponent $(1-n) \approx 0.50$ of Kohlrasuch function used to fit the QENS and the deuteron NMR data. Because of chain connectivity, the correlation function of the primitive α-relaxation in some polymers may have the Hall-Helfand functional form that already deviates from the exponential function. To cover this possibility, we have calculated $\tau_0(T)$ from $\tau(T)$ with $n = 0.4$ and 0.3. The results shown in Figure 5.24 are also consistent with $\tau_{QENS,F}$.

Of great interest here is the insensitivity of the relaxation time, $\tau_{QENS,S}$, of the slow QENS process to compositions of the three blends studied, as seen before in the deuteron NMR experiment and considered as an anomalous behavior of blend dynamics (e.g., Property (i)). But now the fast F- or primitive relaxation is identified by QENS, and its crossover to the slow S- or many-body relaxation has been observed by Sakai et al. Therefore $\tau_{QENS,S}(T)$ and $\tau_{QENS,F}(T)$ obey Eq. (5.2), after substituting τ_α by $\tau_{QENS,S}$ and τ_0 by $\tau_{QENS,F}$. Moreover, the ratio, $\tau_{QENS,F}(T)/t_c$, is not large, less than a factor of about 5.6 in the entire temperature range where $\tau_{QENS,S}(T)$ were available for the 10, 20, and 30% PEO blends. From these two points, it is totally justified to conclude that the CM provides an explanation of Property (i). In the temperature ranges where measurements were made by deuteron NMR or QENS, $\tau_{QENS,F}(T)$ of the 10, 20, and 30% PEO blends is not only insensitive to composition and falls within the range from 10^{-12} to 10^{-11} s, but also is nearly the same as $\tau_{QENS,F}(T)$ of 100% PEO determined experimentally (Garcıa Sakai et al., 2008; Garcıa Sakai et al., 2005) or $\tau_0(T)$ obtained by calculation from $\tau_{QENS,S}(T)$ via Eqs. (5.44) or (5.45) with $n = 0.30$–0.50 and $t_c = 1$ ps (Ngai and Roland, 2004b). The insensitivity of $\tau_{QENS,F}(T)$ to composition is likely because the primitive relaxation is a rather local motion. In conjunction with Eqs. (5.56) or (5.57), this is the crux of the effect observed that $\tau_{QENS,S}(T)$ of the PEO component remains about 12 decades shorter than the relaxation time, $\tau_{\alpha s}$, of the PMMA component at its T_g where $\tau_{\alpha s}$ reaches 100 s in blends with 3, 10, 20, and 30% PEO.

The inset of Figure 5.24 shows once more the $\tau_{QENS,S}(T)$ data of PEO in blends with 30 and 20% PEO from Sakai et al. together with that of 25% PEO from Genix et al. (2005). The insensitivity of $\tau_{QENS,S}(T)$ to composition at high frequencies of measurements (i.e., Property (i)) is demonstrated once more. The red line drawn connecting two data points of Genix et al. at lower temperature is the same as the construction by Colmenero and Arbe (2007) to indicate the crossover to Arrhenius temperature dependence of the PEO relaxation time due to the confinement scenario. The short time scale at the suggested crossover seems inconsistent with the confinement hypothesis.

(i).3 Property (i) Does Not Persist at Lower Temperatures and Longer $\tau(T)$ of the PEO Component

The following question naturally arises: Does Property (i) persist at lower temperatures where the segmental α-relaxation time, $\tau_{\alpha f}(T)$, of the PEO component becomes much longer than that observed by deuteron NMR or QENS? Low frequency ($f < 10^6$ Hz) dielectric relaxation would be ideal for this task, but unfortunately it is extremely difficult to resolve the segmental α-relaxation of the PEO component in the 20% and 10% PEO blends with PMMA as found out by Jin et al. (2004). This is due to interference from the dielectric loss from the intense secondary relaxation of PMMA over broad frequencies. Nevertheless, DSC or calorimetry measurements made by Mpoukouvalas and Floudas (2008) in 20% PEO blends with M_w(PEO) = 10^5 g/mol have detected glass transition of the PEO component with $T_g = 245$ K. Goulart Silva et al. (2000) found DSC $T_g = 229$ K and 237 K of the PEO component in 20% PEO blends with M_w(PEO) = 10^5 and 10^6 g/mol. Lodge et al. (2006) also found DSC $T_g = 237$ K in 25% PEO blends with M_w(PEO) = 300 g/mol. The DSC data show the segmental α-relaxation time of the PEO component reaches typically 10^2 or 10^3 s at the T_g so determined. Despite the absence of $\tau_{\alpha f}(T)$ of the PEO component at lower temperatures than NMR data provide, it should fall somewhere likely as suggested by the VFT dependence, $\log_{10}(\tau_\alpha) = A + B/(T - T_0)$, interpolating the deuteron NMR data and any one of the DSC data, assuming $\tau_{\alpha f}(T_g) = 10^2$ s at T_g of the PEO component. Three possible values of T_g of the PEO component from DSC are 245 K (Mpoukouvalas and Floudas, 2008), 237 K (Goulart Silva et al., 2000), and 229 K (Goulart Silva et al., 2000). Such VFT interpolations have been constructed and shown in Figure 5.31. The parameters for these VFT$_{\alpha i}$ fit for $i = 1$, 2, and 3 are $A_1 = -12.55, B_1 = 251$ K$, T_{0,1} = 227$ K; $A_2 = -12.76, B_2 = 306$ K$, T_{0,2} = 216$ K; $A_3 = -12.96, B_3 = 366$ K$, T_{0,3} = 205$ K, respectively, for $T_{gi} = 245$ K, 237 K, and 229 K, the reciprocal of which are shown by stars of the same color.

The relaxation times of the fast process obtained from QENS by García Sakai et al. (2008) and identified as τ_0 are shown in the main figure by the red closed circles. Fits to τ_0 by two possible VFT temperature dependencies, $\log_{10}(\tau_0) = A + B/(T - T_0)$, are also shown by the black and red dotted lines. The parameters of these two VFT dependencies, VFT1 and VFT2, are, respectively, $A_{VFT1} = -12.36, B_{VFT1} = 154$ K$, T_{0,VFT1} = 176$ K$, A_{VFT2} = -12.27, B_{VFT2} = 122$ K, and $T_{0,VFT2} = 195$ K. From the VFT1 or VFT2 for τ_0 we calculate $\tau_{\alpha f}(T)$ of the PEO component from the CM (Eq. (5.39)) to match the VFT$_{\alpha i}$ fits for $i = 1$, 2, and 3. This is done by letting n be a function of T and used as the sole fitting parameter. The $n(T)$ obtained are shown in Figure 5.32, separately for the two choices, VFT1 and VFT2, for τ_0. The $\tau_{\alpha f}(T)$ obtained this way are shown by the magenta, black, and green symbols in good agreement with the VFT$_{\alpha i}$ lines, $i = 1$, 2, and 3, lines with the same color. At higher T, n is the same for all of them and it stays around 0.65 and 0.67, then n increases on cooling, and it reaches different values $n(T_{gi})$ at T_{gi}, depending on the value of T_{gi}. We find $n(T_{gi}) = 0.82$,0.84, and 0.86 for $i = 1$, 2, and 3, respectively.

We show the deuteron NMR and QENS data and the fits over the short timescale more clearly in the inset of Figure 5.31. In addition to the same symbols and lines of the main figure, it shows QENS data for PEO component in 25% PEO–75% PMMA blend from a backscattering spectrometer at $Q = 1.02$ Å^{-1} (Genix et al., 2005) (asterisks). Also from QENS data at $Q = 1.02$ Å^{-1} (Garcıa Sakai et al., 2008) are open and closed black squares indicating the slower and the faster relaxations for the PEO component in 20% PEO–80% PMMA blend. The red the closed and open triangles stand for the same in the case of the 30% PEO–70% PMMA blend. The blue line is the primitive relaxation time τ_0 of PEO obtained from NMR data of pure PEO. The inset carries the message of good agreement between the faster relaxation observed by QENS is in good agreement with the calculated τ_0 confirming the reality of the primitive relaxation. Also there is good agreement between QENS and deuteron NMR on the slower relaxation time.

The values of $n(T_{gi}) = 0.82$, 0.84, and 0.86 at T_{gi} for $i = 1$, 2, and 3, respectively, shown in Figure 5.32, are needed to account for the VFT$_{\alpha i}$ dependence of $\tau_{\alpha f}(T)$ in Figure 5.31. The full-width at half-maximum (FWHM) of the loss spectra corresponding to KWW functions with $n = 0.65$, 0.80, 0.82, 8.4, and 8.6 are 3.07, 5.0, 5.9, 6.7, and 7.5 decades of frequency, respectively. The large FWHM of loss peak at T_{gi} of the PEO component has support from the dielectric loss spectra of PVME component in 20% PVME/80% PS blends shown before in the lower right inset in Figure 5.25, where the α-loss peak of PVME near T_{gf} of PVME is almost flat. Another example is loss peak of the PI component in 20% PI/80% PtBS blends in Figures 3a and 3b of Arrese-Igor (Arrese-Igor et al., 2010). In this blend, the α-loss peak is already very broad at 6–9° above T_{gf} of PI.

Returning to the question posed at the very beginning of this subsection, the answer to it from Figure 5.31 is as follows. Property (i) does not persist at lower temperatures where $\tau_{\alpha f}(T)$ is much longer than that determined by deuteron NMR. By the way, any interpolation by Arrhenius T-dependence to cover the temperature range between the $\tau_{\alpha f}(T_g)$ at the DSC T_g of the PEO component and the deuteron NMR data, which are extended below T_g of the PMMA component (356 K from Lutz et al. (2003), 345 and 333 K from Goulart Silva et al. (2000)) as possibly suggested by the confinement scenario in the version given by Genix et al. (2005) (mentioned in Section (i).7), will end up with a high activation energy of the order of 2 eV or nearly 200 kJ/mol, which is much too large for the local relaxation predicted by the confinement hypothesis.

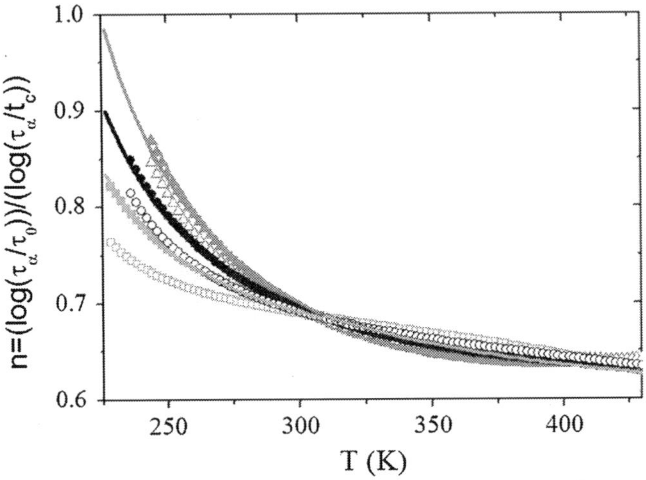

FIGURE 5.32 The coupling parameters $n(T)$ as obtained from the CM (Eq. (5.38)) from the separation between τ_0 and τ_α:τ_0 has been extrapolated according to VFT1 and VFT2 fits in Figure 41a; τ_α are from fitting together the data of Lutz et al. (2003) and three different possibilities for the T_g of PEO component in 20% PEO–80% PMMA blend as obtained from DSC experiments. The solid symbols are from choosing VFT1 for τ_0, the open ones from choosing VFT2 for τ_0. The different symbols are for $n(T)$ with $T_{gi} = 245$ K (Mpoukouvalas and Floudas, 2008), 237 K (Goulart Silva et al., 2000), and 229 K (Goulart Silva et al., 2000), respectively. The lines are guides for the eye.

The impediment to giving unambiguous interpretation of the PEO component dynamics is due to the inability to detect the segmental relaxation of the PEO component in blends with PMMA by dielectric relaxation at frequencies interpolating the high frequency deuteron NMR or the QENS and the very low frequency DSC. The impediment is overcome in the study of another highly asymmetric polymer blend, PI/PtBS. The segmental relaxations of 20% and 35% PI in the blends were resolved by dielectric spectroscopy and detected by DSC, whereby the dielectric and calorimetric T_g of the PI component had been obtained (Arrese-Igor et al., 2010). The dielectric and calorimetric segmental relaxation times, $\tau(T)$, of the PI component in the range, $10^{-7} < \tau(T) < 10^3$ s, increase rapidly on decreasing temperature. This is another example showing Property (i) does not persist at lower temperatures where the segmental α-relaxation time, $\tau(T)$, of the fast component in HAPB becomes much longer than that observed by deuteron NMR or QENS. All the features of $\tau_{\alpha f}(T)$ of the fast PI component in blends with PtBS referred to in the preceding can be seen from Figures 5.30 and 5.31, where the data of $\tau_{\alpha f}(T)$ from Arrese-Igor et al. (2010) and the chain normal mode relaxation time, $\tau_{nf}(T)$, of PI from Arrese-Igor et al. (2011) are reproduced. Not done by Arrese-Igor et al., the display of $\tau_{\alpha f}(T)$ and $\tau_{nf}(T)$ together in the same plot will facilitate the discussion of the relation between segmental α-relaxation and chain relaxation in the next section ((ii)). These figures show $\tau_{\alpha f}(T)$ of the segmental relaxations of 20% and 35% PI in the blends resolved by dielectric relaxation, and also the T_{gf} of PI and T_{gs} of PtBS, the two components by DSC. The dielectric and calorimetric segmental relaxation times, $\tau_{\alpha f}(T)$, of the PI component in the range, $10^{-7} < \tau(T) < 10^{-1}$ s, increase rapidly on decreasing temperature. This is another way to show Property (i) does not persist at lower temperatures where the segmental α-relaxation time, $\tau_{\alpha f}(T)$, of the fast component in HAPB becomes much longer than that observed by deuteron NMR or QENS.

(ii) Explanation of Property (ii)

To explain the observed anomalies in Property (ii), Niedzwiedz et al. (2007) and Brodeck et al. (2010) had proposed the random Rouse model (RRM) in which each monomer has different mobility taken from a broad log-normal distribution. However, Diddens et al. (2011) extracted the distribution of friction coefficients from the MD simulations of a PEO/PMMA blend (Brodeck et al., 2010) and found the distribution is much narrower than expected from the RRM. A simpler explanation of the data and Property (v) without additional assumption of the RRM was proposed in the framework of the CM (Ngai and Wang, 2011) by utilizing alone the observed crossover of PEO chain dynamics at $t_c \approx 1$–2 ns by neutron scattering and simulations (Niedzwiedz et al., 2007; Brodeck et al., 2010). The crossover is just a special case of a general property of relaxation in coupled or interacting systems, which is the crossover from independent relaxation to coupled/constrained many-body relaxation at some t_c determined either by the interaction potential or the strength of the coupling/constraint, depending on the circumstance. The general property is the consequence of classical chaos in phase space originating from the anharmonic interaction potential or coupling/constraint, which causes slowing down of the primitive or independent relaxation rate (Ngai, 1979; Ngai et al., 1988; Ngai and Rendell, 1990; Ngai et al., 1992b; Tsang and Ngai, 1997; Ngai and Tsang, 1999; Ngai, 2011). The onset of chaos starts only at the characteristic time, t_c, and hence the crossover from primitive relaxation to slowed down many-body relaxation occurs at t_c. This is the key result utilized by the Coupling Model. In contrast, the Rouse modes in pure and unentangled PEO are totally entropic and uncoupled (Ferry, 1980; Ngai et al., 1997a), and they should remain the same for all times without showing the crossover because of the absence of PMMA and the constraints imposed. This contrasting behavior has been verified by quasielastic neutron scattering study of pure and unentangled PEO (Niedzwiedz et al., 2008).

Moreover according to the CM (Ngai et al., 1997a), in pure PEO the Rouse modes or its friction factor, $\zeta(T)$, should have the same temperature dependence as that of the primitive relaxation time, $\tau_0(T)$, related to the segmental α-relaxation time, $\tau(T)$, by Eq. (5.39). We can verify this prediction by comparing

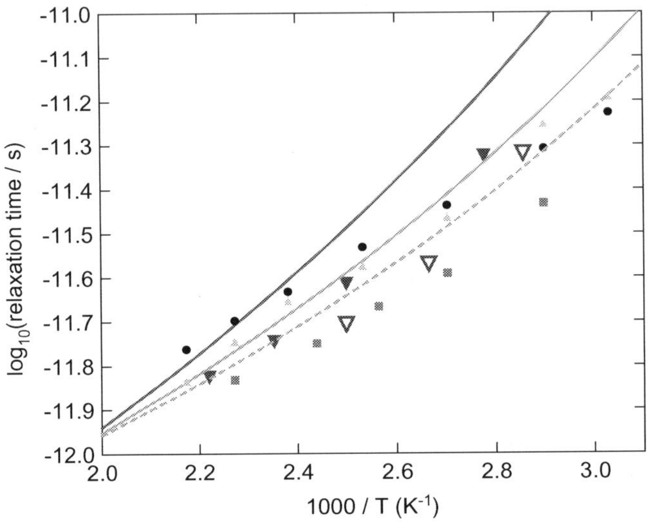

FIGURE 5.33 Temperature dependence of the monomeric friction coefficient of the Rouse chain dynamics obtained experimentally by neutron spin echo for the sample with molecular weights = 2.4 and 81 kg/mol (large open and closed inverted triangles, respectively) from Niedzwiedz et al. (2008) compared with the fast relaxation time, $\tau_{QENS,F}$, from the QENS measurements of the segmental relaxation of PEO in the blends (30% PEO, squares; 20% PEO, triangles, and 10% PEO, circles) shown before in Figure 5.34, or with the temperature dependence of $\tau_0(T)$ of pure PEO calculated by the CM equation with $\tau_{NMR}(T)$ of pure PEO from NMR measurements (line) identified as $\tau(T)$ in the equation with $n = 0.30$ (dashed line), and $n = 0.20$ (line in the middle).

the monomeric friction coefficient, $\zeta(T)$, of the Rouse modes obtained experimentally by neutron scattering by Niedzwiedz et al. (2008) for the sample with molecular weights of 2.4 and 81 kg/mol with $\tau_{QENS,F}$ from the QENS measurements of the segmental relaxation of pure PEO and PEO in the blends shown before in Figure 5.24, or with the temperature dependence of $\tau_0(T)$ calculated from Eq. (5.44) with $\tau_{NMR}(T)$ from NMR measurements identified as $\tau(T)$. The results shown in Figure 5.33 support the prediction. The data of the sample with $M_w = 81$ kg/mol were put together with the data of the other sample by assuming the same distance from the glass transition temperature.

Thus Property (ii) can be derived from the slowing down of the PEO (primitive) Rouse relaxation rate after 1 or 2 ns caused by the constraints imposed on the PEO chains by the much slower PMMA chains. This property of PEO component chain dynamics in blends with PMMA is analogous to Property (i) of segmental dynamics of the PEO component (Ngai and Wang, 2011). In both cases there is crossover from primitive relaxation to coupled (many-body) relaxation, albeit $t_c \approx 2$ ns for chain dynamics compared with 1 ps for segmental relaxation. This is due to weaker coupling of chain modes having longer length scale than segmental relaxation. Crossover properties of the two different relaxation processes are similar, despite the fact that the length scales of the relaxation in these systems are widely different. This is because the physics of the crossover for chain relaxation and segmental relaxation of the PEO component in the blends are the same as far as the CM is concerned. Therefore, Eqs. (5.56) and (5.57) are applicable to explain the chain dynamics properties after replacing τ_0 by $\tau_R(T,Q)$, τ by $\tau(T,Q)$, and n by n_c, where now n_c denotes the interchain coupling parameter. The generality of crossover from (primitive) Rouse dynamics to coupled chain dynamics is brought out by the observation in *entangled* chains relaxation in monodisperse *homopolymers* by neutron scattering at $t_c \approx 1$ ns (Jin et al., 2004; McKenna et al., 1985). Entanglements impose constraints on each chain and modify its dynamics from the primitive Rouse dynamics to coupled many-chain dynamics. Since the crossover exist $t_c \approx 1$ ns, the terminal relaxation time of entangled chains, $\tau(T,M)$, and its dependence on temperature and molecular weight, M, can be obtained from Eqs. (5.44) and (5.45) after replacing τ_0 by $\tau_R(T,M)$, τ by $\tau(T,M)$, and n by n_{ent}, where now n_{ent} denotes the entanglement coupling parameter. This generates the $M^{3.4}$-dependence of $\tau(T,M)$ and larger activation energy of the Arrhenius T-dependence for polyethylene and hydrogenated polybutadiene, which have explained the experimental data as done before (Ngai et al., 1997a; Ngai, 2011; McKenna et al., 1985). Although entanglements in homopolymer are different constraints than that imposed by the less mobile PMMA on unentangled PEO chains in the PEO/PMMA blends, the physical principle behind the Coupling Model applies. For this reason, the crossover from Rousean to non-Rousean dynamics is found in both cases.

(ii).1 The Chain Dynamics of the PEO Component Show Property (i) of Segmental Relaxation When $\tau_R(T,Q)$ Is Close to $t_c = 2$ ns, but Not When $\tau_R(T,Q) \gg t_c$

In Sections (i).1 and (i).3 we have presented two contrasting situations for segmental α-relaxation. Under the condition, $\tau_0(T)$ or $\tau_f(T) \approx t_c$, we have the anomalous Property (i). On the other hand, if $\tau_0(T)$ or $\tau_f(T) \gg t_c$, segmental α-relaxation of the PEO component behaves normally. Analogue of this can be found in the chain dynamics of the PEO component from the results of molecular dynamics simulations of 20% PEO/80% PMMA (Brodeck et al., 2010). Here the primitive Rouse relaxation time is $\tau_R(T,Q)$. Crossover from Rouse dynamics to slowed non-Rousean dynamics was found by neutron scattering at $t_c \approx 1$ ns (Niedzwiedz et al., 2007), and by molecular dynamics simulations at $t_c \approx 2$ ns (Brodeck et al., 2010). We can pick out a case similar to the high frequency NMR measurements of segmental α-relaxation from the data in Figure 4b of Brodeck et al. (2010). When $T = 350$ K and $\log(Q/\text{Å}^{-1}) = -0.4, -0.30$, and -0.22, corresponding to $Q = 0.4, 0.5$, and 0.6 Å^{-1}, it can be seen that $\tau_R(T,Q)$ (open circles in the figure) is less than or equal to $t_c \approx 2$ ns, and the corresponding $\tau(T,Q)$ (closed circles) is merely longer than $\tau_R(T,Q)$ by a factor ranging from 3 to 5, albeit $\beta(T,Q) = 0.34$ at 350 K. However, at $\log(Q/\text{Å}^{-1}) = -1.0$

or $Q = 0.1$ Å^{-1}, $\tau_R(T,Q)$ is more than two orders of magnitude longer than 2 ns, and the corresponding $\tau(T,Q)$ becomes longer by more than 3 orders of magnitude than $\tau_R(T,Q)$. At a lower temperature of $T = 300$ K, $\tau_R(T,Q)$ (open squares) are mostly much longer than $t_c \approx 2$ ns, and the corresponding $\tau(T,Q)$ (closed squares) all become much longer than $\tau_R(T,Q)$. The $Q^{-6.6}$-dependence of $\tau(T,Q)$ deviates greatly from the $Q^{-4.0}$-dependence of the primitive Rouse dynamics, and $\beta(T,Q) = 0.30$ is smaller than 0.50 of pure PEO and the Rouse model.

(iii) Explanation of Property (iii)

The explanation by the Coupling Model of this property has been given in conjunction with its background in Section (i).5 already. It is unnecessary to repeat that here.

(iv) Explanation of Property (iv)

(iv).1 Chain Dynamics versus High Frequency Deuteron NMR Segmental Dynamics of PEO in Blends with PMMA

Like in pure PEO, diffusion of unentangled PEO chains in blends with PMMA is performed by Rouse mode with length scale of the order of the radius of gyration of the PEO chains. The basic motional unit of the Rouse mode is the Gaussian submolecule, which is composed of a number of repeat units of PEO and its length scale is already longer than the segmental α-relaxation, and much longer than the primitive α-relaxation or the JG β-relaxation. The large displacements involved in the global motion of the diffusing PEO chain are now coupled to or constrained by the much slower PMMA chains depending on the composition of the blend. Evidence of this is provided by crossover of the Rousean dynamics of the PEO component to slowed-down and non-Rousean dynamics at $t_c \approx$ 1–2 ns observed by neutron scattering and simulations (Niedzwiedz et al., 2007; Brodeck et al., 2010). Because of the coupling/constraint to PMMA, the friction factor of PEO chain diffusion and rheology at times $t \gg t_c \approx 1$–2 ns is determined in part by that of PMMA, the extent of which depends on the composition.

The extreme case occurs in the study of 0.3 and 1.0% PEO unentangled tracer chain in PMMA matrix (Haley and Lodge, 2005; Zeroni et al., 2008) where the tracer PEO chain is enveloped practically totally at all length scales by PMMA. Thus, the temperature dependence of the diffusion coefficient is governed almost exclusively by the friction factor of PMMA host. This is observed experimentally (Haley and Lodge, 2005; Zeroni et al., 2008) by the T-dependence of the 1% PEO component chain dynamics by tracer diffusion and rheology is only slightly weaker than the friction factor of the PMMA homopolymer (Zeroni et al., 2008), and much stronger than the T-dependence of the segmental α-relaxation time τ_α of PEO obtained by high frequency deuteron NMR (Lutz et al., 2003) and by quasielastic neutron scattering (Garcıa Sakai et al., 2008; Garcıa Sakai et al., 2005). The difference between the temperature dependence of terminal chain relaxation and the high frequency deuteron NMR segmental relaxation decreases significantly on increasing the wt.% of PEO in the blend from 1.0% to 10%, but the effect persists and the difference is still prominent in blends with 20% and 50% PEO blends (Zeroni et al., 2008). The effect is readily understood from the combination of two facts. One fact is the Property (i) exhibited by the very short τ_α of PEO obtained by these high frequency experiments, making τ_α composition insensitive in contrast to terminal chain relaxation. On the other hand, the terminal relaxation of the PEO component is measured at times is much longer than $t_c \approx 2$ ns, and hence the analogue of Property (i) does not apply for chain dynamics. The second fact is the large difference between the T_gs of PEO and PMMA and hence the much stronger T-dependence of the friction factor of the PMMA component. This is not matched by other polymer blends except PI/PtBS, which will be considered in the next subsection. It is also the reason why the effect was not found in the PI/PVE blends where the difference between the T_gs of PI and PVE is not large. Terminal dynamics and the segmental α-relaxation measured by deuteron NMR at high frequencies of the PI component exhibited nearly the same dependences on temperature and composition (Min et al., 2001; Haley et al., 2003). For these blends of PI with PVE, the segmental relaxation times of PI and PVE, and the diffusion coefficient of PI and terminal relaxation of PVE all have about the same temperature and composition dependence (Ngai and Capaccioli, 2007). Therefore, in the PI/PVE blends, the diffusing PI chain sees all chains including PVE with dynamics having friction factors with about the same temperature dependence at all length scales from segmental to terminal. Naturally, the global relaxation time of PI has about the same temperature dependence as its segmental relaxation time in the PI/PVE blends.

(iv).2 Chain Dynamics versus. Segmental Dynamics of PI in Blends with PtBS from Dielectric Relaxation

In the very first statement made on Property (iii) in Section (i).5, we had already mentioned that the relation between the temperature dependences of the terminal relaxation and the segmental α-relaxation of the fast component in HAPB is different when its segmental relaxation time $\tau_{\alpha f}(T)$ is obtained by dielectric relaxation and calorimetry at lower frequencies and lower temperatures than the deuteron NMR measurements. This cannot be substantiated in HAPB where PEO is the fast component, such as 20% PEO in blends with PMMA or PVAc, because $\tau_{\alpha f}(T)$ of PEO cannot be determined. However $\tau_{\alpha f}(T)$ of 20 and 35% PI in blends with PtBS have been determined by dielectric spectroscopy together with the terminal relaxation time, $\tau_{nf}(T)$, of the normal mode (Arrese-Igor et al., 2010; Arrese-Igor et al., 2011). The comparison of $\tau_{\alpha f}(T)$ and $\tau_{nf}(T)$ can be made in Figures 5.30 and 5.31 for the blends. By inspection it is clear that $\tau_{nf}(T)$ no longer has a stronger temperature dependence than $\tau_{\alpha f}(T)$. This happens because at the low frequencies and temperatures of dielectric measurements, the anomaly of the α-relaxation existing at high frequencies (i.e., Property (i)) is no longer present. Under the condition $\tau_{\alpha f} \approx \tau_{nf}$, the published dielectric loss data of 20% PI in PtBS indicates that the width of frequency dispersion of the α-relaxation is larger than that of the normal mode. From this we deduce larger coupling parameter $n_{\alpha f}(T)$ for α-relaxation than $n_{nf}(T)$ for the normal mode, and from the CM we have stronger temperature dependence for $\tau_{\alpha f}$ than τ_{nf} as observed experimentally.

(v) Explanation of Property (v)

The finding of two relaxations of the PVME component in blends of 20% and 30% PVME with PS has invalidated the explanation by the confinement scenario, which has only a single relaxation with its relaxation time changes T-dependence at some temperature. This has been discussed in Section (i).7. On the other hand, this property is consistent with the alternative interpretation of the dielectric relaxation data of the PVME component given before (Ngai and Roland, 2004a; Ngai and Roland, 2004b). In this interpretation, the much faster relaxation with Arrhenius T-dependence is the Johari-Goldstein β-relaxation or the primitive relaxation of the PVME component, and the slower one is the α-relaxation, again of the PVME component in the blend. Pure PVME has a resolved secondary relaxation much faster than the unresolved Johari-Goldstein β-relaxation. The latter is not resolved because it is slower and not well separated from the α-relaxation, but its presence can be construed from the broadened α-loss peak on the high frequency side. This is expected from the weaker intermolecular coupling/constraint of a flexible polymer such as PVME, and the empirical correlation between the separation, ($\log \tau_{\alpha} - \log \tau_{\text{JG}}$), and intermolecular coupling/constraint (Ngai, 1998). In the blend, the presence of the much less mobile PS enhances the intermolecular coupling/constraint of the PVME component. Consequently, ($\log \tau_{\alpha} - \log \tau_{\text{JG}}$) of the PVME component in the blend becomes larger, and the Johari-Goldstein (JG) β-relaxation become resolved in the blend and observed as the much faster of the two relaxations (see Figures 5.27 and 5.28).

A good example of this description of the change of dynamics of the fast component is from molecular dynamics simulations of model miscible polymer blends consisting of chemically realistic 1,4-polybutadiene (CR-PBD) as the slow component (higher T_g) and PBD chains with reduced dihedral barriers as the fast component (LB-PBD) with lower T_g by Bedrov and Smith (2005). The conformational energy of the C(sp^2)-C(sp^3)–C(sp^3)-C(sp^2) (alkyl) dihedrals as a function of dihedral angle for the CR-PBD and LB-PBD models. The simulation was designed to study the influence on the segmental α-relaxation and the JG β-relaxation of the fast component when mixed with the slow component. Before blending, the relaxation times of α- and JG β-relaxations of the neat fast component LB-PBD are too close together, so that the faster but weaker JG β-relaxation cannot be easily resolved like PVME. However, with the addition of the slow CR-PBD component, they found a monotonic increase in the separation between α- and the JG β-relaxations of the fast component, whereby the latter becomes resolved. The increased separation between the two is due to the strong increase of the α-relaxation time, $\tau_{\alpha f}$, but a much smaller change of the relaxation time, $\tau_{\beta f}$, of the JG β-relaxation of the fast component. This is accompanied by a monotonic increase of breadth of the α-dispersion or decrease of the stretch exponent, $\beta_{\alpha f} \equiv (1 - n_{\alpha f})$, of the Kohlrausch function used to fit the correlation function with increased concentration of the slow component. These results from simulations obtained by Bedrov and Smith are fully compatible with the experimental observation of two relaxations of the PVME component in the 30% and 20% PVME blends with PS, and the explanation given earlier.

Another example from polymer blends can be taken from the dynamics of the faster poly(ethyl methacrylate) (PEMA, $T_g = 74$ C) component in blends with poly(4-vinylphenol) (PVPh, $T_g = 171$ C) (Zhang et al., 2002). Neat PEMA has a resolved JG relaxation, which is continued to be observed in the blend and practically unchanged with blending. Since T_g of the PEMA component in the blend is shifted to higher temperature by the slower poly(4-vinylphenol) component, therefore ($\log \tau_{\alpha f}(T_{gf}) - \log \tau_{\beta f}(T_{gf})$) of the PEMA component increase at constant $\tau_{\alpha}(T_g)$.

Since the faster relaxation is the JG β-relaxation of the PVME component in the blends, the crossover of the T-dependence of its relaxation time, $\tau_{\beta f}$, from Arrhenius at temperatures below T_{gf} to a stronger T-dependence above T_{gf} can be understood from the same behavior of JG β-relaxation of neat glass-formers resolved above and below T_g. Examples include m-fluroaniline at elevated pressure (Hensel-Bielowka et al., 2005), copolyesters of poly(ethylene terephthalate) and poly(ethylene isophthalate) (Sanz et al., 2004) and a series of amorphous aromatic polyesters, poly(ethylene terepthalate) (PET), poly(ethylene naphthalene dicarboxylate) (PEN), poly(ethylene isophthalate) (PEI), and poly(butylene isophthalate) (PBI), as found by Nogales et al. (2006). Exactly the same PVME component dynamics, including the simultaneous presence of the α- and the JG β-relaxations, and the crossover of T-dependence of $\tau_{\beta f}$ at T_{gf}, have been found by dielectric relaxation in a number of polar small molecule glass-former when mixed with oligomers of styrene (Capaccioli et al., 2012; Mierzwa et al., 2008; Kessairi et al., 2008). When mixed with oligomer of PS with higher T_{gs}, the JG β-relaxation is resolved and the spectra show the presence of this much faster secondary relaxation together with the slow α-relaxation in mixtures having concentration of small molecules as low as 5–10 mol%. Except that polymer blend has the effect of self-concentration not present in nonpolymeric glass-former mixture, the component dynamics of the two systems are essentially the same (Ngai and Roland, 2004a). Another example from polymers is the poly(n-butyl methacrylate) (PBMA) in copolymer with polystyrene. Like pure PVME, pure PBMA has no resolved JG β-relaxation because of weak intermolecular coupling. On increasing the styrene content from 0 mol% to 66 mol% in a series of PBMA/PS copolymers, a monotonic increase in the separation of the JG relaxation from the segmental relaxation was found (Kahle et al., 1997). The effect was explained by the enhancement of intermolecular coupling of PBMA component in the presence of the less mobile styrene components in the framework of the Coupling Model (Ngai, 1999).

We have reasoned that intermolecular coupling/constraint of the faster component is enhanced by the slower component, leading to increased separation, $(\log \tau_\alpha - \log \tau_{\rm JG})$, between the segmental α- and JG β-relaxation of the faster component in the blend. If this is valid, the reverse must hold for the slower component in the blend. That is the separation, $(\log \tau_\alpha - \log \tau_{\rm JG})$, between the segmental α- and JG β-relaxation of the slower component has to decrease on increasing the concentration in the blend. The amorphous x% PEO/(1 − x)% PMMA blends already discussed offers a test of this expected change of dynamics of the slower PMMA component by the faster PEO component. The calorimetry and dielectric data of Jin et al. (2004) indeed show that $(\log \tau_\alpha(T_g) - \log \tau_{\rm JG}(T_g))$ of the PMMA component at constant $\tau_\alpha(T_g)$ decreases with increasing concentration x of PEO continuously from $x = 0$ to $x = 30$. This conclusion is justified by the fact on increasing x of the PEO component that the T_g of the PMMA component decreases monotonically while its $\tau_{\rm JG}(T)$ remains practically unchanged. For example, log(f/Hz) of the JG β-relaxation at $T = T_g \approx 320$ K of the PMMA component in the $x = 30$ blend (Jin et al., 2004) is 2, corresponding to $\tau_{\rm JG}(T_g) = 1.6 \times 10^{-3}$ s, which is about one and a half decades longer than $\log \tau_{\rm JG}(T_g)$ of neat PMMA.

(vi) Explanation of Property (vi)

The property is the experimental finding by Arrese-Igor et al. (2011) of an increase in separation between the NM relaxation time, $\hat{\tau}_R$, and the segmental α-relaxation time, $\tau_{\alpha f}$, of PI27 at constant $\log(\hat{\tau}_R/\mathrm{s}) \approx -0.5$ on decreasing the concentration of PI in the blends (see Figure 5.30). This is best described in Figure 5.29 by comparing the data of the 20% PI, 35% PI, and 100% PI. The vertical dashed lines giving the separation between $\hat{\tau}_R$ and $\tau_{\alpha f}$ in the logarithmic scale clearly increase on decreasing the PI content. An explanation of this nontrivial relation between $\hat{\tau}_R$ and $\tau_{\alpha f}$ of PI27 at constant $\hat{\tau}_R \approx 3.2 \times 10^{-2}$ s is afforded by the Coupling Model (CM) relations applied to both processes. The applicability of the CM to NM is justified by the observed crossover from Rouse dynamics to slowed-down stretched exponential chain dynamics at $t_{\rm cR} = 1$–2 ns (Property (ii)). The observed crossover from primitive Rouse dynamics to coupled chain dynamics together with the general physical principle of the CM leads to Eq. (5.41) for $\hat{\tau}_R$ given before in Section (i).5. The CM equation for the segmental α-relaxation time $\tau_{\alpha f}$ with $t_{c\alpha} = 1$–2 ps was given earlier in the form of Eq. (5.39). Now we apply these two equations to calculate $\tau_{\alpha f}$ and $\hat{\tau}_R$ for the 35% PI27 ($M_n = 2700$) blended with PtBS23 ($M_n = 2300$) at 260 K, based on the same quantities in pure PI27. The choice of the 35%

PI27 blend is because data are available to determine the coupling parameter, $n_R \equiv (1 - \beta_R)$ of $\hat{\tau}_R$ in this blend but not the others. This opportunity comes from the $M^{3.4}$-dependence of $\hat{\tau}_R$ (Moreno and Colmenero, 2008; Arrese-Igor et al., 2012), from which Eq. (5.41) it follows that $n_R = 0.41$ or $\beta_R = 0.59$ (see Section (i).5). PI29 is not entangled, and hence the measured NM relaxation time can be identified with the primitive Rouse relaxation time, τ_R^0. Assuming that the primitive relaxation time does not change on blending with 65% PtBS, we can use the measured NM relaxation time of pure PI27 in Figure 5.29 as $\tau_R^0(T)$ of the 35% PI27 blend. At 260 K, $\log(\tau_R^0/\mathrm{s}) \approx -4$ and by Eq. (5.43) with $t_{cR} = 1$ ns, the calculated value of $\log(\hat{\tau}_R/\mathrm{s}) = -0.53$ (indicated by the tip of the upper arrow drawn at very near $(1000/260\ \mathrm{K}) = 3.86\ \mathrm{K}^{-1}$) is almost the same as the experimental value of $\log(\hat{\tau}_R/\mathrm{s}) = -0.50$.

To calculate $\tau_{\alpha f}$ at 260 K of the 35% PI27 blend by Eq. (5.32) with $t_{c\alpha} = 2$ ps, we need to know the primitive segmental α-relaxation time, τ_α^0. Assuming it is unchanged on blending with 65% PtBS, τ_α^0 can be obtained from the experimental values of τ_α and the coupling parameter, $n_\alpha \equiv (1 - \beta_\alpha)$, of pure PI29 from Eq. (5.32). The dielectric loss data of PI27 of Arrese-Igor et al. at 260 K have not fully resolved the α-relaxation and the value of β_α has not been determined. Dielectric data of higher molecular weight PI have resolved the α-relaxation and value of $\beta_\alpha = 0.47$ independent of relaxation times in the range from a few hundred seconds to a microsecond has been reported for relaxation times (Roland et al., 2004). Another measurement of PI (Fragiadakis et al., 2011) with molecular weights $M_n = 1100, 9{,}730$, and $20{,}700$ give values of β_α close to 0.47 at longer τ_α, but the reported values at shorter τ_α cannot be taken seriously because it is even smaller than the value at longer τ_α, against the general rule that nonexponentiality decreases at shorter τ_α. Therefore we take $\beta_\alpha = 0.47$ or $n_\alpha = 0.53$ for pure PI29 and calculate τ_α^0 by Eq. (5.32) from the measured τ_α and $t_{c\alpha} = 2$ ps at 260 K. The value of β_α of PI27 in the 35% PI27 blend cannot be determined with certainty from the unresolved α-relaxation at 260 K. By comparing the stronger T-dependence of $\tau_{\alpha f}$ in the blend with the weaker T-dependence of τ_α in pure PI27 at 260 K, we can use again Eq. (5.32) for the two cases to give an estimate of β_α for the 35% PI27 blend in the range $0.30 \leqslant \beta_\alpha \leqslant 0.35$. The values of $\log(\tau_{\alpha f}/\mathrm{s})$ in the 35% PI27 blend at 260 K calculated with the two limiting values of β_α are -4.3 and -5.3. These estimates of $\log(\tau_{\alpha f}/\mathrm{s})$, indicated by the two tips of the pair of arrows slightly displaced to the right of $1000/260\ \mathrm{K}^{-1}$ in Figure 5.29, are consistent with the experimental observed $\log(\tau_{\alpha f}/\mathrm{s})$.

Since our calculated values of $\log(\hat{\tau}_R/\mathrm{s})$ and $\log(\tau_{\alpha f}/\mathrm{s})$ in the 35% PI27 blend are consistent with experimental values, naturally the larger separation between $\log(\hat{\tau}_R/\mathrm{s})$ and $\log(\tau_{\alpha f}/\mathrm{s})$ observed in 35% PI27 blend at 260 K compared with pure PI27 is explained. Nevertheless, this property can be traced to two factors. Blend with lower PI27 content requires a higher temperature for the NM to have relaxation time satisfying the condition, $\log(\hat{\tau}_R/\mathrm{s}) = -0.50$. Consequently, the corresponding τ_α^0 is shorter, and smaller increase of $\tau_{\alpha f}$ from

τ_α^0 results from the CM Eq. (5.32) even for a large $0.65 \leqslant n_\alpha \leqslant 0.70$. This is similar to Property (i) of $\tau_{\alpha f}$ of PEO in blends with PMMA observed by high frequency (short τ_α^0) deuteron NMR experiments and explained in Section (i) by the CM. On the other hand, primitive Rouse relaxation time τ_R^0 of the NM in the 35% PI27 blend is 5 decades longer than $t_{cR} = 1$ ns, and $\hat{\tau}_R$ calculated by the CM is long albeit $n_R = 0.41$ is smaller than n_α.

(vii) Explanation of Property (vii)

The temperature-pressure superposition of the structural α-relaxation at constant f_α or τ_α was found in many nonassociated small molecular glass-formers and amorphous homopolymers (Ngai et al., 2005), which is consistent with the CM equation connecting τ_α with n. It is found for a component in mixtures of two van der Waals liquids, picoline with tri-styrene by Mierzwa et al. (2008), and quinaldine with tri-styrene by Kessairi et al. (2008). In these mixtures as well as some pure small molecular glass-former such as benzoyn isobutylether (BIBE) (Capaccioli et al., 2012; Ngai et al., 2005), the JG β-relaxation shows up together with the α-relaxation. There is good agreement between the calculated τ_0 and τ_{JG}, and the ratio $\tau_\alpha/\tau_{\mathrm{JG}}$ is unchanged on varying P and T while τ_α is kept constant. This means that $\tau_\alpha, \tau_{\mathrm{JG}}$ (or τ_0), and the dispersion (or n) are co-invariants to changes in T and P, which is exactly as predicted by the CM equation in conjunction with $\tau_0 \approx \tau_{\mathrm{JG}}$,

$$\begin{aligned}\tau(T,P) &= [t_c^{-n(T,P)}\tau_0(T,P)]^{1/[1-n(T,P)]}\\ &\approx [t_c^{-n(T,P)}\tau_{\mathrm{JG}}(T,P)]^{1/[1-n(T,P)]}.\end{aligned} \tag{5.46}$$

This equation has been used as the basis to explain the *T-P* superposition of the α-relaxation of a component in mixtures of van der Waals glass-formers and polymer blends as discussed in Capaccioli and Ngai (2005). Concentration fluctuations in the mixture or blend create a distribution of environments $\{i\}$. Each environment, i, has its own coupling parameter, n_i, primitive relaxation time, τ_{0i}, and the corresponding Kohlrausch function with stretch exponent, $(1 - n_i)$, which determines the relaxation time $\tau_{\alpha i}$ by the CM equation $\tau_{\alpha i} = [t_c^{-n_i}\tau_{0i}]^{1/(1-n_i)}$. In the same manner as shown earlier for neat glass-formers, this CM equation ensures co-invariance of $\tau_{\alpha i}$ and n_i (or the frequency dispersion of loss contributed by i) for each i in the distribution for different combinations of T and P at constant $\tau_{\alpha i}$. The co-invariance applies in particular to $\hat{\tau}_\alpha$ and $\hat{n}$ of the most probable $\hat{i}$ in the distribution. The observed segmental relaxation of the component is composed of contributions from all i, and since for each i there is co-invariance $\tau_{\alpha i}$ and n_i, therefore the overall frequency dependence of the α-relaxation of the component is unchanged on varying P and T while maintaining the most probable relaxation time, $\hat{\tau}_\alpha$, or frequency, $\hat{f}_\alpha$, constant. This explains why the shape of the dielectric loss coming from the α-relaxation observed for a component in several polymer blends is independent of thermodynamic conditions (T and P combinations).

None of the models proposed to address the component dynamics of polymer blends (Lodge and McLeish, 2000; Colmenero and Arbe, 2007; Zetsche and Fischer, 1994; Kumar et al., 1996; Kamath et al., 2003), except that based on the Coupling Model (Roland and Ngai, 1991; Ngai and Roland, 2004a; Alegria et al., 1994; Ngai et al., 2006), consider the frequency dispersion of the α-relaxation of a component in the blend (or in its neat state), and naturally these models cannot address the observation of the invariance of the dispersion of the segmental relaxation of the component PVME for different T and P at constant relaxation time. Besides, these models are based on thermodynamic considerations and, if considered, the dispersion will change on changing T and P.

(viii) Explanation of Properties (viii) and (viii′)

Explanation of Property (viii′) of the PVME component in the 25% PVME blend with PS has been offered near the end of the section where these properties are made known. There we have mentioned that this property is shared by some neat glass-formers (Ngai, 2011) including a bead-necklace model for polymer (Bedrov and Smith, 2011) and mixtures of van der Waals liquids (Mierzwa et al., 2008; Kessairi et al., 2008), and the explanation of it readily follows from the co-invariance of τ_α, τ_{JG}, and n to changes of pressure and temperature at either constant τ_α or constant τ_{JG}, found by experiments and computer simulations. The explanation also follows from the CM equation when combined with the relation, $\tau_{JG} \approx \tau_0$, found experimentally valid for many glass-formers between the JG β-relaxation time and the primitive relaxation time, τ_0, of the CM.

For the PVME component in blends with PS, complete experimental data for co-invariance of τ_α, τ_{JG}, and n for different combinations of T and P at constant τ_α and τ_{JG} is not available for any of the blends studied. In the 75% PVME and 50% PVME blends, the data of the α-relaxation demonstrate co-invariance of τ_α and n, which are examples of Property (*vii*) explained previously in Section (vii), but the JG β-relaxation is not resolved in these blends. On the other hand, in blends with less than 25% of PVME, JG β-relaxation is resolved but not the α-relaxation by dielectric spectroscopy. Although not resolved, it is reasonable to assume that τ_α and n are co-invariant, because this was the case in the same blends at higher PVME content, and also in other polymer blends. Experimentally found in the 25% PVME blend is the change of T-dependence of τ_{JG} when it crosses $10^{-4.8}$ s at ambient pressure, $P_a = 0.1$ MPa, and also at elevated pressures up to 300 MPa. At ambient pressure, the temperature at which $\tau_{JG} \approx 10^{-4.8}$ s is the glass transition temperature T_{gf} of the PVME component estimated from TSDC data (Leroy et al., 2002). This observed change of T-dependence of τ_{JG} at T_{gf} at ambient pressure is in accord with that found in pure glass-formers and in mixtures of van der Waals glass-formers. Although the glass transition temperature, $T_{gf}(P)$, of the PVME component at elevated pressures had not been determined, it is reasonable to assume that the change of T-dependence of τ_{JG} observed at elevated pressures is also at the glass transition temperature, $T_{gf}(P)$, where τ_α has the same value independent of pressure.

From the same τ_α and n at $T_{gf}(P)$ for all pressures, the CM (Eq. (58)) explains the remarkable experimental observation of the change of T-dependence of τ_{JG} when it crosses $10^{-4.8}$ s uniformly at ambient and elevated pressure.

(ix) Explanation of Property (ix)

The Property (ix) alone can be explained well by the "fluctuation model" of Richter et al. (1990) and Hall and Helfand (1982) based on concentration fluctuations. As pointed out by Alegrıa et al. (2002), the "fluctuation model" leads to a continuous distribution of compositions in all the different micro-regions of the blend, but not the presence of two micro-environments with distinctly different average mobilities of the two components, as found by experiments. An improved concentration fluctuation model that predicts distinct dynamics of the two components was given by Kumar et al. (1996). Notwithstanding, in their present forms, neither the model of Fischer and coworkers nor the improved version of Kumar et al. can address or explain the other properties (i)–(viii). On the other hand, the Coupling Model (CM) for component dynamics had taken into consideration of concentration fluctuations and the effects engendered (Roland and Ngai, 1991; Alegria et al., 1994) a few years before the models of Fischer et al. and Kumar et al. In this model of A/B polymer blend, each component has its own characteristic primitive relaxation time and its own coupling parameter. This starting point of the CM for blends immediately leads to each component having its own dynamics and relaxation time spectrum, a property called "dynamic heterogeneity" and highlighted by others as one of two important facts on the segmental α-relaxation in polymer blends (Colmenero and Arbe, 2007). The other is the broadening of the α-relaxation function of the component. In the CM, there is enhancement of intermolecular coupling/constraint of the segmental α-relaxation of the faster component A due to the presence of the slower B component. Due to concentration fluctuations, the segmental α-relaxation of component A occurs in environments with different compositions out of a distribution $\{i\}$, and hence the coupling parameter, n_{Ai}, necessarily varies from one environment to another within a distribution $\{n_{Ai}\}$. It readily follows from the CM equation that the relaxation time, τ_{Ai}, is also distributed, thereby the α-relaxation of the A-component is broadened, and the broadening increases at lower temperatures. Larger n_{Ai} gives rise to longer τ_{Ai} in the distribution, and stronger temperature dependence. This explains the broadening of the dielectric loss spectrum of the faster component toward low frequencies, and the breakdown of time-temperature superposition. Conversely, intermolecular coupling/constraints are mitigated and the coupling parameter, n_{Bi}, of the slower B-component is reduced on mixing with the faster A-component. This effect of narrowing the α-relaxation counters the broadening by concentration fluctuations. At sufficiently low concentration of component A where concentration fluctuation is not important, the narrowing of the α-relaxation of the B-component may be observed. In fact, this has been seen in the molecular dynamics simulation of the

LB-PBD blends with CR-PBD by Bedrov and Smith (2005) discussed before. The α-relaxation time of the slower CR-PBD component becomes shorter, and interestingly its frequency dispersion becomes *narrower* when compared with that of neat CR-PBD (Bedrov and Smith, 2005; Ngai and Capaccioli, 2007; Ngai et al., 2006).

5.7.4 Summary

The purpose of this Section IX on polymer blends is to bring back the basic and critical properties of component dynamics in highly asymmetric polymer blends (HAPB) to the attention of the research community. This is a worthwhile task because a number of novel dynamic properties of the terminal chain relaxation and segmental α-relaxation had been found in several of these HAPB, not seen before in any of the conventional synthetic or natural homopolymers. The many properties of the fast component in the highly asymmetric polymer blends are not isolated but related to each other, making the entire collection most interesting. A satisfactory theory must be able to explain all of them together, not just any selected subset, and not offering a separate explanation or rationalization for each of them as currently done by others in the literature.

By and large this purpose is accomplished by demonstrating the general and fundamental nature of the multiple properties discussed in the previous sections. By this exposition, it is hoped that the research community will steer away from the currently common practice of exclusively comparing data to the Lodge-McLeisch (LM) model, and from the interpretation of the dynamics of the fast component as confined local motion by the "frozen" slow component. This is a thankless job, but someone has to do it. Otherwise new physics that could be gained from the study of HAPB are obscured, and advance in the field will stagnate.

The Coupling Model is consistent with all the properties and has fundamental support from quasielastic neutron scattering and simulations. However, the emphasis of the entire section is on the many properties of component dynamics in HAPB that deserve attention and alternative explanation by researchers in glass transition and polymer chain dynamics and viscoelasticity. This is because the new physics found in the segmental and chain dynamics of components in highly asymmetric polymer blends could possibly revolutionize the current understanding of polymer dynamics and viscoelasticity.

REFERENCES

Adachi, K., Kotaka, T., 1990. Polym. Yearbook 6, 43.
Adachi, K., Kotaka, T., 1993. Prog. Polym. Sci. 18, 585.
Adachi, K., Nakamoto, T., Kotaka, T., 1989. Macromolecules 22, 3111.
Alegria, A., Colmenero, J., Ngai, K.L., Roland, C.M., 1994. Macromolecules 27, 4486.
Alegrıa, A., Gomez, D., Colmenero, J., 2002. Macromolecules 35, 2030.

Alig, I., Johari, G.P., 1993. J. Polym. Sci. Part B: Polym. Phys. 31, 299.
Arai, K., Ferry, J.D., 1986a. Rubber Chem. Technol. 59, 592.
Arai, K., Ferry, J.D., 1986b. Rubber Chem. Technol. 59, 605.
Arrese-Igor, S., Alegrıa, A., Colmenero, J., 2010. Macromolecules 43, 6406.
Arrese-Igor, S., Moreno, A.J., Alegrıa, A., Colmenero, J., 2011. Macromolecules 44, 3611.
Arrese-Igor, Silvia, Alegria, Angel, Moreno, Angel J., Colmenero, Juan, 2012. Soft Matter 8, 3739.
Bedrov, D., Smith, G.D., 2005. Macromolecules 38, 10314.
Bedrov, D., Smith, G.D., 2011. J. Non-Cryst. Solids 357, 258.
Beiner, M., Ngai, K.L., 2005. Macromolecules 38, 7033.
Bergman, R., Alvarez, F., Alegria, A., Colmenero, J., 1998. J. Non-Cryst. Solids 235–237, 580.
Berry, G.C., Fox, T.G., 1968. Adv. Polym. Sci. 5, 261.
Brodeck, M., Alvarez, F., Moreno, A.J., Colmenero, J., Richter, D., 2010. Macromolecules 43, 3036.
Capaccioli, S., Ngai, K.L., 2005. J. Phys. Chem. B 109, 9727.
Capaccioli, S., Paluch, M., Prevosto, D., Wang, Li-Min, Ngai, K.L., 2012. J. Phys. Chem. Lett. 3, 735.
Casalini, R., Ngai, K.L., Robertson, C.G., Roland, C.M., 2001. J. Polym. Sci. Polym. Phys. Ed. 38, 1841.
Cendoya, I., Alegria, A., Alberdi, J.M., Colmenero, J., Grimm, H., Richter, D., Frick, B., 1999. Macromolecules 32, 4065.
Cerveny, S., Alegria, A., Colmenero, J., 2008. Phys. Rev. E 77, 031803.
Chang, K., 2008. The Nature of Glass Remains Anything but Clear. The New York Times, July 29.
Chen Quan, Matsumiya, Y., Masubuchi, Y., Watanabe, H., Inoue, T., 2008. Macromolecules 41, 8694.
Choy, I.-C., Plazek, D.J., 1986. J. Polym. Sci. Part B: Polym. Phys. 24, 1303.
Cochrane, J., Harrison, G., Lamb, J., Phillips, D.W., 1980. Polymer 21, 837.
Colby, R.H., 1989. Polymer 30, 1275–1278.
Colmenero, J., 2010. Chain dynamics in dynamically asymmetric polymer blends. In: Multi-Scale Dynamics of Structured Polymeric Materials, The Michelin Materials Science Chair at ESPCI ParisTech Lectures, December 6–7, pdf of lecture available on the web.
Colmenero, J., Arbe, A., 2007. Soft Matter 3, 1474.
Colmenero, J., Arbe, A., Coddens, G., Frick, B., Mijangos, C., Reinecke, H., 1977. Phys. Rev. Lett. 78, 1928.
Colmenero, J., Alegria, A., Arbe, A., Frick, B., 1992. Phys. Rev. Lett. 69, 478.
Correzi, S., Beiner, M., Huth, H., Schröter, K., Capaccioli, S., Casalini, R., Fioretto, D., Donth, E., 2002. J. Chem. Phys. 117, 2435.
Dicke, R.A., Ferry, J.D., 1966. J. Phys. Chem. 70, 2594.
Diddens, D., Brodeck, M., Heuer, A., 2011. Eur. Phys. Lett. 95, 56003.
Doi, M., Edwards, S.F., 1986. The Theory of Polymer Dynamics. Clarendon Press, Oxford.
Ferry, J.D., 1980. Viscoelastic Properties of Polymers, third ed. Wiley, New York.
Ferry, J.D., Manke, R.G., Maekawa, E., Oyanagi, Y., Dicke, R.A., 1964. J. Phys. Chem. 68, 3414.
Flory, P.J., 1953. Principle of Polymer Chemistry. Cornell University Press, Ithaca, NY.
Flory, P.J., 1976. Proc. Roy. Soc. London A 351, 351.
Flory, P.J., 1979a. Rubber Chem. Technol. 52, 110.
Flory, P.J., 1979b. Polymer 20, 1317.
Flory, P.J., 1985. Polym. J. 17, 1.
Flory, P.J., Erman, B., 1982. Macromolecules 15, 800.
Fragiadakis, D., Casalini, R., Bogoslovov, R.B., Robertson, C.G., Roland, C.M., 2011. Macromolecules 44, 1149.
Fytas, G., Ngai, K.L., 1988. Macromolecules 21, 804.
Garcıa Sakai, Victoria, Chowdhuri, Zema, Maranas, Janna K., Peral, I., Copley, John R., 2005. J. Polym. Sci. Part B: Polym. Phys. 43, 2914.
Garcıa Sakai, Victoria, Maranas Inmaculada Peral, Janna K., Copley, John R.D., 2008. Macromolecules 41, 3701.
Genix, A.C., Arbe, A., Alvarez, F., Colmenero, J., Willner, L., Richter, D., 2005. Phys. Rev. E 72, 031808.
Gent, A.N., 2004. In: Science and Technology of Rubber, second ed. Academic Press, New York (Chapter 10).

Goulart Silva, G., Machado, J.C., Song, M., Hourston, D.J., 2000. J. Appl. Polym. Sci. 77, 2043.
Haley, J.C., Lodge, T.P., 2005. J. Chem. Phys. 122, 234914.
Haley, J.C., Lodge, T.P., He, Y., Ediger, M.D., von Meerwall, E.A., Mijovic, J., 2003. Macromolecules 36, 6142.
Hall, C.K., Helfand, E., 1982. J. Chem. Phys. 77, 3275.
He, Y.Y., Lutz, T.R., Ediger, M.D., Ayyagari, C., Bedrov, D., Smith, G.D., 2004. Macromolecules 37, 5032.
He, Y.Y., Lutz, T.R., Ediger, M.D., Pitsikalis, M., Hadjichristidis, N., von Meerwall, E.A., 2005. Macromolecules 38, 6216.
Hensel-Bielowka, S., Paluch, M., Ngai, K.L., 2005. J. Chem. Phys. 123, 014502.
Hodge, I.M., 1983. Macromolecules 16, 898.
Hopkins, I.L., Hamming, R.W., 1957. J. Appl. Phys. 28, 906.
Isono, Y., Ferry, J.D., 1984. Rubber Chem. Technol. 57, 925.
James, H.M., Guth, E., 1947. J. Chem. Phys. 15, 669.
James, H.M., Guth, E., 1953. J. Chem. Phys. 21, 1039.
Jin, X., Zhang, S.H., Runt, J., 2004. Macromolecules 37, 8110.
Johari, G.P., 1976. Ann. N.Y. Acad. Sci. 279, 117.
Johari, G.P., Goldstein, M., 1970. J. Chem. Phys. 53, 2372.
Kahle, S., Korus, J., Hempel, E., Unger, R., Höring, S., Schröter, K., Donth, E., 1997. Macromolecules 30, 7214.
Kamath, A., Colby, R.H., Kumar, S.K., 2003. Phys. Rev. E 67, 010801.
Katana, G., Fischer, E.W., Hack, Th., Abetz, V., Kremer, F., 1995. Macromolecules 28, 2714.
Kessairi, K., Capaccioli, S., Prevosto, D., Lucchesi, M., Sharifi, S., Rolla, P.A., 2008. J. Phys. Chem. B 112, 4470.
Kohlrausch, R., 1847. Pogg. Ann. Phys. 12, 393; Pogg. Ann. Phys., 91, 56, 179 (1854).
Kovacs, A.J., Aklonis, J.J., Hutchinson, J.M., Ramos, A.R., 1979. J. Polym. Sci.: Polym. Phys. Ed. 17, 1097.
Kudlik, A., Benkhof, S., Blochowicz, T., Tschirwitz, C., Rössler, E., 1999. J. Mol. Struct. 479, 210.
Kumar, S.K., Colby, R.H., Anastasiadis, S.H., Fytas, G., 1996. J. Chem. Phys. 105, 3777.
Lartigue, C., Guillermo, A., Cohen-Addad, J.P., 1997. J. Polym. Sci. Part B: Polym. Phys. 35, 1095.
Leaderman, H., 1955. J. Polym. Sci. 16, 261.
Leaderman, H., 1962. J. Polym. Sci. 59 (168), S42.
Leroy, E., Alegria, A., Colmenero, J., 2002. Macromolecules 35, 5587.
Lodge, T.P., McLeish, T.C.B., 2000. Macromolecules 33, 5278.
Lodge, T.P., Wood, E.R., Haley, J.C., 2006. J. Polym. Sci. Part B: Polym. Phys. 44, 756.
Lorthioir, C., Alegrıa, A., Colmenero, J., 2003a. Phys. Rev. E 68, 031805.
Lorthioir, C., Alegrıa, A., Colmenero, J., 2003b. Eur. Phys. J. E 12, S127.
Lutz, T.R., He, Yiyong, Ediger, M.D., Haihui, Cao, Lin, Guoxing, Jones, A.A., 2003. Macromolecules 36, 1724.
Maranas, J.K., 2007. Curr. Opin. Colloid Interface Sci. 12, 29.
Mark, J.E., Erman, B., 1988. Rubberlike Elasticity: A Molecular Primer. Wiley-Interscience, New York.
Mark, J.E. et al., 2004. Physical Properties of Polymers, third ed. Cambridge University Press, Cambridge.
McKenna, G.B., 1989. In: Booth, C., Price, C. (Eds.), Comprehensive Polymer Science, Polymer Properties, vol. 2. Pergamon, Oxford, p. 311.
McKenna, G.B., Ngai, K.L., Plazek, D.J., 1985. Polymer 26, 1651.
Mierzwa, M., Pawlus, S., Paluch, M., Kaminska, E., Ngai, K.L., 2008. J. Chem. Phys. 128, 044512.
Min, B., Qiu, X., Ediger, M.D., Pitsikalis, M., Hadjichristidis, N., 2001. Macromolecules 34, 4466.
Miwa, Yohei, Usami, Kaori, Yamamoto, Katsuhiro, Sakaguchi, Masato, Sakai, Masahiro, Shimada, Shigetaka, 2005. Macromolecules 38, 2355.
Moreno, A.J., Colmenero, J., 2008. Phys. Rev. Lett. 100, 126001.
Moynihan, C.T. et al., 1976. Ann. N.Y. Acad. Sci. 279, 15.
Mpoukouvalas, K., Floudas, G., 2008. Macromolecules 41, 1552.
Mullins, L., 1959. J. Appl. Polym. Sci. 2, 257.
Ngai, K.L., 1979. Comments Solid State Phys. 9, 127.
Ngai, K.L., 1994. In: Richert, R., Blumen, A. (Eds.), Disorder Effects on Relaxational Properties. Springer Verlag, Berlin (Chapter 3).

Ngai, K.L., 1998. J. Chem. Phys. 109, 6982.
Ngai, K.L., 1999. Macromolecules 32, 7140.
Ngai, K.L., 2001. IEEE Trans. Dielectr. Electr. Insul. 8, 329.
Ngai, K.L., 2003. J. Phys.: Condens. Matter 15, S1107.
Ngai, K.L., 2007. J. Non-Cryst. Solids 353, 709.
Ngai, K.L., 2011. Relaxation and Diffusion in Complex Systems. Springer, New York.
Ngai, K.L., Capaccioli, S., 2004. Phys. Rev. E 69, 0315xx.
Ngai, K.L., Capaccioli, S., 2007. J. Phys.: Condens. Matter 19, 205114.
Ngai, K.L., Paluch, M., 2003. J. Phys. Chem. B 107, 6865.
Ngai, K.L., Paluch, M., 2004. J. Chem. Phys. 120, 857–873.
Ngai, K.L., Plazek, D.J., 1986a. J. Polym. Sci. Polym. Phys. Ed. 23, 2169.
Ngai, K.L., Plazek, D.J., 1986b. J. Polym. Sci. B Polym. Phys. Ed. 24, 619.
Ngai, K.L., Plazek, D.J., 1995. Rubber Chem. Technol. Rubber Rev. 68, 376.
Ngai, K.L., Plazek, D.J., 1996. In: Mark, J.E. (Ed.), Physical Properties of Polymers Handbook. American Institute of Physics, Woodbury, NY, p. 341 (Chapter 25).
Ngai, K.L., Plazek, D.J., 2002. Macromolecules 35, 9136.
Ngai, K.L., Rendell, R.W., 1990. In: Campbell, Ian A., Giovannella, C. (Eds.), Relaxation in Complex Systems and Related Topics, NATO ASI Series, vol. 222. Plenum, New York, p. 309.
Ngai, K.L., Rendell, R.W., 1991. J. Non-Cryst. Solids 131–133, 942.
Ngai, K.L., Rendell, R.W., 1997. In: Fourkas, J.T., Kivelson, D., Mohanty, U., Nelson, K. (Eds.), Supercooled Liquids, Advances and Novel Applications, ACS Symposium Series, vol. 676. Am. Chem. Soc., Washington, DC, p. 45.
Ngai, K.L., Roland, C.M., 1994. Macromolecules 27, 2454.
Ngai, K.L., Roland, C.M., 2004a. Rubber Chem. Technol. 77, 579.
Ngai, K.L., Roland, C.M., 2004b. Macromolecules 37, 2817.
Ngai, K.L., Tsang, K.Y., 1999. Phys. Rev. E 60, 4511.
Ngai, K.L., Wang, Limin, 2011. J. Chem. Phys. 135, 194902.
Ngai, K.L., Plazek, D.J., Deo, S.S., 1987. Macromolecules 20, 3047.
Ngai, K.L., Rajagopal, A.K., Teitler, S., 1988. J. Chem. Phys. 88, 5086.
Ngai, K.L., Schönhals, A., Schlosser, E., 1992a. Macromolecules 25, 4915.
Ngai, K.L., Peng, S.L., Tsang, K.Y., 1992b. Phys. A 191, 523.
Ngai, K.L., Roland, C.M., Yee, A.F., 1993a. Rubber Chem. Technol., 66, 817.
Ngai, K.L., Plazek, D.J., Bero, C., 1993b. Macromolecules 26, 1065.
Ngai, K.L., Plazek, D.J., Echeverria, I., 1996. Macromolecules 29, 7937.
Ngai, K.L., Plazek, D.J., Rendell, R.W., 1997a. Rheol. Acta 36, 307.
Ngai, K.L., Plazek, D.J., Rizos, A.K., 1997b. J. Polym. Sci. Polym. Phys. Ed. 35, 599.
Ngai, K.L., Plazek, D.J., O'Rourke, V.M., 1997c. Macromolecules 30, 5450.
Ngai, K.L., Casalini, R., Capaccioli, S., Paluch, M., Roland, C.M., 2005. J. Phys. Chem. B 109, 17356.
Ngai, K.L., Capaccioli, S., Roland, C.M., 2006. Macromolecules 39, 8543.
Ngai, K.L., Capaccioli, S., Ancherbak, S., Shinyashiki, N., 2011. Philos. Mag. 91, 1809.
Niedzwiedz, K., Wischnewski, A., Monkenbusch, M., Richter, D., Genix, A.-C., Arbe, A., Colmenero, J., Strauch, M., Straube, E., 2007. Phys. Rev. Lett. 98, 168301.
Niedzwiedz, K., Wischnewski, A., Pyckhout-Hintzen, W., Allgaier, J., Richter, D., Faraone, A., 2008. Macromolecules 41, 4866.
Nogales, A., Sanz, A., Ezquerra, T.A., 2006. J. Non-Cryst. Solids 352, 4649.
Parthun, M.G., Johari, G.P., 1995. J. Chem. Phys. 103, 7611; 103, 440 (1995).
Payne, A.R., 1960. J. Appl. Polym. Sci. 3, 127.
Payne, A.R., Whittaker, R.E., 1971. Rubber Chem. Technol. 44, 340.
Plazek, D.J., 1960. J. Colloid Sci. 15, 40.
Plazek, D.J., 1966. J. Polym. Sci. A-2 4, 745.
Plazek, D.J., 1980. Polym. J. 12, 43.
Plazek, D.J., 1982. J. Polym. Sci. Polym. Phys. Ed. 20, 729.
Plazek, D.J., 1996. J. Rheol. 40, 987.
Plazek, D.J., Berry, G., 1986. In: Uhlmann, D.R., Kreidl, N.J. (Eds.), Glass: Science and Technology, Viscosity and Relaxation, vol. 3. Academic Press, New York (Chapter 7).
Plazek, D.J., Chay, In-Chul, 1991. J. Polym. Sci. Part B: Polym. Phys. 29, 17.
Plazek, D.J., Choy, I.-C., 1989. J. Polym. Sci. Part B: Polym. Phys. 27, 307.

Plazek, D.J., Frund Jr., Z.N., 1990. J. Polym. Sci. Part B: Polym. Phys. 28, 431.
Plazek, D.J., Ngai, K.L., 1996. In: Mark, J.E. (Ed.), Glass Transition Temperatures, Handbook of Polymer Properties. Am. Inst. Phys. Press, p. 139.
Plazek, D.J., O'Rourke, V.M., 1971. J. Polym. Sci. A-2 9, 209.
Plazek, D.J., Plazek, D.L., 1983. Macromolecules 16, 1469.
Plazek, D.J., Rosner, M., 1998. Rubber Chem. Technol. 71, 679.
Plazek, D.J., Choy, I.-C., Kelley, F.N., von Meerwall, E., Su, L.-J., 1983. Rubber Chem. Technol. 56, 866.
Plazek, D.J., Gu, G.-F., Stacer, R.G., Su, L.-J., von Meerwall, E., Kelley, F.N., 1988. J. Mater. Sci. 23, 1289.
Plazek, D.J., Zheng, X.D., Ngai, K.L., 1992. Macromolecules 25, 4920.
Plazek, D.J., Bero, C., Neumeister, S., Floudas, G., Fytas, G., Ngai, K.L., 1994. J. Colloid Polym. Sci. 272, 1430.
Poh, B.T., Adachi, K., Kotaka, T., 1987. Macromolecules 20, 2563.
Riande, E., Markovitz, H., Plazek, D.J., Raghupathi, N., 1995. J. Poly. Sci. Symp. 50, 405.
Richter, D., Ewen, B., Farago, B., Wagner, T., 1989. Phys. Rev. Lett. 62, 2140.
Richter, D., Farago, B., Fetters, L.J., Huang, J.S., Ewen, B., Lartigue, C., 1990. Phys. Rev. Lett. 64, 1389.
Roland, C.M., Casalini, R., 2007. Macromolecules 40, 3631.
Roland, C.M., Ngai, K.L., 1991. Macromolecules 24, 2261.
Roland, C.M., Ngai, K.L., 1992. Macromolecules 25, 363.
Roland, C.M., Ngai, K.L., 2000. Macromolecules 33, 3184.
Roland, C.M., Ngai, K.L., 1992. J. Rheology 36, 1691.
Roland, C.M., Ngai, K.L., O'Reilly, J.M., Sedita, J.S., 1992. Macromolecules 25, 3906
Roland, C.M., Ngai, K.L., Santangelo, P.G., Qiu, X.H., Ediger, M.D., Plazek, D.J., 2001. Macromolecules 34, 6159.
Roland, C.M., Ngai, K.L., 1992. Macromolecules 25, 363.
Roland, C.M., Schroeder, M.J., Fontanella, J.J., Ngai, K.L., 2004. Macromolecules 37, 2630.
Roland, C.M., McGrath, K.J., Casalini, R., 2005. Macromolecules 38, 8729.
Roland, C.M., Fragiadakis, D., Coslovich, D., Capaccioli, S., Ngai, K.L., 2010. J. Chem. Phys. 133, 124507.
Roland, C.M., Schroeder, M.J., Fontanella, J.J., Ngai, K.L., 2004. Macromolecules 37, 2630.
Ronca, G., Allegra, G., 1975. J. Chem. Phys. 63, 4990.
Sakai, V.G., Chen, C.X., Maranas, J.K., Chowdhuri, Z., 2004. Macromolecules 37, 9975; Liu, J.H., Sakai, V.G., Maranas, J.K., 2006. Macromolecules 39, 2866.
Sakai, V.G., Maranas, J.K., Chowdhuri, Z., Peral, I., Copley, J.R.D., 2005. J. Polym. Sci. B: Polym. Phys. 43, 2914.
Sakai, V.G., Maranas, J.K., Peral, I., Copley, J.R.D., 2008. Macromolecules 41, 3701.
Sanz, A., Nogales, A., Ezquerra, T.A., Lotti, N., Finelli, L., 2004. Phys. Rev. E 70, 021502.
Saunders, J.F., Ferry, J.D., 1974. Macromolecules 7, 681.
Schmidt-Rohr, K., Spiess, H.W., 1991. Phys. Rev. Lett. 66, 3020.
Schönhals, A., 1993. Macromolecules 26, 1309.
Schönhals, A., Schlosser, E., 1993. Phys. Scr. T49A, 233.
Schwartz, Gustavo A., Colmenero, Juan, Alegria, Angel, 2007. Macromolecules 40, 3246.
Schwarzl, F.R., Staverman, A.J., 1952. J. Appl. Phys. 23, 838.
Schweizer, K.S., 1989. J. Chem. Phys. 91, 5802.
Shears, M.F., Williams, G., 1973. J. Chem. Soc., Faraday Trans. 2, 69, 609.
Shi, J.-F., Dickinson, L.C., MacKnight, W.J., Chien, J.C.W., 1993. Macromolecules 26, 5908.
Smith, T.L., 1977. In: Treatise on Materials Science and Technology. Academic Press, New York.
Struik, L.C.E., 1978. Physical Aging in Amorphous Polymers and Other Materials. Elsevier, Amsterdam.
Thirion, P., Chasset, R., 1965. In: Proceedings of the Conference on Physics of Non-Crystalline Solids. North Holland, Amsterdam; Rev. Gen. Caoutchouc 41, 271 (1964).
Tool, A.Q., 1946. J. Am. Ceram. Soc. 29, 240.
Treloar, L.G.R., 1973. Rep. Prog. Phys. 36, 755.
Treloar, L.R.G., 1975. The Physics of Rubber Elasticity, third ed. Clarendon, Oxford.
Tsang, K.Y., Ngai, K.L., 1997. Phys. Rev. E 54, 3067.
Urakawa, O., Fuse, Y., Hori, H., Tran-Cong, Q., Yano, O., 2001. Polymer 42, 765.

Urakawa, O., Sugihara, T., Adachi, K., 2002. Polym. Appl. (Jpn.) 51, 10.
Urakawa, Osamu, Ujii, Takahiro, Adachi, Keiichiro, 2006. J. Non-Cryst. Solids 352, 5042.
Valentine, R.H., Ferry, J.D., Homma, T., Ninomiya, K., 1968. J. Polym. Sci.: Part A-2 6, 479.
Watanabe, H., Matsumiya, Y., Takada, J., Sasaki, H., Matsushima, Y., Kuriyama, A., Inoue, T., Ahn, K.H., Yu, W., Krishnamoorti, R., 2007. Macromolecules 40, 5389.
Zawada, J.A., Ylitalo, C.M., Fuller, G.G., Colby, R.H., Long, T.E., 1992. Macromolecules 25, 2896.
Zeroni, Ilan, Ozair, Sahban, Lodge, Timothy P., 2008. Macromolecules 41, 5033.
Zetsche, A., Fischer, E.W., 1994. Acta Polym. 45, 168.
Zetsche, A., Kremer, F., Jung, W., Schulze, H., 1990. Polymer 31, 1883.
Zhang, S., Jin, X., Painter, P.C., Runt, J., 2002. Macromolecules 35, 3636.
Zhao, Junshu, Zhang, Liang, Ediger, M.D., 2008. Macromolecules 41, 8030.
Zhao, Junshu, Ediger, M.D., Sun, Ye, Yu, Lian, 2009. Macromolecules 42, 6777.
Zorn, R., Arbe, A., Colmenero, J., Frick, B., Richter, D., Buchenau, U., Phys. Rev. E 52, 781 (1995).

Chapter 6

Rheological Behavior and Processing of Unvulcanized Rubber

C.M Roland
Chemistry Division, Naval Research Laboratory, Washington, DC, USA

6.1 RHEOLOGY

6.1.1 Introduction

The mixing and processing of rubber is centrally important to achieving useful products at acceptable cost. Various texts review the general topic of polymer processing (Osswald and Gries, 2006; Baird and Collias, 1998; Morton-Jones, 1989; Tadmor and Gogos, 2006) and the rheology and viscoelasticity of rubbery polymers (Roland, 2011; Macosko, 1994; Ferry, 1980; Watanabe, 1999; Berry, 2003; Graessley, 2008), including some specific to rubber processing (White, 1995; Johnson, 2001). This chapter provides a basic overview of the rheology (from the Greek *rheos*, meaning stream or flow, rheology is the study of the deformation and flow of matter) and processing of uncured rubber, including recent developments concerning the material behavior.

The study of the rheology and processing of rubber had its beginnings with the development of the industry in the early 19th century (Hancock, 1920). A major advance was the Boltzmann superposition principle (Boltzmann, 1874) (Section 6.2.2), enabling prediction of stresses in a material subjected to arbitrary strain histories. Early elastic recoil data of Mooney on natural rubber (Mooney, 1936) are displayed in Figure 6.1, illustrating the Boltzmann superpositioning: A short time delay in the onset of recovery only affects the recovery at short times; the behavior over longer times is essentially the same. The first experimental measurements of viscoelastic behavior were on silk thread, carried out in the mid-19th century (Weber, 1835), although there is an oblique reference to the concept in the biblical book of Judges 5:5,

The Science and Technology of Rubber. http://dx.doi.org/10.1016/B978-0-12-394584-6.00006-6

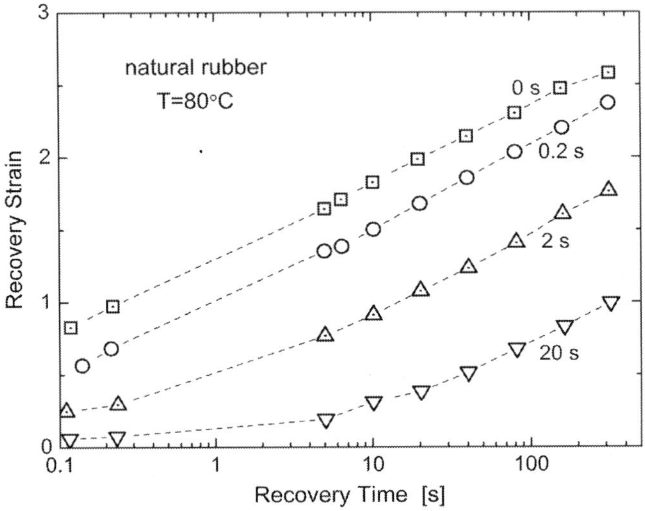

FIGURE 6.1 Elastic recovery of natural rubber following imposition of a strain for the indicated time period. The initial recovery is reduced by maintaining the material in the stretched condition, but the recovery at longer times is unaffected (Mooney, 1936).

"*The mountains flow before the Lord,*" expressing the idea that even solids exhibit viscous behavior over a sufficiently long time frame.

The first recorded processing of rubber is attributed to the Olmec, the earliest major civilization in Mexico. Analysis of excavated rubber balls dating to 1600 BC indicates that the Olmec heated a mixture of rubber latex with liquid extracted from *Ipomoea Alba*, a species of morning glory vine (Hosler et al., 1999). The process removes nonpolymeric contaminants from the 1,4-polyisoprene and probably introduces low levels of crosslinking. Evidently the resulting rubber balls had sufficient shape retention and elasticity to be used for games. This material was possibly used also for waterproofing (Wendt and Cyphers, 2008). About 3400 years later, Macintosh Hancock (1920) coated woven fabrics with solutions of rubber for the same purpose, leading to early manufacture of raincoats bearing the inventor's name. About 20 years later Goodyear invented the process of sulfur vulcanization of rubber, which became universally adopted and enabled most modern applications (Reader's Digest, 1958).

6.1.2 Basic Concepts

The mechanical response of a material is described by constitutive equations, which are geometry-independent, but usually limited to specific conditions of temperature, pressure, and even type of deformation. Hooke's law describes the response to a tensile strain ε:

$$\sigma = E\varepsilon \tag{6.1}$$

in which σ is the stress and E is a material constant known as Young's modulus. For a strictly elastic material, the behavior is reversible. The behavior of sheared fluids can be described by an equation due to Newton:

$$\sigma = \eta\dot{\varepsilon} \tag{6.2}$$

in which σ is the shear stress and $\dot{\varepsilon}$ represents the rate of shear or the shear gradient. The proportionality constant in Eq. (6.2) is the viscosity, which is a material constant for simple liquids.

Before proceeding, some definitions are useful. Stress is the ratio of the force on a body to the cross-sectional area of the body. The true stress refers to the infinitesimal force per (instantaneous) area, while the engineering stress is the force per initial area. Strain is a measure of the extent of the deformation. Normal strains change the dimensions, whereas shear strains change the angle between two initially perpendicular lines. In correspondence with the true stress, the Cauchy (or Euler) strain is measured with respect to the deformed state, while the Green's (or Lagrange) strain is with respect to the undeformed state.

The simple form of Eqs. (6.1) and (6.2) embodies two assumptions. The material is isotropic, so that the properties are the same in all directions. More generally, the stress is a tensor

$$\sigma_{ij} = \begin{pmatrix} \sigma_{11} & \sigma_{12} & \sigma_{13} \\ \sigma_{21} & \sigma_{22} & \sigma_{23} \\ \sigma_{31} & \sigma_{32} & \sigma_{33} \end{pmatrix} \tag{6.3}$$

in which the subscripts refer to the rectilinear coordinates. To avoid rotation or translation of a body, $\sigma_{12} = \sigma_{21}$, $\sigma_{13} = \sigma_{31}$, and so on; thus, only six of the stress components are unique. The set of axes for which one of the normal stresses is maximum defines the principal stresses

$$\sigma_{ij} = \begin{pmatrix} \sigma_{11} & 0 & 0 \\ 0 & \sigma_{22} & 0 \\ 0 & 0 & \sigma_{33} \end{pmatrix}. \tag{6.4}$$

These can be found as roots of the equation

$$\sigma^3 - I_1\sigma^2 - I_2\sigma - I_3 = 0 \tag{6.5}$$

in which I_1, I_2, and I_3 are the so-called strain invariants,

$$I_1 = \lambda_1^2 + \lambda_2^2 + \lambda_3^2,\ I_2 = \lambda_1^2\lambda_2^2 + \lambda_1^2\lambda_3^2 + \lambda_2^2\lambda_3^2,\ I_3 = \lambda_1^2\lambda_2^2\lambda_3^2. \tag{6.6}$$

Strain invariants are independent of the axes used to define the geometry, enabling calculations for inhomogeneous deformations without explicit consideration of the principal directions. During a homogenous deformation,

TABLE 6.1 Common Elastic Constants and Their Interrelationships (Mott and Roland, 2009)

Elastic Constant	Definition	Relation to Poisson's Ratio
$E \equiv$ Young's modulus	$\varepsilon = \varepsilon_{11}$; $\varepsilon_{22} = \varepsilon_{33} = -\nu\varepsilon_{11}$; other $\varepsilon_{ij} = 0$	——
$G \equiv$ shear modulus	$\varepsilon = \varepsilon_{12}$; other $\varepsilon_{ij} = 0$; $\gamma_{12} = \varepsilon_{12} + \varepsilon_{21} = 2\varepsilon_{12}$	$\nu = \frac{E}{2G} - 1$
$B \equiv$ bulk modulus	$\varepsilon = \varepsilon_{11} = \varepsilon_{22} = \varepsilon_{33}$; other $\varepsilon_{ij} = 0$	$\nu = \frac{1}{2} - \frac{E}{6B}$
$M \equiv$ longitudinal modulus	$\varepsilon = \varepsilon_{11}$; other $\varepsilon_{ij} = 0$	$\nu = \frac{1}{4}\left[\frac{E}{M} - 1 \pm \left(\frac{E^2}{M^2} - 10\frac{E}{M} + 9\right)^{1/2}\right]$
$E \equiv$ biaxial modulus	$\sigma = \sigma_{11} = \sigma_{22}$; other $\sigma_{ij} = 0$	$\nu = 1 - \frac{E}{H}$

parallel lines remain parallel, whereas inhomogeneous (heterogeneous) strains are nonuniform and irregular, distorting the body into a more complex form. Most mathematical treatments assume that strains are homogeneous.

For an isotropic body, there are only two stress components that are independent of the other. This means that while different loadings and strains can be imposed, the material constants relating stress and strain are not all unique. There are six common elastic constants (Table 6.1), including Poisson's ratio, defined as the ratio of the lateral strain accompanying a longitudinal strain, $\nu = -\varepsilon_{22}/\varepsilon_{22}$. Since only two are unique for an isotropic body, each elastic constant can be expressed as a function of any other pair; these expressions are tabulated in Table 6.1 in terms of Poisson's ratio (Mott and Roland, 2009).

The second assumption implicit in Eqs. (6.1) and (6.2) is that the material is homogeneous, meaning its properties are the same at all points. (This differs from the idea of a homogeneous, or affine, strain, in which the displacements at all points are linear functions of their original coordinates.) Of course, because of their molecular structure, no material is strictly homogeneous. Continuum mechanics is always an approximation to real materials, with homogeneity being relevant to the type of measurement under consideration. For example, rubber reinforced with filler is homogeneous on the scale of a bulk mechanical measurement, notwithstanding the marked difference in properties between the polymer and the rigid particles, which have dimension on the order of 10–100 nm.

The defining characteristic of rubbery polymers is their macromolecular size, which can be two orders of magnitude larger than the distance between atoms. As a result, internal motions occur over a range of length scales, each associated with a different time scale. This means that whether an external perturbation is "slow" or "fast" depends on the mode of motion it excites. Invariably, some modes move on the time scale of the perturbation, dissipating energy and causing a peak in the out-of-phase, or loss, component of the dynamic response, at a frequency equal to that of the underlying motion. Other modes, such as vibrations, are relatively fast, responding elastically, or they involve the entire molecule, manifested as flow (in the absence of a network structure). If the time scale of the motion is very long, the mode can be unresponsive to the perturbation; a well-known example is the conformational transitions of a polymer backbone below the glass transition temperature. The net effect of this diversity of modes of motion is a frequency-dependent response to mechanical excitation, or equivalently, a time-varying reaction to a transient perturbation. This means that elastic constants such as E or η are not constant for a viscoelastic material, but vary with dynamic frequency or time.

Although the rheology and processing of rubber entails mechanical properties as described earlier, viscoelasticity encompasses other manifestations of the same underlying motions, such as dielectric polarization, dynamic birefringence, and the inelastic scattering of light, X-rays, and neutrons. Viscoelastic materials exhibit a second, related property, the simultaneous storage and dissipation of energy, from which the term originates. The normal forces accompanying shear flow of polymers are due to the elasticity of the material, and hence another manifestation of viscoelasticity.

Viscoelasticity can also be observed in nonpolymeric materials, although usually only over a limited range of conditions (whereas polymers have rate-dependent mechanical properties at almost all temperatures). An example is the viscosity of supercooled molecular liquids (Angell et al., 2000; Roland et al., 2005). Creep of metals under sustained loading at elevated temperature is referred to as viscoelasticity (Kennedy, 1953; McLean, 1966), although the behavior is caused by changes in dislocations or grain size. Strictly speaking, viscoelasticity refers only to the properties of an unchanging material; if the variation in the response with time is due to changes of the material itself (e.g., the drying of concrete or chain scission in a sheared polymer), this is not viscoelasticity.

6.2 LINEAR VISCOELASTICITY

6.2.1 Material Constants

Due to their viscoelastic nature, the response of polymers depends entirely on the loading history, and the simple inverse relationship between moduli and compliances is lost. Thus, Eq. (6.1) for application of a shear strain becomes

$$G(t) = \sigma(t)/\varepsilon_0 \tag{6.7}$$

in which $G(t)$ is the stress relaxation modulus; tensile stresses would yield the corresponding tensile relaxation modulus, $E(t)$. Imposing a constant load yields the compliance, which for shear is

$$J(t) = \gamma(t)/\sigma_0. \tag{6.8}$$

The two response functions are related as

$$\int_0^t G(u)J(t-u)du = t \tag{6.9}$$

but the stress relaxation and compliance are not reciprocals

$$G(t)J(t) < 1. \tag{6.10}$$

For a viscoelastic material, the viscosity in the limit of zero-shear rate, the Newtonian viscosity, can be obtained from the integral of the stress relaxation modulus

$$\eta_0 = \int_0^\infty G(t)dt = \int_0^\infty tG(t)d\ln t. \tag{6.11}$$

A logarithmic time axis is necessary because of the very broad time scale of the dynamics of polymers.

The steady-state recoverable compliance, which is a measure of the elastic strain, can be calculated using

$$J_s^0 = \int_0^\infty tG(t)dt \bigg/ \left(\int_0^\infty G(t)dt\right)^2. \tag{6.12}$$

The compliance function measured in a creep experiment can be decomposed into recoverable and viscous flow components

$$J(t) = J_r(t) + t/\eta, \tag{6.13}$$

where $J_r(t)$ is the recoverable compliance. After steady state is attained, and the viscosity becomes equal to η_0,

$$J_s^0 = \lim_{t\to\infty} J_r(t). \tag{6.14}$$

The steady-state recoverable compliance can be measured by fitting Eq. (6.13) to creep data at long times, or more accurately, J_s^0 is determined from the recovery after removal of the stress subsequent to attainment of steady-state flow.

The Kramers-Kronig formula relates the stress relaxation modulus to dynamic quantities (Kronig, 1926; Kramers, 1927)

$$\int_0^\infty G'(\omega)\frac{\sin\omega t}{\omega}d\omega = \int_0^\infty G''(\omega)\frac{\cos\omega t}{\omega}d\omega = \frac{\pi}{2}G(t) \tag{6.15}$$

or the inverse relations for the storage modulus

$$G'(\omega) = \omega \int_0^\infty G(t) \sin(\omega t) dt \tag{6.16}$$

and the loss modulus

$$G''(\omega) = \omega \int_0^\infty G(t) \cos(\omega t) dt. \tag{6.17}$$

An approximate relation is (Tschoegl, 1989)

$$G''(\omega) \approx \frac{\pi}{2} \frac{dG'(\omega)}{d \ln \omega}. \tag{6.18}$$

An analogous expression for the dielectric loss in terms of the derivative of the permittivity is commonly used to avoid the problem of dielectric loss peaks being masked by dc conductivity (the latter not contributing to the in-phase response) (Wubbenhorst and Van Turnhout, 2002).

The dynamic viscosity, $\eta = G''(\omega)/\omega$, yields the terminal viscosity in the limit of zero frequency

$$\eta_0 = \lim_{\omega \to 0} G''(\omega)/\omega \tag{6.19}$$

and J_s^0 can be obtained from the zero-shear-rate limiting value of the storage modulus

$$J_s^0 = \eta_0^{-2} \lim_{\omega \to 0} \omega^{-2} G'(\omega). \tag{6.20}$$

Alternatively, from steady shearing experiments, which yield η_0 directly in the limit of low shear rate, the steady-state recoverable compliance can be obtained from the first normal stress coefficient, Ψ_0, which is the ratio of the first normal stress difference to the square of the shear rate, measured at low shear rate

$$J_s^0 = \frac{\Psi_0}{2\eta_0^2}. \tag{6.21}$$

From Eq. (6.12) the steady-state recoverable compliance is also given by

$$J_s^0 = \eta_0^{-2} \int_0^\infty t G(t) dt. \tag{6.22}$$

It follows from Eqs. (6.11) and (6.17) that

$$\eta_0 = \lim_{\omega \to 0} G''(\omega)/\omega \tag{6.23}$$

Experimentally steady state in a transient experiment implies

$$J(t) \sim t. \tag{6.24}$$

The steady-state recoverable compliance is an important rheological parameter because it is very sensitive to the high molecular tail of the molecular weight distribution, and thus can be correlated with elastic properties of a rubber.

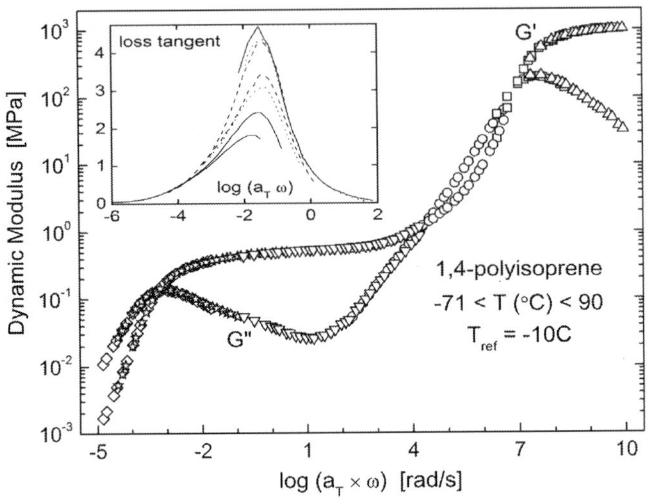

FIGURE 6.2 Dynamic shear modulus master curve for 1,4-polyisoprene. Although the time-temperature shifting is apparently satisfactory, there is a breakdown in the glass transition zone, as seen in the inset, showing the changing shape of the loss tangent peak for different temperatures (Santangelo and Roland, 1998).

The defining feature of the dynamics of rubber comes from the entanglement of the chains, due to their uncrossability. These entanglements are topological and entirely intermolecular, since the volume of a given chain is pervaded by many other chains. Due to entanglements, there is a range of frequencies over which the storage modulus (or equivalently, the relaxation modulus versus time) has a nearly constant magnitude, G_N. This rubbery plateau is evident in the dynamic mechanical spectra for polyisoprene in Figure 6.2 (Santangelo and Roland, 1998). The entanglements cause chain motion to be anisotropic, favoring the chain direction (de Gennes, 1979), and are described by the tube model of Doi and Edwards (1986). The model has two species-dependent parameters, the Rouse friction coefficient, ς, and a parameter characterizing the concentration of entanglements. From Eq. (6.15) (Kronig, 1926; Kramers, 1927),

$$G_N = \frac{2}{\pi}\int_{-\infty}^{+\infty} G''(\omega) d \ln \omega. \tag{6.25}$$

The upper limit in Eq. (6.25) is imposed empirically to avoid inclusion of the glass transition zone. When the material is plasticized with diluent, G_N decreases by roughly the square of the polymer volume fraction.

A determination of how G_N depends on the chemical structure of a polymer requires a definition of an entanglement. Various ideas have been proposed: a fixed number of binary contacts per entanglement (Brochard and De Gennes, 1977) or per entanglement volume (Colby et al., 1992); a fixed number of strands per entanglement volume (Ronca, 1983; Lin, 1987; Kavassalis and

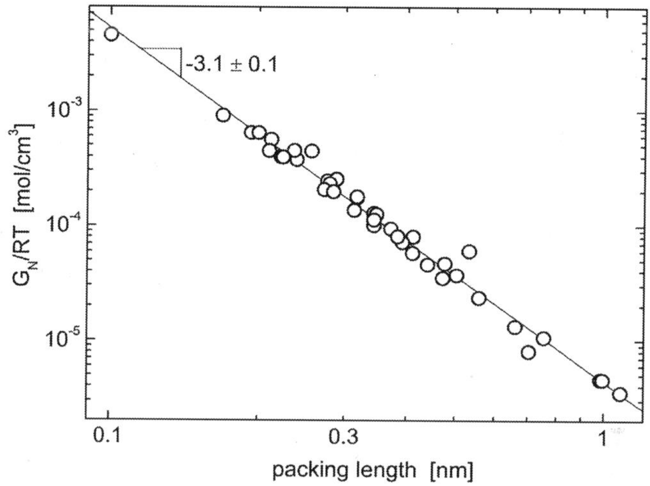

FIGURE 6.3 Entanglement interaction density, defined as the ratio of the plateau modulus to the thermal energy, versus the packing length for different polymers. The power-law slope is indicated (Roland, 2011; Everaers et al., 2004).

Noolandi, 1989; Fetters et al., 1999); or some fraction of the number of interchain contacts (Graessley and Edwards, 1981; Heymans, 2000; Fetters et al., 1994, 2002; Richter et al., 1993). Analysis of experimental data leads to the rough estimate that six binary contacts are necessary to produce one entanglement (Graessley, 2008). The greater the volume pervaded by a chain, the more it encounters neighboring chains; thus, stiffer polymer backbones having smaller, less bulky repeat units are expected to be more entangled. In Figure 6.3 (Roland, 2011; Everaers et al., 2004) the plateau modulus of various polymers is plotted versus the packing length, $l_p \equiv \dfrac{M}{\rho N_A \langle R_g^2 \rangle}$, in which $\langle R_g^2 \rangle$ is the mean square radius of gyration, ρ is the mass density, and N_A is Avogadro's number. The packing length is the ratio of the volume of a chain to the square of its radius. The data in Figure 6.3 indicate that G_N is inversely proportional to the cube of l_p.

A characteristic relaxation time for a rubber can be defined from the ratio of the zero-shear viscosity and the plateau modulus (Roland, 2011; Ferry, 1980)

$$\tau_0 = \eta_0 G_N. \tag{6.26}$$

τ_0 is a terminal relaxation time describing chain motions. Other global relaxation times can be defined experimentally, for example as the inverse of the frequency of the maximum in the terminal dispersion in the loss modulus, τ_{max}; from the time for equilibration following cessation of nonlinear shear flow, $\tau_{R,\eta}$, measured by recovery of the overshoot in the transient viscosity; the corresponding time for recovery of the overshoot in the second normal

TABLE 6.2 Comparison of Global Relaxation and Recovery Times for Entangled 1,4-Polybutadiene Solutions (Roland and Robertson, 2006)

Relaxation Time	$\varphi M_w/M_e$[a]	
	30	47
$\log \tau_0$	-0.64 ± 0.03	-0.10 ± 0.03
$\log \tau_{max}$	-1.49 ± 0.01	-0.944 ± 0.03
$\log \tau_{R,\eta}$	0.84 ± 0.02	1.68 ± 0.02
$\log \tau_{R,G}$	1.59 ± 0.05	1.96 ± 0.05
$\log \tau_{R,\Psi}$	0.53 ± 1	0.93 ± 0.54

[a] *Number of entanglements per chain.*

stress coefficient, $\tau_{R,\Psi}$; and $\tau_{R,G}$, the time for the dynamic storage modulus to return to its equilibrium value after nonlinear flow. Table 6.2 (Roland and Robertson, 2006) compares these different relaxation times for high molecular weight 1,4-polybutadiene. Their magnitudes differ by as much as three orders of magnitude. It has also been found that the temperature dependence of these various relaxation times can be different (Roland et al., 2004).

6.2.2 Boltzmann Superposition Principle

If the rubber exhibits linear mechanical behavior, defined by proportionality between stress and strain and time invariance of the response, the Boltzmann superposition equation can be used to predict the stress for any strain history (Boltzmann, 1874; Larson, 1988). In terms of the shear relaxation modulus, the equation is

$$\sigma(t) = \int_0^t G(t-u)\frac{d\varepsilon}{du}du. \tag{6.27}$$

Rewriting Eq. (6.27) to make explicit the additivity of the stresses

$$\sigma(t) = \int_0^t \frac{dG(t-u)}{du}\big(\varepsilon(t) - \varepsilon(u)\big)\,du + G(t)\varepsilon(t). \tag{6.28}$$

In this equation $\frac{dG(t-u)}{du}du$ represents the "fading memory"; that is, the survival probability at time t of a stress created at time u.

While all polymers obey Eq. (6.27) at sufficiently small strains, for rubbers this linear behavior extends over a very wide range (Yannas, 1974; Chang et al., 1976). When the strains are too large for a linear response, Bernstein (1963, 1964) and, independently, Kaye (1962) developed an integral constitutive equation to describe the nonlinear viscoelastic properties, known as the K-BKZ equation

$$\sigma(t) = \int_0^t [G(t-u, \varepsilon(t-u))] \frac{d\varepsilon}{du} du. \tag{6.29}$$

If the relaxation behavior, *G*(*t*), in Eq. (6.29) is independent of strain, the behavior is referred to as "time invariant." This decoupling of time and strain effects leads to a simpler expression

$$\sigma(t) = \int_0^t [G(t-u) g(\varepsilon)] \frac{d\varepsilon}{du} du. \tag{6.30}$$

The damping function, $g(\varepsilon)$, in Eq. (6.30) accounts for lack of proportionality between stress and strain. The product, $g(\varepsilon)\varepsilon$, quantifies the nonlinear elasticity ($g(\varepsilon) = 1$ for linear viscoelastic behavior). Separability of time and strain is illustrated for 1,4-polyisoprene in Figures 6.4 and 6.5; the time-dependence of the stress relaxation is the same for shear strains of varying amplitude and for different modes of deformation (Fuller, 1988).

Absent mechanisms such as strain crystallization or the finite extension of the chains, uncrosslinked rubbers usually strain-soften, which means that $g(\varepsilon)$ decreases with strain. A damping function that has been shown to successfully describe elastomers for tensile strains of a few hundred percent is (Roland, 1989)

$$g(\varepsilon) = \left(1 + \frac{c_2}{c_1}\lambda^{-1}\right), \tag{6.31}$$

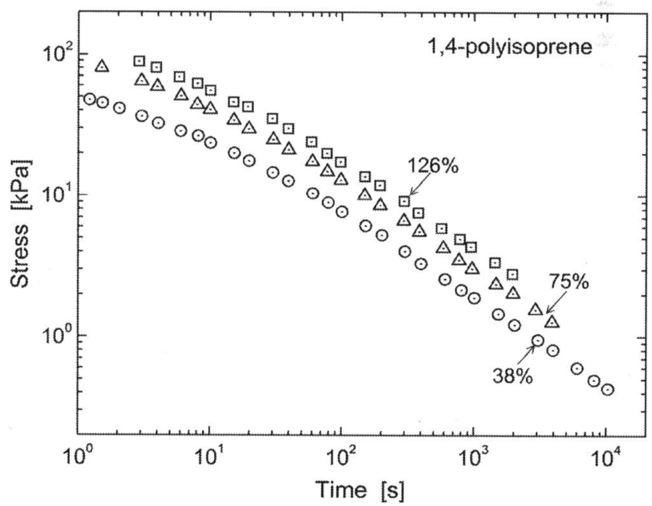

FIGURE 6.4 Stress relaxation of synthetic 1,4-polyisoprene at the indicated shear strains. The parallel nature of the curves reflects the invariance of the time-dependence to strain (Fuller, 1988).

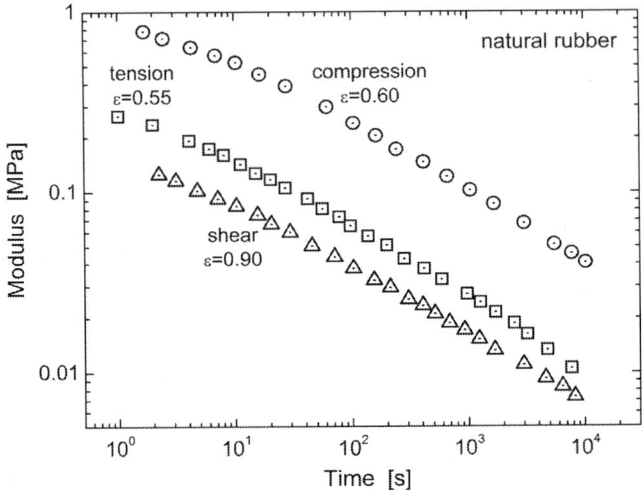

FIGURE 6.5 Stress relaxation modulus of natural rubber at different types and magnitudes of strain. The curves are approximately parallel, indicating that strain and time effects are decoupled (Fuller, 1988).

where c_1 and c_2 are material constants. For a single-step strain, Eq. (6.29) simplifies to

$$\sigma(\varepsilon, t) = G(t)g(\varepsilon)\varepsilon. \tag{6.32}$$

For continuous straining the integral in Eq. (6.30) can be approximated as a series of step strains, each described by Eq. (6.32). From these equations the stress for arbitrary strain history can be predicted, although the calculation requires that both the stress relaxation function and the strain-dependence of the modulus be determined in separate experiments. This general approach to predicting the rheology of polymer melts has met with good success (Tanner, 1988).

The accuracy of K-BKZ predictions can be improved with the addition of correction terms to account for other mechanisms (Vrentas et al., 1991; Coleman and Zapas, 1989; McKenna and Zapas, 1979). Introducing the "independent alignment approximation" to the Doi-Edwards tube model of polymer dynamics yields the K-BKZ form (Doi, 1980). (This independent alignment approximation assumes the relative position of every segment does not change following straining; that is, the "tube" of entanglement constraints does not retract.) The Doi-Edwards damping function for shear can be approximated as

$$g(\varepsilon) = \left(1 + \frac{\varsigma}{3}\varepsilon^2\right)^{-1}, \tag{6.33}$$

where ς is a parameter fitted to experimental data (Larson, 1985; Larson and Valesano, 1986). Using $\varsigma = 0.43$ gives results corresponding to the tube model with constraint release. Constraint release refers to diffusion of the segments comprising the tube of entanglements, whereas in the original tube model, strictly applicable only to a chain dissolved in a network, the segments of the tube are fixed (Milner, 1996).

6.2.3 Time-Temperature Equivalence

Most mechanical test instruments have a limited dynamic range, typically down to ca. 10^{-4} s^{-1} for low strain rates, which almost always requires transient measurements, and up to about 100 s^{-1} on the high frequency end, usually employing dynamic measurements. Servohydraulic test systems can achieve higher speeds, but with limitations on the attainable strains (Bergstrom and Boyce, 1998; Arruda et al., 2001). If only the linear viscoelastic response is to be measured, dielectric spectroscopy (Kremer and Schonhals, 2003; McCrum et al., 1967) can be used. Dielectric spectroscopy can be carried out routinely over 11 decades of frequency, extending to GHz; this range can be extended to as much as 18 decades (Schneider et al., 1999), although this requires corrections be made for effects from the connectors and cables. However, dielectric relaxation only probes the local segmental dynamics of polymers, except for the few cases in which a dipole moment is present parallel to the polymer chain. For the latter the chain modes are dielectrically active, and the dynamics of the chain end-to-end vector can be measured. Although dielectric relaxation times are smaller than the corresponding mechanical relaxations, their respective dependences on temperature (Boese et al., 1989; Colmenero et al., 1991, 1994; Paluch et al., 2002, 2003) and pressure (Bair and Winer, 1980; Roland et al., 2006) are the same. Another advantage of dielectric spectroscopy is the relative ease of obtaining measurements at elevated hydrostatic pressure (Roland et al., 2005; Skorodumov and Godovskii, 1993; Floudas, 2003). Few mechanical studies of the effect of pressure have been carried out, and these are usually limited to studying the viscosity of low molecular weight polymers (Hellwege et al., 1967; Fillers and Tschoegl, 1977; Bair, 2002). The dynamic modulus of polymer melts can be measured in a rheometer pressurized with a gas (Han and Ma, 1983; Gerhardt et al., 1997; Royer et al., 2000), but since gases dissolve in the material, the consequent plasticization can introduce error.

(i) Superposition Principle

The most common means to extend the frequency scale is to invoke time-temperature superpositioning (Ferry, 1980). If all motions of a polymer contributing to a particular viscoelastic response are affected the same by temperature, then changes in temperature only alter the overall time scale; such a material is thermorheologically simple. Thermorheological simplicity means conformance to the time-temperature superposition principle, whereby lower and higher strain rate data can be obtained from measurements at higher and lower temperatures, respectively.

Time-temperature superpositioning was originally derived from free volume models, which assume that the rates of molecular motions are governed by the available unoccupied space. Early studies of molecular liquids led to the Doolittle equation, relating the viscosity to the fractional free volume, $f(\equiv V/(V - V_0)$, where V is the specific volume and V_0 is the occupied volume normalized by the mass) (Doolittle and Doolittle, 1957; Cohen and Turnbull,

1959):

$$\log \eta(T) = \log A_D + B_D/f(T), \tag{6.34}$$

with A_D and B_D constants. However, Eq. (6.34) only approximately fits experimental data (Corezzi et al., 1999; Paluch, 2001; Schug et al., 1998). The assumption that the free volume expands linearly with temperature, $\left.\frac{df}{dT}\right|_P \sim T$, leads to the Williams-Landel-Ferry (WLF) equation (Ferry, 1980; Williams et al., 1955):

$$\log a(T) = \log a(T_R) - \frac{c_1(T - T_R)}{c_2 + T - T_R}, \tag{6.35}$$

where T_R is an arbitrary reference temperature. Assuming the WLF constants c_1 and c_2 are functions of the free volume leads to "universal values" of $c_1 = 8.86$ and $c_2 = 101.6$ (Williams et al., 1955). In Eq. (6.35) $a(T)$ is a time-temperature shift factor, defined for the viscosity as $a(T) = \eta(T)/\eta(T_R)$, or for the relaxation times as $a(T) = \tau(T)/\tau(T_R)$. (η and τ are proportional except near the glass transition (Chang and Sillescu, 1997; Rössler, 1990; Fischer et al., 1992); cf. Eq. (6.26).) Shift factors are used because the actual magnitude of the viscoelastic quantities does not affect $t-T$ superpositioning, only their change with temperature. An ostensibly different, but mathematically equivalent expression to Eq. (6.35) is due to Vogel, Fulcher, Tamman, and Hesse (Ferry, 1980):

$$\log \tau(T) = \log \tau_\infty + \frac{b_{\mathrm{VFTH}}}{T - T_V}, \tag{6.36}$$

in which b_{VFTH} $(=c_1/c_2)$, $T_V (=T_R - c_2)$, and τ_∞ are constants (Zallen, 1981). At the Vogel temperature, T_V, the dynamics would diverge if not for intervention of the glass transition. T_V is often taken to be the Kauzmann temperature at which the extrapolated liquid entropy equals the crystal entropy, ostensibly in violation of the third law of thermodynamics ("Kauzmann paradox") (Angell, 1991).

Cohen and Grest (1979, 1984) and Grest and Cohen (1981) derived a more rigorous free volume model, in which molecules have mobility only when local continuity of empty space exists. Generally free volume models have fallen into disfavor, due to the unphysical results they lead to; for example, free volumes can become negative (Williams and Angell, 1977) or change less with pressure than the occupied volume (Ferry, 1980). Nevertheless, Eqs. (6.35) or (6.36) can still be used to fit experimental data. However, the absence of a firm theoretical foundation means that the validity of time-temperature superpositioning must be established empirically for any material. A major development in the study of polymer viscoelasticity was in 1949, when samples of a high molecular weight polyisobutylene were distributed to laboratories around the world. Both transient, in particular stress relaxation measurements (Andrews and Tobolsky, 1951), and dynamic experiments (Fitzgerald et al., 1953; Ferry et al., 1953) were carried out, with the data spanning 15 decades of frequency. The successful superpositioning of these results led to the acceptance of time-temperature superpositioning as a method of converting viscoelastic data for polymers obtained over a narrow dynamic range to master curves covering many decades of reduced time or frequency (Roland, 2011; Ferry, 1980; Marvin, 1953).

The four requirements for thermorheological simplicity are:

- The local friction coefficient is the same for all motions underlying the viscoelastic property of interest.
- The polymer does not degrade, crystallize, physically age, or undergo other changes over the course of the measurements.
- At high strain rates, the deformation period is not less than the time required for stress waves to travel through the material; this ensures spatial uniformity of the stress.
- Heat build-up, more pronounced for low temperatures and high rates, causes no deviation from isothermal conditions.

A typical master curve is shown in Figure 6.2 (Santangelo and Roland, 1998) for dynamic shear data of a high molecular weight 1,4-polyisoprene measured over about a 160° range, yielding 15 decades of reduced frequency (the actual data spanned less than 5 decades). A tensile stress relaxation master curve for a polyurea rubber that spans 13 decades is displayed in Figure 6.6 (Knauss and Zhao, 2007).

Although time-temperature superpositioning can be very useful, in the glass transition (softening) zone of the viscoelastic response, polymers are thermorheologically complex. This breakdown of time-temperature superpositioning is caused by the weaker temperature dependence of the chain dynamics

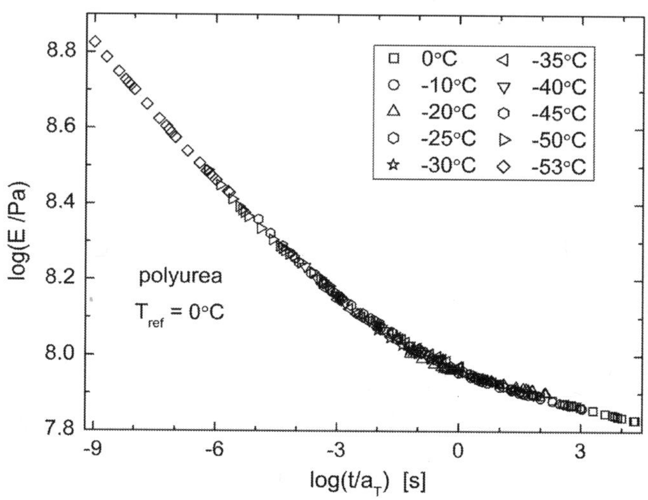

FIGURE 6.6 Tensile stress relaxation modulus of an elastomeric polyurea. Similar to the data in Figure 6.2, there is a breakdown of the time-temperature superpositioning in the glass transition zone (Knauss and Zhao, 2007).

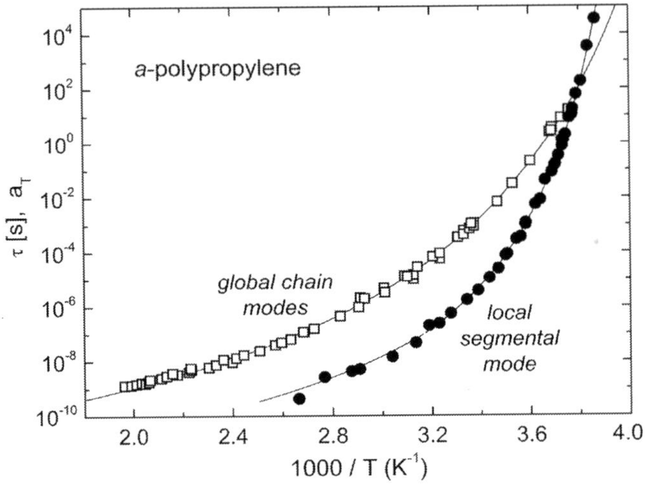

FIGURE 6.7 Atactic polypropylene segmental relaxation times (solid symbols) from mechanical spectroscopy, dynamic light scattering, dielectric relaxation, and ^{13}C NMR, along with the global time-temperature shift factors (open symbols) from dynamic mechanical spectroscopy, creep compliance, and viscosity. Vertical shifts were applied to superpose the data (Roland et al., 2001).

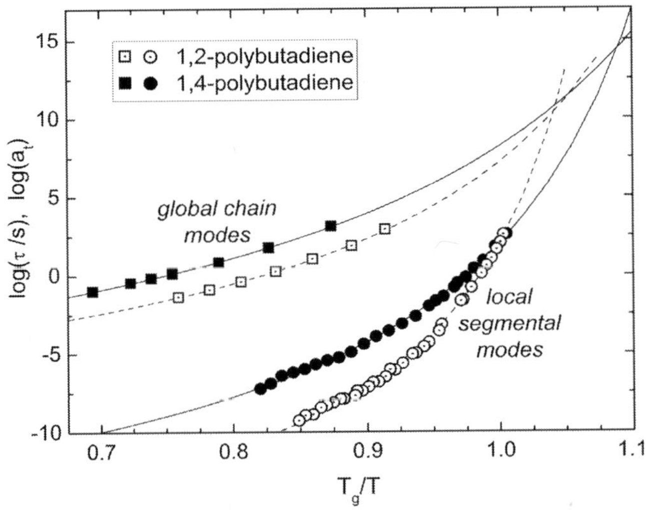

FIGURE 6.8 Local segmental relaxation times (circles) and terminal shift factors (squares) for polybutadienes. The fitted VFTH equations illustrate the marked differences in temperature-dependence of the local and global dynamics (Robertson and Rademacher, 2004).

in comparison to the local segmental modes. Well above the glass transition, relaxation times (or shift factors) for the chain and segmental dynamics are proportional, as shown in Figure 6.7 for *atactic* polypropylene (Roland et al., 2001) and in Figure 6.8 for polybutadiene (Robertson and Rademacher, 2004).

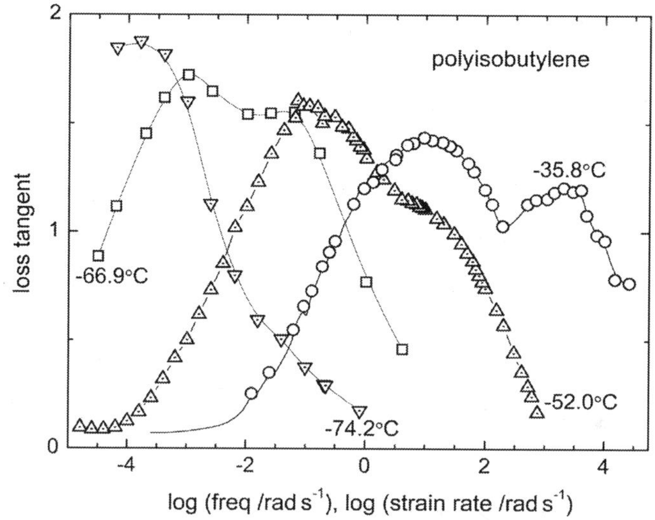

FIGURE 6.9 Loss tangent of polyisobutylene measured by dynamic mechanical spectroscopy and calculated from the recoverable creep compliance. These are not master curves; the abscissa is the actual frequency (Plazek et al., 1995).

However, for measurements within the glass transition zone, where both the chain and the segmental modes contribute, the viscoelastic data do not time-temperature superpose accurately. This breakdown of the time-temperature superpositioning was first discovered for polystyrene (Plazek, 1965) and has since been observed for every polymer for which sufficient data exist (Plazek, 1996; Ngai et al., 1995). Examples of thermorheological complexity include polyisobutylene (Figure 6.9) (Plazek et al., 1995), polyvinylacetate (Plazek, 1980), polypropylene glycol (Ngai et al., 1992), poly(phenylmethylsiloxane) (Plazek et al., 1994), polybutadiene (Robertson and Rademacher, 2004; Palade et al., 1995), atactic polypropylene (Santangelo et al., 1996), poly(alkyl glycidyl ether) (Yamane et al., 2005), polymethylmethacrylate (Plazek et al., 1988), and both the polyisoprene in Figure 6.2 (Santangelo and Roland, 1998) (see inset to figure) and the polyurea in Figure 6.6 (Pathak et al., 2008). The important point when utilizing $t - T$ superpositioning is that high temperature measurements of chain (terminal) properties cannot be combined with low temperature measurements of segmental relaxation times, because the extrapolation of the combined shift factors will be subject to very large errors. This is evident in the crossing of the fits to Eq. (6.36) for the local and global dynamics for the polybutadienes in Figure 6.8.

(ii) Density Scaling

A more realistic alternative to free-volume-derived models is to interpret relaxation and related properties of polymers in terms of the interactions among segments. Thus, in the softening regime of the viscoelastic spectrum, falling intermediate to the rubbery and glassy plateaus, the polymer dynamics entails jumps over potential barriers that are large compared to the available thermal energy (Goldstein, 1969; Stillinger and Weber, 1984; Sampoli et al., 2003). The dense packing at lower temperatures, however, means that the segmental dynamics are correlated—motion of a given segment occurs in cooperative fashion with neighboring segments. This interpretation of the dynamics requires consideration of the nature of the intermolecular potential governing these reciprocal interactions among segments. A simplified potential that considers only two-body interactions is the symmetric Lennard-Jones (LJ) potential energy function

$$U(r) = 4\hat{\varepsilon}\left[(r_0/r)^{12} - (r_0/r)^6\right], \tag{6.37}$$

where $\hat{\varepsilon}$ is a measure of the depth of the potential well and r the separation between segments (which are taken to be spherical particles of radius r_0). The attractive term in this equation has an exponent equal to six because only van der Waals dispersion interactions are considered. The repulsive forces have a stronger dependence on particle separation than the attractive forces; moreover, unlike van der Waals forces, the distance dependence varies with chemical structure (Moelwyn-Hughes, 1961; Bardik and Sysoev, 1998). A more realistic form of the LJ function is then

$$U(r) = 4\hat{\varepsilon}\left[(r_0/r)^{m} - (r_0/r)^6\right], \tag{6.38}$$

where the value of m depends on the species.

If the primary focus is on the segmental dynamics and properties involving local interactions, the attractive forces can be ignored, since they are long ranged and tend to cancel locally (Widen, 1967; Chandler et al., 1983; Stillinger et al., 2001). This follows from the fact that the force on a segment is the vector sum of all forces from neighboring segments, and there are many such neighbors (Widom, 1999). Thus, an approximate form for the potential can be assumed (Hansen, 1970; March and Tosi, 2002):

$$U(r) = \hat{\varepsilon}(r_0/r)^m. \tag{6.39}$$

This inverse power law (IPL) neglects attractive forces, consistent with the fact that the liquid structure is determined primarily by packing effects, which depend only on the repulsions (Hoover and Ross, 1971; Laird and Haymet, 1992). If an IPL accurately represents the local interaction energy between segments, then all thermodynamic properties (such as the energy, volume, and entropy) (March and Tosi, 2002; Hoover and Ross, 1971), as well as dynamic properties (Hiwatari et al., 1974; Ashurst and Hoover, 1975), depend only on the product variable, Tr^m, or in terms of the experimentally measurable specific volume, $V^{m/3}$:

$$\tau = \Im\left(TV^{m/3}\right), \tag{6.40}$$

where $\Im$ is a function. Similar equations can be written for the viscosity, diffusion constant, and so on.

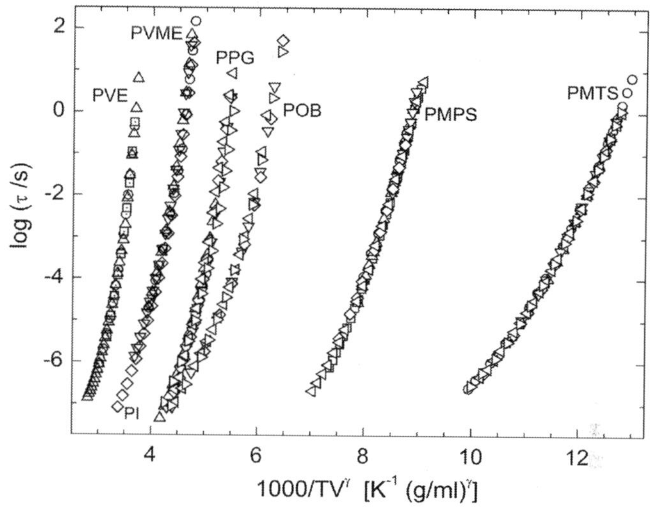

FIGURE 6.10 Density scaling of local segmental relaxation times for polyvinylethylene (PVE, $\gamma = 1.9$); polyvinylmethylether (PVME, $\gamma = 2.55$); polypropylene glycol (PPG, $\gamma = 2.5$); polyoxybutylene (POB, $\gamma = 2.65$); polymethylphenylsiloxane (PMPS, $\gamma = 5.6$); and polymethyl tolylsiloxane (PMTS, $\gamma = 5.0$). For each polymer, every symbol represents a distinct state point, with each symbol type denoting either varying pressure at constant temperature or varying temperature at constant pressure.

The idea is that the polymer dynamics are thermally activated with an activation barrier that depends on volume as $V^{m/3}$. The local segmental relaxation times measured as a function of temperature, pressure, or volume should fall on a single curve when plotted as a function of TV^{γ}, where the scaling exponent $\gamma = m/3$. Note that if free volume models were exact, $\gamma = \infty$.

Figure 6.10 shows segmental relaxation times for six rubbers measured at different conditions of T and P. For any given polymer, the data collapse to a single dependency on TV^{γ}. The value of the scaling exponent depends on the material, reflecting differences in the steepness of the intermolecular repulsive potentials. Generally for polymers γ ranges from 1.9 for 1,4-polyisoprene to 5.6 for polymethylphenysiloxane (Roland et al., 2005). Density scaling has also been applied to viscosity data for polymers (Roland et al., 2006), and to their terminal dispersions (Figure 6.11) (Roland et al., 2004; Casalini and Roland, 2005), although for chain relaxation times the superposition is only approximate (Fragiadakis et al., 2011). In toto, dynamic data for more than 100 polymers and molecular liquids have been shown to conform to density scaling (Roland, 2010, 2011; Fragiadakis and Roland, 2011).

6.2.4 Molecular Weight Dependences

Rheological properties depend strongly on the chain length, but since polymer chains invariably have a distribution of molecular weights, the use of

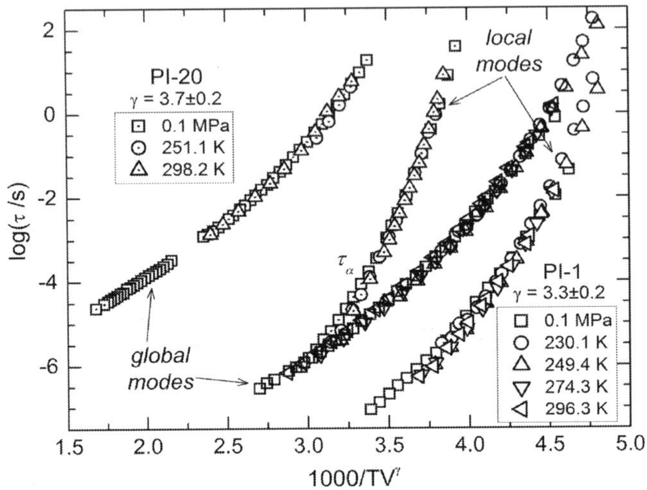

FIGURE 6.11 Density scaling of local segmental and global chain relaxation times for 1,4-polyisoprene having $M_W = 1$ kg/mol (dotted symbols) and 20 kg/mol (open symbols) (Fragiadakis et al., 2011).

averages is required. These include the arithmetic mean or number average molecular weight

$$M_n = \frac{\int_0^\infty n(M) M dM}{\int_0^\infty n(M) dM}, \tag{6.41}$$

in which n is the number of molecules having molecular weight of M and the weight average molecular weight

$$M_w = \frac{\int_0^\infty n(M) M^2 dM}{\int_0^\infty n(M) M dM}. \tag{6.42}$$

Higher order moments can be defined, but they are less useful because they cannot be determined as accurately. Other measures of molecular weight include the peak value, M_p, which corresponds to the maximum in the distribution, and M_v, determined from the intrinsic viscosity, $[\eta]$, using the Mark-Houwink relation (Chapter 3 herein):

$$[\eta] = \hat{K} M_v^{\hat{a}}, \tag{6.43}$$

where $\hat{K}$ and $\hat{a}$ are polymer-specific constants. A metric of the breadth of the chain length distribution is the polydispersity index

$$P_i = M_w / M_n. \tag{6.44}$$

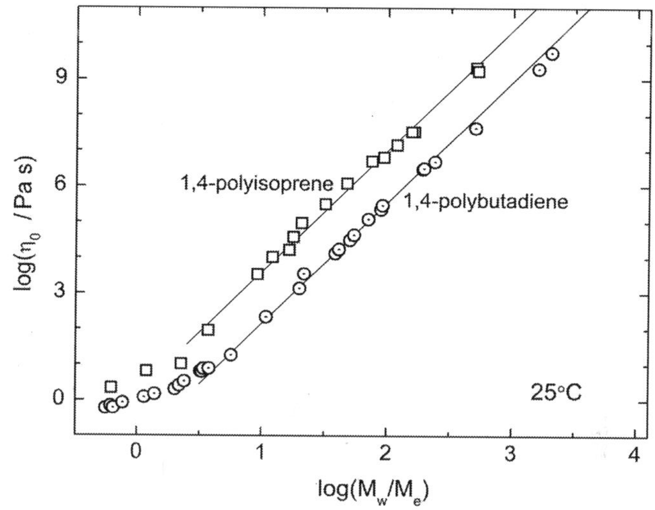

FIGURE 6.12 Terminal viscosity as a function of the molecular weight divided by the entanglement molecular weight; $M_e = 1850$ and 6190 g/mol for 1,4-polybutadiene (circles) and 1,4-polyisoprene (squares), respectively. The solid lines are fits having a slope of 3.4 (Colby et al., 1987; Abdel-Goad et al., 2004).

This index is related to the variance, σ_V^2, in the chain lengths as

$$\sigma_V^2 = (P_i - 1)M_n^2. \tag{6.45}$$

Most often the average molecular weight is used in relation to rheological properties (Berry and Fox, 1968), although this is only an approximation. When model blends are prepared by mixing monodisperse polymers, the molecular weight dependences are complicated. For example, the melt viscosity is less than the value predicted assuming η is determined by M_w (Watanabe et al., 1995; Monfort et al., 1984; Plazek et al., 1991).

The dependence on molecular weight of most rheological properties becomes much stronger for entangled chains (an exception being J_s^0, which becomes constant at high M_w (Roland, 2011)). The most commonly reported data are for the melt viscosity, with zero-shear-rate data shown in Figure 6.12 for 1,4-polybutadiene (Colby et al., 1987) and 1,4-polyisoprene (Abdel-Goad et al., 2004). At high molecular weight, the slope of the double logarithmic plot is 3.4, in accord with data for many other materials (Ferry, 1980; Watanabe, 1999; Doi and Edwards, 1986). Interestingly, when compared at equal number of entanglements per chain, η_0 of 1,4-polyisoprene is larger than for the polybutadiene, although when compared at equal M_w, the viscosity of the former is twofold smaller than η_0 of polybutadiene. From Eq. (6.26) the same M_w-dependence is expected for the terminal relaxation time as for η_0, since J_0 for rubbers is a constant. (Note on cooling toward T_g the recoverable compliance decreases, although the effect can only be observed for low molecular weight polymers

(Roland et al., 2004; Ngai et al., 1995; Santangelo and Roland, 2001).) The diffusion constant of rubbers has a quadratic or somewhat stronger molecular weight dependence (Lodge, 1999; Wang, 2003).

Various characteristic molecular weights can be defined from the molecular weight dependences of the rheological properties, including the molecular weight between entanglements, M_e, determined from the plateau modulus

$$M_e = \frac{\rho RT}{G_N}. \tag{6.46}$$

This equation corresponds to the classic rubber elasticity expression for an affinely deforming network (Roland, 2011). The value of the molecular weight at which the viscosity begins to exhibit a nonlinear dependence is M_c, and since roughly two entanglements are required to affect a chain, there is the approximate relationship, $M_c \sim 2M_e$ (Figure 6.13) (Roland, 2011; Fetters et al., 1999). The molecular weight at which J_s^0 becomes molecular weight invariant is even larger than M_c (Figure 6.14) (Plazek and O'Rourke, 1971; Nemoto et al., 1971, 1972; Roovers, 1985; Roovers and Toporowski, 1990).

The processing of rubber takes place at sufficiently high temperatures that T_g is usually not a relevant parameter, although dynamic properties for different temperatures are often correlated in terms of the difference, $T - T_g$, the ratio T/T_g, or a related quantity (Ferry, 1980; Plazek and Ngai, 1991; Bohmer et al., 1993; Ngai et al., 2008; Ding and Sokolov, 2006). The glass transition defines another characteristic molecular weight, M_∞, the value at which T_g becomes

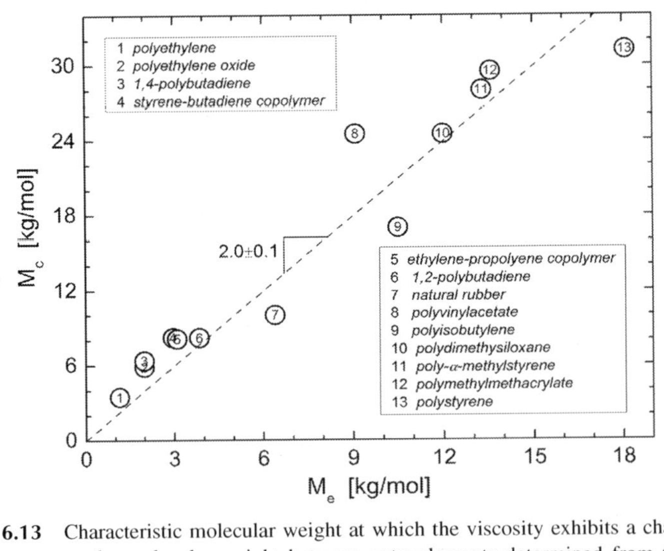

FIGURE 6.13 Characteristic molecular weight at which the viscosity exhibits a change in M-dependence versus the molecular weight between entanglements determined from the plateau modulus (Roland, 2011; Fetters et al., 1999).

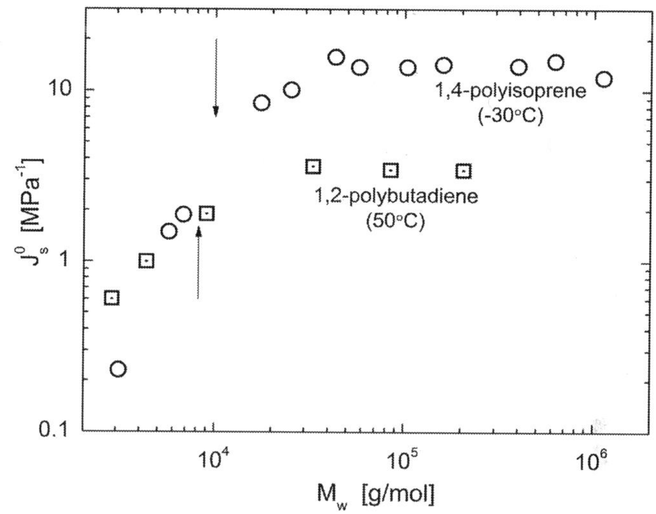

FIGURE 6.14 Steady-state recoverable compliance, which becomes constant at high molecular weight, for 1,4-polyisoprene (circles) and 1,4-polybutadiene (squares). The vertical arrows denote the M_w for entanglements to affect the viscosity, which is smaller than the value of the molecular weight at which J_s^0 become constant (Roland, 2011).

independent of chain length. The general M-dependence of the glass transition can be described empirically (Ueberreiter and Kanig, 1952)

$$T_g = \left(T_{g,\infty}^{-1} + k_{\rm UK}/M_n\right)^{-1}. \qquad (6.47)$$

$T_{g,\infty}$ and $k_{\rm UK}$ are material constants, the latter dependent on the chain end groups (Turner, 1978; Danusso et al., 1993). Figure 6.15 shows the fit of Eq. (6.47) to data for 1,4-polybutadiene (Bogoslovov et al., 2010).

6.2.5 Stress Birefringence

An independent method of determining stress, especially useful for materials under flow, is from the optical birefringence. Of course, it is limited to transparent materials, which precludes application to filled polymers. Because of the need for transparency, stress birefringence has been used more often for plastics than rubber. Residual anisotropy due to unrelaxed orientation can also be assessed using birefringence; this is commonly known as the photoelastic effect. Generally the birefringence is directly proportional to the true stress

$$\Delta n = C\sigma, \qquad (6.48)$$

where Δn represents the difference in refractive index between the strain direction and transversely, and the stress optical coefficient C is a material constant determined by the chemical structure of the chain repeat unit. This

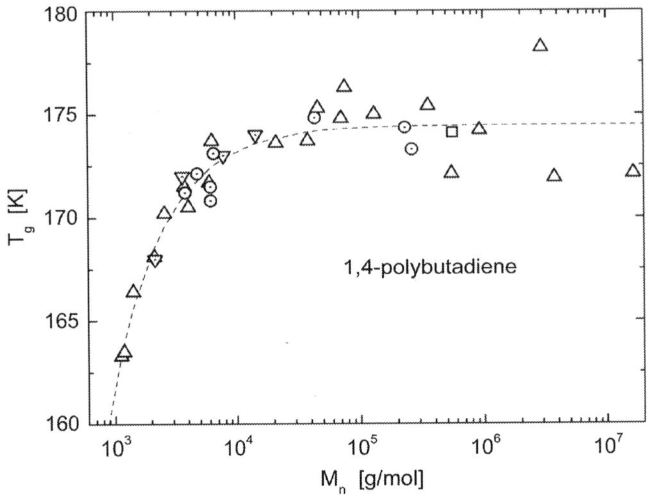

FIGURE 6.15 Glass transition temperatures determined from the thermal expansivity (circles), calorimetry (triangles and squares), Raman scattering (inverted triangles), and as the temperature at which the dielectric relaxation time equals 100 s (star). The solid line is the simultaneous fit of Eq. (6.47) to all data, with the high molecular weight limiting value of $T_{g,\infty} = 174.4$ K (Bogoslovov et al., 2010).

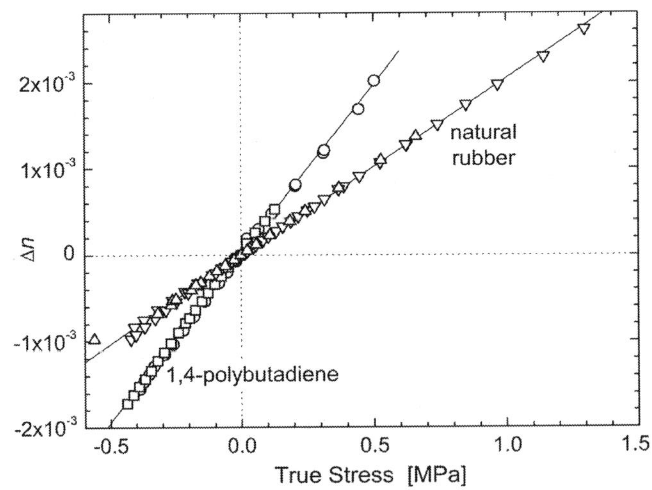

FIGURE 6.16 Optical birefringence for 1,4-polybutadiene and natural rubber networks under tension and compression; the stress optical coefficient is given by the slopes = 3.6 and 2.0 GPa^{-1}, respectively (Mott and Roland, 1996).

stress optical law holds to reasonably high levels of strain, a consequence of the small end-to-end distance of macromolecules compared to their fully extended length. Data for polybutadiene and polyisoprene are shown in Figure 6.16 (Mott

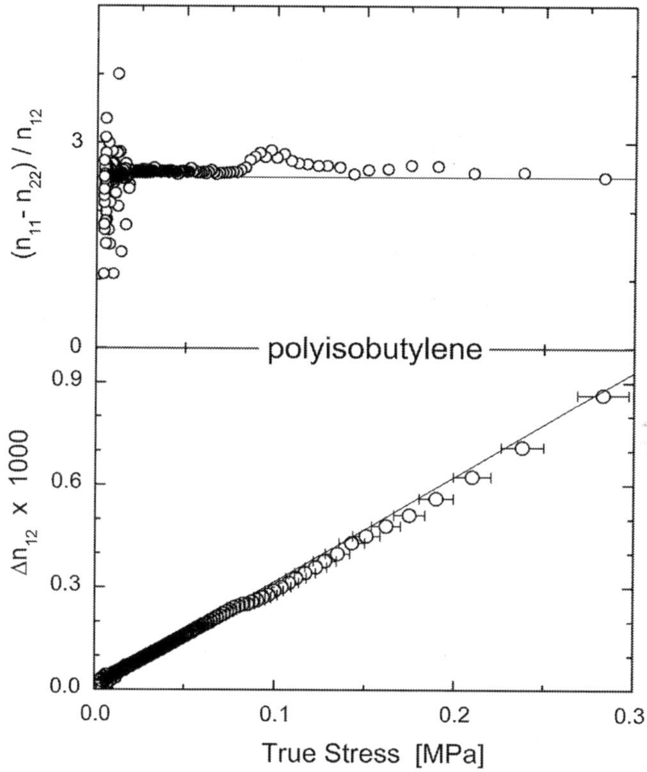

FIGURE 6.17 (Bottom) Birefringence for uncrosslinked PIB after imposition of a shear strain = 2.5. Linear fitting yields 3.07 GPa^{-1} for the stress optical coefficient. (Top) Ratio of refractive index components (1 is the flow direction and 2 the direction of the gradient), which equals the value of the shear strain, in accord with the Lodge-Meissner relation (Eq. (6.66)) (Balasubramanian et al., 2005).

and Roland, 1996). A measure of orientation is the Herman orientation function,

$$f(\theta) = (3\langle\cos^2\theta\rangle - 1)/2, \tag{6.49}$$

which is equal to the ratio of Δn to the value of the birefringence for complete orientation. In Eq. (6.49) the brackets refer to an average over all chains, with θ the angle between the long axis of the repeat unit and the reference axis. Note that in the liquid crystal literature, $f(\theta)$ is known as the order parameter.

The stress optical law is maintained during relaxation of a deformed rubber (Figure 6.17) (Balasubramanian et al., 2005); moreover, the same proportionality to Δn is maintained for the normal components of the stress. And since orientation of rubber also affects heat conduction (Hands, 1980), there is a corresponding proportionality, known as the "stress-thermal rule," between stress and the anisotropy of the thermal conductivity (Venerus et al., 1999);

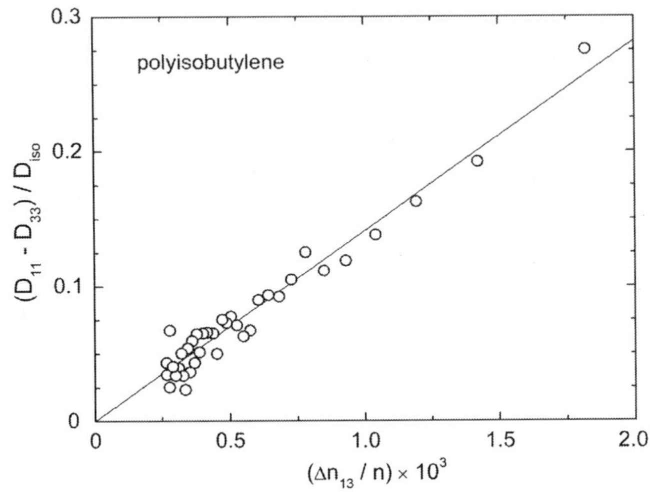

FIGURE 6.18 Normalized difference in thermal diffusivity versus normalized birefringence for polyisobutylene after imposition of a shear strain = 8. The line is the linear fit to the data (Venerus et al., 1999).

consequently, the thermal conductivity is proportional to the birefringence (Figure 6.18) (Balasubramanian et al., 2005; Venerus et al., 1999).

6.3 NONLINEAR VISCOELASTICITY

6.3.1 Shear Thinning Flow

The viscosity of entangled polymers becomes shear-rate dependent when the shear rate exceeds the inverse of the time scale for the terminal dynamics of the chains. This non-Newtonian behavior is a prominent characteristic of rubber, central to processing operations (Figure 6.19). Shear thinning is due to a reversible reduction in the degree of chain entanglement caused by flow (Vinogradov and Belkin, 1965; Simmons, 1968; Tanner and Williams, 1971; Batzer et al., 1980; Isono et al., 1995, 1997; Oberhauser et al., 1998). Since reentanglement of the chains can be slow, the "shear-modified" state of a polymer melt can be exploited to enhance processability of the material. This has been shown to give transient reductions in die swell, for example in polyethylene, polypropylene, and polystyrene melts (Yamaguchi and Gogos, 2001; Rudin and Schreiber, 1983; Leblans and Bastiaansen, 1989; Pohl and Gogos, 1961; Ram and Izrailov, 1986; Kotliar et al., 1990).

Another indication of fewer entanglements is the magnitude of the overshoot in the transient shear stress. For high shear rates, the stress exhibits a maximum, followed by a steady-state value determined by the degree of entanglement under the particular conditions of flow. If the shearing is stopped, the subsequently observed stress overshoot will be smaller. However, its magnitude

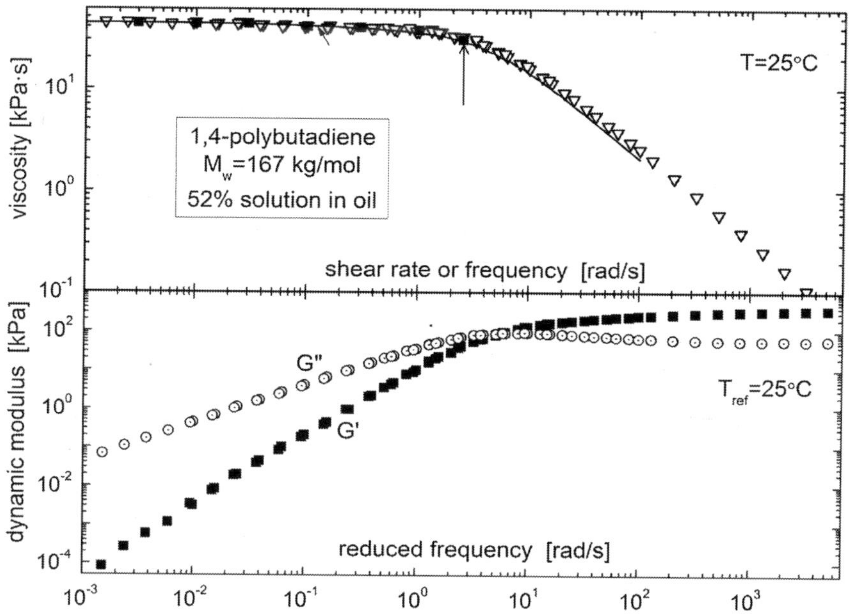

FIGURE 6.19 (Upper panel) Steady-state shear viscosity versus shear rate (solid symbols), dynamic viscosity versus frequency (open symbols), and transient viscosity calculated from Eq. (6.65) versus the inverse of the time of shearing (solid line). (Lower panel) Dynamic storage and loss modulus master curve for the same entangled polybutadiene solution (Roland and Robertson, 2006).

increases with the time prior to restarting the flow, as the chains reentangle (Attane et al., 1985; Dealy and Tsang, 1981; Huppler et al., 1967; Stratton and Butcher, 1973; Menezes and Graessley, 1982; Xu et al., 1995). This effect is illustrated in Figure 6.20 with data for a concentrated polybutadiene solution (Roland and Robertson, 2006; Robertson et al., 2004). The stress overshoot and associated disentanglement requires shear rates in the non-Newtonian regime (Figure 6.21). Note that stress overshoots can occur in any material whose structure is disrupted by flow (Whittle and Dickinson, 1997; DeKee and Fong, 1994), and even in unentangled polymers; however, these effects are usually much weaker (Moore et al., 1999; Santangelo and Roland, 2001) than for entangled polymers. These transient flow phenomena can be described using K-BKZ-type constitutive equations (Eq. (6.29)) (Menezes and Graessley, 1982).

6.3.2 Particulate Fillers

Rubber compounds very often contain small-particle fillers such as carbon black and silica, to improve processability and physical properties, and to reduce material costs. The most common filler is carbon black, use of which as a pigment dates to ca. 4000 BC. Methods to incorporate fillers, and the effect of their distribution and dispersion on properties, are central considerations, with

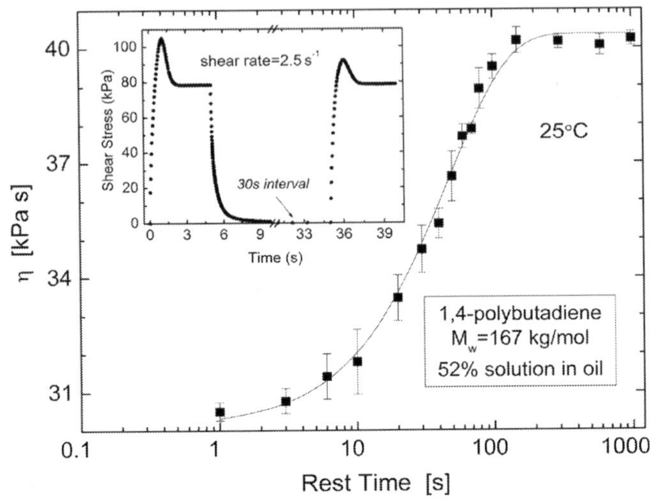

FIGURE 6.20 Maximum in transient viscosity after initiation of shear flow versus the duration of a quiescent interval between shearing. The growth of the overshoot peak is caused by reentanglement of the solution. The inset shows representative transient shear stress data (Roland and Robertson, 2006).

much literature devoted to these topics (Chapter 8 herein; Robertson and Roland, 2008; Carbon Black-Polymer Composites, 1982; Hamed, 2000, 2007; Heinrich et al., 2002; Waddell and Evans, 1996; Medalia, 1987). More recently attention has been directed to nanoparticles, such as carbon nanotubes, nanosilicates, etc., as a replacement for conventional fillers (Nah et al., 2010; Bhattacharya et al., 2008). The expectation is that their enormous surface to volume ratio will enable polymer reinforcement to be achieved with very low levels of a nanofiller.

(i) Payne Effect

In a carbon black reinforced elastomer, the aggregates (fused primary particles) can agglomerate (flocculate) to produce a filler network that gives rise to higher electrical conductivity (Medalia, 1986; Roland et al., 2004; Nikiel et al., 2009) and a modulus that decreases with strain (Chapter 8 herein; Payne, 1962, 1963, 1964; Roland, 1990). Strain breaks up the flocculated network structure, as seen in the strain-dependence of the dynamic modulus in Figure 6.22 (Robertson et al., 2007). This network strain-induced breakup is governed by the strain energy; that is, for different concentrations or types of filler, and for different strain modes and frequencies, the maximum in the loss, associated with network disruption, occurs at the same value of the product of dynamic modulus times strain (Robertson and Wang, 2005; Wang and Robertson, 2005). This phenomenon is illustrated in Figure 6.23 for polybutadiene reinforced with different types of fillers (Robertson et al., 2007).

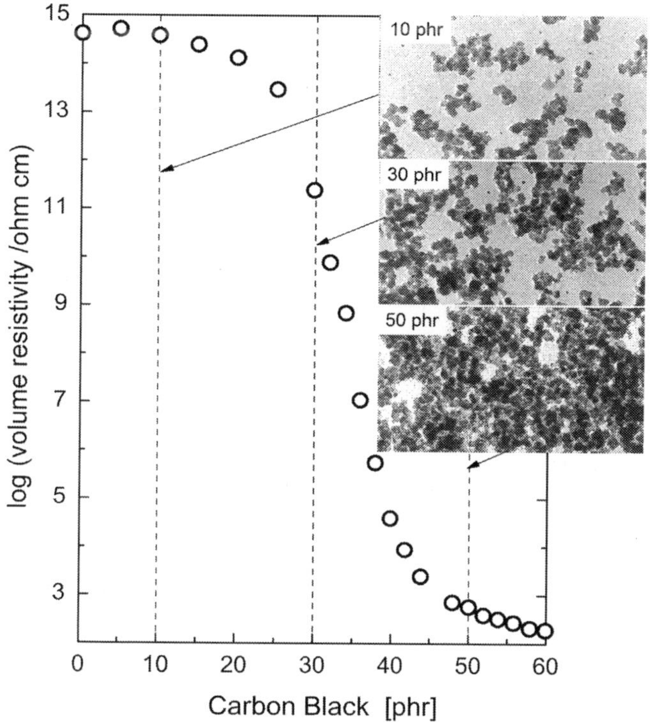

FIGURE 6.21 Change in electrical resistivity of an SBR versus the concentration of N330 carbon black. The property change reflects the degree of particle agglomeration, as seen in the transmission electron micrographs (Nikiel et al., 2009).

At higher strains the network structure is gone, and absent the Payne effect, the main mechanism affecting the processing and flow of filled rubber is the strain amplification in the polymer chains resulting from the inextensibility of the particles. This can be described by the Guth-Gold equation (Guth, 1945; Kohls and Beaucage, 2002):

$$\eta = \eta_0(1 + 2.5\varphi + 14.1\varphi^2), \tag{6.50}$$

where η_0 is the viscosity of the gum rubber and φ is the filler volume fraction. As pointed out by Medalia (1967, 1972), in order for Eq. (6.50) to reproduce experimental results, φ may be augmented by occluded rubber that enhances the hydrodynamic effect of the particles.

(ii) Mullins Effect

It is widely believed that the presence of carbon black or other reinforcing fillers introduces substantial mechanical hysteresis into rubber. Certainly stresses during retraction are lower than during the initial stretching of a rubber, a

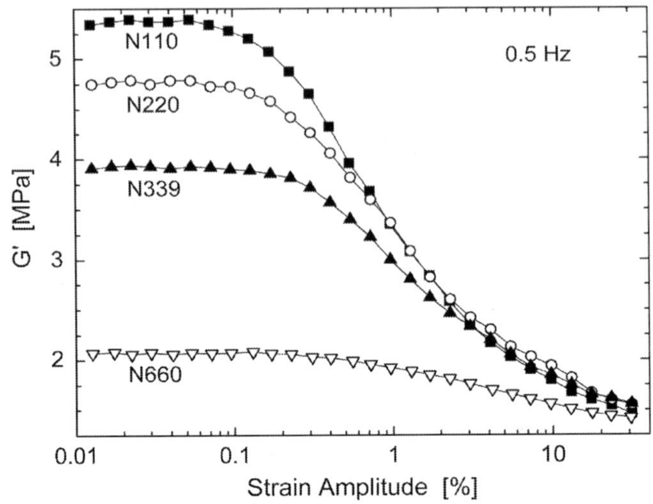

FIGURE 6.22 Storage modulus versus shear strain for 1,4-polybutadiene with 50 phr (18% by volume) of different carbon blacks (Robertson et al., 2007).

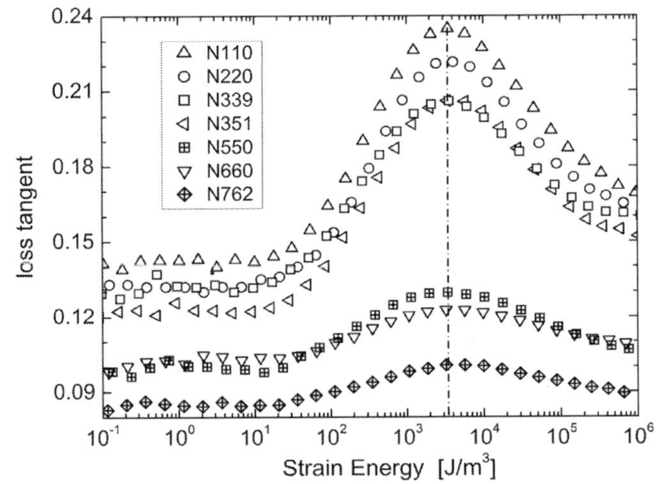

FIGURE 6.23 Loss tangent versus dynamic strain energy for 1,4-polybutadiene with 50 phr of the indicated carbon black. The peak maxima superpose at a strain energy = 3.5 kJ/m^3, that is independent of filler particle size (Robertson et al., 2007).

phenomenon referred to as the Mullins effect (Mullins, 1969). However, this strain-softening is a viscoelastic effect; it does not require filler, or chain scission, strain-crystallization, or other quasi-irreversible material changes. This was shown in the original work of Harwood and Schallamach (1967), who

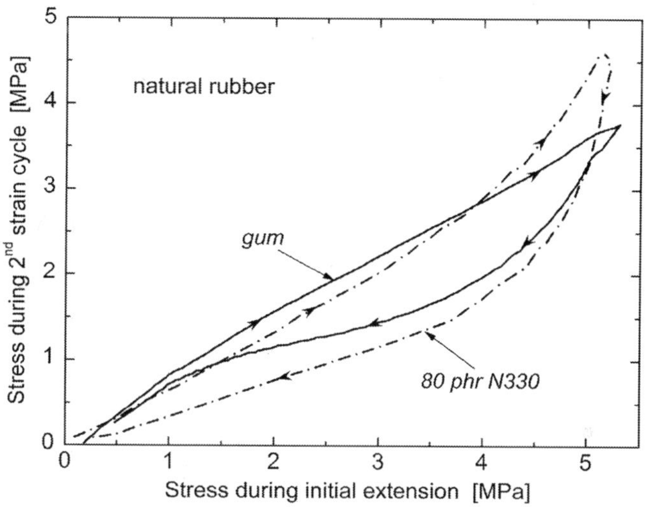

FIGURE 6.24 Hysteresis in unfilled and filled natural rubber networks during the second cycle of tension-retraction. When stretched to the same stress (which requires higher strain in the gum rubber), the softening is roughly equal for the two materials; that is, the softening does not arise primarily due to the reinforcing filler (Harwood and Payne, 1966a,b).

pointed out that when stretched to equivalent stresses, gum and filled elastomers show comparable amounts of softening. Representative data for natural rubber are displayed in Figure 6.24. This strain-softening is not due to the filler network, because disruption of the agglomerated structure transpires at very low strains (<10% in Figure 6.22). Nevertheless, the mechanical hysteresis of gum rubber during retraction is anomalous, since it is always larger than predictions based on Boltzmann superpositioning (e.g., Eq. (6.30)). Various mechanisms have been suggested as the cause of this Mullins softening, including the contribution from network chain ends or relaxation processes too fast to be included in the constitutive description; however, more work is needed to resolve this issue.

(iii) Nanofillers

At a given concentration, smaller filler particles provide more reinforcement due to the greater interfacial area. This is illustrated in Figure 6.25, showing the increase in bound rubber and tensile strength of a 1,4-polybutadiene as the particle size of the carbon black decreases. Interest has been rekindled recently as part of the burgeoning attention to nanoscience and technology, with carbon nanotubes (Bokobza, 2007; Coleman et al., 2006), silica nanoparticles (Bansal et al., 2005; Bogoslovov et al., 2008), nanoclay (Vu et al., 2001; Sadhu and Bhowmick, 2003; Carretero-Gonzalez et al., 2008), graphene (Stankovich et al., 2006; Wakabayashi et al., 2008), and diamond nanoparticles (Behler et al., 2009) investigated as potential fillers for polymers. Substantial improvements in the

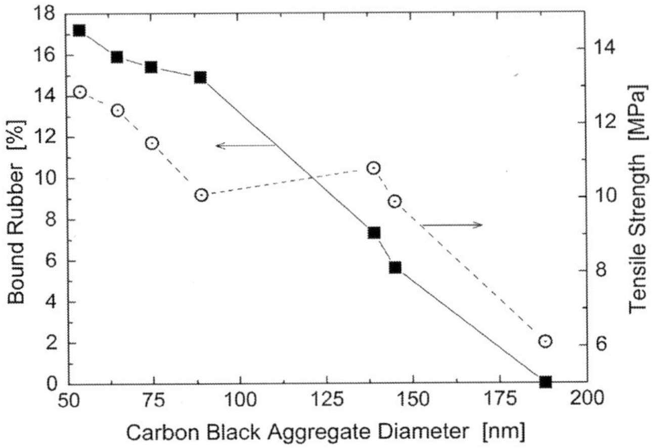

FIGURE 6.25 Bound rubber (squares) and tensile strength (circles) of 1,4-polybutadiene reinforced with 50 phr of carbon black (in order of increasing particle size: N110, N220, N339, N351, N550, N660, and N762) (Robertson et al., 2007).

mechanical properties of polymers can be obtained with a few percent or less of nanofiller (Mark, 2006; Sengupta et al., 2007; Maiti et al., 2008; Vu et al., 2001). An additional factor is the high aspect ratio of particles that are only nm-sized in one or two dimensions. This can augment the reinforcing effect, as seen in Figure 6.26 (Casalini et al., 2012), showing more than two orders

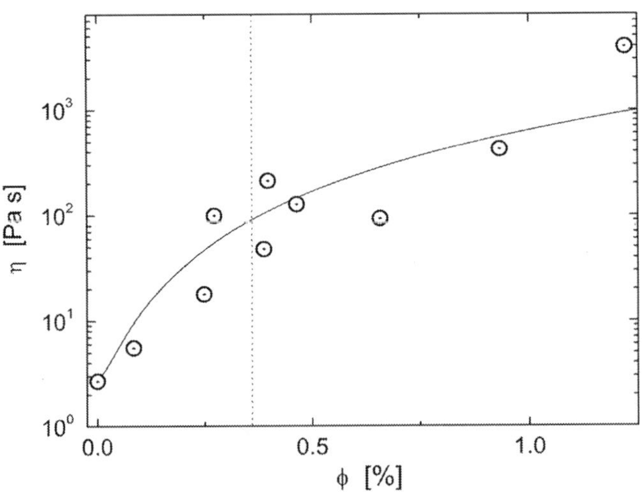

FIGURE 6.26 Dynamic viscosity measured at 0.1 Hz for a polydiamine (M_W = 4 kg/mol) as a function of the volume concentration of multiwall carbon nanotubes. The line is the fit to Eq. (6.51) (Casalini et al., 2012).

of magnitude increase in the viscosity with the addition of just 1% carbon nanotubes. To account for this effect of particle shape, the Guth-Gold relation can be modified by introducing a shape factor, f, into Eq. (6.50) (Kohls and Beaucage, 2002):

$$\eta = \eta_0(1 + 2.5f\varphi + 14.1(f\varphi)^2). \tag{6.51}$$

This shape factor affects the interaction among the particles and thus the onset of the Payne effect. The minimum concentration necessary for formation of a particle network can be estimated from (Hobbie and Fry, 2007; Chatterjee and Krishnamoortim, 2007):

$$\varphi^* = \frac{\pi}{4f^2}. \tag{6.52}$$

These equations assume uniform distribution and random orientation of the particles. However, for nanoparticles good mixing can be difficult to achieve, and often chemical modification of the particles is used to enhance their dispersion in the polymer. Poor mixing causes the property changes for small concentrations of nanofillers to be rather modest (Valadares et al., 2006; Bokobza, 2007), much less than the orders of magnitude enhancement seen in Figure 6.26.

6.3.3 Blends

Since flow can disrupt the morphology of a phase-separated blend, the possibility exists for changes in rheological properties. A striking example of this effect is observed in blends with components of different viscosity; the morphology can rearrange during flow to minimize the viscous dissipation (MacLean, 1973). This phase adjustment can cause differences between dynamic viscosities and viscosities measured by flow through capillary dies (Figure 6.27) (Roland and Nguyen, 1988). Only the latter is affected because enrichment of the surface with one component causes slip at the interface, and thus less resistance to flow. Changes in the phase morphology and surface accumulation of one component can be accentuated if there is an attraction (chemisorption) with the processing equipment. This contamination of the walls by one polymer in a blend can result in slippage at the interface, and consequently large reductions in apparent viscosity of the blend (Figure 6.28) (Shih, 1976). Slippage can also result from surface migration of additives in the rubber compound (Ahn and White, 2003, 2004), and it can be induced physically, for example by reducing the hydrostatic pressure (Brzoskowski et al., 1987) or by introducing air between the flowing rubber and the extruder barrel or die (Montes et al., 1988).

Since the effect of filler is nonlinear (Eqs. (6.50) and (6.51)), its proportioning between the phases of an immiscible blend can affect the viscosity and other rheological properties. Carbon black, for example, has greater affinity for the more polar or unsaturated component (Hess, 1991; Lee, 1981), which

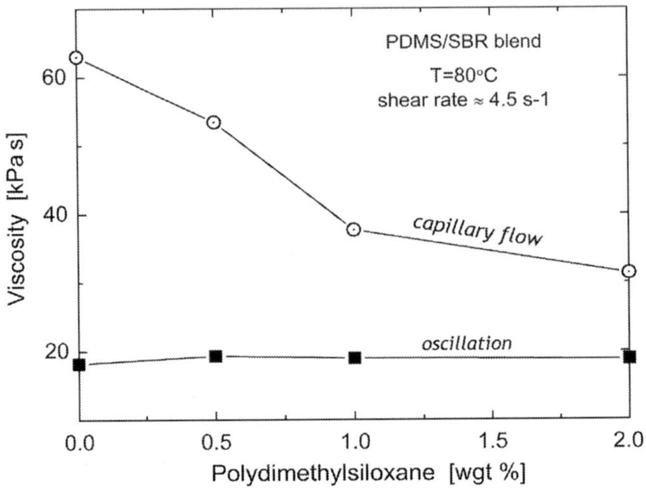

FIGURE 6.27 Viscosity of a blend of styrene-butadiene copolymer and polydimethylsiloxane containing 50 phr N326 carbon black. Presence of the PDMS has no effect on the response to oscillatory strain (squares), but phase segregation and wall slippage occur during flow through a capillary die (circles) (Roland and Nguyen, 1988).

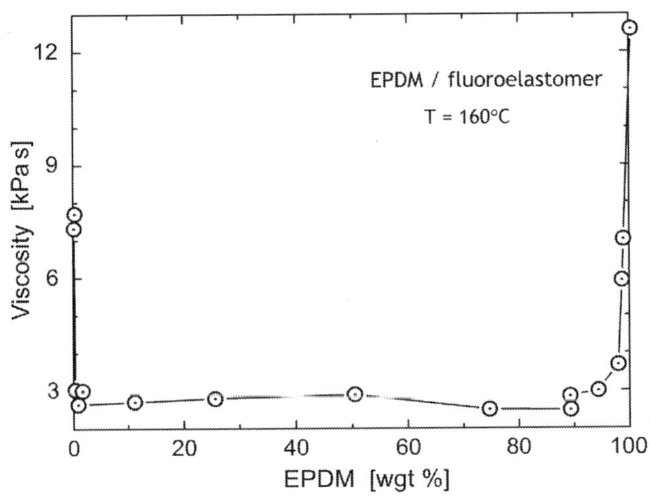

FIGURE 6.28 Apparent viscosity of a blend of EPDM with a fluoroelastomer measured in a capillary rheometer at a shear rate equal to 14 s^{-1} (Shih, 1976).

often leads to a nonuniform distribution. This is seen in Figure 6.29 in the greater uniformity of the respective concentrations of carbon black in the components for SBR blended with a relatively nonpolar component, than when mixed with a polar rubber (Le et al., 2008).

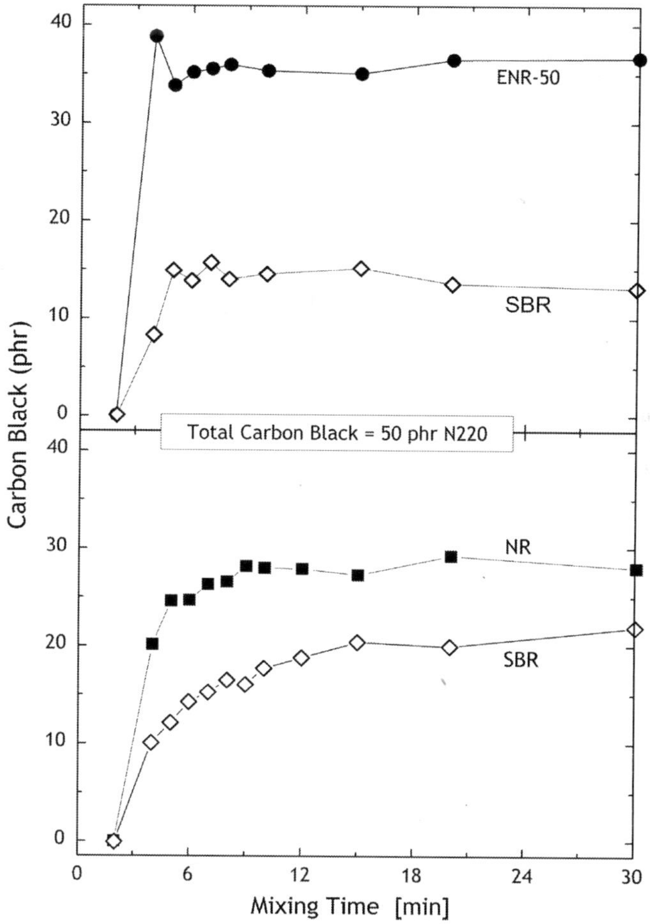

FIGURE 6.29 Respective carbon black contents of the components in immiscible blends of SBR with 50% (top) epoxidized natural rubber and (bottom) natural rubber. The filler preferentially incorporates into the more polar phase (Le et al., 2008).

6.4 ENGINEERING ANALYSIS

6.4.1 Dimensionless Quantities

Dimensional analysis is commonly applied to complex materials and processes to assess the significance of some phenomenon or property regime. It enables analysis of situations that cannot be described by an equation; however, there is no a priori guarantee that a dimensionless analysis will be physically meaningful. Dimensionless quantities are combinations of variables that lack units (i.e., pure numbers), used to categorize the relationship of physical quantities and their interdependence in order to anticipate the behavior. Several dimensionless quantities relevant to polymer rheology and processing are defined in the following. Although these are all ratios, this is not a requirement of dimensionless quantities; they can be characteristic numbers (e.g., quantum numbers) or represent a count (such as a partition function).

(i) Reynolds Number

A very common quantity, this is the ratio of the inertial to viscous forces (Rott, 1990):

$$Re \equiv \frac{\rho v L}{\eta}, \tag{6.53}$$

where ρ is the mass density, v is the velocity, and L is a characteristic dimension. For extensional and shear flows, L represents the dimension parallel and perpendicular, respectively, to the flow. Re can identify the laminar to turbulent flow transition for different fluids and different flow rates, independently of the geometry. For example, for flow in a circular pipe, L is the inside diameter, and the transition from laminar to turbulent flow occurs in the range $2300 < Re < 4000$. A variation on Re is the **Blake Number** (Blake, 1922), which applies to porous media, such as beds of solids.

(ii) Deborah Number

The Deborah number is the ratio of the relaxation time of the material to the experimental timescale:

$$De \equiv \dot{\gamma}\tau. \tag{6.54}$$

A larger Deborah number implies a more elastic response. The term, which comes from the biblical quote in Section 6.1.1, was coined by Reiner (1964).

(iii) Weissenberg Number

Defined as the ratio of the material relaxation time to the processing time during flow, Wi characterizes the amount of molecular orientation induced by the flow. For simple shear flow, Wi is given by the product of τ times the shear rate. Related to the Deborah number, Wi is used for constant straining, whereas De describes deformations with a varying strain history.

(iv) Capillary Number

This quantity is the ratio of the viscous force to the surface tension across an interface,

$$Ca \equiv \frac{\eta v}{\Upsilon}, \tag{6.55}$$

where Υ is the surface tension. Relevant to extensional flow from orifices and domain formation during mixing of immiscible liquids, values of Ca larger than unity imply stretching of dispersed particles or droplets, which is a necessary condition for their dispersion. The ratio of the Deborah and capillary numbers,

$$\frac{De}{Ca} = \frac{\tau \Upsilon}{\eta L}, \tag{6.56}$$

the "elastocapillary number" (Hsu and Leal, 2009), is a measure of the deformation of dispersed phases that is independent of the strain rate. L in Eq. (6.56) is related to the droplet size.

(v) Weber Number

Analogous to the capillary number, *We* is the ratio of the inertial force to the surface tension across an interface,

$$We \equiv \frac{\rho v^2 L}{\Upsilon}. \tag{6.57}$$

(vi) Brinkman Number

This number, used in polymer processing, describes the heat flow between the flowing material and the vessel containing it (e.g., the rotor and walls of an extruder). It is defined as

$$Br \equiv \frac{\eta v^2}{\kappa \Delta T}, \tag{6.58}$$

in which κ is the thermal conductivity of the material and ΔT its temperature difference with the vessel. A large value of $Br (> 1)$ indicates significant viscous heating.

(vii) Bingham Number

Used to characterize a polymer that flows when the stress reaches a characteristic level (Bingham, 1916), this number is the ratio of the yield stress, σ_Y, to the viscous stress,

$$Bn \equiv \frac{\sigma_Y L}{\eta v}. \tag{6.59}$$

(viii) Elasticity Number

The ratio of the elastic forces to the inertial forces,

$$El = \frac{\tau \eta}{\rho L^2}. \tag{6.60}$$

and equal to the Weissenberg number divided by the Reynolds numbers, *El* characterizes the effect of elasticity on the flow. For a polymer solution *El* depends only on the material properties, except for the size of the flow device.

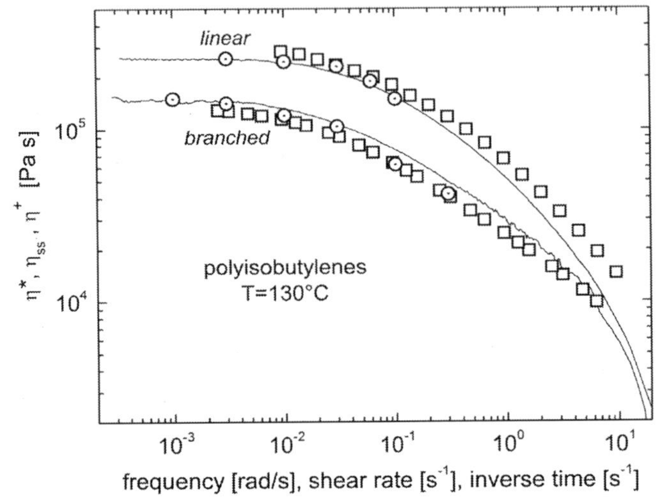

FIGURE 6.30 Dynamic viscosity (squares), steady-state shear viscosity (circles), and the transient viscosity calculated using Eq. (6.65) (solid lines) for a linear ($M_W = 389$ kg/mol) and a highly branched ($M_W = 1080$ kg/mol) with 21 branches per chain and a branch $M_W = 52.7$ kg polyisobutylene (Robertson et al., 2002).

6.4.2 Empirical Rules

(i) Cox-Merz Rule (Cox and Merz, 1958)

The Cox-Merz rule states that the shear-rate dependence of the steady-state viscosity equals the frequency dependence of the complex viscosity,

$$\eta(\dot{\gamma}) = \eta(\dot{\gamma})|_{\omega=\dot{\gamma}}, \tag{6.61}$$

in which $\dot{\gamma}$ is the strain rate. The upper panel of Figure 6.19 (Robertson et al., 2004) shows data for an entangled solution of 1,4-polybutadiene, and in Figure 6.30 (Robertson et al., 2002) are data for linear and branched polyisobutylene; both materials exhibit good compliance with Eq. (6.61). However, the presence of filler degrades the connection between the two viscosities, as shown in Figure 6.31 for a polydimethylsiloxane (Xu et al., 2008). The data for gum silicone rubber agree well with the Cox-Merz rule, but filler causes the steady shear viscosities to be smaller, presumably reflecting breakup of the agglomerated filler structure under continuous shearing conditions.

(ii) Laun Relations (Laun, 1986)

Laun proposed two equations relating the dynamic storage and loss moduli to the first normal stress difference coefficient for steady shearing:

$$\Psi_1(\dot{\gamma}) = 2G'\omega^{-2}\left[1 + \left(G'/G''\right)^2\right]^a\Big|_{\omega=\dot{\gamma}} \tag{6.62}$$

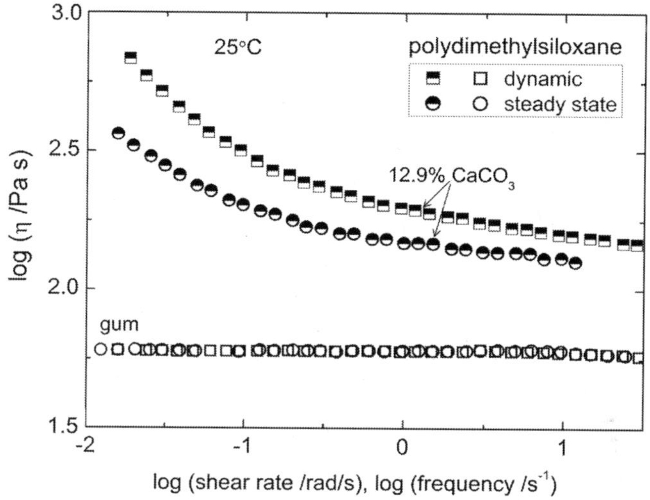

FIGURE 6.31 Dynamic viscosity (squares) and steady-state viscosity (circles) for PDMS with the indicated concentration by volume of calcium carbonate (half-filled symbols) and without filler (open symbols) (Xu et al., 2008).

and to the steady-state recoverable strain,

$$\gamma_{\text{rec}}(\dot{\gamma}) = \left(G'/G''\right)\left[1 + \left(G'/G''\right)^2\right]^{1.5}\Big|_{\omega=\dot{\gamma}}. \tag{6.63}$$

In Eq. (6.62) the exponent a is taken as a fitting parameter having a value around 0.7. The accuracy of these expressions is not quantitative (Roland, 2011). From the definitions of the dynamic moduli in the terminal region (Eqs. (6.23) and (6.24), the first Laun relation reduces to the equation of Coleman and Markovitz (1964),

$$\psi_1^0 = \lim_{\dot{\gamma}\to\infty} \frac{N_1(\dot{\gamma})}{\dot{\gamma}} = 2J_s^0\eta_0^2, \tag{6.64}$$

which is valid only for low shear rates.

(iii) Gleissle Equations (Gleissle, 1980)

This empirical equation connects the steady-state viscosity to the transient viscosity, $\eta^+(t)$, measured from the stress growth during approach to steady state at the limiting (Newtonian) shear rate,

$$\lim_{\dot{\gamma}\to 0} \eta^+(t,\dot{\gamma}) = \eta(\dot{\gamma}) \quad (t^{-1} = \dot{\gamma}). \tag{6.65}$$

As illustrated in Figure 6.19 (upper panel) (Robertson et al., 2004) and Figure 6.30 (Robertson et al., 2002), Eq. (6.65) is only approximate; nevertheless, using it enables estimates for the viscosity over a broad range of rates from a single measurement of stress versus time at one slow shear rate.

(iv) Lodge-Meissner Rule (Lodge and Meissner, 1972)

After imposition of a step strain, the ratio of the first normal stress difference, N_1, to the shear stress is equal to the strain, or

$$N_1(t) = \varepsilon^2 G(t) = \varepsilon\sigma(t). \tag{6.66}$$

This behavior is apparent in the top panel of Figure 6.17, showing that the birefringence perpendicular to the strain direction has the same time-dependence as the stress.

(v) Boyer-Spencer (Boyer and Spencer, 1944) and Simha-Boyer (Simha and Boyer, 1962) Rules

From analysis of data for various polymers, two correlations have been reported concerning properties at the glass transition. The rationalization for both relies on a free volume description of the glass transition in polymers, which means the correlations are strictly empirical. The Boyer-Spencer rule (Boyer and Spencer, 1944) posits that the product of the glass transition temperature and the thermal expansion coefficient at T_g is a universal constant, $\alpha_P T_g = 0.113$. An alternative analysis of Bondi (Van Krevelen, 1990) gives $\alpha_P T_g = 0.116$. The data in Figure 6.32 (Krause et al., 1965) show that this empirical rule is reasonably accurate, with a mean for 51 amorphous polymers yielding $\alpha_P T_g = 0.162 \pm 0.005$. This correlation is affirmed by the density scaling property (Eq. (6.40)), from which a relation between γ and the ratio of the isochoric and isobaric activation energies can be derived (Casalini and Roland, 2004a):

$$\frac{E_V}{E_P} = (1 + \alpha_P T\gamma)^{-1}, \tag{6.67}$$

where the isobaric activation energy (which is more properly called an activation enthalpy) has its usual definition,

$$E_P(T,P) = R\left.\frac{\partial \ln \tau}{\partial T^{-1}}\right|_P, \tag{6.68}$$

and the activation energy at constant density is

$$E_V(T,V) = R\left.\frac{\partial \ln \tau}{\partial T^{-1}}\right|_V. \tag{6.69}$$

E_V/E_P can be calculated from segmental relaxation times for polymers and rotational correlation times of molecular liquids, measured over a range of thermodynamic conditions. Plots of the activation energy ratio versus γ can be fit using a value of $\alpha_P T_g = 0.18 \pm 0.01$ (Casalini and Roland, 2004b), consistent with the Boyer-Spencer rule.

According to the Simha-Boyer rule (Simha and Boyer, 1962), the product of the T_g times the change in thermal expansivity at T_g is a material-independent

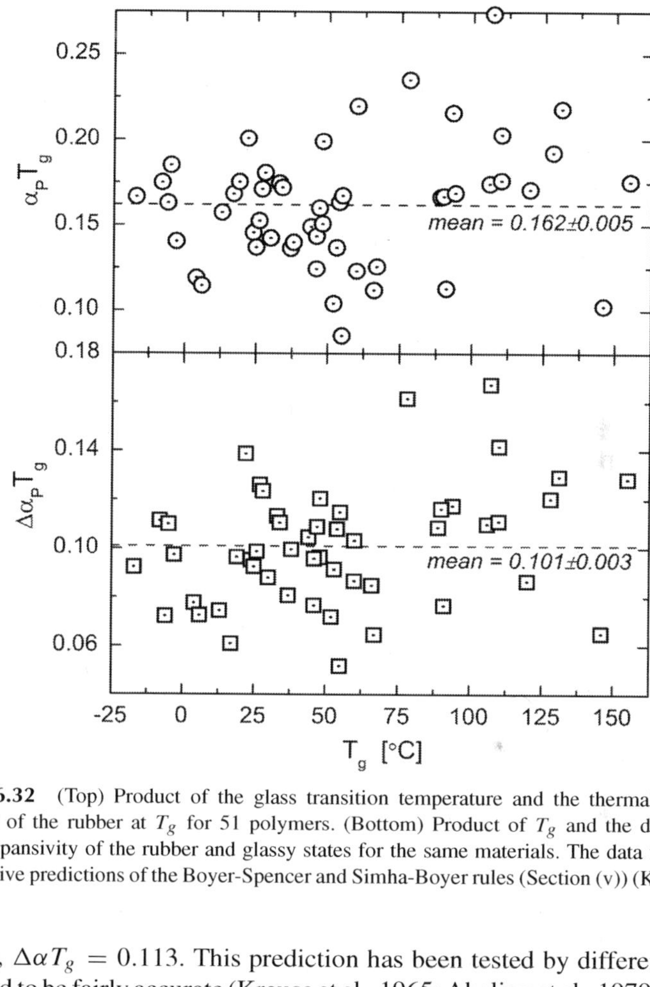

FIGURE 6.32 (Top) Product of the glass transition temperature and the thermal expansion coefficient of the rubber at T_g for 51 polymers. (Bottom) Product of T_g and the difference in thermal expansivity of the rubber and glassy states for the same materials. The data fall close to the respective predictions of the Boyer-Spencer and Simha-Boyer rules (Section (v)) (Krause et al., 1965).

constant, $\Delta\alpha T_g = 0.113$. This prediction has been tested by different groups and found to be fairly accurate (Krause et al., 1965; Abeliov et al., 1979). Results are displayed in Figure 6.32 for 51 polymers, yielding $\Delta\alpha T_g = 0.101 \pm 0.003$.

6.5 PRACTICAL PROCESSING CONSIDERATIONS

6.5.1 Mixing

There are two main aspects to mixing—distributive mixing, to combine the ingredients and obtain a spatially uniform composition, and dispersive mixing, which refers to the breakup of constituents into smaller sizes. The latter presents the greater challenge. The general principle is that the obtained dispersed particle size depends both on the rheological properties of the materials and the type of mixing. Concerning the rheological properties, lower cohesive strength of the dispersed component and a closer matching of its viscosity (assuming fluid particles) to that of the matrix leads to more effective dispersion (Rallison, 1984; Grace, 1982; Nelson et al., 1977). The type of mixing affects the dispersion process primarily because stretching flows are most effective at dispersion and rotational flows are completely ineffective. In order to sustain the stresses and thereby fracture the particle, flow field vorticity should be suppressed, since reorientation of the particles or domains allows the stresses to successively counterbalance each other. Beyond this consideration, the objective is to apply the highest stresses and strain rates, provided degradation (chain scission) is not an issue, and to ensure that all the material passes through the regions of the mixing vessel at which dispersion forces are greatest. Very generally, the strain rate tensor can be written as

$$\Gamma \equiv \nabla_v = \begin{pmatrix} \frac{\partial v_x}{\partial x} & \frac{\partial v_x}{\partial y} & \frac{\partial v_x}{\partial z} \\ \frac{\partial v_y}{\partial x} & \frac{\partial v_y}{\partial y} & \frac{\partial v_y}{\partial z} \\ \frac{\partial v_z}{\partial x} & \frac{\partial v_z}{\partial y} & \frac{\partial v_z}{\partial z} \end{pmatrix}. \tag{6.70}$$

A metric for the strength of the flow field, ζ, can be introduced, where $\zeta = 1$ for pure extensional flow and -1 for rotation (Fuller and Leal, 1981). The strength of the flow field is then given by the ratio of the stretching to the vorticity, equal to $\frac{1+\zeta}{1-\zeta}$. For two-dimensional flow (Fuller and Leal, 1981),

$$\Gamma = \frac{\dot{\gamma}}{2} \begin{pmatrix} (1+\zeta) & (1-\zeta) & 0 \\ (-1+\zeta) & (-1-\zeta) & 0 \\ 0 & 0 & 0 \end{pmatrix}, \tag{6.71}$$

where $\dot{\gamma}$ represents the local velocity gradient. An arbitrary flow field can be expressed as the linear superposition of pure stretching and rotational flow, with shear flow corresponding to $\zeta = 0$. Using their strength criteria, more dispersion will occur on the entrance to a capillary die, associated with ζ close to unity, than for shearing flow through the die. The effect of slippage of the polymer is to reduce the velocity gradient; thus (Fuller and Leal, 1981),

$$\Gamma_{\text{eff}} = \frac{\dot{\gamma}}{2} \begin{pmatrix} (1-\xi)(1+\zeta) & (1-\zeta) & 0 \\ (-1+\zeta) & -(1-\xi)(-1-\zeta) & 0 \\ 0 & 0 & 0 \end{pmatrix}, \tag{6.72}$$

with ξ a measure of the slippage. The strength of the flow field is now given by $\frac{(1-\xi)(1+\zeta)}{1-\zeta}$.

These considerations apply both to mixing of the components of a polymer blend and to dispersion of particulate fillers. For blends the obtained size of the dispersed phase represents the competition between breakup of the particles and their flow-induced coalescence (Roland and Bohm, 1984; Sirisinha et al., 2003; Chen et al., 2006; Peng et al., 2011). Similarly, the final degree of filler dispersion depends not only on the breakup of the aggregates during mixing, but also on the extent of reagglomeration, which occurs, for example, during curing or thermal annealing (Bohm and Nguyen, 1995; Mihara et al., 2009; Schwartz et al., 2003). Chemical treatment of the filler surface is a common method to suppress agglomeration, either with coupling agents that promote interaction with the polymer, or using shielding agents to decrease filler-filler interactions (Robertson et al., 2008; Ye et al., 2012; Lin et al., 2010).

6.5.2 Die Swell

When rubber is deformed during processing operations such as extrusion or calendaring, its inherent elasticity will cause some subsequent recovery. A fundamental material constant characterizing this tendency is J_s^0, defined in Eqs. (6.12) and (6.14). Elastic recovery is also reflected in die swell, $\hat{D}$, defined as the ratio of the final and initial dimensions perpendicular to the flow. Die swell is governed by the normal force, although predictions of $\hat{D}$ for an arbitrary deformation are not possible. However, for steady-state flow there are approximate relations between $\hat{D}$ and the first (primary) normal force (Tanner, 1970),

$$N_1 = 2\sigma(2\hat{D}^6 - 2)^{1/2}, \tag{6.73}$$

and (Bagley and Duffey, 1970),

$$N_1 = 2\sigma(2\hat{D}^4 - \hat{D}^{-2}). \tag{6.74}$$

Substituting for the stress in Eq. (6.66), $\hat{D}$ can be related to the shear strain, or using Eq. (6.62) to the dynamic modulus. Results are shown in Figure 6.33 (Müllner et al., 2008), comparing the die swell measured in a capillary rheometer to values calculated from dynamic measurements. The rough agreement with experimentally measured die swell is satisfactory given the approximate nature of the derivation of Eqs. (6.73) or (6.74) (Tanner, 1970; Bagley and Duffey, 1970).

As mentioned in Section 6.3.3, the affinity of filler for a rubber is expected to be greater for more polar rubbers and those having a higher degree of unsaturation, and this can result in a nonuniform distribution of the filler between the phases of a blend (Hess, 1991; Lee, 1981; Le et al., 2008). Since active fillers such as carbon black interact with the polymer chains, mainly via chemisorption, there is little migration between the phases. This means that a nonuniform distribution can be obtained by design, through the sequence in which the materials are added to the mixing. Properties such as the elasticity depend nonlinearly on the filler content, which means a rubber blend having a nonuniform distribution can behave differently than one in which the filler

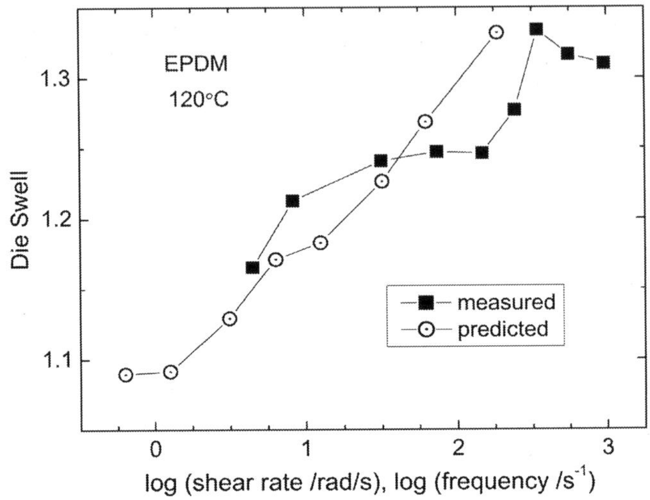

FIGURE 6.33 $\hat{D}$ measured for a filled EPDM rubber (squares) and calculated from dynamic mechanical measurements (circles) (Müllner et al., 2008).

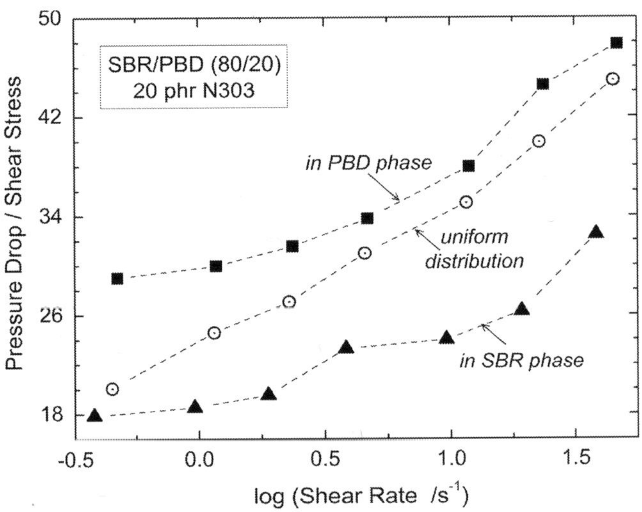

FIGURE 6.34 Exit pressure normalized by the shear stress for an 80:20 blend of styrene-butadiene copolymer with 1,4-polybutadiene, containing 20 phr of carbon black. The polymers and filler were added to the mixer simultaneously (circles), or the filler was first mixed into either the SBR (triangles) or the PBD (squares). The greatest extrudate swell is observed when the major phase lacks filler (Lee, 1981).

is equally partitioned between components. Figure 6.34 (Lee, 1981) shows the pressure drop when a rubber blend exits a die; this exit pressure is a measure of the elasticity and hence of extrudate swell. By incorporating the carbon black into the major component, the exit pressure is smaller than for a uniform distribution of the filler. On the other hand, if the carbon black is mixed into the minor component, the large elasticity of the "gum" phase results in very large exit pressures.

6.5.3 Tack

The ability of uncured rubber to adhere quickly to itself and then be resistant to separation is known as tack or autohesion. Tack can be crucial to the assembly of rubber components that need to remain in place prior to vulcanization. An example is tire assembly, which entails positioning carcasses, belts, and other plies; these layers are held together by virtue of their tack. There are three requirements for "building tack":

- The surfaces must attain microscopic contact to allow interpenetration of chains via diffusion. This surface wetting depends on the surface roughness and the rubber viscosity.
- The ensuing interdiffusion must transpire rapidly, to fuse the layers and yield adequate adhesion.
- The cohesive strength of the rubber must be sufficient, since it serves as an upper bound on the achievable autohesion (Hamed and Shieh, 1986).

Plasticization of the compound facilitates the first two processes; however, it entails loss of green strength and thus it is not usually an effective means to improve tack. Tackifying resins, which are important ingredients in pressure-sensitive adhesives such as Scotch tape, can be employed in rubber formulations to take advantage of the stronger rate dependence the resins impart to the mechanical response (Hamed and Roberts, 1994; Basak et al., 2012). At high strain rates, corresponding to peeling the layers apart, these resins solidify the material, whereas their effect on the viscosity at the slow strain rates relevant to wetting is less significant. Strain-crystallizing rubbers, such as natural rubber, exhibit superior tack due to the dependence of their viscosity and strength on strain. Only at the large strains ($\sim$200% (Choi and Roland, 1997)) that induce crystallization does the material have strong resistance to flow. At low strains relevant to coalescence of the two surfaces, the rubber is amorphous and thus has the compliance necessary for intimate contact to develop between adherends.

The adhesive strength and failure mode can change with both the contact time and the rate at which the surfaces are separated (Hamed and Wu, 1995; Schach and Creton, 2008). Under the usual conditions, the process of interdiffusion causes an increase in the tack with contact time. When the diffusion has transpired over distances on the order of the chain coil size, the interfacial region becomes indistinguishable from the bulk material, and the tack becomes constant, equal to the cohesive strength of the rubber. The coil size of a typical rubber molecule is in the range from 10 to 20 nm. For typical self-diffusion constants of rubbery polymers at room temperature, ca. 0.4–4 nm^2/s

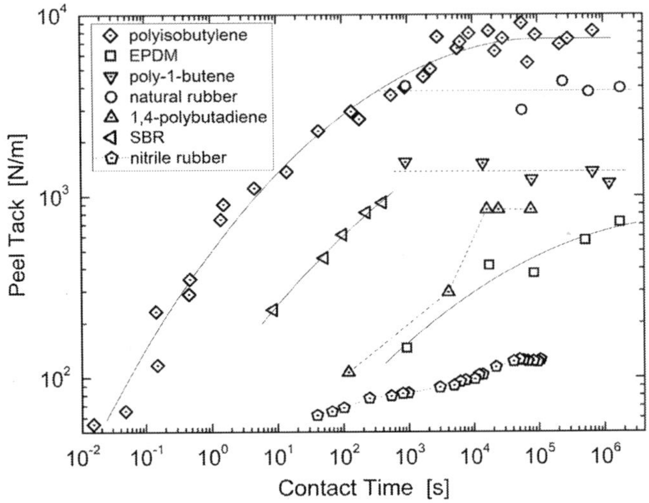

FIGURE 6.35 Peel force per unit length (equal to the detachment energy per unit area) measured for various rubbers versus contact time. The increase is due to the buildup of an interfacial layer, as the chains interdiffuse. Some polymers (natural rubber and poly-1-butene (hydrogenated 1,2-polybutadiene)) show no time-dependence because interdiffusion is fast relative to the measurement time. For polyisobutylene, 1,4-polybutadiene, and an acrylonitrile-butadiene copolymer, the cohesive strength of the materials is attained at longer contact times. For the styrene-butadiene copolymer and ethylene-propylene-diene terpolymer, complete interdiffusion is never complete, and the adhesion continues to increase with time. Of course, the ultimate level of tack and the adhesion kinetics are strongly dependent on molecular weight. The ordinate values for SBR and NBR were shifted for clarity (Skewis, 1966; Roland and Bohm, 1985; Bothe and Rehage, 1982).

(Skewis, 1966; Roland and Bohm, 1985; Bothe and Rehage, 1982), a plateau value of tack is thus expected in time periods ranging from seconds to several minutes. Results for a number of rubbers are in accord with this expectation (Figure 6.35).

ACKNOWLEDGMENT

This work was supported by the Office of Naval Research.

REFERENCES

Abdel-Goad, M., Pyckhout-Hintzen, W., Kahle, S., Allgaier, J., Richter, D., Fetters, L.J., 2004. Macromolecules 37, 8135.
Abeliov, Y.A., Nikolskii, V.G., Kirillov, V.N., Alekseyev, B.F., 1979. Polym. Sci. USSR 20, 2590.
Ahn, S., White, J.L., 2003. J. Appl. Polym. Sci. 90, 1555; Ahn, S., White, J.L., 2004. J. Appl. Polym. Sci. 91, 651.
Andrews, R.D., Tobolsky, A., 1951. J. Polym. Sci. 1, 221.
Angell, C.A., 1991. Pure Appl. Chem. 63, 1387.

Angell, C.A., Ngai, K.L., McKenna, G.B., McMillan, P.F., 2000. J. Appl. Phys. 88, 3113.
Arruda, E., Wang, Y., Przbylo, P., 2001. Rubber Chem. Technol. 74, 560.
Ashurst, W.T., Hoover, W.G., 1975. Phys. Rev. A 11, 658.
Attane, P., Pierrard, J.M., Turrel, G., 1985. J. Non-Newton. Fluid Mech. 18, 295.
Bagley, E.B., Duffey, H.J., 1970. Trans. Soc. Rheol. 14, 545.
Bair, S., 2002. J. Eng. Tribol. 216, 1.
Bair, S., Winer, W.O., 1980. J. Lubricat. Technol. 102, 229.
Baird, D.G., Collias, D.I., 1998. Polymer Processing: Principles and Design. Wiley-Interscience.
Balasubramanian, V., Bush, K., Smoukov, S., Venerus, D.C., Schieber, J.D., 2005. Macromolecules 38, 6210.
Bansal, A., Yang, H.C., Li, C.Z., Cho, K.W., Benicewicz, B.C., Kumar, S.K., Schadler, L.S., 2005. Nat. Mater. 4, 693.
Bardik, V.Y., Sysoev, V.M., 1998. Low Temp. Phys. 24, 601.
Basak, G.C., Bandyopadhyay, A., Bhowmick, A.K., 2012. J. Mater. Sci. 47, 3166.
Batzer, H., Wagner, M.H., Meissner, J., 1980. Makromol. Chem. 181, 1533.
Behler, K.D., Stravato, A., Mochalin, V., Korneva, G., Yushin, G., Gogotsi, Y., 2009. ACS Nano. 3, 363.
Bergstrom, J.S., Boyce, M.C., 1998. J. Mech. Phys. Solids 46, 931.
Bernstein, B., Kearsley, E.A., Zapas, L.J., 1963. Trans. Soc. Rheol. 7, 391.
Bernstein, B., Kearsley, E.A., Zapas, L.J., 1964. J. Res. Natl. Bur. Stand., Sect. B 68, 103.
Berry, G.C., 2003. Polymer Rheology: Principles, Techniques and Applications. In: Brady, J.R.F. (Ed.), Desk Reference of Polymer Characterization and Analysis. American Chemical Society, New York, p. 574.
Berry, G.C., Fox, T.G., 1968. Adv. Polym. Sci. 5, 261.
Bhattacharya, M., Maiti, M., Bhowmick, A.K., 2008. Rubber Chem. Technol. 81, 782.
Bingham, E.C., 1916. US Bureau Stand. Bull. 13, 309.
Blake, F.C., 1922. Trans. AIChE 14, 415.
Boese, D., Momper, B., Meier, G., Kremer, F., Hagenah, I.U., Fischer, E.W., 1989. Macromolecules 22, 4416.
Bogoslovov, R.B., Roland, C.M., Ellis, A.R., Randall, A.M., Robertson, C.G., 2008. Macromolecules 41, 1289.
Bogoslovov, R.B., Hogan, T.E., Roland, C.M., 2010. Macromolecules 43, 2904.
Bohm, G.G.A., Nguyen, M.N., 1995. J. Appl. Polym. Sci. 55, 1041.
Bohmer, R., Ngai, K.L., Angell, C.A., Plazek, D.J., 1993. J. Chem. Phys. 99, 4201.
Bokobza, L., 2007. Polymer 48, 4907.
Boltzmann, L., 1874. Sitzungsber. Kaiserl. Akad. Wiss. Wien 70, 275.
Bothe, L., Rehage, G., 1982. Rubber Chem. Technol. 55, 1308.
Boyer, R.F., Spencer, R.S., 1944. J. Appl. Phys. 15, 398.
Brochard, F., De Gennes, P.G., 1977. Macromolecules 10, 1157.
Brzoskowski, R., White, J.L., Szydlowski, W., Weissert, F.C., Nakajima, N., Min, K., 1987. Rubber Chem. Technol. 60, 945.
Carretero-Gonzalez, J., Retsos, H., Verdejo, R., Toki, S., Hsiao, B.S., Giannelis, E.P., Lopez-Manchado, M.A., 2008. Macromolecules 41, 6763.
Casalini, R., Roland, C.M., 2004a. Phys. Rev. E 69, 062501.
Casalini, R., Roland, C.M., 2004b. Coll. Polym. Sci. 283, 107.
Casalini, R., Roland, C.M., 2005. Macromolecules 38, 1779.
Casalini, R., Bogoslovov, R., Qadri, S.B., Roland, C.M., 2012. Polymer 53, 1282.
Chandler, D., Weeks, J.D., Andersen, H.C., 1983. Science 220, 787.
Chang, I., Sillescu, H., 1997. J. Phys. Chem. B 101, 8794.
Chang, W.V., Bloch, R., Tschoegl, N.W., 1976. Rheol. Acta 15, 367.
Chapter 3 herein.
Chapter 8 herein.
Chatterjee, T., Krishnamoortim, R., 2007. Phys. Rev. E 75, 050403.
Chen, X., Xi, J., Guo, B.H., 2006. J. Appl. Polym. Sci. 102, 3201.
Choi, I.S., Roland, C.M., 1997. Rubber Chem. Technol. 70, 202.
Cohen, M.H., Grest, G.S., 1979. Phys. Rev. B 20, 1077.
Cohen, M.H., Grest, G.S., 1984. J. Non-Cryst. Solids 61–62, 749.
Cohen, M.H., Turnbull, D., 1959. J. Chem. Phys. 31, 1164.

Colby, R.H., Fetters, L.J., Graessley, W.W., 1987. Macromolecules 20, 2237.
Colby, R.H., Rubinstein, M., Viovy, J.L., 1992. Macromolecules 25, 996.
Coleman, B.D., Markovitz, H., 1964. J Appl. Phys. 35, 1.
Coleman, B.D., Zapas, L.J., 1989. J. Rheol. 33, 501.
Coleman, J.N., Khan, U., Blau, W.J., Gunko, Y.K., 2006. Carbon 44, 1624.
Colmenero, J., Alegria, A., Alberdi, J.M., Alvarez, F., Frick, B., 1991. Phys. Rev. B44, 7321.
Colmenero, J., Alegria, A., Santangelo, P.G., Ngai, K.L., Roland, C.M., 1994. Macromolecules 27, 407.
Corezzi, S., Rolla, P.A., Paluch, M., Ziolo, J., Fioretto, D., 1999. Phys. Rev. E 60, 4444.
Cox, W.P., Merz, E.H., 1958. J. Polym. Sci. 28, 619.
Danusso, F., Levi, M., Gianotti, G., Turri, S., 1993. Polymer 34, 3687.
Dealy, J.M., Tsang, W.K.-W., 1981. J. Appl. Polym. Sci. 26, 1149.
de Gennes, P.-G., 1979. Scaling Concepts in Polymer Physics. Cornell University, Ithaca.
DeKee, D., Fong, C.F.C.M., 1994. Polym. Eng. Sci. 34, 438.
Ding, Y.F., Sokolov, A.P., 2006. Macromolecules 39, 3322.
Doi, M., 1980. J. Polym. Sci. Polym. Phys. Ed. 18, 1891; 18, 2055 (1980).
Doi, M., Edwards, S.F., 1986. The Theory of Polymer Dynamics. Clarendon Press, Oxford.
Doolittle, A.K., Doolittle, D.B., 1957. J. Appl. Phys. 28, 901.
Everaers, R., Sukumaran, S.K., Grest, G.S., Svaneborg, C., Sivasubramanian, A., Kremer, K., 2004. Science 303, 823.
Ferry, J.D., 1980. Viscoelastic Properties of Polymers, third ed. Wiley, New York.
Ferry, J.D., Grandine, L.D., Fitzgerald, E.R., 1953. J. Appl. Phys. 24, 911.
Fetters, L.J., Lohse, D.J., Richter, D., Witten, T.A., Zirkel, A., 1994. Macromolecules 27, 4639.
Fetters, L.J., Lohse, D.J., Milner, S.T., Graessley, W.W., 1999. Macromolecules 32, 6847.
Fetters, L.J., Lohse, D.J., Graessley, W.W., 1999. J. Polym. Sci. Polym. Phys. Ed. 3, 1023.
Fetters, L.J., Lohse, D.J., Garcia-Francok, C.A., Brant, P., Richter, D., 2002. Macromolecules 35, 10096.
Fillers, R.W., Tschoegl, N.W., 1977. Trans. Soc. Rheol. 21, 51.
Fischer, E.W., Donth, E., Steffen, W., 1992. Phys. Rev. Lett. 68, 2344.
Fitzgerald, E.R., Grandine, L.D., Ferry, J.D., 1953. J. Appl. Phys. 24, 650.
Floudas, G., 2003. Chapter 8 in Kremer and Schonhals..
Fragiadakis, D., Roland, C.M., 2011. Phys. Rev. E 83, 031504.
Fragiadakis, D., Casalini, R., Bogoslovov, R.B., Robertson, C.G., Roland, C.M., 2011. Macromolecules 44, 1149.
Fuller, K.N.G., 1988. In: Roberts, A.D. (Ed.), Natural Rubber Science and Technology. Oxford, New York.
Fuller, G.G., Leal, L.G., 1981. J. Polym. Sci. Polym. Phys. Ed. 19, 557.
Gerhardt, L.J., Manke, C.W., Gulari, E., 1997. J. Polym. Sci. Polym. Phys. 35, 523.
Gleissle, W., 1980. In: Astarita, G., Marucci, G., Nicolais, L. (Eds.), Rheology, vol. 2. Plenum, New York, p. 457.
Goldstein, M.J., 1969. J. Chem. Phys. 51, 3728.
Grace, H.P., 1982. Chem. Eng. Comm. 14, 225.
Graessley, W.W., 2008. Polymeric Liquids and Networks: Dynamics and Rheology. Taylor & Francis, London.
Graessley, W.W., Edwards, S.F., 1981. Polymer 22, 1329.
Grest, G.S., Cohen, M.H., 1981. Adv. Chem. Phys. 48, 455.
Guth, E., 1945. J. Appl. Phys. 16, 20.
Hamed, G.R., 2000. Rubber Chem. Technol. 73, 524.
Hamed, G.R., 2007. Rubber Chem. Technol. 80, 533.
Hamed, G.R., Roberts, G.D., 1994. J. Adhes. 47, 95.
Hamed, G.R., Shieh, C.H., 1986. Rubber Chem. Technol. 59, 883.
Hamed, G.R., Wu, P.S., 1995. Rubber Chem. Technol. 68, 248.
Han, C.D., Ma, C., 1983. J. Appl. Polym. Sci. 28, 851.
Hancock, T., 1857. Personal Narrative of the Origin and Progress of the Caoutchouc or India-Rubber Manufacture in England. Longman, Brown, Green, Longmans, and Roberts, London; reprinted (1920).
Hands, D., 1980. Rubber Chem. Technol. 53, 80.
Hansen, J.P., 1970. Phys. Rev. A 2, 221.

Harwood, J.A.C., Payne, A.R., 1966a. J. Appl. Polym. Sci. 10, 315.
Harwood, J.A.C., Payne, A.R., 1966b. J. Appl. Polym. Sci. 1203.
Harwood, J.A.C., Schallamach, A., 1967. J. Appl. Polym. Sci. 11, 1835.
Heinrich, G., Kluppel, M., Vilgis, T.A., 2002. Curr. Opin. Sol. State Mater. Sci. 6, 195.
Hellwege, K.H., Knappe, W., Paul, F., Semjonow, V., 1967. Rheol. Acta 6, 165.
Hess, W.M., 1991. Rubber Chem. Technol. 64, 386.
Heymans, N., 2000. Macromolecules 33, 4226.
Hiwatari, Y., Matsuda, H., Ogawa, T., Ogita, N., Ueda, A., 1974. Prog. Theor. Phys. 52, 1105.
Hobbie, E.K., Fry, D.J., 2007. J. Chem. Phys. 126, 124907.
Hoover, W.G., Ross, M., 1971. Contemp. Phys. 12, 339.
Hosler, D., Burkett, S.L., Tarkanian, M.J., 1999. Science 284, 1988.
Hsu, A.S., Leal, L.G., Non-Newt, J., 2009. Fluid Mech. 160, 176.
Huppler, J.D., MacDonald, I.F., Ashare, E., Spriggs, T.W., Bird, R.B., Holmes, L.A., 1967. J. Rheol. 11, 181.
Isono, Y., Obashi, N., Kase, T., 1995. Macromolecules 28, 5145.
Isono, Y., Kamohara, T., Takano, A., Kase, T., 1997. Rheol. Acta 35, 245.
Johnson, P.S., 2001. Rubber Processing: An Introduction. Hanser Publications.
Kavassalis, T.A., Noolandi, J., 1989. Macromolecules 22, 2709.
Kaye, A., 1962. Note to the College of Aeronautics. Cranfield, UK.
Kennedy, A.J., 1953. J. Mech. Phys. Solids 1, 172.
Knauss, W.G., Zhao, J., 2007. Mech. Time-Depend. Mater. 11, 199.
Kohls, D.J., Beaucage, G., 2002. Curr. Opin. Solid State Mater. Sci. 6, 183.
Kotliar, M., Kumar, R., Back, R.A., 1990. J. Polym. Sci., Polym. Phys. Ed. 28, 1033.
Kramers, H.A., 1927. Atti del Congresso Internazionale dei Fisici 2, 545.
Krause, S., Gormley, J.J., Roman, N., Shetter, J.A., Watanabe, W.H., 1965. J. Polym. Sci. A 3, 3573.
Kremer, F., Schonhals, A. (Eds.), 2003. Broadband Dielectric Spectroscopy. Springer-Verlag, Berlin.
Kronig, R. de L., 1926. J. Opt. Soc. Am. 12, 547.
Laird, B.B., Haymet, A.D.J., 1992. Mol. Phys. 75, 71.
Larson, R.G., 1985. J. Rheol. 29, 823.
Larson, R.G., 1988. Constitutive Equations for Polymer Melts and Solutions. Buttersworth, Boston.
Larson, R.G., Valesano, V.A., 1986. J. Rheol. 30, 1093.
Laun, H.M., 1986. J. Rheol. 30, 459.
Leblans, P.J.R., Bastiaansen, C., 1989. Macromolecules 22, 3312.
Lee, B.L., 1981. Polym. Eng. Sci. 21, 294.
Le, H.H., Ilisch, S., Kasaliwal, G.R., Radusch, H.-J., 2008. Rubber Chem. Technol. 81, 767.
Lin, Y.-H., 1987. Macromolecules 20, 3080.
Lin, J.C., Hergenrother, W.L., Hilton, A.S., 2010. J. Appl. Polym. Sci. 115, 655.
Lodge, T.P., 1999. Phys. Rev. Lett. 83, 3128.
Lodge, A.S., Meissner, J., 1972. Rheol. Acta 11, 351.
MacLean, D.L., 1973. Trans. Soc. Rheol. 17, 385.
Macosko, C.W., 1994. Rheology: Principles, Measurements, and Applications. Wiley-VCH.
Maiti, M., Bhattacharya, M., Bhowmick, A.K., 2008. Rubber Chem. Technol. 81, 384.
March, N.H., Tosi, M.P., 2002. Introduction to the Liquid State. World Scientific, Singapore.
Mark, J.E., 2006. Accts. Chem. Res. 39, 881.
Marvin, R.S., 1953. In: Harrison, V.G.W. (Ed.), Proc. 2nd Inter. Congress on Rheology. Buttonworths Sci. Pub., London.
McCrum, N.G., Read, B.E., Williams, G., 1967. Anelastic and Dielectric Effects in Polymer Solids. Wiley, London.
McKenna, G.B., Zapas, L.J., 1979. J. Rheol. 23, 151.
McLean, D., 1966. Rep. Prog. Phys. 29, 1.
Medalia, A.I., 1967. J. Coll. Inter. Sci. 24, 393.
Medalia, A.I., 1972. Rubber Chem. Technol. 45, 1171.
Medalia, A.I., 1986. Rubber Chem. Technol. 59, 432.
Medalia, A.I., 1987. Rubber Chem. Technol. 60, 45.
Menezes, E.V., Graessley, W.W., 1982. J. Polym. Sci. Polym. Phys. Ed. 20, 1817.
Műllner, H.W., Ernst, G., Eberhardsteiner, J., 2008. J. Appl. Polym. Sci. 110, 76.

Mihara, S., Datta, R.N., Noordermeer, J.W.M., 2009. Rubber Chem. Technol. 82, 524.
Milner, S.T., 1996. J. Rheol. 40, 303.
Moelwyn-Hughes, E.A., 1961. Physical Chemistry, second ed. Pergamon Press, NY.
Monfort, J.P., Mann, G., Monge, P., 1984. Macromolecules 17, 1551.
Montes, S., White, J.L., Nakajima, N., 1988. J. Non-Newton. Fluid Mech. 28, 183.
Mooney, M., 1936. Physics 7, 413.
Moore, J.D., Cui, S.T., Cochran, H.D., Cummings, P.T., 1999. Phys. Rev. E 60, 6956.
Morton-Jones, D.H., 1989. Polymer Processing. Chapman & Hall.
Mott, P.H., Roland, C.M., 1996. Macromolecules 29, 8492.
Mott, P.H., Roland, C.M., 2009. Phys. Rev. B 80, 132104.
Mullins, L., 1969. Rubber Chem. Technol. 42, 339.
Nah, C., Lim, J.Y., Cho, B.H., Hong, C.K., Gent, A.N., 2010. J. Appl. Polym. Sci. 118, 1574.
Nelson, C.J., Avgeropoulos, G.N., Weissert, F.C., Bohm, G.G.A., 1977. Ange. Makro. Chem. 60, 49.
Nemoto, N., Moriwaki, M., Odani, H., Kurata, M., 1971. Macromolecules 4, 215.
Nemoto, N., Odani, H., Kurata, M., 1972. Macromolecules 5, 531.
Ngai, K.L., Schonhals, A., Schlosser, E., 1992. Macromolecules 25, 4915.
Ngai, K.L., Plazek, D.J., Roland, C.M., 2008. Macromolecules 41, 3925.
Ngai, K.L., Plazek, D.J., 1995. Rubber Chem. Technol. 68, 376.
Nikiel, L., Wampler, W., Neilsen, J., Hershberger, N., 2009. Rubber Plast. News, p. 12.
Oberhauser, J.P., Leal, L.G., Mead, D.W., 1998. J. Polym. Sci. Polym. Phys. Ed. 36, 265.
Osswald, T.A., Gries, T., 2006. Polymer Processing: Modeling and Simulation. Hanser Gardner Publications.
Palade, L.I., Verney, V., Attane, P., 1995. Macromolecules 28, 7051.
Paluch, M., 2001. J. Chem. Phys. 115, 10029.
Paluch, M., Roland, C.M., Best, A., 2002. J. Chem. Phys. 117, 1188.
Paluch, M., Roland, C.M., Gapinski, J., Patkowski, A., 2003. J. Chem. Phys. 118, 3177.
Pathak, J.A., Twigg, J.N., Nugent, K.E., Ho, D.L., Lin, E.K., Mott, P.H., Robertson, C.G., Vukmir, M.K., Epps, T.H., Roland, C.M., 2008. Macromolecules 41, 7543.
Payne, A.R., 1962. J. Appl. Polym. Sci. 6, 57.
Payne, A.R., 1963. J. Appl. Polym. Sci. 7, 873.
Payne, A.R., 1964. J. Appl. Polym. Sci. 8, 2661.
Peng, X., Huang, Y., Xia, T., Kong, M., Li, G., 2011. Eur. Polym. J. 47, 1956.
Plazek, D.J., 1965. J. Phys. Chem. 6, 612.
Plazek, D.J., 1980. Polymer J. 12, 43.
Plazek, D.J., 1996. J. Rheol. 40, 987.
Plazek, D.J., Ngai, K.L., 1991. Macromolecules 24, 1222.
Plazek, D.J., O'Rourke, V.M., 1971. J. Polym. Sci. B2 (9), 209.
Plazek, D.J., Rosner, M.J., Plazek, D.L., 1988. J. Polym. Sci. Polym. Phys. 26, 473.
Plazek, D.J., Seoul, C., Bero, C.A., Non-Cryst, J., 1991. Solids 131–133, 570.
Plazek, D.J., Bero, C., Neumeister, S., Floudas, G., Fytas, G., Ngai, K.L., 1994. Colloid. Polym. Sci. 272, 1430.
Plazek, D.J., Chay, I.-C., Ngai, K.L., Roland, C.M., 1995. Macromolecules 28, 6432.
Pohl, H.A., Gogos, C.G., 1961. J. Appl. Polym. Sci. 5, 67.
Rallison, J.M., 1984. Ann. Rev. Fluid Mech. 16, 45.
Ram, A., Izrailov, L., 1986. J. Appl. Polym. Sci. 31, 85.
Reader's Digest, 1958. The Reader's Digest Assoc. Inc., Pleasantville, NY, January.
Reiner, M., 1964. Phys. Today 17, 62.
Richter, D., Farago, B., Butera, R., Fetters, L.J., Juang, J.S., Ewen, B., 1993. Macromolecules 26, 795.
Robertson, C.G., Rademacher, C.M., 2004. Macromolecules 37, 10009.
Robertson, C.G., Roland, C.M., 2008. Rubber Chem. Technol. 81, 506.
Robertson, C.G., Wang, X., 2005. Phys. Rev. Lett. 95, 075703.
Robertson, C.G., Roland, C.M., Puskas, J.E., 2002. J. Rheol. 46, 307.
Robertson, C.G., Warren, S., Plazek, D.J., Roland, C.M., 2004. Macromolecules 37, 10018.
Robertson, C.G., Bogoslovov, R., Roland, C.M., 2007. Phys. Rev. E 75, 051403.
Robertson, C.G., Lin, C.J., Rackaitis, M., Roland, C.M., 2008. Macromolecules 41, 2727.
Roland, C.M., 1989. J. Rheol. 33, 659.

Roland, C.M., 1990. J. Rheol. 34, 25.
Roland, C.M., 2010. Macromolecules 43, 7875.
Roland, C.M., 2011. Viscoelastic Behavior of Rubbery Materials. Oxford University Press.
Roland, C.M., Bohm, G.G.A., 1984. J. Polym. Sci. Pt. B–Polym. Phys. 22, 79.
Roland, C.M., Bohm, G.G.A., 1985. Macromolecules 18, 1310.
Roland, C.M., Nguyen, M., 1988. J. Appl. Polym. Sci. 35, 2141.
Roland, C.M., Robertson, C.G., 2006. Rubber Chem. Technol. 79, 267.
Roland, C.M., Ngai, K.L., Santangelo, P.G., Qiu, X.H., Ediger, M.D., Plazek, D.J., 2001. Macromolecules 34, 6159.
Roland, C.M., Paluch, M., Casalini, R., 2004. J. Polym. Sci. Polym. Phys. Ed. 42, 4313.
Roland, C.M., Robertson, C.G., Nikiel, L., Puskas, J.E., 2004. Rubber Chem. Technol. 77, 372.
Roland, C.M., Ngai, K.L., Plazek, D.J., 2004. Macromolecules 37, 7051.
Roland, C.M., Hensel-Bielowka, S., Paluch, M., Casalini, R., 2005. Rep. Prog. Phys. 68, 1405.
Roland, C.M., Bair, S., Casalini, R., 2006. J. Chem. Phys. 125, 124508.
Ronca, G., 1983. J. Chem. Phys. 79, 1031.
Roovers, J., 1985. Macromolecules 8, 1359.
Roovers, J., Toporowski, P.M., 1990. Rubber Chem. Technol. 63, 734.
Rössler, E., 1990. Phys. Rev. Lett. 65, 1595.
Rott, N., 1990. Ann. Rev. Fluid Mech. 22, 1.
Royer, J.R., Gay, Y.J., Desimone, J.M., Khan, S.A., 2000. J. Polym. Sci. Polym. Phys. 38, 3168.
Rudin, A., Schreiber, H.P., 1983. Polym. Eng. Sci. 23, 422.
Sadhu, S., Bhowmick, A.K., 2003. Rubber Chem. Technol. 76, 860.
Sampoli, M., Benassi, P., Eramo, R., Angelani, L., Ruocco, G., 2003. J. Phys. Condens. Matter. 15, S1227.
Santangelo, P.G., Roland, C.M., 1998. Macromolecules 31, 3715.
Santangelo, P.G., Roland, C.M., 2001. J. Rheol. 45, 583.
Schach, R., Creton, C., 2008. J. Rheol. 52, 749.
Schneider, U., Lunkenheimer, P., Brand, R., Loidl, A., 1999. Phys. Rev. E 59, 6924.
Schug, K.U., King, H.E., Böhmer, R., 1998. J. Chem. Phys. 109, 1472.
Schwartz, G.A., Cerveny, S., Marzocca, A.J., Gerspacher, M., Nikiel, L., 2003. Polymer 44, 7229.
Sengupta, R., Chakraborty, S., Bandyopadhyay, S., Dasgupta, S., Mukhopadhyay, R., Auddy, K., Deuri, A.S., 2007. Polym. Eng. Sci. 47, 1956.
Shih, C.K., 1976. Polym. Eng. Sci. 16, 742.
Sichel, E.K. (Ed.), 1982. Carbon Black-Polymer Composites. Marcell Dekker, New York.
Simha, R., Boyer, R.F., 1962. J. Chem. Phys. 37, 1003.
Simmons, J.M., Rheol. Acta 1968. 7, 184.
Sirisinha, C., Saeoui, P., Guaysomboon, J., 2003. J. Appl. Polym. Sci. 90, 4038.
Skewis, J.D., 1966. Rubber Chem. Technol. 39, 217.
Skorodumov, V.F., Godovskii, Y.K., 1993. Polym. Sci. 35, 562.
Stankovich, S., Dikin, D.A., Dommett, G.H.B., Kohlhaas, K.M., Zimney, E.J., Stach, E.A., Piner, R.D., Nguyen, S.T., Ruoff, R.S., 2006. Nature 442, 282.
Stillinger, F.H., Weber, T.A., 1984. Science 225, 983.
Stillinger, H., Debenedetti, P.G., Truskett, T.M., 2001. J. Phys. Chem. B 105, 11809.
Stratton, R.A., Butcher, A.F., 1973. J. Polym. Sci. Polym. Phys. Ed. 11, 1747.
Tadmor, Z., Gogos, C.G., 2006. Principles of Polymer Processing, second ed. Wiley-Interscience.
Tanner, R.I., 1970. J. Polym. Sci. Polym. Phys. Ed. 8, 2067.
Tanner, R.I., 1988. J. Rheol. 32, 673.
Tanner, R.I., Williams, G., 1971. Rheol. Acta 10, 528.
Tschoegl, N.W., 1989. The Phenomenological Theory of Linear Viscoelastic Behavior. Springer-Verlag, Berlin.
Turner, D.T., 1978. Polymer 19, 789.
Ueberreiter, K., Kanig, G., 1952. J. Colloid Sci. 7, 569.
Valadares, L.F., Leite, C.A.P., Galembeck, F., 2006. Polymer 47, 672.
Van Krevelen, D.W., 1990. Properties of Polymers. Elsevier, Amsterdam.
Venerus, D.C., Schieber, J.D., Iddir, H., Guzman, J.D., Broerman, A.W., 1999. Phys. Rev. Lett. 82, 366.
Vinogradov, G.V., Belkin, I.M., 1965. J. Polym. Sci. A3, 917.
Vrentas, J.S., Venerus, D.C., Vrentas, C.M., 1991. J. Polym. Sci. Polym. Phys. Ed. 29, 537.

Vu, Y.T., Mark, J.E., Pham, L.H., Engelhardt, M., 2001. J. Appl. Polym. Sci. 82, 1391.
Waddell, W.H., Evans, L.R., 1996. Rubber Chem. Technol. 69, 377.
Wakabayashi, K., Pierre, C., Dikin, D.A., Ruoff, R.S., Ramanathan, T., Brinson, L.C., Torkelson, J.M., 2008. Macromolecules 41, 1905.
Wang, S.Q., Poly, J., 2003. Sci. Poly. Phys. Ed. 41, 1589.
Wang, X., Robertson, C.G., 2005. Phys. Rev. E 72, 031406.
Watanabe, H., 1999. Prog. Polym. Sci. 24, 1253.
Watanabe, H., Sakamoto, T., Kotaka, T., 1995. Macromolecules 18, 1008.
Weber, W., 1835. Ann. Phys. Chem. 34, 247; 54, 1 (1841).
Wendt, C.J., Cyphers, A., 2008. J. Anthropol. Archaeol. 27, 175.
White, J.L., 1995. Rubber Processing: Technology, Materials, and Principles. Epic Press, Hanser Publications.
Whittle, M., Dickinson, E., 1997. Mol. Phys. 90, 739.
Widen, B., 1967. Science 157, 375.
Widom, B., 1999. Physica A 263, 500.
Williams, E., Angell, C.A., 1977. J. Phys. Chem. 81, 2323.
Williams, M.L., Landel, R.F., Ferry, J.D., 1955. Am. J. Chem. Soc. 77, 3701.
Wubbenhorst, M., Van Turnhout, J., J. Non-Cryst. 2002. Solids 305, 40.
Xu, Y.Z., Dekee, D., Fong, C.F.C.M., 1995. J. Appl. Polym. Sci. 55, 779.
Xu, X., Tao, X., Gao, C., Zheng, Q., 2008. J. Appl. Polym. Sci. 107, 1590.
Yamaguchi, M., Gogos, C.G., 2001. Adv. Polym. Test. 20, 261.
Yamane, M., Hirose, Y., Adachi, K., 2005. Macromolecules 38, 10686.
Yannas, I.V., J. Polymer Sci, 1974. Macro. Rev. 9, 163.
Ye, X., Tain, M., Xin, Y., Zhang, L.-Q., 2012. J. Appl. Polym. Sci. 124, 927.
Zallen, R., 1981. The Physics of Amorphous Solids. Wiley, New York.

Chapter 7

Vulcanization

A.Y. Coran
A.Y. Coran Consulting, 1465 Gulf of Mexico Drive, Ste b106, Longboat Key, FL 34228, USA

7.1 INTRODUCTION

The vulcanization process is necessary to produce most useful rubber articles, like tires and mechanical goods. Unvulcanized rubber is generally not strong, does not retract essentially to its original shape after a large deformation, and it can be very sticky. In short, unvulcanized rubber can have the same consistency as chewing gum.

The first commercial method for vulcanization has been attributed to Charles Goodyear. His process (heating natural rubber with sulfur) was first used in Springfield, Massachusetts, in 1841. Thomas Hancock used essentially the same process about a year later in England. However, Hancock filed his patent on November 21, 1843, 8 weeks before Goodyear filed his US Patent on January 30, 1844.

Since those early days, there has been continued progress toward the improvement of the process and in the resulting vulcanized rubber articles. In addition to natural rubber, over the years, many synthetic rubbers have been introduced. Also, in addition to sulfur, other substances have been introduced as components of curing (vulcanization) systems. This chapter is an overview of the science and technology of vulcanization. Emphasis is placed on general-purpose "high-diene" rubbers; for example, natural rubber (NR), styrene-butadiene rubber (SBR), and butadiene rubber (BR), vulcanized by sulfur in the presence of organic accelerators.

The accelerated-sulfur vulcanization of these rubbers along with the vulcanization of other rubbers, which are vulcanized by closely related technology, comprises more than 90% of all vulcanization. These rubbers include ethylene-propylene-diene-monomer rubber (EPDM), butyl rubber (IIR), halobutyl rubbers, and nitrile rubber (NBR). Nevertheless, we give some consideration to vulcanization by the action of other vulcanization agents such as organic peroxides, phenolic curatives, and quinoid curatives.

The Science and Technology of Rubber. http://dx.doi.org/10.1016/B978-0-12-394584-6.00007-8

Dynamic vulcanization (DV) is also considered. DV is the crosslinking of one polymer in a blend of polymers during its mixing, all polymers of the blend being in the molten state. The process is used in the preparation of thermoplastic elastomers from rubber-plastic blends.

7.2 DEFINITION OF VULCANIZATION

Vulcanization is a process generally applied to rubbery or elastomeric materials. These materials forcibly retract to their approximately original shape after a rather large mechanically imposed deformation. Vulcanization can be defined as a process that increases the retractile force and reduces the amount of permanent deformation remaining after removal of the deforming force. Thus, vulcanization increases elasticity while it decreases plasticity. It is generally accomplished by the formation of a crosslinked molecular network (Figure 7.1).

According to the theory of rubber elasticity (Flory, 1953), the retractile force to resist a deformation is proportional to the number of network-supporting polymer chains per unit volume of elastomer. A supporting polymer chain is a linear polymer molecular segment between network junctures. An increase in the number of junctures or crosslinks gives an increase in the number of supporting chains. In an unvulcanized linear high polymer (above its melting point) only molecular chain entanglements constitute junctures.

Vulcanization, thus, is a process of chemically producing network junctures by the insertion of crosslinks between polymer chains. A crosslink may be a group of sulfur atoms in a short chain, a single sulfur atom, a carbon-to-carbon bond, a polyvalent organic radical, an ionic cluster, or a polyvalent metal ion.

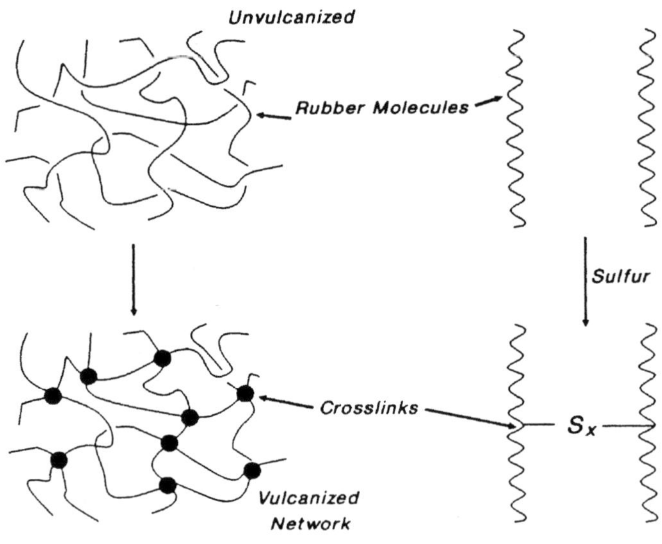

FIGURE 7.1 Network formation.

The process is usually carried out by heating the rubber, mixed with vulcanizing agents, in a mold under pressure. We note that, for extruded rubber goods, the compounded rubber extrudate can be heated in a salt bath or even in hot air.

7.3 EFFECTS OF VULCANIZATION ON VULCANIZATE PROPERTIES

Vulcanization causes highly significant changes at the molecular level. The long rubber molecules (molecular weight usually between 100,000 and 500,000 daltons) become linked together with junctures (crosslinks) spaced along the polymeric chains, with the average distance between junctures corresponding to a molecular weight between crosslinks of about 4000 and 10,000 daltons. Because of network formation, the rubber becomes essentially insoluble in any solvent, and it cannot be processed by any means that requires it to flow, like processing in a mixer or extruder; on a mill or calendar; or during shaping, forming, or molding. Thus, it is essential that vulcanization occur only after the rubber article is in its final geometric form.

Effects of vulcanization (Bateman et al., 1963; Hofmann, 1967; Coran, 1978) on end-use properties are illustrated in Figure 7.2. It should be noted that static modulus increases with vulcanization to a greater extent than does the dynamic modulus. (Here, static modulus is more correctly the equilibrium modulus, approximated by a low-strain, slow-strain-rate modulus. Dynamic modulus is generally measured with the imposition of a sinusoidal, small strain at a frequency of 1–100 Hz.) The dynamic modulus is a composite of viscous

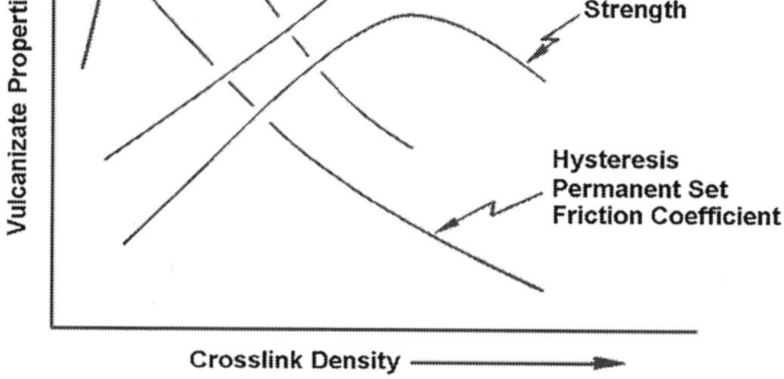

FIGURE 7.2 Vulcanizate properties as a function of the extent of vulcanization.

and elastic behavior, whereas static modulus is largely a measure of only the elastic component of rheological behavior.

Crosslink formation greatly reduces hysteresis. Hysteresis is the ratio of the rate dependent or viscous component to the elastic component of deformation resistance. It is also a measure of deformation energy, which is not stored (or borne by the elastic network), but is converted to heat. Vulcanization then causes a trade-off of elasticity for viscous or plastic behavior. Tear strength, fatigue life, and toughness are related to the breaking energy. Values of these properties increase with small amounts of crosslinking but they are reduced by further crosslink formation. Properties related to the energy to break increase with increases in both the number of network chains and hysteresis. Since hysteresis decreases as more network chains are developed, the energy-to-break related properties are maximized at some intermediate crosslink density.

It should be noted that the properties given in Figure 7.2 are not functions only of crosslink density. The type of crosslink, the type of polymer, and type and amount of filler, and such also affects them.

Reversion is a term generally applied to the loss of network structures by nonoxidative thermal aging. It is usually associated with isoprene rubbers vulcanized by sulfur. It can be the result of too long of a vulcanization time (overcure) or of hot aging of thick sections. It is most severe at temperatures above about 155°C. It occurs in vulcanizates containing a large number of polysulfidic crosslinks. Though its mechanism is complex, a good deal about the chemical changes that occur during the reversion of natural rubber has been deduced (Morrison and Porter, 1984).

Sometimes the term reversion is applied to other types of nonoxidative degradation, especially with respect to rubbers not based on isoprene. For example, thermal aging of SBR (styrene-butadiene rubber), which can cause increased crosslink density and hardening, has been called reversion, since it can be the result of overcure and can also degrade a product's usefulness.

7.4 CHARACTERIZATION OF THE VULCANIZATION PROCESS

The time elapsed before crosslinking starts, the rate of crosslink formation once it starts, and the extent of crosslinking at the end of the process, are important characteristics related to the vulcanization process. There must be sufficient delay or scorch resistance (resistance to premature vulcanization) to permit mixing, shaping, forming, and flowing (e.g., in the mold) before vulcanization. After this necessary delay, or induction period, the formation of crosslinks should be rapid and the extent of crosslinking must be controlled (Figures 7.3 and 7.4).

Scorch resistance is usually measured by the time required, at a given temperature, for the onset of crosslink formation, as indicated by an abrupt increase in viscosity. The Mooney viscometer has been used for this

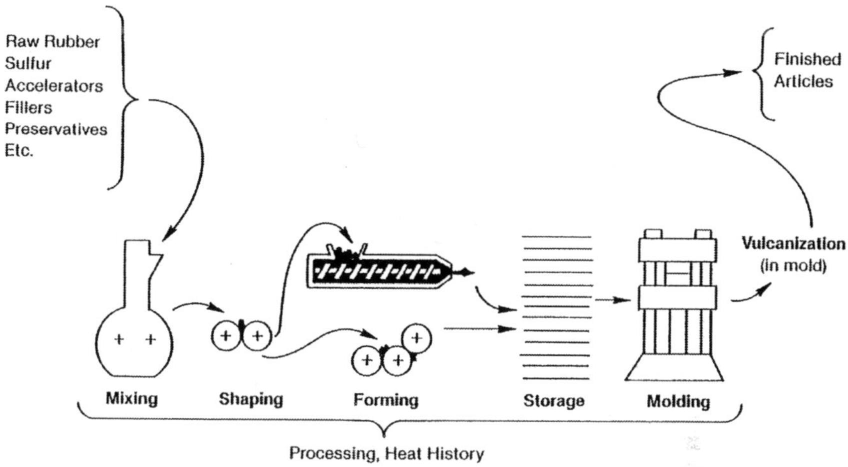

FIGURE 7.3 The effect of processing on heat history.

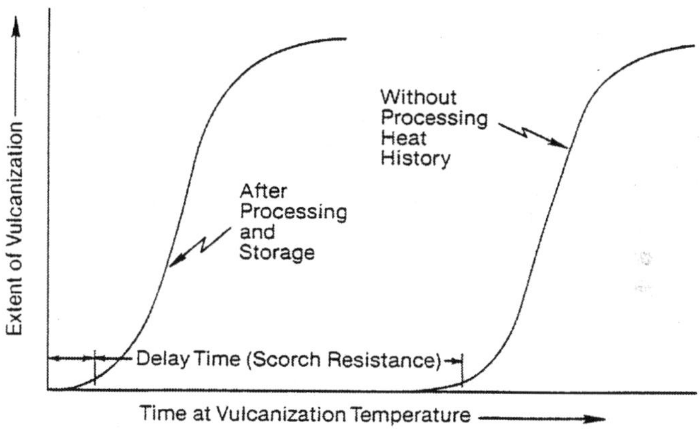

FIGURE 7.4 The effect of heat history (processing) on scorch safety.

(Coran, 1978). During this test, fully mixed but unvulcanized rubber is contained in a heated cavity. Imbedded in the rubber is a rotating disc. Viscosity is continuously measured (by the torque required to keep the rotor rotating at a constant rate) as a function of time. The temperature is selected to be characteristic of rather severe processing (extrusion, calendaring, etc.).

Both the rate of vulcanization after the scorch period and the final extent of vulcanization are now measured by using devices called cure meters. Many researchers have contributed to this development (Decker et al., 1963; Juve et al., 1964). Widely used cure meters are oscillating disc rheometers of the type introduced by the Monsanto Company around 1965. The development of the oscillating disc rheometer, largely through the efforts of Wise, was the beginning of modern vulcometry, which has become standard practice in the industry. Before the development of the cure meter, it was necessary to measure

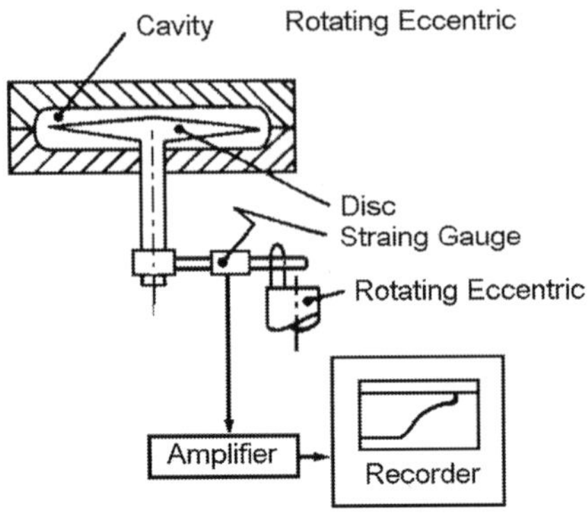

FIGURE 7.5 Oscillating disc rheometer.

mechanical properties of many specimens of a rubber sample, each vulcanized for a different length of time at a given temperature.

In order to measure the vulcanization characteristics the rubber is enclosed in a heated cavity (Figure 7.5). Imbedded in the rubber is a metal disc that oscillates sinusoidally in its plane about its axis. Vulcanization is measured by the increase in the torque required to maintain a given amplitude (e.g., degrees of arc) of oscillation at a given temperature. The torque is proportional to a low-strain modulus of elasticity. Since this torque is measured at the elevated temperature of vulcanization, the portion of it due to viscous effects is minimal. Thus, it has been assumed that the increase in torque during vulcanization is proportional to the number of crosslinks formed per unit volume of rubber. The torque is automatically plotted against time to give a so-called rheometer chart, rheograph, or cure curve.

Newer versions of the cure meter have been introduced (e.g., Figure 7.6). The cavity is much smaller and there is no rotor. In this type of cure meter, one-half of the die (e.g., the upper half) is stationary and the other half oscillates. These instruments are called moving-die rheometers. The sample is much smaller and heat transfer is faster. Also, because there is no rotor, the temperature of the cavity and sample can be changed more rapidly. In either case (oscillating disc or moving die), torque is automatically plotted against time. Such a chart is shown in Figure 7.7.

The cure curve gives a rather complete picture of the overall kinetics of crosslink formation and even crosslink disappearance (reversion) for a given rubber mix. In some cases, instead of reversion, a long plateau or marching cure can occur. The cure meter is, therefore, extensively used to control the quality and uniformity of rubber stocks (also called rubber compounds).

Vulcometry started as a research tool to study vulcanization. It was then used to control uniformity of rubber mixed in the factory. Also, programmed temperature-profile vulcometry has been used to develop recipes for industrial use. The cure temperature-time profile of an industrial mold can be imposed on the curing cavity of the cure meter. The test sample can then be vulcanized in

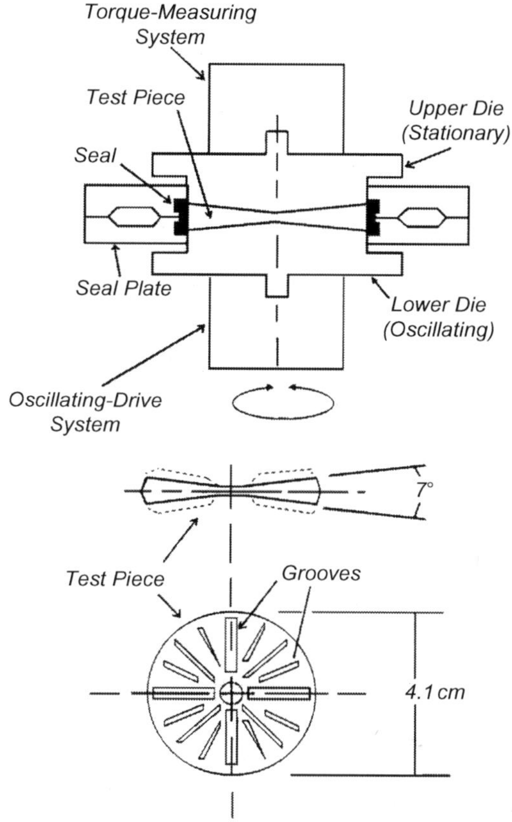

FIGURE 7.6 Moving-die rheometer.

the cure meter under the same conditions as those encountered in the factory. Both the extent of cure and temperature can be simultaneously displayed as functions of time.

7.5 VULCANIZATION BY SULFUR WITHOUT ACCELERATOR

Initially, vulcanization was accomplished by using elemental sulfur at a concentration of 8 parts per 100 parts of rubber (phr). It required 5 h at 140°C. The addition of zinc oxide reduced the time to 3 h. The use of accelerators, in concentrations as low as 0.5 phr, has since reduced the time to as short as 1–3 min. As a result, elastomer vulcanization by sulfur without accelerator is no longer of much commercial interest. (An exception to this is the use of about 30 or more phr of sulfur, with little or no accelerator to produce molded products of hard rubber or "ebonite.") Even though unaccelerated-sulfur vulcanization is not of commercial significance, its chemistry has been the object of much of the early research and study.

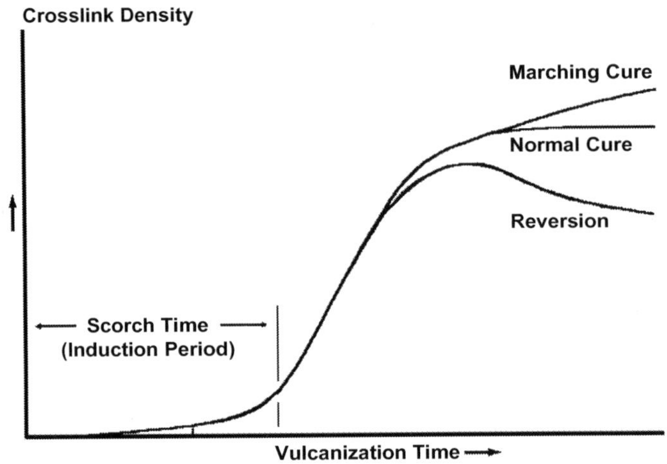

FIGURE 7.7 Rheometer cure curve.

The chemistry of unaccelerated vulcanization is controversial. Many slow reactions occur over the long period of vulcanization. Some investigators have felt that the mechanisms involved free radicals (Farmer and Shipley, 1946; Farmer, 1947a,b):

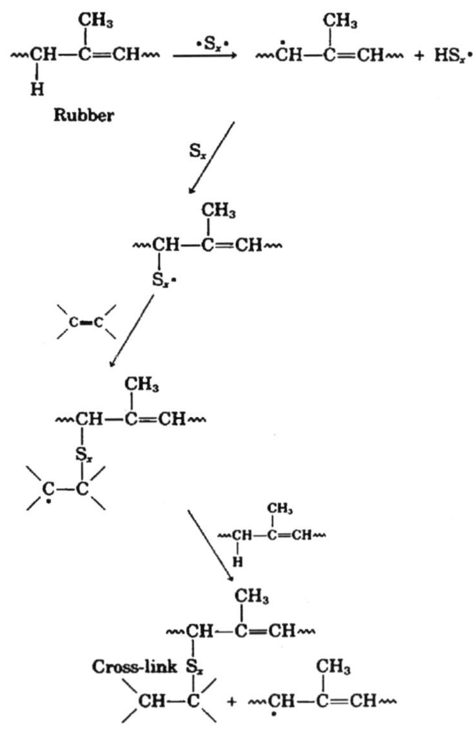

Other investigators have promoted ionic mechanisms (Bateman et al., 1958):

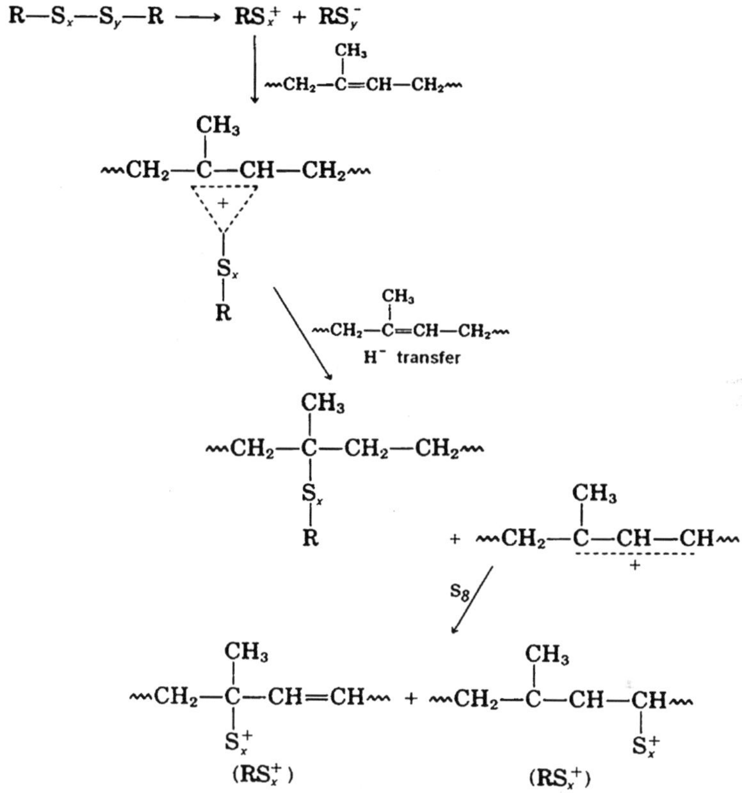

The intermediates

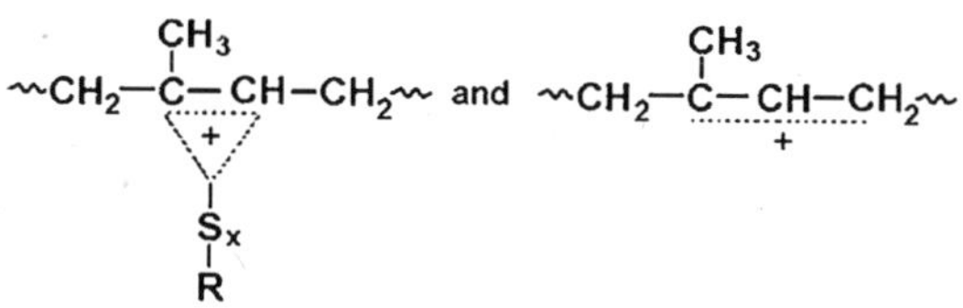

were proposed in Scheme 2 to explain the fact that model compound reactions gave both unsaturated and saturated products, sulfur atoms being connected to both secondary and tertiary carbon atoms.

7.6 ACCELERATED-SULFUR VULCANIZATION

Organic-chemical accelerators were not used until 1906 (65 years after the Goodyear-Hancock development of unaccelerated vulcanization; Figure 7.8),

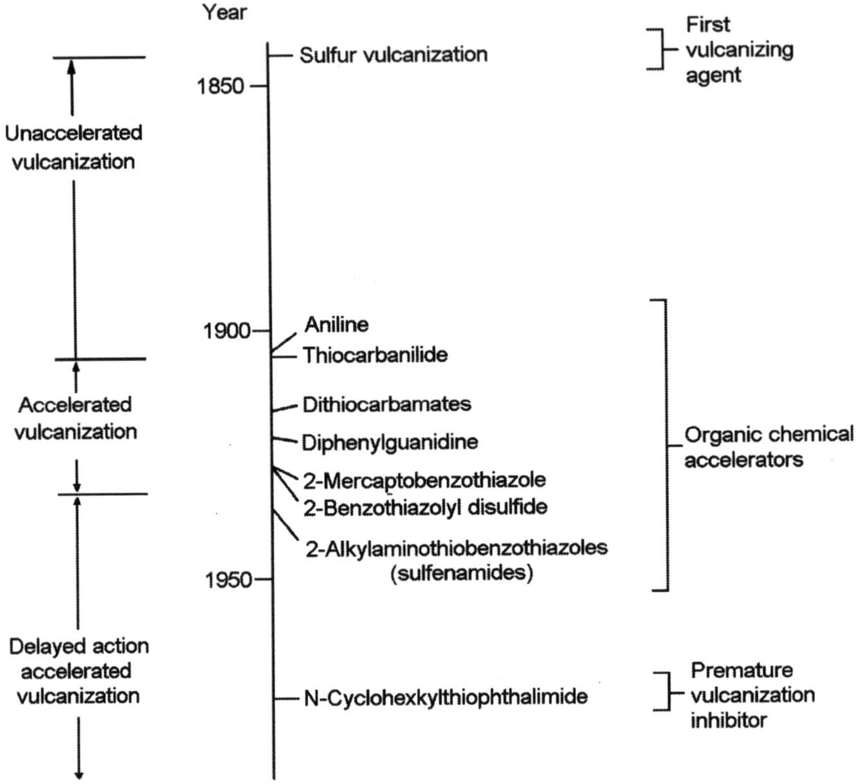

FIGURE 7.8 The history of vulcanization by sulfur.

when Oenslager discovered the effect of aniline on sulfur vulcanization (Oenslager, 1933). This could have been, at least partially, in response to the development of pneumatic tires and automobiles near the turn of the century. Aniline, however, is too toxic for use in rubber products. Its less toxic reaction product with carbon disulfide (i.e., thiocarbanilide) was introduced as an accelerator in 1907. Further developments led to guanidine accelerators (Weiss, 1922). Reaction products formed between carbon disulfide and aliphatic amines (dithiocarbamates) were first used as accelerators in 1919 (Malony, 1920). These were and are still the most active accelerators with respect to both crosslinking rate and extent of crosslink formation. However, most of the dithiocarbamate accelerators give little or no scorch resistance and their use is impossible in many factory-processing situations. The first delayed-action accelerators were introduced in 1925 with the development of 2-mercaptobenzothiazole (MBT) and 2-benzothiazole disulfide (or 2,2′-dithiobisbenzothiazole) (MBTS) (Bedford, 1921; Sebrell and Bedford, 1925; Bruni and Romani, 1921). This nearly coincided with the deployment of cord-ply construction (1920–1930), which enabled mass production of automobile tires. Even more delayed-action and yet faster curing vulcanization were possible in 1937 with the introduction of the first commercial benzothiazolesulfenamide accelerator (Zaucker et al., 1934; Harmon, 1937). Still more delay became possible in 1968, with the availability of an extremely effective premature vulcanization inhibitor (PVI). This compound was N-(cyclohexylthio)phthalimide (CTP), small concentrations of which were used along with benzothiazolesulfenamide accelerators. Figures 7.8–7.10 illustrate the history of the progress toward faster vulcanization, but with better control of premature vulcanization or scorch resistance.

Accelerated-sulfur vulcanization is the most widely used method. For many applications, it is the only rapid crosslinking technique that can, in a

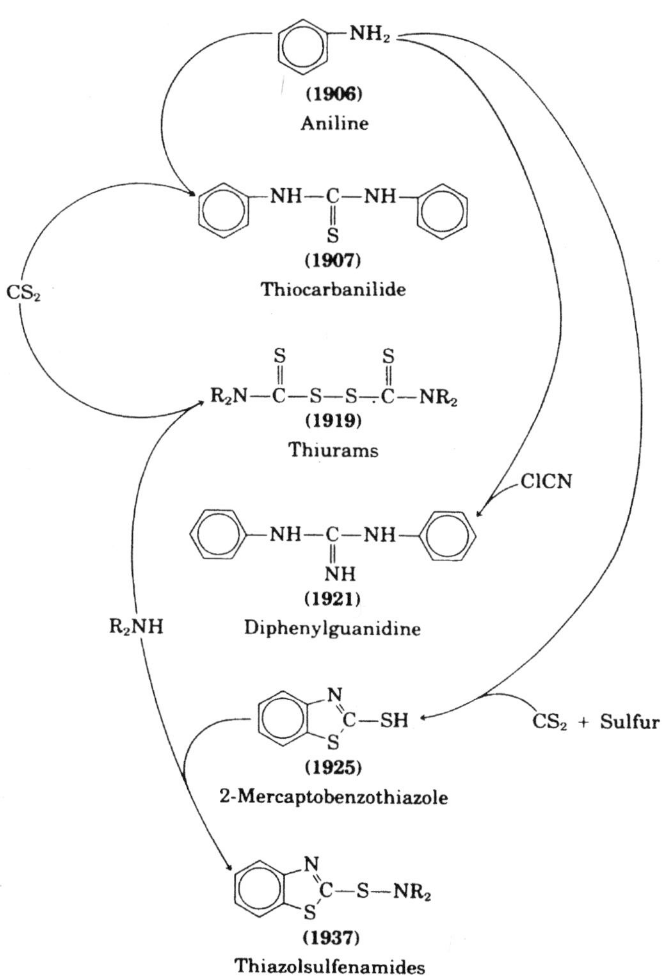

FIGURE 7.9 The chemistry of accelerator synthesis.

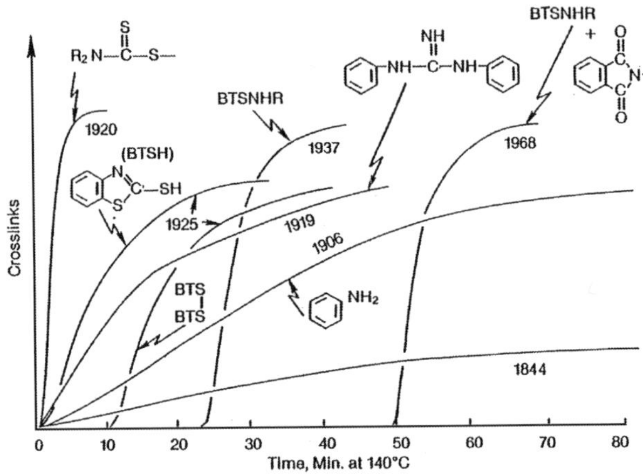

FIGURE 7.10 Improvements in the accelerated-sulfur vulcanization of natural rubber.

practical manner, give the delayed action required for processing, shaping, and forming before the formation of the intractable vulcanized network. It is used to vulcanize natural rubber (NR), synthetic isoprene rubber (IR), styrene-butadiene rubber (SBR), nitrile rubber (NBR), butyl rubber (IIR), chlorobutyl rubber (CIIR), bromobutyl rubber (BIIR), and ethylene-propylene-diene-monomer rubber (EPDM). The reactive moiety for all of these elastomers can be represented by

H
|
>C-C=CH-

Typically a recipe for the vulcanization system for one of these elastomers contains 2–10 phr of zinc oxide, 1–4 phr of fatty acid (e.g., stearic), 0.5–4 phr of sulfur, and 0.5–2 phr of accelerator. Zinc oxide and the fatty acid are vulcanization-system activators. The fatty acid with zinc oxide forms a salt, which can form complexes with accelerators and reaction products, formed between accelerators and sulfur. Accelerators are classified and illustrated in Table 7.1.

Frequently, mixtures of accelerators are used. Typically, a benzothiazole type is used with smaller amounts of a dithiocarbamate (thiuram) or an amine type. An effect of using a mixture of two different types of accelerators can be that each activates the other and better-than-expected crosslinking rates can be obtained. Mixing accelerators of the same type gives intermediate or average results.

We should note here that there is a need to reduce the use of accelerators based on secondary amines, which can react with nitrogen oxides to form suspected carcinogenic nitrosamines. This is especially a problem

TABLE 7.1 Accelerators for Sulfur Vulcanization

Compound	Abbreviation	Structure
Benzothiazoles		
2-Mercaptobenzothiazole	MBT	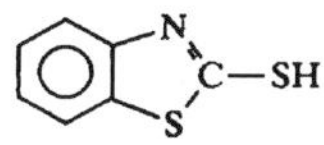
2,2′-Dithiobisbenzothiazole	MBTS	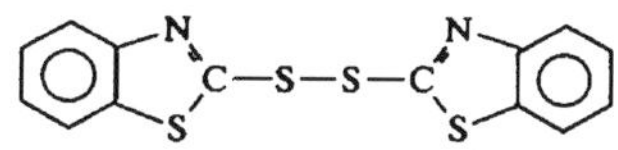
Benzothiazolesulfenamides		
N-Cyclohexylbenzothiazole-2-sulfenamide	CBS	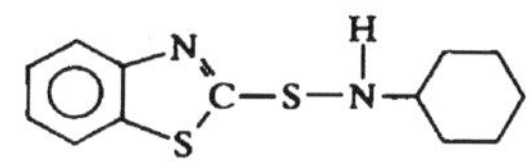
N-t-butylobenzothiazole-2-sulfenamide	TBBS	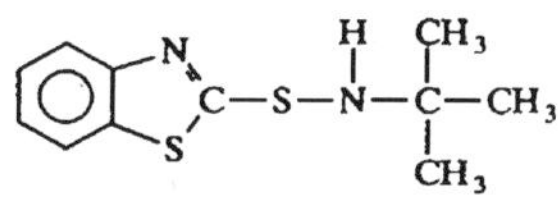
2-Morpholinothiobenzothiazole	MBS	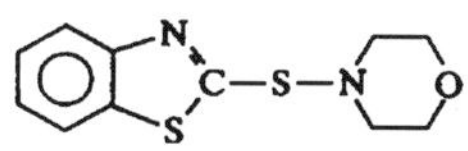
N-Dicyclohexylbenzothiazole-2-sulfenamide	DCBS	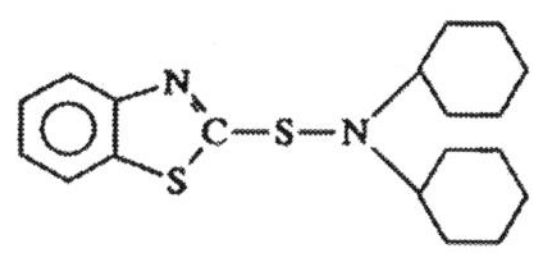
Dithiocarbamates		
Tetramethylthiuram monosulfide	TMTM	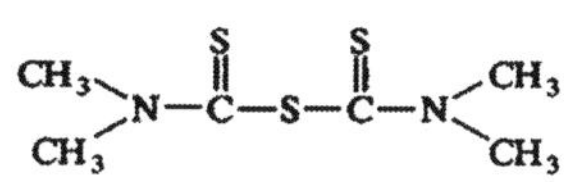
Tetramethylthiuram disulfide	TMTD	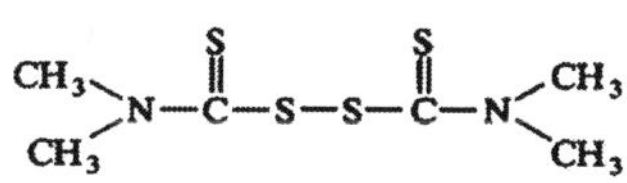
Zinc diethyldithiocarbamate	ZDEC	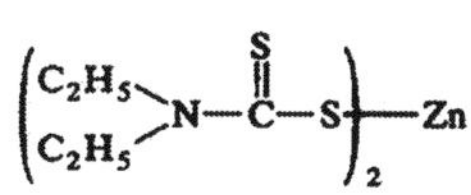
Amines		
Diphenylguanidine	DPG	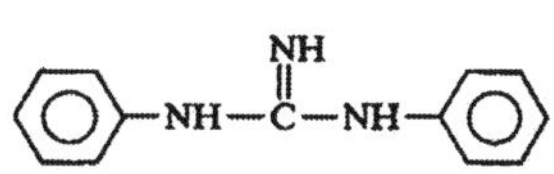
Di-o-tolylguanidine	DOTG	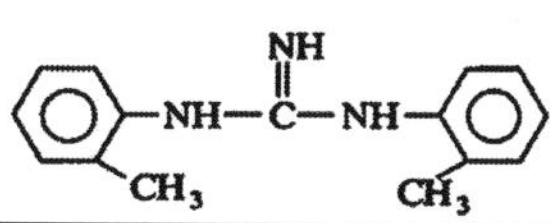

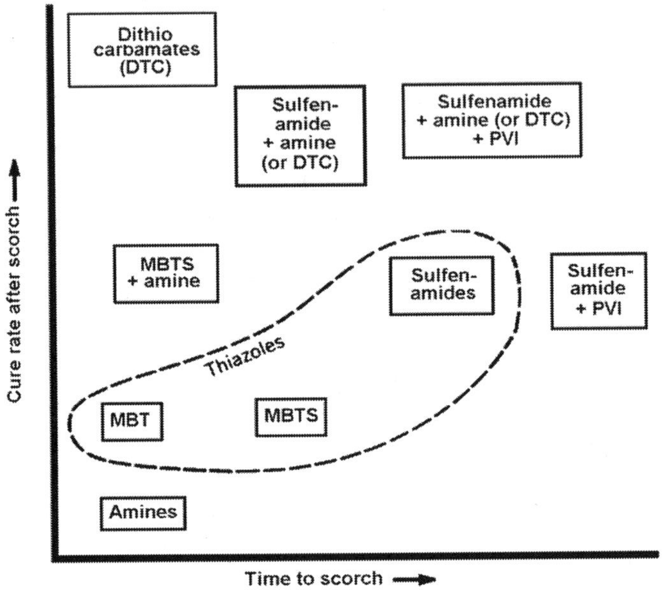

FIGURE 7.11 Vulcanization characteristics given by various accelerators and combinations.

with dithiocarbamate-type accelerators. Proposed accelerators, which do not give carcinogenic nitrosamine derivatives, include dibenzylamine-derived dithiocarbamates and those based on sterically hindered amines.

Different types of accelerators impart vulcanization characteristics that differ with respect to both scorch resistance and crosslinking rate. Figure 7.11 is a map of accelerator system characteristics. Within groups or types, differences can be obtained by choosing the individual accelerators. In the group of benzothiazolesulfenamides, the scorch resistance and vulcanization time increase in the order: TBBS or CBS, MBS, DCBS.

The effect of the addition of small concentrations of the premature vulcanization inhibitor (PVI), N-(cyclohexylthio)phthalimide, is also given by Figure 7.11. This retarder (Coran and Kerwood, 1970) is frequently used to independently control scorch resistance with little effect on the rate of crosslinking (Coran, 1978). Before the development of N-(cyclohexylthio)phthalimide as a PVI, acidic retarders like salicylic acid, acetylsalicylic acid, phthalic anhydride, and benzoic acid were used. These additives improved scorch resistance but also gave greatly reduced rates of crosslink formation after the delay. Another retarder of the past was N-nitrosodiphenylamine, which is less active and not now used because of toxicological concerns.

7.6.1 The Chemistry of Accelerated-Sulfur Vulcanization

The general reaction path of accelerated-sulfur vulcanization is thought to be as follows (Coran, 1978; Campbell and Wise, 1964a,b; Hu and Scheele, 1962; Morrison and Porter, 1984; Ghosh et al., 2003): Accelerator reacts with sulfur to give monomeric polysulfides of the structure Ac-S_x-Ac where Ac is an organic radical derived from the accelerator (e.g., benzothiazyl-). The monomeric polysulfides interact with rubber to form polymeric polysulfides (e.g., rubber-S_x-Ac). During this reaction, 2-mercaptobenzothiazole (MBT) is formed if the accelerator is a benzothiazole derivative and if the elastomer is natural rubber. (In SBR the MBT becomes bound to the elastomer molecular chain probably as the thioether rubber-S_x-Ac.) When MBT itself is the accelerator in natural rubber, it first disappears then reforms with the formation of rubber-S_x-Ac. Finally, the rubber polysulfides react, either directly or through an intermediate, to give crosslinks, rubber-S_x-rubber. A reaction scheme can be written as follows:

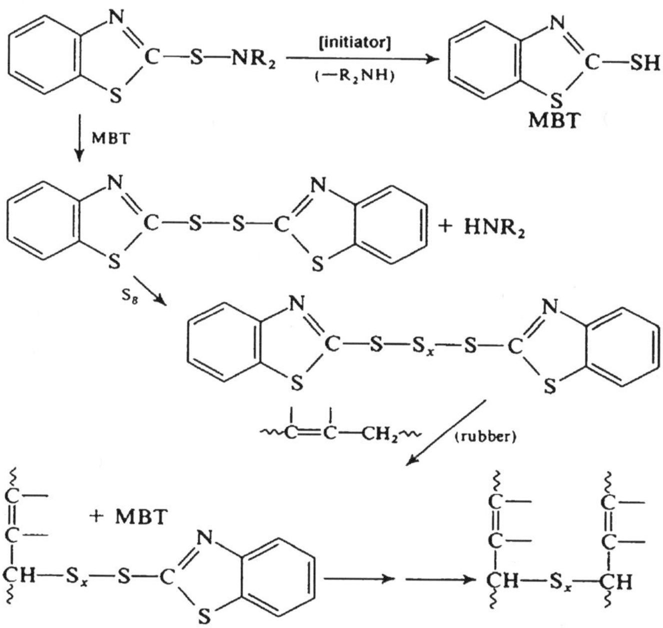

There are obvious differences between accelerated vulcanization and unaccelerated vulcanization. (Greater crosslinking efficiencies and greater crosslinking rates are obtained with accelerated vulcanization.) But there are more subtle differences. Results from model reactions with curing ingredients indicate that sulfur becomes attached to the rubber hydrocarbon almost exclusively at allylic positions (Skinner, 1972). This is not the case with unaccelerated-sulfur vulcanization, thus:

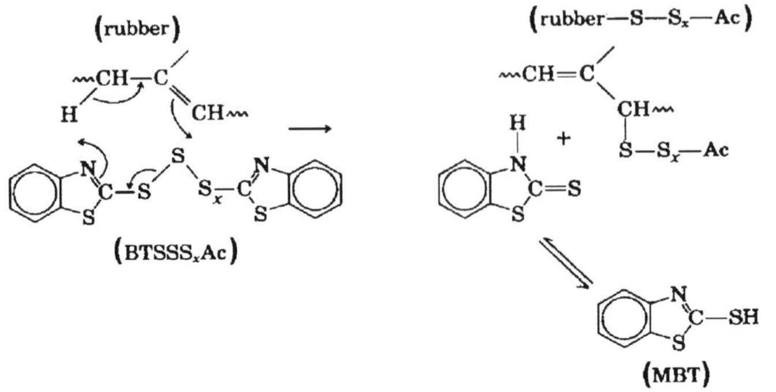

Rather than the eight-membered-ring intermediate shown above, one could propose a six-membered-ring sulfurization intermediate:

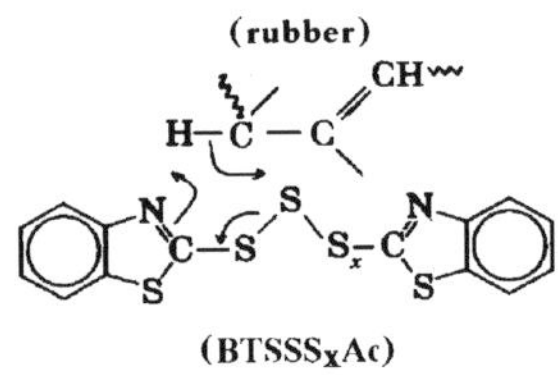

This is similar to what was suggested by Nieuwenhuizen et al. with respect to their work on dithiocarbamate acceleration (Nieuwenhuizen et al., 1997):

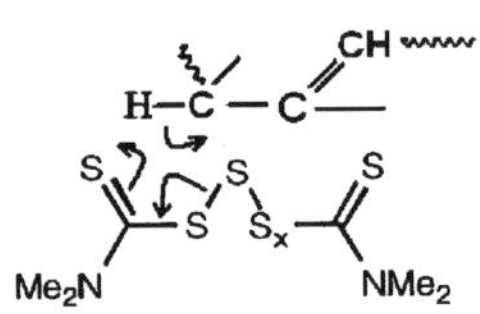

Others have proposed that zinc must be present for sulfuration (Ignatz-Hoover et al., 1999). At any rate, crosslinks could then form, in a number of ways; for example:

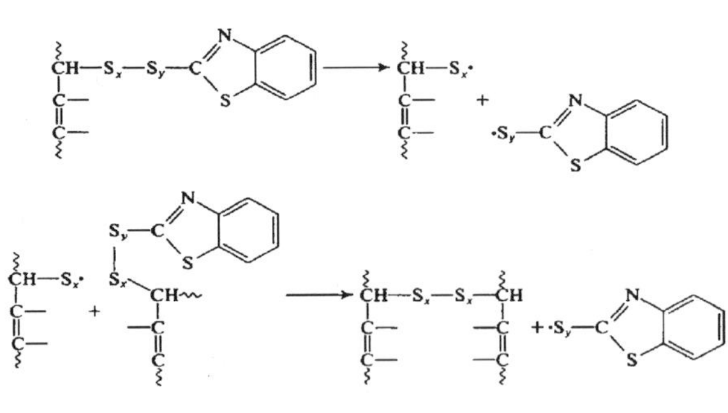

By using solid-state C-13 NMR spectroscopy, Koenig and his group (Koenig, 2000; Pellicioli et al., 2002) have added much detail to the chemical structure of the sulfurated network backbone. Following are the types of structures that have been assigned to the attachment points of the sulfur atoms to the rubber molecular backbone:

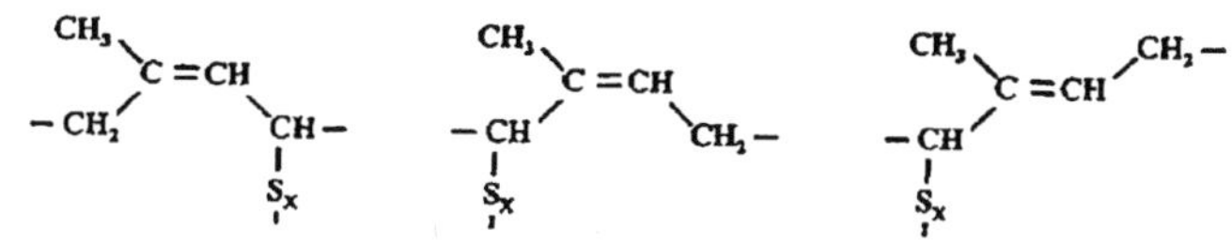

A1c B1c B1t

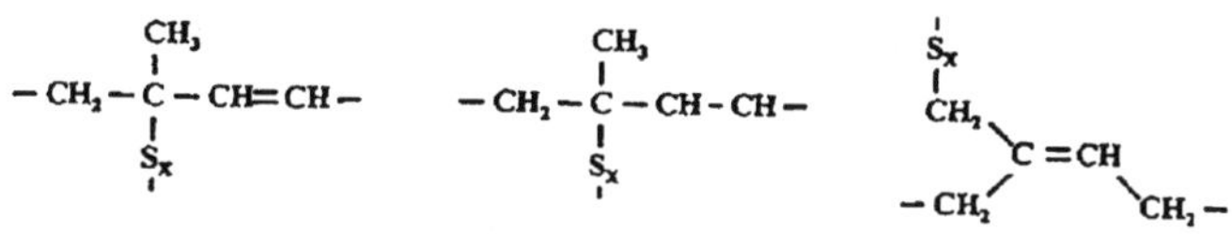

Unsaturated A2 Saturated A2 C1c

Koenig's group has done much work on conditions that change the relative amounts of the various types of attachments. For example, concentrations of both the B1c- and B1t-type polysulfides increase with the level of carbon black loading (for types N110, N220, N326, N330, N550, and N765) (Mori and Koenig, 1995).

7.6.2 Delayed-Action Accelerated Vulcanization

If crosslink formation is by a free radical mechanism, delayed action could be the result of a quenching action by the monomeric polysulfides formed by reactions between accelerator and sulfur. If the polymeric polythiyl radicals (crosslink precursors) are rapidly quenched by an exchange reaction before they are able to form crosslinks, crosslink formation would be impeded until substantial depletion of the monomeric polysulfides (Coran, 1978). This is illustrated as follows:

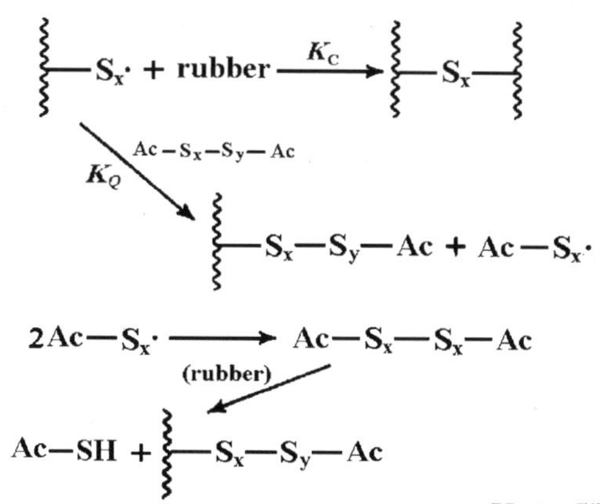

$$K_Q \gg K_C$$

Thus, one theory for delayed action is the quenching of free radical crosslink precursors by monomeric polysulfides. It has been found that, if bisalkylpolysulfides are mixed with uncured rubber stocks, more delay results. It is also been shown that the early reaction products formed by the interaction between accelerator and sulfur (Ac-S_x-Ac) are inhibitors of crosslink formation. The very substances that give rise to the formation of the crosslink precursor (rubber-S_x-Ac) inhibit the formation of the crosslinks. We note that other mechanisms for delayed action have been proposed. In the case of acceleration by benzothiazolesulfenamides, the accelerator is depleted in an autocatalytic fashion with the formation of 2-mercaptobenzothiazole (MBT). The rate of this depletion is about proportional to the amount of MBT present. There is strong evidence, which indicates that the following reactions occur in sulfenamide-accelerated systems:

BT-S-NHR + **BT-SH** ⟶ **BT-S-S-BT** + **RNH_2**

sulfenamide accelerator | MBT | MBTS

If MBT could be taken out of the system as fast as it forms, substantial increases in processing safety would result. Such is the case when the premature vulcanization inhibitor, N-(cyclohexylthio)phthalimide (CTP), is present. This compound (Coran and Kerwood, 1970) and others like it react rapidly with MBT to form 2-(alkyldithio)benzothiazoles, R-S-S-BT, which are active accelerators but which do not interact rapidly with the sulfenamide accelerator:

L-S-R + **BT-SH** ⟶ **BT-S-S-R** + **L-H**

premature vulcanization inhibitor | MBT | 2-(alkyldithiyl)-benzothiazole | inactive compound

where L is a "leaving group" of the premature vulcanization inhibitor (e.g., N-phthalimido- for CTP).

Gradwell and Stephenson have also studied prevulcanization inhibition by using CTP (Gradwell and Stephenson, 2004).

The importance of scorch control cannot be overemphasized. Present-day tire plants could not compete without good control of scorch resistance or processing safety as it is commonly called. Such safety is necessary in order to rapidly process rubber mixes at high temperatures (through extrusion, calendaring, etc.) into preforms for molding (e.g., tire components).

Delayed-action mechanisms and reaction kinetics have also been discussed and reviewed elsewhere (Ignatz-Hoover et al., 1999; Koenig, 2000;

Pellicioli et al., 2002; Mori and Koenig, 1995; Fan et al., 2002; Coran, 1983; Leib et al., 1188; Son, 1973; Sullivan et al., 1061; Ding et al., 1996; Choi, 2006; Gradwell and Stephenson, 2004; Arem et al., 1999; Heideman et al., 2004).

7.6.3 The Role of Zinc in Benzothiazole-Accelerated Vulcanization

An increase in the concentration of fatty acid and hence increases in the concentration of available Zn++ causes an increased overall rate in the early reactions (during the delay period), which lead to the formation of rubber-S_x-Ac. However, it gives rise to a decrease in the rate of crosslink formation but an increase in the extent of crosslinking (Coran, 1965). The increase in the rates of the early reactions has been explained by the interaction:

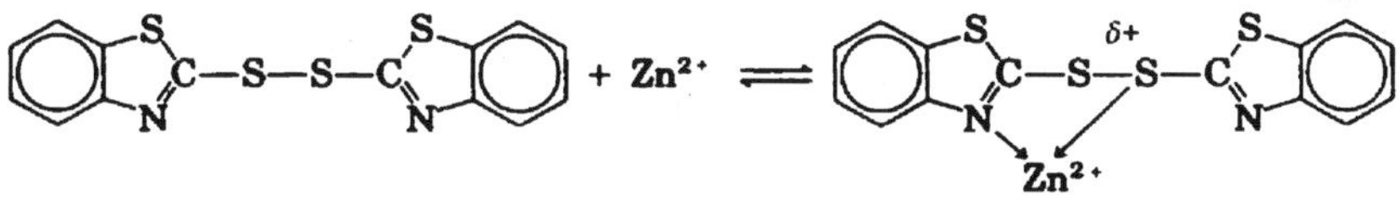

where the chelated form of the accelerator is more reactive than the free accelerator during the early reactions:

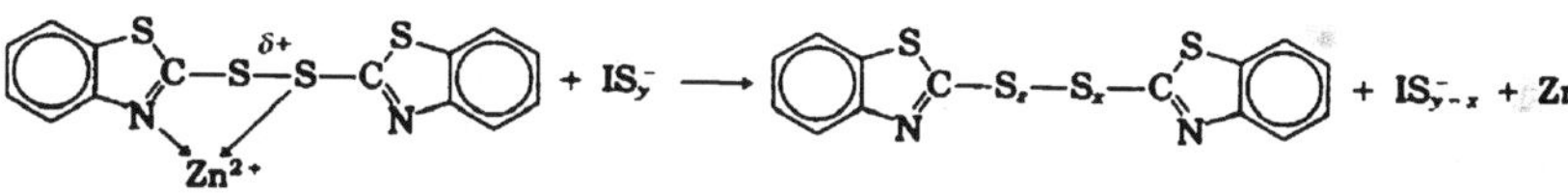

Here, I-S_y is an ionized form of linear sulfur. It could be rapidly formed in a reaction between sulfur and any number of initiating species. Others have proposed that the presence of Zn++ can increase the rate of sulfurization through the formation of complexes of the type:

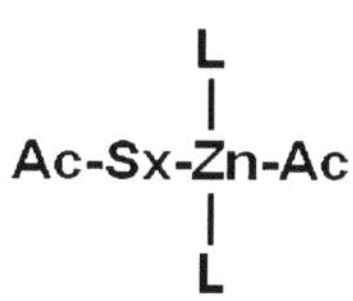

where L is a ligand such as an amine molecule (Ignatz-Hoover et al., 1999; Choi, 2006).

The decreased specific rate of crosslink formation, and the increased extent of crosslinking due to the presence of Zn++ in benzothiazole-accelerated vulcanization, have been explained by the following scheme (Coran, 1965):

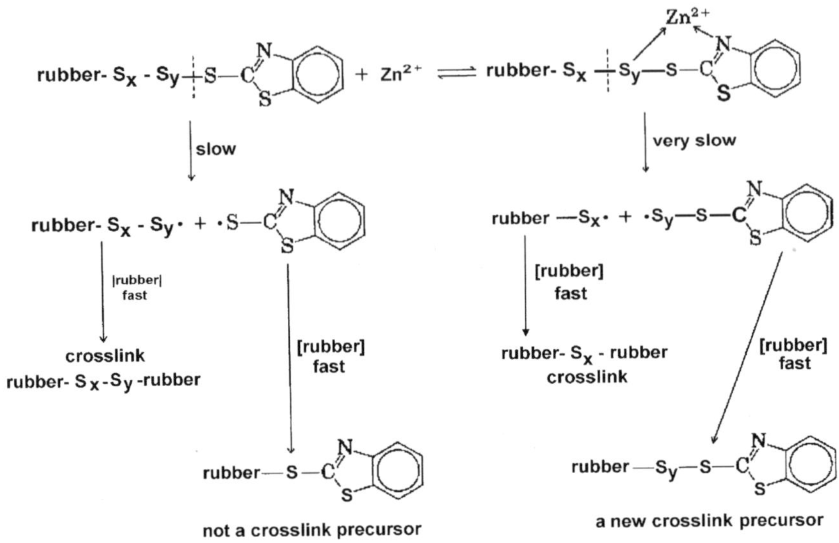

Zinc chelation changes the position of the S-S bond most likely to break. Since a stronger bond must break, the rate is slower. Though the rate of crosslinking is slower, the extent of crosslink formation is increased since less sulfur is used in each crosslink. That is, the crosslinks are of lower sulfidic rank.

The presence of zinc compounds can also promote the reduction of the sulfur rank of crosslinks during high-temperature aging of the vulcanizate, for example, during reversion (Layer, 1992). In some cases zinc compounds actually promote the decomposition of crosslinks (Morrison and Porter, 1984).

More recently, Heideman et al. have particularly discussed the role of the zinc activator system (ZnO + fatty acid) (Heideman et al., 2004).

7.6.4 Achieving Specified Vulcanization Characteristics

It has been difficult, for many years, to independently control the two main vulcanization characteristics, scorch resistance (processing safety) and rate of crosslink formation. In the case of natural rubber (NR), if one had chosen a fast accelerator system in order to obtain short curing times in the press, then process safety would have suffered greatly. If one had chosen a delayed-action acceleration system, then the rate of vulcanization in the press would have been limited. The development of the highly efficient premature vulcanization inhibitor N-(cyclohexyl-thio)-phthalimide (CTP) changed all of that, since great improvements in scorch resistance with little or no change in crosslinking rate became possible (Trivette et al., 1977). Thus, the rate of crosslink formation can be adjusted by the selection of accelerators. For example, the moderately fast delayed-action accelerator t-butylbenzothiazolesulfenamide (TBBS) can be partially replaced by a small amount of a coaccelerater (e.g., 0.1–0.2 phr of tetramethylthiuram disulfide [TMTD] or tetramethylthiuram monosulfide [TMTM]) to obtain a greatly increased cure rate; however, the scorch resistance will be significantly reduced. In such a case, the scorch resistance can be regained by the addition of 0.05–0.25 phr of CTP, without a noticeable decrease in the rate of crosslinking (Coran, 1978; Rodger and Roebuck, 1976).

It is true that merely increasing the concentration of TBBS will give an increase in cure rate with only a small change in scorch resistance. However, the increase in accelerator concentration will generally be rather large and the concentration of sulfur will be adjusted downward to keep the hardness and stiffness constant (maintaining constant crosslink density). The relatively large change in the concentrations of sulfur and accelerator will cause changes in vulcanizate-performance properties. (See the later section on vulcanizate properties.)

In rubber mixes containing only a synthetic rubber, such as styrene-butadiene rubber (SBR) or butadiene rubber (BR), the effects of cure-system changes may not be as pronounced as they are in the case of NR. However, if even a relatively small amount of NR is present, the effects of cure-system changes on the vulcanization process parameters resemble those obtained with NR alone.

One of the curing characteristics that we would like to control is reversion, which can occur in compounds containing natural rubber. There is more than one approach to reducing the amount of reversion. We can use sulfur donors or increase the ratio of accelerator concentration to sulfur concentration, or we could carry out the vulcanization at a reduced temperature for a longer period. However, these approaches give rise to effects that will have to be compensated. Another approach is use of additives such as certain bisimides, e.g., N,N′-*m*-phenylene-biscitraconimide and N,N′-*m*-phenylene-bismaleimides (Datta et al., 1997, 1998; Datta and Ivan, 1995) or trialkoxysilylalkylpoly-sulfides such as bis-(3-triethoxisilylpropyl)-tetrasulfide (Tan and Wolff, 1985).

7.6.5 Effects on Adhesion to Brass-Plated Steel

The adhesion between rubber and brass-plated steel (e.g., steel tire cords for belted radial tires) has been the subject of much study and speculation. Brass plating is the major method of obtaining adhesion between natural rubber and the steel of tire cords. Over the years there has been much speculation about its mechanism, but there is agreement on one aspect of the adhesion of natural rubber to brass-plated steel: the actual adhesion between the natural rubber and the brass-plated cord, formed *in situ* during the vulcanization process, is an interfacial layer of sulfides and oxides of copper (Buchan, 1959; van Ooij, 1979, 1984).

The adhesive layer between the rubber and cord is generally considered to be formed by the interaction between the copper and the vulcanization system. As a result of this, optimization of the vulcanization system with respect to adhesion is critical. Also, a change in the composition of the brass coating on the steel wires, or a change in the thickness, can require a change in the vulcanization system in order to maintain the optimum level of adhesion.

van Ooij, who has done much of the work in the field, has written reviews on the subject of brass-plated steel cord/natural rubber adhesion. van Ooij (1979, 1984) has given a model for rubber-brass adhesion in which a copper sulfide layer forms on the brass before the onset of crosslink formation. The thin film of copper sulfide has good adhesion and cohesion. In addition, the film is so porous that rubber molecules can become entangled with it. It is not required that the film forms simultaneously with the formation of crosslinks during vulcanization; but, rather, it is required that the copper sulfide film be completely formed before crosslinking starts. Indeed, adhesion between brass-plated steel and natural rubber can frequently be improved by the use of the retarder, CTP (Coran, 1978), or by using a more delayed-action accelerator such as N-dicyclohexylbenzothiazole-2-sulfenamide (DCBS) (Albrecht, 1973).

Failure rarely occurs between the rubber and the copper sulfide film. It generally occurs cohesively within the sulfide film or adhesively in a layer below the sulfide film.

Sulfidation of the brass surface is not due to its interaction with elemental sulfur, but it is the result of the interaction between the brass surface and accelerator-sulfur reaction products, which can be represented by the general structure, Ac-S_y-Ac and Ac-S_y-H, where Ac is an accelerator-derived moiety (e.g., benzothiazolyl group). The value of the subscript, y, increases with the ratio of the concentration of sulfur to the concentration of accelerator used in the curing system. Generally, high sulfur levels and high ratios of sulfur concentration to accelerator concentration favor good rubber-to-brass adhesion.

Recently, van Ooij et al. have reviewed adhesion of steel tire cord to rubber (van Ooij et al., 2009). The authors reviewed the literature extensively and provided an updated model for adhesion to brass-plated tire cord, which incorporated observations made by many techniques. They discussed the effects of different compounding ingredients and the possible alternatives to the current brass coatings. They note that the use of cobalt compounds improves the adhesion between rubber and brass-coated cords, but new adhesion promoters have been developed as replacements for Co, or for combined use with Co. They also discussed the use of phenolic-resin adhesion promoters. They describe the various techniques that have been developed to study the rubber-brass interface and its aging mechanism.

7.6.6 The Effect on Vulcanizate Properties

Increases in sulfur and accelerator concentrations give higher crosslink densities and, therefore higher moduli, stiffness, hardness, and so on. However, as the ratio of the concentration of accelerator to the concentration of sulfur increases, the proportion of monosulfidic crosslinks increases in natural rubber

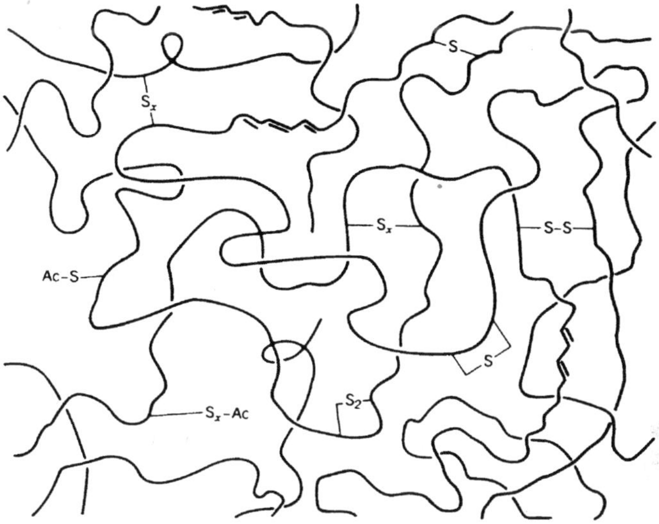

FIGURE 7.12 Crosslink types and chain modifications.

stocks (also called rubber compounds). Greater amounts of accelerator (with respect to sulfur) also give an abundance of pendant groups of the type, -S_x-Ac, which are attached to and "dangle" from the rubber molecular chains. Higher ratios of sulfur concentration to accelerator concentration give both more polysulfide crosslinks and more sulfur combined with the rubber chains to form sulfur-containing six-membered heterocyclic rings along the rubber molecular chains. In addition, conjugated olefinic double bonds appear along the polymer backbone chain. Figure 7.12 indicates these features. Such changes in the vulcanizate network structure, no doubt, are responsible for changes, which occur in vulcanizate properties as a result of changes made in the curing-system recipe (Studebaker, 1966; Moore et al., 1961; Moore and Trego, 1961; Skinner and Watson, 1969; Russell et al., 1969; Lewis, 1986).

Effects of changes in the concentrations of accelerator and sulfur on vulcanizate properties have been studied by using the following recipe (parts by weight): natural rubber, 100; N330 carbon black, 50; N-isopropyl-N′-phenyl-*p*-phenylenediamine (IPPD antidegradant) 2; zinc oxide, 5; stearic acid, 3; plasticizer, 3; sulfur, variable; N-cylohexylbenzothiazolesulfenamide (CBS), variable (Coran, 1978).

The effects of changes in the accelerator concentration on 300% modulus (jargon for stress at 300% tensile strain), thermal-oxidative aging and fatigue life (DeMattia flex crack) are given in Figure 7.13. The effects on 300% modulus are indicated by the diagonal contours of negative slope. They are parallel and show that the stress at 300% strain increases with an increase in either sulfur or accelerator concentration. The contours for % retention of ultimate

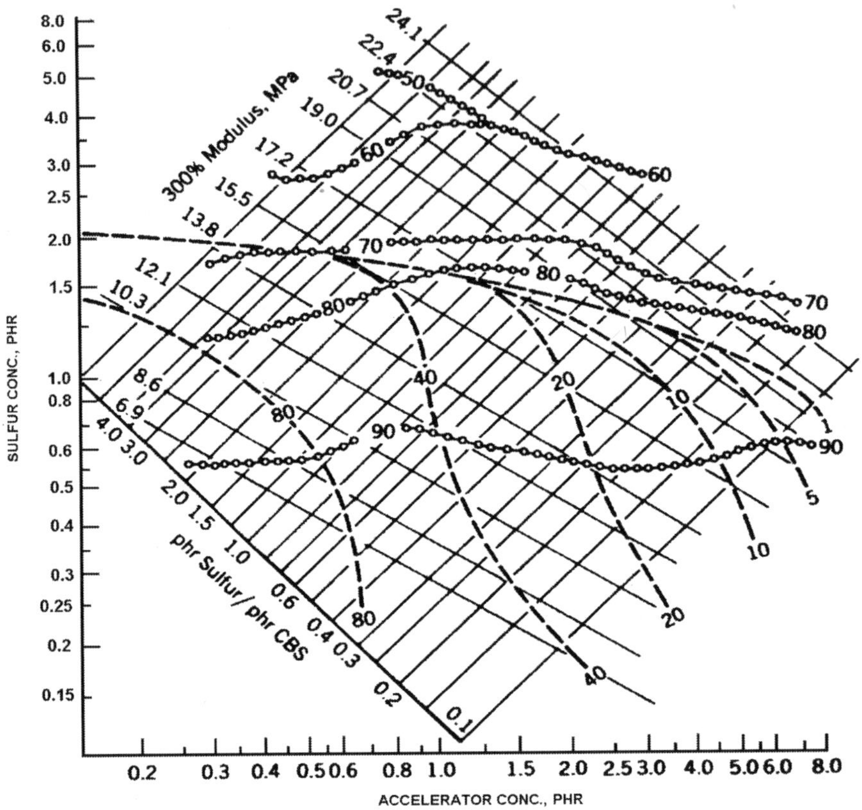

FIGURE 7.13 Vulcanizate properties: —, 300% modulus (Mpa); - - -, DeMattia flex fatigue life (kHz × 10^-1); -o-o-o-, % retention of ultimate elongation after 2 days at 100°C.

elongation after hot air aging (at 100°C for 2 days) indicate that oxidative aging, in the presence of IPPD, depends only on the concentration of sulfur. Higher concentrations of sulfur give poor aging characteristics in correlation with the higher number of points of chain sulfuration. This suggests that sulfur substitution along the chain can activate chain scission by reactions with oxygen (Lorenz, 1963). Another view is that sulfur interferes with the antidegradant activity (in this case with IPPD) (Bell and Cuneen, 1967).

The contours for flex fatigue life are complex. The test is run such that the specimens are about equally strained; however, there is some question as to whether the tests should be run at equal strain or at equal strain energy (Gent et al., 1964; McCall, 1969). For some cases, where strain is restricted by fabric reinforcement, fatigue test data should be compared at equal strain amplitude. For other applications, where the strain is not limited, the tests should be run at equal strain energy. The contours as presented here can be interpreted in terms of either constant strain or constant strain energy: All points on the chart can be compared at an approximately equal strain per cycle; however, if we interpolate

TABLE 7.2 Fatigue Life at Constant 300% Modulus

phr Sulfur/phr Sulfenamide	Flex Life, kc
6	400
5	500
4	530
3	550
2	550
1	400
0.6	350
0.4	270
0.3	250
0.2	190

between the flex-life contours but only along a constant modulus contour, we can extract values corresponding to approximately equal strain energy per cycle. By choosing higher modulus contours, we are considering higher strain energies.

Considering the group of flex-life contours as a whole, or at approximately constant strain energy per cycle, we may conclude the following: a high level of either sulfur or accelerator gives poor flex life. However, by the selection of the proper ratio of sulfur concentration to accelerator concentration, higher modulus vulcanizates can be obtained with at least some optimization of fatigue life. The flex life at approximately equal strain energy per cycle can be illustrated by extracting values along the 13.8 Mpa modulus contour line. Table 7.2 can then be constructed.

For the strain energy corresponding to a 300% modulus of 13.8 Mpa, the maximum flex life (as measured by the DeMattia flex test) is obtained when 2.5 times as much sulfur as accelerator is used. Other optimum ratios for various 300% moduli can be obtained from the contours. Some of these are given in Table 7.3.

These optimum ratios and fatigue life data should not be considered to be universal values. Different recipes, different types of antidegradants, different types of fillers, different concentrations of antidegradants and fillers, different base polymers, different types of fatigue tests, and such give rise to different optimum sulfur concentration/accelerator concentration ratios and different optimum fatigue life values. Nevertheless, the trends given here have been generally noted.

The effects of concentrations of sulfur and accelerator on network structure and vulcanizate properties have been studied (McCall, 1969; Datta et al., 2007; Fan et al., 2001; Dijkhuis et al., 2009; Campbell, 1409; Cooper, 1958). The low values for fatigue life at low levels of sulfur, but high levels of accelerator,

TABLE 7.3 Optimized Fatigue Life

Stress at 300% Strain, 300% Modulus, MPa	Optimum Sulfur Conc./ Accelerator Conc. Ratio	Optimized Flex Life, kc
6.9	3.50	800
10.4	3.00	800
13.8	2.50	550
15.5	1.00	300
17.2	0.45	120
19.0	0.27	70

have been attributed to high concentrations of accelerator-terminated appended groups (Campbell, 1409) and high concentrations of monosulfidic crosslinks. Monosulfidic crosslinks are not able to exchange, rearrange, or break to relieve stresses without the breakage of main chains.

On the other hand, polysulfidic crosslinks are able to rearrange under stress (Bateman et al., 1963; Coran, 1978; Cooper, 1958). The rearrangement of a crosslink occurs in two steps: (1) breaking and (2) reforming. Work by Brown et al. (1987) indicates that only the breaking of the weak polysulfide crosslinks is required for the strengthening of the vulcanizate network. It is only required that enough of the crosslinks be weak (in comparison to backbone bonds) for the rubber to be strong. At any rate, when moderately high concentrations of sulfur (with respect to accelerator) are used, flex life improves, presumably due to the presence of enough weak or rearrangeable polysulfidic crosslinks.

When even higher concentrations of sulfur are used (with the maintenance of constant modulus), flex life decreases. It is possible that this is due to the large amount of cyclic chain modification associated with high levels of sulfur.

Natural rubber compositions, vulcanized by high levels of accelerator and low levels of sulfur, have been called EV and semi-EV vulcanizates. Here, "EV" means efficient vulcanization, since sulfur is used efficiently in the production of crosslinks. On the average, the crosslinks are shorter than in the case of conventional vulcanization; they contain more monosulfidic crosslinks and less polysulfidic ones, or their average sulfur rank is lower. Though EV vulcanizates suffer with respect to fatigue resistance, they are frequently used because of their excellent reversion resistance (resistance to nonoxidative thermal aging or overcure) and good resistance to thermal-oxidative aging. The resistance to reversion or thermally induced loss of crosslinks is thought to be the result of the greater intrinsic stability of the lower rank (disulfidic and especially monosulfidic) crosslinks. Semi-EV vulcanizates (wherein the sulfur concentration to accelerator-moiety concentration ratio is at an intermediate value) are an advantageous compromise in which fairly good unaged fatigue life is obtained, but maintained after aging.

Rather than using high levels of accelerator to obtain EV and semi-EV vulcanizates, it is sometimes advantageous to replace some of the sulfur with a so-called sulfur donor. Examples of these are tetramethylthiuram disulfide (TMTD) and 4,4′-dithiodimorpholine (DTDM).

This type of vulcanization-system design was reported by McCall (1969). He found that by judiciously balancing the levels of accelerator, sulfur, and DTDM, he could obtain good vulcanization characteristics, good thermal stability, good flex life, and superior retention of flex life. Others have reported on more recent work on the effects of crosslink type on reversion (Datta et al., 2007; Fan et al., 2001).

7.6.7 Accelerated-Sulfur Vulcanization of Various Unsaturated Rubbers

Over the years, much of the research on accelerated-sulfur vulcanization was done by using natural rubber as a model substrate. Natural rubber was the first elastomer and therefore the search for the understanding of vulcanization originated with work on natural rubber. Most of the work cited in the previous sections is related to natural rubber. However, some rather early studies have been directed to the vulcanization of butadiene 1,4-polymers (Skinner and Watson, 1969; Wolfe et al., 1329; Gregg and Katrenick, 1970). More recent is the work of Pellicioli and coworkers. Early basic work on the vulcanization of ethylene-propylene-diene-monomer rubber (EPDM) has been carried out (van den Berg et al., 1984a,b). Recently, Kuno and coworkers did basic work on EPDM networks. They found that, essentially, the vulcanizate properties depend only on the crosslink density, not on the type of curing system (Dijkhuis et al., 2009).

The chemistry of the accelerated vulcanization of BR, SBR, and EPDM appears to have much in common with the vulcanization of natural rubber: Before the formation of crosslinks the rubber is first sulfurated by accelerator-derived polysulfides (Ac-S_x-Ac) to give macromolecular, polysulfidic intermediates (rubber-S_x-Ac). However, whereas in the case of MBTS- or benzothiazolesulfenamide-accelerated sulfur vulcanization of natural rubber, MBT is given off during the formation of rubber-S_x-BT from the attack of rubber by BT-S_x-BT, in the case of BR and SBR, MBT is not eliminated and remains unextractable presumably because it becomes bound as the macromolecular thioether rubber-S_x-BT. (BT is a 2-benzothiazolyl group.) As in the case of natural rubber, the average length of a crosslink (its sulfidic rank, the value of x in the crosslink, rubber-S_x-rubber) increases with the ratio of sulfur concentration to accelerator concentration (S/Ac) used in the compounded rubber mix. However, in the case of BR or SBR, the crosslink sulfidic rank is not nearly as sensitive to S/Ac as it is in the case of natural rubber. Model compound studies of the vulcanization of EPDM (e.g., wherein ethylidenenorbornane was used as a model for EPDM) (van den Berg et al., 1984a,b) indicate that the polysulfidic rank of the EPDM crosslinks probably responds to changes in

S/Ac in a natural rubber-like fashion. A difference here is that evidence for model monosulfidic crosslinks was lacking while model disulfidic crosslinks were more apparent than in the case of natural rubber vulcanization.

Reversion (when defined as the loss of crosslinks during nonoxidative thermal vulcanizate aging) is a problem associated mainly with natural rubber or synthetic isoprene polymers. It can occur only under severe conditions in butadiene rubber; in SBR, instead of the softening associated with the nonoxidative aging of natural rubber, we can observe hardening (the so-called marching modulus) during extensive overcure. In natural rubber and synthetic isoprene-polymer rubbers, the crosslinks tend to be more polysulfidic than in the case of BR or SBR. The highly polysulfidic crosslinks are more heat-labile than their lower rank cousins in BR and SBR; they are more likely to break and then form cyclic chain modifications. But the reason for the formation of the crosslinks of higher polysulfidic rank in isoprene rubbers than in butadiene polymers is grandly elusive, though it almost has to be related to the methyl groups, which are substituents along the isoprene-polymer chains but which are absent from butadiene-polymer chains.

The effect of zinc is much greater in the vulcanization of isoprene rubbers than it is in the vulcanization of BR and SBR. Again, the reason for the difference is not known, but a strong speculation is that this difference is also related to the presence of methyl groups only in the case of the isoprene rubbers.

7.6.8 Selected Accelerated-Sulfur System Recipes

Examples of recipes are given in Table 7.4. These recipes are not intended as ultimate solutions to compounding problems. Variations will undoubtedly be necessary to meet particular requirements.

7.7 VULCANIZATION BY PHENOLIC CURATIVES, BENZOQUINONE DERIVATIVES, OR BISMALEIMIDES

Diene rubbers such as natural rubber, SBR, and BR can be vulcanized by the action of phenolic compounds (van der Meer, 1943; Thelamon, 1963; Giller, 1966; Van Duin and Souphanthong, 1995), which are (usually di-)substituted by $-CH_2$-X groups where X is an –OH group or a halogen atom substituent. A high-diene rubber can also be vulcanized by the action of a dinitrosobenzene, which forms *in situ* by the oxidation of a quinonedioxime (Rehner and Flory, 1946; Martell and Smith, 1962; Sullivan, 1966; Baker et al., 1962; Gan and Chew, 1983), which had been incorporated into the rubber along with the oxidizing agent, lead peroxide.

The attack upon rubber molecules by the vulcanization system Can be visualized in a way similar to that which was postulated for the sulfurization of the rubber molecules by the action of accelerated-sulfur vulcanization systems. Reaction schemes for these two types of vulcanization can be written as follows:

TABLE 7.4 Recipes for Accelerated-Sulfur Vulcanization Systems[a]

			Nitrile (NBR)		Butyl (IIR)	EPDM
	NR	SBR	1	2		
Zinc oxide	5.00	5.00	3.00	2.00	3.00	5.00
Stearic acid	2.00	2.00	0.50	0.50	2.00	1.00
Sulfur	2.50	1.80	0.50	0.25	2.00	1.50
DTDM[b]	–	–	–	1.00	–	–
TBBS[b]	0.60	1.20	–	–	–	–
MBTS[b]	–	–	2.00	–	0.50	–
MBT[b]	–	–	–	–	–	0.50
TMTD[b]	–	–	1.00	1.00	1.00	1.50
Vulcanization conditions[c]						
Temperature (°C)	148	153	140	140	153	160
Time (min)	25	30	60	60	20	20

[a]*Concentrations in phr.*

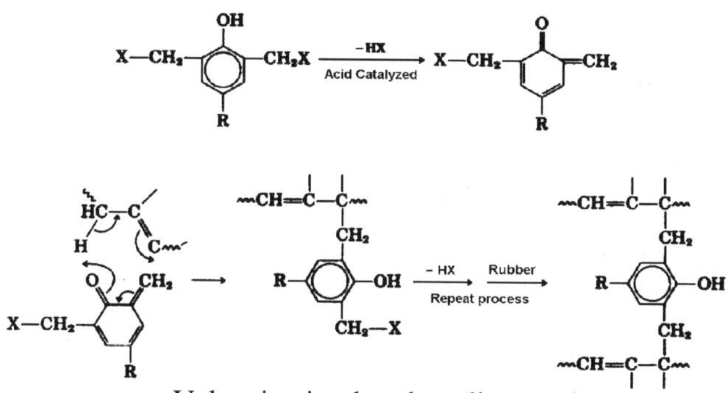

Vulcanization by phenolic curatives

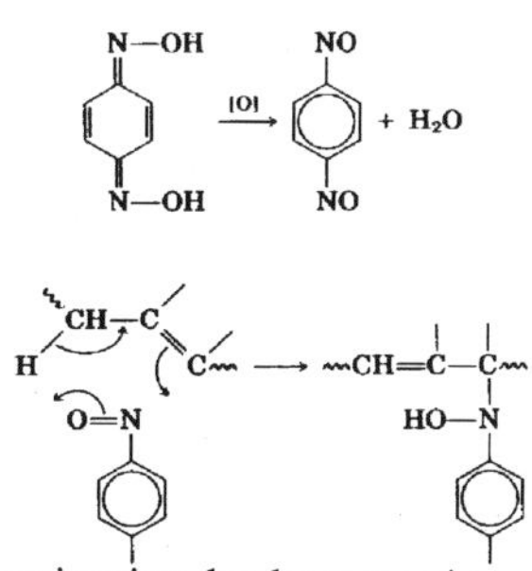

Vulcanization by benzoquinonedioxime

As shown above, the chemical structural requirements for these types of vulcanization are that the elastomer molecules contain allylic hydrogen atoms. The attacking species from the vulcanization system must contain sites for proton acceptance and electron acceptance in proper steric relationship. This will then permit the following rearrangement, where A is the proton acceptor site and B is the electron acceptor site:

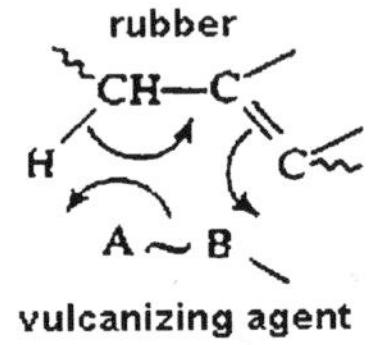

This is an explanation for the fact that this type of vulcanization is not enabled by double bonds per se, without allyic hydrogens in the elastomer molecules (Coran, 1978; Baldwin et al., 1970). (It should be pointed out that the phenolic curative could also act by a slightly different mechanism to give crosslinks, which contain chromane structural moieties (Greth, 1951) the allylic hydrogens still being required.) Another vulcanizing agent for high-diene rubbers is *m*-phenylenebismaleimide. A catalytic free radical source such as dicumyl peroxide or benzothiazyl disulfide (MBTS) is usually used to initiate the reaction. A reaction scheme for this type of vulcanization, adapted from Kovacic and Hein (1962), is as follows:

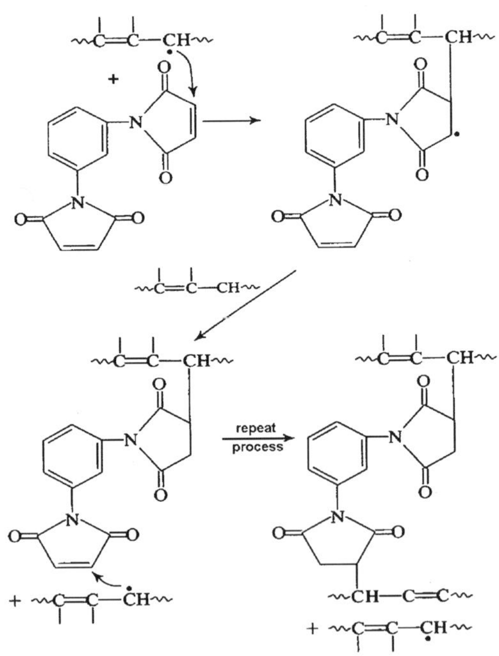

Although a free radical source is frequently used with a maleimide vulcanizing agent, at high enough vulcanization temperatures, the maleimides react with the rubber without the need for a free radical source. This could occur as shown here:

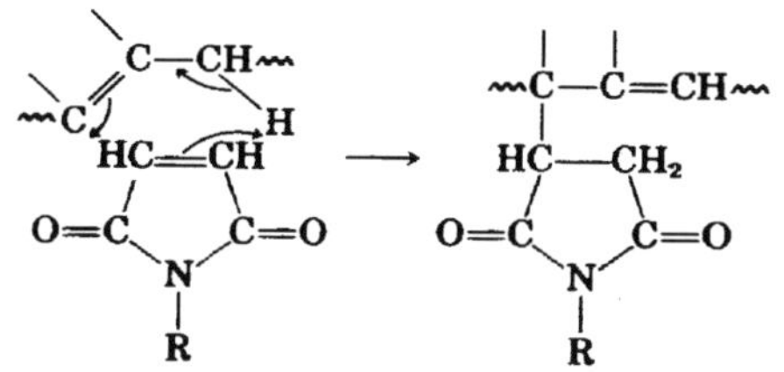

This is another example of what has variously been called a pseudo-Diels-Alder, ene, or "no-mechanism" reaction (Hoffmann, 1969). It is similar to the reaction written for the attack of rubber molecules by phenolic curatives or the *in situ* formed nitroso derivative of the quinoid (e.g., benzoquinonedioxime) vulcanization system. It is also closely related to the sulfurization scheme written for accelerated-sulfur vulcanization. Comparisons between accelerated sulfur, phenolic, quinoid, and maleimide vulcanization can then be visualized as follows:

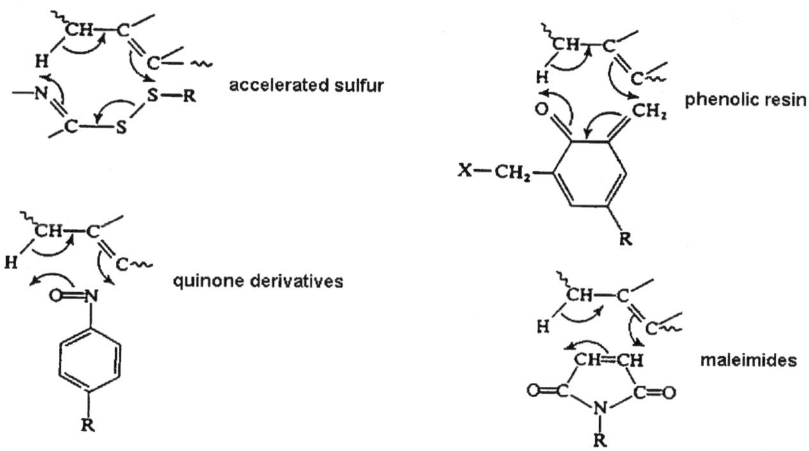

Table 7.5 gives selected recipes for vulcanization by phenolic curatives, benzoquinone-dioxime, or *m*-phenylenebismaleimide. Vulcanizates based on these types of curatives are particularly useful in cases where thermal stability is required.

TABLE 7.5 Recipes for Vulcanization by Phenolic Curatives, Quinone Derivatives, or Maleimides[a]

	IIR		SBR		NBR
	1	2	1	2	
Zinc oxide	5.00	5.00	–	–	–
Lead peroxide (Pb_3O_4)	–	10.00	–	–	–
Stearic acid	1.00	–	–	–	–
Phenolic curative (SP-1056)[b]	12.00	–	–	–	–
Benzoquinonedioxime (GMF)	–	2.00	–	–	–
m-Phenylenebismaleamide (HVA-2)[c]	–	–	0.85	0.85	3.00
2-Benzothiazyl disulfide (MBTS)	–	–	2.00		
Dicumyl peroxide	–	–	–	0.30	0.30
Vulcanization condition[d]					
Temperature (°C)	180	153	153	153	153
Time (min)	30	20	25	25	30

[a]*Concentrations in phr.*

[b]*Schenectady Chemicals.*

[c]*DuPont.*

[d]*Conditions change depending on other aspects of the compositions.*

7.8 VULCANIZATION BY THE ACTION OF METAL OXIDES

Chlorobutadiene or chloroprene rubbers (CR), also called neoprene rubbers, are generally vulcanized by the action of metal oxides. CR can be represented by the structure:

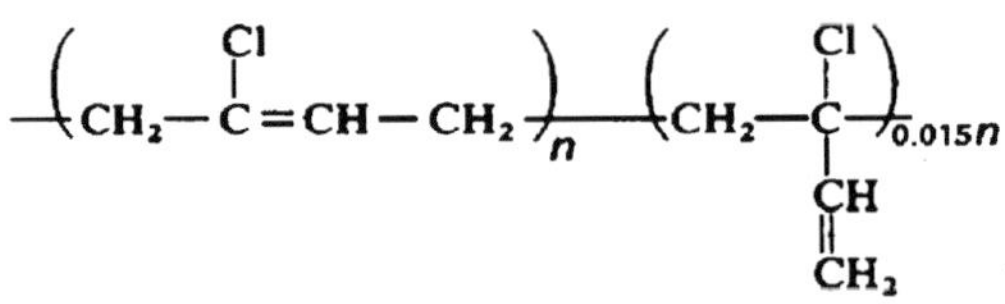

The crosslinking agent is usually zinc oxide, which is used along with magnesium oxide. CR can be vulcanized in the presence of zinc oxide alone; however, magnesium oxide is necessary to give scorch resistance. The reaction is thought to involve the allylic chlorine atom, which is the result of the small amount of 1,2-polymerization:

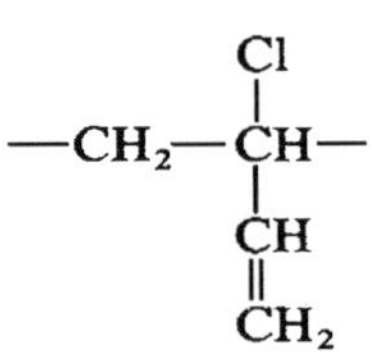

A mechanism, which has been written for the vulcanization of CR by the action of zinc oxide and magnesium oxide, is as follows (Hofmann, 1967):

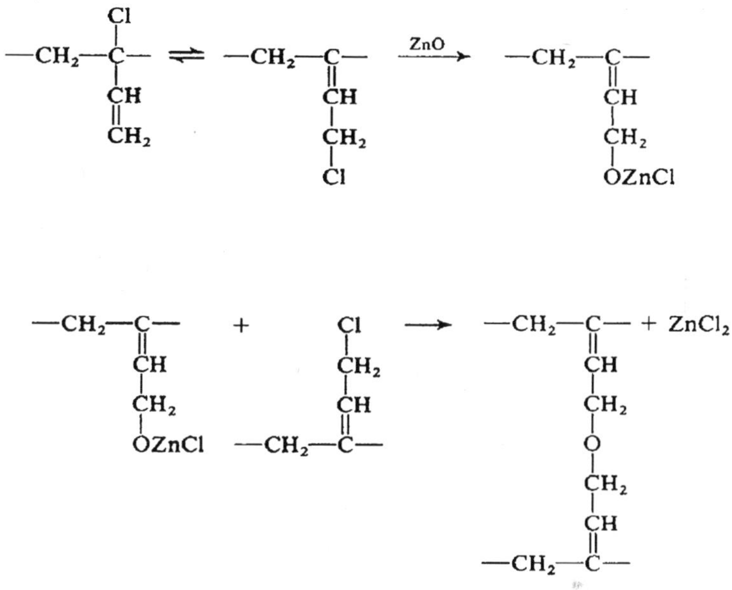

$$ZnCl_2 + MgO \longrightarrow ZnO + MgCl_2$$

Most accelerators used in the accelerated-sulfur vulcanization of other high-diene rubbers are not applicable to the metal oxide vulcanization of neoprene rubbers. An exception to this is in the use of the so-called mixed curing system for CR, in which metal oxide vulcanization is combined with accelerated-sulfur vulcanization. Along with the metal oxides, tetramethylthiuram disulfide (TMTD), N,N′-di-o-tolylguanidine (DOTG), and sulfur are used. This may be desirable for high resilience or for good dimensional stability. The accelerator, which has been most widely used with metal oxide cures, is ethylenethiourea (ETU) or 2-mercaptoimidazoline. Further extensive use of ETU in the vulcanization of CR is somewhat in doubt since it is a suspected carcinogen. The related compound, thiocarbanalide, an old accelerator for sulfur vulcanization, has been revived for CR vulcanization. Other substitutes for ETU have been proposed (Mori and Nakamura, 1984; Kato and Jujita, 1982). A mechanism for ETU acceleration has been given by Pariser (1960):

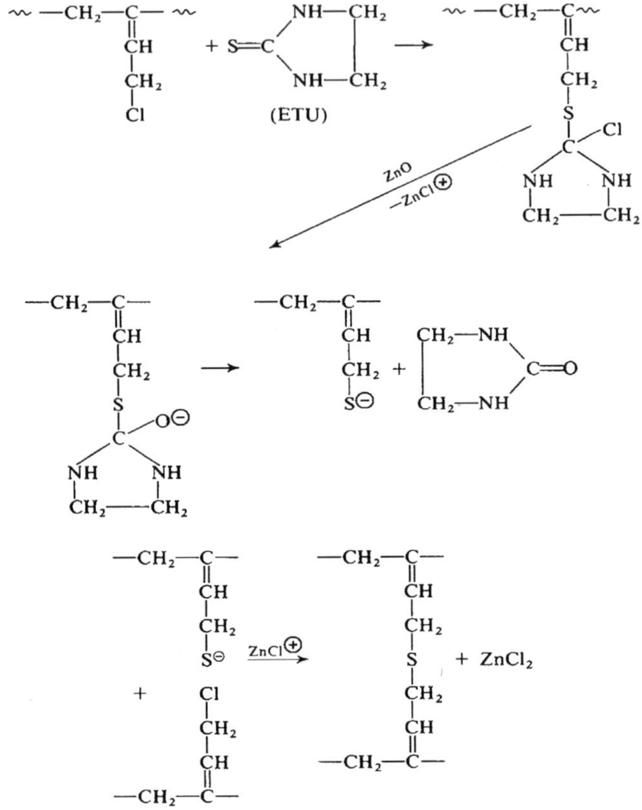

Examples of recipes for metal oxide vulcanization are given in Table 7.6. It should be noted that in one case, calcium stearate was used instead of magnesium oxide to obtain better aging characteristics (Becker, 1964).

7.9 VULCANIZATION BY THE ACTION OF ORGANIC PEROXIDES

Most elastomers can be vulcanized by the action of organic peroxides (Loan, 1967; Dluzneski, 2001). Diacyl peroxides, dialkyl peroxides, and peresters have been used. Dialkyl peroxides and t-butyl perbenzoate give efficient crosslinking. Di-t-butyl peroxide and dicumyl peroxide give good vulcanizates, but the former is too volatile for general use. Dicumyl peroxide is widely used, however its vulcanizates have the odor of acetophenone, which is a byproduct of the vulcanization process. Other nonvolatile peroxides of the same class, which give vulcanizates free of the odor of acetophenone, are 1,1-bis(t-butylperoxy)-3,3,5-trimethylcyclohexane and 2,5-dimethyl-2,5-bis(t-butylperoxy)hexane. This latter compound is particularly good for vulcanization at higher temperatures (as high as 180°C) since it is more thermally stable than the others.

TABLE 7.6 Vulcanization Systems for Chloroprene Rubber[a]

ZnO	5.00	5.00	5.00
MgO	4.00		4.00
Calcium stearate	–	5.50	–
Stearic acid	–	–	1.00
TMTM	–	–	1.00
DOTG	–	–	1.00
ETU	0.5	0.5	–
Sulfur	–	–	1.0
Vulcanization condition[b]			
Temperature (°C)	153	153	153
Time (min)	15	15	15

[a]*Concentrations in parts by weight per 100 parts of neoprene W.*

[b]*Conditions change depending on other aspects of the compositions.*

It should be noted that acidic compounding ingredients (fatty acids, certain carbon blacks, and acidic silicas) can catalyze nonradical-generating, wasteful decomposition of peroxides. Other compounding ingredients such as antidegradants can reduce crosslinking efficiency by quenching or altering the free radicals before they can react with the polymeric substrate.

Peroxides are vulcanizing agents for elastomers, which contain no sites for attack by other types of vulcanizing agents. They are useful for ethylene-propylene rubber (EPR), ethylene-vinyl acetate copolymers (EAM), certain millable urethane rubbers, and silicone rubbers. They generally are not useful for vulcanizing butyl rubber (poly[isobutylene-co-isoprene]) because of a tendency toward chain scission, rather than crosslinking, when the polymer is subjected to the action of peroxide.

Elastomers derived from isoprene and butadiene are readily crosslinked by peroxides; but many of the vulcanizate properties are inferior to those of accelerated-sulfur vulcanizates. However, peroxide vulcanizates of these diene rubbers may be desirable in applications where improved thermal aging and compression set resistance are required.

7.9.1 Peroxide Vulcanization of Unsaturated Hydrocarbon Elastomers

The initiation step in peroxide-induced vulcanization is the decomposition of the peroxide to give free radicals. Thus

$$\textbf{Peroxide} \longrightarrow \textbf{2 R}\bullet$$

where R is an alkoxyl, alkyl, or acyloxyl radical, depending on the type of peroxide used. (Dibenzoyl peroxide gives benzoyloxyl radicals but dicumyl peroxide gives cumyloxyl and methyl radicals (Scanlan and Thomas, 1963).) If the elastomer is derived from butadiene or isoprene, the next step is either the abstraction of a hydrogen atom from an allylic position on the polymer molecule or the addition of the peroxide-derived radical to a double bond of the polymer molecule (Farmer and Michael, 1942; Farmer and Moore, 1951a,b).

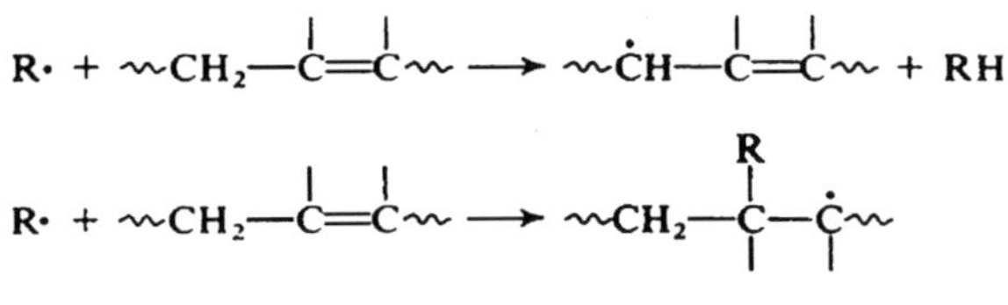

For isoprene rubber, the abstraction route predominates over radical addition. Two polymeric free radicals then unite to give a crosslink.

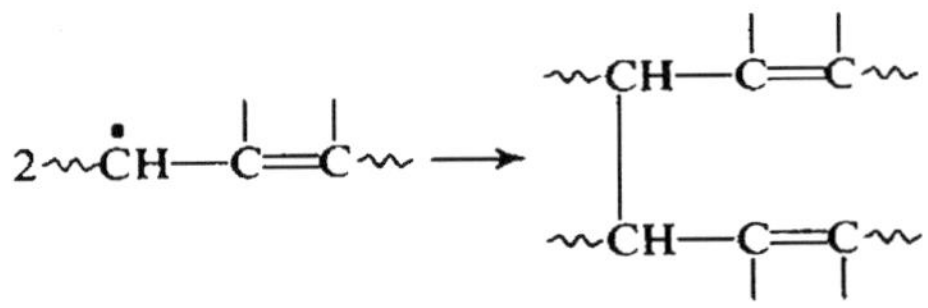

Crosslinks could also form by a chain reaction, which involves the addition of polymeric free radicals to double bonds (Loan, 1967, 1963).

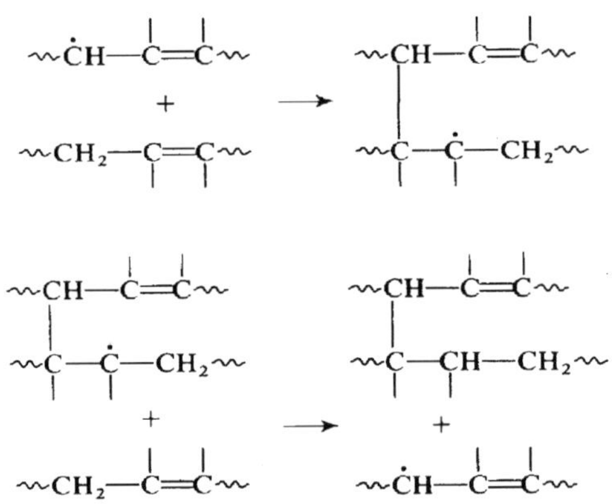

In this case crosslinking occurs without the loss of a free radical, so that the process can be repeated until termination by radical coupling. Coupling can be between two polymeric radicals to form a crosslink or by an unproductive process: a polymeric radical can unite with a radical derived from the peroxide. If a polymeric radical decomposes to give a vinyl group and a new polymeric radical, a scission of the polymer chain is the result.

Few monomeric radicals are lost by coupling with polymeric radicals when dialkyl peroxides are used as the curative. Also, if the elastomer is properly chosen, the scission reaction is not excessive (Loan, 1967; Dluzneski, 2001; Scanlan and Thomas, 1963; Farmer and Michael, 1942; Farmer and Moore, 1951a,b; Loan, 1963). For dicumyl peroxide in natural rubber, the crosslinking efficiency has been estimated at about 1.0. One "mole" of crosslinks is formed for each mole of peroxide; crosslinking is mainly by the coupling of two polymeric radicals (Thomas, 1962; Scott, 1962). One peroxide moiety gives two monomeric free radicals, which react with rubber to give two polymeric radicals that couple to form one crosslink.

In the case of BR or SBR, the efficiency can be much greater than 1.0, especially if all antioxidant materials are removed. A chain reaction is indicated here. It might be explained by steric considerations. In butadiene-based rubbers, double bonds are quite accessible. Radical addition to double bonds could give highly reactive radicals, which would be likely to add to other polymer double bonds. A chain of additions might be more likely in butadiene rubber than in the presence of hindering methyl groups in isoprene rubbers.

We might expect that nitrile rubber would also be vulcanized with efficiencies greater than 1.0; however, though the double bonds in nitrile rubber are highly accessible, the crosslinking efficiency is somewhat less than 1.0 (Loan, 1963).

7.9.2 Peroxide Vulcanization of Saturated Hydrocarbon Elastomers

Saturated hydrocarbon polymers are also crosslinked by the action of organic peroxides, though branching reduces the efficiency. Polyethylene is crosslinked by dicumyl peroxide at an efficiency of about 1.0, saturated EPR gives an efficiency of about 0.4, while butyl rubber cannot be cured at all. For polyethylene, the reaction scheme is similar to that of the unsaturated elastomers.

$$\text{peroxide} \longrightarrow 2\text{R}\cdot$$

$$\text{R}\cdot + -\text{CH}_2-\text{CH}_2- \longrightarrow \text{RH} + -\text{CH}_2-\dot{\text{C}}\text{H}-$$

$$2\ -\text{CH}_2-\dot{\text{C}}\text{H}- \longrightarrow \begin{array}{c} -\text{CH}_2-\text{CH}- \\ | \\ -\text{CH}_2-\text{CH}- \end{array}$$

However, branched polymers undergo other reactions.

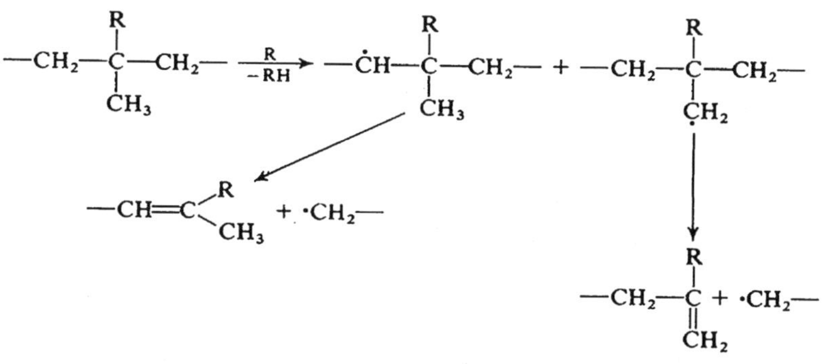

Here, though the peroxide has been depleted, no crosslinks have been formed between polymer chains, and the average molecular weight of the polymer has even been reduced by scission.

Sulfur, or the so-called coagents (Loan, 1963; Lenas, 1964), can be used to suppress scission. Examples of coagents are *m*-phenylenebismaleimide, high-1,2 (high-vinyl)polybutadiene, triallyl cyanurate, diallyl phthalate, and ethylene diacrylate. Their mechanism of action may be as follows:

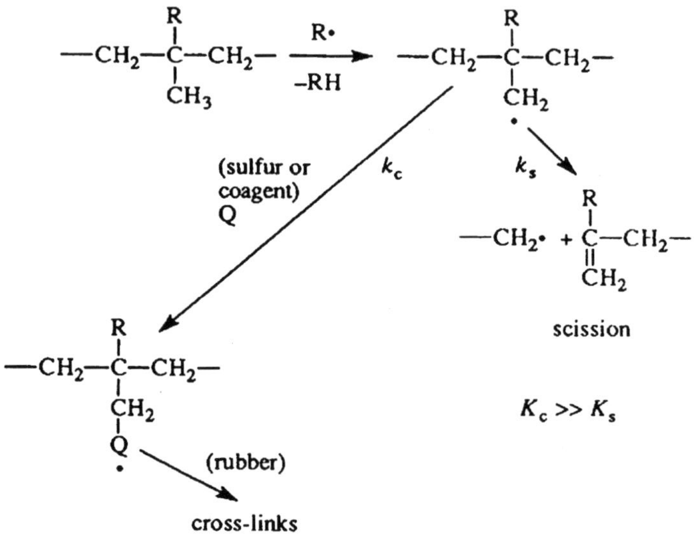

Recent work with coagents has been reported by Alvarez-Grima and coworkers (Alvarez Grima et al., 2006). They point out that scorch is a common problem in peroxide cure, especially for injection molding and extrusion applications. Several additives can help to improve scorch safety; however, they always result in a lower peroxide efficiency, thus inferior vulcanizate properties. These authors presented work on the use of a combination of a bismaleimide-type coagent, like N,N′-*m*-phenylenedimaleimide (BMI-MP), and a sulfur-containing compound like dipentamethylenethiuram tetrasulfide (DPTT). This combination provides scorch safety and at the same time improves the mechanical properties of the vulcanizates. The concentrations of coagent and sulfur-containing compound have a big influence on the scorch time and on the mechanical properties. Optimum concentrations were 4 phr of coagent and 0.7–0.96 phr of the sulfur-containing compound. The chemistry of this type of vulcanization was also studied (Alvarez Grima et al., 2009).

7.9.3 Peroxide Vulcanization of Silicone Rubbers

Silicone rubbers can be represented by

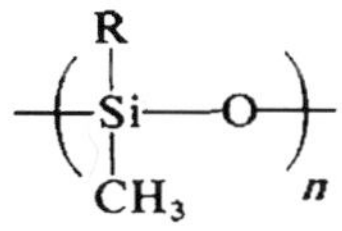

where R can be methyl, phenyl, vinyl, trifluoropropyl, or 2-cyanoethyl. Dialkyl peroxides such as dicumyl peroxide can cure silicone rubbers, which contain vinyl groups. Saturated silicone rubbers require diacyl peroxides such as bis-(2,4-dichlorobenzoyl)peroxide. In the case of saturated siloxane rubbers, the mechanism is hydrogen atom abstraction followed by polymeric radical coupling to give crosslinks. Nonproductive use of peroxide results from the coupling of the polymeric radicals with the lower-molecular-weight free radicals formed by the decomposition of the peroxide curative. The incorporation of vinyl groups improves the crosslinking efficiency (Lewis, 1962; Thomas, 1966).

Vulcanization is frequently done in two steps. After a preliminary vulcanization in a mold, a high-temperature (e.g., 180°C) postcure is carried out in air. The high-temperature postcure removes acidic materials that can catalyze hydrolytic decomposition of the vulcanizate (Roush et al., 1964). Also the high temperature enables the formation of additional crosslinks of the following type (Hofmann, 1967):

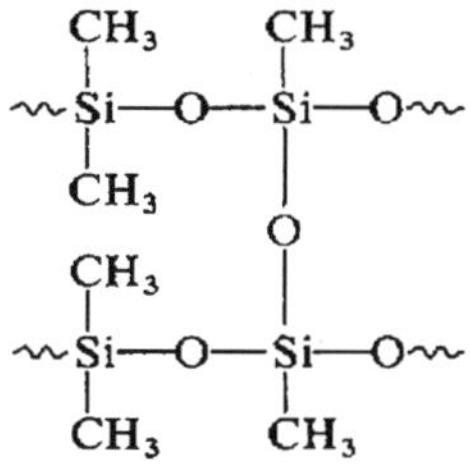

7.9.4 Peroxide Vulcanization of Urethane Elastomers

Urethane elastomers suitable for peroxide vulcanization typically are prepared from a hydroxyl-group-terminated oligomeric adipate polyester and 4,4′-methylenediphenylisocyanate (MDI). A typical structural representation is as follows:

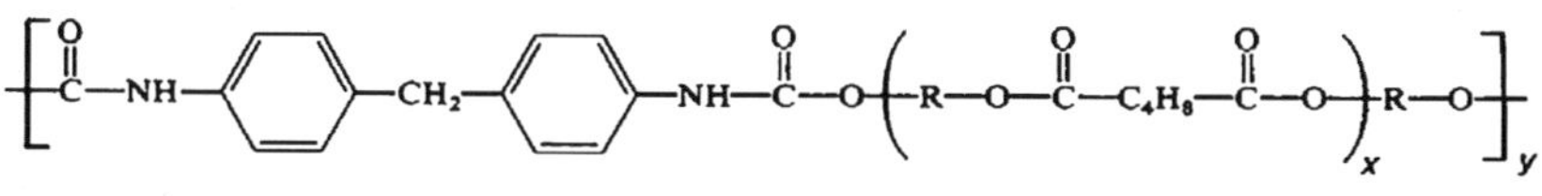

Hydrogen atoms can be abstracted from arylated methylene groups, but hydrogen atoms may also be abstracted from α-methylene groups of the adipate moieties. Though they are usually sufficient, vulcanization efficiencies can be increased by the incorporation of urea structures into the polymer chain (Bork and Roush, 1964).

TABLE 7.7 Recipes for Peroxide Vulcanization[a]

	NR	SBR	EPR	Silicone Rubber	Millable Urethane
Dicumyl peroxide	1.0	1.0	2.7		2
Bis(2,4-dichlorobenzoyl peroxide)	–			1.0	
Triallyl cyanurate			1.5		
Vulcanization conditions[b]					
Temperature (°C)	150	150	160	115, 250[c]	153
Time (min)	45	45	30	14,1440[c]	45

[a]*Concentrations in phr.*

[b]*Conditions change depending on other aspects of the compositions.*

[c]*Temperature and time of postcure in air.*

7.9.5 Recipes for Peroxide Vulcanization

Examples of starting-point recipes are given in Table 7.7. Outstanding characteristics of peroxide vulcanizates are low permanent set and thermal stability of the network.

7.10 DYNAMIC VULCANIZATION

Dynamic vulcanization (DV) is the vulcanizing or crosslinking of one polymer during its molten-state mixing with another polymer or with other polymers. The polymers are first thoroughly mixed and then, during further mixing, *one* of the polymers is obliged to become crosslinked, whereas the remaining other polymeric material remains uncrosslinked. The process produces a dispersion of crosslinked polymer in a matrix or continuous phase of uncrosslinked polymer. If the dispersed crosslinked material is elastomeric and the continuous or matrix material is of a melt-processable plastic, then the composition can be used as an impact resistant thermoplastic resin, or if there is a large enough proportion of rubber in the composition, it might be suitably used as a thermoplastic elastomer (TPE). Such materials are referred to as thermoplastic vulcanizates (TPVs).

TPVs combine the rapid molding of thermoplastics with the elastomeric properties of thermoset rubber. They are useful in a broad range of applications. TPVs comprise the second largest group of soft, rubbery thermoplastic elastomers, after styrenic-based block copolymers.

Fischer (1973) used the DV process to prepare compositions containing *partially* vulcanized rubber. It has since been found that improved, very strong elastomeric compositions of EPDM and polypropylene could be prepared by dynamic vulcanization provided that the rubber was completely vulcanized (Coran and Patel, 1980).

The DV process for thermoplastic elastomers can be described as follows: After sufficient melt-mixing of plastic and rubber, vulcanizing agents are added. Vulcanization of the rubber phase occurs as mixing continues. After removal from the mixer, the cooled blend can be chopped, extruded, pelletized, injection molded, and so on. Such a composition is described as a dispersion of very small particles of vulcanized rubber in a thermoplastic resin matrix. Such compositions are prepared commercially by a continuous process by using a twin-screw extruder.

Dynamic vulcanization gives the following improvements, in comparison with blends which have not been dynamically vulcanized: reduced set, improved ultimate properties, improved fatigue resistance, improved resistance to attack by hot oils, greater stability of melt-phase morphology, greater melt strength, and so on.

Recent developments in dynamic vulcanization have been reviewed by Babu and Naskar (2011).

7.10.1 EPDM-Polyolefin Compositions

The dynamic vulcanization of blends of EPDM rubber with polyolefins (PP or PE) has been described (Coran and Patel, 1980). The rubber-plastic proportions and the extents of vulcanization were varied. In a few instances the rubber was first press-cured and then ground to various particle sizes. The ground rubber particles were then mixed with molten polypropylene. It was found that the ultimate properties (UE and UTS) varied inversely with rubber particle size. Since the smallest particle sizes of vulcanized rubber were obtained by dynamic vulcanization (not by grinding of cured rubber), the more durable compositions were obtained by dynamic vulcanization.

Only a small amount of crosslink formation is required for a large improvement in tension set. However, tensile strength improves rather continuously as the crosslink density of the rubber phase is increased. Compositions can be vulcanized by accelerated sulfur, methylolphenolic materials (e.g., catalyzed by $SnCl_2$), or other curatives (Coran and Patel, 1996).

As the concentration of the polyolefin resin increases, the compositions become less like rubber and more like plastic. Modulus, hardness, tension set, and strength increase.

7.10.2 NBR-Nylon Compositions

Excellent elastomeric NBR-nylon compositions have also been prepared by dynamic vulcanization during the melt-mixing of intimate blends of NBR with various nylons. In this case, the effect of curatives was complicated by the fact that some nitrile rubbers tend to self-cure at temperatures of mixing. Sulfur, phenolic, maleimide, or peroxide curatives can be used. The thermoplastic elastomeric compositions prepared by the dynamic vulcanization of nylon-NBR blends are highly resistant to hot oil. As in the case of the EPDM-polyolefin blends, increases in the amount of rubber in the composition reduce stiffness but increase resistance to permanent set.

7.10.3 Other Elastomeric Compositions Prepared by Dynamic Vulcanization

In addition to EPDM-polyolefin and NBR-nylons combinations, a large number of other rubber-plastic combinations have been used to prepare thermoplastic vulcanizates by dynamic vulcanization (Coran et al., 1982).

The best compositions are prepared when the surface energies of the rubber and plastic material are matched, when the entanglement molecular length of the rubber molecule is small (i.e., molecular entanglement density of the rubber molecule is greatest), and when the plastic material is crystalline. It is also necessary that neither the plastic nor the rubber decomposes in the presence of the other at temperatures required for melt-mixing. Also, in each case, a curing system appropriate for the rubber under the conditions of melt-mixing is required.

7.10.4 Technological Applications

The lower cost of thermoplastic processing is the motivation for the development of thermoplastic elastomers. However, failure in the achievement of truly

TABLE 7.8 Commercially Available High-Performance TPVs[a]

Grade Name	Elastomer Type	Matrix Phase	Supplier
TPSiV	Silicone	Polyamide (PA), thermoplastic polyurethane (TPU)	Dow Corning
Zeotherm	Polyacrylate (ACM)	Polypropylene (PP), PA, polyester	Zeon
E-TPV	Ethylene acrylate	Thermoplastic copolyester (COPE)	DuPont
Uniprene-XL	Hydrogenated-styrene block copolymer (H-SBC)	PP	Teknor-Apex
Serel	Styrene-butadiene (SSBR)	PP	Goodyear
Septon V	H-SBC, reactive hard block	SBC triblock	Kuraray

[a]*Robert Eller Associates Inc. (2004). http://www.robertellerassoc.com/.*

rubber-like properties has impeded the acceptance of thermoplastic-elastomer technology. Nevertheless relatively recently commercialized compositions based on polypropylene and *completely* vulcanized EPDM have many of the excellent properties of the polyurethane and copolyester-type thermoplastic elastomers and even improved set and fatigue resistance. Applications of these materials can be listed as follows: caster wheels, convoluted bellows, diaphragms, gaskets, seals, tubing, mounts, bumpers, glazing seals, shields, suction cups, torque couplings, vibration isolators, plugs, connectors, rollers, oil-well injection lines, handles, grips, hose covers, vacuum tubing, bushings, grommets, protective sleeves, shock isolators, ducts, various hoses (e.g., hydraulic, agricultural spray, paint spray, plant air-water, mine hose, etc.), wire and cable insulation and strain relief, jacketing, and more.

7.10.5 Extra-High-Performance TPVs

TPVs have been developed that are more resistant to high temperatures, hot oils, oxidation fatigue, and so on. There is little open, published work on these materials. Some of these materials were discussed by Babu and Naskar (2011). This new TPV technology is expected to bridge the gap between PP/EPDM TPVs and the more costly engineering TPVs. Examples of commercialized extra-high-performance TPVs are given in Table 7.8.

REFERENCES

Albrecht, K.D., 1973. Rubber Chem. Technol. 46, 981.
Alvarez Grima, M.M., Talma, A.G., Datta, R.N., Noordermeer, J.W.M., 2006. Rubber Chem. Technol. 79, 694.
Alvarez Grima, M.M., Eriksson, J.G., Talma, A.G., Datta, R.N., Noordermeer, J.W.M., 2009. Rubber Chem. Technol. 82, 442.
Arem, A.B., Joseph, K., Thomas, S., 1999. Rubber Chem. Technol. 72, 458.
Babu, R.R., Naskar, K., 2011. Adv. Polym. Sci. 239, 219.
Baker, C.S.L., Barnard, D., Porter, M., 1962. Rubber Chem. Technol. 35, 141.
Baldwin, F.P., Borzel, P., Cohen, C.A., Makowski, H.S., Castle, J.F., 1970. Rubber Chem. Technol. 43, 522.
Bateman, L., Moore, C.G., Porter, M., 1958. J. Chem. Soc. 2866.
Bateman, L., Moore, C.G., Porter, M., Saville, B., 1963. In: Bateman, L. (Ed.), The Chemistry and Physics of Rubber-Like Substances. John Wiley & Sons, Inc., New York (Chapter 19).
Becker, R.O., 1964. Rubber Chem. Technol. 37, 76.
Bedford, C., 1921. US Patent 1 371, pp. 922–924.
Bell, C.L.M., Cuneen, J.I., 1967. J. Appl. Polym. Sci. 11, 2201.
Bork, P.G., Roush, C.W., 1964. In: Alliger, G., Sjothun, I.J. (Eds.). Vulcanization of Elastomers. Reinhold, New York, p. 366.
Brown, P.S., Porter, M., Thomas, A.G., The role of crosslink breakage in determining the strength of vulcanizates. In: A Paper Presented at the International Rubber Conference, Harrogate, 18a, 1987 (preprint).
Bruni, G., Romani, E., 1921. India Rubber J. 62, 63.
Buchan, S., 1959. Rubber to Metal Bonding. Crosby Lockwood and Sons, London.
Campbell, D.S., 1970. J. Appl. Polym. Sci. 14, 1409.
Campbell, R.H., Wise, R.W., 1964a. Rubber Chem. Technol. 37, 635.

Campbell, R.H., Wise, R.W., 1964b. Rubber Chem. Technol. 37, 650.
Choi, W., 2006. e-J. Soft Mater. 2, 47.
Cooper, W., 1958. J. Polym. Sci. 28, 195.
Coran, A.Y., 1965. Rubber Chem. Technol. 38, 1.
Coran, A.Y., 1978. In: Eirich, F.R. (Ed.), Science and Technology of Rubber. Academic Press, New York (Chapter 7).
Coran, A.Y., 1983. Chemtech 23, 106.
Coran, A.Y., Kerwood, J.E., 1970. US Patent 3 546 185.
Coran, A.Y., Patel, R., 1980. Rubber Chem. Technol. 53, 141.
Coran, A.Y., Patel, R., 1996. In: Holden, G., Legg, N.R., Quirk, R., Schroeder, H.E. (Eds.), Thermoplastic Elastomers, second ed., Hanser, Munich, p. 153 (Chapter 7).
Coran, A.Y., Patel, R., Williams, D., 1982. Rubber Chem. Technol. 55, 116.
Datta, R.N., Ivan, M.S., 1995. Rubber World 212 (5), 24.
Datta, R.N., Schotman, A.H.M., Weber, A.J.M., Van Wijk, F.G.H., Van Haeren, P.J.C., Hofstraat, J.W., Talma, A.G., Bovenkamp-Bouwman, A.G.V.D., 1997. Rubber Chem. Technol. 70, 129.
Datta, R.N., Talma, A.G., Schotman, A.H.M., 1998. Rubber Chem. Technol. 71, 1073.
Datta, R.N., Huntink, N.M., Datta, S., Talma, A.G., 2007. Rubber Chem. Technol. 80, 436.
Decker, G.E., Wise, R.W., Guerry, D., 1963. Rubber Chem. Technol. 36, 451.
Dijkhuis, K.A.J., Noordemeer, J.W.M., Dierkes, W.K., 2009. Eur. Polym. J. 45, 3302.
Ding, R., Leonov, A.I., Coran, A.Y., 1996. Rubber Chem. Technol. 69, 81.
Dluzneski, P.R., 2001. Rubber Chem. Technol. 74, 451.
Fan, R., Zhang, Y., Huang, C., Zhang, Y., Fan, Y., Sun, K., 2001. J. Appl. Polym. Sci. 81, 710.
Fan, R., Zhang, Y., Huang, C., Gong, P., Zhang, Y., 2002. Rubber Chem. Technol. 43, 287.
Farmer, E.H., 1947a. J. Chem. Soc. 1519.
Farmer, E.H., 1947b. J. Soc. Chem. Ind. 66, 86.
Farmer, E.H., Michael, S.E., 1942. J. Chem. Soc. 513.
Farmer, E.H., Moore, C.B., 1951a. J. Chem. Soc. 131.
Farmer, E.H., Moore, C.B., 1951b. J. Chem. Soc. 142.
Farmer, E.H., Shipley, F.W., 1946. J. Polym. Sci. 1, 293.
Fischer, W.K., 1973. US Patent 3 758 643.
Flory, P.J., 1953. Principles of Polymer Chemistry. Cornell University Press, Ithaca, New York (Chapter 11).
Gan, L.M., Chew, C.H., 1983. Rubber Chem. Technol. 56, 883.
Gent, A.N., Lindley, P.B., Thomas, A.G., 1964. J. Appl. Polym. Sci. 8, 455.
Ghosh, P., Katare, S., Patkar, P., Caritjers, J.M., Venkatasubramanian, V., Walker, K.A., 2003. Rubber Chem. Technol. 76, 592.
A. Giller, 1966. Kaut. Gummi Kunstst. 19.
Gradwell, M.H.S., Stephenson, N.R., 2004. Rubber Chem. Technol. 77, 931.
Gregg Jr., E.C., Katrenick, S.E., 1970. Rubber Chem. Technol. 43, 549.
Greth, A., 1951. Kunststoffe 31, 345.
Harmon, M.W., 1937. US Patent 2 100 692.
Heideman, G., Datta, R.N., Noordermeer, J.W.M., van Baarle, B., 2004. Rubber. Chem. Technol. 77, 512.
Hoffmann, H.M.R., 1969. Angew. Chem. Int. Ed. Engl. 8, 556.
Hofmann, W., 1967. Vulcanization and Vulcanizing Agents. Maclaren and Sons Ltd., London, p. 242.
Hu, P.L., Scheele, W., 1962. Kautsch. Gummi 15, 440.
Ignatz-Hoover, F., Katritzky, A.R., Lobanov, V.S., 1999. Rubber Chem. Technol. 72, 318.
Juve, A.I., Karper, P.W., Schroyer, L.O., Veith, A.G., 1964. Rubber Chem. Technol. 37, 434, and references therein.
Kato, H., Jujita, H., 1982. Rubber Chem. Technol. 55, 949.
Koenig, J.L., 2000. Rubber Chem. Technol. 73, 385.
Kovacic, P., Hein, P.W., 1962. Rubber Chem. Technol. 35, 528.
Layer, R.W., 1992. Rubber Chem. Technol. 65, 211.
Leib, R.T., Sullivan, A.B., Trivette, C.D., 1970. Rubber Chem. Technol. 43, 1188.
Lenas, L.P., 1964. Rubber Chem. Technol. 37, 229.
Lewis, F.M., 1962. Rubber Chem. Technol. 35, 1222.
Lewis, P.M., 1986. NR Technol. 17, 57.

Loan, L.D., 1963. J. Appl. Polym. Sci. 7, 2259.
Loan, L.D., 1967. Rubber Chem. Technol. 40, 149.
C.R., Lorenz, O., 1963. Ind. Eng. Chem. Prod. Res. Dev. 2 (4), 279.
Malony, S., 1920. US Patent 1 343 222.
Martell, R.F., Smith, D.E., 1962. Rubber Chem. Technol. 35, 141.
McCall, E.B., 1969. J. Rubber Res. Inst. Malaya 22, 354.
Moore, C.G., Trego, B.R., 1961. J. Appl. Polym. Sci. 5, 299.
Moore, C.G., Mullins, L., McL, P., Swift, 1961. J. Appl. Polym. Sci. 5, 293.
Mori, M., Koenig, J.L., 1995. Rubber Chem Technol. 68, 551.
Mori, K., Nakamura, Y., 1984. Rubber Chem. Technol. 57, 34.
Morrison, N.J., Porter, M., 1984. Rubber Chem. Technol. 57, 63.
Nieuwenhuizen, P.J., Reedijk, J., Van Duin, M., McGill, W.J., 1997. Rubber. Chem. Technol. 70, 368.
Oenslager, G., 1933. Ind. Eng. Chem. 23, 232.
Pariser, R., 1960. Kunststoffe 50, 623.
Pellicioli, L., Mowdood, S.K., Negroni, F., Parker, D.D., Koenig, J.L., 2002. Rubber Chem. Technol. 75, 65.
Rehner, J., Flory, P.J., 1946. Rubber Chem. Technol. 19, 900.
Rodger, E.R., Roebuck, H.S., 1976. J. Elastom. Plast. 8, 81.
Roush, C.W., Kosmider, J., Baufer, R.L., 1964. Rubber Age 94, 744.
Russell, R.M., Skinner, T.D., Watson, A.A., 1969. Rubber Chem. Technol. 42, 418.
Scanlan, J., Thomas, D.K., 1963. J. Polym. Sci. Part A 1, 1015.
Scott, K.W., 1962. J. Polym. Sci. 58, 517.
Sebrell, L., Bedford, C., 1925. US Patent 1 522 687.
Skinner, T.D., 1972. Rubber Chem. Technol. 45, 182.
Skinner, T.D., Watson, A.A., 1969. Rubber Chem. Technol. 42, 404.
Son, P.N., 1973. Rubber Chem. Technol. 46, 999.
Studebaker, M.L., 1966. Rubber Chem. Technol. 39, 1359.
Sullivan, A.B., 1966. J. Org. Chem. 31, 2811.
Sullivan, A.B., Davis, L.H., Maender, O.W., 1983. Rubber Chem. Technol. 56, 1061.
Tan, E.-W., Wolff, S., 1985. US Patent 4 517 336.
Thelamon, C., 1963. Rubber Chem. Technol. 36, 268.
Thomas, D.K., 1962. J. Appl. Polym. Sci. 6, 613.
Thomas, D.K., 1966. Polymer 7, 243.
Trivette Jr., C.D. Morita, E., Maender, O.W., 1977. Rubber Chem. Technol. 50, 570.
van den Berg, J.H.M., Beulen, J.W., Duynstee, E.F.J., Nelissen, H.L., 1984a. Rubber Chem. Technol. 57, 265.
van den Berg, J.H.M., Beulen, J.W., Hacking, J.M.H., Duynstee, E.F.J., 1984b. Rubber Chem. Technol. 57, 725.
van der Meer, S., 1943. Rev. Gen. Caoutch. Plast. 20, 230.
Van Duin, M., Souphanthong, A., 1995. Rubber Chem. Technol. 68, 717.
van Ooij, W.J., 1979. Rubber Chem. Technol.52, 605.
van Ooij, W.J., 1984. Rubber Chem. Technol. 57, 421, 1984.
van Ooij, W.J., Harakuni, P.B., Buytaert, G., 2009. Rubber Chem. Technol. 82, 315.
Weiss, M., 1922. US Patent 1 411 231.
Wolfe, J.R., Pugh, T.L., Killian, A.S., 1968. Rubber Chem. Technol. 41, 1329.
Zaucker, E., Bogemann, M., Orthner, L., 1934. US Patent 1 942 790.

Chapter 8

Reinforcement of Elastomers by Particulate Fillers

Jean-Baptiste Donnet* **and Emmanuel Custodero**†

**EnscMu-Uha, Mulhouse Cedex, France*

†Manufacture Française des Pneumatiques Michelin, Clermont-Ferrand Cedex, France

8.1 INTRODUCTION

The reinforcement of elastomers by particulate fillers has been extensively studied in the past, particularly in the 1960s and 1970s. The first reason is naturally the drastic changes in mechanical properties that induce fillers reinforcement: Many of the usual applications of elastomers could not be envisaged without the use of particulate fillers. The other reason seems to us to be of a very different nature, and probably resides in the "mystery" of the reinforcement mechanism that has fascinated many scientists and remains, despite their efforts, mainly not understood today.

It is necessary to define precisely what is *reinforcement*, because this word covers very different meanings when applied to thermoplastics, thermosets, or elastomers. Confusion is mainly due to the fact that reinforcement qualifies an increase in mechanical properties, but what is expected as mechanical properties is very different considering the different matrices and applications.

For plastics, reinforcement results in an increase in modulus and hardness. The effect of particulate fillers is quite clear—they replace a part of the matrix. So modulus becomes higher, but deformation at break decreases in the same time.

The situation is very different for elastomers: the use of reinforcing fillers induces a simultaneous increase modulus and deformation at break. Curiously, the replacement of a part of the deformable matrix by solid objects doesn't reduce its deformability. The increase of these two antagonistic properties characterizes elastomer reinforcement. This fascinating paradox, despite not being fully understood, explains the ability of reinforced elastomers to provide

The Science and Technology of Rubber. http://dx.doi.org/10.1016/B978-0-12-394584-6.00008-X

unique material properties and applications and justify their success in different technological fields.

8.2 PREPARATION OF FILLERS

8.2.1 Nonreinforcing Fillers

As it will be discussed later, the size of the filler is probably one of the most important properties for reinforcement. So, particulate fillers obtained by grinding of minerals or by coarse precipitation are *usually* nonreinforcing fillers because of their size: they are too big. Such fillers can even be used in elastomers but just confer them a very slight increase in modulus and a very significant drop in break properties occurs.

8.2.2 Reinforcing Fillers

(i) Carbon Black

- Historical Processes.
- Furnace Process.
- Posttreatments, Surface Modification.

The great majority of carbon black posttreatment studies have been conducted to increase strength/quality of reinforcement. So the chemical modifications that have been tested are strongly linked to the different theories envisaged for reinforcement.

In the 1960s, carbon black-elastomer interaction was considered as the result of a *chemical bonding* (Bueche, 1961, 1960) between acidic surface functions and natural rubber alkaline moieties (Donnet and Heinrich, 1960; Le Bras and Papirer, 1983). So many studies have been conducted to increase carbon black activity by surface oxidation (Le Bras and Papirer, 1983): oxygen at high temperatures, H_2O_2, ozone, nitric acid. The type of oxidation used determines the number and the type of functions obtained; it is interesting to underline that such chemical modifications are used at industrial scale for specialty carbon blacks (inks, pigments).

In the early 1980s, Danneberg proposed the mechanism of molecular slippage (Dannenberg, 1975) and posttreatments turned to chemical grafting of polymeric chains onto carbon black surface (Donnet et al., 1975).

More recently, the need for low hysteresis compounds has reactivated chemical modification studies. Many modification processes have been proposed: functionalization (Tsubokawa and Hosoya, 1991), surface coating of carbon black by silica (Wang et al., 1997), and alumina (EP1034222 et al., 1997).

(ii) Silicas

The use of silica in rubber mixes cannot be considered as new at all, because this filler has been used in rubber formulations since the beginning of the 20th century (Voet et al., 1977). Silicas are not reinforcing fillers in the proper sense, because silica-reinforced mixes exhibit much lower mechanical properties, particularly considering modulus at break and abrasion resistance. So silicas weren't used as reinforcing fillers but mainly in association with carbon black.

Two major breakthroughs have transformed this facility product into a reinforcing filler that can achieve all carbon black mixes properties with, in addition, a decreased hysteresis of major interest for tire applications. The first step was made in the 1970s by Wolff, who proposed a specific silane coupling agent, the TESPT (Wolff, 1975, 1981). The second step occurred in the 1990s with Rauline's patent, which introduced the use of specific precipitation silica, elastomers, and mixing conditions to achieve reinforcement (EP0501227 and Rauline, 1991).

(i) Precipitated Silicas

Silicas used as reinforcing fillers are mainly obtained by precipitation (Legrand, 1998; EP0157703 et al., 1984; Chevallier and Morawski, 1986). The process basically consists in the preparation of a silica glass by alkaline fusion of pure sand and an alkaline salt. Then this glass is solubilized in water at high temperature and acid precipitated. The silica suspension obtained is then filtered, washed, and dried.

In order to obtain reinforcing silicas, much care must be taken in precipitation recipes (to obtain small rigid objects) and drying conditions (to maintain high dispersibility) (EP0157703 et al., 1984; Chevallier and Morawski, 1986; Shabanova and Silos, 1996; EP0890602 et al., 1997).

It is also interesting to underline that silicas can be very easily chemically modified by doping or grafting of species during or at the end of their preparation (Bomal et al., 1995).

(ii) Fumed Silicas

Fumed silicas are obtained by high temperature oxydecomposition of SiH_4, or other methyl hydride precursors ($SiHMe_3, SiH_2Me_2 \cdots$):

$$n\,SiH_4 + 2n\,O_2 \rightarrow n\,SiO_2 + 2n\,H_2O.$$

Coming out of the furnace, fumed silicas are obtained in a fluffy form, and because of their high temperature of formation, they present a very stable morphology and few surface silanols compared to precipitation silicas. This confers a high dispersibility and reactivity to fumed silicas but, because of their higher price, they are rarely used in the rubber industry.

(iii) New Reinforcing Fillers

Very recently, many studies have been conducted to identify new reinforcing systems. These systems are similar to silica compounds and characterized by the use of a coupling agent to chemically bond elastomer chains to filler surface. Many reinforcing systems have been patented: alumina oxyhydroxide and oxide (WO9928376 et al., 1997; EP0810258 et al., 1996), titanium oxides (EP1114092 et al., 1999), and silicon nitride/carbide (WO2002053634 et al., 2002; WO2004003067 et al., 2002).

8.3 MORPHOLOGICAL AND PHYSICOCHEMICAL CHARACTERIZATION OF FILLERS

As will be demonstrated later, morphology and physicochemical properties of reinforcing fillers are of crucial importance because they directly define their reinforcement ability. Their characterization formerly was based essentially on morphological properties (surface area and structure), but because of the use of silicas as reinforcing filler, there is now a strong need for dispersibility and surface chemistry characterization.

8.3.1 Filler Morphology Characterization

(i) Filler Morphology

It is important to emphasize that the actual morphology of carbon blacks has remained unknown for decades, even if it was commonly used in the rubber industry. This is due to the very small size of its constituting objects; they are smaller than 0.1μm and can only be resolved by transmission electron microscopy.

(i) MET

As observed by transmission electron microscopy, carbon blacks appear as irregular, chainlike, branched aggregates of partially fused spheres (Hess et al., 1973).

Aggregates constitute the smallest dispersible unit of carbon black and are virtually unbreakable in usual conditions of use; therefore, aggregates must be considered as the actual reinforcing objects.

The chainlike, branched structure of aggregates makes them very bulky, and their effective volume is much higher than the volume of the aggregate itself. This observation is of primary importance because the effective volume of the aggregate will be more or less its volume in the mixes and define which part of rubber can be deformed and not. This bulkiness is usually called *structure* and generally measured by other methods (see later section on structure); some very interesting studies have been conducted in the past to classify carbon black aggregates in different shape classes (bulk, ellipsoid, linear, etc.) (Hess, 1991).

Even if they are usually called *primary particles*, spheres that constitute aggregate are partially fused together and never exist by themselves. Anyway, their size is of great importance because it defines the actual surface of interaction between carbon black and elastomeric phase: the lower the size of primary particles, the higher the interface extension. Primary particle size distribution has been estimated by TEM image analysis, but carbon black surface area is usually more efficiently obtained by adsorption methods (see later section on surface area).

At very high magnification, it is possible to observe directly the internal structure of carbon black primary particles. They are constituted by overlapping graphitic layers that locally present a quasi-crystalline *turbostratic* structure with an approximately 0.35 nm interlayer spacing, close to pure graphite (~0.332 nm).

(ii) STM, AFM

Scanning tunneling microscopy (STM) (Custodero et al., 1992; Probst and Donnet, 1993; Donnet and Custodero, 1992) has been applied to carbon black characterization. As suggested by TEM, carbon black surface morphology consists in the overlapping of graphitic sheets in an onionlike structure (Figure 8.1).

Carbon black surfaces appear surprisingly ordered, and graphitic edges should be identified with chemically reactive zones that were previously assigned to "amorphous" zones (Donnet and Custodero, 1992).

Atomic-force microscopy (AFM) has also been used for three-dimensional characterization of carbon black aggregates (Donnet and Wang, 1994).

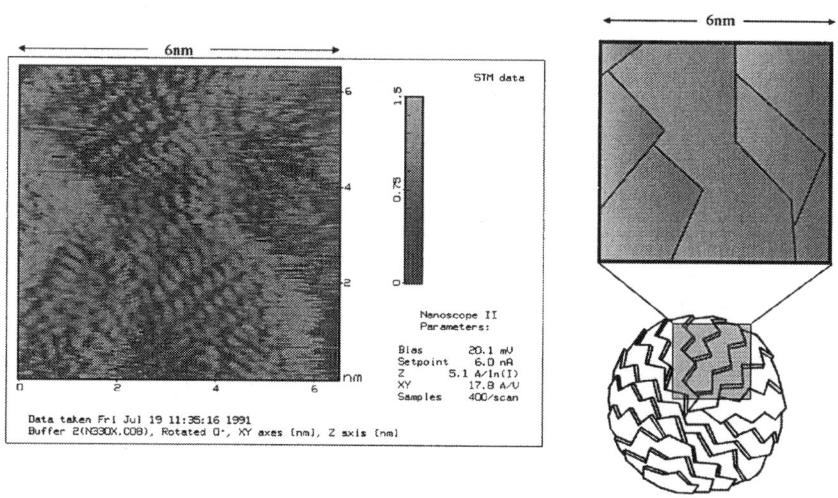

FIGURE 8.1 STM observation of carbon black surface at atomic scale resolution (from Bueche, 1960).

(iii) Silicas

Electron transmission microscopy of silicas is much more difficult, particularly for precipitation silicas, which tend to reagglomerate during preparation and are fairly often unstable under high magnification.

Reinforcing silicas observed by TEM present a morphology very close to carbon black; they are constituted by small, chainlike, branched aggregates. However, the identification of silicas' primary particles is much more difficult because they are significantly more fused.

(ii) Surface Area

Surface area is probably the most important morphological characterization of reinforcing fillers because it corresponds to the extension of the interface; that is, the interaction between elastomer and filler surface.

As evidenced by transmission electron microscopy (Hess, 1991; Herd et al., 1992), surface area is directly linked to the size of primary particles so that American Society of Testing Materials (ASTM) has chosen this parameter for carbon black nomenclature. More precisely, ASTM nomenclature includes four digits, the first one relates to vulcanization speed (N as *normal* or S as *slow*), then tree numbers, which first correspond to the primary particle diameter.

ASTM Numbers	Primary Particle Diameter (nm)	Previous Nomenclature
900–999	201–500	MT: Medium Thermal
800–899	101–200	FT: Fine Thermal
700–799	61–100	SRF: Semi-Reinforcing Furnace
600–699	49–60	GPF: General Purpose Furnace
500–599	40–48	FEF: Fine Extrusion Furnace
400–499	31–39	FF: Fine Furnace
300–399	26–30	HAF: High Abrasive Furnace
200–299	20–25	ISAF: Intermediate Super Abrasive Furnace
100–199	11–19	SAF: Super Abrasive Furnace
000–099	1–10	–

Particle size diameter can be done only by TEM characterization and is difficult and costly, so surface area of fillers usually is obtained by different adsorption methods. (From Janzen and Kraus (1971) and Janzen (1982).)

(i) Nitrogen Adsorption/BET

Adsorption of nitrogen and surface area determination by BET method is probably the most widely used method for surface area characterization of reinforcing fillers (Kraus and Jansen, 1978; Wampler, 1997; Magee, 1995a,b).

This method is very sensitive, reliable, and can be applied to all reinforcing fillers because it is either not or very weakly influenced by surface chemistry.

The main drawback of BET characterization is that the surface area obtained includes micropores whose surface cannot be reached by elastomeric chains, which are much bigger than nitrogen (Bradley et al., 1996). So, many are now using the *t-plot* method that allows the determination of the net surface excluding micropores (Wampler, 1997).

(ii) CTAB Adsorption

Cetyl triethyl ammonium bromide (CTAB) adsorption is widely used in carbon black industry (Janzen and Kraus, 1971; Kraus and Jansen, 1978; Bele et al., 1998). This method consists of making a suspension of a known mass of carbon black into a water CTAB solution of known concentration. Carbon black is then filtered and the quantity of adsorbed CTAB is determined by titration of the remaining CTAB in the filtrate. Surface area of carbon black is then deducted from the amount of CTAB adsorbed, using a previously determined calibration with a reference carbon black.

CTAB surface area characterization is not influenced by micropores because of the size of CTAB molecule, but it can be influenced by surface chemistry and impurities (Wampler et al., 1997). Hence, it tends to disappear and to be replaced by BET/*t-plot*, which is much more reliable.

(iii) Iodine Adsorption

Iodine surface characterization proceeds in exactly the same way as CTAB, except that the adsorbed species is I_2 instead of CTAB. This method can only be applied to carbon black characterization (Puri and Bansal, 1965).

Because iodine probably partly adsorbs and partly reacts with double bonds on the surface, the iodine method is extremely sensitive to any surface functions, modifications, or contaminations (Medalia and Kraus, 1994). Hence, this method, formerly very widely used in carbon black industry, is now replaced by CTAB or BET/*t-plot*.

(iii) Structure

If structure is rather easy to define on the basis of MET images, it is much more difficult to measure it quantitatively. Nevertheless, the determination of structure is of primary importance because structure defines the actual volume of the filler in the mix and therefore the level of strain amplification of the deformable phase (see "Mechanical Properties of Filled Rubbers").

(i) TEM Measurements

Some attempts have been made to use TEM measurements to determine structure of fillers. But in spite of the well-constructed studies (Hess and

McDonald, 1983; Gruber et al., 1993), the qualification of TEM images that are two dimensional do not lead to a three-dimensional image.

(ii) DBP Absorption

Practically, structure determination of fillers is obtained by a very simple and easy method: dibutyl phthalate absorption (DBPA) (Creeden, 1997). This method consists of filling up the voids in and between aggregates with DBP, which is a viscous liquid. Historically, the DPB volume was determined by making a solid pellet of carbon black and DBP with a spoon. Now, this delicate measurement is made automatically with a couple-monitored Banbury, in which a well-known quantity of dry carbon black is placed. DBP is then added drop by drop, and the structure value is obtained when torque reaches a given value.

DBP measurements are surprisingly reliable and can be easily and quickly obtained. Nevertheless, DBP measurement can be sensitive to surface chemistry, and values obtained with fillers of a different nature cannot be directly compared. For silicas, dioctyl phthalate can also be used instead of DBP (EP0157703 et al., 1984; Chevallier and Morawski, 1986).

DBP values also depend on pelletization/granulation of the filler. This makes sense because DBP measures the *total* void volume: intraaggregate voids, which are the pertinent parameter, but also interaggregate voids, which essentially reflect pelletization/granulation. Thus, fluffy or loosely pelletized black will have higher DBP value; in the same way, spray-dried silicas will have greater DBP value than granulated ones, even if their aggregates are exactly the same.

(iii) CDBP Absorption

Very frequently, DBP adsorption is made with carbon black previously submitted to very high pressure into a cylinder; for example, 24,000 psi four times (Creeden, 1997). Crushed DBP absorption or CDBPA is equal to DBP (for *low* structure carbon blacks) or significantly lower than DBP (for *high* structure carbon blacks).

The high pressure crush procedure is supposed to reproduce aggregate breakage during mixing, but because DBP measures either intraaggregate or interaggregate voids, it is difficult to settle if the decrease in DBP absorption is due to the compaction of the filler induced by the pressure (Dollinger et al., 1967) or to an actual breakage of aggregates.

(iv) Mercury Porosimetry

Filler structure can also easily be determined by mercury porosimetry (Pirard et al., 1999; Moscou et al., 1971; Evans and Waddell, 1995). Filler is placed in a small chamber and mercury is forced into the voids by increasing pressure. The intrusion curve gives the volume of mercury intruded in pores for each applied pressure. Usually, intrusion curves present a well-defined step at high pressure, which corresponds to the filling of the smallest pores (intraaggregate and interprimary-particles voids). Step's high gives a structure index, which excludes interaggregate voids and so is more representative of the intrinsic structure of the filler. In addition, mercury porosimetry allows a direct determination of filler surface area, excluding micropores (Milburn et al., 1988), and can be applied to any particulate filler (Paul et al., 1982).

(iv) Aggregate Size Distribution

Aggregate size distribution is the last morphological characterization of reinforcing fillers. This measurement is very rarely used, despite the great interest in knowing the size of aggregates, which directly influences distances between reinforcing objects in the mix, and therefore the strain amplification. This surprising situation is mainly due to the fact that this determination is particularly difficult to make (Janzen, 1982).

(i) TEM/AI Measurement

Transmission electron microscopy/image analysis (TEM/AI) has been used for a long time to determine aggregate size distribution of carbon black and silicas (Göritz et al., 1997). Such studies are very costly because they need at least a few thousand aggregate size measurements to determine precisely the size distribution. Nevertheless, using TEM/AI aggregates are measured as two-dimensional projections, which probably maximizes their sizes.

(ii) Disk Centrifugation

One better method to access aggregate size distribution is probably to use disk centrifugal photo or X-ray sedimentometers (Allen, 1987). This apparatus consists of a transparent void disk that can be rotated at very high speed. A sedimentation medium is first injected into the rotating disk, and then a very small quantity of filler suspension is injected at its surface. Sedimentation is registered by light or X-ray transmission.

Aggregate size distribution of carbon blacks (Patel and Lee, 1990a,b) and silicas can be easily obtained by this method. Its main drawback is that, unlike with EM-AI, aggregate sizes are probably underestimated because they settle following their lowest project area.

(iii) Tint

Coming from the ink and pigment industry, the tint measurement is a very simple way to evaluate the mean aggregate size of carbon black (Medalia and Richards, 1972; Medalia and Richards, 1972). This characterization consists of making a paste of known amounts of carbon black, white solid powder (titanium or zinc oxide), and oil (DBP, for example). Then the reflectivity of the paste is measured; it has been demonstrated that tint values will roughly correlate with mean aggregate size and can be considered as an indicator of aggregate size distribution broadness (Janzen, 1982; Medalia et al., 1976).

8.3.2 Dispersibility

Even if the relationship between filler dispersion and abrasion resistance is well established, relatively few studies have been done on the characterization of filler dispersibility. This is mainly due to the fact that carbon black dispersibility was commonly judged satisfactory, partly because it is indeed high, but more probably because all mixing apparatuses were designed for dispersing carbon blacks.

The use of silica has given a new light to this domain, because, contrary to carbon black, dispersibility is one of the key properties for achieving silica-reinforced mixes (EP0157703 et al., 1984; Chevallier and Morawski, 1986; Evans and Waddell, 1995; Cochet et al., 1993, 1994).

Obviously, filler dispersibility is mainly influenced by interactions between agglomerates and/or aggregates; in other words, the force/energy needed in order to separate two objects. For carbon black, these interactions are mainly due to van der Waals forces, which are very low compared to the hydrogen bonding existing between silica objects.

In addition, pelletization process has a great influence on dispersibility: Any action leading to a higher compaction of the filler increases interaction between filler objects and so decreases its dispersibility.

(i) Reflectivity

The most commonly used technique to qualify filler dispersibility is to study light reflectivity of clean-cut mixes. Some apparatuses have been developed to evaluate filler dispersion using a calibrated set of reference mixes (Dispergrader). However, such characterization mainly detects dispersion defects of a few tens of microns, and direct comparison of carbon black and silica mixes has to be done cautiously. In any case, it is necessary to make a mix, which means choosing a formula, a mixer, and mixing conditions; thus the result cannot be considered as an intrinsic dispersibility measurement of the filler, but just reflects the dispersibility of the filler in one mix with a set of mixing conditions.

(ii) Laser Granulometry

Recently, a method has been patented to determine filler dispersibility. It consists of measuring continuously the size of the filler by laser granulometry during an ultrasonic desagglomeration (WO9928376 et al., 1997). This characterization can be applied to any filler and is an intrinsic property; however, the use of water as a desagglomeration medium can be a problem because of its high polarity compared to elastomers.

8.3.3 Filler Physicochemistry

Compared to morphology, filler chemistry has been only slightly studied, partly because of the difficulty of such characterizations and more probably because since the 1970s reinforcement is broadly considered as a physical interaction between elastomer and filler. So carbon black chemical characterizations mainly date from the 1960s, and few new technical methods have been applied to carbon black surface characterization since then. The situation is somewhat different for silicas, because silica reinforcement is the consequence of a chemical reaction of silane with silica surface. Few studies have been published in the elastomer reinforcement area, probably because silica surface was already well characterized for other applications.

Concerning physicochemical characterization, the studies are limited to surface energy distribution determination, which will be discussed first.

(i) Surface Energy

Elastomer reinforcement by carbon black is generally considered as the consequence of the adsorption of polymeric chains onto carbon black surface. Therefore carbon black surface energy knowledge is of primary importance in carbon black characterization. However, very few carbon black surface energy measurements have been published; this can be easily understood considering the difficulty of such measurements on a highly heterogeneous and tortuous surface.

Immersion calorimetry allows carbon black (Beebe et al., 1947; Gonzalez-Martin et al., 1997) or silica (Donnet et al., 2000) surface energy determination. This technique can be used with different liquids or solutions of low mass elastomers, but it presents the main drawback of giving a mean surface energy value ($\sim$50–70 mJ/m^2), when surface adsorption of polymeric trains probably preferentially occurs on the highest surface energy sites.

This justifies the use of inverse gas chromatography (IGC) for filler surface energy characterization. This technique can be used in two very different modes: "infinite" or "finite" dilution.

In "infinite" dilution, very small quantities of alkanes of growing numbers of carbons are adsorbed onto carbon black (Wolff et al., 1994; Donnet and Lansinger, 1992; Wolff and Wang, 1992) or silica surface (Donnet et al., 2000); from their retention times, it is possible to calculate the dispersive component of surface energy $\gamma_{s,d}$. Because of the very low surface coverage during characterization, this value corresponds to the highest energetic sites (Custodero et al., 1992). According to this technique, surface energy grows with carbon black surface area; because high surface area carbon blacks are highly reinforcing, this result should be considered encouraging. Surface energy values obtained can reach values of 300–500 mJ/m^2, which can't be considered as realistic for carbonaceous surfaces. In addition, this determination seems to be also sensitive to chemical modification of carbon black surface even if the probes used should only characterize the dispersive part of surface energy.

"Finite" dilution is a more powerful technique in that it is possible to obtain the complete energetic site distribution for carbon black (Custodero et al., 1992; Wang and Wolff, 1991a,b) or silica (Wang et al., 1991; Jagiello et al., 1990). In this technique, surface is fully covered by the probe and the distribution is calculated by a specific posttreatment of desorption signal. Using this technique, carbon blacks present approximately the same surface energy distribution differing only in the number of adsorption sites. The energetic site distribution is particularly broad, with sites of high ($\sim$100 mJ/m^2) to low ($\sim$10 mJ/m^2) energy. Mean values are consistent with those obtained by immersion calorimetry.

Finally, it should be mentioned that a procedure similar to IGC, a "finite" dilution, can be applied to nitrogen adsorption isotherm and allow surface nanoroughness characterization of any filler (Lapra et al., 2004).

(ii) Surface Chemistry

(i) Carbon Black

(i) Impurities Because of its manufacturing process, carbon black surface includes some organic and mineral impurities (Schubert et al., 1969; Deviney, 1969).

Organic impurities are mainly polyaromatic hydrocarbons (PAHs) (Taylor et al., 1980). They correspond to partially unconverted fuel that has been readsorbed onto carbon black. These PAHs are present at a very low content and, because of their firm adsorption on carbon black, the extraction must be conducted in a Soxhlet apparatus with a strong solvent (toluene) and at high temperature (80°C). However, it has been demonstrated that organic impurities have no significant effect on carbon black reinforcement (Custodero et al., 1992).

Mineral impurities come from quench and pelletization steps in the carbon black production process. As presented before, the decrease in temperature of carbon black and exhaust gases is mainly obtained by injection of a great mass of water. Additional water is also added to carbon black during pelletization. Even if this water is purified, the remaining mineral salts precipitate onto the carbon black surface and, because of the high temperature, are reduced to basic salts. Mineral impurities of carbon blacks can easily be extracted by solubilization in water, as in the so-called "pH of carbon black," in which carbon black is suspended in water and the pH then filtered and the pH of the filtered water measured. Mineral impurities don't seem to alter carbon black reinforcement properties but they have a significant effect on vulcanization speed, which increases with the pH value of carbon black.

(ii) Oxygenated Functions Oxygenated functions on carbon black surface were observed in the early 1950s (Villars, 1947) and completely characterized by Boehm (1966). At this time, interaction between carbon black and natural rubber was considered the consequence of chemical reactions between the

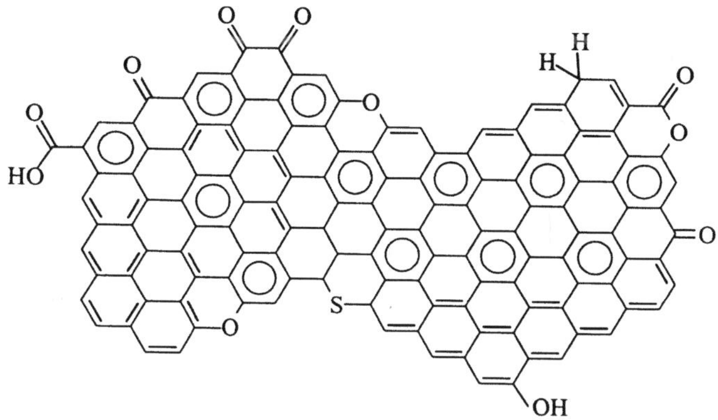

FIGURE 8.2 Chemical functions on carbon black surface (from Bueche, 1960).

carbon black surface's acidic groups and basic moieties present in the natural rubber structure (Donnet et al., 1961).

The carbon black surface function characterization consists of suspending a given amount of carbon black in solutions of known normality of basis of different strength: $NaHCO_3$, Na_2CO_3, NaOH in water, and EtONa in ethanol (Boehm, 1994). Then carbon black is filtered and the number of reacted acidic groups obtained by titrating the remaining basis in filtrate (Figure 8.2).

In the 1960s, carbon blacks were mainly prepared by *channel* processes, and their acidic functions were present at about 10^{-3}eq/g, which allows relatively easy determinations; but now, with *furnace* blacks, the surface acidic functions are generally of about 10^{-5} eq/g, and specific techniques (Bertrand and Weng, 1998) or drastic reaction conditions must be used (Custodero et al., 1992). Obviously, such delicate determination must be conducted on previously extracted black, in order to eliminate basic mineral impurities that would hinder any characterization of the rare acidic groups present on carbon black.

This observation allows us to believe that acidic groups on carbon blacks are mainly produced by surface oxidation in the production process, probably during drying following pelletization. Therefore, acidic groups could be considered an alteration of carbon black surface.

This point of view is supported by the fact that oxidized blacks, which have an acidic surface group content of about 10^{-2} eq/g, exhibit a very low reinforcement ability, even if their slow vulcanization is corrected.

(iii) Double Carbon Bonds, Hydrogen Content All chemical studies done in the past on carbon black have focused on chemical impurities or on functions produced by partial surface oxidation of carbon black, and not on its *own* surface reactivity.

Now carbon blacks cannot be considered as chemically inert surfaces. Their reaction with iodine or oxygen, their structure evidenced by STM (Custodero et al., 1992), demonstrates the presence of a great number of reactive double bonds on their surface.

Following Medalia and Kraus (1994), such double bonds could react with sulfur (Rivin, 1963, 1971; Lewis et al., 1970; Papirer et al., 1978), olefins (Rivin et al., 1968), and radicals (Donnet and Heinrich, 1960; Watson, 1955) to provide chemical bonding between carbon black surface and polymer, but their direct quantitative determination has never been obtained. Several authors have observed a strong correlation between hydrogen content of carbon blacks and their reinforcement ability. The content and reactivity of hydrogen present on graphitic edges have been determined by isotopic exchange and correlated to carbon black reinforcement ability (Papirer et al., 1969, 1971). This characterization is particularly difficult and usually hydrogen mass content is used; these values also are surprisingly well correlated to reinforcement ability of carbon blacks (Hess et al., 1988).

(ii) Silica

(i) Silanols Because of its numerous and longstanding uses in other applications, silica surface chemistry is clearly better known than that of carbon black (Legrand, 1998).

Silica surface chemistry is mainly defined by the surface content in silanols Si—OH; silanols can be "isolated," O≡Si—OH, or "geminated," $O{=}Si{=}(OH)_2$. They are generally highly associated by hydrogen bonds (Figure 8.3).

For *fumed* silicas, silanol content is about 2 Si–OH/nm^2 (Läufer, 1980; Humbert, 1995), but for *precipitation* silicas, silanol content can reach values as high as 6 to more than 10 Si–OH/nm^2 (Zaborski et al., 1989). It is essential to emphasize that such high silanol content cannot be considered as true per se: considering bond length and silicon-oxygen arrangement, a content of 2–3 Si–OH/nm^2 corresponds to a full coverage of the surface with silanols (Legrand, 1998). Such high values are generally attributed to the existence of polysilicic acid chains, $[Si(OH)_2]_n$–O–$Si(OH)_3$, and to the fact that BET surface area doesn't take into account pores of very small size in which some silanols are located.

Silanol surface content cannot reasonably be considered an indicator for silica reinforcement ability; indeed, because of its size, the coupling agent used to bond silica surface and elastomer cannot react with more than two or three

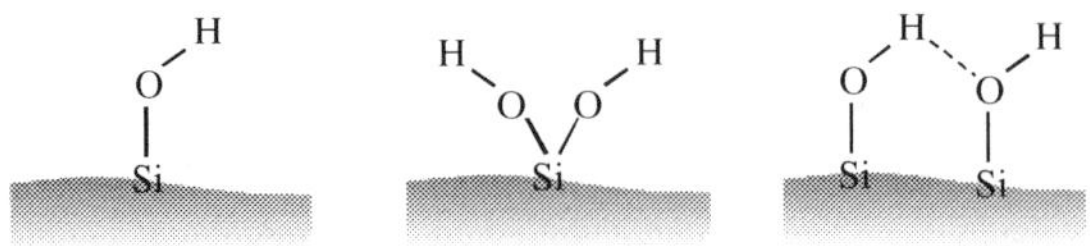

FIGURE 8.3 "Isolated," "geminated," and associated silanols.

silanols of the surface (Hunsche et al., 1997; Görl et al., 1997). Moreover, a high number of silanols will induce more associations by hydrogen bonding and so decrease their chemical reactivity.

(ii) Other Surface Functions Silica surface chemistry cannot be used to determine the silanol surface content. Particularly, for silicas synthesized by precipitation, some other chemical species can modify silica's surface reactivity.

For example, because of an incomplete hydrolysis of silicate, ≡Si–O–Na can be observed; because of an incomplete washing of silica after filtering, sulfate can be adsorpted onto silica surface. In addition, some salts used for silica processing can also be added and modulate its reactivity (Bomal et al., 1995).

Obviously, as discussed earlier in the silica production process, silica surface can also be modified by other chemical species added during and after its preparation. In any case, this modifier only changes silanol reactivity by enhancing (Cardona-Martinez and Dumesic, 1990; Uytterhoeven and Fripiat, 1962) or decreasing its acidity.

8.4 THE MIX: A NANOCOMPOSITE OF ELASTOMER AND FILLER

Even if the term *nanocomposite* is usually not used in reinforcement by particulate fillers, it would be particularly adapted: mixing of reinforcing solids and elastomers is not limited to the arithmetic "sum" of the properties of both taken independently but gives a synergetic alliance that achieves new properties.

Moreover, the term *filler* is more or less inadequate, because the particulate solid is not used to fill a *void*; that is, to diminish the cost of the elastomeric product. Indeed, elastomer and reinforcing filler should be considered as two inseparable parts of equal merit in the composite. As expressed nicely by Papirer, "carbon black and polymer is the wedding of the century."

This introduction is much more than a semantical debate; it underlines the importance of a *global* approach including reinforcing filler *and* elastomeric matrix relationships.

8.4.1 Dispersion, Aggregate Sizes, and Distances

(i) Dispersion

(i) Dispersion of Filler in Rubber Matrix

The quality of particulate filler dispersion in the elastomer matrix is of primary importance for compound mechanical and use properties (Cotten, 1983; Funt, 1986; Gerspacher and O'Farrell, 1993; Bomo and Morawski, 1983; Richmond et al., 1993).

In very recent years, filler dispersion characterization has been brought again into light because of the difficulties encountered to disperse silica in rubber (Bomal et al., 1998, 1993).

Usually, filler dispersion is achieved in a Banbury mixer and is presented to proceed in two different steps. In the first step, filler is distributed in pure polymer in the form of pellets or subpellets (i.e., *agglomerates*) (Boonstra and Medalia, 1963) and then, in the second step, subpellets are eroded into aggregates by an "onion peeling" process (Shiga and Furuta, 1985). This second phase, which eliminates agglomerates and determines interactual-aggregate distances, is much longer and necessitates higher mechanical energetic input.

As discussed earlier, dispersibility and so dispersion is mainly controlled by the strength of interaction between agglomerates and/or aggregates, which is a direct consequence of their surface energy. The influence of surface energy on dispersion has been clearly demonstrated by the use of different matrices; when matrix surface energy increases and becomes closer to filler surface energy, dispersion is facilitated.

(ii) Characterization

Filler dispersion characterization is particularly difficult because it must be conducted on a very broad range of scales (Nikiel et al., 2000): from microscopic undispersed agglomerates, which are defects and will decrease a product's life, to nanoscopic distances between aggregate, which will greatly influence reinforcement level (Coran and Donnet, 1993). Indeed, filler dispersion characterization has been conducted not only with a large number of analytical techniques: optical, electronic (Hess et al., 1994), and atomic-force microscopy (Maas and Gronski, 1994; Lapra, 1999), but also X-rays (Hess, 1991; Ehrburger-Dolle et al., 2001; Legrand et al., 1992) and neutron diffraction. The main difficulty is to recompose a global image from very different data, because any of these techniques gives directly a complete description of dispersion.

(iii) Influence of Filler's Properties

In addition to filler's surface energy, which is of major importance, dispersion can also be influenced by filler's morphological properties.

Dispersion is highly influenced by filler surface area: the higher the surface area, the lower the dispersion (Sone et al., 1992). This result is probably due to the fact that high surface area usually has smaller aggregates, which will develop more interactions with their neighbors in the dry state.

Filler structure also has a neat influence on dispersion: the higher the structure, the higher the dispersion. This result is well established and likened to the fact that more "open" aggregate structures develop a lower number of contact with their neighbors in the dry state.

(ii) Object Sizes in the Mix

As discussed earlier, filler occurs as a distribution of different aggregate sizes. This characterization is interesting as a potential, but this final aggregate size distribution could not be achieved in mixes. Indeed, dispersion can be incomplete and some agglomerates can be left in the mix, shifting the actual size of reinforcing objects to a higher value. On the other hand, because of the high shear strengths developed during mixing, some aggregate breakage can also occur and produce an actual size of reinforcing objects lower than expected on the basis of filler characterization. Therefore, considering the possibility of filler in the mix, we will use the term *object*, which can refer to either aggregates *or* agglomerates.

Obviously, aggregate size distribution characterization in the mix is very delicate. Some transmission electron microscopy observations have been conducted on microtome thin cuts, but such characterizations are restricted to a small number of aggregates and can only lead to qualitative conclusions (Hess, 1991). Direct characterization of object distribution in the mix has also been conducted using X-ray (Young et al., 1986) or neutron diffraction, but such approaches are strongly limited by the high concentration of filler objects and their refraction index, which is relatively close to that of rubber. One other way to characterize object size distribution is to extract the filler from the mix by thermal or catalyzed polymer decomposition; these procedures probably greatly affect object size, because of possible reagglomerations.

The characterization of filler object size distributions in the mix mainly remains a domain to develop. In any case, it is generally accepted that aggregate size of carbon black decreases during mixing (Herd et al., 1992; McDonald and Hess, 1977), even if any of the methods used can eliminate possible artifact. About this, it is interesting to recall that crushed DBP, as previously discussed, has been developed to take into account possible aggregate breakage. Considering silica, explicit data has been published considering this problem, but the same possible aggregate breakage seems also possible.

(iii) Distances

Characterization of interobject distances in the mix is the reciprocal problem of object size determination and involves the same difficulties—and the same lack of experimental data.

However, it should be noted that usual filler loadings used in elastomers (around 20% in volume) are very close to the maximal fraction that can be incorporated into the elastomeric matrix. This fraction is very low compared to compact spheres arrangement, but, as discussed earlier, filler's aggregates present a highly open structure; moreover, maximum loading is highly dependent on the filler's structure.

Based on simple models, it has been demonstrated that interobject distances are in the range of a few tens of nanometers; this result is consistent with the fact that carbon black mixes are electrically conducting, which implies that inter-object distances are low enough to allow tunnel conduction (Medalia, 1986).

Moreover, it has been demonstrated that optimum filler loading corresponds to very similar interaggregate distances, without regard to aggregate size or structure (Sone, 1999). This very interesting result underlines the decisive influence of interaggregate distances in elastomer reinforcement.

8.4.2 Filler-Elastomer Interactions

(i) Carbon Black

(i) Elastomer Adsorption

(i) "Filler Network" Because of carbon black high surface energy, elastomeric chains are strongly adsorbed onto its surface. This adsorption, even if it is limited to a small part of the elastomeric chains, called "trains," drastically slows down their mobility (Legrand et al., 1992; O'Brien et al., 1976; Vilgis and Heinrich, 1994).

As a simplified—but slightly incorrect—picture, it can be considered that "trains" have a lowered transition temperature (Kaufmann et al., 1971; De Candia et al., 1996). The exact thickness of this layer remains disputed (O'Brien et al., 1976; Leisen et al., 1998; Semaan et al., 2001), but values of 1–5 nm are usually considered (Kraus, 1971). Anyway, it is very noticeable that such thickness corresponds at least to 3–15% of total elastomeric phase (Rigbi, 1993). Taking into account carbon black structure/tortuosity, values as high as 30% have been proposed.

A more accurate approach considers that elastomeric chains present a gradient of mobility coming from carbon black surface to bulk.

The high surface areas and loadings of carbon blacks used in elastomer reinforcement induce such small distances between reinforcing objects that almost any elastomeric chain contacts at least one aggregate (Sone, 1999). In addition, because the statistical size of polymeric chains is in the range of interaggregate distances, close neighboring objects are probably bounded together by chains adsorbed onto both aggregates.

The bonding of carbon black aggregates constitutes the *filler network*.

(ii) Bound Rubber The filler network is clearly evidenced by *bound rubber* measurements (Stickney and Falb, 1964; Kraus, 1965; Ban et al., 1974a,b; Villars, 1996). *Bound rubber* is a very specific measurement done on green mixes; it consists of determining the part of rubber that cannot be extracted by a good solvent (Dannenberg, 1986). A small part of rubber, previously weighted, is put in toluene and submitted to extraction at a room temperature. Samples swell but usually do not delicate; the surrounding solvent is regularly renewed to ensure an optimal extraction, and samples are weighted to follow extraction progression. After 1 or 2 weeks, extraction is completed and samples are dried and weighted. The weight difference corresponds to the soluble part of elastomer; that is, to the chains that were weakly adsorbed on the carbon black surface.

When the extraction temperature increases, the bound rubber decreases and, above about 80°C, samples completely delitate, indicating the disappearance of the continuous networking of carbon black aggregates by elastomer chains.

Because *bound rubber* measures elastomer adsorption onto filler surface, it is highly dependent on filler loading, specific surface, and structure, which are parameters that can be measured independently (Meissner, 1974). However, at given loading and carbon black surface area and structure, it has been demonstrated that *bound rubber* is also dependent on carbon black surface energy (Wolff et al., 1993).

(ii) Chemical Surface Bonding

Before the 1970s, carbon black reinforcement of elastomers was generally considered chemical by nature (Wiegand, 1925). It was supposed that carbon black surface acidic groups were reacting with natural rubber basic moieties conducting to a strong covalent bond that was responsible for carbon black reinforcement ability.

In the 1970s, *furnace* gradually replaced *channel* technology. But even if *furnace* carbon blacks present 10 times less surface acidic groups, their reinforcing ability remains unchanged or increased. On the other hand, synthetic elastomers, which obviously have more basic moieties than natural rubber, were also perfectly reinforced by carbon black. In addition, the preparation of surface-oxidized carbon or grafted blacks (Le Bras and Papirer, 1983) leads to a decreased reinforcement ability.

It was then obvious that chemical reaction of carbon black surface acidic groups with natural rubber basic moieties was not responsible for reinforcement. So the newly discovered mechanism of "molecular slippage," proposed by Dannenberg and based on molecular adsorption, was quickly and fully adopted (Dannenberg, 1975; Dannenberg and Brennan, 1966; Dannenberg, 1960).

These observations do not allow the full refutation of chemical reinforcement theory. Chemical bonding by acidobasic reaction is clearly rejected, but, following Medalia and Kraus (1994), other chemical reactions could occur in carbon black mixes. For example, elastomeric chain breaking during mixing can lead to radicals that could react with carbon black surface (Le Bras and Papirer, 1979); sulfur-direct bonding of elastomer and carbon black could also be envisaged (Rivin et al., 1968).

(ii) Silica

(i) Silica-Silane Reaction

In contrast to carbon black, it is necessary to use a coupling agent to achieve silica elastomers reinforcement. TESPT, triethoxysilylpropyltetrasulfide, is the most widely used coupling agent.

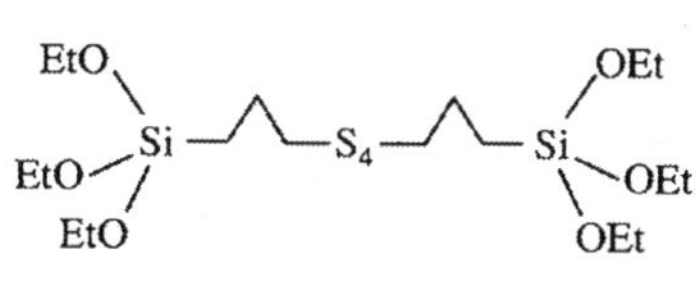

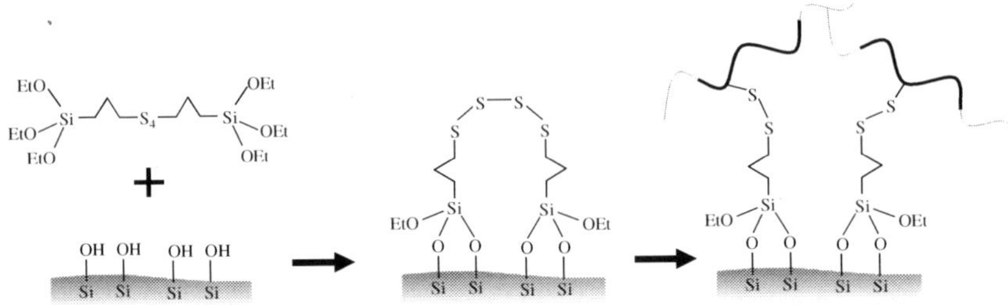

FIGURE 8.4 TESPT loadings are about one $Si(OEt)_3$ per nm^2, which roughly corresponds to a complete coverage of silica surface. So, for precipitation silicas with content 6–10 Si–OH/nm^2, many surface silanols will remain unreacted.

TESPT is a bifunctional molecule with a triethoxysilyl moiety reactive toward silica's silanols and a polysulfidic moiety that reacts with elastomeric chains.

Because reaction temperatures are somewhat different, the reaction of TESPT with silica surface mainly occurs during mixing when the reaction of polysulfidic moiety takes place during curing (Hunsche et al., 1997; Görl et al., 1997) (Figure 8.4).

(ii) Polymer Adsorption

When a coupling agent is used to generate a covalent bond between silica surface and elastomeric chains, it also limits polymer adsorption because of its shielding effect. So in silica-silane-elastomer compounds, the "filler network" will be much lower than in carbon black-elastomer systems. However, it remains *qualitatively* the same, and elastomer chain mobility is also limited in the close neighboring of silica surface (Hommel et al., 1993).

Obviously, if any coupling agent is used, polymer adsorption will naturally occur (Bomo, 1989; Killian et al., 1987); in addition, because of the high polarity of silica, some direct interaction between silica aggregates will also take place and constitute an additional filler-filler network. These effects will not happen in silica-reinforced systems when an appropriate amount of coupling agent is used.

Bound rubber determination is also applied to silica compounds, even if the numerous possible interactions naturally limit the interpretation of the values (Ashida et al., 1977; Strauss et al., 1994).

8.5 MECHANICAL PROPERTIES OF FILLED RUBBERS

8.5.1 Mechanical Properties in Green State

The increased lifetime expected from reinforcement by particulate fillers naturally refers to cured pieces. The incorporation of reinforcing fillers greatly changes the viscosity of green compounds, conducting to a mainly plastic behavior that allows their processing.

(i) Viscosity

It is generally reported that elastomers filled with a volume fraction ϕ present a viscosity following (Guth and Gold, 1938):

$$\eta = \eta_0 \cdot (1 + 2.5 + \phi + 14.1 \cdot \phi^2).$$

It is interesting to remember that this relationship has been established for solid spherical objects having any interaction among them and/or with the surrounding medium. As widely discussed earlier in this chapter, reinforcing systems are very far from this: carbon black or silica aggregate is highly structured, and elastomeric chains strongly adsorb onto carbon black surfaces.

(ii) Occluded Rubber

Elastomer interaction with carbon black or silica is very difficult to estimate and correct. On the other hand, it is much easier to take into account the actual volume of aggregate in the mix, and it has been proposed that a corrected volume fraction ϕ_c be used, which integrates the influence of filler structure as represented by DBP (Medalia, 1974; Kraus, 1970; Medalia, 1973).

$$\phi_c = \frac{\phi}{2} \cdot \left[1 + \frac{1 + 0.02139 \cdot \text{DBP}}{1.46}\right] \quad \text{and} \quad \eta = \eta_0 \cdot (1 + 2.5 \cdot \phi_c + 14.1 \cdot \phi_c^2).$$

The ϕ_c value represents the "actual" size of filler aggregates in the mix; it includes, naturally, the filler object itself plus a significant volume of polymer that is shielded from deformation by aggregate tortuosity. This part of the polymer that will not be deformed is usually called *occluded rubber* (Kraus, 1970; Medalia, 1970;1974).

Nevertheless, occluded rubber must not be confused with the polymer part whose molecular mobility is changed by adsorption. *Occluded rubber*, which is mainly trapped in aggregate fractal sites, only represents a part of the volume of elastomer whose molecular motion is slowed down.

Occluded rubber and viscosity increases with filler structure and loading; on return, specific surface area of the filler has an influence on green mix viscosity.

(iii) Shear Dependence of Viscosity, Non-Newtonian Behavior

The presence of reinforcing fillers also increases the non-Newtonian behavior of elastomers. This effect is mainly due to the fact that the incorporation of fillers in elastomers decreases the volume of the deformable phase. As discussed in the following text, this decrease is not limited to the actual volume of the filler, but must also include the existence of occluded rubber. So, when filled mixes are submitted to shear forces, because of the lower deformable volume, the actual deformation and speed of deformation are much higher than in unfilled mixes (Bueche, 1961; Medalia, 1974). This phenomenon is usually called *strain amplification effect*; obviously *strain amplification* is not specific to reinforced systems but to any filled polymer.

The influence of filler is not limited to this enhancement of the non-Newtonian behavior of elastomers. At very small shear rates, filled green compounds also exhibit an additional increase of viscosity that cannot be explained by *strain amplification*. This effect is usually attributed to the existence of the filler network: the direct bonding of reinforcing objects by adsorbed chains implies a increased force to be broken. Obviously this influence can be observed only at very low strain, because a very small increase of interaggregate distances immediately implies a desorption of the bridging elastomeric chains.

8.5.2 Mechanical Properties in Vulcanized State

As for pure elastomers, the vulcanization step provides sulfur bridges between elastomeric chains and connects them into an infinite network. Vulcanization is supposed to be mainly unaffected by the presence of the reinforcing fillers and transforms the roughly plastic green mixes into viscoelastic vulcanizates.

For silica mixes, the high temperature of the vulcanization step allows the reaction of the polysulfidic moiety of the coupling agent with elastomer, ensuring the chemical covalent bonding of polymer to silica surface.

As in the green state, the *strain amplification*, due to the limited volume of the actually deformable phase, remains the first-order result of filler incorporation. For a given macroscopic deformation, the actual deformation of the polymeric matrix will always be much higher, obviously depending on the filler volume and its structure, which defines occluded rubber volume.

The viscosity equation is usually generalized to Young or shear modulus G^* (Guth and Gold, 1938; Medalia, 1973; Smallwood, 1944):

$$G^* = G_0^* \cdot (1 + 2.5 \cdot \phi_c - 14.1 \cdot \phi_c^2),$$

where G_0^* is the shear modulus of the unfilled vulcanized matrix at the same shear strain. Obviously, it is possible to vary the modulus by changing vulcanization conditions, providing more or fewer sulfur bridges and so higher or lower G^* values. Anyway, as evidenced by the equation, vulcanization changes G_0^* but roughly does not affect reinforcement by itself.

(i) Small-Strain Properties, Dynamic Viscoelastic Measurements

(i) Payne Effect

Reinforced vulcanized samples generally present a marked viscoelastic behavior that is usually studied by dynamic viscoelastic measurements. In this experiment, a sample is subjected to periodic sinusoidal shear strain γ

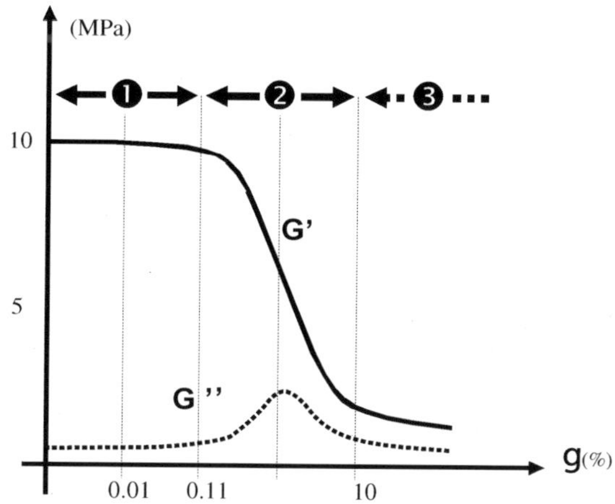

FIGURE 8.5 Schematic illustration G′ and G″ variations.

(at defined frequency ω and temperature T). Its dynamic shear modulus G* is complex and can be written as the sum of the storage modulus G′, and the loss modulus G″.

$$G^*(\gamma) = G'(\gamma) + i \cdot G''(\gamma).$$

The dynamic storage modulus G′ presents an interesting variation when γ increases: at very low shear strain, G′ is constant (phase 1 in Figure 8.5), then strongly decreases (phase 2 in Figure 8.5), and reaches a plateau value (phase 3 in Figure 8.5). This evolution of G′ is usually described as the *Payne effect* (Harwood et al., 1965).

This change in G′ also corresponds to an important variation of G″ that passes through a maximum value. δ, the phase angle between stress and strain, is given by:

$$\tan(\delta[\gamma]) = \frac{G''(\gamma)}{G'(\gamma)}.$$

The evolution of G′ and G″ in the range of 0.1–0.5 strain amplitude is of major importance because this domain corresponds to the most common solicitations of filled rubber compounds, for example in tire tread applications (Medalia, 1991).

G″, the loss modulus, must not be confused with hysteretic losses of which the expression naturally depends on solicitation mode: hysteretic losses are proportional to G″ (constant strain), or $G''/G^{*2} \sim \tan(\delta)G'$ (constant stress) or to tan (δ) (constant energy).

It is also important to stress that filled elastomers are a very complex thermorheological system: particularly, G′ and G″ variations do not follow the same laws in frequency and temperature (Duperray and Leblanc, 1982).

(ii) Mechanism

The Payne effect is widely accepted as the mechanical consequence of the progressive destruction of the "filler network" under shear strain.

The attribution of the Payne effect to the filler network is strongly supported by the fact that carbon black pastes, made with carbon black and low molecular weight oils, present very similar G* levels at very low shear strain (Payne, 1965). Obviously, when strain increases, G* drops drastically for carbon black pastes and much more slowly for filled rubber compounds, because of the *progressive* desorption of elastomeric chains.

Dannenberg's molecular slippage model, which will be discussed in detail in the next section, gives a good schematic view of the molecular mechanism that is responsible for the Payne effect. Numbers ❶, ❷, and ❸ refer to Figure 8.5.

❶ At equilibrium, elastomer chains are adsorbed onto filler surface (state ⓪). When strain increases, it induces a progressive extension of elastomer chain segments that bridges filler particles (state ❶). Obviously, this extension is much greater than macroscopic deformation because of *strain amplification*. At very low strain, the macroscopic deformation energy is stored in elongated chains as elastic energy and so can be fully recovered when strain decreases: G″ is low and constant.

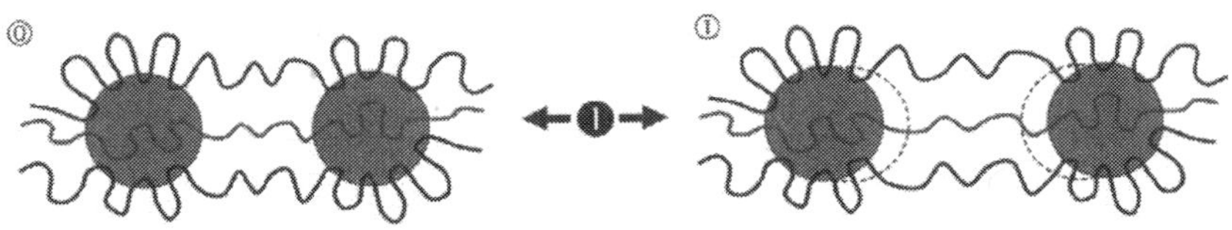

❷ At higher deformation, it is necessary to decompose the solicitation cycle. During the first extension at higher rate, stored elastic energy overpasses adsorption energy, and elastomer chains progressively desorb from filler surface (state ②, ★ sites desorbed).

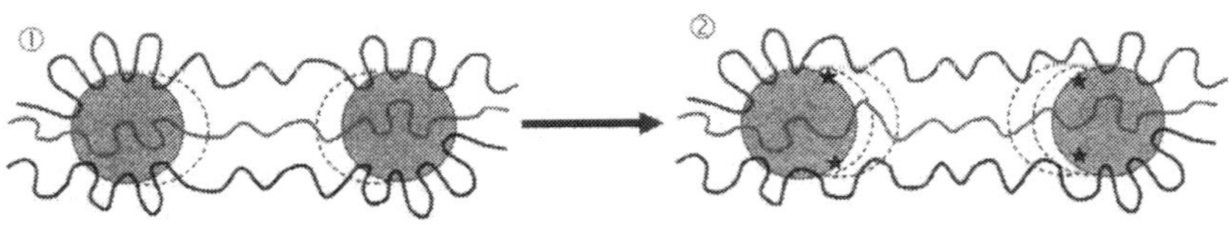

This desorption lengthens the bridging polymer segments that have to direct impacts. First, G′, which is roughly inversely dependent on the length of bridging chains, decreases; second, a part of the initially elastic energy stored in deformed chains is converted into molecular mobility and mechanically lost, which corresponds to an increase in G″.

The decrease of deformation that immediately follows the first extension at the higher rate does not exactly lead to initial state ①; indeed, in the short time of the dynamic deformation cycle, adsorption cannot reach equilibrium state and remains imperfect, as illustrated by state ③.

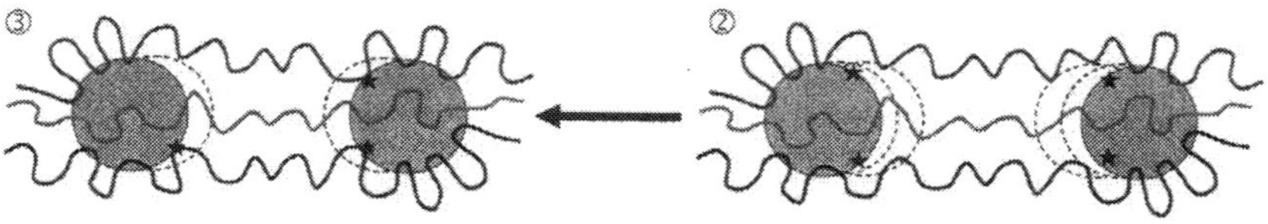

Thus during phase ❷, bridging elastomeric chains undergo adsorption-desorption cycles between "pseudo" equilibrium states ② and ③:

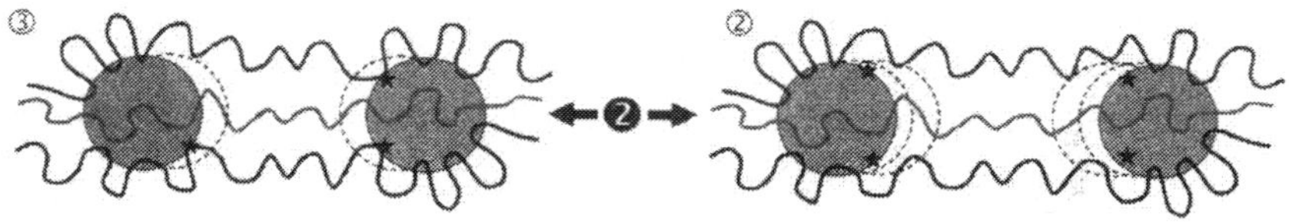

Obviously, elastomer desorption occurs gradually, because of the very broad interaggregate distance distribution that induces an also broad distribution of bridging elastomer segments. This explains the smooth decrease of G′ for mixes and its step shape for shorter molecules such as oils.

❸ Progressively, desorption induces a homogenization of bridging elastomer segment lengths. This homogenization and the stabilization of the modulus will be discussed in detail in the next section.

(iii) Hysteresis

It appears from the previous mechanism that G′ and G″ variations will be directly influenced by interaggregate distances (Sone, 1999) and the strength of elastomeric chain adsorption on the filler surface.

(iv) Interaggregate Distances

Any change in mix composition or processing that influences interaggregate distance distribution will change hysteresis: the lower the average distance, the higher the hysteresis and vice versa (Harris and Wise, 1965; Wang et al., 1993).

Filler loading evidently decreases interaggregate distances. When filler mean aggregate size decreases, at the same loading level, the number of reinforcing objects increases and diminishes mean interaggregate distance. Increased structure provides, at the same loading ratio, lower interaggregate distances. Thus, increasing loading, surface area (i.e., decreasing aggregate size), or structure induces higher hysteresis of mixes.

On the other hand, low hysteresis mixes can be achieved by dispersion methods that increase interaggregate distances; prolonged or two-stage mixing or master-batch techniques have been used in order to decrease hysteresis (Hess, 1991; Sone et al., 1992).

(v) Adsorption Strength

Coupling agents that mask filler surface strongly reduce elastomer adsorption and thus hysteresis of mixes. This partially explains the low hysteresis value of silica mixes (Harris and Wise, 1965).

It is also possible to use functionalized elastomers to reduce mixes hysteresis; such elastomers have specific chemical moieties that react with the filler surface. These reactions lower hysteresis because of the decrease of polymer dangling chains and by shielding of the filler surface (Nagata et al., 1987; Tsutsumi et al., 1990).

(vi) Large-Strain Properties

(i) Observations At large strain, dynamic viscoelastic measurements cannot be made accurately because of the important self-heating of the sample during the experiment. Therefore large-strain properties are usually determined by uniaxial extension (Soos et al., 1982; Soos, 1984) (Figure 8.6).

As was observed for G* in dynamic shear-strain measurements, a clear decrease of Young's modulus $E + \sigma/\varepsilon$ is observed at low strain ($\varepsilon < 1$). This corresponds to the previously described phases ❶ and ❷, even if phase ❶ is obviously not observable (see Figure 8.5).

At higher extension rates ($\varepsilon > 1$), Young's modulus increases and reaches a "pseudo maximum" just before the sample break.

A very significant observation is the *stress softening effect*, also called the *Mullins effect* (Mullins and Tobin, 1965; Mullins, 1969). In this experiment, a compound sample is stretched to ε_1 and returned to zero strain, then stretched again. For strain below ε_1, its stress-strain curve is significantly below the first one but rejoins it at ε_1. *Stress softening* is dependent on the initial strain level; it can be partially reduced by thermal treatment but not be totally effaced (Figure 8.7).

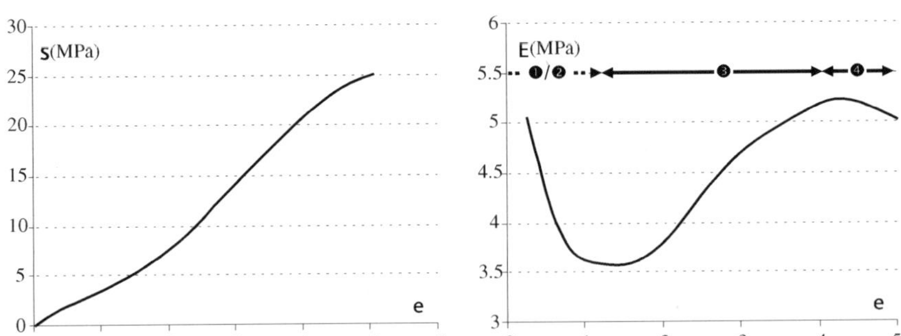

FIGURE 8.6 Stress and Young's modulus of reinforced compound. (Data from Medalia and Kraus, 1994.)

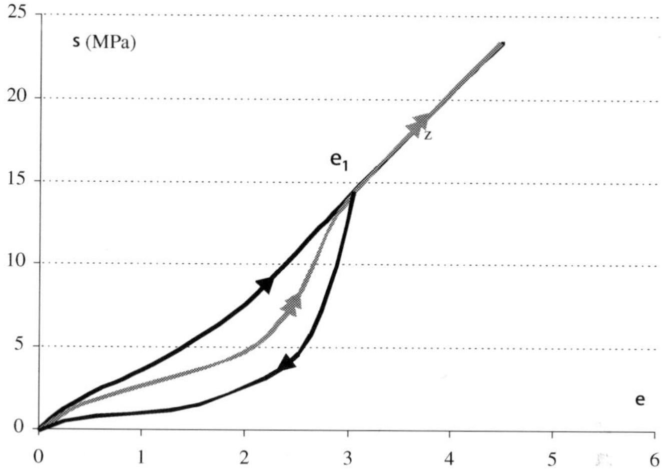

FIGURE 8.7 Schematical illustration of *stress softening effect*.

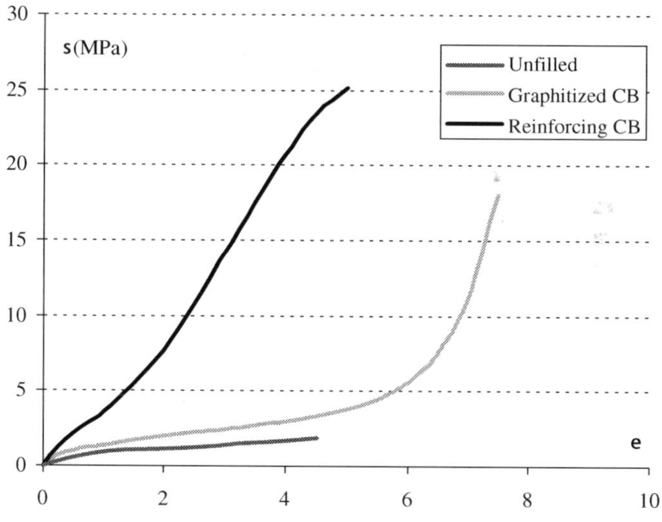

FIGURE 8.8 Stress-strain curve of unfilled, graphitized, and reinforcing carbon black samples. (Data from Medalia and Kraus, 1994.)

(ii) Interpretation The paradox of reinforcement by particulate fillers is that there is a simultaneous increase of modulus *and* elongation at break. This fact is clearly illustrated by the comparison of stress-strain curves of pure and carbon-black-filled elastomer (Figure 8.8).

The modulus increase is the logical consequence of *strain amplification* due to the replacement of a part of elastomeric deformable phase by a particulate rigid filler: for a macroscopic deformation ε, the *local* deformation of bridging chains is much higher. Strain amplification should also induce a neat decrease in elongation at break, which is not observed: here is the paradox.

Even if this point clearly remains mainly undecided, one way to surpass this paradox is to consider that fillers allow a locally more cooperative sharing of the stress.

In the early 1960s, Bueche was probably the first to consider carbon black as a part of a polyfunctional network (Bueche, 1961;1960). In his model, carbon black aggregates were chemically linked by chains and constitute what Medalia and Kraus describe as a "giant multifunctional crosslink" network (Medalia and Kraus, 1994).

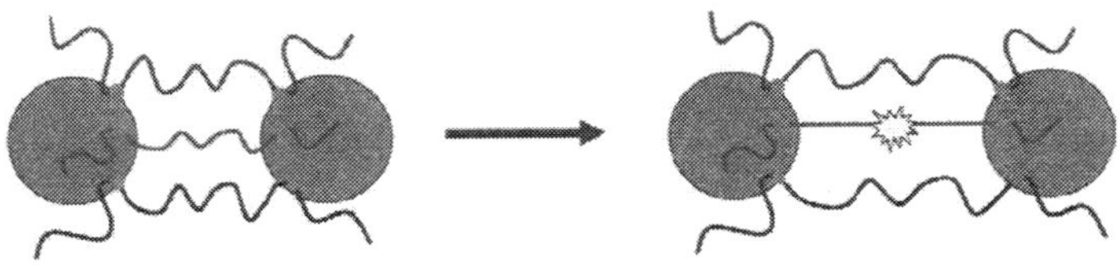

Even if Bueche's model tried to give a molecular origin of reinforcement, it remains difficult to consider that it ensures a massive local sharing of the stress: when the shortest chain reaches its finite extensibility, it is really not evident that a large part of the stress is shared by other bridging chains. Moreover, Bueche's model supposes chemical bonding of elastomeric chains, which is, *a minimo*, debatable (see previous text). In any case, it must be mentioned that chain breaks during extension have been demonstrated using ESR (Sullivan and Wise, 1967).

At the end of the 1960s, Dannenberg completely renewed reinforcement understanding by proposing the "molecular slippage" model that we have previously used to illustrate Payne's effect (Dannenberg, 1975, 1986). In contrast to Bueche, Dannenberg suggested that interaction between elastomer and carbon black was mainly caused by adsorption and not by chemical bonding. Because of its low energy and reversibility, adsorption permits elastomer-filler contacts to change continuously and so allows the homogenization of bridging segment lengths, which ensures the local sharing of stress (Medalia and Kraus, 1994; Dannenberg and Brennan, 1966; Dannenberg, 1966).

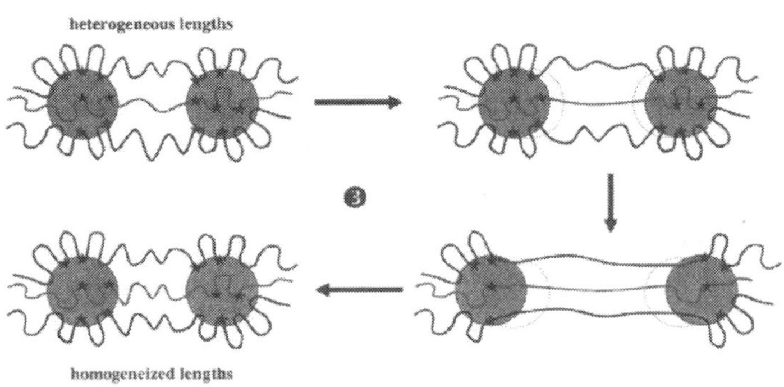

It is noteworthy that the concept of "giant multifunctional crosslink" network associated with the "molecular slippage" model proposes a possible mechanism for length homogenization of bridging chains and gives a rather satisfactory answer to the reinforcement paradox. So the increase of modulus in phase ❸ is due to this stress sharing between bridging chains.

Obviously, the obtained homogenization directly depends on the maximum strain at which the compound has been accommodated. In other words, chain segment lengths are homogenized for any strain below the maximum strain; at higher strains, a new homogenization should occur (Gent, 1974). This naturally corresponds to the so-called stress softening effect previously described (Mullins, 1969).

The stress sharing by segment homogenization is naturally limited by the number of bridging segments between reinforcing objects; when all connecting chains reach approximately the same length, modulus is at its maximum. A further increase of strain produces total chain slippage of the shortest chain; then the number of bridging segments decreases (phase ❹). Each remaining bridging chain must support an increased force that produces their massive dewetting and macroscopic break. The filler dewetting under strain has been evidenced by TEM direct observation (Hess et al., 1967).

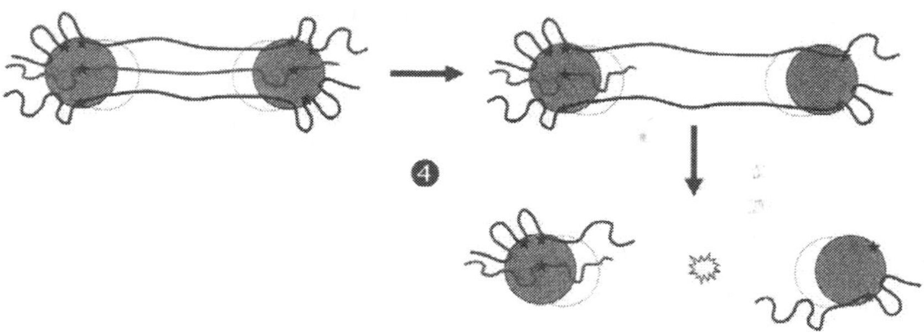

8.5.3 Applications

As discussed in the introduction, reinforcement of elastomers can be considered only for a specific application, because it corresponds to an increase of product service life. Hence, in order to give some practical illustration of the different topics discussed earlier, we will present some results about carbon black reinforcement for materials subjected to wear; for example, reinforced compounds used for tire treads.

Carbon black loading, surface area, and structure basically increase wear resistance; for very high loadings or surface area, a significant decrease in wear resistance is observed (Figure 8.9); this effect can be attributed to deficient dispersion (Sone, 1999).

Carbon black surface activity, as revealed by hydrogen content, has also a significant influence on wear resistance (Hess et al., 1988) (Figure 8.10).

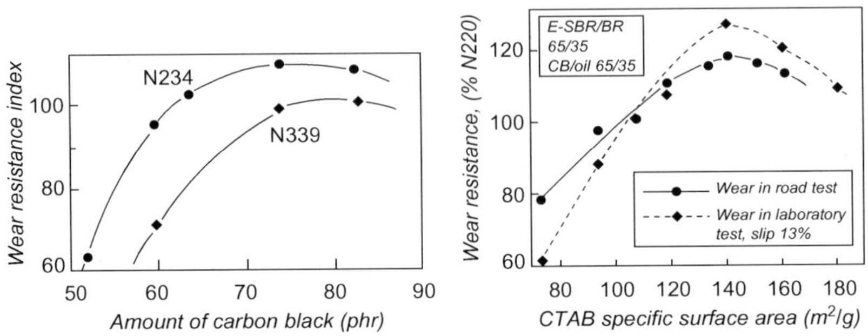

FIGURE 8.9 Wear resistance index versus carbon black loading and surface area (from Sone, 1999).

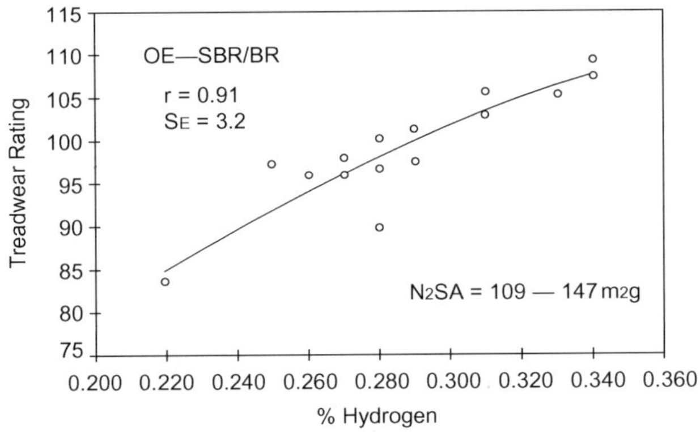

FIGURE 8.10 Treadwear versus carbon black hydrogen content (Hess et al., 1988).

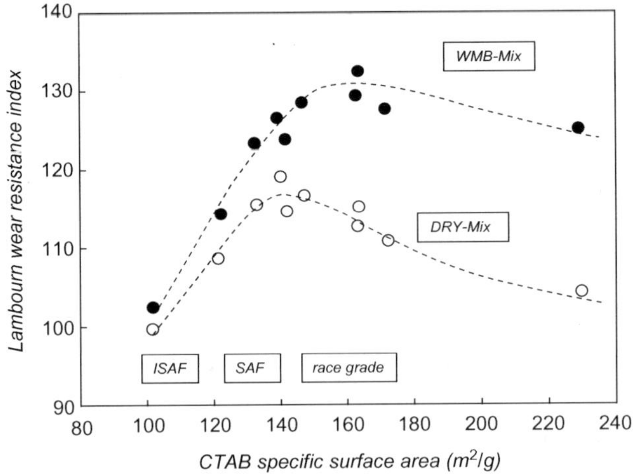

FIGURE 8.11 Comparison of abrasion resistance for dry mixes and masterbatches (from Sone et al., 1992).

Carbon black dispersion also influences abrasion resistance, and the maximum observed for high carbon black loadings and surface area can be significantly shifted by using specific dispersion techniques like masterbatching (Sone et al., 1992; Wang et al., 2002) (Figure 8.11).

REFERENCES

Allen, T., 1987. Photocentrifuges. Powder Technol. 50, 193.
Ashida, M., Abe, K., Watanabe, T., 1977. Int. Polym. Sci. Technol. 4, T42.
Ban, L.L., Hess, W.M., Papazian, L.A., 1974a. Rubber Chem. Technol. 47, 585.
Ban, L.L., Hess, W.M., Papazian, L.A., 1974b. Rubber Chem. Technol. 47, 858.
Beebe, R.A., Polley, M.H., Smith, W.R., Wendell, C.B., 1947. J. Am. Chem. Soc. 69, 2294.
Bele, M., Kodre, A., Grdadolnik, J., Pejovnik, S., Besenhard, J.O., 1998. Carbon 36, 1207.
Bertrand, P., Weng, L.T., 1998. Rubber Chem. Technol. 72, 384.
H.P. Boehm, 1966. Angew. Chem. 78, 617; Boehm, H.P., 1973. Farbe Lak 79, 419.
Boehm, H.P., 1994. Carbon 32, 759.
Bomal, Y., Cochet, P., Dejean, B., Machurat, J., 1993. Rubber World 6, 33.
Bomal, Y., Chevallier, Y., Cochet, P., 1995. Novel method for preparing precipitated silica, novel aluminium-containing precipitated silicas, and use thereof for reinforcing elastomers. Patent WO9630303.
Bomal, Y., Cochet, P., Dejean, B., Gelling, I., Newell, R., 1998. Kautsch. Gummi Kunstst. 51, 259.
Bomo, F., 1989. Makromol. Chem. Macromol. Symp. 23, 321.
Bomo, F., Morawski, J.C., 1983. In: Rubber Division American Chemical Society, Houston, October.
Boonstra, B.B., Medalia, A.I., 1963. Rubber Age 92, 892; Boonstra, B.B., Medalia, A.I., 1963. Rubber Age 93, 82.
Bradley, R.H., Sutherland, I., Sheng, E., 1996. J. Colloid Interf. Sci. 179, 561.
Bueche, F., 1960. J. Appl. Polym. Sci. 4, 107.
Bueche, F., 1961. J. Appl. Polym. Sci. 5, 271.
Cardona-Martinez, N., Dumesic, J.A., 1990. J. Catal. 125, 427.
Chevallier, Y., Morawski, J.C., 1986. EUCHEM.
Cochet, P., Barruel, P., Barriquant, L., Grobert, J., Bomal, Y., Prat, E., 1993. Meeting Rubber Div. Am. Chem. Soc. Orlando, 26–29 October, 1–43.
Cochet, P., Barruel, P., Barriquand, L., Grobert, J., Bomal, Y., Prat, E., 1994. Rubber World 135, 20–24.
Coran, A.Y., Donnet, J.B., 1993. Rubber World 65, 1016 (1993); A.Y. Coran, J.B. Donnet. Rubber Chem. Technol. 65, 998.
Cotten, G.R., 1983. In: Meeting Rubber Division American Chemical Society, Houston, vol. 25–2 (8), p. 38.
Creeden, D., 1997. The Carbon Aggregate 4, 1.
Custodero, E., 1992. Caractérisation de la surface des noirs de carbone, nouveau modèle de surface et implications pour le renforcement. Ph.D. Thesis, Université de Haute Alsace.
Dannenberg, E.M., 1975. Rubber Chem. Technol. 48, 410.
Dannenberg, E.M., 1986. Rubber Chem. Technol. 59, 512.
Dannenberg, E.M., 1996. Trans. Inst. Rubber Ind. 42, T26.
Dannenberg, E.M., Brennan, J.J., 1966. Rubber Chem. Technol. 39, 597.
De Candia, F., Carotenuto, M., Gargani, L., Guadagno, L., Lauretti, E., 1996. Kautsch. Gummi Kunstst. 49, 99.
Deviney, M.L., 1969. Adv. Colloid Interf. Sci. 2, 237.
Dollinger, R.E., Kallenberg, R.H., Studebaker, M.L., 1967. Rubber Chem. Technol. 40, 1311.
Donnet, J.-B., Custodero, E., 1992. C. R. Acad. Sci. Ser. 2 314, 579.
Donnet, J.-B., Custodero, E., 1992. Carbon 30, 813.
Donnet, J.-B., Heinrich, G., 1960. Bull. Soc. Chim. Fr. 135, 1609.
Donnet, J.-B., Lansinger, C.M., 1992. Kautsch. Gummi Kunstst. 45, 459.

Donnet, J.-B., Wang, T.K., 1994. Analusis 22, M24; Donnet, J.-B., Custodero, E., Wang, T.K., 1996. Kautsch. Gummi Kunstst. 49, 274.
Donnet, J.-B., Heinrich, G., Riess, G., 1961. Rev. Gen. Caoutch. Plast. 38, 1803; Donnet, J.-B., Heinrich, G., Riess, G., 1964. Rev. Gen. Caoutch. Plast. 41, 519.
Donnet, J.-B., Papirer, E., Vidal, A., 1975. Grafting of macromolecules onto carbon blacks. In: Walker Jr., P.L., Thrower, P.A. (Eds.), Chemistry and Physics of Carbon. Marcel Dekker, New York.
Donnet, J.-B., Wang, T.K., Li, Y.J., Balard, H., Burns, G.T., 2000. Rubber Chem. Technol. 73, 634.
Duperray, B., Leblanc, J.-L., 1982. Kautsch. Gummi Kunstst. 35, 298.
Ehrburger-Dolle, F., Hindermann-Bischoff, M., Livet, F., Bley, F., Rochas, C., 2001. Langmuir 17, 329.
EP0157703, Chevallier, Y., Morawski, J.C., 1984. Precipitated silica with morphological properties, process for producing it and its application, especially as a filler.
EP0501227, Rauline, R., 1991. Rubber compound and tires based on such a compound.
EP0810258, Custodero, E., Tardivat, J.-C., 1996. Diene rubber composition containing alumina as reinforcing filler and use in tire treads.
EP0890602, Hergenrother, W.L., Oziomek, J.C., William, M., 1997. Addition of salts to improve the interaction of silica with rubber.
EP1034222, Custodero, E., Simonot, L., Tardivat, J.-C., 1997. Carbon black coated with an aluminous layer and method for obtaining same.
EP1114092, Custodero, E., Simonot, L., Tardivat, J.-C., 1999. Rubber composition for tyre, based on diene elastomer and a reinforcing titanium oxide.
Evans, L.R., Waddell, W.H., 1995. Kautsch. Gummi Kunstst. 48, 718.
Funt, J.M., 1986. Rubber World 21 (February).
Gent, A.N., 1974. J. Appl. Polym. Sci. 18, 1397.
Gerspacher, M., O'Farrell, C.P., 1993. Rubber Plast. News 23, 85.
Gonzalez-Martin, M.L., Janczuk, B., Labajos-Broncano, L., Bruque, J.M., 1997. Langmuir 13, 5991.
Göritz, D., Böhm, J., Freund, B., 1997. In: International Rubber Conference 1997, Kuala Lumpur, October 6–9, pp. 201–205.
Görl, U., Hunsche, A., Müller, A., Koban, H.G., 1997. Rubber Chem. Technol. 70, 608.
Gruber, T.C., Zerda, T.W., Gerspacher, M., 1993. Carbon 37, 1209.
Guth, E., Gold, O., 1938. Phys. Rev. 53, 322.
Harris, J.O., Wise, R.W., 1965. In: Kraus, G. (Ed.), Reinforcement of Elastomers. Wiley Interscience, New York.
Harwood, J.A.C., Mullins, L., Payne, A.R., 1965. J. Appl. Polym. Sci. 9, 3011; Harwood, J.A.C., Payne, A.R., 1966. J. Appl. Polym. Sci. 10, 315; Harwood, J.A.C., Payne, A.R., Smith, J.F., 1970. Rubber Chem. Technol. 43, 687.
Herd, C.R., McDonald, G.C., Hess, W.M., 1992. Rubber Chem. Technol. 65, 107.
Hess, W.M., 1991. Rubber Chem. Technol. 64, 386.
Hess, W.M., McDonald, G.C., 1983. Rubber Chem. Technol. 56, 892.
Hess, W.M., Lyon, F., Burgess, K.A., 1967. Kautsch. Gummi Kunstst. 20, 135.
Hess, W.M., McDonald, G.C., Urban, E., 1973. Rubber Chem. Technol. 46, 204.
Hess, W.M., Ayala, J.A., Vegvari, P.C., Kistler, F.D., 1988. Kautsch. Gummi Kunstst. 41, 1215; Soeda, M., Kurata, Y., 1995. Nippon Gomu Kyokaishi 68, 616.
Hess, W.M., Herd, C.R., Sebok, E.B., 1994. Kautsch. Gummi Kunstst. 47, 328.
Hommel, H., Legrand, A.-P., Balard, H., Papirer, E., 1993. Makromol. Chem. 194, 879.
Humbert, B., 1995. J. Non-Crystalline Solids 191, 29.
Hunsche, A., Görl, U., Müller, A., Knaack, M., Göbel, T., 1997. Kautsch. Gummi Kunstst. 50, 881.
Jagiello, J., Ligner, G., Papirer, E., 1990. J. Colloid Interf. Sci. 137, 128.
Janzen, J., 1982. Rubber Chem. Technol. 55, 669.
Janzen, J., Kraus, K., 1971. Rubber Chem. Technol. 44, 1287.
Kaufmann, S., Slichter, W.P., Davis, D.D., 1971. J. Polym. Sci. A 9, 829.
Killian, H.G., Schenk, H., Wolff, S., 1987. Colloid Polym. Sci. 265, 410.
Kraus, G., 1965. Rubber Chem. Technol. 38, 1070.
Kraus, G., 1970. J. Polym. Sci. B 8, 601.
Kraus, G., 1971. Fortschr. Hochpolym. Forsch. 8, 155.
Kraus, G., Jansen, J., 1978. Kautsch. Gummi Kunstst. 31, 569.

Lapra, A., 1999. Caractérisation moléculaire et propriétés mécaniques des réseaux élastomères SBR renforcés par la silice. Université Paris VI.

Lapra, A., Custodero, E., Simon, N., 2004. Kautsch. Gummi. Kunstst. 57, 52.

Läufer, S., 1980. J. Mol. Struct. 60, 409.

Le Bras, J., Papirer, E., 1979. Rubber Chem. Technol. 52, 43.

Le Bras, J., Papirer, E., 1983. J. Appl. Polym. Sci. 22, 525.

Legrand, A.P., 1998. In: Legrand, A.P. (Ed.), The Surface Properties of Silicas. vol. 1. John Wiley and Sons, New York, pp. 1–18.

Legrand, A.P., Lecomte, N., Vidal, A., Haidar, B., Papirer, E., 1992. J. Appl. Polym. Sci. Appl. Polym. Symp. 46, 2223.

Leisen, J., Breidt, J., Kelm, J., 1998. Rubber Chem. Technol. 72, 1.

Lewis, J.E., Deviney Jr., L.R., McNabb, C.F., 1970. Rubber Chem. Technol. 43, 449.

Maas, S., Gronski, W., 1994. Kautsch. Gummi Kunstst. 47, 409.

Magee, R.W., 1995a. Rubber Chem. Technol. 68, 590.

Magee, R.W., 1995b. The Carbon Aggregate. vol. 2, p. 1.

McDonald, G.C., Hess, W.M., 1977. Rubber Chem. Technol. 50, 842.

Medalia, A.I., 1970. J. Colloid Interf. Sci. 32, 115.

Medalia, A.I., 1973. Rubber Chem. Technol. 46, 877.

Medalia, A.I., 1974. Rubber Chem. Technol. 47, 411.

Medalia, A.I., 1986. Rubber Chem. Technol. 59, 432.

Medalia, A.I., 1991. Rubber Chem. Technol. 64, 481.

Medalia, A.I., Kraus, G., 1994. Reinforcement of elastomers by particulate fillers. In: Mark, J.E., Erman, B., Eirich, F.R. (Eds.), Science and Technology of Rubber, 2nd ed. Academic Press, San Diego.

Medalia, A.I., Richards, L.W., 1972. J. Colloid Interf. Sci. 40, 233.

Medalia, A.I., Dannenberg, E.M., Heckmann, F.A., Cotten, G.R., 1976. Rubber Chem. Technol. 46, 1239.

Meissner, B., 1974. J. Appl. Polym. Sci. 18, 2483; Meissner, B., 1993. J. Appl. Polym. Sci. 50, 285.

Milburn, D.R., Adkins, B.D., Davis, B.H., 1988. In: Unger, K.K. (Ed.), Characterization of Porous Solids. Elsevier, Amsterdam, p. 501.

Moscou, L., Lub, S., Bussemaker, O.K.F., 1971. Rubber Chem. Technol. 44, 805.

Mullins, L., 1969. Rubber Chem. Technol. 42, 339.

Mullins, L., Tobin, N.R., 1965. J. Appl. Polym. Sci. 9, 2993.

Nagata, N., Kobatake, T., Watanabe, H., Ueda, A., Yoshioka, A., 1987. Rubber Chem. Technol. 60, 837.

Nikiel, L., Gerspacher, M., Yang, H., O'Farrell, C.P., 2000. In: 157th Meeting Rubber Division American Chemical Society Dallas, 4–6 April , Paper 31.

O'Brien, J., Cashell, E., Wardell, G.E., McBrierty, V.J., 1976. Macromolecules 9, 653; McBrierty, V.J., Kenny, J.C., 1994. Kautsch. Gummi Kunstst. 47, 342.

Papirer, E., Voet, A., Given, P.H., 1969. Rubber Chem. Technol. 42, 1200.

Papirer, E., Donnet, J.B., Heinkele, J., 1971. J. Chim. Phys. 68, 581.

Papirer, E., Nguyen, V.T., Donnet, J.B., 1978. Carbon 16, 141.

Patel, A.C., Lee, K.W., 1990a. Elastomerics 122, 14.

Patel, A.C., Lee, K.W., 1990b. Elastomerics 122, 22.

Paul, I.B., Roy, T.N., Mukherjee, P.N., 1982. Indian J. Technol. 20, 441.

Payne, A.R., 1965. In: G. Kraus (Ed.), Reinforcement of Elastomers. Wiley Interscience, New York (Chapter 3).

Pirard, R., Sahouli, B., Blacher, S., Pirard, J.P., 1999. J. Colloid Interf. Sci. 217, 216.

Probst, N., Donnet, J.B., 1993. Differentiation of carbon black particularities by scanning tunneling microscopy. In: Second International Conference on Carbon Black, pp. 261–263.

Puri, B.R., Bansal, R.C., 1965. Carbon 3, 227.

Richmond, B.R., 1993. In: Conference Proceedings of the International Relations Committee 1993, Orlando, 26–29 October, Paper 158.

Rigbi, Z., 1993. Kautsch. Gummi Kunstst. 46, 36.

Rivin, D., 1963. Rubber Chem. Technol. 36, 729.

Rivin, D., 1971. Rubber Chem. Technol. 44, 307.

Rivin, D., Aron, J., Medalia, A.I., 1968. Rubber Chem. Technol. 41, 330.

Schubert, B., Ford, F.P., Lyon, F., 1969. Encyclopedia Ind. Chem. Anal. 8, 1.
Semaan, M.E., Nikiel, L., Quarles, C.A., 2001. Carbon 39, 1379.
Shabanova, N.A., Silos, I.V., 1996. Colloid J. 58, 256.
Shiga, S., Furuta, M., 1985. Rubber Chem. Technol. 58, 1.
Smallwood, H.M., 1944. J. Appl. Phys. 15, 758.
Sone, K., 1999. Int. Polym. Sci. Technol. 26, 60.
Sone, K., Ishiguro, M., Akimoto, H., Ishida, M., 1992. Rubber World 206, 29.
Soos, I., 1984. Int. Polym. Sci. Technol. 11, T4.
Soos, I., 1982. Int. Polym. Sci. Technol. 9, T84; Soos, I., 1983. Int. Polym. Sci. Technol. 10, T77.
Stickney, P.B., Falb, R.D., 1964. Rubber Chem. Technol. 37, 1299.
Strauss, M., Killian, H.G., Freund, B., Wolff, S., 1994. Colloid Polym. Sci. 272, 1208.
Sullivan, A.B., Wise, R.W., 1967. In: Proceedings of Fifth International Rubber Conference, pp. 235.
Taylor, G.T., Redington, T.E., Bayley, M.J., 1980. Am. Ind. Hyg. Assoc. J. 41, 819.
Tsubokawa, N., Hosoya, M., 1991. Reactive Polym. 14, 33; Tsubokawa, N., Hosoya, M., 1991. Reactive Polym. 14, 95.
Tsutsumi, F., Sakakibara, M., Oshima, N., 1990. Rubber Chem. Technol. 63, 8.
Uytterhoeven, J.J., Fripiat, J.J., 1962. Bull. Soc. Chim. Fr. 788.
Vilgis, T.A., Heinrich, G., 1994. Macromolecules 27, 7846.
Villars, D.S., 1947. J. Am. Chem. Soc. 69, 214.
Villars, D.S., 1996. J. Polym. Sci. 21, 257.
Voet, A., Morawski, J.C., Donnet, J.B., 1977. Rubber Chem. Technol. 50, 342.
Wampler, W.A., 1997. The Carbon Aggregate. vol. 5, p. 2.
Wampler, W.A., Gerspacher, M., O'Farrell, C.P., 1997. The Carbon Aggregate 4, 3.
Wang, M.J., Wolff, S., 1991a. Meeting Rubber Div. Am. Chem. Soc., Detroit, 8–11 October.
Wang, M.J., Wolff, S., Donnet, J.B., 1991b. Rubber Chem. Technol. 64, 714.
Wang, M.J., Wolff, S., Donnet, J.B., 1991. Rubber Chem. Technol. 64, 559.
Wang, M.J., Wolff, S., Tan, E.H., 1993. Rubber Chem. Technol. 66, 178.
Wang, M.J., Brown, T.A., Patterson, W.J., Francis, R.A., October 1997. In: International Rubber Conference 1997, Kuala Lumpur, pp. 6–9.
Wang, M.J., Zhang, P., Mahmud, K., Lanoye, T., Vejins, V., 2002. Tire Technol. Int. 58, 54.
Watson, W.F., 1955. Ind. Eng. Chem. 47, 1281.
Wiegand, W.B., 1925. Trans. Inst. Rubber Ind. 1, 141.
WO2002053634, Simonot, L., Chartier, T., Custodero, E., 2002. Rubber composition made with diene elastomer and a reinforcing silicon carbide.
WO2004003067, Simonot, L., Lapra, A., Veyland, A., Custodero, E., 2002. Rubber composition based on diene elastomer and a reinforcing silicon nitride.
WO9928376, Custodero, E., Simonot, L., Tardivat, J.-C., 1997. Reinforcing aluminous filler and rubber composition comprising such a filler.
Wolff, S., 1975. In: International Rubber Conferences, 14–17 October, p. 295.
Wolff, S., 1981. Kautsch. Gummi Kunstst. 34, 280.
Wolff, S., Wang, M.J., 1992. Rubber Chem. Technol. 65, 329.
Wolff, S., Wang, M.J., Tan, E.H., 1993. Rubber Chem. Technol. 66, 163.
Wolff, S., Tan, E.H., Donnet, J.B., 1994. Kautch. Gummi Kunstoffe 47, 485.
Young, R.J., Al-Khudhairy, D.H.A., Thomas, A.G., 1986. J. Mater. Sci. 21, 1211.
Zaborski, M., Vidal, A., Ligner, G., Balard, H., Papirer, E., Burneau, A., 1989. Langmuir 5, 447.

Chapter 9

The Science of Rubber Compounding

Brendan Rodgers and Walter Waddell
ExxonMobil Chemical Company, Houston, TX, USA

9.1 INTRODUCTION

Compounding, a term that has evolved within the tire and rubber industry, is the materials science of modifying a rubber or elastomer or a blend of polymers and other materials to optimize properties to meet a given service application or set of performance parameters. Compounding is therefore a complex multidisciplinary science necessitating knowledge of materials physics, organic and polymer chemistry, inorganic chemistry, and chemical reaction kinetics. The materials scientist, when designing a rubber formulation, has a range of objectives and restrictions within which to operate. Product performance requirements will dictate the initial selection of formula ingredients. These materials must be environmentally safe, meet occupational health and safety requirements, be processable in the product manufacturing facilities, and be cost effective.

Compounded rubber has many unique characteristics not found in other materials, such as dampening properties, high elasticity, and abrasion resistance. Hence rubber has found use in applications such as tires, conveyor belts, large dock fenders, building foundations, automotive engine components, and a wide range of domestic appliances. The ingredients available to the materials scientist for formulating a rubber compound can be divided into five categories:

- *Polymers*: Natural rubber, synthetic polymers.
- *Filler systems*: Carbon blacks, clays, silicas, calcium carbonate.
- *Stabilizer systems*: Antioxidants, antiozonants, waxes.
- *Vulcanization system components*: Sulfur, accelerators, activators.
- *Special materials*: Secondary components such as pigments, oils, resins, processing aids, and short fibers.

Each class of materials is reviewed in this chapter.

The Science and Technology of Rubber. http://dx.doi.org/10.1016/B978-0-12-394584-6.00009-1

9.2 POLYMERS

World rubber usage of around 25.8 million metric tons is split between natural rubber, which constitutes about 43% of global consumption, and synthetic rubber, of which styrene-butadiene rubber (SBR) accounts for 21%. The balance of synthetic rubbers (36%) consists of polybutadiene rubber (BR) and a range of specialty polymers such as polyurethanes, halogenated polymers, silicones, and acrylates. Traditionally, the growth of synthetic and natural rubber consumption is virtually in line with the change in gross domestic product of, collectively, North America, Europe, Japan, China, and India.

9.2.1 Natural Rubber

Global natural rubber consumption is split among tires (75%), automotive mechanical products (5%), nonautomotive mechanical products (10%), and miscellaneous applications such as medical and health-related products (10%). Since the 1960s, the quality and consistency of natural rubber has improved, primarily because of the implementation of standard specifications defining a range of grades of rubber. Natural rubber is available in three basic types: technically specified rubbers, visually inspected rubbers, and specialty rubbers.

The American Society for Testing and Materials (ASTM) describes six basic grades of coagulated technically specified natural rubber, which is processed and compacted into 34-kg blocks (ASTM D2227-96, 2002) (Table 9.1). These six general grades of technically specified natural rubber are defined in more detail by the respective producing countries. Standard Malaysian Rubber (SMR), Standard Indonesian Rubber (SIR), and Thai Technical Rubber (TTR) expand the range of rubbers available. For example, two constant-viscosity Standard Malaysian Rubber CV grades are available, SMR CV50 and CV60

TABLE 9.1 Specifications for Technically Graded Natural Rubber

Property	Rubber Grade					
	L	CV	5	10	20	50
Dirt (% maximum)	0.050	0.050	0.050	0.100	0.200	0.500
Ash (% maximum)	0.60	0.60	0.60	0.750	1.00	1.50
Volatile matter (%)	0.80	0.80	0.80	0.80	0.80	0.80
Nitrogen (%)	0.60	0.60	0.60	0.60	0.60	0.60
Plasticity	30	–	30	30	30	30
Plasticity retention index	60	60	60	50	40	30
Color index	6.0	–	–	–	–	–
Mooney viscosity	–	60	–	–	–	–

TABLE 9.2 International Natural Rubber Type and Grade Specification

Type	Natural Rubber	Description
1	Ribbed smoked sheet	Coagulated sheets, dried, and smoked latex. Five grades available (RSS1–5)
2	White and pale crepe	Coagulated natural liquid latex milled to produce a crepe
3	Estate brown crepe	Fresh lump and other high-quality scrap generated on the plantation
4	Compo crepe	Lump, tree scrapes, and smoked sheet cuttings are milled into a crepe
5	Thin brown crepe	Unsmoked sheets, wet slab, lump, and other scrap from estates and small holdings
6	Thick blanket crepe	Wet slab, lump, and unsmoked sheets milled to give a crepe
7	Flat bark crepe	All types of scrap natural rubber including earth scrap
8	Pure smoked blanket crepe	Milled smoked rubber derived exclusively from ribbed smoked sheets

(Datta, 2004). SMR 10 and SMR 20 grades are also available as viscosity stabilized (SMR 10CV and SMR 20CV).

The Rubber Manufacturers Association has a further set of standards for quality and packing of latex natural rubber grades. Table 9.2 defines the eight types of rubber covered in their specifications. Here, coagulated latex is sheeted, dried, and packed into bales of up to 113.5 kg. Grading is by visual inspection. Quality assurance laboratories have sets of visual standards for inspections (The Rubber Manufacturers Association, 1979).

The third category of natural rubbers are the specialty materials, which include liquid low molecular weight rubber, methyl methacrylate grafted polymers, oil-extended natural rubber, deproteinized natural rubber, epoxidized natural rubber, and superior-processing natural rubber.

Natural rubber usage has increased substantially in modern radial tires. Bernard et al. (1985) compared the natural rubber levels of heavy-duty radial truck tires to those of the equivalent bias tire and noted the following increase:

Natural Rubber (%)	Bias	Radial
Tread	47	82
Skim coat	70	100
Sidewall	43	58

The reasons for the increase have been attributed to improved green strength, increase in component-to-component adhesion, improved tear strength, lower tire temperatures generated under loaded dynamic service conditions, and lower tire rolling resistance to improve vehicle fuel efficiency.

The increase in natural rubber usage translates into approximately 21 kg per tire for a radial construction compared with approximately 9 kg found in a bias truck tire. Natural rubber compounds also tend to find use in covers of high-performance conveyor belts where a similar set of performance parameters such as those of a truck tire tread compound are found. Low hysteretic properties, high tensile strength, and good abrasion resistance are required for both products.

9.2.2 Synthetic Elastomers

Classification of synthetic rubber is governed by the International Institute of Synthetic Rubber Producers (IISRP). In the case of styrene-butadiene rubber, polyisoprene rubber, and polybutadiene, a series of numbers have been assigned that classify the general properties of the polymer (The Synthetic Rubber Manual, 1999). For example, the IISRP 1500 series defines cold emulsion-polymerized (i.e., below 10°C), nonpigmented SBR. The 1700 series of polymers describes oil-extended cold emulsion SBR. Table 9.3 illustrates the general numbering used by IISRP. The numbering system for solution-polymerized stereo elastomers is given in Table 9.4.

Tire production consumes approximately 60% of the global synthetic rubber production. Of this, SBR is the largest-volume polymer, representing over 65% of the synthetic rubber used in tires. Polybutadiene (BR) ranks second in production output (Worldwide Rubber Statistics, 2010; Datta, 2004). Tables 9.5–9.7 illustrate the consumption of synthetic rubber by-product group. Styrene-butadiene rubber finds extensive use in tire treads because it offers wet skid and traction properties while retaining good abrasion resistance. Polybutadiene (BR) is frequently found in treads, sidewalls, and some casing components of the tire because it offers good abrasion resistance and tread wear

TABLE 9.3 Classification of Synthetic Rubbers by IISRP

Class Number	Description
1000 series	Hot nonpigmented emulsion SBR (polymerized above 38°C)
1500 series	Cold nonpigmented emulsion SBR (polymerized below 10°C)
1600 series	Cold polymerized/carbon black master batch/14 phr oil (max) SBR
1700 series	Oil-extended cold emulsion SBR
1800 series	Cold emulsion-polymerized/carbon black master batch/more than 14 phr oil SBR
1900 series	Emulsion resin rubber master batches

TABLE 9.4 IISRP Solution-Polymerized Stereo Elastomers

	Butadiene and Copolymers	Isoprene and Copolymers
Dry polymer	1200–1249	2200–2249
Oil extended	1250–1299	2250–2299
Black master batch	1300–1349	2300–2349
Oil-black master batch	1350–1399	2350–2399
Latex	1400–1449	2400–2449
Miscellaneous	1450–1499	2450–2499

TABLE 9.5 Synthetic Rubber Consumption

Tires	60%
Automotive parts	10%
Nonautomotive mechanical goods	9%
Thermoplastic elastomer composites	6%
Footwear	4%
Construction	3%
Wire and cable	2%
Adhesives	1%
Miscellaneous products	5%

TABLE 9.6 US Consumption of SBR

Product	Percent of Total Consumption
Passenger tires	55
Retread rubber	13
Truck tires	1
Special tires (aircraft, earthmover, farm, racing)	5
Automotive mechanical goods	6
Miscellaneous applications (domestic appliances, medical equipment, construction)	20

TABLE 9.7 US Consumption of Polybutadiene

Product	Percent of Total Consumption
Passenger tires	45
Truck tires	25
Retread rubber	3
Special tires (aircraft, earthmover, farm, racing)	5
Automotive mechanical goods	2
Miscellaneous applications (polymer blends, modifiers with polystyrene or styrene acrylonitrile butadiene terpolymers)	15

performance and enhances resistance to cut propagation. BR can also be blended with natural rubber, and many authors have reported that such compositions give improved fatigue and cut-growth resistance (Mezynski and Rodgers, 1993).

Before reviewing elastomer characteristics required to meet any given set of tire performance parameters, it is appropriate to identify two means by which the materials scientist may describe a polymer: polymer macrostructure and polymer microstructure. The macrostructure of a polymer defines the molecular weight and crosslink distribution, polymer chain branching, and crystallite formation. The arrangement of the monomers within a polymer chain constitutes its microstructure. Butadiene can adopt one of three configurations as illustrated in Figure 9.1. These molecular configurations or stereochemistry can be described as follows:

vinyl-(1,2): The third and fourth carbon atoms are pendant; the first and second carbon atoms participate in the polymer backbone.
trans-(1,4): The hydrogen atoms attached to the carbon-carbon double bond on the polymer backbone are on opposite sides.
cis-(1,4): The two hydrogen atoms attached to the carbon-carbon double bond in the polymer are on the same side of the double bond.

Table 9.8 illustrates the effect of the catalyst system on polymer microstructure (Rodgers et al., 2004).

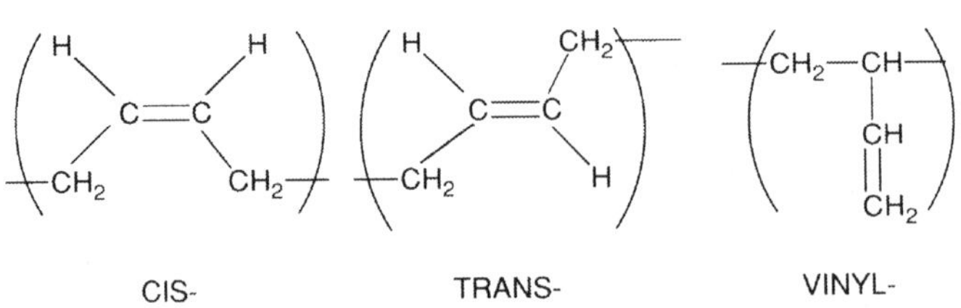

FIGURE 9.1 Polymer microstructure: possible configurations for butadiene in SBR and BR.

TABLE 9.8 Polybutadiene Microstructure

Catalyst	Isomer Level to +/−1%		
	cis-%	*trans*-%	*vinyl*-%
Li	35	55	10
Ti	91–94	2–4	4
Co	96	2	2
Nd	98	1	1
Ni	96–98	0–1	2–4

The relative levels of each of the three isomers in a polymer such as BR can have a dramatic effect on the material's performance. For example, lithium-catalyzed solution polymers, with approximately 36% *cis* content, tend to process easily, whereas high-*cis* Ti and Ni polymers (92% *cis*) are more difficult to process at factory processing temperatures but show better abrasion resistance. High-*trans* BR (93% *trans*) tends to be a tough, crystalline material at room temperature. High vinyl-butadiene BR polymers in tire treads tend to show good wet skid and wet traction performance (Mezynski and Rodgers, 1993; Simpson, 1978; Nordsiek, 1985; Rodgers et al., 2004).

Nordsiek (1985) documented a series of empirical guidelines that might be used in designing a polymer for a set of tire performance targets. By preparing various blends of BR and SBR, Nordsiek produced a series of compounds in which the T_g increased from −100 to −30°C. He noted the following points:

- As the T_g increases there is a near-linear drop in abrasion resistance.
- Wet grip or traction improves nearly linearly with the increase in compound T_g.
- Addition of a catalyst modifier during the preparation of solution-polymerized, lithium-catalyzed BR results in an increase in the vinyl-1,2-butadiene level in the polymer and causes an increase in T_g. There is a corresponding drop in abrasion resistance and an increase in wet traction.
- Inclusion of styrene leads to an increase in traction performance and loss in abrasion resistance. There is a linear relationship between styrene and vinyl-1,2-butadiene. Approximately two vinyl-1,2-butadiene units gave a tire traction performance equivalent to that of one styrene unit.
- Inclusion of 3,4-isoprene in polyisoprene leads to an increase in T_g and a corresponding increase in traction, and an increase in the percentage incorporation of 1,2- or 3,4-piperylene in polypiperylenes results in a T_g increase, causing a loss in abrasion resistance and an increase in grip.

This allowed the tan δ temperature curve of a tread compound run from −100 to +100°C to be segmented into zones that would characterize that

TABLE 9.9 Characterization of an Idealized Tread Compound: tan δ Temperature Curve[a]

Temperature Zone (°C)	Feature	Performance Parameter
−60 to −40	T_g	Abrasion
−20		Low-temperature properties
+20	–	Wet traction
+40 to +60	–	Rolling resistance
+80 to +100	–	Heat buildup

[a]Data taken from Nordsiek (1985).

tire tread compound's performance (Table 9.9). Such property targets enabled development of the concept of "integral rubber"; that is, a polymer can be designed to meet rolling resistance, traction, and tread wear targets without a drop in overall tire performance.

Day and Futamura (1986) evaluated the impact of variation in 1,2-butadiene and styrene content in SBR on the properties of a compounded formulation. Briefly, (1) an increase in styrene produced an increase in tensile strength, (2) an increase in vinyl-1,2-butadiene resulted in a drop in both tear strength and ultimate elongation, and (3) at equal T_g, neither vinyl-1,2-butadiene nor styrene level affected the formulation's hysteretic properties.

Brantley and Day then conducted a study to compare the tire performance of emulsion- and solution-polymerized SBR (Brantley and Day, 1986). The authors noted that solution-polymerized polymers, which tend to have a narrower molecular weight distribution and lower T_g than equivalent emulsion-polymerized polymers, have lower hysteretic properties. They then showed that a solution SBR with the same bound styrene as an emulsion SBR will give lower rolling resistance, improved dry traction, and better tread wear. Emulsion SBR, however, tends to show better wet skid, wet traction, and wet handling performance. Kern and Futamura later elaborated on this work by evaluating the impact of vinyl-1,2-butadiene level in a solution SBR and again comparing this with an emulsion SBR (Kern and Futamura, 1987). Though this work was conducted with passenger tires, many of the principles should be applicable to the range of tires such as light truck and heavy-duty truck tires.

The authors collected the test data shown in Table 9.10. From these data it can be noted that the number-average molecular weight, or M_n, of a commercial emulsion SBR such as IISRP 1500 or 1712 is typically 90,000–175,000. The primary molecular weight of a solution-polymerized polymer produced with an anionic lithium catalyst can, in contrast, be increased toward 250,000 without gelation. In addition, emulsion-polymerized SBR contains only about 92% rubber hydrocarbon as a result of the presence of residues from the production process; solution polymers tend to be near 100% hydrocarbon. As a consequence, the authors concluded that the number-average molecular weight

TABLE 9.10 Comparison of Emulsion and Solution-Polymerized SBR

Property	Emulsion SBR	Solution SBR
Viscosity ($M_L 1+4$ at 100°C)	50	57
Time to optimum cure (min at 150°C)	40	25
Tensile strength (MPa)	26	21
Ultimate elongation (%)	400	300
Rebound (%)	48	61

TABLE 9.11 Effect of Polymer Butadiene Vinyl Level on Tire Performance

Vinyl level	10%	50%
Glass transition temperature	−90°C	−60°C
Tire properties[a]		
Wet traction	100	120
Rolling resistance rating	100	95
Tread wear rating	100	90

[a]*Higher rating is better.*

can be considered the key parameter of polymer macrostructure, particularly with respect to the hysteretic characteristics of a tread formulation. Hence the differences in macrostructure between emulsion- and solution-polymerized polymers will dictate many of their properties in a tire tread compound.

When considering only solution polymers, polymer microstructure has a greater effect on tire tread compound performance. Table 9.11 illustrates the impact on tire traction, rolling resistance, and tread wear of a polybutadiene tread on which the vinyl-1,2-butadiene level had been increased from 10% to 50% (Brantley and Day, 1986). The corresponding drop in wear and increase in tire rolling resistance are in agreement with the empirical rules presented by Nordsiek (1985), who attributed such tire property trends to the polymer T_g.

Table 9.12 shows how polybutadiene microstructure and macrostructure (i.e., molecular weight, Mw, Mn, polydispersity, and branching) can affect the processability of a polymer (Kumar et al., 1996). A study with both cobalt- and neodenium-catalyzed polybutadiene showed the relationship between polydispersity or molecular weight distribution and increases in stress relaxation. Increases in stress relaxation, as measured by the Mooney viscometer, will infer greater difficulty in compound processing, gauge control, "nerve," and extrudate or calendered sheet shrinkage (Waddell et al., 2004).

Halobutyl rubber (HIIR) is used primarily in tire innerliner and white sidewalls. These elastomers are best for tire air retention owing to lower air permeability as well as aging and fatigue resistance. The chlorinated (CIIR)

TABLE 9.12 Macrostructure and Mooney Viscometry (Kumar et al., 1996)

Polymer Catalyst	*Sample*	M_w	M_n	M_w/M_n	Mooney Stress $M_L 1+4$	*Relaxation*
Cobalt	1	338	156	2.17	47	4.50
	2	318	131	2.43	45	7.50
	3	321	125	2.57	46	9.00
	4	303	108	2.81	44	14.00
Neodenium	1	353	186	2.10	50	5.00
	2	381	103	3.70	42	8.00
	3	347	87	3.99	44	9.00
	4	368	86	4.28	42	10.00

and brominated (BIIR) versions of isobutylene-isoprene rubber (IIR) can be blended with other elastomers to improve adhesion between HIIR compounds and those based on general-purpose elastomers, and improve vulcanization kinetics (Waddell et al., 2004).

Attempts at using halogenated isobutylene-based polymers in tread compounds have been limited, even though such tread compounds display good performance in winter applications and have good traction performance. A new isobutylene polymer modified with *p*-methylstyrene and then brominated is also available, which offers a fully saturated backbone to resist aging while improving compatibility with general-purpose elastomers such as natural rubber and styrene-butadiene rubbers.

It is common to blend more than one type of rubber within a given tire tread compound. An example of this is the truck drive axle tire tread compounds that must not only possess high strength but must also have good fatigue resistance. In passenger car tires, as many as four different polymers may be used for the tread compound totaling 100 phr; for example, 25 phr emulsion SBR, 25 phr solution SBR, 30 phr BR, and 20 phr NR. If solution SBR categories can be considered as each one falling with a 10°C T_g range, there are at least nine groups of specialty SSBR polymers commercially available, in addition to the range of proprietary polymers chemical operations produce to support tire manufacturing (Mezynski and Rodgers, 1993; Waddell et al., 2004).

Figure 9.2 shows the effect of T_g on wet skid. If an increase in wet grip is required with minimum impact on rolling resistance, then a change in T_g is best accomplished via an increase in the vinyl-butadiene level rather than in the bound styrene content. Alternatively, if wear is of higher importance, T_g should be adjusted by a change in the bound styrene level. The optimum

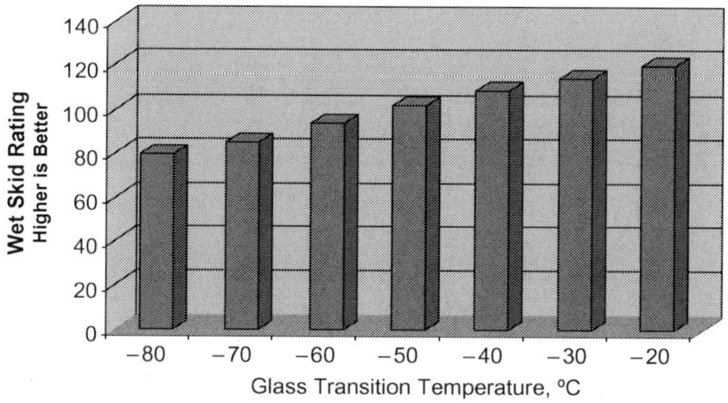

FIGURE 9.2 Effect of T_g on tire traction performance (from Saito, 1999).

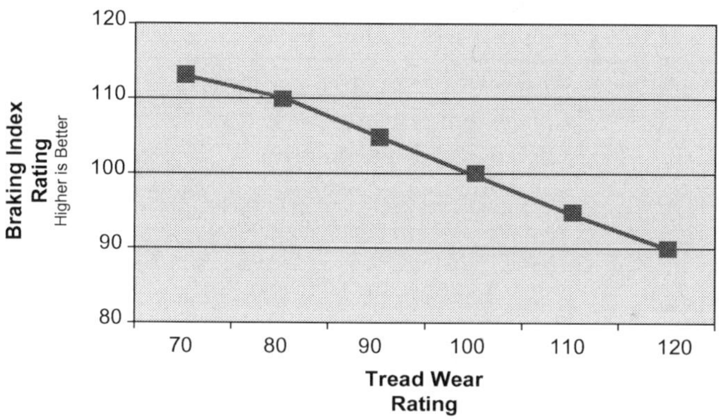

FIGURE 9.3 Relationship between wear and braking qualities (from Oberster et al., 1973).

T_g could therefore be obtained by adjustment of either the vinyl-butadiene or styrene contents to obtain the required wet grip, rolling resistance, and wear performances. It has been demonstrated (Mezynski and Rodgers, 1989; Saito, 1999; Oberster et al., 1973) that an increase in wear performance would lead to a trade-off in traction performance (Figure 9.3).

High-molecular-weight commercial polymers are oil extended to facilitate processing and to enable the production of polymers that will yield compounds with better mechanical properties than those with lower molecular weight polymers of corresponding structure (Schneider et al., 1989). Table 9.13 displays a selection of emulsion SBR grades. Aromatic oils can raise the glass transition temperature of the corresponding oil-free polymer. Naphthenic oils will tend to shift the transition temperature below the value of the oil-free rubber.

TABLE 9.13 Oil-Extended Emulsion SBR (Schneider et al., 1989)

IISRP Polymer	Mooney Nominal Viscosity (M_L1+4)	Styrene (%)	Oil Type	Oil Level (phr)
1707	50	23.5	Naphthenic	37.5
1712	50	23.5	Aromatic	37.5
1720	40	23.5	Naphthenic	50.0
1721	55	40.0	Aromatic	37.5

TABLE 9.14 ASTM and IISRP Classification of Oils for Oil-Extended Elastomers (The Synthetic Rubber Manual, 1999)

Type	Asphaltenes	Polar Compound Content (%)	Saturated Hydrocarbon Content (%)	Category	Viscosity Gravity Constant
101	0.75	25.0	20.0	Highly aromatic	>0.900
102	0.50	12.0	20–35	Aromatic	0.900
103	0.30	6.0	35–65	Naphthenic	0.875
104A	0.10	1.0	65.0	Paraffinic	>0.820
104B	0.10	1.0	65.0	Paraffinic	0.820 Max

The primary function of oil in rubber is to facilitate improvement in processing; that is, the ease of mixing in an internal mixer, to improve mixed compound uniformity such as viscosity, and to improve downstream processing such as in extrusion. The specific oils used in oil-extended elastomers have been categorized into essentially five groups, which are summarized in Table 9.14.

Though natural rubber, SBR, and BR represent the largest consumption of elastomers, several additional polymers merit a brief discussion because of their economic significance—nitriles, polychloroprene, butyl, and ethylene-propylene-diene monomer (EPDM) elastomers (Datta, 2004; The Synthetic Rubber Manual, 1999).

Nitrile rubber (NBR) is a copolymer of acrylonitrile and butadiene. Its most important property is resistance to oil absorption; it therefore finds extensive use in such products as hydraulic hose and automotive engine components, where oil resistance is essential. Figure 9.4 illustrates the effect of acrylonitrile level on oil absorption (IRM 903 oil). Conversely, NBR polymers have poor cold flex properties, which prohibits their use on equipment operating in cold climates.

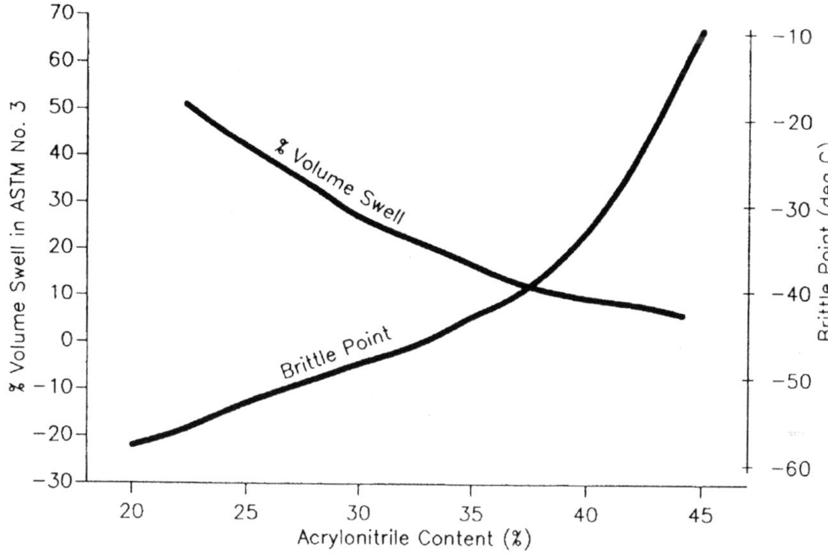

FIGURE 9.4 Acrylonitrile content and NBR oil absorption.

NBR tends to break down readily on a mill or Banbury. Peptizers are not normally required, though antigel agents are needed if mixing temperatures exceed 140°C. Because of the polymer's low green strength, sufficient shear during mixing is not achieved to enable use of SAF or ISAF carbon blacks. It can also result in poor processing qualities such as mill bagging. Antioxidants are essential in NBR compounds since NBR will oxidize readily in hot air. Polymerized 2,2,4-trimethylhydroquinolene is the most effective antioxidant. Antiozonants and waxes are ineffective with NBR compounds.

Polychloroprene is made either from acetylene:

$$\underset{\textbf{Acetylene}}{HC{\equiv}CH} \rightarrow HC{\equiv}C{-}CH{=}CH_2 \xrightarrow{HCl} \underset{\textbf{Chloroprene}}{H_2C{=}\underset{|\atop Cl}{C}{-}CH{=}CH_2}$$

$$H_2C{=}\underset{|\atop Cl}{C}{-}CH{=}CH_2 \xrightarrow{\text{Polymerization}} \underset{\textbf{Polychloroprene}}{[-CH_2{-}CH{=}\underset{|\atop Cl}{C}{-}CH_2-]_n} \qquad (9.1)$$

or butadiene:

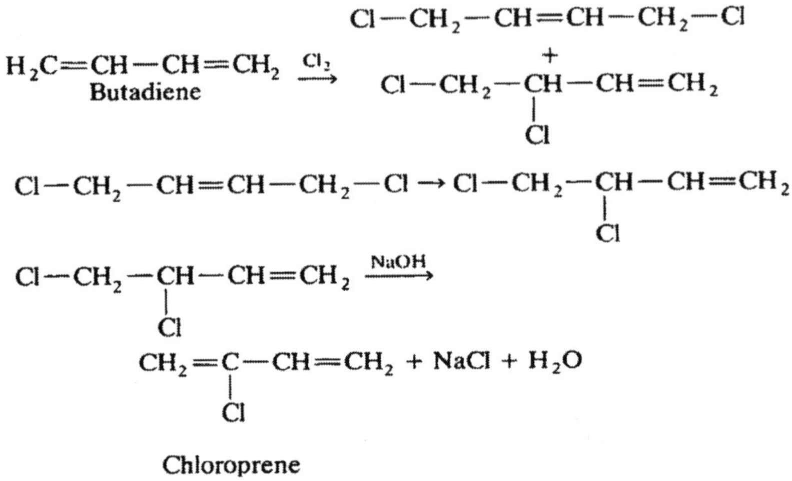

(9.2)

Acetylene is reacted to form vinyl acetylene, which is then chlorinated to form chloroprene. This can then be polymerized to polychloroprene. Polychloroprene contains approximately 85% *trans*-, 10% *cis*-, and 5% *vinyl*-chloroprene. Because of its high-*trans* content, polychloroprene tends to crystallize readily.

Depending on the grade of polymer, polychloroprene can be vulcanized by zinc oxide or magnesium oxide. Tetramethylthiuram disulfide can serve as a retarder. Polychloroprene is inferior to NBR for oil resistance but is still significantly better than natural rubber, SBR, or BR. Like NBR it also finds extensive use in such products as oil seals, gaskets, hose linings, and automotive engine transmission belts where resistance to oil absorption is important.

Butyl rubbers are a copolymer of isobutylene and isoprene:

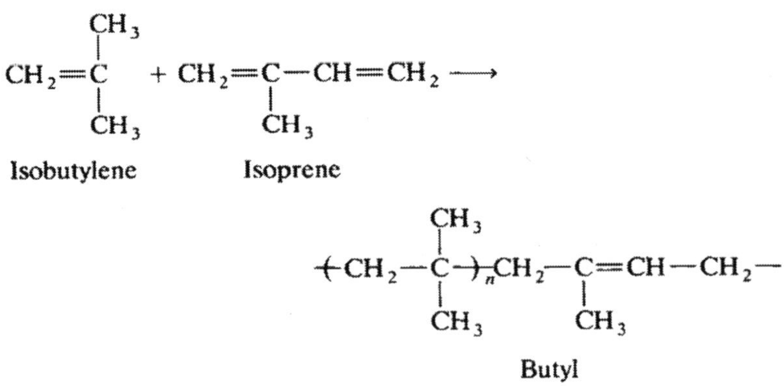

(9.3)

Isobutylene and isoprene are in a ratio of approximately 50:1. Chlorobutyl rubber and bromobutyl rubber are produced by the halogenation of butyl rubber. Butyl rubber and halobutyl rubber are highly impermeable to air and show very low water absorption, and good heat, oxygen and ozone resistance. As noted earlier, they therefore find extensive use in liners of radial tires, covers and insulation of high-voltage electric cables, and automobile engine and radiator hoses.

High-tensile-strength butyl compounds generally use FEF- or GPF-grade carbon blacks. Vulcanization systems tend to be based on thiazole accelerators such as mercaptobenzothiazole disulfide (MBTS) and thiuram accelerators such as tetramethylthiuram disulfide (TMTD). Low-tensile-strength compounds will use a clay or silica reinforcing filler in place of carbon black.

Copolymerization of ethylene and propylene produces an elastomeric polymer that is virtually inert because of the absence of carbon-carbon double bonds (EPM). Such polymers thus tend to be crosslinked with peroxides or by radiation. To improve the reactivity of ethylene-propylene copolymers, 1–10% of a third monomer can be added to give a terpolymer or ethylene-propylene-diene monomer (EPDM). The primary diene monomers used in EPDM are 1,4-hexadiene, dicyclopentadiene, and ethylidene norbornene. Introduction of an unsaturated monomer such as ethylidene norbornene will enable use of sulfur-based crosslinking systems.

EPDM tends to show good resistance to ozone attack, oxidation resistance, and moisture resistance. It is therefore used in applications that require good weather resistance and heat stability. Roofing materials, outer covers of high-voltage electric cables, and selected automotive hoses use EPDM.

See Table 9.15 for the abbreviations of selected elastomers.

9.3 FILLER SYSTEMS

Fillers, or reinforcement aids, such as carbon black, clays, and silicas are added to rubber formulations to meet material property targets such as tensile strength and abrasion resistance. Carbon black technology is as complex as polymer science, and an extensive range of blacks are available, each imparting specific sets of properties to a compound. The correct choice of carbon black is therefore as important as the development of a formulation's polymer system in meeting a product performance specification. Table 9.16 displays the general classes of rubber-grade carbon blacks as defined in ASTM Standard D1765-04 (ASTM D1795-04, 2004).

9.3.1 Carbon Black Properties

Carbon black can be described qualitatively by a series of properties: particle size (and surface area); particle size distribution; structure (particle aggregates); and surface activity (chemical functional groups such as carboxyl and ketones).

TABLE 9.15 Nomenclature for Selected Elastomers (The Synthetic Rubber Manual, 1999)

AU	Polyester urethane
BIIR	Brominated isobutylene-isoprene (bromobutyl)
BR	Polybutadiene
CIIR	Chlorinated isobutylene-isoprene (chlorobutyl)
CPE	Chlorinated polyethylene
CR	Chloroprene rubber
CSM	Chlorosulfonyl polyethylene
EAM	Ethylene-vinyl acetate coploymer
EPDM	Terpolymer of ethylene, propylene, and a diene with a residual unsaturated portion in the chain
EPM	Ethylene-propylene copolymer
EU	Polyether urethane
HNBR	Hydrogenated acrylonitrile-butadiene rubber (highly saturated nitrile rubber)
IIR	Isobutylene-isoprene rubber (butyl)
IR	Synthetic polyisoprene
NBR	Acrylonitrile-butadiene rubber
X-NBR	Carboxylated nitrile-butadiene rubber
SBR	Styrene-butadiene rubber
E-SBR	Emulsion styrene-butadiene rubber
S-SBR	Solution styrene-butadiene rubber
X-SBR	Carboxylated styrene-butadiene rubber
YSBR	Block copolymers of styrene and butadiene

TABLE 9.16 Types of Carbon Blacks

Type	ASTM Designation	Particle Size (nm)	General Use
SRF	N 762	61–100	Nontread components
GPF	N 660	49–60	Nontread components
FEF	N 550	40–48	Nontread components
FF	N 475	31–39	Nontread components
HAF	N 330	26–30	Tread and other components
ISAF	N 220	20–25	Tread
SAF	N 110	11–19	Tread

TABLE 9.17 Carbon Black Properties

ASTM Designation	Iodine Number	DBP	Compressed DBP	NSA Multipoint	STSA	Tint Strength
N 110	145	113	97	127	115	123
N 115	160	113	97	137	124	123
N 120	122	114	99	126	113	129
N 121	121	132	111	122	114	119
N 125	117	104	89	122	121	125
N 134	142	127	103	143	137	131
N 219	118	78	75			123
N 220	121	114	98	114	106	116
N 231	121	92	86	111	107	120
N 234	120	125	102	119	112	123
N 299	108	124	104	104	97	113
N 326	82	72	68	78	76	111
N 330	82	102	88	78	75	104
N 339	90	120	99	91	88	111
N 343	92	130	104	96	92	112
N 347	90	124	99	85	83	105
N 351	68	120	95	71	70	100
N 358	84	150	108	80	78	98
N 375	90	114	96	93	91	114
N 472	250	178	114	270	145	
N 550	43	121	85	40	39	
N 630	36	78	62	32	32	
N 650	36	122	84	36	35	
N 660	36	90	74	35	34	
N 762	27	65	59	29	28	
N 772	30	65	59	32	30	
N 990		43	37	8	8	
N 991		35	37	8	8	

Key properties describing a carbon black can be listed as follows (see Table 9.17):

- *Iodine number:* Measure of surface area (particle size). The higher the iodine number, the smaller the particle size.
- *DBP:* Measure of structure or size of carbon black aggregate. The higher the DBP number, the higher the structure.
- *Tint:* Optical absorbance, which increases with smaller particles.
- *CTAB:* Specific surface area measurement corrected for the effect of micropores.

TABLE 9.18 Definitions of Carbon Black Terms

Furnace carbon black	Class of carbon blacks produced by injection of defined grades of petroleum feedstock into a high-velocity stream of combustion gases under a set of defined processing conditions; for example, N 110 to N 762
Thermal carbon black	Type of carbon black produced by thermal decomposition of hydrocarbon gases; for example, N 990, N 991
Microstructure	Carbon black microstructure describes the arrangement of carbon atoms within a carbon black particle
Particle	Small spherical component of a carbon black aggregate produced by fracturing the aggregate. Particle size is measured by electron microscopy
Aggregate	Distinct, colloidal mass of particles in its smallest dispersible unit
Agglomerate	Arrangement or cluster of aggregates
Structure	Measure of the deviation of the carbon black aggregate from a spherical form
Iodine number	Weight in grams of iodine absorbed per kilogram of carbon black. Measure of particle surface area. The smaller the particle size, the greater the iodine number
Carbon black DBP	Volume of dibutyl phthalate in cubic centimeters absorbed by 100 g of carbon black. DBP number is a measure of the structure of the carbon black aggregate
Tint	Tint is a ratio of the reflectance of a reference paste to that of a sample paste consisting of a mixture of zinc oxide, plasticizer, and carbon black
CTAB	Measure of the specific surface area corrected for the effect of micropores. CTAB (cetyltrimethylene ammonium bromide) is excluded from the smaller interstices and thus better represents the portion of a particle surface area in contact with the polymer
Nitrogen surface area	Measure of total particle surface area, due to nitrogen gas being able to cover the full surface including pores without interface from surface organic functional groups
Compressed DBP	The DBP test, but where the sample undergoes a series of compressions (four times to 24,000 lb) before testing. This enables a measure of changes the carbon black will undergo during compound processing
Pellet	Mass of compressed carbon black formed to reduce dust levels, ease handling, and improve flow
Fines	Quantity of dust present in a pelletized carbon black; should be at the minimum level possible
Pellet hardness	Measure of the load in grams to crush a defined number of pellets. It is controlled by the quantity of pelletizing agent. For best pellet durability and compound mixing, pellet hardness range should have a narrow distribution. Examples of pelletizing agents are lignosulfonates and molasses
Ash	Residue remaining after burning carbon black at 550°C for 16 h; primarily a measure of the quality of plant cooling water

(continued)

TABLE 9.18 (Continued)

Toluene discoloration	Hydrocarbons extractable in toluene from carbon black; can be used as a measure of the residence time in a furnace
Hydrogen and oxygen content	Residual hydrogen and oxygen remaining after carbon black is produced; will be in the form of phenolic lactonic, carboxylic, quinonic, and hydroxyl functional groups. Such groups can have significant effects on vulcanization kinetics and reinforcement potential of the carbon black

Carbon black terms are defined in Table 9.18. Further reference can also be made to ASTM Standards D1566-04 on general compounding terms (ASTM D1566-04, 2004) and D3053-04 specifically for carbon blacks (ASTM D3053-04, 2004).

As an empirical guide, an increase in a carbon black aggregate size or structure will result in an improvement in cut growth and fatigue resistance. A decrease in particle size results in an increase in abrasion resistance and tear strength, a drop in resilience, and an increase in hysteresis and heat buildup. The impact of carbon black type and loading on tread compound performance has been studied by Hess and Kemp (1983), who evaluated 16 types of carbon black in three tread formulations with varying oil levels. The authors documented a number of criteria relating carbon black to the hysteretic properties of rubber compounds. These included loading, aggregate size, surface area, aggregate size distribution, aggregate irregularity (structure), surface activity, dispersion, and phase distribution within a heterogeneous polymer system.

From tire testing of the selected carbon black types, the following points were noted:

- Reduction of carbon black loading lowers tire rolling resistance. At a constant black loading, an increase in oil level will increase rolling resistance but also improve traction (at low oil levels, an increase in oil level may decrease compound hysteresis by improving carbon black dispersion).
- Increasing black fineness raises both rolling resistance and traction.
- An increase in the broad aggregate size distribution decreases the tire rolling resistance with constant surface area and DBP.
- Tread-grade carbon blacks can be selected to meet defined performance parameters of rolling resistance, traction, wear, etc.

Figure 9.5 illustrates the general trends for tread-grade carbon black loading and the effect on compound physical properties. As carbon black level increases, there are increases in compound heat buildup and hardness and, in tires, an increase in rolling resistance and wet skid properties. Tensile strength, compound processability, and abrasion resistance, however, go through an optimum after which these properties deteriorate.

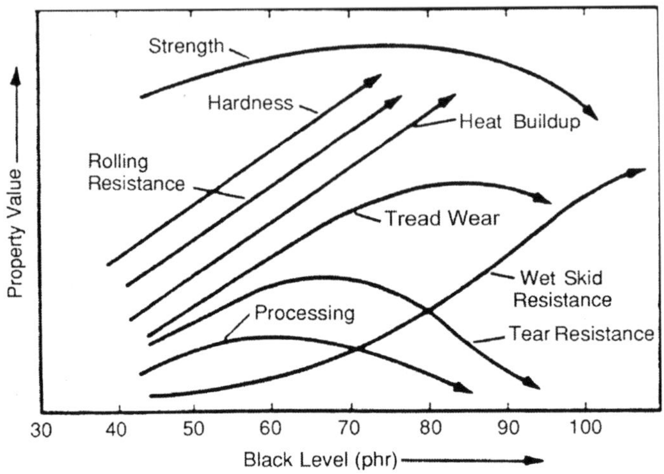

FIGURE 9.5 Effect of carbon black level on compound properties.

To exploit the results of work such as that of Hess and Klamp in improving tire rolling resistance, Swor and coworkers have developed new-technology N 200 series carbon blacks that they claim will give a better balance of tire performance properties, and the general principles will be applicable to a range of tire designs (Hess and Kemp, 1983).

In attempting to predict the direction that future research in carbon black technology will follow, a review of the literature suggests that carbon black-elastomer interactions will provide the most potential to enhance compound performance. Le Bras demonstrated that carboxyl, phenolic, quinone, and other functional groups on the carbon black surface react with the polymer and provided evidence that chemical crosslinks exist between these materials in vulcanizates (LeBras and Papirer, 1979). Ayala et al. (1990, 1990) determined a rubber-filler interaction parameter directly from vulcanizate measurements. The authors identified the ratio σ/n, where σ = slope of the stress-strain curve that relates to the black-polymer interaction, and n = the ratio of dynamic modulus E' at 1 and 25% strain amplitude and is a measure of filler-filler interaction. This interaction parameter emphasizes the contribution of carbon black-polymer interactions and reduces the influence of physical phenomena associated with networking. Use of this defined parameter enabled a number of conclusions to be made:

- The σ/n values obtained provided a good measure of black-polymer interaction for a range of polymers including SBR, IIR, NR, and NBR.
- Higher σ/n values were obtained for SBR and NBR, the aromatic structure in SBR and the polar—CN group in NBR clearly influencing black-filler interaction.

- Analysis of dry carbon black surface indicated the presence of a range of hydrocarbon groups, which is in line with earlier work (LeBras and Papirer, 1979). These groups are capable of reacting with other functional groups.

Given the establishment of organic functional groups on the carbon black surface, Wolff and Gorl investigated the reactivity of organosilane such as bis(3-triethoxysilylpropyl)tetrasulfane with furnace blacks (Wolff and Gorl, 1991). The authors deduced that such groups as carboxyl, lactol, quinone, and ketone will react with the ethoxy group of bis(3-triethoxysilylpropyl)tetrasulfane, which then become pendant on the carbon black surface:

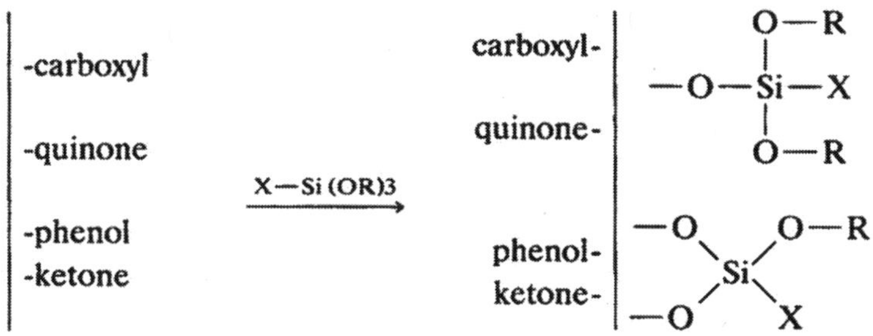

Carbon black surface **Activated carbon black surface**

(9.4)

On the basis of extract analysis and compound properties of organosilane-treated carbon black, Wolff and Gorl concluded:

- Carbon black is able to bind with a specific amount of trialkoxysilane.
- The quantity of bound organosilane correlates with carbon black particle surface area and level of oxygen-containing functional groups.
- The triethoxysilyl group constitutes the reactive part of the silane, forming a covalent bond with the carbon black.
- Reaction of bis(3-triethoxysilylpropyl)tetrasulfane with carbon black allows a reduction of compound hysteresis.

This work laid the foundation for many of the newer technology carbon blacks. These fall into two categories: postprocess modification, where the surface of the carbon black is treated to improve its properties; and in-process modification, where another material is introduced to again enhance the basic properties of the filler (Wampler et al., 2004).

Examples of postprocess systems under evaluation include surface oxidation using ozone, hydrogen peroxide, or nitric acid. Such approaches are used in the production of conductive blacks. Reaction with diazonium salts, plasma treatment, and polymer grafting are also under investigation.

In-process modification includes metal addition, development of inversion blacks or nanostructure blacks, and carbon black-silica dual phase fillers.

9.3.2 Silica and Silicates

Addition of silica to a rubber compound offers a number of advantages such as improvement in tear strength, reduction in heat buildup, and an increase in compound adhesion in multicomponent products such as tires. Two fundamental properties of silica and silicates influence their use in rubber compounds: ultimate particle size and the extent of hydration. Other physical properties such as pH, chemical composition, and oil absorption are of secondary importance.

Silicas, when compared to carbon blacks of the same particle size, do not provide the same level of reinforcement, though the deficiency of silica largely disappears when coupling agents are used with silica. Wagner reported that addition of silica to a tread compound leads to a loss in tread wear, even though improvements in hysteresis and tear strength are obtained (Wagner, 1976). The tread wear loss can be corrected by the use of silane coupling agents (Blume et al., 2004).

The chemistry of silica can be characterized as follows:

- Silica, which is amorphous, consists of silicon and oxygen arranged in a tetrahedral structure of a three-dimensional lattice. Particle size ranges from 1 to 30 nm and surface area from 20 to 300 m^2/g. There is no long-range crystal order, only short-range ordered domains in a random arrangement with neighboring domains.
- Surface silanol concentration (silanol groups –Si–O–H) influences the degree of surface hydration.
- Silanol types fall into three categories: isolated, geminal (two –OH hydroxyl groups on the same silicon atom), and vicinal (on adjacent silicon atoms), as illustrated in Figure 9.6.
- Surface acidity is controlled by the hydroxyl groups on the surface of the silica and is intermediate between those of P–OH and B–OH. This intrinsic acidity can influence peroxide vulcanization, although in sulfur curing, there is no significant effect. Rubber-filler interaction is affected by these sites.
- Surface hydration caused by water vapor absorption is affected by surface silanol concentration. High levels of hydration can adversely affect final

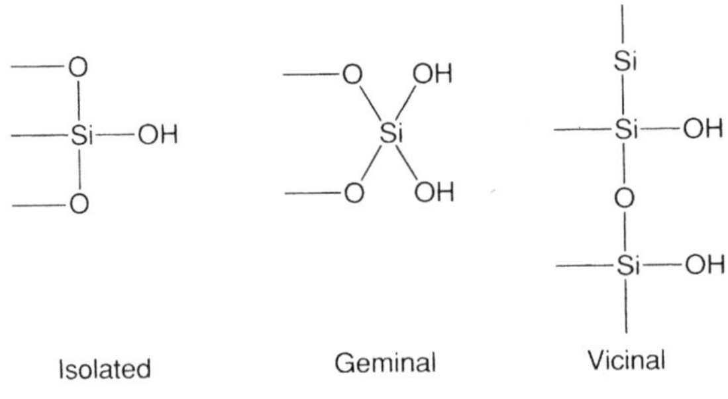

FIGURE 9.6 Typical silanol groups on silica.

compound physical properties. Silicas are hydroscopic and thus require dry storage conditions.

To illustrate the influence of surface hydroxyl groups and hydration levels on rubber properties, Wagner (1976) took a series of silicas of different surface areas, hydroxylated to different extents, and then added them to an SBR compound at 50 phr (Table 9.19). The author concluded that a reduction in silanol level as a result of an increase in absorbed water will decrease cure time, tensile strength, and abrasion resistance.

In general, silicas produce relatively greater reinforcement in more polar elastomers such as NBR and CR than in nonpolar polymers such as SBR and NR. The lack of reinforcement properties of silica in NR and SBR can be corrected through the use of silane coupling agents. An essential prerequisite for a coupling agent is that the molecule be bifunctional; that is, capable of reacting chemically with both the silica and either directly or indirectly with the polymer via participation in the vulcanization reaction or sulfur crosslinking process.

Use of silicas in rubber compounds offers two advantages: reduction in heat buildup when used as a part for part replacement of carbon black and improvement in tear strength, cut, chip, and chunking resistance. When loadings approach 20%, however, the drop in abrasion resistance of, for example, a tread compound renders the formulation no longer practical. Silane coupling agents offer the potential to overcome such drops in compound performance. Therefore, to compound silica effectively, a discussion of the properties and chemistry of coupling agents, and specifically silane coupling agents, is pertinent.

Silicas can be divided into three groups or classes. These include standard or conventional silicas, semi-highly dispersible (semi-HD) or easily dispersible silica, and the latest group developed, termed highly dispersible silica or HDS

TABLE 9.19 Effect or Surface Hydration on Silica Properties

	Silica A	Silica B
Surface area (m^2/g)	152.0	152.0
Loss at 105°C (%H_2O)	6.8	0.5
Mooney scorch	14.0	16.0
Rheometer t_{-90}	27.0	47.0
Tensile strength (MPa)	18.3	21.7
Elongation at break	480	480
300% modulus (MPa)	7.0	9.5
Pico abrasion	67.0	103.0

TABLE 9.20 Silica Groups

Surface Area (m^2/gm)	90–130	130–180	180–220
Conventional	Tire casing. Nontire internal and external components	Tire treads. Tire casings and nontire products external components	Tire treads and nontire external components for abrasion resistance
Semi-highly dispersible	Tire casing. Nontire products external components	Tire treads. Tire casings	Tire treads
Highly dispersible	Tire casings. Tire treads	Tire treads	High-performance tire treads

(Table 9.20). The silanol composition on the surface of three types of silicas remains to be elucidated, but it would be anticipated that the HDS silicas would have higher concentrations of geminal groups, whereas the conventional silica would have a greater amount of isolated silanols (Rodgers et al., 2004; Blume et al., 2004).

9.3.3 Chemistry of Silane Coupling Agents

There are three silane coupling agents of commercial significance and these have similar properties: mercaptopropyltrimethoxysilane (A189), bis(triethoxysilylethyltolylene)polysulfide (Y9194), and bis(3-triethoxisilylpropyl) tetrasulfane (TESPT). Commercial designations are given in parentheses. The coupling agent TESPT has been covered more extensively in the literature than other silane coupling agents; however, the following discussion on the use of silane coupling agents is applicable to all three materials (Blume et al., 2004).

TESPT, a bifunctional polysulfidic organosilane, was introduced as a coupling agent to improve the reinforcement properties of silicas in rubbers. Use of coupling agents offers the following advantages:

- Lowers heat buildup and hysteresis in silica-loaded compounds.
- Increases 300% modulus and tensile strength, again, in silica-loaded compounds.
- Improves reinforcing effect of clays and whiting.

- Serves as a reversion resistor in equilibrium cure systems.
- Improves DIN abrasion resistance.

The mechanism of silane coupling agent reinforcement comprises two phases: (1) the hydrophobation reaction in which coupling agent reacts with silica, and (2) the formation of crosslinks between the modified silica and polymer. Silanization of the silica surface can occur quite readily, though with TESPT systems, the reaction is generally carried out *in situ* at between 150 and 160°C in an internal mixer. Though an excess of silanol groups are present on the silica surface and reaction rates are fast, this high temperature is required because of the steric hindrance around the silylpropyl group in TESPT.

As noted earlier, three types of functional silanol groups exist on the silica surface: isolated hydroxyl groups, geminal groups (two –OH groups on one Si atom), and vicinal groups (see Figure 9.6). The silanization reaction is illustrated in Eq. (9.5).

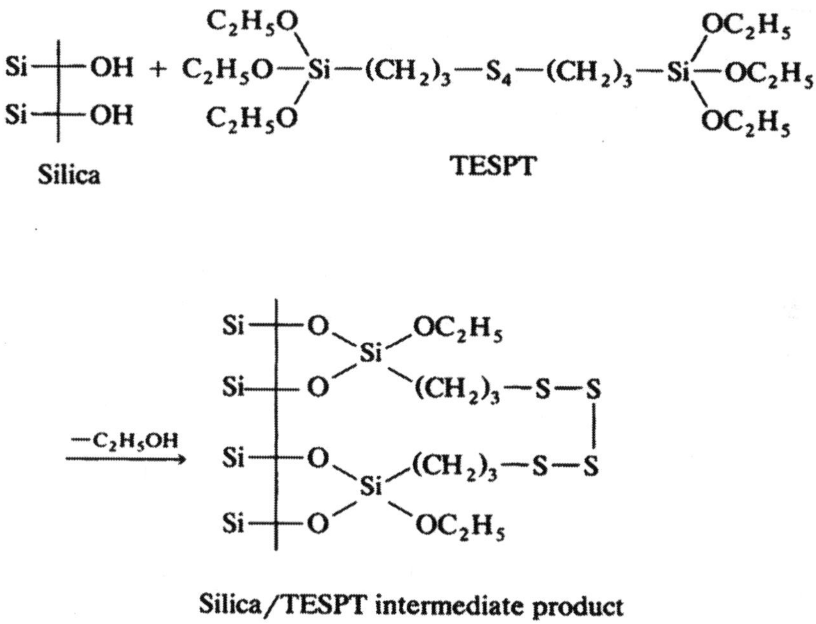

(9.5)

The filler/silane intermediate can now react with the allyl position of unsaturated sites on the polymer chain. The vulcanization of rubber is known to proceed via reaction of an accelerator, such as a sulfenamide, with sulfur, zinc oxide, and stearic acid, to generate a sulfurating agent (Bateman et al., 1963).

Equation (9.6) gives a somewhat simplistic schematic of the vulcanization reaction. On completion of the reaction, the pendant accelerator will cleave off (i.e., Captax) after generation of a crosslink. This accelerator residue, Captax, is an accelerator in its own right and continues to participate in further crosslinking as vulcanization continues.

In silica reinforcement systems containing TESPT, Wolff has suggested that the reaction is similar when the TESPT/silica intermediate is present instead of sulfur, in which case the crosslinking agent is the polysulfidic sulfur chain. Wolff showed that mercaptobenzothiazyl disulfide (MBTS) reacts with the tetrasulfane group, thus forming 2 mol of the polysulfide:

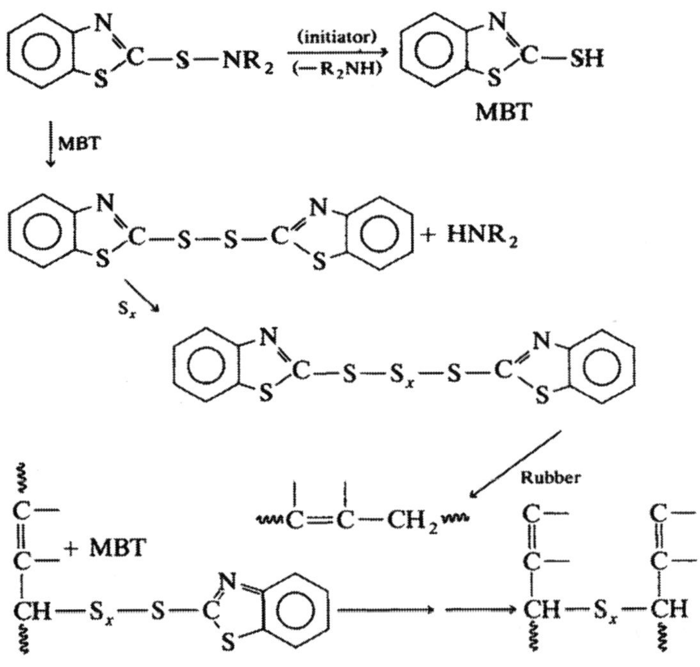

(9.6)

The silica particle is on one side and the mercaptobenzthiazolyl on the other. This polysulfidic pendant group on the silica surface will now undergo crosslink formation with the polymer in much the same way as occurs in rubber-bound intermediates that convert to crosslinks. Wolff (1983) suggested that the MBT entity reacts with the allyl position of a double bond of the rubber, thus releasing MBT and forming the rubber-silica bond.

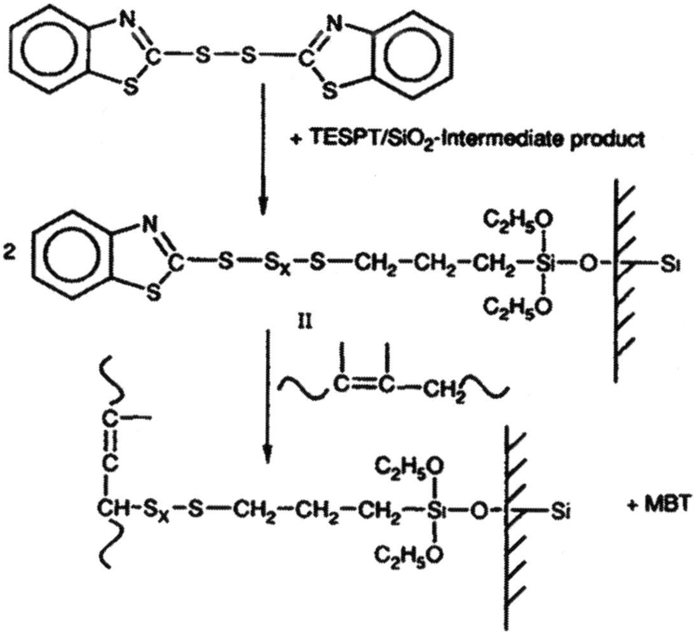

(9.7)

Proper compounding of silica with coupling agents has permitted the use of such filler systems in applications including shoe soles; engine mounts in which coupling agent/silica NR compounds provide the necessary hysteretic properties; tire treads in which, again, hysteretic properties are important; and a range of other applications such as golf balls (Blume et al., 2004).

9.3.4 Other Filler Systems

A series of additional filler systems merit brief discussion, not because of their reinforcement qualities but because of their high consumption. These include kaolin clay (hydrous aluminum silicate), mica (potassium aluminum silicate), talc (magnesium silicate), limestone (calcium carbonate), and titanium dioxide.

As with silica, the properties of clay can be enhanced through treatment of the surface with silane coupling agents. Thioalkylsilanes can react with the surface to produce a pendant thiol group, which may react with the polymer through either hydrogen bonding, van der Waal forces, or crosslinking with other reactive groups:

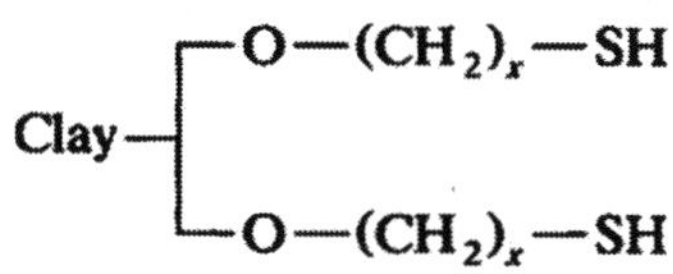

(9.8)

Such clays show improved tear strength, an increase in modulus, improved component-to-component adhesion in multicomponent products, and improved aging properties.

Calcium carbonate is used as a low-cost filler in rubber products for static applications such as carpet underlay. Titanium dioxide finds extensive use in white products such as white tire sidewalls where appearance is important.

9.4 STABILIZER SYSTEMS

The unsaturated nature of an elastomer accounts for its unique viscoelastic properties. However, the presence of carbon-carbon double bonds renders elastomers susceptible to attack by oxygen, ozone, and thermal degradation. A comprehensive review of elastomer oxidation and the role of antioxidants and antiozonants is available (Hong, 2004).

9.4.1 Degradation of Rubber

Oxidation of elastomers is accelerated by a number of factors including heat, heavy metal contamination, sulfur, light, moisture, swelling in oil and solvents, dynamic fatigue, oxygen, and ozone. Three variables in the compound formulation can be optimized to resist degradation: polymer type, cure system, and antidegradant system.

Thermo-oxidative stability is primarily a function of the vulcanization system. Peroxide vulcanization or cure systems tend to perform best for reversion resistance as a result of the absence of sulfur and use of carbon-carbon crosslinks. Efficient vulcanization (EV) systems that feature a low sulfur level (0.0–0.3 phr), a high acceleration level, and a sulfur donor similarly show good heat stability and oxidation resistance. Such systems do, however, have poor resistance to fatigue because of the presence of predominantly monosulfidic crosslinks. Conventional cure systems that feature a high sulfur level and low accelerator concentration show poor heat and oxidation resistance because the polysulfidic crosslinks are thermally unstable and readily oxidized. Such vulcanization systems do, however, have better fatigue resistance. Semi-EV cure systems, which are intermediate between EV and conventional systems, are a compromise between resistance to oxidation and required product fatigue performance.

Oxidation proceeds by two fundamental mechanisms.

- *Crosslinking:* A predominantly di- or polysulfidic crosslink network breaks down into monosulfidic crosslinks. Compound hardness increases, fatigue resistance decreases, and the compound becomes much stiffer. SBR, EPDM, NBR, and polychloroprene tend to show this behavior.
- *Chain scission:* The polymer chain breaks, causing a softening of the compound and decreased abrasion resistance. Natural rubber tends to show such degradation.

The degradation of unsaturated elastomers is an autocatalytic, free radical chain reaction, which can be broken into three steps:

1. Initiation

$$\text{RH (elastomer)} \xrightarrow[\text{heat}]{\text{O}_2,\text{light}} \text{R}_{\bullet}\text{HOO}_{\bullet}\text{(free radical)}. \tag{9.9}$$

2. Propagation

$$\begin{aligned} \text{R}_{\bullet} + \text{O}_2 &\longrightarrow \text{ROO}_{\bullet}, \\ \text{ROO}_{\bullet} + \text{R}'\text{H} &\longrightarrow \text{ROOH} + \text{R}'_{\bullet}, \\ \text{ROOH} &\longrightarrow \text{RO}_{\bullet} + {}_{\bullet}\text{OH}, \\ \text{RO}_{\bullet} + \text{R}''\text{H} &\longrightarrow \text{ROH} + \text{R}''_{\bullet}, \\ {}_{\bullet}\text{OH} + \text{R}'' &\longrightarrow \text{R}''_{\bullet} + \text{H}_2\text{O}. \end{aligned} \tag{9.10}$$

3. Termination

$$\left.\begin{aligned} &\text{R}_{\bullet} + {}_{\bullet}\text{R}, \\ &\text{R}_{\bullet} + \text{ROO}_{\bullet} \longrightarrow \text{Nonradical products}, \\ &\text{ROO}_{\bullet} + {}_{\bullet}\text{OOR}. \end{aligned}\right. \tag{9.11}$$

Like any chemical process, the rate of reaction will increase with temperature. Increase in service temperature will thus accelerate the degradation of rubber, the rate of reaction with oxygen being governed by the Arrhenius equation.

Ultraviolet light initiates free radical oxidation at the exposed surface of an elastomeric product to generate a layer of oxidized rubber. Heat, moisture, or high humidity can then initiate crazing of the surface, which subsequently can be abraded off. Such degradation of the surface is more severe with nonblack stocks than with black compounds. Nonblack compounds such as white tire sidewalls thus require higher levels of nonstaining antioxidants than carbon black-loaded formulations.

Heavy transition metal ions such as iron, manganese, and copper catalyze oxidation of elastomers. Compounds of manganese or copper such as oleates and stearates are readily soluble in rubber, enabling rapid oxidation of the polymer. *para*-Phenylenediamine antidegradants are used to hinder the activity of such metal ions.

A major cause of failure in rubber products is surface crack development. The growth of such cracks under cyclic deformation results in fatigue failure. Fatigue-related cracks are initiated at high-stress zones. Attack by ozone can induce crack initiation at the surface, which then propagates as a result of flexing. Ozone-initiated cracking can be seen as crazing on the sidewalls of old tires. Ozone readily reacts with the carbon-carbon double bonds of unsaturated elastomers to form ozonides. Under strain, ozonides readily decompose, resulting in chain cleavage and a reduction in polymer molecular weight.

Such polymer molecular weight reduction becomes apparent as surface crazing and cracking.

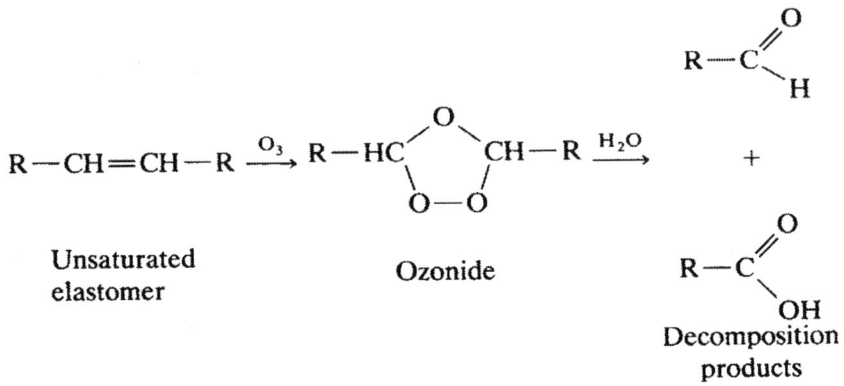

(9.12)

Polymer blends, in which the constituent polymers are incompatible, tend to improve fatigue resistance. For example, natural rubber and polybutadiene show good resistance to fatigue, crack initiation, and growth because of the formation of heterogeneous polymer phases; a crack growth in one polymer phase is arrested at the boundary with the adjacent polymer phase.

Natural rubber and polybutadiene blends tend to be used in tire sidewalls, which undergo flexing, and in tire treads, which have a lug pattern and contain high-stress zones at the base of the tread blocks.

In summary, the addition of antidegradants becomes important in order to protect the elastomeric compound from this broad range of environmental, chemical, and service-related aging phenomena.

9.4.2 Antidegradant Use

The selection criteria governing the use of antidegradants can be summarized as follows:

1. *Discoloration and staining:* In general, phenolic antioxidants tend to be nondiscoloring and amines are discoloring. Thus for elastomers containing carbon black, more active amine antioxidants are preferred since discoloration is not important.
2. *Volatility:* As a rule, the higher the molecular weight of the antioxidant, the less volatile it will be, though hindered phenols tend to be highly volatile compared with amines of equivalent molecular weight. Thus, correct addition of antioxidants in the compound mix cycle is critical if loss of material is to be avoided.
3. *Solubility:* Low solubility of an antidegradant will cause the material to bloom to the surface, with consequent loss of protection of the product. Therefore, solubility of antidegradants, particularly antiozonants, controls

their effectiveness. The materials must be soluble up to 2.0 phr, must be able to migrate to the surface, but must not be soluble in water or other solvents such as hydraulic fluid so as to prevent extraction of the protectant from the rubber.

4. *Chemical stability:* Antidegradant stability against heat, light, oxygen, and solvents is required for durability.
5. *Concentration:* Most antidegradants have an optimum concentration for maximum effectiveness after which the material solubility becomes a limiting factor. *para*-Phenylenediamines offer good oxidation resistance at a loading of 0.5–1.0 phr and antiozonant protection in the range of 2.0–5.0 phr. Above 5.0 phr *para*-phenylenediamines tend to bloom.
6. *Environment, health, and safety:* For ease of handling and avoidance of dust and inhalation, antidegradants should be dust-free while free flowing.

9.4.3 Antidegradant Types

Antidegradant types include:

1. *Nonstaining antioxidants:* This class of antioxidants is subdivided into four groups: phosphites, hindered phenols, hindered bisphenols, and hydroquinones. Hindered bisphenols such as 4,4′-thiobis(6-*t*-butyl-*m*-cresol) are the most persistent of the four classes of material. Because of their lower molecular weight, hindered phenols tend to be volatile. Phosphites tend to be used as synthetic rubber stabilizers, and hydroquinones such as 2,5-di-*tert*-amylhydroquinone are used in adhesives:

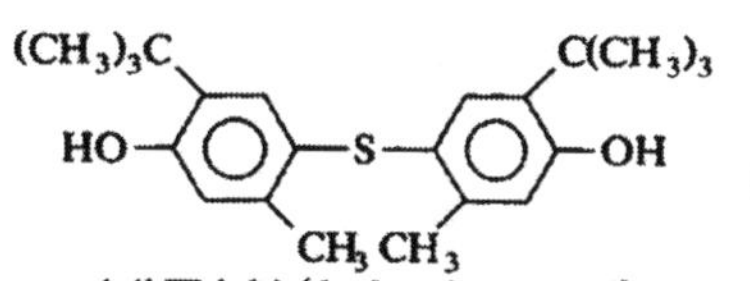

4,4′-Thiobis(6-*t*-butyl-*m*-cresol)

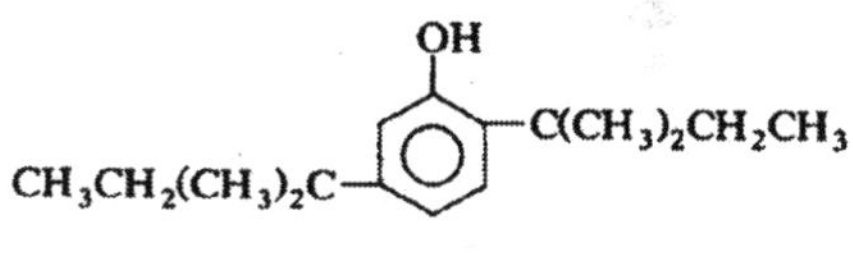

2,5-di-*tert*-amylhydroquinone

2. *Staining antioxidants:* Two classes of staining or discoloring antioxidants find extensive use, polymerized dihydroquinolines and diphenylamines:

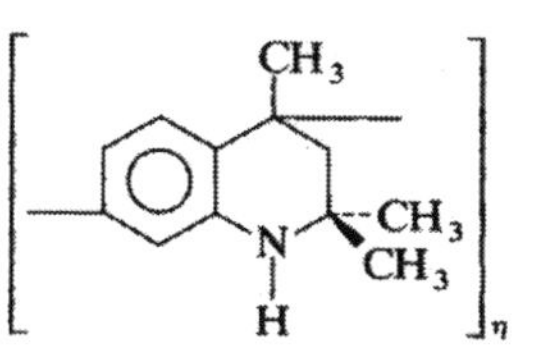

Polymerized 1,2-dihydro-2,2,4-trimethylquinoline

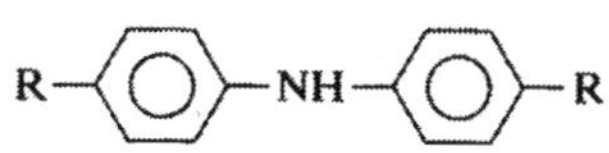

Diphenylamine class antioxidants

Dihydroquinolines differ in the degree of polymerization, thus influencing migratory and long-term durability properties. They are good general antioxidants and are effective against heavy metal pro-oxidants such as nickel and copper ions. The polymeric nature of dihydroquinolines results in low volatility and migratory properties in a vulcanizate. Thus, there is minimum loss of protectant through extraction or diffusion, durability is improved, and high-temperature stability is improved. Diphenylamine antioxidants tend to show a directional improvement in compound fatigue resistance.

3. *Antiozonants: para*-Phenylenediamines (PPDs) are the only class of antiozonants used in significant quantities. The general structure is:

They not only serve to protect rubber products from ozone but also improve resistance to fatigue, oxygen, heat, and metal ions. The three general categories of paraphenylenediamines are:

   a. *Dialkyl PPDs:* The substituent R groups are both alkyls, as in diisopropyl-*p*-phenylenediamine. The R group can range from C3 up to C9. Dialkyl PPD antidegradants tend to induce higher levels of scorch in a compound than other classes of PPD antidegradants, and tend to migrate faster than other PPD because of their low molecular weight. They lack persistence.
   b. *Alkyl-aryl PPDs:* One R group is an aromatic ring; the other is an alkyl group. The most widely used PPD in this class is *N*-1,3-dimethylbutyl-*N'*-phenyl-*p*-phenylenediamine. This antiozonant offers good dynamic protection, good static protection when combined with wax, better compound processing safety and scorch safety, and slower migratory properties, allowing it to be more persistent and suitable for long product life.
   c. *Diaryl PPDs:* The third class of PPDs contains two aromatic pendant groups, as in diphenyl-*p*-phenylenediamine or di-β-naphthyl-*p*-phenylenediamine. They are less active than alkyl-aryl PPDs and tend to bloom, thus rendering them unsuitable for many applications.

4. *Waxes:* Waxes are an additional class of materials used to improve rubber ozone protection primarily under static conditions. Waxes used in elastomeric formulations fall into two categories: microcrystalline and paraffin. Microcrystalline wax has a melting point in the range of 55–100°C and is extracted from residual heavy lube stock of refined petroleum. Paraffin wax has melting points in the range of 35–75°C and is obtained from the light lube distillate of crude oil.

The properties of waxes are listed in Table 9.21. Wax protects rubber against static ozonolysis by forming a barrier on the surface. Wax migrates from the bulk of the rubber continuously, maintaining an equilibrium concentration at

TABLE 9.21 Composition of Paraffin and Microcrystalline Waxes

	Microcrystalline	Paraffin
Molecular weight	500–800	340–430
Melting point (°C)	55–100	35–75
Mean carbon chain length	C-25	C-60
Features	Branched molecules	Linear molecules

the surface. Microcrystalline waxes migrate to the rubber surface at a slower rate than paraffins because of the higher molecular weight and branching. Furthermore, microcrystalline waxes tend to perform best at high service temperatures, whereas paraffin waxes protect best at low temperatures. This is related to the rate of migration of the wax to the product surface.

It should be noted that under dynamic conditions, the protective wax film breaks down, after which the antiozonant system in the rubber formulation will serve as the primary stabilizer or protection mechanism. Waxes are used to ensure protection against ozone for products in storage, such as tires in a warehouse.

In summary, a number of empirical guidelines can be used to develop an antidegradant system for an elastomeric formulation:

1. Short-term static protection is achieved by use of paraffin waxes.
2. Microcrystalline waxes provide long-term ozone protection while the finished product is in storage.
3. A critical level of wax bloom is required to form a protective film for static ozone protection.
4. Optimized blends of waxes and PPDs provide long-term product protection under both static and dynamic applications and over a range of temperatures.
5. Excess levels of wax bloom can have a detrimental effect on fatigue resistance, because the thick layer of wax can crack under strain and the crack can propagate into the product.

9.5 VULCANIZATION SYSTEM

Vulcanization, named after Vulcan, the Roman God of Fire, describes the process by which physically soft, compounded rubber materials are converted into high-quality engineering products. The vulcanization system constitutes the fourth component in an elastomeric formulation and functions by inserting crosslinks between adjacent polymer chains in the compound. A typical vulcanization system in a compound consists of three components: (1) activators; (2) vulcanizing agents, typically sulfur; and (3) accelerators.

The chemistry of vulcanization has been reviewed elsewhere in this text. It is appropriate, however, to review each of these components within the context of developing a compound for a defined service application.

9.5.1 Activators

The vulcanization activator system consisting of zinc oxide and stearic acid has received much less research effort than other components in the rubber compound. Stearic acid and zinc oxide levels of 2.0 and 5.0 phr, respectively, are accepted throughout the rubber industry as being adequate for achievement of optimum compound physical properties when in combination with a wide range of accelerator classes and types and accelerator-to-sulfur ratios. To clarify why zinc oxide is selected over the other metal oxides, a comparative study was conducted with magnesium oxide, calcium oxide, titanium dioxide, lead oxide, and zinc oxide. All the metal oxides were evaluated in ASTM D3184 (ASTM D3184-89, 1989); compound numbers 1A (gum stock) and 2A (which contains carbon black) are also referred to as American Chemical Society (ACS) compounds 1 and 2, respectively (Table 9.22). Test data are presented in Tables 9.23 and 9.24 (Barbin and Rodgers, 1994).

A plot of the electronegativity of the six metals of the oxides evaluated in the study versus rheometer torque ($M_H - M_L$) indicates that outside a given electronegativity range of 1.6–1.8, optimum vulcanizate properties will not be obtained (see Figures 9.7–9.8). Electronegativity is a measure of the metal atom's affinity for electron attraction. Viewing Figures 9.7–9.8 it can be concluded that for metals of electronegativity less than 1.55, a consequent shift to ionic bonding with sulfur induces a reduction in electrophilicity in the penultimate sulfur atoms of complexes:

$$\mathrm{X{-}S}_x\mathrm{{-}S^{-}{-}{-}{-}M^{+2}{-}{-}{-}S^{-}{-}S}_x\mathrm{{-}X}$$
$$\mathrm{R{-}H{-}{-}{-}S}_y\mathrm{{-}X}$$

TABLE 9.22 ASTM D3184 Formulations 1A and 2A (ACS 1 and 2)

ACS 1		ACS 2	
Natural rubber	100	Natural rubber	100
Metal oxide	6	Metal oxide	5
Stearic acid	0.5	Stearic acid	2
Sulfur	3.5	Carbon black (IRB 5)	35
MBT	0.5	Sulfur	2.25
		TBBS	0.7

TABLE 9.23 Effect of Metal Oxide Type on Compound Properties of ACS 1 Base Formulation

Compound:	**1**	**2**	**3**	**4**	**5**	**6**
Metal oxide:	***MgO***	***CaO***	***TiO_2***	***FeO***	***ZnO***	***PbO***
Cation electronegativity	1.2	1.0	1.5	1.9	1.6	1.8
% Free sulfur	2.69	2.51	2.60	2.74	1.43	1.03
Monsanto rheometer at 150°C, deta torque [$M_H - M_L$ (dN m)]	14.0	17.0	19.00	24.0	29.0	43.0
Tensile strength (MPa)	4.84	6.37	3.09	4.40	14.80	20.0
Elongation (%)	731	695	817	530	667	634
300% modulus (MPa)	0.90	1.14	0.58	1.85	2.96	2.2
Shore A hardness at 21°C	32	34	26	24	38	42
ASTM tear strength, die B (kN/m)	38	53	11	21.5	68	67

TABLE 9.24 Effect of Metal Oxide Type on Compound Properties of ACS 2 Base Formulation

Compound:	**1**	**2**	**3**	**4**	**5**	**6**
Metal Oxide:	***MgO***	***CaO***	***TiO_2***	***FeO***	***ZnO***	***PbO***
Cation electronegativity	1.2	1.0	1.5	1.9	1.6	1.8
% Free sulfur	0.39	0.35	0.72	0.68	0.15	0.15
Monsanto rheometer at 150°C, deta torque [$M_H - M_L$ (dN m)]	24.0	23.5	18.50	29.5	61.5	54.0
Tensile strength (MPa)	14.21	20.28	12.03	15.41	25.63	26.38
Elongation (%)	631	592	595	565	492	502
300% modulus (MPa)	2.94	4.86	2.91	2.53	11.04	8.9
Shore A hardness at 21°C	39	46	40	39	57	57
ASTM tear strength, die B (kN/m)	30	61	22	31.5	161	140

Conversely, with metals of electronegativity greater than 1.85, such as iron, the greater covalent character of the M + · · S− linkage with reduced charge separation would adversely affect generation of amine or carboxylate ligands

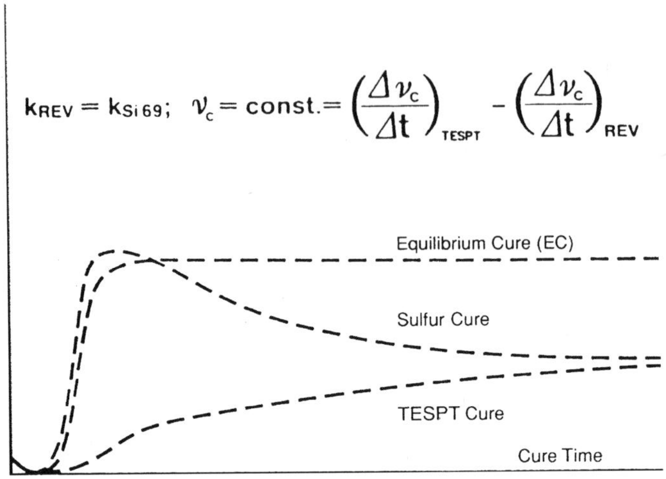

FIGURE 9.7 Rheometer profile of the EC system.

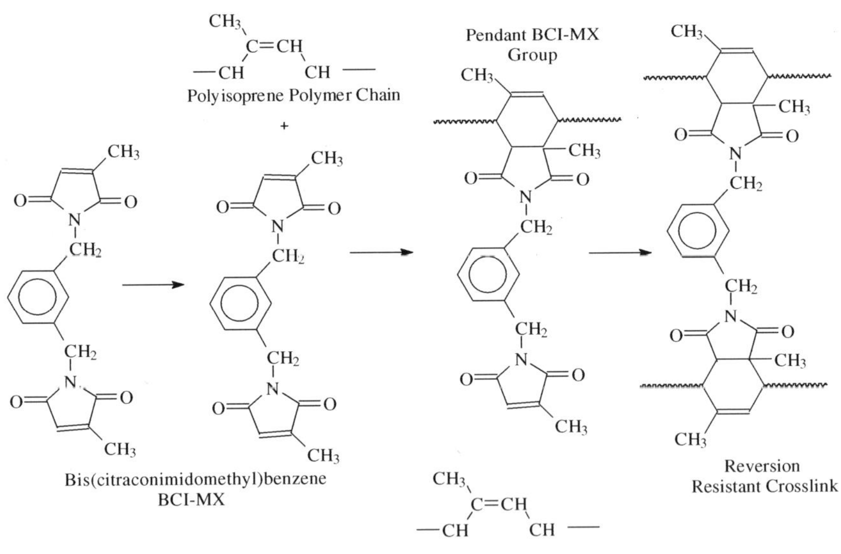

FIGURE 9.8 Proposed reversion resistance mechanism of bis(citraconimidomethyl)benzene (BCI-MX).

to the metal ion as in

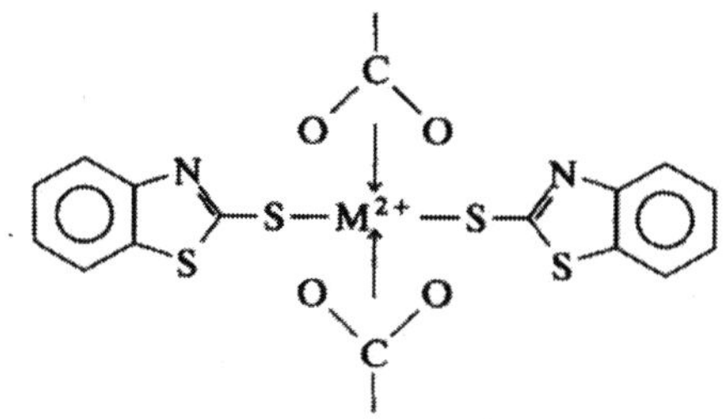

which in turn will reduce the solubility of the sulfurating reagent, consequent drop in sulfurating agent activity, and resultant drop in vulcanizate properties.

In summary, zinc is most suited to participate in formation of the sulfurating complex. Coordination of external ligands (ROO—,R′2NH:) of the zinc atom causes the bonding between $XS{-}S_x \ldots$ and $\ldots S_y{-}SX$ groups to weaken, thereby increasing the contribution of the polar canonical form:

$$XS{-}S_x^-{-}Zn^{2+}{-}S_y^-{-}SX$$

This effect is induced by ligands satisfying vacant 4*p* orbitals and distributing positive charge from the metal. The result will be increased nucleophilicity of XSS_{x^-} but decreased electrophilicity of XS_{y^+} in the sulfurating complex. The same is true for Cd^{2+} and Pb^{2+} complexes, which have vacant *p* orbitals to accommodate coordination ligands. In the case of Mg^{2+} and Ca^{2+} complexes, coordination will not readily occur, the reduced ease of formation being further influenced by the inability of the metal to achieve an inert gas configuration as in more stable organometallics. Toxicity of CdO and PbO prohibits their use, thus ZnO has found virtually universal use in the rubber industry, the ultimate loading in a compound being dependent on the product application.

As part of the metal oxide study, a comparative study of oleic acid and stearic acid, each at 1.0, 2.0, and 3.0 phr, was conducted on ASTM No. 2A (ACS 2) compound.

The data outlined in Table 9.25 illustrate a number of points:

1. An increase in the fatty acid level reduces vulcanization activation energy, the effect being greater for stearic acid.
2. Stearic acid/ZnO-activated compounds show higher crosslink densities compared with oleic acid systems.
3. Aging and tear strength properties of stearic acid/ZnO compounds are superior to those of oleic acid systems.

The effectiveness of stearic acid in activating vulcanization is a function of its solubility in the elastomer, molecular weight, and melting point.

TABLE 9.25 Influence of Fatty Acid Level in Vulcanizates

Compound:	1	2	3	4	5	6
Fatty Acid:	***(Stearic)***			***(Oleic)***		
phr:	***1.0***	***2.0***	***3.0***	***1.0***	***2.0***	***3.0***
Crosslink density (rating)	100	94	106	75	80	89
Activation energy (kJ mol^{-1})	131.5	101.5	97.6	135.1	114.2	110.5
Tensile strength (MPa)	27.50	26.8	26.9	28.5	28.0	26.4
Elongation (%)	545	535	538	591	576	551
Shore A hardness	52	53	50	50	52	52
Tear strength, ASTM die B (kN/m)	72	112	103	72	94	80
Aged tensile strength (MPa)	17.50	18.1	21.3	15.8	16.8	17.3

9.5.2 Vulcanizing Agents

Three vulcanizing agents find extensive use in the rubber industry: sulfur, insoluble sulfur, and peroxides. The chemistry of peroxides has been reviewed in Chapter 7. Rhombic sulfur is the most common form of sulfur used in the rubber industry and, other than normal factory hygiene and operational procedures, does not require any special handling or storage. Sulfur is soluble in natural rubber at levels up to 2.0 phr. Above this concentration, insoluble sulfur must be used to prevent migration of sulfur to the compound surface; that is, sulfur bloom.

9.5.3 Accelerators

Accelerators are products that increase both the rate of sulfur crosslinking in a rubber compound and crosslink density. Secondary accelerators, when added to primary accelerators, increase the rate of vulcanization and degree of crosslinking, with the terms primary and secondary being essentially arbitrary. A feature of such binary acceleration systems is the phenomenon of synergism. Where a combination of accelerators is synergistic, its effect is always more powerful than the added effects of the individual components.

Accelerators can be readily classified by one of two techniques:

1. *Rate of vulcanization:* Ultra-accelerators include dithiocarbamates and xanthates. Semiultra-accelerators include thiurams and amines. Fast accelerators are thiazoles and sulfenamides. A medium-rate system is diphenylguanidine. A slow accelerator is thiocarbanilide.

2. *Chemical classifications:* Most accelerators fall into one of eight groups: aldehydeamines, sulfenamides, thioureas, dithiocarbamates, guanidines, thiurams, thiazoles, and xanthates.

Factors involved in the selection of vulcanization systems must include the type of elastomer, type and quantity of zinc oxide and fatty acid, rate of vulcanization, required resistance to fatigue, and service conditions. It is also recommended that use of nitrosamine-generating accelerators be avoided.

The type of elastomer will influence the rate of cure and the resultant crosslink network. Natural rubber tends to cure faster than SBR. Cure systems containing thiuram accelerators such as tetramethylthiuram disulfide will show short induction times and fast cure rates compared with a system containing diphenylguanidine.

Sulfenamide accelerators represent the largest class of accelerators consumed on a global basis:

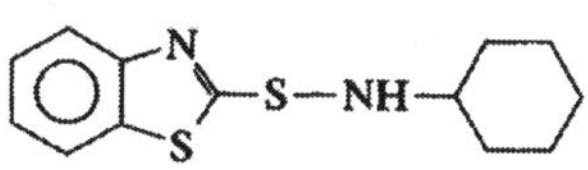

Cyclohexylbenzothiazole sulfenamide (CBS)

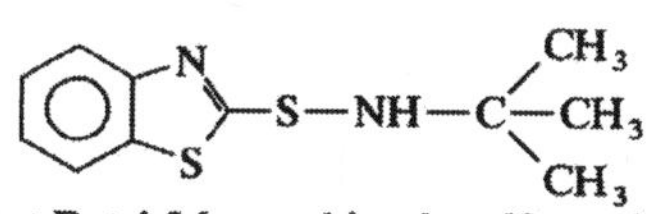

***tert*-Butyl-2-benzothiazole sulfenamide (TBBS)**

The mechanism and chemistry of vulcanization have been reviewed earlier. It is therefore more appropriate to define the general principles governing the activity of an accelerator such as a sulfenamide. Three parameters merit elucidation:

1. *Bond strength of the sulfur-nitrogen bond:* Sulfenamides are cleaved into mercaptobenzothiazole and amine fragments during formation of the sulfurating complex, and the amine forms ligands with the zinc ion. Bond energy must be sufficiently low so as not to prevent generation of active accelerator species or sulfurating reagent.
2. *Stereochemistry of the amine fragment:* The steric bulk of the amine ligand coordinated with the zinc ion, if too large, can hinder the formation of an active sulfurating agent. This is seen as an increase in induction times, change in vulcanization rate, and, ultimately, change in physical properties.
3. *Basic strength of the amine fragment:* An increase in the basicity of the amine fragment of the sulfenamide results in an increase in the rate of vulcanization. More basic amines also tend to induce poor scorch resistance (Table 9.26). Further reference should be made to Chapter 7 on vulcanization.

9.5.4 Retarders and Antireversion Agents

The induction time or scorch resistance of a compound can be improved by addition of a retarder. *N*-Cyclohexylthiophthalimide (CTP) is by far the largest-tonnage retarder used in the rubber industry. See the review by Morita for

TABLE 9.26 Effect of Sulfenamide Amide Fragment Basicity (pK_b) on Compound Scorch Activity[a,b]

R Radical	pK_b of Free Amine	Mooney Scorch t^{-10} at 135°C	Cure Rate Index
	3.3	29.8	1.00
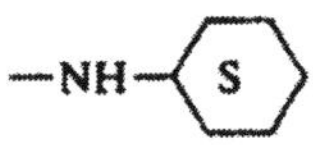	3.7	32.7	2.05
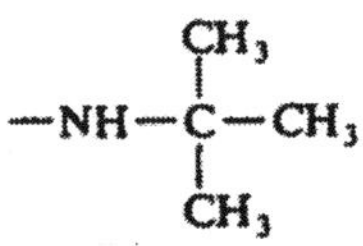	4.2	31.8	2.10
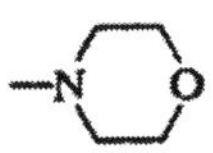	6.2	42.4	2.32
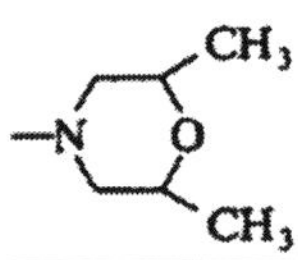	6.2	45.2	2.57

[a]*Compound: 100 phr SBR-1500, 50 HAF, 4.0 ZnO, 2.0 stearic acid, 10.0 oil, 1.5 antioxidant, 1.75 sulfur, equimolar accelerator levels.*

[b]*Cure rate index is $t_{35} - t_{10}$ at 135° C on a Mooney plastimeter (Barbin and Rodgers, 1994).*

discussion of the mechanism of CTP reactivity and the chemistry of other special retarders such as the thiosulfonamide class of materials (Morita, 1980).

Resistance to compound reversion, particularly of natural rubber compounds, has received more recent attention because of the broad range of requirements including faster processing of compounds in production, processing at higher temperatures, and, perhaps more important, extension of product service life. Three antireversion agents have been used commercially:

- Reviewed earlier, a semi-EV system is a compromise designed to produce, in structural terms, a vulcanizate containing a balance of monosulfidic and polysulfidic crosslinks at a defined optimum cure state. If polysulfidic crosslinks are to persist over extended periods, new ones must be created to replace those lost through reversion. With use of normal acceleration systems, there is limited opportunity for such events. Maintenance of a polysulfidic network through the curing process thus dictates utilization of a dual-cure system, in which both cures are independent of each other. This is the principle of the equilibrium cure (EC) system. Here, bis(3-triethoxysilylpropyl)tetrasulfane (TESPT) is added as a slow sulfur donor (Wolff, 1990) (Figure 9.7).

- Bis(citraconimidomethyl)benzene, commercial name Perkalink 900, has been introduced and functions exclusively as a reversion resistor. It is understood to react via a Diels-Alder reaction to form a six-membered ring on the polymer chain (Figure 9.8). The ultimate crosslink is thermally stable and replaces sulfur crosslinks that disappear during reversion (Rubber Chemicals, 1998).
- Sodium hexamethylene-1,6-bisthiosulfide dihydrate, when added to the vulcanization system, breaks down and inserts a hexamethylene-1,6-dithiyl group within a disulfide or polysulfide crosslink. This is termed a hybrid crosslink. During extended vulcanization periods or accumulated heat history due to product service, polysulfidic-hexamethylene crosslinks shorten to produce thermally stable elastic monosulfidic crosslinks. At levels up to 2.0 phr, there is little effect on compound induction or scorch times, nor on other compound mechanical properties (Rubber Chemicals, 1998).

$$^{+}Na^{-}SO_3{-}S{-}(CH_2)_6{-}S{-}SO_3^{-}Na^{+} \cdot 2H_2O$$

Hexamethylene-1,6-bis(thiosulfate)
disodium salt dihydrate

Table 9.27 shows industry-recognized abbreviations for various accelerators.

9.6 SPECIAL COMPOUNDING INGREDIENTS

In addition to the four primary components in a rubber formulation—the polymer system, fillers, stabilizer system, and vulcanization system—there are a range of secondary materials such as processing aids, resins, and coloring agents (e.g., titanium dioxide used in tire white sidewalls). These are briefly discussed to establish a guideline for the use of the materials in practical rubber formulations.

9.6.1 Processing Oils

Processing oils in a rubber formulation serve primarily as a processing aid. Oils fall into one of three primary categories: paraffinic, naphthenic, and aromatic. The proper selection of oils for inclusion in a formulation is important. If the oil is incompatible with the polymer, it will migrate out of the compound with consequent loss in required physical properties, loss in rubber component surface properties, and deterioration in component-to-component adhesion, as in a tire. The compatibility of an oil with a polymer system is a function of

TABLE 9.27 Comparison of the Scorch Activity of a Range of Selected Commercial Accelerators[a,b]

Accelerator	Abbreviation	Scorch Time (min)	
		ACS 1	ACS 2
Diisopropylbenzothiazole	DIBS	63	21
Cyclohexyl-2-benzothiazolesulfenamide	CBS	26	15
tert-Butyl-2-benzothiazolesulfenamide	TBBS	60	23
Oxydiethylene benzothiazole-2-sulfenamide	MBS	60	24
Dimethylmorpholine benzothiazole-2-sulfenamide	—	60	27
Benzothiazyl-*N,N*-diethylthiocarbamyl sulfide	—	26	15
Mercaptobenzothiazole	MBT	12	9
Benzothiazyl disulfide	MBTS	73	12
Tetramethylthiuram monosulfide	TMTM	25	12
Tetramethylthiuram disulfide	TMTD	13	6
Tetraethylthiuram disulfide	TETD	18	9
Zinc dibutyldithiocarbamate	ZDBC	5	4
Zinc dimethylthiocarbamate	ZDMC	6	4
Diphenylguanidine	DPG	16	10

[a]*Data obtained from the formulation ACS 2 (ASTM D3184-89, 2a) but containing 50 phr HAF.*
[b]*Use of nitrosamine-generating accelerators should be avoided.*

the properties of the oil such as viscosity, molecular weight, and molecular composition. Table 9.28 defines the physical properties of three typical classes of oils.

Aniline point is a measure of the aromaticity of an oil. It is the point at which the oil becomes miscible in aniline. Thus the lower the aniline point, the higher the aromatic content. All three classes of oils contain high levels of cyclic carbon structures; the differences are in the number of saturated and unsaturated rings. Oils can therefore be described qualitatively as follows:

TABLE 9.28 Physical Properties of Three Classes of Oils Used in the Rubber Industry

Physical Property	ASTM Methods	Paraffinic	Naphthanic	Aromatic
Specific gravity	D1250	0.85–0.89	0.91–0.94	0.95–1.0
Pour point (°F)	D97	0.0 to +10	−40 to +20	+40 to +90
Refractive index	D1747	1.48	1.51	1.55
Aniline point	D611	200–260	150–210	95.0–150.0
API gravity	D287	28.0–34.0	19.0–28.0	10.0–19.0
Molecular weight	D2502	320–650	300–460	300–700
Aromatic content (%)		15.0	44.0	68.0

- Aromatic oils contain high levels of unsaturated rings, unsaturated naphthanic rings, and pendant alkyl and unsaturated hydrocarbon chains. The predominant structure is aromatic.
- Naphthenic oils have high levels of saturated rings and little unsaturation.
- Paraffinic oils contain not only high levels of naphthenic rings but also higher levels of alkyl pendant groups, unsaturated hydrocarbon pendant groups, and, most important, fewer naphthenic groups per molecule. Pure paraffins from refined petroleum condense out as wax.

The selection of an oil for a given polymer depends on the presence of polar groups in the polymer, such as –CN groups in NBR and –Cl in CR. Hydrogen bonding and van der Waals forces impact on the effectiveness of an oil in a compound. Table 9.29 presents a general guide for selection of an oil for a given polymer. This selection guide is necessarily brief and there are many exceptions. The key parameters to be noted though are the oil's tendency to discolor the product, the oil's tendency to stain adjacent components in a product, and the solubility of the oil in the polymer (Barbin and Rodgers, 1994).

9.6.2 Plasticizers

Though processing oils, waxes, and fatty acids can be considered as plasticizers, within the rubber industry the term *plasticizer* is used more frequently to describe the class of materials that includes esters, pine tars, and low-molecular-weight polyethylene.

Phthalates are the most frequently used esters. Dibutylphthalate (DBP) tends to give soft compounds with tack; dioctylphthalate (DOP) is less volatile and tends to produce harder compounds because of its higher molecular weight. Polymeric esters such as polypropylene adipate (PPA) are used when low volatility is required along with good heat resistance.

TABLE 9.29 Oil Selection Guide for Range of Commercial Elastomers

Oil	Polymer	Examples of Product Applications
Naphthenic	Ethylene-propylene rubber	Sealants, caulking
	EPDM	Adhesives
	Polychloroprene	General rubber products
	SBR	
	PBD	
Paraffinic	Natural rubber	Textile application
	Polyisoprene	Caulking
	Butyl	Sealants
	SBR	
	Polychloroprene	
Aromatic	Natural rubber	Tires
	SBR	Automotive components
	Polybutadiene	

Though total consumption is tending to fall, pine tars are highly compatible with natural rubber, give good filler dispersion, and can enhance compound properties such as fatigue resistance and component-to-component adhesion, which is important in tire durability. Other low-volume plasticizers include factice (sulfur-vulcanized vegetable oil); fatty acid salts such as zinc stearate, which can also act as a peptizer; rosin; low-molecular-weight polypropylene; and organosilanes such as dimethylpolysiloxane.

9.6.3 Chemical Peptizers

Peptizers serve as either oxidation catalysts or radical acceptors, which essentially remove free radicals formed during the initial mixing of the elastomer. This prevents polymer recombination, allowing a consequent drop in polymer molecular weight, and thus the reduction in compound viscosity. This polymer softening then enables incorporation of the range of compounding materials included in the formulation. Examples of peptizers are pentachlorothiophenol, phenylhydrazine, certain diphenylsulfides, and xylyl mercaptan. Each peptizer has an optimum loading in a compound for most efficiency.

Peptizers such as pentachlorothiophenol are generally used at levels between 0.1 and 0.25 phr. This enables significant improvement in compound processability, reduction in energy consumption during mixing, and improvement in compound uniformity. High levels can, however, adversely affect the compound properties, as excess peptizer continues to catalyze polymer breakdown as the product is in service.

9.6.4 Resins

Resins fall into one of three functional categories: (1) extending or processing resins, (2) tackifying resins, and (3) curing resins. Resins have been classified in an almost arbitrary manner into hydrocarbons, petroleum resins, and phenolic resins.

Hydrocarbon resins tend to have high glass transition temperatures so that at processing temperatures they melt, thereby allowing improvement in compound viscosity mold flow. They will, however, harden at room temperature, thus maintaining compound hardness and modulus. Within the range of hydrocarbon resins, aromatic resins serve as reinforcing agents, aliphatic resins improve tack, and intermediate resins provide both characteristics. Coumarone-indene resin systems are examples of such systems. These resins provide:

- Improved tensile strength as a result of stiffening at room temperature.
- Increased fatigue resistance as a result of improved dispersion of the fillers and wetting of the filler surface.
- Retardation of cut growth by dissipation of stress at the crack tip (as a result of a decrease in compound viscosity).

Petroleum resins are a by-product of oil refining. Like hydrocarbon resins, a range of grades are produced. Aliphatic resins, which contain oligomers of isoprene, tend to be used as tackifiers, whereas aromatic resins, which also contain high levels of dicyclopentadiene, tend to be classed more as reinforcing systems.

Phenolic resins are of two types, reactive and nonreactive. Nonreactive resins tend to be oligomers of alkyl-phenyl formaldehyde, where the *para*-alkyl group ranges from to C4 to C9. Such resins tend to be used as tackifying resins. Reactive resins contain free methylol groups. In the presence of methylene donors such as hexamethylenetetramine, crosslink networks will be created, enabling the reactive resin to serve as a reinforcing resin and adhesion promoter.

9.6.5 Short Fibers

Short fibers may be added to compounds to further improve compound strength. They can be processed just as other compounding ingredients. Short fibers include nylon, polyester, fiberglass, aramid, and cellulose. The advantages of adding short fibers to reinforce a compound depend on the application for which the product is used; however, general advantages include improved tensile strength, improvement in fatigue resistance and cut growth resistance, increase in stiffness, increased component or product stiffness, and improved cutting and chipping resistance as in tire treads.

9.7 COMPOUND DEVELOPMENT

The preceding discussion reviewed the range of materials that are combined in an elastomeric formulation to generate a defined set of mechanical properties. Elastomeric formulations can be developed by one of two techniques. Model formulations can be obtained from raw material supplier literature or other industry sources such as trade journals. Such formulations approximate the required physical properties to meet the product performance demands. Further optimization might then include, for example, acceleration level determination to meet a required compound cure induction time, and carbon black level evaluation to match a defined tensile strength target. Where more complex property targets must be met and no model formulations are available, a more efficient technique is to use either Taguchi analysis or multiple regression analysis.

A series of components in a formulation can be optimized simultaneously through use of a computer optimization. A number of models are suitable for use in designed experiments (DelVecchio, 1997). Regardless of the technique or model selected, a series of simple steps is still pertinent before the experimental work is initiated:

1. Define the objective of the work.
2. Identify the variables in the formulation to be analyzed.
3. Select the appropriate analysis for the accumulated experimental data.
4. Analyze the data within the context of previously published data and knowledge of the activity and characteristics of the raw materials investigated.
5. Assess the statistical significance of the data (data scatter, test error, etc.).

The designed experiment will then entail these steps:

1. Define the key property targets, such as tensile strength, fatigue resistance, and hysteretic properties.
2. Select an appropriate design, for example a two-variable factorial or three-, four-, or five-variable multiple regression.
3. Calculate multiple regression coefficients from the accumulated experimental data. The coefficients can be computed from the regression equations, which can be either a linear equation,

$$\text{property} = aX + bY + C$$

in the case of a simple factorial design, or a second-order polynomial where interactions between components in a formulation can be viewed:

$$\text{property} = aX + bY + cZ + dX^2 + eY^2 + fZ^2 + gXY + hXZ + jYZ + C$$

Here, a property or the dependent variable might be modulus, and X, Y, and Z are independent variables such as oil level, carbon black level, and sulfur level. The terms a, b, c, d, and so on are coefficients for the respective dependent variables, and C is a constant for the particular model. Clearly, other equations are possible but depend on the objective of the study in question.

4. Construct appropriate contour plots to visualize trends in the data and highlight interaction between components in the formulation.
5. Compute an optimization of the ingredients.
6. Run a compound confirmation study to verify the computed compound optimization.

A wide range of experimental designs are available, and it is recommended that the attached reference be reviewed for further information (DelVecchio, 1997).

Tables 9.30–9.35 display a series of model formulations on which further compound optimization can be based. Additional formulations are available in industry publications such as those from the Malaysian Rubber Producer's Research Association (Malaysian Rubber Producers' Research Association, 1984; Waddell et al., 1990).

9.8 COMPOUND PREPARATION

In a modern tire or general products production facility, rubber compounds are prepared in internal mixers. Internal mixers consist of a chamber to which the compounding ingredients are added. In the chamber are two rotors that generate high shear forces, dispersing the fillers and other raw materials in the polymer. The generation of these shear forces results in the production of a uniform, quality compound. After a defined mixing period, the compound is dropped onto a mill or extruder where mixing is completed and the stock sheeted out for ease of handling. Alternatively, the compound can be passed into a pelletizer.

Depending on the complexity of the formulation, size of the internal mixer, and application for which the compound is intended, the mix cycle can be divided into a sequence of stages. For an all-natural-rubber compound containing 50 phr carbon black, 3 phr of aromatic oil, an antioxidant system, and a semi-EV vulcanization system, a typical Banbury mix cycle will be as follows:

Stage 1: Add all natural rubber; add peptizer if required. Drop into a mill at 165°C.

Stage 2: Drop in carbon black, oils, antioxidants, zinc oxide, stearic acid, and miscellaneous pigments such as flame retardants at 160°C.

Stage 3: If required to reduce compound viscosity, pass the compound once again through the internal mixer for up to 90 s or 130°C.

TABLE 9.30 Examples of Tire Compounds

	Tread	Sidewall	Wire Coat	Ply Coat	Inner liner
Polymer system	NR BR S-SBR E-SBR	NR BR	NR	NR BR E-SBR	IIR CIIR BIIR NR
Filler system	SAF ISAF HAF	FF FEF GPF	HAF	HAF	GPF
Vulcanization system	Semi-EV	Adapted to polymer system	Conventional system	Conventional system Semi-EV	Adapted to polymer system
Miscellaneous components	Oils Waxes	Antidegradants Waxes	Adhesion promoters	Adhesion promoters	–

TABLE 9.31 Model Truck Tire Tread Formulation

Natural rubber	100.00
Carbon black (N 220)	50.00
Peptizer	0.20
Paraffin wax	1.00
Microcrystalline wax	1.00
Aromatic oil	3.00
Polymerized dihydrotrimethylquinoline	1.00
Stearic acid	2.00
Zinc oxide	5.00
Sulfur	1.20
TBBS	0.95
DPG	0.35
Retarder (if required)	0.25

Stage 4: Add the cure system to the compound and mix it up to a temperature not exceeding 115°C.

Computer monitoring of the internal mixer variables such as power consumption, temperature gradients through the mixing chamber, and mix times enables modern mixers to produce consistent high-quality compounds in large volumes. The mixed compound is then transported to either extruders for production of extruded profiles, calenders for sheeting, or injection molding.

Depending on the compound physical property requirements, compounds can be prepared on mills. Mill mixing takes longer, consumes larger amounts of energy, and gives smaller batch weights. The heat history of the compound is reduced, however, and this can be advantageous when processing compounds with high-performance fast acceleration systems. Two-roll mills function by shear created as the two rolls rotate at different speeds (friction ratio). This ratio of roll speeds is variable and is set dependent on the particular type of compound. The higher the friction ratio, the greater the generated shear and intensity of mixing.

9.9 ENVIRONMENTAL REQUIREMENTS IN COMPOUNDING

In addition to developing products to satisfy customers, the environmental implications of the technology must be taken into consideration. The environmental impact on compound development must be viewed in two parts: (1) product use and long-term ecological implications; (2) health and safety, in both product service and product manufacture.

TABLE 9.32 Model Sidewall Formulation

Natural rubber	50.00
Polybutadiene	50.00
Carbon black (N 330)	50.00
Peptizer	0.20
Paraffin wax	1.00
Microcrystalline wax	1.00
Aromatic oil	10.00
Polymerized dihydrotrimethylquinoline	1.00
Dimethylbutylphenyl-*p*-phenylenediamine	3.00
Stearic acid	2.00
Zinc oxide	4.00
Sulfur	1.75
TBBS	1.00
DPG	0.35
Retarder (if required)	0.25

An example of the impact of product usage and the environmental implications is tire rolling resistance and its effect on vehicle fuel consumption. Reduction in tire rolling resistance results in a drop in vehicle fuel consumption. This has an immediate impact on the generation of exhaust gases such as carbon monoxide, carbon dioxide, and nitrous oxide.

The crown area of the tire, which includes the tread and belts, accounts for approximately 75% of the radial passenger tire rolling resistance. Improvements in the hysteretic properties of the tread compound will therefore enable a reduction in tire rolling resistance and consequent improvements in vehicle fuel economy. The crown area and particularly the tread compound also affect the life cycle of the tire. Longer-wearing tires (including retreading) delay the point in time when used tires must enter the solid waste disposal system.

Critical to a tire's life cycle performance is the ability to maintain air pressure. Tire innerliners composed of halobutyl-based compounds exhibit very low air and moisture permeability. Therefore, tires built with the proper selection of compounds can reduce the rate of premature failure, again delaying entry into the scrap tire and solid waste streams.

Table 9.36 illustrates how the incorrect selection of a tire innerliner polymer will lead to more rapid deterioration in tire performance properties. Replacement of chlorobutyl with natural rubber or reclaimed butyl will lead to a more rapid loss in tire air pressure and loss in overall tire performance (Niziolek, 2000).

Improved tire designs have enabled reduction in noise levels. This has become an important environmental consideration. Optimum footprint

TABLE 9.33 Model Tire Casing Ply Coat

Natural rubber	70.00
Polybutadiene	30.00
Carbon black (N 660)	55.00
Peptizer	0.20
Paraffin wax	1.00
Microcrystalline wax	1.00
Aromatic oil	7.00
Dimethylbutylphenyl-*p*-phenylenediamine	2.50
Stearic acid	2.00
Zinc oxide	4.00
Sulfur	2.50
MBTS	0.80

pressures reduce damage to highway pavements and bridges. All of these improvements in tire rolling resistance, life cycle duration, noise generation, and tire footprint pressure have been incorporated into the full range of tires, from small automobile to heavy truck to large earthmover equipment tires. Today's radial tires use 60–80% natural rubber as the polymer portion of compounds. Because natural rubber is obtained from trees, it is an ideal renewable resource, and thus as a biotechnology material is preferred to petroleum-based synthetic polymers, when equivalent compound properties can be attained.

Tires are one of the most durable technological products manufactured today. They are a resilient, durable composite of fabric, steel, carbon black, natural rubber, and synthetic polymers. The qualities that make tires or other engineered rubber products a high-value item create a special challenge of disposal. Tires and other rubber products, such as conveyor belts and hydraulic hoses, are not biodegradable and cannot be recycled like glass, aluminum, or plastic. Four potential applications for such products entering the solid waste stream have been identified:

- The calorific energy of tires is higher (35 MJ/kg) than that of coal (24 MJ/kg). With properly designed equipment, tires can be burned to produce heat in cement kilns.
- Tires can be burned in furnaces at power-generating facilities to produce electrical energy.
- Ground up scrap tires are beginning to find use in some special asphalt applications.
- Tires with the proper installation technology can serve a variety of applications in the construction industry as marine reefs, energy-efficient house construction, highway bank reinforcement, and erosion control.

TABLE 9.34 Model Conveyor Belt Cover

Natural rubber	100.00
Carbon black (N 330)	50.00
Peptizer	0.20
Paraffin wax	1.00
Microcrystalline wax	1.00
Aromatic oil	5.00
Polymerized dihydrotrimethylquinoline	1.00
Dimethylbutylphenyl-*p*-phenylenediamine	2.50
Stearic acid	2.00
Zinc oxide	4.00
Sulfur	2.50
TBBS	1.00

These four methods of disposal represent the best options for scrap tire and rubber products disposal. It is anticipated that a variety of new applications for disposal of scrap rubber products will emerge in the future.

In summary, materials scientists must consider the implications of their materials choices, from the quantity of energy to manufacture the product, to the performance during its useful life cycle, and finally to disposal methods. US Environmental Protection Agency (EPA) also provides constraints that the materials scientists must consider in the design of compounds. As most rubber compounds contain approximately 6–20 different materials, not only the materials themselves must be clean and harmless, but any by-products that form during product tire manufacturing must also be harmless to humans and the environment.

The aromatic content of carbon blacks and oils was once considered hazardous. Data were generated that showed that carbon black was stabilized and did not represent a hazard to workers. Resins for cure or tack, antioxidants, antiozonants, and cure accelerators also must be investigated to ensure that the materials and any impurities meet changing health and safety standards. Materials safety data sheets and chemical health and toxicity data must be maintained on all materials.

Nitrosamine-generating chemicals represent an area where suspect materials have been removed from rubber products, even though no governing legislation has yet been drafted. Nitrosamines can be formed when secondary amine accelerators are used to cure rubber. These accelerator changes have a very significant effect on the total rubber industry.

Solvent composition and volatility limits can have significant effects on synthetic rubber production and tire manufacturing. Limits of exposure to some

TABLE 9.35 Carpet Underlay

SBR	100.00
Calcium carbonate	150.00
Reclaim (as filler)	20.00
Clay	25.00
Iron oxide	4.00
Sodium bicarbonate	10.00
Blowing agent	1.00
Peptizer	0.20
Paraffin wax	1.00
Microcrystalline wax	1.00
Aromatic oil	60.00
Stearic acid	1.50
Zinc oxide	3.00
Sulfur	3.00
MBT	1.50
TMTD	0.50
DPG	0.50

trace impurities defined in the US Federal Clean Air Act are to be based on the hazard represented, not simply the best available measurement capability.

In conclusion, the materials scientist must continue to adjust to the changes in both the environment and health and safety standards.

9.10 SUMMARY

This chapter has reviewed both the types and the properties of elastomers, compounding with a range of filler or reinforcement systems such as carbon black, and enhancement of filler performance by novel use of compounding ingredients such as silane coupling agents. Other issues such as antioxidant systems and vulcanization systems were also discussed. The role of the modern materials scientist in the tire and rubber industry is to use materials to improve current products and develop new products. Four key parameters govern this development process:

- *Performance:* The product must satisfy customer expectations.
- *Quality:* The product must be durable and have a good appearance, and appropriate inspection processes must ensure consistency and uniformity.
- *Environment:* Products must be environmentally friendly in manufacturing, use, and disposal.
- *Cost:* The systems must provide a value to the customer.

TABLE 9.36 Relationship Between HIIR/IIR Content and Permeability, ICP, and Step-Load Endurance

Rubber Hydrocarbon Content in Liner	HIIR/IIR, % vol	Liner Permeability (O × 10^8) at 65 C+	Equilibrium Intracarcass Pressure (IOF) (MPa)	FMVSS 109 Step-Load Hours to Failure
100 BIIR	65.2	3.0	0.032	61.5
75 BIIR/25 NR	48.3	4.2	0.063	56.9
65 BIIR/35 NR/20 IIR[a]	42.2	5.9	0.063	40.2
60 SBR/40 NR/20 IIR[a]	16.1	6.9	0.090	31.5

[a]*IIR via whole tube reclaim, containing ~50% by weight IIR. Treated as filler.*

In meeting these goals rubber compounding has evolved from a "black art," as it was at the start of the 20th century, to a complex science necessitating knowledge in advanced chemistry, physics, and mathematics.

REFERENCES

ASTM D1566-04, 2004. Standard Terminology Relating to Rubber.
ASTM D1795-04, 2004. Standard Classification System for Carbon Blacks Used in Rubber Products.
ASTM D2227-96, 2002. Standard Specification for Natural Rubber Technical Grades.
ASTM D3053-04, 2004. Standard Terminology Relating to Carbon Black.
ASTM D3184-89, 1989. Standard Methods for Evaluation and Test Formulations for Natural Rubber.
Ayala, J.A., Hess, W.M., Kistler, F.D., Joyce, G., 1990. Paper 42 Presented at a Meeting of the Rubber Division. American Chemical Society, Washington, DC.
Ayala, J.A., Hess, W.M., Dotson, A.O., Joyce, G., 1990. Rubber Chem. Technol. 63, 747.
Barbin, W.W., Rodgers, M.B., 1994. In: Mark, J.E., Erman, B., Eirich, F.R. (Eds.), Science and Technology of Rubber, 2nd ed. Academic Press, New York.
Bateman, L., Moore, C.G., Porter, M., Saville, B., 1963. In: Chemistry and Physics of Rubber-Like Substances. John Wiley, New York.
Bernard, D., Baker, C.S.L., Wallace, I.R., 1985. NR Technol. 16, 19–26.
Blume, A., Luginsland, H.D., Meon, W., Uhrlandt, S., 2004. In: Rodgers, B. (Ed.), Rubber Compounding, Chemistry and Applications. Marcel Dekker Inc., New York.
Brantley, H.L., Day, G.L., 1986. In: Paper 33 Presented at a Meeting of the Rubber Division. American Chemical Society, New York.
Datta, S., 2004. In: Rodgers, B. (Ed.), Rubber Compounding, Chemistry and Applications. Marcel Dekker Inc., New York.
Day, G.L., Futamura, S., 1986. Paper 22 Presented at a Meeting of the Rubber Division. American Chemical Society, New York.
DelVecchio, R.J., 1997. Understanding Design of Experiments. Hanser, Cincinnati, OH.
Hess, W.M., Kemp, W.K., 1983. Rubber Chem. Technol. 56, 390.
Hong, S.W., 2004. In: Rodgers, B. (Ed.), Rubber Compounding, Chemistry and Applications. Marcel Dekker Inc., New York.
Kern, W.J., Futamura, S. 1987. Solution SBR as a Tread Rubber. ACS Rubber Division, Quebec, Paper 78.

Kumar, N., Chandra, A., Mukhopadhyay, R., 1996. Int. J. Polym. Sci. 34, 91.
LeBras, J., Papirer, E., 1979. Rubber Chem. Technol. 52, 43.
Malaysian Rubber Producers' Research Association, 1984. The Natural Rubber Formulary and Property Index. Imprint of Luton, England.
Mezynski, S.M., Rodgers, M.B., 1989. Radial Medium Truck Tire Performance and Materials. ACS Rubber Division, Akron Rubber Group.
Mezynski, S.M., Rodgers, M.B., 1993. Heavy Duty Truck Tire Materials and Performance. vol. 46. Kautschuk Gummi Kunststoffe, Frankfurt, 718–726.
Morita, E., 1980. Rubber Chem. Technol. 53, 393.
Niziolek, A., Nelsen, Jones, R., 2000. Kautsch. Gummi Kunstst. 53, 358.
Nordsiek, K.H., 1985. The Integral Rubber Concept—An Approach to an Ideal Tire Tread Rubber. Kautschuk Gummi Kunstsoffe, Frankfurt, vol. 38, 178–185.
Oberster, A.E., Bouton, T.E., Valaites, J.K., 1973. Angew. Mackromol. Chem. 29/30 (Nr 367), 291.
Rodgers, M.B., Waddell, W.H., Klingensmith, W., 2004. Rubber compounding. In: Kirk-Othmer Encyclopedia of Chemical Technology, 5th ed. John Wiley & Sons, New York.
The Rubber Manufacturers Association, 1979. The International Standards of Quality and Packaging for Natural Rubber Grades. The Green Book, The International Rubber Quality and Packaging Conference, Office of the Secretariat, Washington, DC (January).
Rubber Chemicals, Flexsys Technical Bulletin, Flexsys Rubber Chemicals Ltd., 1998.
Saito, A., 1999. Int. Polym. Sci. Technol. 26, T/19.
Schneider, W.A., Huybrechts, F., Nordesik, K.H., 1989. Kaut. Gummi Kunstst. 44, 528.
Simpson, B.D., 1978. In: R. Babbit (Ed.), The Vanderbilt Rubber Handbook, 12th ed. R.T. Vanderbilt Co., Inc., Norwalk, CT.
The Synthetic Rubber Manual, 1999. 14th ed., International Institute of Synthetic Rubber Producers, Houston, TX.
Waddell, W.H., Bhukuni R.S., Barbin, W.W, Sandstrom, P.H., 1990. In: Ohm, R.F. (Ed.), The Vanderbilt Rubber Handbook, 13th ed., R.T. Vanderbilt Company, Norwalk, CT.
Waddell, W.H., Rodgers, M.B., Tracey, D.S., 2004. Natural Rubber. ACS Rubber Division Meeting, Grand Rapids, MI, Paper A.
Waddell, W.H., Rodgers, M.B., Tracey, D.S., 2004. Tire Applications of Elastomers, Part 1. Treads, Rubber Division Meeting, Grand Rapids, MI, Paper H.
Wagner, M.P., 1976. Rubber Chem. Technol. 55, 703.
Wampler, W.A., Carlson, T.F., Jones, W.R., 2004. In: Rodgers B. (Ed.), Rubber Compounding, Chemistry and Applications. Marcel Dekker Inc., New York.
Wolff, S., 1983. Rubber Chem. Technol. 55, 967.
Wolff, S., 1990. In: Presented at a meeting of the Rubber Division. American Chemical Society, Las Vegas, NV.
Wolff, S., Gorl, U., 1991. The influence of modified carbon blacks on viscoelastic compound properties. In: International Rubber Conference.
Worldwide Rubber Statistics, 2010. International Institute of Synthetic Rubber Producers, Houston, TX.

Chapter 10

Strength of Elastomers

A.N. Gent* **and W.V. Mars**†

*The University of Akron, Akron, Ohio, USA †Endurica LLC, Findlay, Ohio, USA

10.1 INTRODUCTION

Fracture is a highly selective process: only a small number of those molecules making up a test piece or a component actually undergo rupture; the great majority are not affected. For example, of the 10^{26} chain molecules per cubic meter in a typical elastomer, only those crossing the fracture plane, about $10^{18}/m^2$, will definitely be broken. Moreover, these will not all break simultaneously but successively as the fracture propagates across the specimen at a finite speed.

Thus, the first questions posed in studying the strength of elastomers (and other materials as well) are: where and under what conditions does fracture begin? Also, what laws govern the growth of a crack once it has been initiated? This chapter seeks to answer such questions, first in a general way and then with particular reference to important modes of failure of elastomers in service. It does not deal with the rather complex problem of the strength of composite structures, such as a pneumatic tire, which involves failure of adhesive bonds at interfaces between the components as well as fracture of the components themselves.

We consider first the initiation of fracture from crack precursors. These are features of the material's microscopic structure that magnify applied stresses. The rate of development of cracks after initiation is treated next. Naturally, this depends on the local stress levels but also on the way in which these stresses vary with time. For example, rapid crack growth may take place if stresses are applied and removed frequently, whereas the crack may grow quite slowly, if at all, when the same stresses are held constant and never removed. This phenomenon of accelerated growth under dynamic stressing is termed *mechanical fatigue* or *dynamic crack growth*. It is treated in Sections 10.4.5 and 10.6.

The Science and Technology of Rubber. http://dx.doi.org/10.1016/B978-0-12-394584-6.00010-8

Because rubber is viscoelastic, or more generally anelastic, to varying extents and because the mechanical properties depend on rate of deformation and temperature, it is not surprising to find that the strength is also dependent on the rate at which stresses are applied and on the temperature of measurement. These effects are discussed in Sections 10.4.2 and 10.5.1. Other effects of the environment, notably the destructive action of ozone, are discussed in Section 10.8. Finally, a brief survey is given of abrasive wear.

10.2 INITIATION OF FRACTURE

10.2.1 Flaws and Stress Raisers

Under sufficient magnification, every solid body contains heterogeneities of composition or structure, each of which may serve as the precursor to a crack. In addition to features intrinsic to the material, extrinsic features may also be present in the form of sharp corners, nicks, cuts, scratches, and embedded dirt particles or other sharp inclusions. Because of such features, applied stresses are concentrated in certain regions of the body so that they greatly exceed the mean applied stress. Fracture will begin at any such site where the local stress exceeds a critical level, causing the crack precursor to develop (Choi and Roland, 1996; Roland and Smith, 1985). The stress concentration factor—that is, the ratio of the stress at the tip of a sharp flaw, σ_t, to the applied tensile stress σ—is given by Inglis's relation for elastic solids that obey a direct proportionality between stress and strain (Inglis, 1913):

$$\sigma_t/\sigma = 1 + 2(l/r)^{1/2}, \tag{10.1}$$

where l is the depth of an edge flaw and r is the radius of the tip in the unstressed state. If the flaw is totally enclosed, it is roughly equivalent to an edge flaw of depth $l/2$ (Figure 10.1). Thus, edge flaws are more serious stress raisers than enclosed flaws of the same size, and they are more usual sources of fracture than inclusions, but not exclusively so. Heterogeneities of composition have been

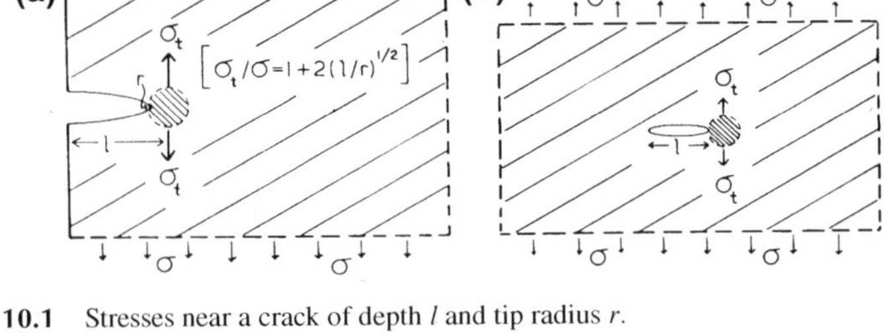

FIGURE 10.1 Stresses near a crack of depth l and tip radius r.

shown to nucleate fatigue cracks internally (Eldred, 1972). Also, some types of cracks cannot form near a free surface (see Section 10.7).

When the tip radius is much smaller than the depth of the flaw, as seems probable for the severe stress raisers responsible for fracture, Eq. (10.1) can be approximated by

$$\sigma = (\sigma_t r^{1/2})/2l^{1/2}. \tag{10.2}$$

Thus, the breaking stress, denoted by σ_b, is predicted to vary inversely with the depth of the flaw, l, in proportion to $1/l^{1/2}$. This prediction has been tested for brittle polymers (i.e., in the glassy state (Berry, 1972)) and for rubbery materials cracked by ozone (Figure 10.2). In both cases the breaking stress σ_b was found to vary in accordance with Eq. (10.2) with the depth of a crack or razor cut made in one edge of the test piece. For elastomers broken by mechanical stress alone, however, elongations at break are generally much too large for the assumption of linear stress-strain behavior to be valid, and Eq. (10.2) becomes a relatively poor approximation. Even so, by extrapolating measured values of the breaking stress for different depths of edge cut to the breaking stress for a test piece having no cuts introduced at all, the depth of flaw characteristic of the material may be inferred. (Actually, the value obtained is the depth of a cut equivalent in stress-raising power to natural flaws, which may be smaller and sharper, or larger and blunter, than knife cuts of equivalent stress-concentration power.)

For both rubbers and glasses the value obtained in this way is about 40 ± 20 μm. The same value is also obtained by extrapolating measured stresses for ozone cracking (see Figure 10.2) back to that value observed for a test piece having no initial cut in it, and by extrapolating the fatigue lives of test pieces with cuts introduced in them back to the fatigue life for test pieces with no cuts (Figure 10.3). In all these cases, substantially the same value of the natural flaw size is obtained. Moreover, it is largely independent of the particular elastomer

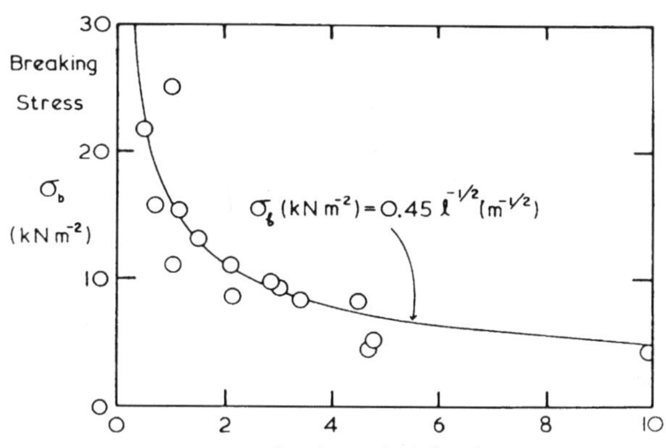

FIGURE 10.2 Fracture stresses for test pieces having cuts of depth l, exposed to ozone. (From Braden and Gent (1961).)

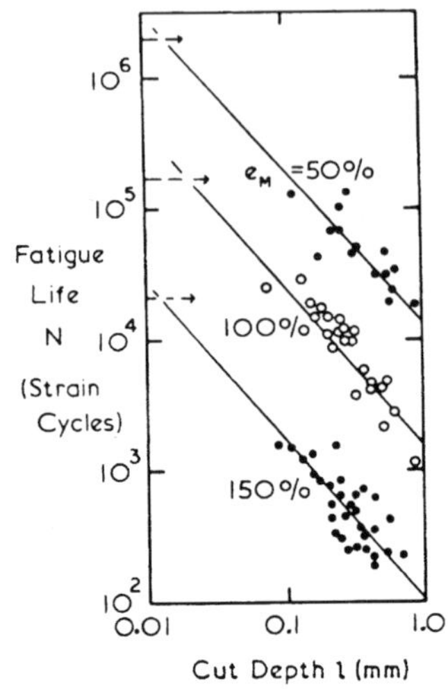

FIGURE 10.3 Fatigue lives N for test pieces having initial cuts of depth l, subjected to repeated extensions to the indicated strains.

or mix formulation used, even though these factors greatly alter the way in which the breaking stress or fatigue life changes with cut size, as discussed later. Thus, a variety of fracture processes appear to begin from a natural flaw equivalent to a sharp edge cut 40 μm deep.

The exact nature of these failure initiation sites is difficult to ascertain. They may consist of accidental nicks in molded or cut surfaces, but even if great care is taken in preparing test pieces (e.g., by molding against polished glass), the breaking stress is not greatly increased. Dust or dirt particles or other heterogeneities nearly as effective as mold flaws seem to be present in a sufficient amount to initiate fracture. Only when the test piece size is reduced to about 10^{-8} m^3 or less is a significant increase in strength observed, suggesting that powerful stress raisers are present only in concentrations of $10^8/m^3$ or less. Of course, if a way could be found to eliminate them, or at least reduce the effective sharpness of these natural flaws, substantial increases in strength, and even more striking increases in fatigue life, might be achieved, as discussed later. At present, however, they appear to be an inevitable consequence of the processes used in making elastomeric materials and components.

10.2.2 Stress and Energy Criteria for Rupture

Equations (10.1) and (10.2) raise several other questions. What is the radius r of a natural flaw? What is the magnitude of the breaking stress at the tip σ_t when the flaw starts to grow as a crack? From a comparison of experimental relations such as that shown in Figure 10.2 with the predictions of Eq. (10.2), only the product $\sigma_t r^{1/2}$ can be determined and not the two quantities separately.

The value of r is, however, unlikely to exceed 1 μm for a sharp cut, and hence the tip stress σ_t may be inferred to be greater than 200 MPa, taking a value for the product $\sigma_t r^{1/2}$ of 0.2 M N m$^{-3/2}$ as representative of fracture under mechanical stress. For ozone cracking this product takes the value 900 N m$^{-3/2}$ (see Figure 10.2), and hence the tip stress in this case is presumably 1 MPa or greater.

We must recognize, however, that a tear that begins to propagate from an initial cut or flaw will soon develop a characteristic tip radius r of its own, independent of the sharpness of the initiating stress raiser (Thomas, 1958). It is therefore more appropriate to treat the product $\sigma_t r^{1/2}$ as a characteristic fracture property of the material. Indeed, Irwin (1948, 1957) proposed that fracture occurs for different shapes of test piece and under varied loading conditions at a characteristic value of a "stress intensity factor," K_c, defined as

$$K_c = (\pi^{1/2}/2)\sigma_t r^{1/2} = \pi^{1/2}\sigma_b l^{1/2} \tag{10.3}$$

when expressed in terms of the applied stress σ_b by means of Eq. (10.2).

An alternative but equivalent view of the critical stress criterion for fracture was proposed by Griffith (1921, 1924) for elastic solids, and applied by Irwin (1948, 1957) and Orowan (1949) to solids that are globally elastic, even when they exhibit plastic yielding around the crack tip. Griffith suggested that a flaw would propagate in a stressed material only when, by doing so, it brought about a reduction in elastically stored energy, W, more than sufficient to meet the free energy requirements of the newly formed fracture surfaces. Irwin and Orowan recognized that in practice the energy expended in local plastic deformation during crack growth generally far exceeds the true surface energy; however, provided that the total energy expended is proportional to the amount of surface created by fracture, Griffith's relations may still be employed.

Griffith's fracture criterion takes the form

$$-(\partial W/\partial A) \geqslant G_c/2, \tag{10.4}$$

where A is the surface area of the specimen, which increases as the crack grows, and G_c is the amount of energy required to tear through a unit area of the material. The factor 2 arises on changing from the area torn through to the area of the two newly formed surfaces. The derivative is evaluated at constant length of the sample, so that the applied forces do no work as the crack advances. (An example where this is not appropriate, because the crack will advance only when the applied forces do the necessary work, is given in Section 10.2.4.)

In Griffith's original treatment, the surface free energy per unit area of fracture plane was employed in place of the generalized fracture energy, $G_c/2$. His results therefore carried the implication of thermodynamic reversibility. In contrast, G_c merely represents energy dissipated during fracture. Nevertheless, provided that it is dissipated in the immediate vicinity of the crack tip and is independent of the overall shape of the test piece and the way in which forces are applied to its edges, the magnitude of G_c can be employed as a characteristic fracture property of the material, independent of the test method. This expectation has been borne out by critical experiments on a variety of materials, including elastomers, using test pieces for which the relation between the breaking stress, σ_b, and the rate of release of strain energy on fracture, G_c, defined by Eq. (10.4), can be either calculated or measured experimentally (Rivlin and Thomas, 1953; Thomas, 1960). Two important cases are considered here.

10.2.3 Tensile Test Piece

As shown in Figure 10.4, a thin strip of thickness t with a cut in one edge of depth l is placed in simple extension until it breaks. The effect of the cut in diminishing the total stored elastic energy at a given extension may be calculated approximately by considering a small triangular region around the cut (shown shaded in Figure 10.4) to be unstrained and the remainder of the test piece to be unaffected by the presence of the cut, with stored strain energy U per unit volume. The reduction in strain energy due to the presence of the cut is thus kl^2tU, where k is a numerical constant whose value depends on the applied strain (k is given approximately by $\pi/(1+e)^{1/2}$, where e is the tensile strain (Greensmith, 1963)). Thus, as the tensile strain increases, k decreases from a value of π at small strains, to about 1 at large strains.

For a tensile test piece, therefore, $-(\partial W/\partial A) = kltU$ and $(\partial A/\partial l) = 2t$. Equation (10.4) becomes (Rivlin and Thomas, 1953)

$$2klU \geqslant G_c. \tag{10.5}$$

We note that the breaking stress, σ_b, does not appear explicitly in this fracture criterion. σ_b is the stress at which the strain energy density, U, satisfies Eq. (10.5). It therefore depends on the elastic properties of the material and the length of the initial cut, as well as on the fracture energy, G_c. For a material

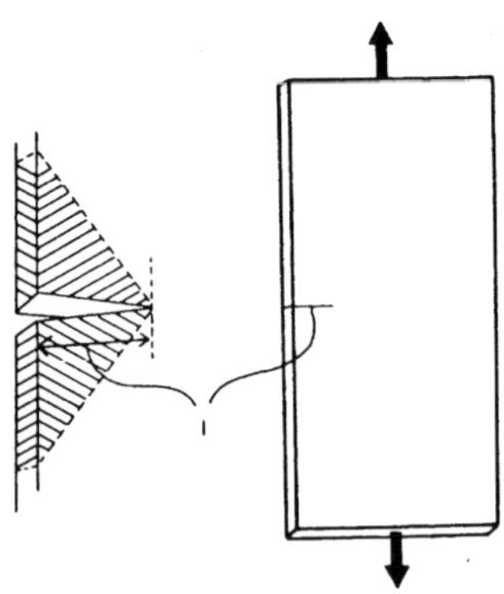

FIGURE 10.4 Tensile test piece.

obeying a linear relation between tensile stress, σ, and extension, e, the stored energy, U, is given by $Ee^2/2$ or $\sigma^2/2E$, where E is Young's modulus. The stress and extension at break are therefore given by

$$\sigma_b = (G_c E/\pi l)^{1/2}, \tag{10.6}$$

$$e_b = (G_c/\pi l E)^{1/2}, \tag{10.7}$$

where k has been given the value π appropriate to linearly elastic materials. Equations (10.6) and (10.7) were obtained by Griffith (1924).

On comparing Eqs. (10.3) and (10.6), we see that the critical stress intensity factor, K_c, and the fracture energy, or critical strain energy release rate, G_c, are related to each other and to the breaking stress at the crack tip, as follows:

$$K_c^2 = EG_c = (\pi/4)\sigma_t^2 r = (\pi/2)U_t E r, \tag{10.8}$$

where U_t is the strain energy density at the crack tip. Hence,

$$G_c = (\pi/2)U_t r. \tag{10.9}$$

The fracture energy, G_c, is thus a product of the energy required to break a unit volume of material at the crack tip; that is, in the absence of nicks or external flaws, and the effective diameter of the tip, as pointed out by Thomas (1955). These two factors can be regarded as independent components of the fracture energy: an "intrinsic" strength, U_t, and a characteristic roughness or bluntness of a developing crack (Papadopoulos et al., 2008), represented by r. Because K_c also involves the elastic modulus, E, it is not considered as suitable a measure of the fracture strength as G_c for materials, like elastomers, of widely different moduli.

Equation (10.5) is more generally applicable than Eq. (10.6) because it is not restricted to linearly elastic materials. It constitutes a criterion for tensile rupture of a highly elastic material having a cut in one edge of length, l, in terms are of the fracture energy, G_c. Two important examples of test pieces of this type are (1) the ASTM "tear" test piece for vulcanized rubber (ASTM D624-54) and (2) a typical tensile test piece that has accidental small nicks caused, for example, by imperfections in the surface of the mold or die used to prepare it.

Several features of Eqs. (10.6) and (10.7) are noteworthy. For a given value of fracture energy, G_c, stiffer materials with higher values of Young's modulus, E, will have higher breaking stresses and lower extensions at break than softer materials. These correlations are well known in the rubber industry. Less well known is the effect of the size of an initial cut or flaw on both the breaking stress and elongation at break. Finally, if the fracture criterion, Eq. (10.5), is met for an initial flaw of depth, l, it will be greatly exceeded as fracture proceeds. As a consequence, a crack will accelerate across the specimen catastrophically.

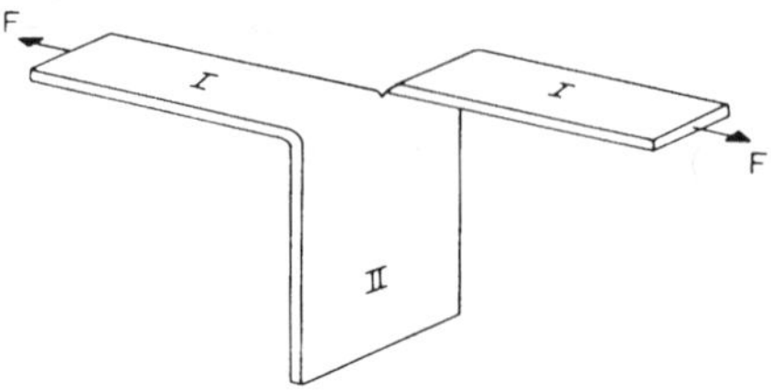

FIGURE 10.5 Tear test piece.

10.2.4 Tear Test Piece

This test piece, shown in Figure 10.5, has regions I in the arms that are in simple extension and a region II that is virtually undeformed. If the arms are sufficiently wide, or if they are reinforced with inextensible tapes, their extension under the tear force, F, will be negligibly small. The work of fracture, $G_c \Delta A$, is then provided directly by the applied force, F, acting through a distance $2\Delta l$, where Δl is the distance torn through. The corresponding area torn through is $t\Delta l$, where t is the thickness of the sheet.[1] On equating the work supplied to that required for tearing, the fracture criterion becomes (Rivlin and Thomas, 1953)

$$F \geqslant G_c t/2. \tag{10.10}$$

This relation can also be derived by considering the balance of material forces (Verron, 2010). Because the tear force in this case is a direct measure of the fracture energy, G_c, and is independent of the elastic properties of the material and of the length of the tear, this test piece is particularly suitable for studying the effects of composition and test conditions on G_c (Greensmith and Thomas, 1955; Greensmith, 1956; Veith, 1965; Ahagon and Gent, 1975; Gent and Tobias, 1982; Bhowmick et al., 1983). An additional feature of this test geometry is that it offers control of the rate of tearing, since the tearing rate is related to the rate of extension of the clamps (1 mm growth of the tear is equal to 2 mm extension of the clamps). In cases involving knotty or unsteady tearing, Sakulkaew et al. (2011) have developed a procedure that gives the dependence of G_c on the rate of the tearing energy, dG/dt.

It is important to recognize that the fracture energy, G_c, is not a constant value for a particular material; it depends strongly on the temperature and rate of tear; that is, the rate at which material is deformed to rupture at the tear tip, as discussed in Section 10.5.2. Nevertheless, several critical values may be distinguished. The smallest possible value is, of course, twice the surface free energy, about 50 mJ/m^2 for hydrocarbon liquids and polymers (Tarkow, 1958).

[1] Actually, the tear tends to run at 45° to the thickness direction, i.e., at right angles to the principal tensile stress, and thus the tear path has a width instead of about t (Papadopoulos et al., 2008; Ahagon et al., 1975).

Values of this order of magnitude are indeed observed for fracture induced by ozone, when the function of the applied forces is merely to separate molecules already broken by chemical reaction, as discussed in Section 10.8.

Another critical value is that necessary to break all the molecules crossing a plane, in the absence of any other energy-absorbing processes. This minimum energy requirement for mechanical rupture is found to be about 50 J/m^2; it is treated in the following section. Finally, there are the considerably larger values found in normal fracture experiments, ranging from 100 to 100,000 J/m^2. These are described in Section 10.5.

10.3 THRESHOLD STRENGTHS AND EXTENSIBILITIES

A threshold value for the fracture energy of elastomers was first pointed out by Lake and Lindley (1965) from studies of fatigue crack growth. By extrapolation, they found that a minimum amount of mechanical energy, about 50 J/m^2 of torn surface, was necessary for a crack to propagate at all. Mueller and Knauss (1971) measured extremely low tearing energies directly, by employing low rates of tear, high temperatures, and a urethane elastomer composition swollen highly with a mobile fluid. Under these near-equilibrium conditions, they obtained a lower limit of about 50 J/m^2 for the tear energy, similar to Lake and Lindley's extrapolated value. More recently, threshold tear strengths have been measured for several elastomers, crosslinked to varying degrees (Ahagon and Gent, 1975; Gent and Tobias, 1982; Bhowmick et al., 1983). Again, the values are about 20–100 J/m^2, much smaller than tear energies obtained in conventional tearing experiments, which range from about 10^3 to about 10^5 J/m^2, depending on tear rate, test temperature, and elastomer composition (Greensmith et al., 1963). Indeed, they amount to only about 1 lb of force to tear through a sheet several inches thick.

Nevertheless, they are much larger than would be expected on the basis of C–C bond strengths alone. For example, about 2×10^{18} molecules cross a randomly chosen fracture plane having an area of 1 m^2, and the dissociation energy of the C–C bond is about 5×10^{-19} J. Thus, a fracture energy of only about 1 J/m^2 would be expected on this basis, instead of the observed value of about 50 J/m^2. This large discrepancy has been attributed by Lake and Thomas

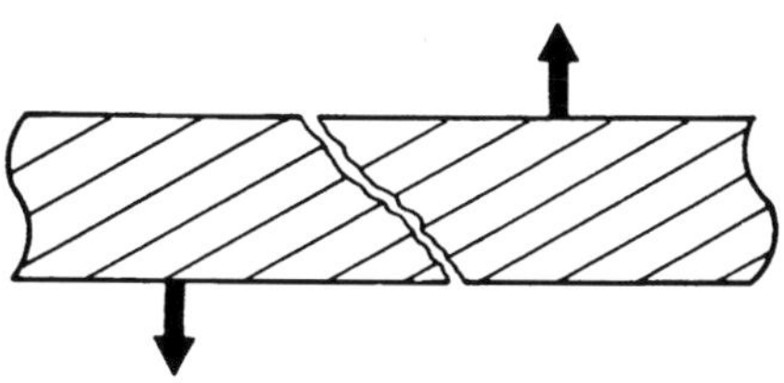

FIGURE 10.6 Tearing under shear stresses (schematic). (From Knauss (1970) and Ahagon et al. (1975).)

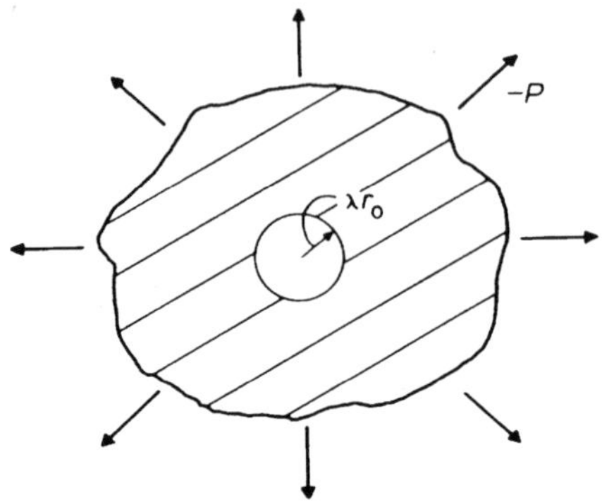

FIGURE 10.7 Expansion of a cavity under a triaxial tension.

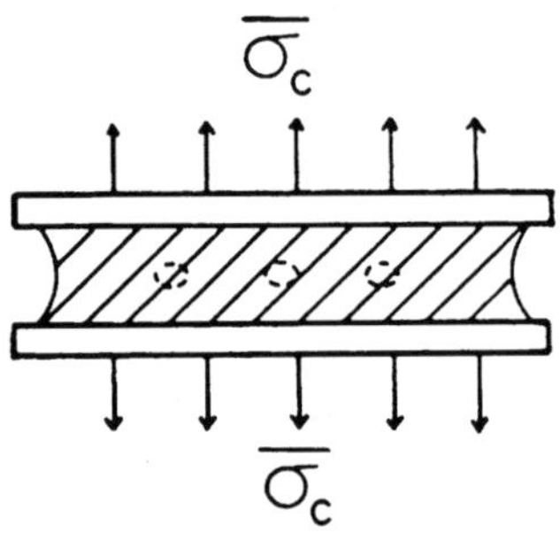

FIGURE 10.8 Cavitation in a bonded block (schematic).

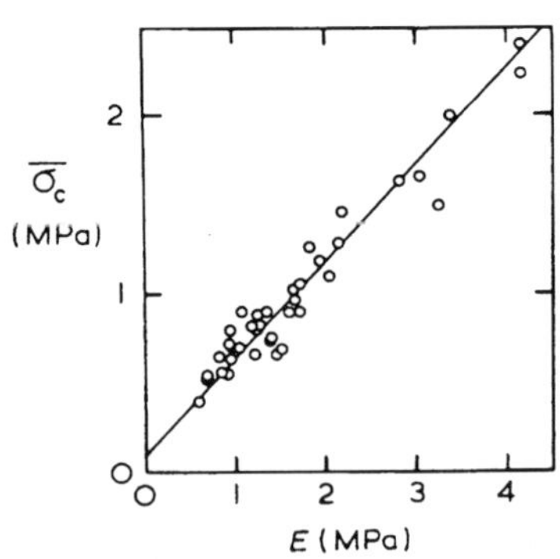

FIGURE 10.9 Critical applied stress, σ_c, for cavitation in bonded blocks (see Figure 10.8) versus Young's modulus, E, of the elastomer. (From Gent and Lindley (1958).)

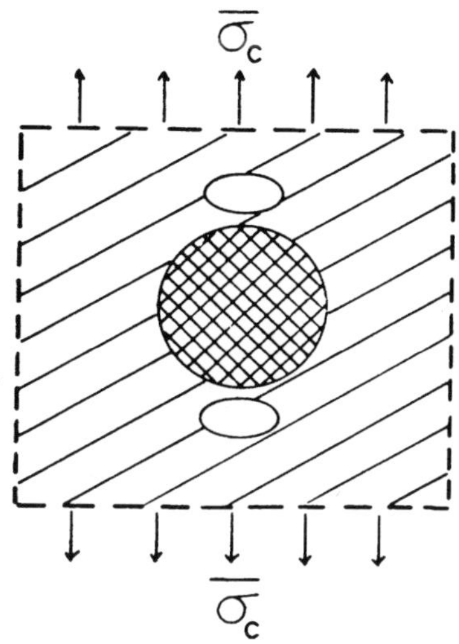

FIGURE 10.10 Cavitation near a rigid inclusion (schematic).

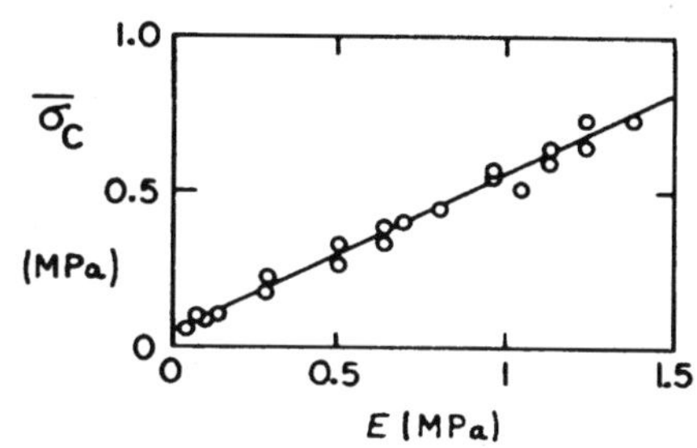

FIGURE 10.11 Critical applied stress, σ_c, for cavitation near rigid inclusions (see Figure 10.10) versus Young's modulus, E, of the elastomer. (From Oberth and Bruenner (1965).)

(1967) to the polymeric character of elastomers: many bonds in a molecular chain must be stressed equally to break one of them. Thus, the greater the molecular length between points of crosslinking, the greater the energy needed to break a molecular chain. On the other hand, when the chains are long, a smaller number of them cross a randomly chosen fracture plane. These two factors do not cancel out; the net effect is a predicted dependence of the threshold fracture energy, G_0, on the average molecular weight, M_c, of chains between points of crosslinking, of the form (Lake and Thomas, 1967)

$$G_0 = \alpha M_c^{1/2}. \tag{10.11}$$

Two other features of molecular networks can be taken into account, at least in an approximate way: the presence of physical entanglements between chains, at a characteristic spacing along each chain of molecular weight, M_e, and the presence of molecular ends that do not form part of the load-bearing network. Equation (10.11) then becomes

$$G_0 = \alpha[(1/M_c) + (1/M_e)]^{-1/2}[1 - 2(M_c/M)], \tag{10.12}$$

where M is the molecular weight of the polymer before crosslinking (Lake and Thomas, 1967). The constant α involves the density of the polymer, the mass, length, and effective flexibility of a monomer unit, and the dissociation energy of a C–C bond, assumed to be the weakest link in the molecular chain. If reasonable values are taken for these quantities (Lake and Thomas, 1967), α is found to be about 0.3 $J/m^2(g/g\text{-mol})^{-1/2}$. Thus, for a representative molecular network, taking $M_c = M_e = 15{,}000$ and $M = 300{,}000$, the threshold fracture energy obtained is about 25 J/m^2, in reasonable agreement with experiment in view of uncertainties and approximations in the theory. Moreover, the predicted increase in fracture energy with molecular weight, M_c, between crosslinks appears to be correct; increased density of crosslinking (shorter network chains) leads to lower threshold fracture energies (Ahagon and Gent, 1975; Gent and Tobias, 1982; Bhowmick et al., 1983). Because, however, the tensile strength, σ_b, also involves the elastic modulus, E [Eq. (10.6)], and E is increased by crosslinking, the threshold tensile strength shows a net increase with increased crosslinking.

Threshold values of tensile strength and extensibility may be calculated by means of Eq. (10.6), using an average threshold fracture energy of 50 J/m^2, a "natural" flaw size of 40 μm (assumed to be independent of composition), and a typical value for Young's modulus, E, for rubber of 2 MPa (corresponding to a Shore A hardness of about 48°). The results are $\sigma_{b,0} = 0.9$ MPa and $e_{b,0} = 0.45$. These values are indeed close to experimental "fatigue limits"; that is, stresses and strains below which the fatigue life is effectively infinite in the absence of chemical attack (Lake and Lindley, 1965).

A surprisingly large effect has been found on the tensile strength and, by inference, on the tear strength of rubber as a result of the specific *distribution* of molecular weights, M_c, in the network (Mark and Tang, 1984). When a small proportion of short chains, about 5 mol%, is combined in a network of long chains, the tensile strength is considerably higher than for other mixtures. It seems likely that the threshold strength is also higher. This remarkable enhancement of strength may be the result of strain redistribution within the network; that is, the ability to undergo nonaffine deformation, so that internal stress concentrations are minimized. Whatever the cause, the effect has been demonstrated in several material systems (Buckley et al., 2011), and the phenomenon is clearly of both scientific and practical interest.

Low values of fracture energy, only about one order of magnitude greater than the threshold level, have been obtained by measuring the resistance of rubber to cutting with a sharp knife or a razor blade. Frictional effects were minimized by stretching the sample as it was being cut. By adding the energy supplied by the stretching force to that supplied by the cutter, the total fracture energy was found to be rather constant, and low, of the order of 300 J/m^2

(Lake and Yeoh, 1978). The value was affected by the sharpness of the blade used, being lower for sharper blades, as would be expected from Eq. (10.9).

10.4 CRACK PROPAGATION

10.4.1 Overview

Whereas the initiation of fracture appears to be a similar process for all elastomers, the propagation of a crack is widely different. Three basic patterns of crack propagation, or tearing, can be distinguished corresponding to three characteristic types of elastomeric compounds:

1. Amorphous elastomers like SBR.
2. Elastomers, like natural rubber and Neoprene, that crystallize on stretching, even if only at the crack tip where local stresses are particularly high.
3. Reinforced elastomers containing large quantities, about 30% by volume, of a finely divided reinforcing particulate filler such as carbon black.

Elastomers in the first category show the simplest tearing behavior and are therefore described first. For these materials, once fracture has been initiated, a tear propagates at a rate dependent on two principal factors: the strain energy release rate, G, and the temperature, T. The former quantity represents the rate at which strain energy is converted into fracture energy as the crack advances. It is defined by a relation analogous to Eq. (10.4):

$$G = -2(\Delta W/\Delta A). \tag{10.13}$$

Here W denotes the total strain energy of the specimen and A denotes the surface area (which, of course, increases as the crack advances).

Even if a crack is stationary, because the critical value, G_c, at which fracture takes place has not been attained, Eq. (10.13) is still a useful definition of the rate, G, at which energy would be available from the strained specimen. For a tensile strip with an edge cut, it yields

$$G = 2\pi l/U \tag{10.14}$$

by analogy with Eq. (10.5), and for a tear test piece,

$$G = 2F/t \tag{10.15}$$

from Eq. (10.10).

10.4.2 Viscoelastic Elastomers

Experimental relations between the fracture energy G, the rate of tearing, and the temperature of test are shown as a three-dimensional diagram in Figure 10.12 for

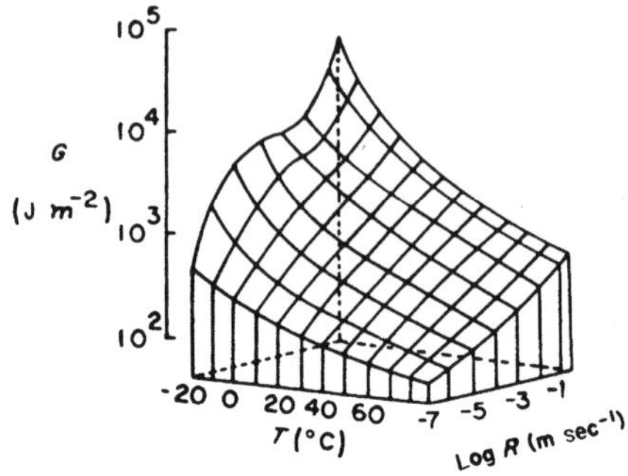

FIGURE 10.12 Fracture energy, G, for an unfilled SBR material as a function of temperature, T, and rate of tearing, R. (From Greensmith and Thomas (1955).)

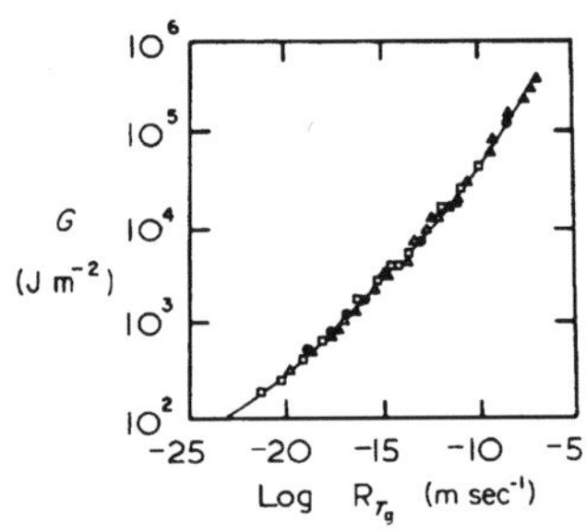

FIGURE 10.13 Fracture energy, G, versus rate of tearing, R, reduced to T_g for six unfilled amorphous elastomers. (From Mullins (1959).)

an SBR material. The fracture energy is seen to be high at high rates of tearing and at low temperatures, and vice versa, in a manner reminiscent of the dependence of energy dissipation in a viscous material on rate of deformation and temperature. Indeed, when the rates of tear are divided by the corresponding molecular segmental mobility, ϕ_T, at the temperature of test, the relations at different temperatures superpose to form a single master curve, as shown in Figure 10.13 (Mullins, 1959). In this figure, the rates have been multiplied by the factor

$$a_T = \phi_{T_g}/\phi_T \tag{10.16}$$

to convert them into equivalent rates of tearing at the glass transition temperature, T_g, of the polymer, $-57°C$ for the SBR material of Figure 10.12.

Furthermore, values of fracture energy G for five other amorphous elastomers, two butadiene-styrene copolymers of lower styrene content ($T_g = -72$ and $-78°C$) and three butadiene-acrylonitrile copolymers having T_g values of -30, -38, and $-56°C$, all fall on a single curve in this representation, increasing with rate of tearing in a similar way to the dissipation of energy internally by a viscous process (Mullins, 1959). We conclude that the fracture energy, G, is approximately the same for all unfilled amorphous lightly crosslinked elastomers under conditions of equal segmental mobility, and that the dependence of tear strength on temperature arises solely from corresponding changes in segmental mobility.

Thus, internal energy dissipation determines the tear resistance of such elastomers: the greater the dissipation, the greater the tear strength. This point will also emerge in connection with variations in tensile strength of elastomers (see Section 10.6). It is demonstrated strikingly in the present case by a proportionality between fracture energy, G, and a direct measure of energy dissipation, namely, the shear loss modulus, G'', for the same six elastomers (Figure 10.14) (Mullins, 1959).

For simple C–C crosslinked elastomers (Gent and Lai, 1994), the reduction factors, a_T, used to transform tear energy results at different temperatures as in Figure 10.12 to yield a master curve as in Figure 10.13 are found to correspond closely to the universal form of the WLF rate–temperature equivalence relation (Ferry, 1970):

$$\log a_T = -17.5(T - T_g)/(52 + T - T_g). \tag{10.17}$$

At low rates of tearing, about 10^{-20} m/s at T_g, the contribution from viscous losses becomes vanishingly small and the tear strength is reduced to the small threshold value, G_0. As the tear rate increases, the tear strength rises, approximately in proportion to $(Ra_T)^{0.24}$, until it becomes about $1000 \times G_0$; that is, by about three orders of magnitude. Thus, over wide ranges of rate and temperature the strength of simple rubbery solids can be expressed as:

$$G/G_0 = 1 + 2.5 \times 10^4 (Ra_T)^{0.24}, \tag{10.18}$$

where G_0 is determined by the structure of the molecular network (Eqs. (10.11) and (10.12)) (Gent and Lai, 1994).

At the highest rates of tear, about 1 m/s at T_g, the tear strength is extremely high, approaching 10^6 J/m^2. Simultaneously, elastomers become leathery and, eventually, glasslike. Indeed, at still higher rates and lower temperatures they would fracture like typical polymeric glasses, by a failure process in which a narrow craze band forms and propagates, and is followed by a running

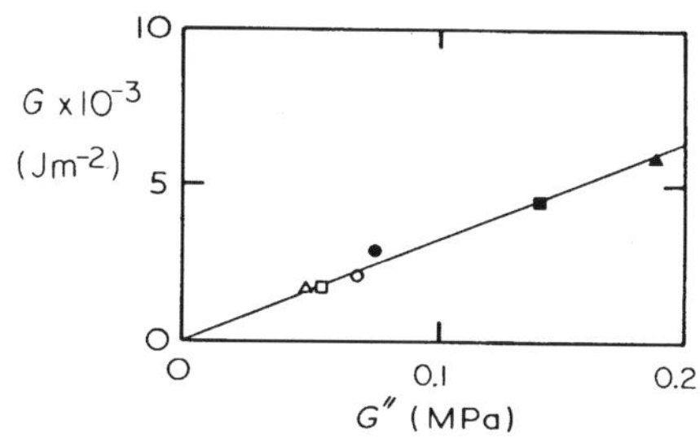

FIGURE 10.14 Fracture energy, G, versus shear loss modulus, G'', for six unfilled amorphous elastomers. (From Mullins (1959).)

crack (Kambour, 1973). The fracture energy for this process is relatively small, about 500–1000 J/m^2, in comparison with that for highly viscous but highly deformable elastomers, so that the curve shown in Figure 10.13 turns sharply down at higher rates to level off at this value.

Why is the tear strength so strongly dependent on tear rate and temperature; that is, on the viscoelastic response of the polymer? This striking feature has been explained by Knauss (1992) as a consequence of retarded elasticity. His treatment assumes that the stress intensity factor, K_c, given by Eq. (10.8), is largely unchanged by changes in rate and temperature, so that

$$K_c^2 = E(t)G_0 = E(t = \infty)G, \quad (10.19)$$

where $E(t)$ denotes the tensile modulus, E, at a time, t, after straining. $E(t = \infty)$ denotes the equilibrium modulus. Energy, G, obtained from a quasi-equilibrium solid far from the crack tip, is expended in breaking material with a time-dependent modulus, $E(t)$, at the crack tip. Thus, the fracture energy, G, is expected to increase with rate of tearing, R, approximately in proportion to the increase in elastic modulus, E, with rate of deformation,

$$G/G_0 = E(t = \delta/R)/E(t = \infty), \quad (10.20)$$

where δ is a characteristic distance ahead of the crack tip over which the high stress concentration at the crack tip is built up. This simple picture gives a good representation of the observed increase in fracture energy with rate and temperature, but the value of δ obtained by a direct comparison of G and E is too small, about 1 Å, to be physically reasonable (Gent and Lai, 1994). The discrepancy may arise because tearing is discontinuous on a microscopic scale; that is, it may take place in a stick-slip fashion. When rubber is cut at a controlled rate with a sharp knife, the variation of cutting resistance with rate of cutting and temperature is found to be accounted for correctly by Eq. (10.20) using a value for δ about equal to the blade tip diameter (Lake and Yeoh, 1978, 1987; Gent et al., 1994).

The tear strength of sulfur-crosslinked elastomers is higher and its dependence on temperature is greater than would be expected from Eq. (10.17) (Gent and Lai, 1994). This has been attributed to fracture of weak polysulfide crosslinks rather than the stronger C–C bonds in network chains. As a result, a second temperature-dependent process is introduced, because the strength of S–S bonds falls as the temperature is raised. The strength of practical rubber compounds thus reflects not only internal dissipation of energy from viscous processes (Mullins, 1959) but also in detachment from filler particles, from changes in tear tip radius (Lake and Yeoh, 1987), and from fracture of weak crosslinks (Gent and Lai, 1994).

10.4.3 Strain-Crystallizing Elastomers

As shown in Figure 10.15, the tear strength of strain-crystallizing elastomers is greatly enhanced over the range of tear rates and temperatures at which

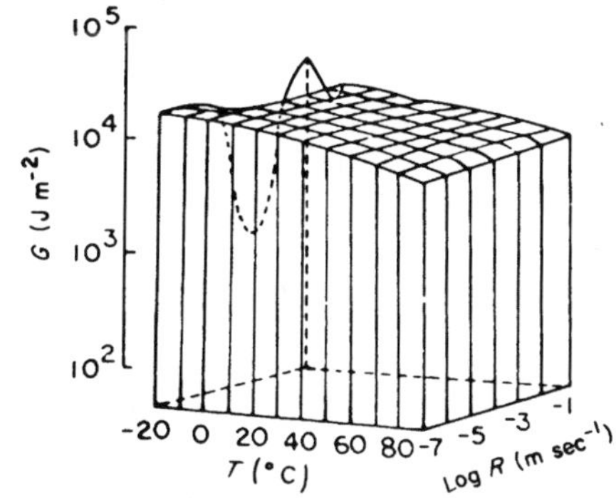

FIGURE 10.15 Fracture energy, G, for a strain-crystallizing elastomer, natural rubber, as a function of temperature, T, and rate of tearing, R. (From Greensmith and Thomas (1955).)

crystallization occurs on stretching, at the tear tip. At high temperatures, however, crystallization becomes thermodynamically prohibited because even the high melting temperatures of crystallites in highly stretched elastomers have been exceeded. Conversely, at low temperatures and high rates of tear, molecular reorganization into crystallites cannot take place in the short times of stretching as the crack tip advances. Thus, the strengthening effect of strain-induced crystallization is limited to a particular range of tear rates and temperatures, as seen in Figure 10.15. Outside this range, the material has only the strength associated with its viscous characteristics, dependent on $T - T_g$.

The high strength of strain-crystallizing materials has been attributed to pronounced energy dissipation on stretching and retraction, associated with the formation and melting of crystallites under nonequilibrium conditions (Andrews, 1961; Huneau, 2011). Reinforcing particulate fillers have a similar strengthening action, as discussed later, and they also cause a marked increase in energy dissipation. Whether this is the sole reason for the strengthening effect of crystallites and fillers, and other strengthening inclusions such as hard regions in block copolymers, hydrogen-bonded segments, is not clear, however.

10.4.4 Reinforcement with Fillers

A remarkable reinforcing effect is achieved by adding fine particle fillers such as carbon black or silica to a rubber compound. They cause an increase in tear strength and tensile strength by as much as 10-fold when, for example, 40% by weight of carbon black is included in the mix formulation. But this strengthening action is restricted to a specific range of tear rates and test temperatures—ranges that depend on both the type of filler and the elastomer (Greensmith and Thomas, 1955; De and Gent, 1996) (Figure 10.16). Outside this range of effectiveness, the filler does not enhance the observed strength to nearly the same degree.

The marked enhancement of tear strength in certain circumstances is associated with a pronounced change in the character of the tear process, from relatively smooth tearing with a roughness of the torn surface of the order of

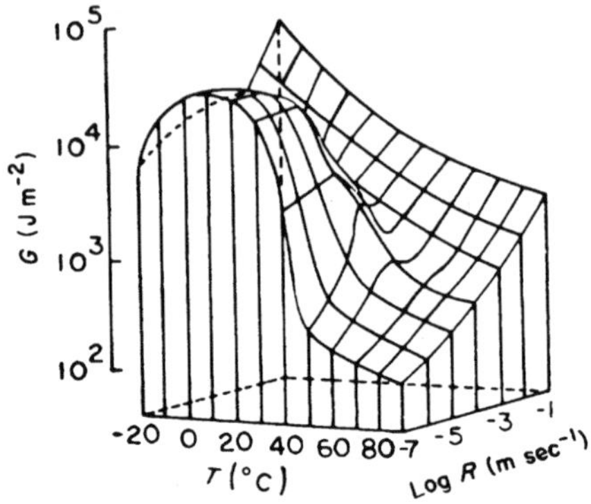

FIGURE 10.16 Fracture energy, G, for an amorphous elastomer (SBR) reinforced with 30% by weight FT carbon black. (From Greensmith (1956).)

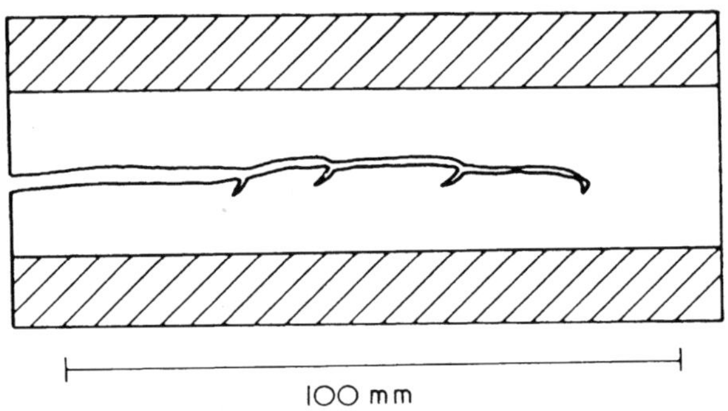

FIGURE 10.17 "Knotty" tear in a carbon-black-reinforced elastomer. (From Gent and Henry (1968).)

0.1–0.5 mm to discontinuous stick-slip tearing, where the tear deviates from a straight path and even turns into a direction running parallel to the applied stress, until a new tear breaks through. This form of tearing has been termed *knotty tearing* (Greensmith and Thomas, 1955); an example is shown in Figure 10.17. A typical tear force relation is shown in Figure 10.18a; it may be compared with the corresponding relation for an unfilled material in Figure 10.18c. The peak tearing force at the "stick" position reaches high values, but the force during catastrophic "slip" tearing drops to a much lower level, only about twice as large as that for continuous tearing of the unfilled elastomer. Indeed, when the tear is prevented from deviating from a linear path by closely spaced metal guides, or is made to propagate in a straight line by stretching (Gent and Henry, 1968) or prestretching (Houwink and Janssen, 1956) the sample in the tearing direction, then the tear force is much smaller, only two to three times that for the corresponding unfilled material (see Figure 10.18b).

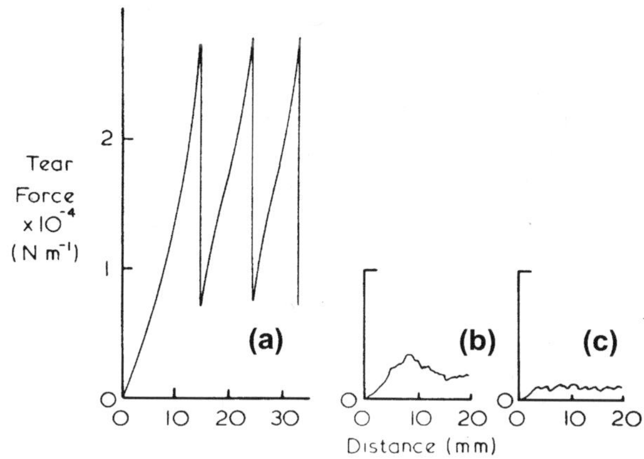

FIGURE 10.18 Tear force relations for (a) a filled elastomer without constraints; (b) the same material with the tear confined to a linear path; and (c) the corresponding unfilled elastomer, with and without constraints. (From Gent and Henry (1968).)

Thus, reinforcement of tear strength by fillers is of two kinds: a small (no more than two- to threefold) increase in intrinsic strength, and a major deviation of the tear path on a scale of several millimeters under special conditions of rate of tearing, temperature, and molecular orientation. The first effect may be attributed to enhanced energy dissipation in filled materials, as discussed in the previous section. The second is attributed to a *lowering* of the tear resistance sideways, parallel to the stretching direction. If the tear resistance is reduced sufficiently in the sideways direction, then the tear will be deflected sideways and rendered relatively harmless. Paradoxically, rubber is reinforced if its strength is lowered in a certain direction—one that does not result in catastrophic fracture. This mechanism of reinforcement is supported by two observations: measurements of tear strength in the stretching direction show a pronounced decrease as the stretch is increased, and calculations reveal that the energy available to turn a crack into this direction is surprisingly large (40% or more of that for continuing in the straight-ahead direction when the sample is highly stretched) (Gent et al., 2003). Thus, when the tear strength in the stretching direction falls to 40% or less of that in the straight-ahead direction, the crack is expected to turn sideways.

10.4.5 Repeated Stressing: Dynamic Crack Propagation

Although amorphous elastomers are found to tear steadily, at rates controlled by the available energy for fracture, G (as shown in Figures 10.13 and 10.14), strain-crystallizing elastomers do not tear continuously under small values of G, of less than about 10^4 J/m^2 for natural rubber, for example (see Figure 10.15). Nevertheless, when small stresses are applied repeatedly, a crack will grow in

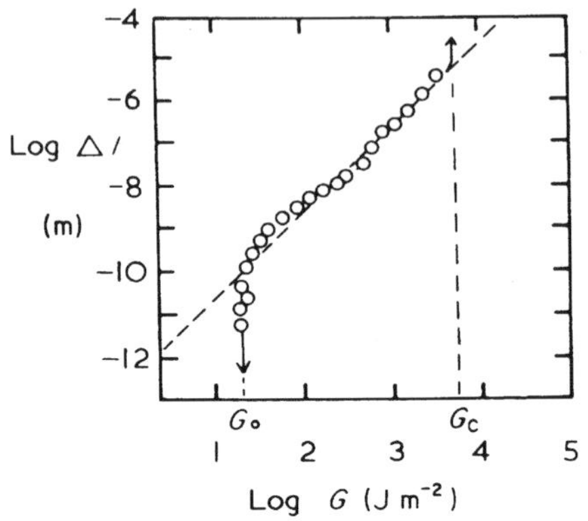

FIGURE 10.19 Crack growth step, Δl, per stress application versus energy, G, available for fracture, for a natural rubber vulcanizate. (From Lake and Lindley (1965).)

a stepwise manner by an amount, Δl, per stress application, even though the corresponding value of G is much below the critical level (Thomas, 1958).

Experimentally, four distinct growth laws have been observed (Figure 10.19) corresponding to four levels of stressing (Lake and Lindley, 1964, 1966):

1. $G < G_0$: no crack growth occurs by tearing, but only by chemical (ozone) attack.
2. $G_0 < G < G_1$: the growth step, Δl, is proportional to $G - G_0$.
3. $G_1 < G < G_c$: the growth step, Δl, is proportional to G^α.
4. $G \sim G_c$: catastrophic tearing.

The transitional value of G between one crack growth law and another, denoted G_1 above, is found to be about 400 J/m^2. No explanation has yet been advanced either for the form of these experimental growth laws or for the transition between them. They must therefore be regarded for the present as empirical relations for the growth step, Δl, per stress application.

In practice it is customary to approximate crack growth over a wide range of G values (but greater than the threshold value, G_0) by a "Paris Law" relation that can be put in the form:

$$\Delta l = B'(G/G_0)^\alpha, \tag{10.21}$$

where the constant, B', is found to be about 1 Å per stress application for many rubber compounds and the exponent, α, takes different values for different elastomers, ranging from 2 for natural rubber compounds, represented by the broken line in Figure 10.19, to values of 4–6 for noncrystallizing elastomers such as SBR and *cis-/trans*-polybutadiene.

Crack growth in natural rubber compounds is brought about only by imposing the deformation; if the deformation is maintained, the crack does not grow further under forces insufficient to cause catastrophic tearing. The reason for this is that a crystalline region develops in the highly stressed material at the crack tip and effectively precludes further tearing. This explains a striking feature of crack growth in strain-crystallizing elastomers: the growth steps under repeated stressing become extremely small if the test piece is not relaxed completely between each stress application (Lake and Lindley, 1964). In these circumstances, the crystalline region does not melt; it remains intact to prevent further crack growth when the stresses are reimposed. As a result, the mechanical fatigue life (discussed in the following section) becomes remarkably prolonged if the component is never relaxed to the zero-stress state (see Figure 10.33). Indeed, failure in these circumstances is a consequence of chemical attack, usually by atmospheric ozone (Lake and Lindley, 1964), rather than mechanical rupture. The retardation of fatigue crack growth rate under nonrelaxing cycles caused by strain crystallization was studied by Lindley (1973). Mars has demonstrated that the prolonged fatigue life under nonrelaxing cycles shown in Figure 10.33 can be computed as a direct consequence of the behavior measured by Lindley. Mars introduced (Mars and Fatemi, 2003; Mars, 2009) the phenomenological relation of Eq. (10.22), in which the power-law slope, α, of the fatigue law depends on the degree of nonrelaxation, quantified by the ratio $R = G_{\min}/G_{\max}$:

$$\Delta l = B''(G/G_c)^{\alpha(R)}. \tag{10.22}$$

Here, B'' is the amount of crack growth apparently occurring at the point where the fatigue crack growth curve is extrapolated to intersect with the asymptote associated with the fracture strength, G_c. A typical value is roughly 0.01 mm/cyc. A simple empirical form for $\alpha(R)$ is

$$\alpha(R) = \alpha_0 \exp(CR). \tag{10.23}$$

α_0 is the power-law exponent for the fully relaxing (i.e., $R = 0$) crack growth relationship. C is an empirical parameter that describes the sensitivity of the power-law slope to increases in R ratio. For natural rubber, $C \approx 4$. Note that, as used here, $0 \leqslant R \leqslant 1$. Negative values of R are prohibited because the energy release rate of a crack cannot be negative.

Amorphous elastomers show more crack growth under intermittent stressing than under a steady stress, and the additional growth step per stress cycle is found to depend on the available energy for fracture, G, in substantially the same way as for natural rubber. The principal difference is that over region 3, the exponent, α, in Eq. (10.21) is about 4 for SBR in place of 2 for NR (Lake and Lindley, 1966).

Andrews has put forward a general explanation for the slowing down of a crack in an amorphous elastomer (and the complete cessation of tearing in a strain-crystallizing elastomer), after the stresses have been applied, in terms of time-dependent stress changes at the tip of the crack (Andrews, 1963; Andrews and Fukahori, 1977). As a result, the stress concentration at the growing tip is smaller for a viscoelastic material, or for one that is energy-dissipating, than would be expected from purely elastic considerations. Crack growth is correspondingly slowed. This concept has features in common with that of Knauss, described briefly in Section 10.5.2. The same considerations may also be applied to explain a phenomenon observed by Harbour et al. (2007b) and later modeled by Näser et al. (2009), in which an unloaded dwell period was introduced between load applications. In this case, time-dependent stress increases upon reloading were responsible for significant increases of the crack growth rate. Longer dwell periods were associated with faster rates of crack growth, up to a factor of 30×, for a filled styrene-butadiene rubber.

Oxygen in the surrounding atmosphere is found to increase crack growth, presumably by an oxidative chain scission reaction catalyzed by mechanical rupture. The minimum energy, G_0, is found to be somewhat larger for experiments carried out in vacuo (Lake and Lindley, 1964; Gent and Hindi, 1990; Gent et al., 1991). When antioxidants are included in the elastomer formulation, then the results in an oxygen-containing atmosphere approach those obtained in vacuo.

10.4.6 Thermoplastic Elastomers

Thermoplastic elastomers derive their physical characteristics from the fundamental immiscibility of different polymers. They consist of triblock molecules having the general structure A–B–A, where A denotes a glassy polymeric strand (e.g., of polystyrene) and B denotes a flexible polymeric strand (e.g., of polybutadiene). The end sequences A are immiscible in polymer B and hence they tend to cluster together to form small domains of a glassy polymer isolated within an elastomeric matrix. Moreover, because the sequences A at each end of one triblock molecule generally become part of different glassy domains, a network of elastomeric strands is formed, linked together by small hard domains, 10–30 nm in diameter. Materials of this kind behave in a characteristically rubberlike way, at least at temperatures below the glass temperature of polymer A and above that of polymer B.

The tear strength of a representative thermoplastic elastomer, Kraton 1101 (Shell Chemical Company), is quite comparable to that of a well-reinforced amorphous or strain-crystallizing elastomer, about 20 kJ/m^2 at room temperature. This remarkably high value is attributed to plastic flow of the hard domains under high local stresses, approaching the breaking point. Indeed, such a deformation process seems essential for these materials to have the capacity to dissipate strain energy, as any tough material must do.

10.5 TENSILE RUPTURE

10.5.1 Effects of Rate and Temperature

In Figure 10.20, several relations are shown for the breaking stress, σ_b, of an unfilled vulcanizate of SBR as a function of the rate of elongation at

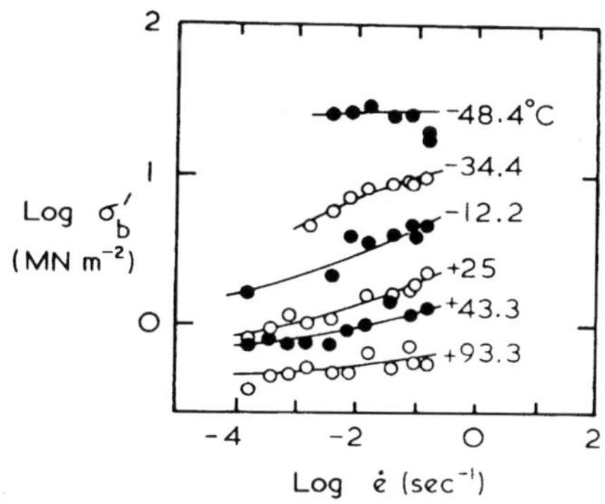

FIGURE 10.20 Breaking stress, σ_b, for an SBR vulcanizate versus rate of elongation, e. (From Smith (1958).)

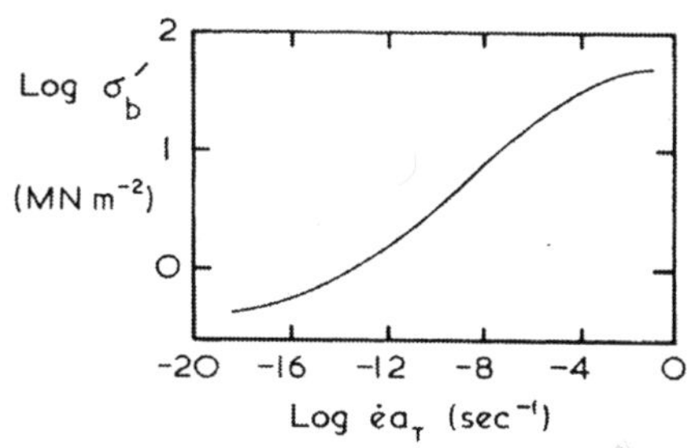

FIGURE 10.21 Master relations for breaking stress, σ_b, as a function of rate of elongation, e, reduced to T_g (−60°C) by means of the WLF relation, Eqs. (10.16) and (10.17). (From Smith (1958).)

different temperatures (Smith, 1958). A small correction factor (T_g/T) has been applied to the measured values to allow for changes in the elastic modulus with temperature. The corrected values are denoted σ_b'.

The experimental relations appear to form parallel curves, superimposable by horizontal displacements. The strength at a given temperature is thus equal to that at another temperature provided that the rate is adjusted appropriately, by a factor depending on the temperature difference. (Using a logarithmic scale for rate of elongation, a constant multiplying factor is equivalent to a constant horizontal displacement.) As in the case of fracture energy, G (see 10.4.2), this factor is found to be the ratio, $/\phi_T$, of segmental mobilities at the two temperatures (Smith, 1958). It is readily calculated from the WLF relation (Eq. (10.17)). A master curve may thus be constructed for a reference temperature, T_s, chosen here for convenience as T_g, by applying the appropriate shift factors to relations determined at other temperatures. The master curve for tensile strength, obtained from the relations shown in Figure 10.20, is given in Figure 10.21.

The variation of tensile strength with temperature, like the variation in fracture energy, is thus due primarily to a change in segmental mobility.

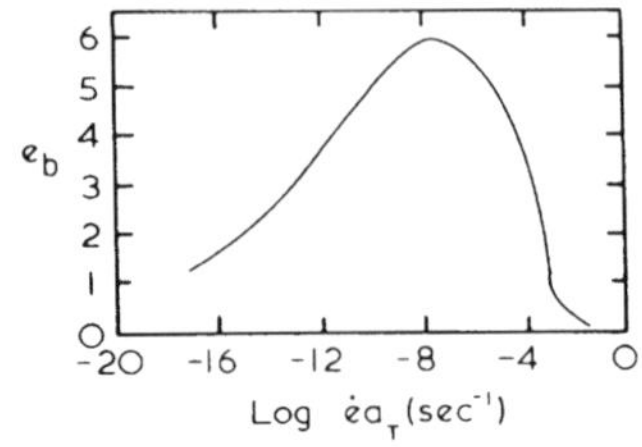

FIGURE 10.22 Master relations for breaking elongation, e_b, as a function of rate of elongation, e, reduced to T_g (−60°C). (From Smith (1958).)

Moreover, the master curve has the form expected of a viscosity-controlled quantity: it rises sharply with increased rate of elongation to a maximum value at high rates when the segments do not move and the material breaks as a brittle glass (Bueche, 1955). The breaking elongation at first rises with increasing rate of elongation, reflecting the enhanced strength, and then falls at higher rates as the segments become unable to respond sufficiently rapidly (Figure 10.22).

Rupture of a tensile test piece may be regarded as catastrophic tearing at the tip of a chance flaw. The success of the WLF reduction principle for fracture energy, G, in tearing thus implies that it will also hold for tensile rupture properties. Indeed, σ_b and e_b may be calculated from the appropriate value of G at each rate and temperature, using relations analogous to Eqs. (10.6) and (10.7). The rate of extension at the crack tip will, however, be much greater than the rate of extension of the whole test piece, and this discrepancy in rates must be taken into account (Bueche and Halpin, 1964).

In addition, it is clear from the derivation of Eq. (10.5) that U represents the energy obtainable from the deformed material rather than the energy put into deforming it. For a material with energy-dissipating properties, the energy available for fracture is only a fraction of that supplied. Such a material will therefore appear doubly strong in a tensile test or in any other fracture process in which the tear energy is supplied indirectly by the relief of deformations elsewhere.

10.5.2 The Failure Envelope

An alternative representation of tensile rupture data over wide ranges of temperature and rate of elongation is obtained by plotting the breaking stress, σ_b, against the corresponding breaking extension, e_b (Smith, 1963). Tensile strength results given in Figures 10.20–10.22 are replotted in this way in Figure 10.23. They yield a single curve, termed the *failure envelope*, which has a characteristic parabolic shape. Following around the curve in an anticlockwise sense corresponds to increasing the rate of extension or to decreasing the temperature, although these two variables do not appear explicitly. Thus, at the lower extreme, the breaking stress and elongation are both small. These conditions are found at low rates of strain and at high temperatures. Conversely, the upper extreme corresponds to a high breaking stress and low extensibility.

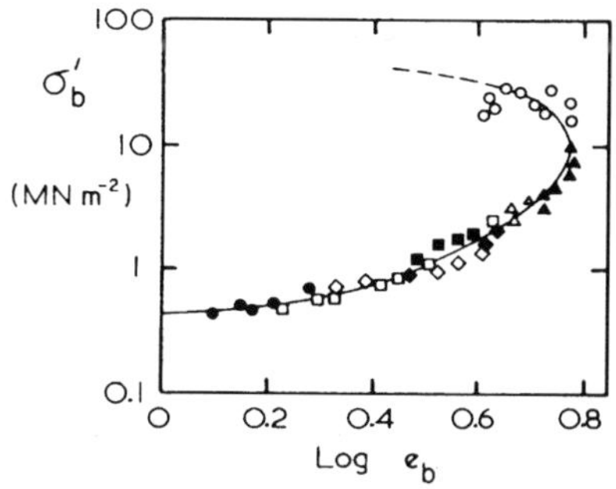

FIGURE 10.23 Failure envelope for an SBR vulcanizate. (From Smith (1963).)

These conditions obtain at high rates of strain and low temperatures, when the material responds in a glasslike way.

The principal advantages of the "failure envelope" representation of data seem to be twofold. First, it clearly indicates the maximum possible breaking elongation, $e_{b,\max}$, for the material. This is found to be well correlated with the degree of crosslinking, specifically with the molecular weight between crosslinks, M_c, as predicted by elasticity theory:

$$\lambda_{\max} = 1 + e_{b,\max} \propto M_c^{1/2}. \tag{10.24}$$

Second, the failure envelope can be generalized to deal with different degrees of crosslinking, as discussed later. It has therefore been employed to distinguish between changes in crosslinking and other changes that affect the response to rate of elongation and temperature but do not necessarily affect M_c. Examples of this are plasticization and addition of reinforcing fillers (Smith, 1969).

10.5.3 Effect of Degree of Crosslinking

The breaking stress is usually found to pass through a sharp maximum as the degree of crosslinking is increased from zero. An example is shown in Figure 10.24. This maximum is due primarily to changes in viscoelastic properties with crosslinking and not to changes in intrinsic strength. For example, it is much less pronounced at lower rates of extension (Bueche and Dudek, 1963), and it is not shown at all by swollen specimens (Epstein and Smith, 1958). Bueche and Dudek (1963) and Smith and Chu (1972) therefore conclude that it would not exist under conditions of elastic equilibrium.

Two different reduction schemes have been employed to construct failure envelopes for materials having different degrees of crosslinking. The first, shown in Figure 10.25, consists of scaling the breaking elongation, e_b, in terms of its maximum value, which is, of course, dependent on the degree of crosslinking (Eq. (10.24)). Also, the breaking stress, σ_b, is converted into a true stress at break, rather than the nominal stress employed until now.

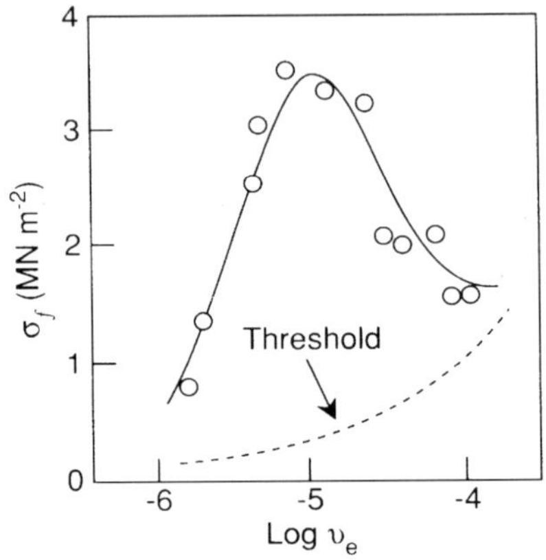

FIGURE 10.24 Tensile strength of SBR vulcanizates versus degree of crosslinking, represented by υ_e. (From Bueche and Dudek (1963).) Broken curve: author's estimate of threshold strength under nondissipative conditions.

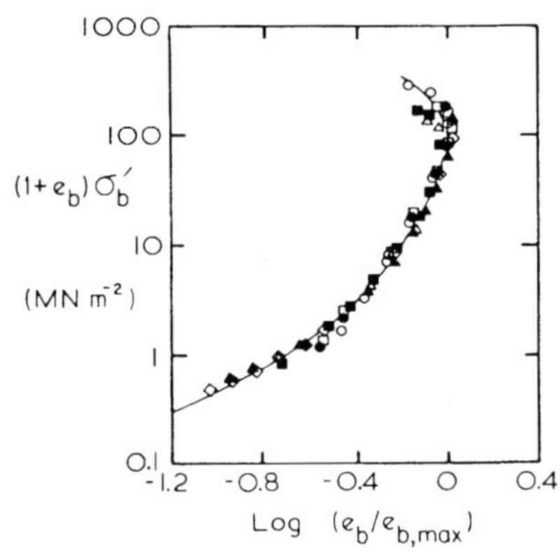

FIGURE 10.25 Failure envelope for Viton A-HV materials crosslinked to various extents ($e_{b,\max}$ ranging from 2.5 to 18). (From Smith (1969).)

(The nominal tensile stress is given by the tensile force divided by the unstrained cross-sectional area of the specimen. It has been commonly used in the literature dealing with deformation and fracture of elastomers.) This reduction scheme is clearly quite successful in dealing with a wide range of crosslinking (see Figure 10.25) (Smith, 1969).

The second method consists of scaling the stress axis by dividing the nominal stress at break by a measure of the density of crosslinking (Landel and Fedors, 1966). This method also appears to bring data from differently crosslinked materials into a common relationship.

10.5.4 Strain-Crystallizing Elastomers

Whereas amorphous elastomers show a steady fall in tensile strength as the temperature is raised (see Figure 10.20), strain-crystallizing elastomers show a rather sudden drop at a critical temperature, T_c (Figure 10.26) (Boonstra, 1949; Harwood et al., 1970; Thomas and Whittle, 1970; Gent and Zhang, 2002). This temperature depends strongly on the degree of crosslinking, as shown in Figure 10.26. It is clearly associated with failure to crystallize at high temperatures; however, although the bulk of the specimen is amorphous above T_c, the highly strained material at the flaw tip probably continues to crystallize. Thomas and Whittle (1970) draw a parallel between the drop in strength at the critical temperature, T_c, and the similar sharp drop at a critical depth, l_c, of an edge cut (Figure 10.27), for strength measurements made at room temperature.

Two other aspects of the critical temperature are noteworthy. First, it is substantially the same for compounds reinforced with fillers. Second, it depends strongly on the type of crosslinking, being highest for long, polysulfidic crosslinks and lowest for carbon–carbon crosslinks. Apparently, labile crosslinks are an important factor in promoting crystallization.

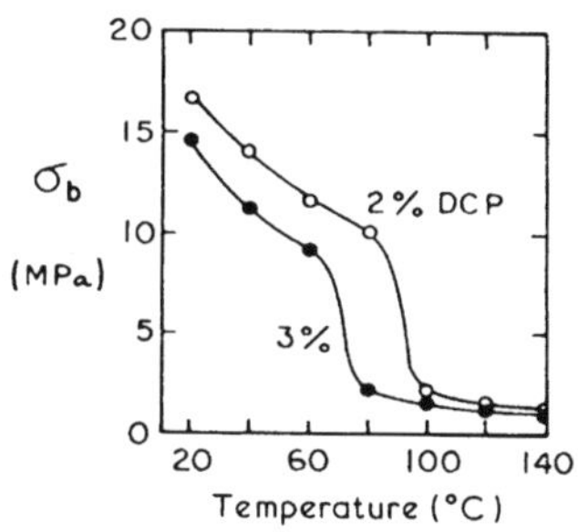

FIGURE 10.26 Tensile strength of natural rubber crosslinked with dicumyl peroxide (DCP) versus temperature. (From Thomas and Whittle (1970).)

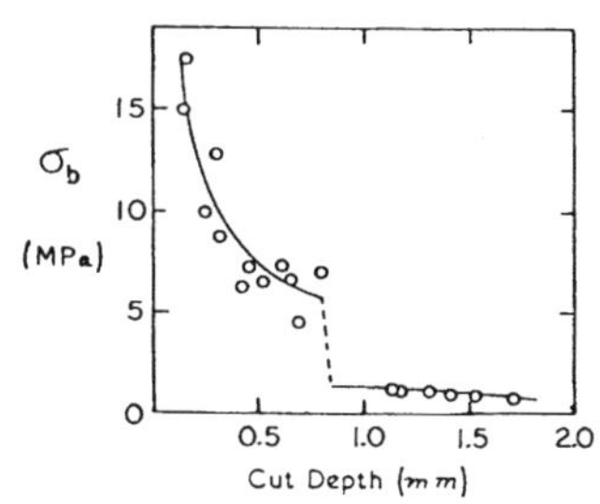

FIGURE 10.27 Tensile strength of natural rubber crosslinked with 2% dicumyl peroxide versus depth of initial edge cut. (From Thomas and Whittle (1970).)

10.5.5 Energy Dissipation and Strength

A general correlation between tensile strength and the temperature interval ($T - T_g$) between the test temperature, T, and the glass transition temperature, T_g, has been recognized for many years, as discussed in Section 10.5.2 (Borders and Juve, 1946). An example is shown in Figure 10.28, where the strengths of polyurethane elastomers with T_g values ranging from −67 to −17°C are plotted against $T - T_g$ (Smith, 1969). All the results fall on a single curve in this representation, indicating once more that segmental viscosity governs the observed strength.

A more striking demonstration of the close connection between energy dissipation and strength has been given by Grosch et al. (1966). They showed that a direct relationship exists between the energy density required to break elastomers, W_b, and the energy density dissipated on stretching them almost to the breaking elongation, W_d. This relationship held irrespective of the mechanism of energy loss; that is, for filled and unfilled, strain-crystallizing,

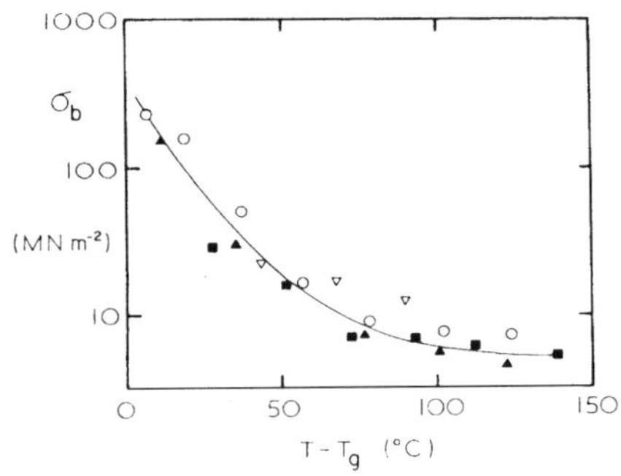

FIGURE 10.28 Tensile strength of polyurethane elastomers versus $T - T_g$ (T_g ranging from −67 to −17°C). (From Smith (1969).)

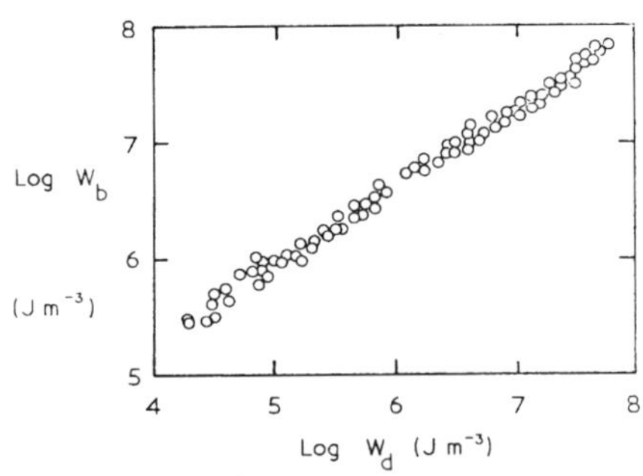

FIGURE 10.29 Work-to-break (W_b) versus energy dissipated (W_d) on stretching almost to the breaking elongation. (From Grosch et al. (1966).)

and amorphous elastomers (Figure 10.29). Their empirical relation is

$$W_b = 410 W_d^{2/3}. \tag{10.25}$$

W_b and W_d are expressed in joules per cubic meter. Those materials that require the most energy to bring about rupture—the strongest elastomers—are precisely those in which the major part of the energy is dissipated before rupture.

10.6 REPEATED STRESSING: MECHANICAL FATIGUE

Under repeated tensile deformations, cracks appear, generally in the edges of the specimen, and grow across it in an accelerating way. This process is known as fatigue failure. It has been treated quantitatively in terms of stepwise tearing from an initial nick or flaw (Lindley and Thomas, 1962; Gent et al., 1964), as follows: Every time a deformation is imposed, energy, G, becomes available in the form of strain energy to cause growth by tearing of a small nick in the edge of the specimen. The value of G for tensile test pieces is given by Eq. (10.5). The corresponding growth step, Δl, is assumed to obey Eq. (10.20); that is, to be proportional to G^a, so that the crack growth law becomes

$$\Delta l / l^\alpha = (2kU)^\alpha B' \Delta n, \tag{10.26}$$

where n is the number of times the deformation is imposed and k is a numerical constant, about 2 (see Section 10.2.3). The depth of the crack after N strain cycles is then obtained by integration,

$$l_0^{(1-\alpha)} - l^{(1-\alpha)} = (2kU)^\alpha B' N, \tag{10.27}$$

where l_0 is the initial depth of the nick. An example of crack growth is shown in Figure 10.30; it conforms closely to Eq. (10.27) with $\alpha = 2.0$.

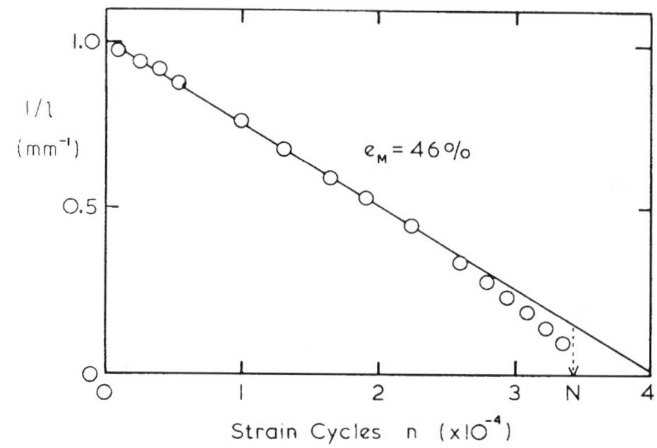

FIGURE 10.30 Growth of an edge crack in a test piece of a natural rubber vulcanizate stretched repeatedly to 46% extension. (From Greensmith et al. (1963).)

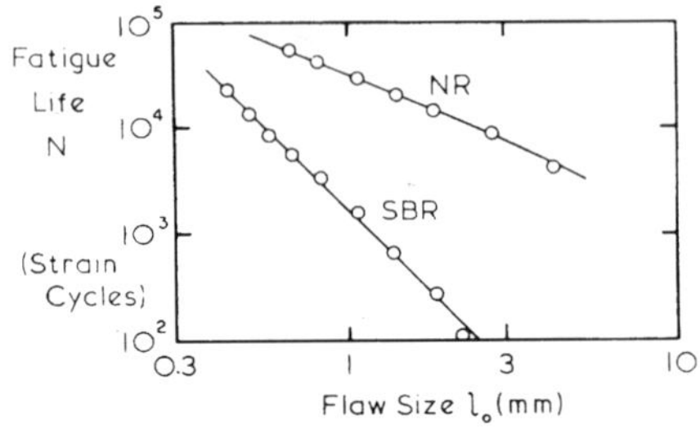

FIGURE 10.31 Fatigue life versus depth of initial cut for test pieces of natural rubber and SBR stretched repeatedly to 50% extension. (From Lake and Lindley (1964).)

If the crack grows to many times its original depth, so that $l \gg l_0$ before fracture ensues, then the corresponding fatigue life may be obtained by setting $l = \infty$ in Eq. (10.27) yielding

$$1/N = (2kU)^{\alpha} B' l_0^{(\alpha-1)}. \tag{10.28}$$

This is a quantitative prediction for the fatigue life, N, in terms of the strain energy, U, and two material properties, the crack growth exponent, α, and the characteristic dimension, B', which can be determined in a separate experiment as described earlier (see Section 10.4.5). Measured fatigue lives for specimens with initial cuts of different length (see Figure 10.3) and for imposed deformations of different magnitude have been found to be in good agreement with the predictions of Eq. (10.28) (Lake and Lindley, 1964; Lindley and Thomas, 1962; Gent et al., 1964).

Examples of the dependence of fatigue life on initial cut size are shown in Figures 10.3 and 10.31. Lives for test pieces that contain no deliberately introduced cuts, represented by horizontal broken lines in Figure 10.3, may be interpreted as stepwise tearing from a hypothetical nick or flaw, about 20μm deep, as discussed previously. It is particularly noteworthy that closely similar sizes are deduced for natural flaws for both strain-crystallizing and noncrystallizing elastomers by such extrapolations, because for a noncrystallizing elastomer (SBR), the crack growth law is quite different over the main tearing region. The exponent, α, in the fatigue life relation (Eq. (10.28)), becomes 4 in place of 2. Measured fatigue lives for an unfilled SBR compound have been found to be in good accord with this relation for a wide range of initial cut depths, l_0, and deformation amplitudes (Lake and Lindley, 1964).

The different crack growth laws for strain-crystallizing and noncrystallizing elastomers thus lead to quite different fatigue life relations. For a noncrystallizing elastomer, the fatigue life is much more dependent on the size of the initial flaw (see Figure 10.31) and the magnitude of the imposed deformation, so that such elastomers are generally longer lived at small deformations and with no accidental cuts, but much shorter lived under more severe conditions. The fatigue life is also drastically lowered at high temperatures as a result of the sharp increase in cut growth rate as the internal viscosity is decreased (Figure 10.32). In contrast, the hysteresis associated with strain-induced crystallization is retained, provided that the temperature does not become so high (about 100°C for natural rubber) that crystallization no longer occurs. The fatigue life for natural rubber is therefore not greatly affected by a moderate rise in temperature.

A more striking difference is found between strain-crystallizing and noncrystallizing elastomers when the stress is not relaxed to zero during each cycle. As shown in Figure 10.33, the fatigue life of a natural rubber vulcanizate is greatly increased when the minimum strain is raised from zero to, say, 100% because the crystalline barrier to tearing at the tip of a crack does not then disappear in the minimum-strain state. As a result, the growth of flaws is virtually stopped unless the total applied strain is very large, about 400%. No comparable strengthening effect on raising the minimum-strain level is found for noncrystallizing elastomers.

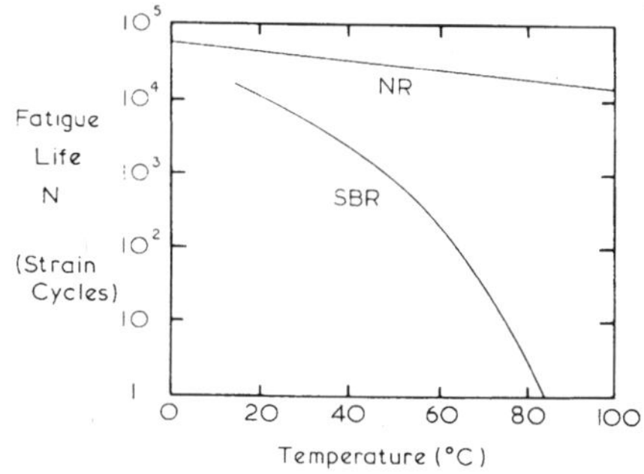

FIGURE 10.32 Fatigue life versus temperature for test pieces of natural rubber and SBR stretched repeatedly to 175% extension. (From Greensmith et al. (1963).)

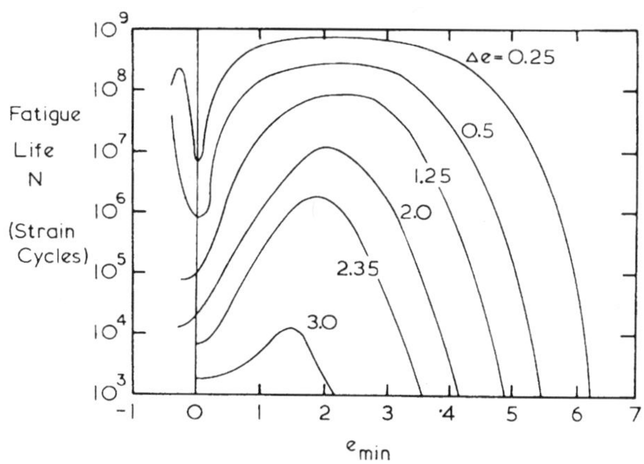

FIGURE 10.33 Fatigue life for test pieces of natural rubber versus minimum extension, e_{min}. Δe, denotes the additional strain imposed repeatedly. (From Cadwell et al. (1940).)

Corresponding to the threshold value, G_0, of tearing energy, below which no crack growth occurs by mechanical rupture, there is a minimum tensile strain, e_0, below which normal-sized flaws do not grow under fatigue conditions. For typical elastomers, this mechanical fatigue limit is found to be about 50–100% extension, by calculation from Eq. (10.7) and by direct observation (Lake and Lindley, 1965). At extensions below this level, the fatigue life would be infinite in the absence of chemical attack, for example, by ozone in the surrounding atmosphere, or by aging.

Reinforcing fillers greatly enhance the tear strength and tensile strength of elastomers but do not cause an equivalent improvement in the crack growth and fatigue properties. At a given strain energy input, the measured lives are appreciably longer, but if compared at equal available energy levels, they are not much increased. The initial flaw size and threshold tear energy, G_0, are therefore deduced to be similar to those for unfilled materials. The growth steps are apparently too small for pronounced deviation of the tear, and hence "reinforcement" against fatigue failure by this mechanism is not so pronounced.

Apart from a rubber compound's fatigue properties, stiffness can also greatly influence suitability for a given fatigue application, through its effect on the strain energy release rate. In strain-controlled applications, softer compounds will tend to exhibit greater durability. In stress-controlled applications, stiffer compounds will tend to exhibit greater durability. The mode of test control is therefore an essential feature of any fatigue test that must be specified and controlled when evaluating materials or products (Mars, 2011).

10.7 FAILURE UNDER MULTIAXIAL STRESSES

While lab tests often employ simple uniaxial loads, rubber components in service commonly experience states that combine loads from multiple directions. Such loads are said to be *multiaxial*. Fracture and fatigue processes occurring under such loads are governed by the same general principles as for uniaxial loads, but additional consideration must be made to account for the way in which local effects on crack precursors depend on orientation.

10.7.1 Critical Plane Hypothesis

Under the action of multiaxial loads, crack precursors will develop in a manner that reflects their particular loading experiences. It may be assumed that a crack precursor is present at every point, and that the crack precursor may occur in any orientation. By computing the rate at which each possible precursor will develop, it can be determined (Mars, 2002) which particular precursor(s) will develop the fastest. The orientation of this precursor identifies the critical plane on which cracks will develop, which is needed in order to accurately estimate fatigue life.

For simple duty cycles consisting of in-phase combinations of tensile and shear loads (i.e., proportional loading), the most critical plane is the plane perpendicular to the direction of maximum principal tensile stress. For more general duty cycles—those involving compression loads, or duty cycles involving out-of-phase combinations of loading components (i.e., nonproportional loading)—a search for the critical plane must be made in order to identify the most highly damaged orientation.

Several critical plane analysis methods have been independently proposed and studied in the last decade (Mars, 2002; Mars and Fatemi, 2006a,b; Saintier et al., 2006a,b; Harbour et al., 2008b; Andriyana and Saintier, 2010; Zine, 2011; Ayoub et al., 2012). They are noteworthy for their ability to provide a detailed and accurate account of the effects of multiaxial loading on fatigue life, and for their ability to predict the plane in which cracks will develop.

10.7.2 Energy Density Available for Driving Growth of Crack Precursors

The energy required to drive the growth of a small crack precursor comes from strain energy stored in the nearest material surrounding the precursor. For example, if the precursor has a size, l, then the largest decreases in strain energy occur in the close neighborhood of the crack, a radius which may very roughly be estimated at several times the crack size l. In the case of simple tension, growth of the stress-relieved region adjacent to crack faces is capable of releasing all the strain energy in the region.

In multiaxial loading scenarios, however, it is more common that growth of the crack and the associated stress-relieved region releases only a fraction of the material's strain energy. For example, under the equibiaxial tension loading state, Yeoh (2002) showed that a small through-crack at large strain releases only a fraction of what would have been released under simple tension at the same nominal strain, the fraction being roughly 80% at small strains, and approaching 50% as the strain approaches 200%. This result is in agreement with the results of equibiaxial fatigue experiments that indicated the available energy density was only half of the total strain energy density (Roberts and Benzies, 1977). Note that, in general, the part of the strain energy density that is available for stress relief of crack faces depends strongly on the orientation of the crack considered. An approach for estimating the available energy density for any multiaxial loading state, for any crack orientation, has been proposed and tested successfully for a broad range of multiaxial loading conditions (Mars, 2002).

The fact that the energy release rate of a small crack surrounded by homogenously strained material scales linearly with the size of the crack, for all multiaxial loading states, has been established from experience (Gough and Muhr, 2005), and can be established mathematically by considering the balance of configurational stresses (Aït-Bachir et al., 2012).

10.7.3 Compression and Shear

Elastomers tend to fracture at 45° to the direction of shear (Figure 10.6), at right angles to the corresponding principal tensile stress (Knauss, 1970; Ahagon et al., 1975), at a shear stress theoretically equal to the tensile strength (Griffith, 1924). Indeed, the general condition for rupture appears to be the attainment of a specific tensile stress, σ_t, at the tip of an existing flaw, and this circumstance can arise even when both applied stresses are compressive, provided that they are unequal (Griffith, 1924). When all the compressive stresses are equal (i.e., under a uniform triaxial compression), the elastomer will merely decrease in volume. No case of fracture under such a loading condition is known.

Under a uniaxial compressive stress, the theory of brittle fracture predicts a breaking stress eight times as large as in tension (Eq. (10.6)) by growth of a crack in an oblique direction (Griffith, 1924). A uniform compressive stress is not, however, readily achieved. Instead, friction at the loaded surfaces of a thin compressed block generally prevents the elastomer from expanding freely in a lateral direction, and a complex stress condition is set up. The outwardly bulging surfaces may split open when the local tensile stress is sufficiently high, but this local fracture does not propagate inward very far because the interior is largely under triaxial compression. Instead, the tear curves around and eventually causes a ring of rubber to break away from the outside of the block, leaving the remainder of the block intact but with a narrower central cross-section (Stevenson, 1983). Thus, a rubber block in compression is remarkably resistant to fracture, but its stiffness may be seriously reduced after many load cycles by loss of rubber from the outer regions.

Failure in simple shear is still more complex. An approximate treatment for an interfacial crack, starting at one edge, yields a relation analogous to Eq. (10.5):

$$G = kUt. \tag{10.29}$$

Here it has been found through numerical analysis that the constant k is 0.4 initially and then varies between 0.2 and 1.0 as the crack length increases (Lindley and Teo, 1979).

10.7.4 Equibiaxial Tension

Quite surprisingly, the breaking stress in equibiaxial tension has been found to be significantly greater than in uniaxial tension (Dickie and Smith, 1969; Extrand and Gent, 1991), by about 20–30%. The breaking elongation is lower but the stored elastic energy at fracture is greater. It should be noted that test sheets put into a state of biaxial extension do not have a cut edge at the desired point of failure, in the central region of the sheet, whereas specimens for uniaxial tests are usually cut from sheets in the form of thin strips.

Experiments with rather brittle rubber sheets that contained deliberately introduced initial cracks of the same size and type in both uniaxial and biaxial specimens have shown that the breaking stress is still about 20–30% higher in equibiaxial tension (Extrand and Gent, 1991). The results were, however, consistent with a single value for the fracture energy, G_c (about 150 J/m^2). The difference between the two tests is that when a crack grows in a sheet stretched equibiaxially, only about one-half of the strain energy stored in that area is released, whereas for a crack growing in a uniaxially stretched specimen, all the energy is released. As a consequence, the strain energy needs to be considerably larger, about twice as large, to cause fracture in equibiaxial stretching.

10.7.5 Triaxial Tension

A small spherical cavity within a block of rubber will expand elastically from its original radius, r_0, to a new radius, λr_0, under the action of an inflating pressure, P. In the same way, it will expand to an equal degree when the faces of the block are subjected to a uniform triaxial tension of $-P$—that is, to a negative hydrostatic pressure (Figure 10.7)—provided that the rubber is itself undilatable. When the expansion is small, it is proportional to P and given by $\lambda = 1+3P/4E$, where E is Young's modulus of the rubber. When the expansion is large, it increases more rapidly than in direct proportion. Indeed, for rubber obeying the neo-Hookean constitutive relation for elasticity (see Chapter 1), the expansion becomes indefinitely large at a finite value of the applied tension, given by (Green and Zerna, 1954; Gent and Lindley, 1958; Lindsey, 1967)

$$-P = 5E/6. \quad (10.30)$$

Of course, the original cavity will burst when the expansion of its wall reaches the breaking extension of the rubber, and a large tear will form, governed by an energy requirement for growth. For large precursor voids, the tensile stress for bursting open is much smaller than what Eq. (10.30) predicts (Williams and Schapery, 1965). And for small precursor voids, there are two further complexities: the actual surface energy of the void needs to be taken into account (Williams and Schapery, 1965; Gent and Tompkins, 1969), and the bursting stresses become so large that the rubber around the cavity will cease to follow elasticity relations valid only for low and moderate strains (Gent and Wang, 1991). However, over a surprisingly wide range of initial radius, r_0, from about 0.5 μm to about 1 mm, Eq. (10.30) is found to be a close approximation to the predicted fracture stress (Gent and Wang, 1991).

Rubber samples are almost invariably found to undergo internal cavitation at the triaxial tensions given by Eq. (10.30). This phenomenon must therefore be regarded as the consequence of an elastic instability, namely, the unbounded elastic expansion of preexisting cavities, too small to be readily detected, in accordance with the theory of large elastic deformations. It does not generally involve the fracture energy, because it is principally a transformation of potential energy (from the loading device) into strain energy. Apparently, rubber contains many precursor voids lying in the critical range, 0.5 μm to 1 mm (not larger because they would break open at lower stresses).

The critical stress predicted by Eq. (10.30) depends only on the elastic modulus and not at all on the strength of the elastomer. In agreement with this, cavitation stresses in bonded rubber blocks under tension (Figures 10.8 and 10.9) (Gent and Lindley, 1958), and near rigid inclusions, at points where a triaxial tension is set up (Figures 10.10 and 10.11) (Oberth and Bruenner, 1965), are found to be accurately proportional to E and independent of the tear strength of the elastomer, in accordance with the dominant role of an elastic rather than a rupture criterion for failure.

Cavitation near small rigid inclusions is more difficult to induce (Gent and Park, 1984), probably because the volume of rubber subjected to a critical triaxial tension is too small to contain relatively large precursor voids. And larger stresses are necessary to expand small voids less than about 0.5 μm in diameter.

If elastomers could be prepared without any microcavities greater than, say, 10 nm in radius, they would be much more resistant to cavitation. This seems an unlikely development, however, so Eq. (10.30) remains an important general fracture criterion for elastomers. It predicts a surprisingly low critical triaxial tension, of the order of only a few atmospheres, for soft, low-modulus elastomers. Conditions of triaxial tension should probably be avoided altogether in these cases.

Cavitation is an important practical issue when elastomers are used for containing high-pressure gases (Briscoe and Zakaria, 1990). If the gas dissolves in the rubber and migrates to fill the precursor voids, it will be at the (high) external pressure. Then, when the outside pressure is released suddenly, the voids break open in accordance with Eq. (10.30).

10.8 SURFACE CRACKING BY OZONE

In an atmosphere containing ozone, stretched samples of unsaturated elastomers develop surface cracks that grow in length and depth until they eventually sever the test piece. Even when they are quite small, they can cause a serious reduction in strength and fatigue life. Quite small concentrations of ozone will induce cracking, given sufficient time. The effect is so sensitive, in fact, that it is the basis for an ozone detection system (Mott and Roland, 1999).

The applied tensile stress necessary for an ozone crack to appear may be calculated approximately from Eq. (10.6). The fracture energy G is only about 0.1 J/m^2 (Braden and Gent, 1961), representing the small amount of energy needed for "fracture" of a liquid medium; that is, about twice the surface energy for a hydrocarbon liquid (Tarkow, 1958). Molecular scission apparently occurs readily by reaction with ozone and does not require mechanical energy to be induced. Taking a representative value for E for a soft rubber of 2 MPa and a value of 40 μm for the effective depth, l, of a chance surface flaw, Eq. (10.6) yields a critical tensile stress for ozone cracking of about 50 kPa and a critical tensile strain of about 5%. These predictions are in reasonably good agreement with experimentally observed minimum values for ozone attack.

As the stress level is raised above the minimum value, numerous weaker stress raisers become effective and more cracks form. Actually, a large number of small, mutually interfering cracks are less harmful than a few widely separated cracks that develop into deep cuts, so that the most harmful condition is just above the critical stress.

The rate at which a crack grows when the critical energy condition is satisfied depends on two factors: the rate of incidence of ozone at the crack tip and the rate of segmental motion in the tip region. When either of these processes is sufficiently slow, it becomes rate controlling. The overall rate, R, of crack growth is thus given approximately by

$$R^{-1}\ (\mathrm{s/m}) = 8 \times 10^{13}\phi_T^{-1} + 1.2 \times 10^{5}C^{-1}, \qquad (10.31)$$

where ϕ_T (s^{-1}) is the natural frequency of Brownian motion of molecular segments at the temperature, T, given by the WLF relation (Eqs. 10.16 and 10.17), in terms of (where $=$ 0.1 s^{-1} (Ferry, 1970)), and C (mg/L) is the concentration of ozone in the surrounding air (Gent and McGrath, 1965).

For a typical outdoor atmosphere, C is of the order of 10^{-4} mg/L, and the second term in Eq. (10.32) is then dominant for values of ϕ_T greater than about $10^4\mathrm{s}^{-1}$; that is, at temperatures more than 25°C above T_g (Gent and McGrath, 1965).

10.9 ABRASIVE WEAR

10.9.1 Mechanics of Wear

Abrasive wear consists of the rupture of small particles of elastomer under the action of frictional forces, when sliding takes place between the elastomer surface and a substrate. A suitable measure of the rate of wear is provided by the ratio, A/μ, where A is the volume of rubber abraded away per unit normal load and per unit sliding distance, and μ is the coefficient of friction. This ratio, termed *abradability*, represents the abraded volume per unit of energy dissipated in sliding. Master curves for the dependence of abradability on the speed of sliding, reduced to a convenient reference temperature by means of the WLF relation (Eqs. (10.16) and (10.17)), are shown in Figure 10.34. The abradability is seen to decrease with increasing speed, pass through a minimum, and then rise again at high speeds as the material becomes glasslike in response. This behavior resembles the variation of the reciprocal of the breaking energy, U_b, with rate of deformation (a reciprocal relationship because high abradability corresponds to low strength). Indeed, Grosch and Schallamach (1965) found a general parallel between A/μ and $1/U_b$. For this comparison values of the breaking energy, U_b, were determined at high rates of extension, about 10,000%

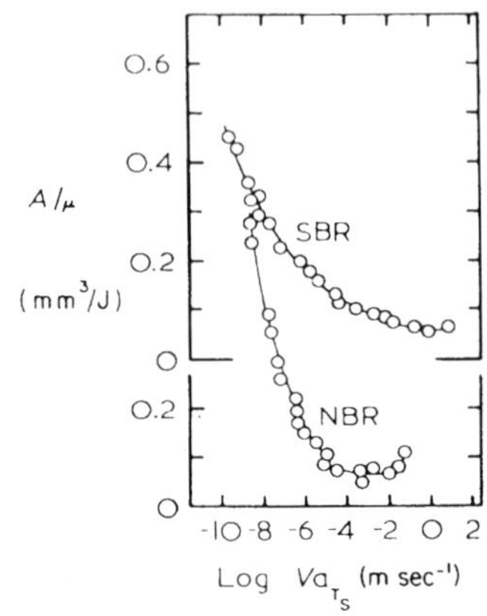

FIGURE 10.34 Abradability A/μ versus speed of sliding V, reduced to 20 °C, for SBR and NBR vulcanizates. (From Grosch and Schallamach (1965).)

per second, to bring them into agreement with measurements of abradability carried out at a sliding speed of 10 mm/s. This scaling relation indicates that the size of the rubber elements involved in deformation and wear was of the order of 0.1 mm, comparable to the size of the abrasive asperities on the particular track employed in the experiments. Moreover, the coefficient of proportionality, C, between abradability and breaking energy was found to be similar, about 10^{-3}, for all the unfilled elastomers examined. The magnitude of C represents the volume, A, of rubber abraded away by unit energy applied frictionally to a material for which unit energy, U_b, per unit volume is necessary to cause tensile rupture. It may be regarded as a measure of the inefficiency of rupture by tangential surface tractions; large volumes are deformed but only small volumes are removed. Apparently the ratio is similar throughout the rubber-to-glass transition and for a variety of elastomers.

The abradabilities, A/μ, were found to be generally about twice as large for carbon black-filled elastomers as for corresponding unfilled materials. This surprising observation that "reinforced" materials wear away faster can be partially accounted for in terms of the tear strength measurements referred to in a previous section. Under conditions of relatively smooth tearing, it was concluded that the intrinsic strength of reinforced materials is not particularly high; instead, it was found to be comparable to that of unfilled elastomers. The measurements of abradability considered here suggest that it is actually somewhat lower under abrasion conditions.

Southern and Thomas (1978) have related the rate of wear, A, to the crack growth resistance of the rubber by a simple theoretical treatment. A pattern of lateral ridges is generated in steady-state wear, known as the Schallamach abrasion pattern. A single ridge is shown in Figure 10.35. The frictional force, F, pulls laterally on the ridge crest and tends to tear the rubber in the direction indicated by a broken line, at an angle, θ, to the surface. Fracture energy, G, is made available for tearing in this direction, given by $F(1+\cos\theta)$. The crack will therefore advance by a distance, Δl, given by Eq. (10.21). This leads to a loss in thickness of rubber of $\Delta l(\sin\theta)$. Thus,

$$A = B'/G_0^{\alpha} F^{\alpha}(1+\cos\theta)^{\alpha}\sin\theta, \tag{10.32}$$

where α is either 2 or 4, depending on elastomer type. The angle, θ, may be estimated by direct inspection of the way in which abrasion patterns move over the surface during wear. All other terms in Eq. (10.32) can be determined from tear growth measurements. Thus, the theory does not involve any fitting constants. In these circumstances, it is remarkably successful in accounting for the rate of wear of several unfilled elastomers under severe conditions of pattern abrasion (Figure 10.36).

Two difficulties must be mentioned, however. The agreement is unsatisfactory for unfilled natural rubber, which wears away much more rapidly than crack growth measurements would predict (see Figure 10.36). It has been suggested that this material may not undergo strain-induced crystallization under abrasive conditions (i.e., under rapidly applied compressive and shearing stresses), and therefore does not show the high resistance to crack growth associated with crystallization. Liang et al. used a detailed numerical treatment of the Southern and Thomas theory to study these difficulties (Liang et al., 2009, 2010). Also, the wear of reinforced rubber is much slower than would be

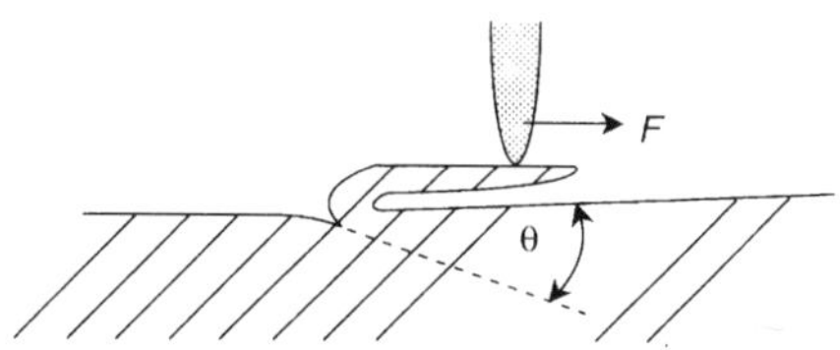

FIGURE 10.35 Sketch of a single surface ridge subjected to a frictional (tearing) force F. (From Southern and Thomas (1978).)

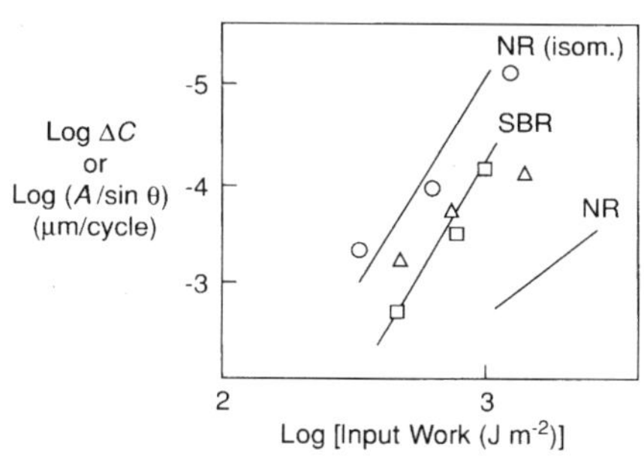

FIGURE 10.36 Rate of wear A versus frictional work input for NR, n; SBR, h; and isomerized (noncrystallizing) NR, s. Solid lines represent crack growth properties Δl $(=\Delta C)$ of the same materials under repeated stressing. (From Southern and Thomas (1978).)

predicted on the basis of crack growth measurements. Further work is needed to clarify this point, which is of great practical importance.

10.9.2 Chemical Effects

Under mild abrasion conditions, chemical changes within the elastomer become important in wear (Brodskii et al., 1960; Schallamach, 1968; Gent and Pulford, 1983). The scale of wearing remains small, but the particles of debris are often sticky and agglutinate to form larger particles, several millimeters in size. Indeed, *cis*-polyisoprene and poly(ethylene–co-propylene), for which molecular rupture under shearing conditions is particularly pronounced, both develop a tarry liquid surface during abrasion. In contrast, *cis*-polybutadiene shows no signs of structural deterioration. The debris appears to be unchanged chemically and nonadhering. Evidently, different chemical changes are undergone by different elastomers and are responsible for the different types of wear.

Two reactions can occur during wear: oxidative degradation as a result of frictional heating in the contact zone and mechanochemical degradation initiated by shear-induced rupture of chemical bonds. Present evidence favors the latter process. For example, in the absence of oxygen, the wear of *cis*-polyisoprene changes to resemble that of *cis*-polybutadiene, whereas the wear of poly(ethylene-co-propylene) is unaltered (Gent and Pulford, 1983). These results are in accord with the response of these materials to free radical reactions.

Surprisingly, the products of chemical changes within the elastomer are capable of causing rapid wear of metals used as abraders (Gorokhovskii et al., 1972; Evdokimov et al., 1973; Gent and Pulford, 1979a,b). For example, when a knife blade is used as a scraper to abrade rubber, the blade itself wears away and becomes blunted, and the volume of metal removed is substantially greater when the broken elastomer molecules form relatively stable free radicals (Gent and Pulford, 1979a,b). Thus, wear of the metal is attributed to direct chemical attack by reactive polymeric species, probably free radicals, during frictional contact (Gent and Pulford, 1979a,b). Similar wear has been observed with polymers in other physical states (Vinogradov et al., 1965); it appears to be a quite general phenomenon when molecular rupture takes place during sliding.

10.10 COMPUTATIONAL APPROACHES TO FAILURE MODELING

The rapid advance of computing technology over the past few decades has opened new paths to the numerical simulation of rubber's fracture and fatigue behavior.

Perhaps the first application of such an approach was Lindley's Finite Element Analysis (Lindley, 1972) of the tearing energy of a crack in several plane stress cases. The widespread adoption of Finite Element Analysis now enables the energy release rate of a static crack to be evaluated for rubber components under load (Busfield et al., 2005; Asare and Busfield, 2011). The approach has been implemented through various schemes, including a path-independent integral (Rice, 1968), virtual crack extension or closure (Parks, 1977; Shih et al., 1986; Shivakumar et al., 1988), and material configurational forces (Steinmann, 2000; Mueller and Maugin, 2002). Approaches have also been developed that permit growth of the crack during the simulation. The so-called eXtended Finite Element Method (XFEM) offers this capability, for example Belytschko et al. (1999), Melenk et al. (1996), Remmers et al. (2008), Song et al. (2006), Gurvich (2011), and Kaliske et al. (2010).

The influence of inelastic stress-strain behavior on crack tip conditions and failure is often neglected in numerical analysis, but may be considered using some computational approaches. For this task, approaches based on configurational mechanics (Mueller and Maugin, 2002) have proven quite powerful (Näser et al., 2007, 2009; Timmel et al., 2009).

When cracks grow from naturally occurring, microscopic precursors, it is often advantageous to focus analysis efforts on the questions of where and when a crack might first reach a critical size. Crack nucleation analysis approaches have been developed for this purpose (Mars, 2002; Mars and Fatemi, 2006a,b; Saintier et al., 2006a,b; Harbour et al., 2008b; Andriyana and Saintier, 2010; Zine, 2011; Ayoub et al., 2012) and are capable of providing accurate estimates of the effects of material, geometry, and duty cycle variations. When a critical plane analysis approach is adopted, these methods are particularly well-suited for dealing with duty cycles involving multiaxial and variable amplitude loading (Harbour et al., 2007a, 2008a,b).

ACKNOWLEDGMENTS

The original review was prepared in the course of a program of research on fracture supported by a grant from the Office of Naval Research. It is based largely on earlier reviews (Andrews, 1968; Gent, 1972; Eirich and Smith, 1972; Gent and Pulford, 1979a,b; Lake and Thomas, 2001; Mars and Fatemi, 2002; Mars, 2007). The first author is indebted to E.H. Andrews, T.L. Smith, A.G. Thomas, and other authors cited in the text for many helpful discussions, and to R.A. Paden for preparing many of the diagrams. The second author acknowledges the assistance of M. Sarkar in preparing this updated manuscript.

FURTHER READING

Halpin 1965; Lake 1970; Bowen and Knauss 1992; Lake and Yeoh 1978Halpin (1965), Lake (1970), Bowen and Knauss (1992), and Lake and Yeoh (1980).

REFERENCES

Ahagon, A., Gent, A.N., 1975. J. Polym. Sci., Polym. Phys. Ed. 13, 1903.
Ahagon, A., Gent, A.N., Kim, H.-W., Kumagai, Y., 1975. Rubber Chem. Technol. 48, 896.

Aït-Bachir, M., Mars, W.V., Verron, E., 2012. Int. J. Non-Linear Mech. 47, 22.
Andrews, E.H., 1961. J. Appl. Phys. 32, 542.
Andrews, E.H., 1963. J. Mech. Phys. Solids 11, 231.
Andrews, E.H., 1968. Fracture in Polymers, American Elsevier, New York.
Andrews, E.H., Fukahori, Y., 1977. J. Mater. Sci. 12, 1307.
Andriyana, A., Saintier, N., Verron, E., 2010. Int. J. Fatigue 32, 1627.
Asare, S., Busfield, J.J.C., 2011. Plast. Rubber Compos. Macromol. Eng. 40, 194.
Ayoub, G., Naït-Abdelaziz, M., Zaïri, F., Gloaguen, J.M., Charrier, P., 2012. Mech. Mater. 52, 87.
Belytschko, T., Black, T., 1999. Int. J. Numer. Meth. Eng. 45, 601.
Berry, J.P., 1972. Fracture: an advanced treatise. In: Liebowitz, H. (Ed.), Fracture of Non-metals and Composites, vol. 7. Academic Press, New York (Chapter 2).
Bhowmick, A.K., Gent, A.N., Pulford, C.T.R., 1983. Rubber Chem. Technol. 56, 226.
Boonstra, B.B.S.T., 1949. India Rubber World 121, 299.
Borders, A.M., Juve, R.D., 1946. Ind. Eng. Chem. 38, 1066.
Bowen, J.M., Knauss, W.G., 1992. J. Adhesion 39, 43.
Braden, M., Gent, A.N., 1961. Kautsch. Gummi 14, WT157; Andrews, E.H., Barnard, D., Braden, M., Gent, A.N., 1963. In: Bateman L. (Ed.), The Chemistry and Physics of Rubberlike Substances, Wiley, New York, (Chapter 12).
Briscoe, B.J., Zakaria, S., 1990. Polymer 31, 440.
Brodskii, G.I., Sakhnovskii, N.L., Reznikovskii, M.M., Evstratov, V.F., 1960. Sov. Rubber Technol. 18, 22.
Buckley, G.S., Fragiadakis, D., Roland, C.M., 2011. Rubber Chem. Technol. 84, 520.
Bueche, F., 1955. J. Appl. Phys. 26, 1133.
Bueche, F., Dudek, T.J., 1963. Rubber Chem. Technol. 36, 1.
Bueche, F., Halpin, J.C., 1964. J. Appl. Phys. 35, 36.
Busfield, J.J.C., Jha, V., Liang, H., Papadopoulos, I.C., Thomas, A.G., 2005. Plast. Rubber Compos. 34, 349.
Cadwell, S.M., Merrill, R.A., Sloman, C.M., Yost, F.L., 1940. Ind. Eng. Chem. Anal. Ed. 12, 19.
Choi, I.S., Roland, C.M., 1996. Rubber Chem. Technol. 69, 591.
De, D., Gent, A.N., 1996. Rubber Chem. Technol. 69, 834.
Dickie R.A., Smith, T.L., 1969. J. Polym. Sci. A- 2 7, 687.
Eirich, F.R., Smith, T.L., 1972. Fracture: an advanced treatise. In: Liebowitz, H. (Ed.), Fracture of Non-metals and Composites, vol. 7. Academic Press, New York (Chapter 7).
Eldred, R.J., 1972. J. Polym. Sci. B 10, 391.
Epstein, L.M., Smith, R.P., 1958. Trans. Soc. Rheol. 2, 219.
Evdokimov, Y.A., Sanches, S.S., Sukhorukov, N.A., 1973. Polym. Mech. 9, 460.
Extrand, C.W., Gent, A.N., 1991. Int. J. Fracture Mech. 48, 281.
Ferry, J.D., 1970. Viscoelastic Properties of Polymers, second ed. Wiley, New York.
Gent, A.N., 1972. Fracture: an advanced treatise. In: Liebowitz, H. (Ed.), Fracture of Non-metals and Composites, vol. 7. Academic Press, New York (Chapter 6).
Gent, A.N., Henry, A.W., 1968. In: Proceedings of International Rubber Conference, London, 1967. Maclaren, London, pp. 193–204.
Gent, A.N., Hindi, M., 1990. Rubber Chem. Technol. 63, 123.
Gent, A.N., Lai, S.-M., 1994. J. Polym. Sci.: Part B: Polym. Phys. 32, 1543.
Gent, A.N., Lindley, P.B., 1958. Proc. R. Soc. (London) A 249, 195.
Gent, A.N., McGrath, J.E., 1965. J. Polym. Sci. A 3, 1473.
Gent, A.N., Park, B., 1984. J. Mater. Sci. 19, 1947.
Gent, A.N., Pulford, C.T.R., 1979a. J. Mater Sci. 14, 1301.
Gent, A.N., Pulford, C.T.R., 1979b. In: Andrews, E.H. (Ed.), Developments in Polymer Fracture—1. Applied Science Publishers, London (Chapter 5).
Gent, A.N., Pulford, C.T.R., 1983. J. Appl. Polym. Sci. 28, 943.
Gent, A.N., Tobias, R.H., 1982. J. Polym. Sci., Polym. Phys. Ed. 20, 2051.
Gent, A.N., Tompkins, D.A., 1969. J. Polym. Sci. A- 2 7, 1483.
Gent, A.N., Wang, C., 1991. J. Mater. Sci. 26, 3392.
Gent, A.N., Zhang, L.-Q., 2002. Rubber Chem. Technol. 75, 923.
Gent, A.N., Lindley, P.B., Thomas, A.G., 1964. J. Appl. Polym. Sci. 8, 455.
Gent, A.N., Liu, G.L., Sueyasu, T., 1991. Rubber Chem. Technol. 64, 96.
Gent, A.N., Lai, S.-M., Nah, C., Wang, C., 1994. Rubber Chem. Technol. 67, 610.

Gent, A.N., Razzaghi-Kashani, M., Hamed, G.R., 2003. Rubber Chem. Technol. 76, 122.
Gorokhovskii, G.A., Chernenko, P.A., Smirnov, V.A., 1972. Sov. Mater. Sci. 8, 557.
Gough, J., Muhr, A.H., 2005. In: Austrell, Kari (Eds.), Constitutive Models for Rubber IV. Taylor & Francis, London, pp. 51–57.
Green, A.E., Zerna, W., 1954. Theoretical Elasticity, Section 3.10 Oxford Univ. Press, London.
Greensmith, H.W., 1956. J. Polym. Sci. 21, 175.
Greensmith, H.W., 1963. J. Appl. Polym. Sci. 7, 993.
Greensmith, H.W., Thomas, A.G., 1955. J. Polym. Sci. 18, 189.
Greensmith, H.W., Mullins, L., Thomas, A.G., 1963. In: Bateman, L. (Ed.), The Chemistry and Physics of Rubberlike Substances. Wiley, New York (Chapter 10).
Griffith, A.A., 1921. Philos. Trans. R. Soc. (London), Ser. A 221, 163.
Griffith, A.A., 1924. In: Proceedings of 1st International Congress on Applied Mechanics, Delft, pp. 55–63.
Grosch, K.A., Schallamach, A., 1965. Trans. Inst. Rubber Ind. 41, 80.
Grosch, K.A., Harwood, J.A.C., Payne, A.R., 1966. Nature 212, 497.
Gurvich, M.R., 2011. Rubber Chem. Technol. 84, 354.
Halpin, J.C., 1965. Rubber Chem. Technol. 38, 1007.
Harbour, R.J., Fatemi, A., Mars, W.V., 2007a. Fatigue Fract. Eng. Mater. Struct. 30, 640.
Harbour, R.J., Fatemi, A., Mars, W.V., 2007b. Rubber Chem. Technol. 80, 838.
Harbour, R.J., Fatemi, A., Mars, W.V., 2008a. J. Mater. Sci. 43, 1783.
Harbour, R.J., Fatemi, A., Mars, W.V., 2008b. Int. J. Fatigue 30, 1231.
Harwood, J.A.C., Payne, A.R., Whittaker, R.E., 1970. J. Appl. Polym. Sci. 14, 2183.
Houwink, R., Janssen, H.H.J., 1956. Rubber Chem. Technol. 29, 4.
Huneau, B., 2011. Rubber Chem. Technol. 84, 425.
Inglis, C.E., 1913. Trans. Inst. Nav. Archit. (London) 55, 219.
Irwin, G.R., 1948. Fracturing of metals, Am. Soc. Met. Cleveland.
Irwin, G.R., 1957. J. Appl. Mech. 24, 361.
Kaliske, M., Dal, H., Fleischhauer, R., Jenkel, C., Netzker, C., 2010. Comput. Mech. 122, 1.
Kambour, R.P., 1973. J. Polym. Sci. Macromol. Rev. 7, 1.
Knauss, W.G., 1970. Int. J. Fracture Mech. 6, 183.
Knauss, W.G., 1974. In: Kausch, H.H., Hassell, J.A., Jaffee, R.I. (Eds.), Deformation and Fracture of Polymers. Plenum Press, New York, 501–540.
Lake, G.J., 1970. In: Proceedings of International Conference on Yield Deformation Fracture Polymers, Cambridge. Institute of Physics, London, p. 53.
Lake, G.J., Lindley, P.B., 1964. Rubber J. Int. Plast. 146, 24; 146, 30.
Lake, G.J., Lindley, P.B., 1965. J. Appl. Polym. Sci. 9, 1233.
Lake, G.J., Lindley, P.B., 1966. J. Appl. Polym. Sci. 10, 343.
Lake, G.J., Thomas, A.G., 1967. In: Proc. R. Soc. (London) A 300, 108.
Lake, G.J., Thomas, A.G., 2001. Strength. In: Gent, A.N. (Ed.), Engineering with Rubber, second ed. Hanser Publishers, Munich (Chapter 5).
Lake, G.J., Yeoh, O.H., 1978. Int. J. Fracture Mech. 14, 509.
Lake G.J., Yeoh, O.H., 1980. Rubber Chem. Technol. 53, 210.
Lake, G.J., Yeoh, O.H., 1987. J. Polym. Sci. 25, 1157.
Landel, R.F., Fedors, R.F. 1966. In: Proceedings of the 1st International Conference on Fracture, Sendai, vol. 2. Japan. Soc. Strength and Fracture of Materials, Tokyo, p. 1247.
Liang, H., Fukahori, Y., Thomas, A.G., Busfield, J.J.C., 2009. Wear 266, 288.
Liang, H., Fukahori, Y., Thomas, A.G., Busfield, J.J.C., 2010. Wear 268, 756.
Lindley, P.B., 1972. J. Strain Anal. 7, 132.
Lindley, P.B., 1973. Int. J. Fracture, 9, 449.
Lindley, P.B., Teo, S.C., 1979. Plast. Rubber Mater. Appl. 4, 29.
Lindley, P.B., Thomas, A.G., 1962. In: Proceedings of the Fourth International Rubber Conference, London, pp. 428–442.
Lindsey, G.H., 1967. J. Appl. Phys. 38, 4843.
Mark, J.E., Tang, M.-Y., 1984. J. Polym. Sci., Polym. Phys. Ed. 22, 1849.
Mars, W.V., 2002. Rubber Chem. Technol. 75, 1.
Mars, W.V., 2007. Rubber Chem. Technol. 80 (3), 481.
Mars, W.V., 2009. Rubber Chem. Technol. 82, 51.
Mars, W.V., 2011. Rubber Chem. Technol. 84, 178.

Mars, W.V., Fatemi, A., 2002. Int. J. Fatigue 24 (9), 949.
Mars, W.V., Fatemi, A., 2003. Rubber Chem. Technol. 76, 1241.
Mars, W.V., Fatemi, A., 2006a. Rubber Chem. Technol. 79, 589.
Mars, W.V., Fatemi, A., 2006b. J Mater Sci, 41, 7324.
Melenk, J., Babuska, I., 1996. Comput. Method. Appl. Mech. Eng. 39, 289.
Mott, P.H., Roland, C.M., 1999. Rubber Chem. Technol. 72, 769.
Mueller, H.K., Knauss, W.G., 1971. Trans. Soc. Rheol. 15, 217.
Mueller, R., Maugin, G.A., 2002. Comput. Mech. 29, 52.
Mullins, L., 1959. Trans. Inst. Rubber Ind. 35, 213.
Näser, B., Kaliske, M., Müller, R., 2007. Comput. Mech. 40, 1005.
Näser, B., Kaliske, M., Dal, H., Netzker, C., 2009. Z. Angew. Math. Mech. 89, 677.
Oberth, A.E., Bruenner, R.S., 1965. Trans. Soc. Rheol. 9, 165.
Orowan, E., 1949. Rep. Progr. Phys. 12, 185.
Papadopoulos, I.C., Thomas, A.G., Busfield, J.J.C., 2008. J. Appl. Polym. Sci. 109, 1900.
Parks, D.M., 1977. Comput. Method. Appl. Mech. Eng. 12, 353.
Remmers, J.J.C., de Borst, R., Needleman, A., 2008. J. Mech. Phys. Solids. 56, 70.
Rice, J.R., 1968. J. Appl. Mech. 35, 379.
Rivlin, R.S., Thomas, A.G., 1953. J. Polym. Sci. 10, 291.
Roberts, B.J., Benzies, J.B., 1977. In: Proc. Rubbercon '77, 2, 1.
Roland, C.M., Smith, C.R., 1985. Rubber Chem. Technol. 58, 806.
Saintier, N., Cailletaud, G., Piques, R., 2006a. Int. J. Fatigue 28, 61.
Saintier, N., Cailletaud, G., Piques, R., 2006b. Int. J. Fatigue 28, 530.
Sakulkaew, K., Thomas, A.G., Busfield, J.J.C., 2011. Polym. Test. 30, 163.
Schallamach, A., 1968. J. Appl. Polym. Sci. 12, 281.
Shih, C.F., Moran, B., Nakamura, T., 1986. Int. J. Fract. 30, 79.
Shivakumar, K.N., Tan, P.W., Newman, J.C., 1988. Int. J. Fract. 36, R34.
Smith, T.L., 1958. J. Polym. Sci. 32, 99.
Smith, T.L., 1963. J. Polym. Sci. A 1, 3597.
Smith, T.L., 1969. In: Eirich F.R. (Ed.), Rheology, vol. 5. Academic Press, New York, (Chapter 4).
Smith, T.L., Chu, W.H., 1972. J. Polym. Sci. A-2 10, 133.
Song, J.H., Areias, P.M.A., Belytschko, T., 2006. Int. J. Numer. Meth. Eng. 67, 868.
Southern, E., Thomas, A.G., 1978. Plast. Rubber Mater. Appl. 3, 133.
Steinmann, P., 2000. Int. J. Solids Struct. 37, 7371.
Stevenson, A., 1983. Int. J. Fracture Mech. 23, 47.
Tarkow, H., 1958. J. Polym. Sci. 28, 35.
Thomas, A.G., 1955. J. Polym. Sci. 18, 177.
Thomas, A.G., 1958. J. Polym. Sci. 31, 467.
Thomas, A.G., 1960. J. Appl. Polym. Sci. 3, 168.
Thomas, A.G., Whittle, J.M., 1970. Rubber Chem. Technol. 43, 222.
Timmel, M., Kaliske, M., Kolling, S., 2009. Z. Angew. Math. Mech. 89, 698.
Veith, A.G., 1965. Rubber Chem. Technol. 38, 700.
Verron, E., 2010. Rubber Chem. Technol. 83, 270.
Vinogradov, G.V., Mustafaev, V.A., Podolsky, Y.Y., 1965. Wear 8, 358.
Williams, M.L., Schapery, R.A., 1965. Int. J. Fracture Mech. 1, 64.
Yeoh, O.H., 2002. Mech. Mater. 34, 459.
Zine, A., Noureddine B., Naït Abdelaziz, M., 2011. Int. J. Fatigue 33, 1360.

Chapter 11

The Chemical Modification of Polymers

A.F. Halasa[*], **Jean Marie Massie**[†] **and R.J. Ceresa**[‡]

[*] *Research and Development, The Goodyear Tire & Rubber Company, Akron, OH, USA*
[†] *Lexmark International, Lexington, KY, USA*
[‡] *Chemistry and Polymer Technology Department, Polytechnic of South Bank, London, UK*

11.1 INTRODUCTION

The terms *rubber* and *elastomer* embrace those polymers that have useful rubberlike, highly elastic properties at ambient temperatures; however, many polymers that are nonrubbers by themselves can be chemically modified to a relatively small extent to give products with very useful viscoelastic properties. For example, the introduction of a few chlorine atoms and sulfoxide groups into polyethylene changes the macromolecules so that they no longer tend to form crystalline regions, but become elastomers over a wide temperature range. These groups also allow vulcanization so that reversibly deforming, solvent-resistant products can be formulated. On the other hand, starting with a conventional rubber, it is possible, by means of chemical reactions, to convert the macromolecular chains so that they lose their viscoelasticity and become thermoplastic at ambient temperatures (e.g., the cyclization of polyisoprenes). Quite apart from the question of the temperature at which observations are carried out, the borderline between the viscoelastic and the plastic states is a relatively ill-defined one. The common factor of rubbers and plastics is, of course, their macromolecular nature. Now that we have a better understanding of their structure at a molecular level (not forgetting rubber-modified plastics and "plasticized" rubbers), we are able to say that, with some exceptions, a modification that can be carried out on one polymer species may, under suitable conditions, be carried out on a polymer of a different species. Whether the same type of chemical modification will give us the properties we are seeking without the loss of properties we would wish to retain is a matter, first, for conjecture and, subsequently, for experimental verification.

In the early days of polymer chemistry and technology, a new chemical modification successfully applied to one polymer was quickly evaluated with a

The Science and Technology of Rubber. http://dx.doi.org/10.1016/B978-0-12-394584-6.00011-X

whole range of chemically similar reagents (e.g., the esterification of cellulose), and the same type of reaction was attempted with the available range of similar material. Today, the number of homopolymers available runs into the hundreds, and the number of random copolymers runs into the thousands; therefore, the number of possible chemical modifications, including block and graft copolymers, runs into astronomical figures. Only a fraction of these potential systems has been evaluated (and then frequently only in a superficial way to obtain patent coverage) so that the field is wide open for research and development. Because of its breadth and depth, the field of the chemical modification of polymers can be treated only in outline in a single chapter, and only the more important reactions can be described. To serve the interest of the reader, however, this broad survey is punctuated by discussions in greater depth of areas that are of interest to rubber chemists and that, in the opinion of the writer, have considerable potential for further development. The emphasis is on the principles underlying the chemical modification of polymers; specific details of reactions conditions are deliberately omitted. A number of aspects of the chemical modification of rubbers have been covered in detail elsewhere in this book, and the reader is referred to the appropriate chapters on chemistry, vulcanization, characterization,block copolymers, and so on.

11.2 CHEMICAL MODIFICATION OF POLYMERS WITHIN BACKBONE AND CHAIN ENDS

Polymer properties are dependent on many factors, including chain end interactions with substrates such as carbon black or silica fillers, as well as clay and calcium carbonate. At this point, there is a large volume of work that has been done on chain end modification, particularly those made by anionic polymerization with group I or group II metals (Bielinsk et al., 1995; Yamato and Oahu, 1996; Schulz et al., 1974), as seen later.

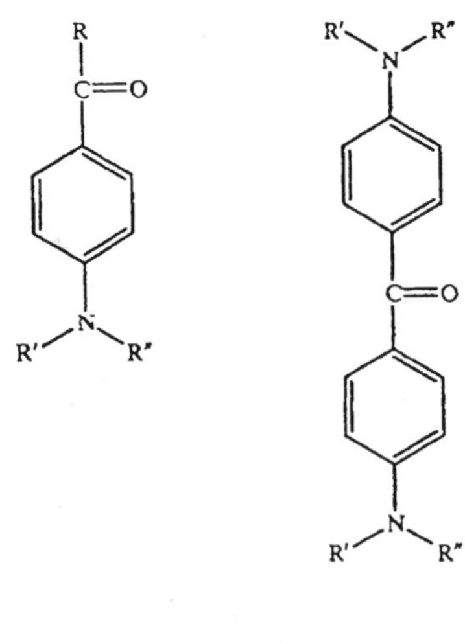

This approach was made possible via the reaction of metallic chain ends with active metal halide as shown on the following page, such as tin tetrachloride and silica terachlorides. These types of reactions led to increased molecular weight as well as improvements in polymer filler interactions that result in improved properties of the polymers (Urneck and Short, 1970).

Modification of polymers through the introduction of polar moieties such as amine siloxy groups made by anionic catalysts or so-called "living polymerizations" is made by either functional initiators or by chain functional monomers, as has been reported by several groups. The functional initiators are cyclic amines attached to alkyl groups and the functional monomers are based on styrene and disoproenyl benzene (Halasa, 1981b; Yoshioka et al., 1987; Uda Brain, 1992; Bohm and Graves, 1988; Halasa and Hsu, 2004).

More recently, a US Patent has been issued to the Goodyear Tire & Rubber Company, which claimed that polar functional monomers could be copolymerized with conjugated dienes and vinyl aromatics to chemically modify the polymer chain. Functional monomers, such as 3-(2-pyrrolidinoethyl) styrene:

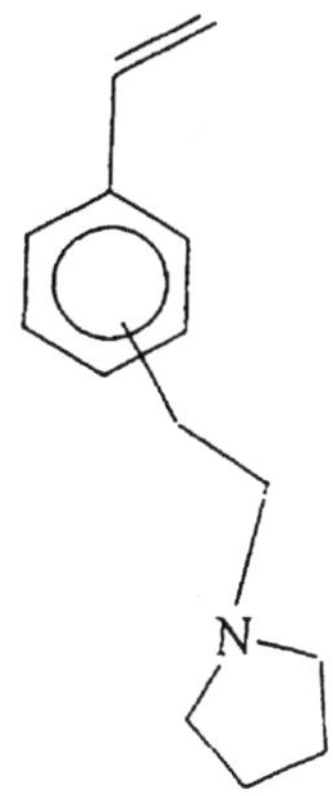

or 3-(pyrrolidino-2-methylethyl) α-methylstyrene:

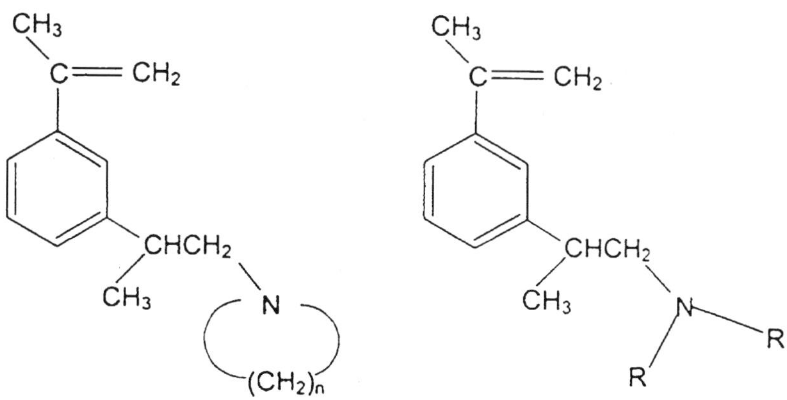

have been copolymerized with solution and emulsion styrene butadiene copolymers in ratios from 1% to 10% and have imparted major improvements in polymer properties resulting in lower hysteresis and better wet grip of tires (Hsieh and Quirk, 1996; Hall and Antikowiak, 1955).

11.3 ESTERIFICATION, ETHERIFICATION, AND HYDROLYSIS OF POLYMERS

The chemical modifications discussed in this section are historically and scientifically so closely linked to one polymer, cellulose, that although the latter occurs primarily as a fiber and not an elastomer, a discussion of this group of cellulose modifications seems appropriate. Apart from the fact that some cellulose derivatives, like ethyl cellulose, when plasticized, can be quite elastomeric, the effects of modification of a basic polymer are particularly well demonstrable on a substance as stiff and highly crystalline as cellulose. Moreover, in view of the expected hydrocarbon shortage, cellulose may soon gain a new role as a polymeric starting commodity.

Cellulose, identified chemically as β-1,4-glucan, is the most widely found natural polymer, constituting the permanent structure of plant cell walls. For the general properties and chemistry of cellulose itself the reader is referred to standard textbooks and recent reviews.

Much of the early history of the chemical modification of cellulose is related to the attempts to find a solvent for it, as its macromolecular structure was not understood at that time. In 1844, Mercer discovered and commercialized the interaction of alkali with cellulose fibers, a process still in use under the name *mercerization*. The initial product of the reaction, alkali cellulose, is not a chemical modification but a physical form in which water and sodium ions penetrate the macromolecular structure and reduce the hydrogen bonding, with consequent swelling of the fibers. The initial product, cellulose I, is converted to cellulose II, a complex physicochemical modification, in the final washing stage. The degree and rate of swelling in this process are dependent on the source of the cellulose, and if the fibers are stretched prior to and during the reaction, optimum interaction is achieved. Many other inorganic salt solutions swell cellulose (Haskell, 1965), and of these, zinc chloride has found the widest application. Aqueous solutions of thiourea, resorcinol, chloral hydrate, and benzene sulfonates also lead to limited swelling of cellulose. In all cases, the reduction in physical crosslinking can be followed by a study of the X-ray diffraction diagrams of the crystalline content.

The complete solubility of cellulose in cuprammonium solutions, discovered in 1857 by Schweizer, led to the development of the rayon industry, but, as in the case of alkali cellulose, the regenerated polymer is chemically the same as the precursor. Regeneration via cellulose xanthate solutions, invented by Cross and Bevan in 1893, is another process still in use; it forms the basis for the manufacture of Cellophane.

The first "chemical" modification of cellulose was achieved by Braconnot in 1833 with the production of cellulose nitrate from a wide range of cellulosic materials. The products were highly inflammable powders that could be dissolved in concentrated acetic acid to give clear tough varnishes. (Note the conversion of a fiber to a film by chemical modification.) In 1847, highly nitrated cellulose, guncotton, was discovered by Bottger and Schonbein, and in 1870, Schutzenberger produced acetylated cellulose using hot acetic anhydride as the reaction medium. These reactions have a common mechanism (Bruxelles and Grassie, 1965), namely, the esterification of the hydroxyl groups in the basic cellulose moiety:

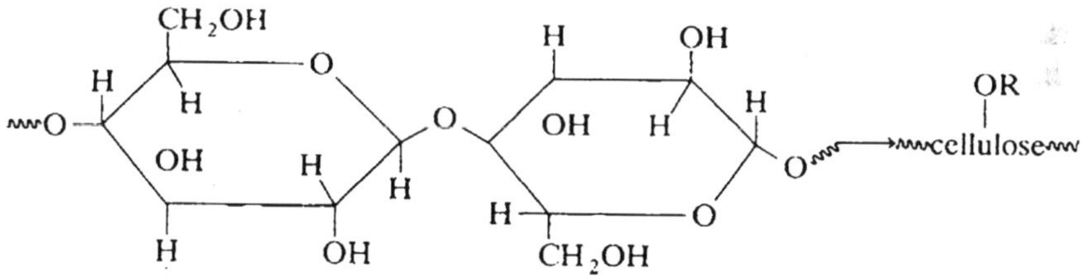

(11.1)

Since this early work, a very large range of organic acids has been used to prepare cellulose esters, mixed esters, and ether esters (Rouse, 1965). A typical example of considerable commercial importance is the acetylation of cellulose. As in all esterifications of macromolecular materials, the accessibility of the hydroxyl groups to the esterifying acid is of prime importance. Reaction (11.1) represents complete esterification, a process that is probably never fully achieved. The identification of the esterified products is, therefore, dependent not only on the content of acetyl groups but also on the location of these groups on the macromolecular backbone. Both factors are affected by the method of preparation and the esterification conditions.

Although many esterification reactions (Bruxelles and Grassie, 1965) are based on inorganic acids, for insoluble hydroxyl compounds like cellulose, xanthation is more important. Sodium hydroxide normally is used to produce the swollen alkali cellulose, which (after aging) is reacted with carbon disulfide to form the sodium salt of cellulose xanthate:

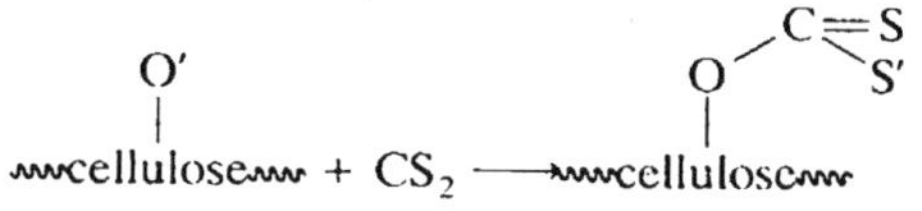

(11.2)

The cellulose can be regenerated by spinning (or extruding as a film) into an acid bath containing salts such as sodium and zinc sulfate (Haskell, 1965).

During spinning or extrusion, the macromolecules are oriented in the direction of flow to give high strength to the viscose fiber or the Cellophane film. The occurrence of macromolecular orientation during spinning is very important, and it is used in the chemical modification of many polymers.

The cellulose ethers constitute another important group of cellulose derivatives prepared from alkali cellulose by standard etherification reactions between the hydroxyl groups and an alkyl halide. The properties of the ethers depend on the extent of the reaction; that is, the degree of etherification. In general, the ethyl celluloses are water-insoluble thermoplastic materials, whereas methyl ether, ethyl hydroxyethyl cellulose, and carboxymethyl cellulose are soluble in cold water and are used as viscoelastic thickeners and adhesives.

For the preparation of synthetic hydroxy polymers, hydroxyl groups can be introduced by copolymerization of the base monomer with a hydroxy monomer. These groups can then be used for esterification or etherification (Jarowenko, 1965), but the relatively high cost of hydroxy monomers detracts from the widespread use of direct copolymerization. Instead, one introduces the groups required by the complete or partial hydrolysis of the ester groups in an appropriately hydrolyzable polymer, such as poly(vinyl acetate). Complete hydrolysis yields poly(vinyl alcohol) (PVA), a water-soluble polymer with considerable utility as a stabilizer and viscosity modifier for aqueous systems:

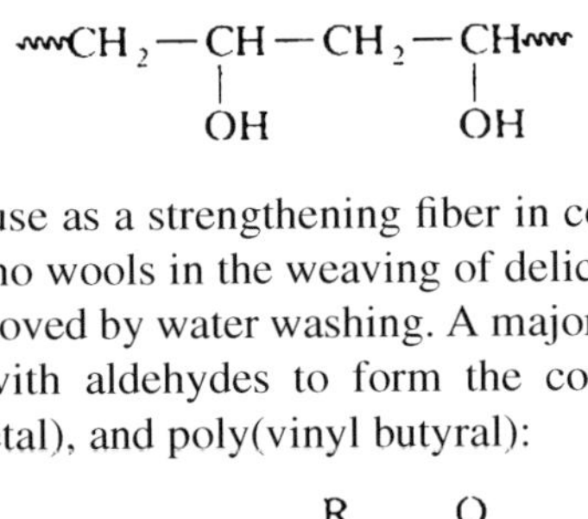
(11.3)

PVA has a unique use as a strengthening fiber in conjunction with weaker materials such as merino wools in the weaving of delicate fabrics, from which it can afterward be removed by water washing. A major portion of the polymer produced is reacted with aldehydes to form the corresponding poly(vinyl formal), poly(vinyl acetal), and poly(vinyl butyral):

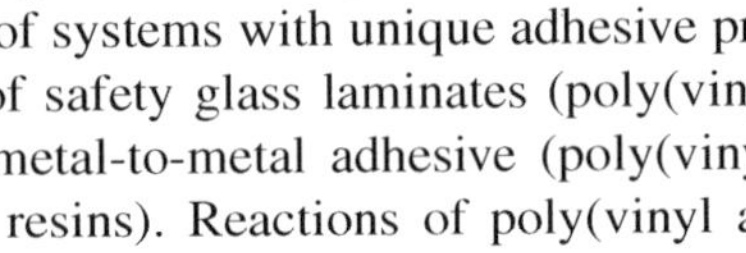
(11.4)

There are different grades of each of these materials according to the overall molecular weight and the degree of substitution. These polymers are used as components of systems with unique adhesive properties; for example, in the manufacture of safety glass laminates (poly(vinyl butyral) and mixed derivatives) and of metal-to-metal adhesive (poly(vinyl formal) cured with phenolics and other resins). Reactions of poly(vinyl alcohol) with acids or anhydrides occur as normal esterifications, a route used to synthesize polymers and copolymers that cannot be readily formed by conventional polymerization (e.g., when the reactivity ratios of the monomers are not suitable).

Natural rubber and synthetic rubbers in general do not have hydroxyl groups in sufficient numbers for them to be used for esterification reactions. Terminal hydroxyl groups may be introduced into synthetic rubbers as terminal catalyst or initiator fragments and used for coupling or extension reactions.

11.4 THE HYDROGENATION OF POLYMERS

Any polymer with unsaturated hydrocarbon groups, either in the main chain or as side groups, can be hydrogenated. Early research on the hydrogenation of elastomers focused on destructive hydrogenation with consequent loss of the macromolecular structure. This is beyond the scope of this chapter; refer to several references on the subject (Hall and Antikowiak, 1955; Wichlatz, 1964; Yakubchik et al., 1962). The most recent work in hydrogenation has produced excellent products, such as linear polyethylene from the hydrogenation of poly(1,4-butadiene) (Rachapudy et al., 1979; Halasa, 1981a) and poly(ethylene-co-propylene) rubber from the hydrogenation of polyisoprene

$$\text{+}CH_2-C=C-CH_2\text{+} \xrightarrow{H_2} \text{+}CH_2-CH_2CH_2-CH_2\text{+}_n \quad (11.5)$$

$$\left(CH_2-\underset{\displaystyle CH_3}{\underset{|}{C}}=C-CH_2\right) \xrightarrow{H_2} \left(CH_2-\underset{\displaystyle CH_3}{\underset{|}{C}}-CH_2-CH_2\right)_n \quad (11.6)$$

These reactions were carried out using a heterogenous catalyst.

Homogeneous soluble transition metal catalysts for hydrogenation have been used to create novel polymers. Homogeneous hydrogenation catalysts usually are generated from Ziegler-type catalysts based on nickel or cobalt organic salts reduced in the presence of organoaluminum or organolithium compounds. These catalysts are used to form saturated elastomers by hydrogenating unsaturated elastomers. The resulting polymers have vastly different viscoelastic properties than their unsaturated parent polymers. For example, the hydrogenation of a 99% poly(1,2-butadiene) has resulted in the formation of polybutene (Halasa and Massie, 1993), which has a lower glass transition temperature than its parent elastomer. It is interesting to note that hydrogenation does not affect the polymer molecular weight or backbone architecture.

The ease of hydrogenation and the resultant degree of saturation achieved reflect the microstructure of the polymer. Hydrogenation of unsaturated elastomers usually proceeds in a blocky way. This is due to the different reactivities of the various double bonds. In general, double bonds of 1,2 structure are four times more reactive than double bonds of 1,4 structure. *cis*-1,4 units are more reactive than *trans*-1,4 units. Chamberline et al. (1981) have shown that hydrogenation of 1,2 units is statistically random, whereas the hydrogenation of 1,4 units is not.

The complete hydrogenation of poly(1,4-butadiene), *cis* or *trans* structure, forms a polyethylene with a low melting point of about 115°C. It is believed that this linear polyethylene is a low-density material. Unpublished data obtained on the partial hydrogenation of a 99% *cis* poly(1,4-butadiene) showed that the hydrogenation proceeded in a blocky fashion. The 40–50% hydrogenation of poly(1,4-butadiene) *(cis* content 98%) made by nickel catalyst gives a polymer with a melting point of +98°C and a crystallization temperature of −13°C as measured by differential scanning calorimetry (DSC). This confirms the fact that hydrogenation of *cis*-poly-1,4-butadiene proceeds in a blocky fashion to produce a block of polyethylene and a block of poly(*cis*-1,4-butadiene).

Polybutadienes made by anionic catalysts in the presence of polar modifiers contain a mixed microstructure of *cis*-1,4, *trans*-1,4, and 1,2 units. Hydrogenation of these polymers leads to interesting products. As mentioned previously, hydrogenation favors the 1,2 units over the 1,4 units by a 3- (or 4)-to-1 ratio. Because of this mismatch in reactivity, hydrogenation of a polybutadiene containing 40–50% 1,2 units produces a polymer containing a polyethylene portion with a T_m of 85–95°C and a rubbery portion with a T_g of −62°C.

Block copolymers can also be hydrogenated to produce unique products. Hydrogenated triblock copolymers of poly(styrene-co-butadiene-co-styrene) (SBS) are commercially available from the Shell Company under the trade name Kraton G. The middle block is usually a mixed microstructure of poly(1,2-butadiene) and poly(1,4-butadiene) units. The resulting product is a hydrogenated unsaturated polymer, which exhibits greater thermal and oxidative properties than the parent SBS triblock.

Similar procedures have been used by several workers (Halasa et al., 1982) to hydrogenate poly(1,4-butadiene-co-1,2-butadiene) diblocks (Halasa, 1985) and poly(l,4-butadiene-co-1,4-isoprene-co-1,4-butadiene) triblocks. Hydrogenation of these diblock and triblock copolymers forms thermoplastic elastomers with crystalline and amorphous segments. All these materials exhibit crystallinity, glass transition, solubility, and dynamic mechanical loss spectra different from those of their unsaturated counterparts.

Another method of preparing saturated rubbers was developed (Harwood et al., 1973) using the diimide reduction. This method can be used to produce a high degree of saturation dependent on the type of reagent used; however, side reactions can occur in this method. Generation of the diimide from *p*-toluenesulfonyl hydrides leaves an acidic fragment that may cause cyclization in some unsaturated elastomers.

11.5 DEHALOGENATION, ELIMINATION, AND HALOGENATION REACTIONS IN POLYMERS

11.5.1 Dehydrochlorination of Poly(vinyl chloride)

The dehydrochlorination of poly(vinyl chloride) has been the subject of much investigation, particularly with the view of developing greater stability in PVC polymers and copolymers. Like many polymeric reactions, dehydrochlorination is a complex process. The vinylene groups, created by the elimination of HCl from adjacent carbon atoms in the chain,

$$\sim\sim CH_2{-}CHCl{-}CH_2{-}CHCl \sim\sim \longrightarrow HCl + \sim\sim CH{=}CH{-}CH_2CHCl\sim\sim \quad (11.7)$$

may be the result of free radical ionic, or ion-radical steps. The presence of a small proportion of head-to-head, tail-to-tail, and other configurational irregularities in the backbone structure of poly(vinyl chloride) leads to more complex elimination steps by thermal degradation (alone or in the presence of catalysts such as aluminum chloride). The introduction of ring structures, a major process during dehydrochlorination, is likewise affected by the distribution of the chlorine atoms along the polymeric backbone. Hydrogen bromide can be effectively eliminated thermally from poly(vinyl bromide), as dehydrohalogenation is a universal thermal reaction, the complexities of which increase from chloride to iodide.

11.5.2 Thermal Elimination

The thermal elimination process can be applied to most "substituted" groups in vinyl polymers by controlled pyrolysis at 600–700°C, producing polyvinylene compounds, for example, by the splitting off of acetic acid from poly(vinyl acetate). By careful temperature control, one can achieve bifunctional reactions and/or intramolecular cyclizations. This has been developed commercially at relatively high temperatures, in the case of the polymerization of methacrylamide above 65°C, to yield a polymer with a substantial proportion of imide groups:

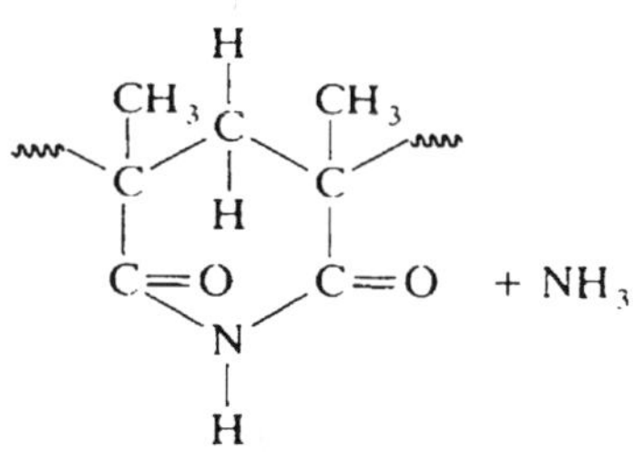

(11.8)

Polymers and copolymers of multifunctional vinyl monomers (using the term to cover the presence of a halogen or other reactive group in addition to the vinyl group, rather than in the sense of more than one polymerizable group), such as α-chloroacrylic acid, often undergo partial lactonization and hydrolysis during polymerization. Heating in alcohol solution or electrolyzing alcoholic solutions, one obtains, for example, the introduction of double lactam rings during the acid hydrolysis of poly(α-acetamineoacrylic acid)

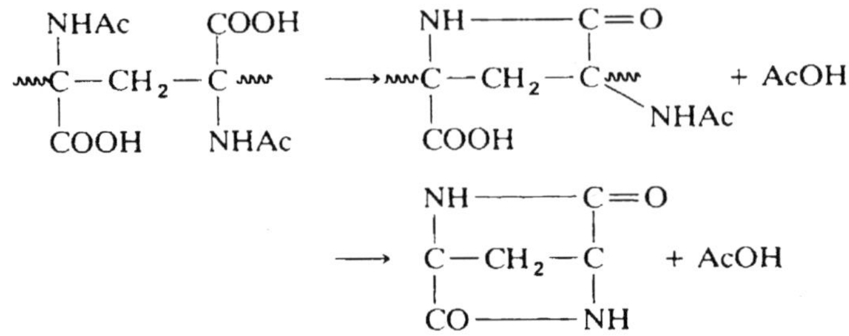

(11.9)

The discoloration of polyacrylonitrile is due to a similar type of elimination reaction, which in this case occurs intra- as well as intermolecularly to give crosslinked insoluble ring products:

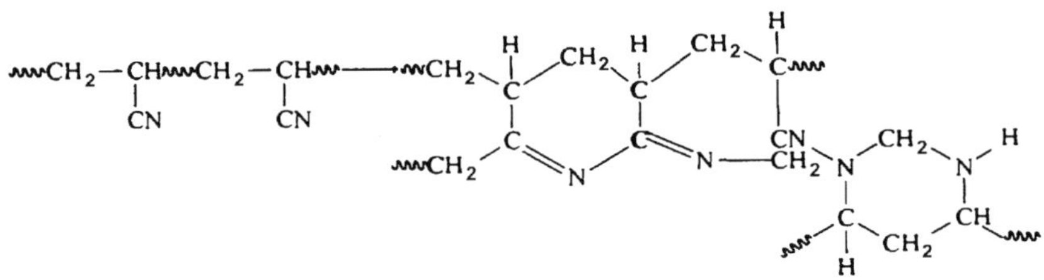

(11.10)

The controlled heating of polyacrylonitrile fibers under tension also causes an elimination of nitrogenous products to leave a "carbon fiber" of high tensile strength that can be considered as the end product of the line of chemical elimination reactions. Carbon fibers from cellulosic materials, lignin, and various interpolymers and blends have been developed. The structures of these products consist largely of three-dimensional carbon networks, partially crystalline and partially graphitic or amorphous.

11.5.3 Halogenation of Polymers

Halogenation and hydrohalogenation of elastomers have been reported extensively in the literature (Poutsma, 1969). The main problems with these reactions are the cyclization and chain scission that occur parallel to the halogenation reaction. These introduce difficult problems in the characterization of the resulting products. Despite these problems, several products have been prepared and commercialized. Chlorination of poly(1,4-butadiene) to prepare a product similar to poly(vinyl chloride) has been reported by several workers (Bevington and Ratt, 1975). This process had extensive side reactions and chain degradation. The chlorination of butyl rubber and conjugated diene-butyl rubbers gives end products that are used in the tire industry as inner liners for air retention.

Ethylene-propylene copolymers (EPDM) are, by their random copolymerization, amorphous in structure and therefore easily halogenated. EPDM has been chlorinated to improve its properties and cocurability with other rubbers. The chlorination was directed toward the termonomer dicyclopentadiene to form the allylic chloride (Schoen et al., 1975). In this manner, EPDM was chlorinated, and the resulting products had improved properties.

EPDM rubbers are modified by 1,2-addition of *N*-chlorothiosulfonamides to their olefinic sites (Hopper, 1975, 1976; Hopper et al., 1984). Such additions may be carried out in solution or without solvent in an internal mixer or extruder. The solventless reactions are facilitated by added carboxylic acids (Hopper, 1989, 1990) or by certain metal salts of weak acids (White et al., 1990). The modified EPDMs are of interest because of their ability to covulcanize in ozone-resistant blends with polydiene rubbers (Hopper, 1975, 1976; Hopper et al., 1984). Although less fully explored, *N*-chlorothiocarboxamides and imides also react with EPDM to produce modified products that covulcanize in blends with polydienes (Hopper, 1977).

In the absence of oxygen, the chlorination of polyethylene, with or without a catalyst, can be controlled to provide products with varying chlorine content. The chlorination process is statistically random so that chlorination of polyethylene to the same chlorine content as poly(vinyl chloride) (60%) gives a product that is chemically different from PVC yet fully compatible with it. This random chlorination of polyethylene destroys its crystallinity. At a degree of chlorination corresponding to the loss of all its crystallinity, the chlorinated product becomes soluble at room temperature. The *p*-bromination of polyethylene follows a similar course to yield a rubberlike polymer at 55% bromine content.

Both chlorination and bromination of polypropylene and isobutylene lead to degradation of the main chain, with the loss of many useful properties. Degradation during chlorination can, however, be avoided at low temperatures by limiting the reaction to a maximum of about 2%. This procedure forms a useful commercial product.

The addition of hydrogen chloride to unsaturated elastomers has also received considerable attention. Extensive work has been done on the hydrochlorination of *Hevea* [poly(*cis*-1,4-1,4-isoprene)] and *Balata* [poly(*trans*-1,4-isoprene)] rubbers since 1940 (Staudinger, 1944; Gordon and Tyler, 1953). Both *cis*-1,4 and *trans cis*-1,4-polyisoprenes readily add hydrogen chloride following Markovnikov's rules with only a small amount of cyclization.

11.5.4 Cyclization of Polymers

Cationic cyclization of unsaturated elastomers such as poly(*cis*-1,4-isoprene), poly(3,4-isoprene), poly(1,2-butadiene), and poly(1,4-butadiene) usually leads to the formation of cyclized resinous products of no commercial value. An extensive review on the subject has been published by Schults et al. (1983). Cyclization of unsaturated elastomers, such as polyisoprene, can be carried out in the solid state, in solution, or even in the latex. The process involves the transformation of linear macromolecular chains into much shorter ones consisting of mono-, di-, tri-, and tetrapolycyclic groups distributed randomly along the chain. Cyclization of polyisoprene increases its glass transition temperature by 20–30°C. The mechanism of cyclization of elastomers depends on the catalyst employed in the process.

11.6 OTHER ADDITION REACTIONS TO DOUBLE BONDS

11.6.1 Ethylene Derivatives

Besides the addition of halogens and hydrohalogens "across the double bond" just covered, there are many other reagents that will react similarly with unsaturated polymers by free radical, ionic, or radical-ion mechanisms. Of prime importance is the addition of ethylene derivatives to polydienes. One of the earliest reactions of natural rubber to be studied in detail was the combination with maleic anhydride (Cunneen and Porter, 1965). Depending on the reaction conditions and the presence or absence of free radical initiators, one or more of four basic reactions may take place, with the products shown (the arrows indicate where the addition has taken place and the new bonds formed).

1. Intramolecular addition to the double bond within polyisoprene chains:

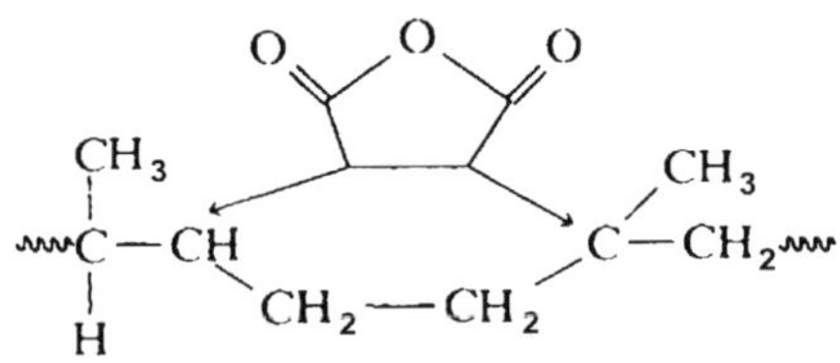

(11.11)

2. Intermolecular addition to double bonds in different polymer chains. In this group should be included the statistically possible reaction between widely separated double bonds within the same molecule:

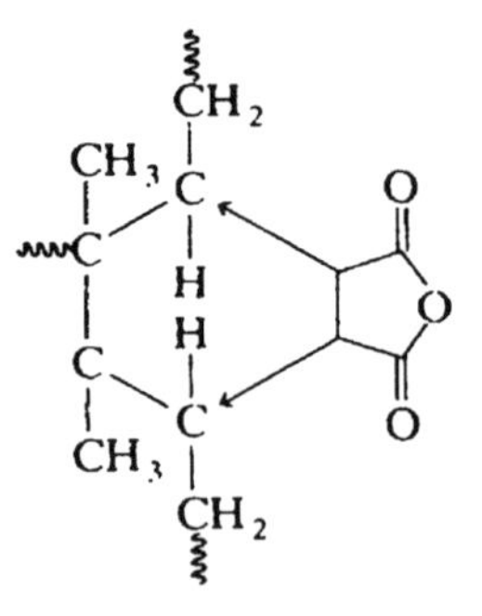

(11.12)

3. Addition to α-methylenic carbon atoms of a polyisoprene chain:

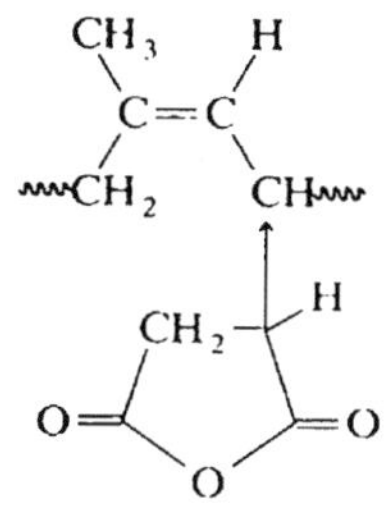

(11.13)

4. Intermolecular addition to α-methylenic carbon atoms in adjacent chains (or widely spaced α-methylenic carbon atoms in the same macromolecule):

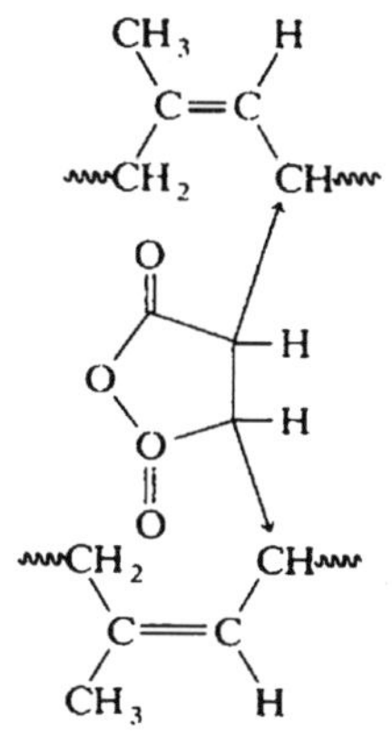

(11.14)

In general, the overall reaction rates increase with rising temperature and in the presence of oxygen or free radical initiators, but these same conditions promote intermolecular reactions leading to gel formation. Similar reactions take place with gutta-percha, synthetic poly(*cis*-1,4-isoprene), and poly(*cis*-1,4-butadiene). Many workers used two-roll mills and other mastication techniques as convenient ways of blending the maleic anhydride with rubbers at elevated temperatures, but where these techniques have been used, mechanochemical reactions have complicated the overall process. Reaction products of natural rubber containing 5–10% combined maleic anhydride can be vulcanized by conventional sulfur cures: of greater interest is the possibility of creating crosslinking by the use of oxides of calcium, magnesium, and zinc.

Other compounds reacting similarly via activated double bonds (excluding here block or graft copolymerization) include maleic acid, *N*-methyl-maleimide, chloromaleic anhydride, fumaric acid, γ-crotonolactone, *p*-benzoquinone, and acrylonitrile. Other polymers with unsaturated backbones, such as polybutadiene, copolymers of butadiene with styrene and with acrylonitrile, and butyl rubber, react in similar ways, but the recorded reaction with poly(vinyl chloride) is largely mechanochemical in nature (discussed later).

11.6.2 The Prins Reaction

Another addition to polymers with main-chain unsaturation is the Prins reaction between ethylenic hydrocarbons and compounds containing aldehydic carbonyl groups. Kirchof in 1923, described the reaction of natural rubber in benzene solution with aqueous formaldehyde in the presence of concentrated sulfuric acid. The general reaction of an aldehyde, RCHO, with a polyisoprene in the presence of an inorganic or organic acid or an anhydrous metal salt, is represented by

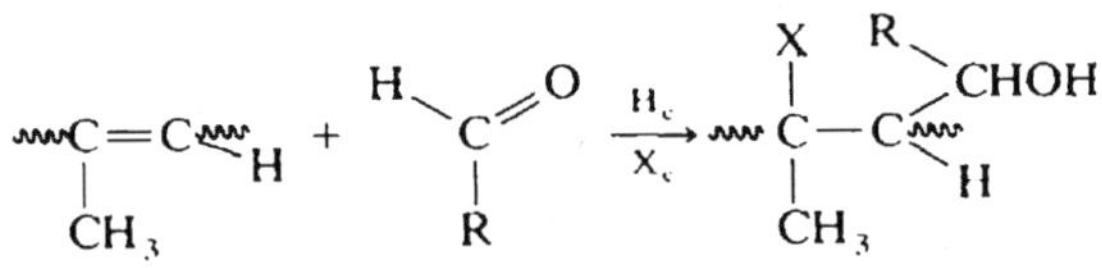

(11.15)

In the absence of such catalysts, the reaction leads to a shift in the double bond rather than its elimination:

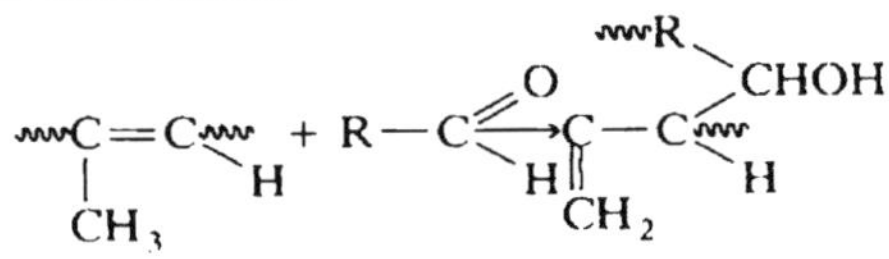

(11.16)

These reactions can be carried out in solution or in dispersion or by reaction in the solid phase (Cunneen and Porter, 1965); in the last case it is again difficult to differentiate the Prins reaction from mechanochemical reactions initiated by chain rupture during mastication.

Other aldehydic compounds, such as glyoxal and chloral, also react in a similar way with polyisoprenes and unsaturated rubbers (e.g., poly(*cis*-1,4-isoprenes), poly(*cis*-1,4-butadiene), and copolymers of isobutylene and isoprene). The use of strong acids, or Lewis acids, causes complications, as the acids themselves, under suitable conditions, catalyze cyclization and *cis-trans* isomerization, and these reactions may occur simultaneously with the addition reactions.

11.7 OXIDATION REACTIONS OF POLYMERS

Uncontrolled oxidation of rubber is detrimental to its physical properties. Oxidation reactions take place readily at unsaturated groups in polymers and are often referred to collectively as *epoxidation*; however, oxidation under controlled conditions can lead to useful products such as the epoxidized natural rubber introduced by the Malaysian Rubber Producers Association (Schults et al., 1983; Cunneen and Porter, 1965; Ceresa, 1965; Avery and Watson, 1956). Natural rubber in the latex form is treated with hydrogen peroxide dissolved in acetic acid. This gives 50% epoxidized natural rubber. This rubber shows very interesting physical properties and excellent carbon black dispersion. Similarly, nonaqueous epoxidizations of synthetic polyisoprene can be achieved using either hydrogen peroxide or hypochloride in t-butanol. The controlled degree of epoxidation usually leads to some interesting products. For example, the 25% epoxidized synthetic polyisoprene is an elastomer with viscoelastic properties similar to those of the unepoxidized material, but has better carbon black dispersion. This gives high modulus and tensile strength; however, higher degrees of epoxidation (60–75%) produce a resinous material that is not rubbery.

11.8 FUNCTIONALIZATION OF POLYMERS

Polymers with stable backbones such as polystyrene, polyethylene, and polypropylene can be functionalized. Functionalization of polystyrene has received considerable attention, because it is a unique polymer with aromatic rings capable of undergoing many nucleophilic as well as electrophilic reactions. A resin recently introduced on the market is based on sulfonated polystyrene. Applications for this resin include ion-exchange material and catalyst binding materials.

Electrophilic substitution on polystyrene through a chlorometallation reaction yields chlorine functionality. This has opened up the possibilities of making many derivatives of polystyrene. Starting with chlorometallated polystyrene, derivatives such as quaternary, ammonium, or phosphonium salts have been made. Similarly, ethers, esters, sulfonamides, silanes, and ketone derivatives have been made by replacing the chlorine atom on chlorometallated polystyrene. In the case of polystyrene, however, it was discovered that chain end functionalization can be realized if the chain ends were terminated by group I metals such as lithium and potassium.

Both the Japanese Synthetic Rubber Company and Nippon Zeon have reported that anionically prepared elastomers that are functionally terminated by active lithium can be chain terminated with Michler ketone, benzophenone, and a variety of enamide groups. Moreover, these chains can be terminated with silicone or tin metals. Chain end functionalization did not change the viscoelasticity of the polymer chains but rather dramatically improved the elastomer-filler interaction and, therefore, reduced its hysteretic properties.

11.9 MISCELLANEOUS CHEMICAL REACTIONS OF POLYMERS

Direct replacement of the hydrogen atoms of aromatic rings such as styrene or the allylic hydrogen in poly(1,4-butadiene) or poly(1,4-isoprene) can be carried out via metallation with organometallic compounds of group I such as lithium, sodium, and potassium. Usually, the yield tends to be low and the product is insoluble; however, the use of chelating diamines with organolithium compounds has increased the yield, and the products are soluble in cyclohexane. For example, polystyrene has been metallated in high yields to give polylithiated polystyrene in which several functional groups have been successfully introduced. Similarly, polyisoprene and polybutadiene have been successfully metallated with either s-butyllithium or t-butyllithium in the presence of tetramethylethylenediamine (TMEDA) at 50°C. In the case of the polyisoprene, chain scission and reduction in molecular weight resulted at longer metallation temperatures and times. In many cases, these lithiated polymers have been used to prepare graft and block copolymers. These are discussed in more detail in Section 11.9.

11.10 BLOCK AND GRAFT COPOLYMERIZATION

11.10.1 Effects on Structure and Properties of Polymers

Some of the most significant changes in structure and properties of polymers can be brought about by either block or graft copolymerization (see Chapter 13). The term *block copolymer* is applied to macromolecules made up of sequences with different chemical (or physical, i.e., tactic) structures, usually represented by A, B, and C. The sequences are of a molecular weight that would give them polymeric features even if separated. The manner in which these sequences are arranged defines the type of block copolymer prepared. A diblock copolymer is represented by AB, indicating that a segment with chemical composition A is connected to a segment with composition B. Other possible types of block copolymers include triblocks ABA and ABC, for example. They may be linear or branched; the linear structures are called block copolymers,

$$\text{AAAAAA} \cdot \text{BBBB} \cdots \text{BBBBB} \cdot \text{AAAAA} \tag{11.17}$$

and the branched structures graft copolymers (11.18),

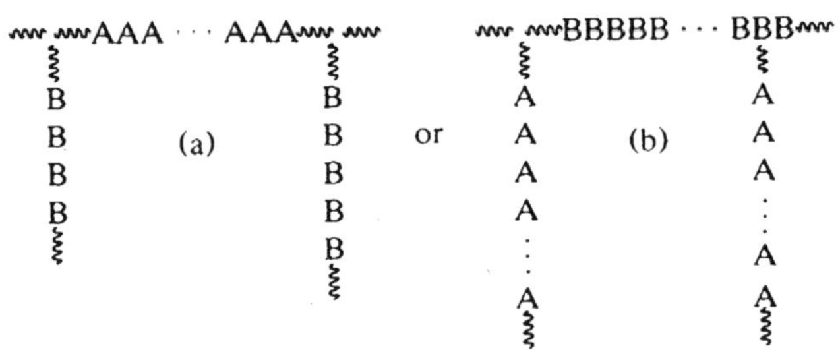

(11.18)

Thus the same polymeric sequences may be put together as block or as graft copolymers, with differing properties, though in the author's experience, the major differences between the properties of block and graft copolymers of the same constituent polymers are pronounced only in solution or in the melt. For example, natural rubber may be block copolymerized with poly(methyl methacrylate), or methyl methacrylate monomer may be grafted onto natural rubber. In an attempt to distinguish by nomenclature one structure from the other, insertion of the letters *b* and *g*, for block and graft, respectively, between the names of the specific sequences was introduced; for example, natural rubber-*b*-poly(methyl methacrylate) and natural rubber-*g*-poly(methyl methacrylate) in the case of the examples cited. The structures of these two macromolecules would be represented by (11.17) and (11.18a), where the ...AAAA... sequences represent natural rubber and the ...BBBBB... sequences represent poly(metyl methacrylate).

The number and order of sequences may be more complicated. Block copolymers are usually made by free radical or living polymerizations. These processes can produce polymers that consist of a pure A block connected to a pure B block, with no interphase zone of mixed A and B structure. The preparation of block copolymers is not limited to monomers A and B, but can also encompass segments of random copolymers. For example, a block of a random copolymer AB can be connected to a block of polymer A or B. Moreover, the point of attachment of the blocks can be either at the end or the middle of the polymer chain. Several examples of the various types of block copolymers possibly follow:

```
~ AAAAAA BBBB AAAAA ~      or      ~ AAAAAAAAAAAA ~
                                       B       B
                                       B       B
                                       B       B
                                       B       B          (11.19)

~ ABABAB   AAAAAA ~      or     ~ ABABBABB  AAAAAAAA ~
                                     A               B
                                     A               B
~ ABABA    BBBBBBB ~                 A               B
                                     A               B    (11.20)
```

When the sequences making up the segments are random copolymers, the prefix *co* may be introduced, with the major component monomer preceding the minor constituent. A backbone polymer of butadiene-styrene rubber grafted with styrene containing a small percentage of acrylic acid would be described as poly[(butadiene-co-styrene)-(styrene-co-acrylic acid)] and could be schematically represented as

```
~AABABAAABBABAAAABBA~
  ~          ~
  B          B
  B          B
  B          C
  C          B
  B          B
  B          B
  B          C
  B          B
  C          B
  B          B
  ~          B
             ~                (11.21)
```

where A represents butadiene, B, styrene, and C, acrylic acid.

11.10.2 Block Copolymer Synthesis

Several methods can be used to synthesize block copolymers. Using living polymerization, monomer A is homopolymerized to form a block of A; then monomer B is added and reacts with the active chain end of segment A to form a block of B. With careful control of the reaction conditions, this technique can produce a variety of well-defined block copolymers. This ionic technique is discussed in more detail in a later section. Mechanicochemical degradation provides a very useful and simple way to produce polymeric free radicals. When a rubber is mechanically sheared (Ceresa, 1965), as during mastication, a reduction in molecular weight occurs as a result of the physical pulling apart of macromolecules. This chain rupture forms radicals of A and B, which then recombine to form a block copolymer. This is not a preferred method because it usually leads to a mixture of poorly defined block copolymers.

Using living polymerizations, the Shell Company was able to commercialize several poly(styrene-co-butadiene) and poly(styrene-co-isoprene) block copolymers known in the industry as Kraton 1101 and Kraton G. These block copolymers have found many uses in the shoe sole and adhesive industries. The physical properties were dependent on the macrostructure and microstructure of these block copolymers.

11.10.3 Examples

As major examples, let us consider the three monomers butadiene, styrene, and acrylonitrile, and see how they can be block copolymerized together by mechanochemical means. From the large number of theoretical possibilities, 11 have been selected for discussion; these may be prepared by mastication of the following:

1. A butadiene-styrene copolymer rubber with acrylonitrile monomer.
2. Polyacrylonitrile (plasticized) with a mixture of butadiene and styrene monomers.
3. A butadiene-acrylonitrile copolymer rubber with styrene monomer.
4. Polystyrene with a mixture of butadiene and acrylonitrile monomers.
5. A styrene-acrylonitrile resin with a mixture of styrene and butadiene monomers.
6. Polybutadiene with a mixture of styrene and acrylonitrile monomers.
7. A butadiene-styrene rubber with polyacrylonitrile (best plasticized).
8. A styrene-acrylonitrile resin with polybutadiene.
9. A butadiene-acrylonitrile rubber with polystyrene.
10. A high styrene-butadiene resin with acrylonitrile monomer.
11. A high styrene-butadiene resin with polyacrylonitrile (plasticized).

(All the foregoing reactions except 2 and 11 have been reported in the patent literature or are known to have been commercially evaluated.) In each example, the products would be chemically and physically different in terms of the makeup of the structural sequences, and all properties would also depend on the relative proportions of the initial components. In all mechanochemical reactions, some of the starting polymer or copolymer remains unchanged, mainly the low-molecular-weight fraction, which is not effectively sheared, and some homopolymer may be formed from the polymerizing monomers by chain transfer reactions. Varying the mastication conditions greatly influences the yield and rate of reaction; the chemical nature of the products is less affected, except that the presence of butadiene as one of the constituents (either polymer or monomer) will cause increasing gel contents with continued mastication. Processes 3, 5, 6, 8, and 9 are known to give products in which a rubber phase is dispersed in a resinous matrix; that is, they are alternative methods for producing an A-B-S-type copolymer.[1] The presence of a proportion of block or graft copolymer in the system assists in stabilizing the dispersion of the rubber phase in the resin matrix by acting at the phase boundary as a "soap"; that is, a compatibilizing agent at the phase boundary.

11.10.4 Other Methods of Effecting Mechanicochemical Reactions

Mechanicochemical degradation is the term used in describing chain scission of polymer backbones through the application of shear during a processing operation. It was previously believed that this type of process led to carbon-carbon chain scission, which usually causes a dramatic change in rheological properties. In the early 1950s, Watson and coworkers (Avery and Watson, 1956; Ceresa and Watson, 1959) showed that, in the absence of oxygen, radicals produced by mechanical shear can be used to initiate the polymerization of vinyl monomers to form block copolymers. For example, vibromilling of natural rubber below its glass transition temperature has enabled block copolymerization of natural rubber with methyl methacrylate to be carried out on a small scale, with conversions as high as 86%. Similar results were achieved with styrene and with acrylonitrile.

This type of approach has also been used to attach antioxidants to unsaturated polymers. The novel approach of Scott in the 1970s (Scott, 1984) using the technique employed by Watson enabled the attack of substituted allyl mercaptans and disulfides to olefinic double bonds employing the Kharasch reactions.

Mechanicochemical reactions can occur during processing, when the polymer is converted to a finished product. The chain scission can occur in both saturated and unsaturated polymer backbones. For example, during processing in a screw extruder, backbone scission in polypropylene produces long-tail

---

[1] It has been found in practice that a number of monomers that normally do not polymerize by free radical processes in the temperature range 10–50°C can be block copolymerized by cold mastication techniques, indicating ionic initiation via heterolytic scission.

free radicals, which can form macroalkyl radical peroxides. These peroxides are responsible for the observed decrease in melt viscosity. In the absence of oxygen, these monoallyl macroradicals can be used to graft new monomers or such polymers as polypropylene and polyethylene to the backbone. In this manner, maleic adducts of polypropylene have been prepared, giving improved dyeability, hydrophilicity, and adhesion.

11.10.5 Ionic Mechanisms

Ionic mechanisms for the preparation of block copolymers are a very important tool of the synthetic polymer chemist. A feature of many homogeneous anionic polymerizations in solution is that termination can be avoided by careful control of experimental conditions. In fact, an infinite life of the active chain end is theoretically possible, and this has led to the term *living polymers*. Polymer carbanions can resume growth after the further addition of monomer. By changing the monomer composition, block copolymerization is readily initiated, and this process can be repeated. A major advantage of this type of synthesis over most free radical processes is the ability to control the chain length of the sequences by adjusting the concentrations of initiating sites and of monomer at each stage of the block copolymerization.

Anionic block copolymerization employs organolithium initiators, which have wider use because of their extended range of solubility, which includes hydrocarbons (Janes, 1973). Organolithium compounds act as initiators by direct attack of the organic anion on the monomer species, again a fast reaction and, in the absence of compound with active hydrogen atoms, without transfer or termination steps. If carefully executed, the reaction permits one to have precise control over molecular weight and (the narrow) molecular weight distribution.

The convenience of this technique has led to the development of many commercial products, including thermoplastic elastomers based on triblocks of styrene, butadiene, and isoprene. The initiator used in these systems is based on hydrocarbon-soluble organolithium initiators. In some cases, a hydrocarbon-soluble dilithio initiator has been employed in the preparation of multiblock copolymers. Several techniques are used to prepare thermoplastic elastomers of the ABA type. All these are discussed in detail in Chapter 2. A short summary of these techniques is given here.

(i) Three-Stage Process with Monofunctional Initiators

In this technique—used, for example, for the synthesis of block copolymers of poly(styrene-*b*-butadiene-*b*-styrene) (SBS)—a polystyrene block is formed by employing *n*-butyllithium as the initiator in an aromatic solvent. Butadiene monomer is then added to react with the polystyrene-lithium chain end to form the poly(butadiene) block. If the reaction was terminated at this stage, a poly(styrene-*b*-butadiene) copolymer would result, which has no thermoplastic properties. Therefore, styrene monomer is added to produce the triblock SBS. The process for the preparation of SBS is very carefully controlled to avoid the formation of a diblock, as the presence of any appreciable amount of SB dramatically reduces the thermoplastic properties of SBS.

(ii) Two-Stage Process with Difunctional Initiators

Several commercial processes using difunctional initiators based on soluble organolithium compounds have been developed. These compounds can polymerize at both ends. Difunctional initiators are useful in the cases of ABA block copolymers where B can initiate A, but A cannot initiate B. These difunctional initiators are useful in the preparation of SBS. The elastomeric butadiene block is polymerized with hexane as a solvent. The added styrene monomer is also soluble in hexane. This method is also useful in preparing triblocks with hydrocarbon middle blocks and polar end blocks such as poly(methacrylonitrile-*b*-isoprene-*b*-methacrylonitrile).

(iii) Monofunctional Initiation and Coupling

In this two-stage process, B is sequentially polymerized onto A, and then the two chains are coupled to yield an ABBA block copolymer. Triblocks of SBS have been prepared using this method, with methylene dichloride as the coupling agent. The disadvantage is the formation of radical anions, which can lead to contamination of the triblock with multiblock species.

(iv) Tapered Block Copolymers

This method is used to form a block copolymer, which consists of two segments of essentially homopolymeric structure separated by a block of a "tapered" segment of random copolymer composition. These are usually prepared by taking advantage of the differences in reaction rates of the component monomers. When polymerized individually in hexane, butadiene reacts six times more slowly than styrene; however, when styrene and butadiene are copolymerized in a hydrocarbon solvent such as hexane, the reaction rates reverse, and the butadiene becomes six times faster than the styrene. This leads to a tapering of the styrene in a copolymerization reaction. For more details on the synthesis techniques, refer to Chapters 2 and 13.

11.10.6 Graft Copolymer Synthesis

The synthesis of graft copolymers is much more diverse but can nevertheless be divided into groups of related processes: (1) polymer transfer, (2) copolymerization via unsaturated groups, (3) redox polymerization, (4) high-energy radiation techniques, (5) photochemical synthesis, and, most importantly, (6) metallation using activated organolithium with chelating diamines.

(i) Polymer Transfer

In a free radical polymerization, chain transfer is an important reaction. Chain transfer to a monomer, solvent, mercaptan, or other growing chain can take place. When a chain transfer reaction to another chain takes place, it creates a radical, which acts as a site for further chain growth and grafting (see Chapter 2 for additional details):

$$\begin{aligned} P + P &\rightarrow PH + P^{\bullet} \\ R^{\bullet} + P &\rightarrow RH + P^{\bullet} \\ P^{\bullet} + P^{\bullet} &\rightarrow PH + P^{\bullet} \end{aligned} \qquad (11.22)$$

The reaction proceeds by the transfer of a hydrogen or halogen (in the case of halogenated polymers) atom from a macromolecule P to the growing chains $P^{\bullet}$ (or to an excess initiator free radical $R^{\bullet}$, thereby "terminating" them). The reactivity is now located on the transfer molecule, which in turn initiates copolymerization; that is, the growth of a grafted side chain of a newly introduced second monomer. A measure of grafting occurs with most monomer-polymer systems, especially those initiated by benzoyl peroxide, if the concentrations of polymer and initiator are high.

The simplest technique is to dissolve the polymer in the appropriate solvent; add the peroxide initiator, which abstracts a hydrogen radical and generates a radical on the polymer chain; and then add fresh monomer for grafting onto this site. This technique has been employed in grafting methylacrylate onto natural rubber and synthetic polyisoprene. In this manner, several commercially useful products such as ABS resins have been prepared; however, tire elastomers are not made in this manner because of the generation of micro and macro gel particles, which are detrimental to physical properties. In many cases when latex grafting has been used, the product has usually been targeted toward thermoplastic applications rather than rubber applications.

(ii) Copolymerization via Unsaturated Groups

Other methods (e.g., most car body repairs) are based on the polyester-styrene copolymerization process (reinforced with various types of inert mesh or glass fabric), a graft copolymerization of styrene onto backbone unsaturated polymer of relatively low molecular weight. In general, for high grafting yields, a reasonably high concentration of pendant vinyl groups is required on the backbone polymer. For glass-reinforced plastics, the polyester resins are selected with this in view. In natural rubber, a few such groups per molecule [2] are always present and these undoubtedly participate during normal grafting. The content of pendant vinyl groups can be increased by mastication of unsaturated rubbers under nitrogen, because the resonance structures

[2] About 0.4% of the unsaturated groups are pendant vinyl groups in an average sample of acetone-extracted pale crepe rubber.

recombine as

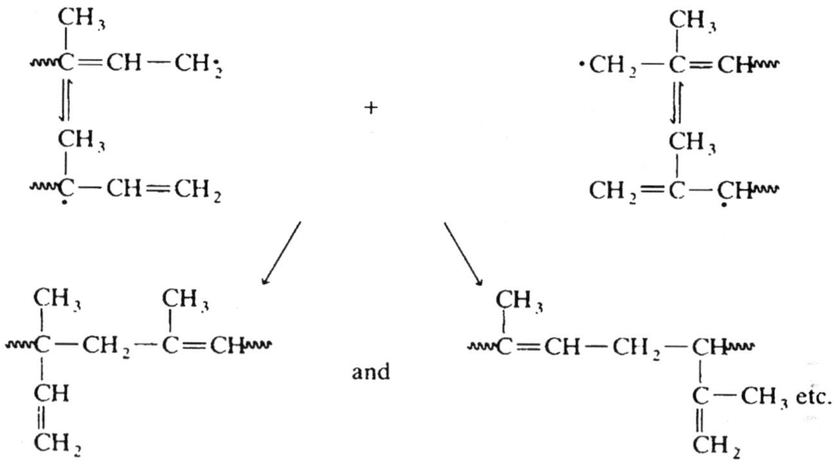

(11.23)

The direct introduction of peroxide groups into the backbone of polymers, such as poly(methyl methacrylate), has been used to produce macromolecular initiators for the synthesis of block copolymers; for example, poly(methyl methacrylate-*b*-acrylonitrile) and poly(methyl methacrylate-*b*-styrene). Ozonization can also be used, with careful control of the degree of ozonolysis, to introduce epoxy ring structures into natural rubber:

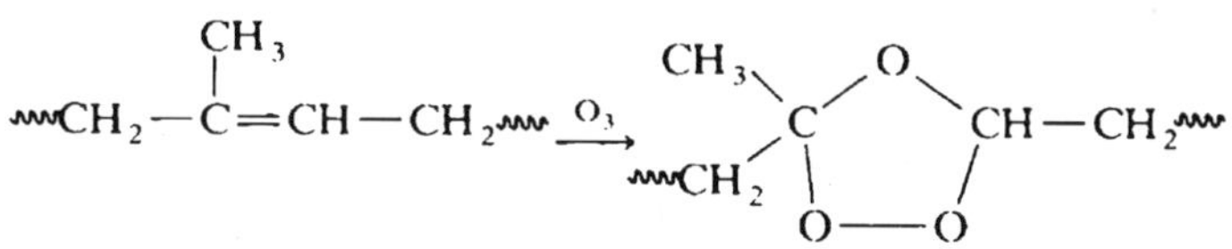

(11.24)

By carrying out the reaction to about 4% of the available double bonds in a solvent such as toluene at a low temperature followed by a nitrogen purge, grafting can be effected by addition under nitrogen of methyl methacrylate (MMA) monomer (reacting at 80°C in sealed ampules) and formation of two MMA chains attached to the oxygens of the opened –O–O– bridge. This technique should be applicable to isoprene and butadiene copolymers.

(iii) Redox Polymerization

Redox polymerizations are among the most popular techniques for grafting reactions, and of the possible initiator systems, ferrous ion oxidation and those based on ceric ion reduction are widely used. In a redox polymerization, a hydroperoxide or similar group is reduced to a free radical plus an anion, while the metal ion is oxidized to a higher valency state, and at the same time a monomer is added. When the reducible group is attached to a polymeric chain, the free radical grafting sites thus formed on the macro-molecular backbone act as initiators for graft copolymerization

$$\sim\sim CH_2-\underset{\displaystyle O-OH}{\underset{|}{CH}}\sim\sim + Fe^{2-} \xrightarrow[\text{aldehyde}]{\text{amine of}} \sim\sim CH_2-\underset{\displaystyle \dot{O}}{\underset{|}{CH}}\sim\sim + Fe^{3+} + OH^- \quad (11.25)$$

This method has been used to graft methyl methacrylate to natural rubber latex. (Actually, fresh latex contains a few hydroperoxide groups per macromolecule, which can take part in grafting reactions.) Recentrifuged latex concentrate is mixed with methyl methacrylate and a solution of tetraethylene pentamine is added, followed by a small quantity of ferrous sulfate solution. The homogenized blend is allowed to stand, often overnight (Morris and Sekher, 1959). The graft copolymer is isolated by coagulation. As practically all free radical sites are formed on the rubber backbone, there is very little free poly(methyl methacrylate) in the grafted system; on the other hand, some rubber chains are without grafts, as not all chains have hydroperoxy groups. Higher yields of graft copolymer are obtained by allowing the monomer to dissolve in, and equilibrate with, the latex particles before adding the amine and ferrous ion initiator. It has been claimed that passing oxygen (air) through the latex for several hours reduces the free rubber content of the polymerization product, but nitrogen purging is then necessary to prevent dissolved oxygen from acting as a polymerization inhibitor.

Hydroxy polymers can be grafted by redox polymerization by using a water-soluble peroxide, such as hydrogen peroxide, in conjunction with ferrous ions. The OH radicals thus produced abstract H atoms from the hydroxy groups in the polymer, giving free radical grafting sites on the backbone. This method has been used with starch and cellulose derivatives, but considerable quantities of homopolymer are formed from the initial hydroxyl radicals in parallel with the H abstraction. By introducing a few hydroxyl groups into a copolymeric synthetic rubber, grafting can be effected, provided the presence of homopolymer can be tolerated. Mixtures of ferrous ammonium sulfate and ascorbic acid are suitable redox initiation systems. Many patents claim the preferred use of ceric ions, which easily oxidize hydroxyl groups by a radical-ion reaction:

$$R-OH + Ce^{3+} \rightarrow R-\underset{\bullet}{\overset{+}{O}}H + Ce^{2\bullet}$$

$$R-\underset{\bullet}{\overset{+}{O}}H + OH \rightarrow R-O^{\bullet} + H_2O \quad (11.26)$$

The advantage of this reaction lies in the fact that only hydroxyls on the polymer are converted into R–O free radicals, so that no homopolymer can be produced and pure graft is obtained.

(iv) High-Energy Radiation Techniques

During high-energy irradiation in vacuo, for example, from a ^{60}Co source, some main-chain degradation of natural rubber and other polyisoprenes occurs:

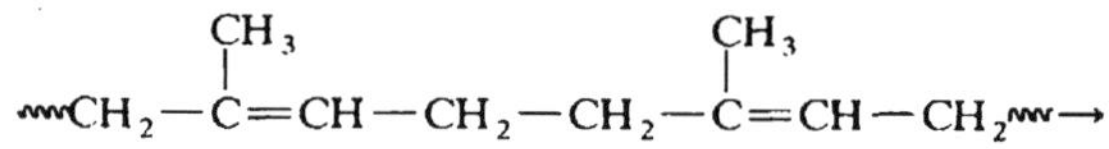

(11.27)

Much of the irradiation energy is also adsorbed by the removal of hydrogen atoms:

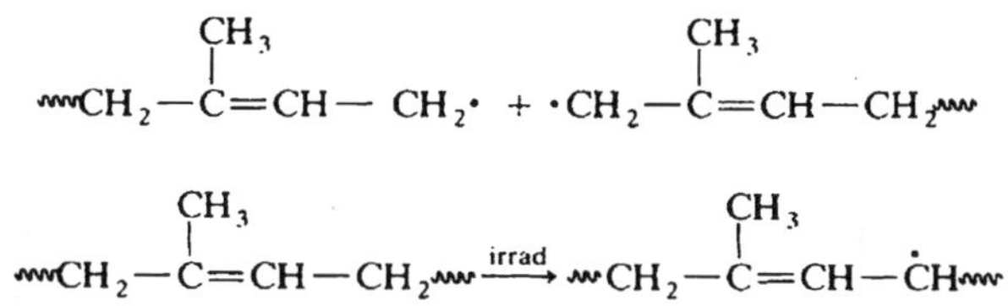

(11.28)

The irradiation of natural rubber in the presence of a vinyl monomer thus leads primarily to a synthesis of graft copolymers, but some block copolymer is certainly always present. Irradiation syntheses may be carried out in solution, either in contact with liquid monomer (with or without a diluent) or in contact with monomer in the vapor phase, or in emulsion or suspension. The rubber may be preirradiated in the absence of air to produce free radicals for later monomer addition, but the life of these radicals is short as a result of mobility within the rubber matrix. Irradiation at very low temperatures makes it possible to use the trapped radicals technique for a variety of natural and synthetic rubbers. Plastics and polymers with a crystalline phase are more readily preirradiated to initiate later grafting by trapped radicals. Irradiation may also be carried out in air to introduce peroxide groupings:

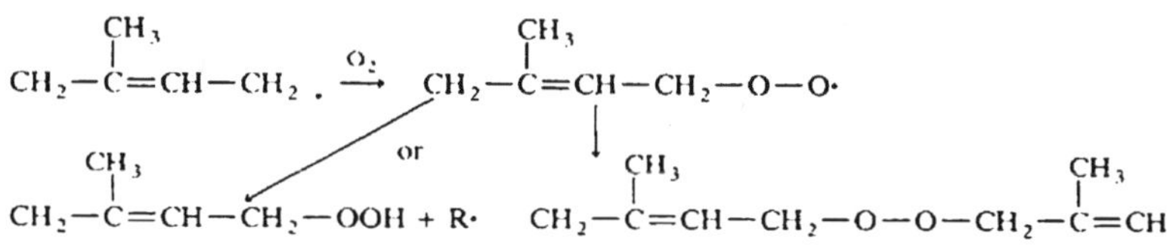

(11.29)

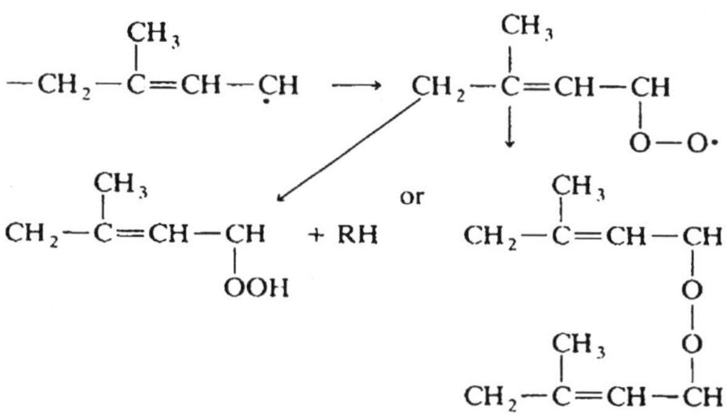

(11.30)

These groups can then be used to initiate grafting by any of the methods already discussed. Latex phase grafting is generally favored for its simplicity; natural rubber grafts with methyl methacrylate styrene, acrylonitrile, and vinyl chloride have been made in this way (Cockbain et al., 1959).

The irradiation of mixed lattices for subsequent combination of the ruptured chains is another approach; it has been carried out with natural rubber and poly(vinyl chloride) lattices to prepare graft (and block) copolymers in fairly high yields without the problem of monomer recovery. The same method has been used to graft polychloroprene onto synthetic polyisoprene dispersions and onto polybutadiene lattices of various compositions.

(v) Photochemical Synthesis

Macromolecules containing photosensitive groups that absorb energy from ultraviolet frequencies often degrade by free radical processes. The degradative process as a rule is fairly slow, but by the addition of photosensitizers, such as xanthone, benzyl, benzoin, and 1-chloroanthraquinone, the rate can be sped up to enable graft copolymerization to take place in the presence of methyl methacrylate or other monomers. This can be done in the case of natural rubber in the latex phase with reasonably high yields of graft copolymer. Natural rubber-*y*-polystyrene and poly(butadiene-*y*-styrene) have both been prepared by ultraviolet irradiation of sensitized latex-monomer dispersions. A combination of photochemical synthesis and redox-type initiation can also be carried out—a process known as one-electron oxidation—to achieve grafting with minimal homopolymer formation.

Bromine atoms on the backbone of a polymer can be liberated readily by ultraviolet irradiation to give free radical sites for grafting reactions. The bromination can be photochemically induced

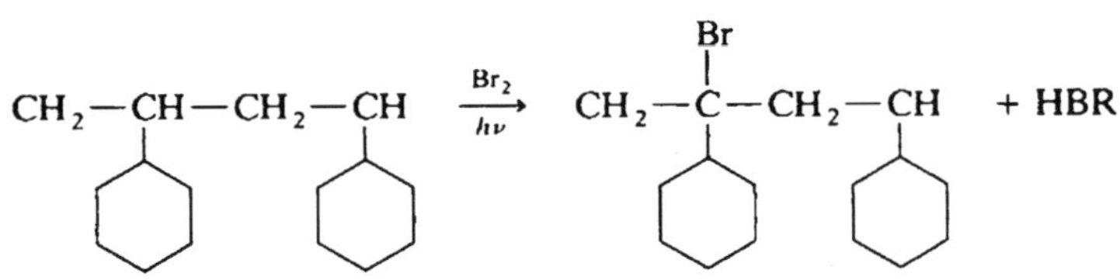

(11.31)

or a chain transfer agent such as carbon tetrabromide may be used in the polymerization step to introduce the labile groups,

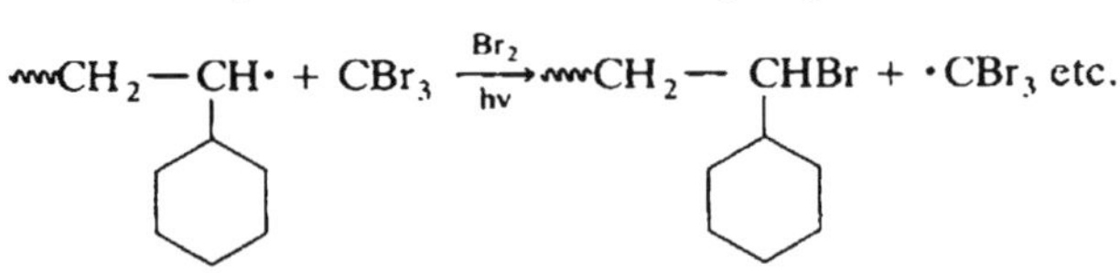

(11.32)

With the aid of suitable sensitizers, polymers such as brominated butyl rubber, valuable because of their flame retardancy, may act as backbone polymers for a variety of grafting reactions.

An early synthesis of block copolymers was based on the ultraviolet irradiation of poly(methyl vinyl ketone) in the presence of acrylonitrile. The initial degradative step is

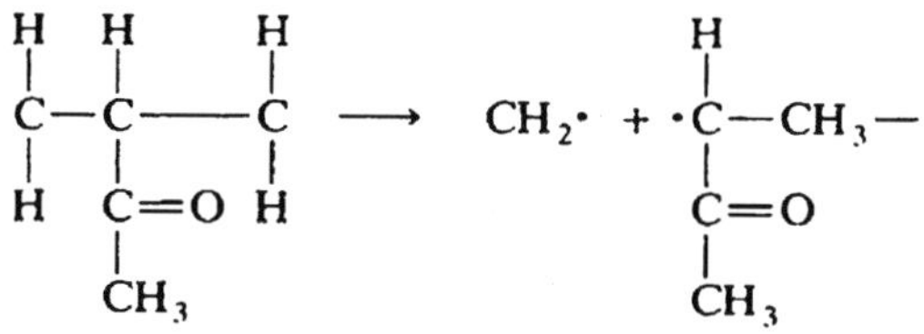

(11.33)

This degradation reaction, supplemented by various subsequent oxidation steps, has found renewed interest in the form of the introduction of photodegradable plastics as part of the campaign to reduce plastic litter from throwaway packaging. Although as yet there has been no demand for photodegradable rubbers, the incorporation of a small percentage of a vinyl ketone into a rubber copolymer or homopolymer would open the way to a useful synthesis of block copolymers.

Many other syntheses of block and graft copolymers have been reported, but enough has been said to indicate the scope of these reactions and to indicate a potential that has still to be thoroughly explored. Many grafting and block copolymerization systems have only been evaluated for plastic materials but are capable of extension to rubbers.

(vi) Metallation Using Activated Organolithium with Chelating Diamines

Unsaturated elastomers can be readily metallated with activated organolithium compounds in the presence of chelating diamines or alkoxides of potassium or sodium. For example, polyisoprene, polybutadiene, styrene-butadiene copolymers, and styrene-isoprene copolymers can be metallated with *n*-butyllithium ·TMEDA complexes (1/1 or 1/2 ratio) to form allylic or benzylic anions. The resulting allylic anion can be employed as an initiator site to grow certain branched or comb polymer species. These polymers can include polystyrene, which would form hard domains, or polybutadiene, which forms soft domains.

Research in this area has resulted in the preparation of several comb polymers (Halasa, 1974; Folk et al., 1044). The metallation technique is a useful and versatile method as it can be used with any polymeric material that contains a few double bonds. For example, ethylene-propylene was successfully grafted with norbornene. Similar reactions were performed on polymeric materials that contain aromatic rings, such as polystyrene, poly-α-methyl styrene, and polyphenylene oxide (PPO).

In general, polymeric materials that contain either side groups or main-chain allylic or acidic hydrogens can be metallated with organolithium compounds in the presence of chelating diamines. They can also be grafted with ionically polymerizable monomers to produce comblike materials (Halasa et al., 1976).

11.10.7 Base Polymer Properties

Though the properties of block and graft copolymers are discussed in Chapters 3, 5, and 13, some properties of the copolymer particularly germane to this discussion are briefly mentioned here.

The properties of, say, natural rubber grafted with poly(methyl methacrylate) cannot be evaluated unless the copolymer is isolated from either homopolymer species. The methods used are based on fractional precipitation, selective solution, or a combination of these basic techniques. For details, refer to Chapter 3. In many cases, though, technologists are concerned with the materials as manufactured, so we consider in this context also the properties of the block and graft copolymers without homopolymer removed.

The presence of two chemically different polymeric sequences in the same chain causes that macromolecule to act as a soap; that is, it helps to compatibilize two species of homopolymer in a blend by accumulating at their interface, assisting a more gradual transition from one phase to the other, and thus reducing the interfacial energy. Microphase separation, of course, still occurs, the predominant case in practice, but macrophase separation is thereby usually prevented. In high-impact polystyrene and in A-B-S copolymers prepared by grafting reactions, the dispersed rubber phase in the glassy matrix and the dispersed glassy phase within the rubbery particles are both prevented from forming separate phases by the graft copolymer chains, which on a molecular scale have their rubbery segments associated with the rubber particles and their plastic segments with the glass phase. In this respect, there is little difference in properties between a graft and a linear block copolymer—the essential feature is the presence of the two types of sequences in the same macromolecule.

The block copolymeric thermoelastic polymers owe their properties to this very structure, whereby the polystyrene end blocks (along with any homopolymeric polystyrene) form the microphase, which is dispersed within a continuous phase of polybutadiene formed from the polybutadiene segments of the central sequences in S-B-S-type block copolymers. (For this to happen, the total volume of the polybutadiene segments must exceed the total volume of the polystyrene segments. When the reverse is the case, the product exhibits the properties of a high-impact polystyrene; see again Chapter 13.) Although the polystyrene "structures" act as physical crosslinks at low temperature, at processing temperatures above the softening temperature of polystyrene both segment types exhibit viscoelasticity, allowing the material to be extruded, injection molded, and so on. On cooling, the polystyrene domains become rigid again and assert their influence on the material properties.

When block and graft copolymers are dispersed in solvents, the solutions have properties that depend on whether or not the copolymer is eventually fully solvated. If the solvent is a "good" solvent for both sequences—for example, chloroform in the case of natural rubber graft copolymerized with poly(methyl methacrylate) (Halasa et al., 1976)—then both segment types are expanded and films cast from dilute solutions will usually be intermediate in properties to the two homopolymers (in this example the properties of a reinforced rubber film). If the solvent is a good solvent for the rubber but a poor solvent or nonsolvent for poly(methyl methacrylate) (e.g., petroleum ether), then the solutions show the typical turbidity of a block or graft copolymer and the cast film is highly elastic. When the solvent is acetone, a good solvent for poly(methyl methacrylate) but a nonsolvent for rubber, the cast films are plastic with high tear strengths. See again Chapter 10.

When grafting is carried out on a polymer under conditions such that the physical form of the substrate polymer is maintained, then the original properties of the substrate usually predominate, and supplementary properties accrue as a result of the grafting. This is invariably the case when the substrate is in fibrous form; for example, cellulose, nylons, and terylene grafted with various monomer systems. The nature of the grafting reaction to these fibers is usually such as to form a surface coating over the substrate polymer; the surface characteristics, such as dyeing, are therefore usually those of the grafting system.

It is very doubtful that any blends of two polymers, or of chemically different copolymers, from a thermodynamic point of view can ever be fully compatible. Even most block or graft copolymers systems therefore show microphase separation, which will be typical for the properties of a given system. Chemical modification, as discussed in the earlier part of this chapter, will in general lead to the formation of polymeric single phases, provided the reaction has been carried out homogeneously. The choice need not be restricted, however, to just these two approaches, since chemical modification can be carried out after block or graft copolymerization or vice versa. Very little has been published on such a consecutive use of these two physically different ways of modifying polymers chemically, so there is considerable scope for developing new modifications of long-established rubbers, as well as generally for changing old into new polymers.

REFERENCES

Avery, C., Watson, W., 1956. J. Appl. Polym. Sci. 19, 1.
Bevington, J., Ratt, L., 1975. Polymer 16, 66.
Bielinsk, D.M., Sulsarki, L.A., Stanley, H.M., Pethrick, R.A.J., 1995. Appl. Polym. Sci. 56, 853.
Bohm, G., Graves, D., 1988. US Patent 4 788 229.
Bruxelles, G.N., Grassie, N.R., 1965. In: Mark, H.S. et al. (Eds.), Encyclopedia of Polymer and Technology, vol. 3. Wiley-Interscience, New York, p. 307.
Ceresa, R.J., 1965. In: Mark, H.S. et al. (Eds.), Encyclopedia of Polymer Science and Technology, vol. 2. Wiley-Interscience, New York, p. 502.
Ceresa, R.J., Watson, W., 1959. J. Appl. Polym. Sci. 101, 1.
Chamberline, Y., Pascoult, J., Razzouk, H., Cheradem, H., 1981. Makromol. Chem. Rapid Commun. 2, 322.
Cockbain, E.G., Pendle, T.D., Turner, D.T., 1959. J. Polym. Sci. 39, 419.
Cunneen, J.I., Porter, M., 1965. In: Mark, H.S. et al. (Eds.), Encyclopedia of Polymer Science and Technology, vol. 2. Wiley-Interscience, New York, p. 502.
Folk, J., Sclott, R., Hoey, D., Pendeltor, J., 1973. Rubber Chem. Technol. 46, 1044.
Gordon, M., Tyler, J., 1953. J. Polym. Chem. 3, 537.
Halasa, A.F., 1974. Adv. Chem. Ser. 130, 77.

Halasa, A.F., 1981a. Rubber Chem. Technol. 54, 627.
Halasa, A.F., 1981b. J. Polym. Sci. Polym. Ed. 19, 1357.
Halasa, A.F. (to The Goodyear Tire & Rubber Co.), 1985. US Patent 3 872 072 and US Patent 4 237 245.
Halasa, A.F., Hsu, Wen-ling., (to The Goodyear Tire & Rubber Co.), 2004. US Patents 6 693 160 and 6 753 447.
Halasa, A.F., Massie, J.M., 1993. Kirk-Othemer Encyclopedia of Chemical Technology. fourth ed., vol. 8. John Wiley and Sons, Inc., New York.
Halasa, A.F., Mitchell, G., Stayer, M., Tate, D.P., Koch, R., 1976. J. Polym. Sci. Chem. Ed. 14, 297.
Halasa, A.F., Vescelius, L.E., Futamura, S., Hall, J., 1982. IUPAC Symposium on Macromolecules, vol 28. Amherst, MA.
Hall, J.E., Antikowiak, T.A., 1955. European Patent 0693 505.
Harwood, H., Russell, D., Verthe, J., Zymons, J., 1973. Macromol. Chem. 163, 1.
Haskell, V.C., 1965. In: Mark, H.S. et al. (Eds.), Encyclopedia of Polymer Science and Technology. vol. 3. Wiley-Interscience, New York, p. 60.
Hopper, R.J. (to The Goodyear Tire & Rubber Co.), 1975. US Patent 3 915 907.
Hopper, R.J., 1976. Rubber. Chem. Technol. 49, 341.
Hopper, R.J. (to The Goodyear Tire & Rubber Co.), 1977. US Patent 4 017 468.
Hopper, R.J. (to The Goodyear Tire & Rubber Co.), 1989. US Patent 4 820 780.
Hopper, R.J. (to The Goodyear Tire & Rubber Co.), 1990. US Patent 4 910 266.
Hopper, R.J., Mcquateand, R.D., Hutchin, T.G., 1984. In: International Conference on Advances in the Stablization and Controlled Degradation of Polymers, Lucerne, Switzerland, May 23–25, (Preprints).
Hsieh, H.L., Quirk, R.P., 1996. Anionic Polymerization: Principles and Practical Applications. Marcel Dekker, New York.
Janes, W.H., 1973. In: Allport, D.C., Janes, W.H. (Eds.), Block Copolymers. Applied Science Publishers, London, p. 62.
Jarowenko, W., 1965. In: Mark, H.S. et al. (Eds.), Encyclopedia of Polymer Science and Technology, vol. 3. Wiley-Interscience, New York, p. 787.
Morris, J.E., Sekher, B.C., 1959. In: Proceedings of the International Rubber Conterence, Washington, DC, 277.
Poutsma, M., 1969. In: Huyser, S. (Ed.), Methods of Free Radical Chemistry, vol. 1. Dekker, New York (Chapter 3).
Rachapudy, H., Smith, G., Raju, V., Graessley, W., 1979. J. Polym. Sci. Phys. Ed. 17, 1211.
Rouse, B.P., 1965. In: Mark, H.S., et al. (Eds.), Encyclopedia of Polymer and Technology, vol. 3. Wiley-Interscience, New York, p. 325.
Schoen, L., Raajien, W., Van'twout, W., 1975. Br. Polym. J. 7, 165.
Schults, D.N., Turner, S., Golub, M., 1983. Rubber Chem. Technol. 85, 809.
Schulz, D.N., Halasa, A.F., Oberster, A.E., 1974. J. Polym. Sci. Chem. Ed. 12, 153.
Scott, G., 1984. Polym. Eng. Sci. 24, 1007.
Staudinger, A., 1944. Rubber Chem. Technol. 17, 15.
Uda Brain, A., 1992. In: 33rd Annual Meeting Proceedings of the International Institute of Synthetic Rubber Producers, pp. 65–83.
Urneck, C.A., Short, J.N., 1970. J. Appl. Polym. Sci. 14, 1421.
White, D.A., Auda, R.S., Davis, W.M., Ferrughlli, D.T., (to Exxon Chemical Co.), 1990. US Patent 4 956 420.
Wichlatz, J., 1964. In: Fetters, M. (Ed.), Chemical Reaction of Polymers. Interscience, New York (Chapter 2).
Yakubchik, A., Tikhominov, B., Sumilov, V., 1962. Rubber Chem. Technol. 35, 1063.
Yamato, M., Oahu, M., 1996. Progress in Organic Coating 27, 277.
Yoshioka, A., Ueda, A., Watanabe, H., Nagata, N., 1987. Nippon Kagaka Kaisha 341.

Chapter 12

Elastomer Blends

Sudhin Datta
ExxonMobil Chemical Company, Baytown, TX, USA

12.1 INTRODUCTION

Blending of elastomers is a commercially feasible and frequently practiced method to access a combination of properties in a final compounded and vulcanized rubber which is commercially desirable but otherwise inaccessible from a single elastomer. In addition, blends inherently provide a continuous range of properties by an adjustment of the ratios of the individual components. Thus the potential utility of blends is significantly larger than a single new elastomer. The economic driver for the use of blends arises from a combination of either lack of suitable synthetic pathway to the single elastomer or, more commonly, a considerable economic penalty for the synthesis of specialty elastomer compared to the blend or two more common ones. A suitable example of this effort is the attempt to use blends of EPDM and NBR to attain both thermal oxidative resistance as well as durability in contact with liquid, aliphatic hydrocarbons. These are the predominant attributes of the two blend components, respectively. These efforts continue principally because the single elastomer that combines the properties, such as HNBR or Kalrez, commands a substantial financial premium compared to the components of the blend. In theory, blends are more versatile since adjustment of the ratio of EPDM to NBR, in our example, should lead to a corresponding variation of the properties of the blend, reflecting the change in the amount of the two components. This continuous variability allows, in theory, an exact match of the property profile of the blend to the desired application. A single elastomer, in contrast, does not have this capability. The potentially improved property profile of elastomer blends can include chemical, physical, and processing benefits. In reality, the properties of blends of elastomers rarely follow a linear or predictable correlation to the individual components. Hence the technology of elastomer blends is largely focused on the choice of component elastomers and the procedures for the formation of the blends, followed by vulcanization to achieve a set of final properties. This chapter shows some of these choices of elastomers and the

The Science and Technology of Rubber. http://dx.doi.org/10.1016/B978-0-12-394584-6.00012-1

procedures for the formation of blends. We conclude with empirical guidelines for the creation of novel blends of elastomers.

A further conceptual advantage of elastomer blends, compared to the use of the component elastomers, is the apparent simplicity of use and the lack of new processing steps in commercial practice. Since elastomers are compounded with fillers, plasticizers, curatives, and other necessary additives prior to fabrication and vulcanization, it is reasonable to expect that a binary blend of elastomers could be processed similarly with the same ingredients to form a final vulcanized elastomer blend. At a first glance the technology of blending elastomers is quite simple since the blend is formed in situ during compounding of the elastomer and the blend is then treated by conventional fabrication and vulcanization. In contrast, new elastomers would require, inevitably, new compounding ingredients and compounding procedures along with new methods of vulcanization. In practice these expectations of using established processing protocols with known ingredients are rarely met since very specific mixing procedures (e.g., phase blending) are followed to avoid undesirable secondary reactions in the formation of blends. These secondary reactions include redistribution of curatives and fillers.

The vast majority of elastomer blends, including the exemplary EPDM/NBR blends, are largely immiscible. For these blends this manifests as two thermodynamically stable phases that are very close to the parents in composition. The morphology of these blends is biphasic with domains of these compositions. We define domains to be aggregations of principally compositionally similar or identical molecules. Domains are uniform in composition across any macroscopic region of several hundred contiguous atoms. Domains are separated from other domains by a distinct boundary or interface. In immiscible elastomer blends the different domains contain elastomers, fillers, plasticizers, and other additives in different amounts. In addition, the domains may have different levels of crosslinking. The size and the distribution of the elastomer domains is dictated by the contrary factors of mixing intensity, which favors smaller domains and interfacial energy between dissimilar elastomers, which favor aggregation and thus larger domains. The typical high viscosity of the elastomers (compared to the typical thermoplastics) favors persistence of either domain segregated morphology by slowing down the formation or collapse of the dispersed domains. An alternate visualization for immiscible elastomer blends is that the thermodynamic steady state is segregation, which is somewhat disrupted by the kinetics of dispersive actions such as mixing. The resultant morphology is thus an arbitrary transit point between these points, dictated by mixing parameters and time.

In addition to changes and variation in morphology, immiscible blends show additional, more complex changes due to different chemical properties of the two-component elastomer. This difference in the chemical structure manifests as three distinct but interrelated properties. First, the dissimilar elastomers differ in the retention of the fillers (e.g., carbon black) and plasticizers (e.g., process oils) because of the inherent differences in the solubility parameters. The distribution of fillers is influenced by the thermodynamic partitioning but is slower to reach equilibrium. Several procedures, referred to as phase blending (Hess and Chirico, 1977), seek to ensure some filler retention in each phase by initially adding all the fillers to the elastomer that is less favored to retain it, and then blending this mixture with other elastomer and the remaining ingredients. These procedures are industrially quite successful, though a rational academic investigation (Wang et al., 1991a,b; Wang and Wolff, 1992a,b) is feasible but challenging. Second, the chemical reactants for the vulcanization and later usage of the elastomers including curatives, cure accelerators, oxidation retarders—all of which are polar, organic small molecules—distribute between the two phases of elastomers according to thermodynamic solubility parameter. This distribution will determine the vulcanization and the subsequent utility of each elastomer phase. These additives are added during compounding to the elastomers and distribute to minimize free energy by preferentially residing in the more polar elastomer. Since these classes of materials easily diffuse, the final distribution is principally a thermodynamic equilibrium and unlike the dispersion of elastomers themselves, kinetic parameters of diffusion are relatively less important. Third, during vulcanization the different elastomer phases consume vulcanizing agents at different rates. This gradient in concentration leads to flux of curatives from regions of higher to lower concentrations. The effect of the diffusion is most pronounced at the boundaries of the elastomer phases where the diffusion distances are the least. This complex set of phenomenon dictated by the difference of chemistry of the elastomers and the diffusion of additives leads not only to different vulcanization of the elastomer phases but also to differences between the bulk and the periphery of each elastomer phase. In practice the use of elastomer blends accounts for the changing properties of elastomers by uneven distribution of fillers and vulcanization in blends of immiscible elastomers (Robeson, 2007).

The engineering properties of elastomers (i.e., tensile strength, hysteresis) in vulcanized compounds depend not only on the elastomer itself but also on the amount and identity of the fillers and plasticizers, as well as on the extent of cure. In an immiscible blend, the amount of these additives in any phase can be modulated by changes in the viscosity and chemical identity of the elastomer, the surface chemistry of the filler, the chemical nature of the plasticizer, and the sequence of addition of the components as well as the details of the mixing procedure. A large body of experimental procedures (*vide infra*) (Wang et al., 1991a,b; Wang and Wolff, 1992a,b) has been developed to attain a metastable, interphase distribution of additives in blends. On vulcanization, this phase morphology is rendered essentially immobile with a trapped filler distribution. This in conjunction with the distribution of the plasticizers and other mobile phases lead to final engineering properties of the blend. It is noteworthy that the interphase migration of plasticizers does not stop on vulcanization and this process leads to dilation and contraction of the individual elastomer domains within the invariant morphology.

Immiscible elastomer blends, which comprise the vast majority, have most of the considerations outlined earlier. The primary expectations of an economic advantage, wide and continuous variability in properties, and inherently simple processing are largely met is the cases the two elastomers are miscible and mix to form a single phase. The key is the formation of a single phase with intermixing at a molecular level. Using our earlier definitions there is only a single domain, and the thermal and mechanical properties of the blend such as glass transition temperature and tensile properties are a composition weighted function of the components (Leroy et al., 2002; Roland et al., 1994; Alegria et al., 1995; Roland and Casalini, 2007). This miscibility could be because of a close match of the solubility parameters and a corresponding match of the constitutive thermodynamic parameters of the two polymers or due to specific interactions between the component polymers. Both circumstances are rare. Most elastomers are high molecular weight, which minimize the favorable entropic effects of molecular mixing. Further the small number of commercial elastomers limits availability of pairs of elastomers with specific favorable attractive interaction.

These concepts for formation of miscible blend of elastomers with similar or near equivalence of solubility parameters require the components to be similar in properties. Thus a wide variation in the properties of the elastomer blends by changing the relative amounts of the two elastomers is not typical since it is unlikely that, for example, a nonpolar polyolefin elastomer and a polar elastomer like acrylate would be similar in solubility parameters. This relative invariance in the properties of the blend compared to the components is an inherent limitation on the basic, economic, and technological need for elastomer blends, which is to generate new properties by blends of existing materials. Similar or near equivalence of solubility parameters can be difficult to predict from chemical structure. For example, chemically distinct 1,4-polyisoprene and 1,2-polybutadiene are miscible, but isomeric 1,2-polybutadiene and 1,4-polybutadiene are immiscible. It is illustrative of this concept that an apolar hydrocarbon elastomer and a highly polar elastomer such as an acrylate cannot have, under any practical structural manifestation for either, a similar solubility parameter and thus be miscible.

Miscible blends of elastomers persist through the process of fabrication and crosslinking with the desirable properties of the miscible blend essentially intact and manifest in the mechanical and thermal properties of the vulcanized fabrications. The final vulcanizates will have uniform crosslinking across any microscopic region larger than the dimensions of a few polymer chains and have, in addition, uniform distribution of reinforcing agents. The properties of the constituents are, to a first approximation, a good measure of the properties of the final blend. Miscible blends are commonly used though they have been very rarely recognized since an analytical deconstruction of a homogeneously vulcanized elastomer is challenging. The current analytical techniques are only slightly more capable than the classical techniques of selective precipitation of the components of an unvulcanized elastomer blend from solution (German et al., 1967), though some key advances using solid state NMR (Senake Perera, 2001; Voelkel, 1988; Klei and Koenig, 1997; Legrand et al., 1992; Brus et al., 2006) are being introduced.

The foregoing shows that the benefits of achieving miscible blends of elastomers, which are different in solubility parameters, are a sufficiently attractive target. Various synthetic strategies have been adopted to achieve some degree of miscibility in blends of elastomers that are ordinarily not miscible. These blends are not thermodynamically miscible but they are compatible in the polymers technology usage of the concept. The common synthetic strategy is the addition of low molecular weight but involatile solvent during compounding, which produces a single, ternary phase in both elastomers and solvent. The solvent reduces unfavorable interactions between the polymer molecules. This strategy is useful when the solubility parameter differences between the elastomers are not very large. It also has a deficiency that the elastomer blend is diluted with a solvent with consequent reduction of entanglements and thus dependent physical properties. Another conceptual procedure is for both elastomers to interact with a single particulate filler. These procedures use the stabilizing interaction of binary elastomer mixture on a filler to form a compatible blend with pairs of elastomers. Another synthetic strategy is to use complementary interacting functionality such as acid and base functionality on each of the elastomers to form a single miscible blend. A deficiency of this procedure is the need for complementary functionality on each of the elastomers: the generation of this pair of structural feature may require additional synthetic steps. Another utilitarian deficiency of this last procedure is the unavoidable rise in viscosity and consequent loss of processing and fabrication when elastomers interact chemically.

In spite of these difficulties, miscible blends of different elastomers do exist. Miscible blends are easier to achieve for polyolefin elastomers like EPR, where the components differ in the relative ethylene content, than for polar NBR elastomers, which differ in their relative acrylonitrile content. The difference between these two cases is that a common change in the relative weight amount of the monomer indicating the change in the solubility parameter is much less for the polyolefin elastomer than for NBR. It is illustrative of the concepts we have described earlier that saturated polyolefin elastomers with a minimum of intermolecular enthalpic interactions, compared to more polar (NBR) or polarizable (NR) elastomers, form a large and probably predominant family of miscible blends. The general principle of achieving miscible elastomer blends between copolymer elastomers, where the inserted monomers units are either structural isomers of each other (such as 1,2 and 1,4 insertion isomers of BR), or closely related such as ethylene and propylene, where the small difference of solubility parameters required for miscibility can be maintained over a wide range of difference in relative ethylene to propylene ratios, is quite general and widely used. Common examples of miscible blends are ethylene-propylene copolymers of different composition that result in an elastomer comprising a semicrystalline, higher ethylene content and an amorphous, lower ethylene content component (Dikland and Van Duin, 2003). These blends combine the higher tensile strength of the semicrystalline, higher ethylene content polymers and the favorable low temperature properties of amorphous polymers.

Another pathway to changing the solubility parameter of elastomers that are copolymers (e.g., styrene-butadiene rubber) is a topological change in the enchainment of the constituent monomers from random to blocky at the same average composition. This is illustrated by styrene-butadiene copolymers and the comparable styrene-butadiene block copolymer. When the block length is sufficiently large each block may be miscible with polymers that are immiscible with the corresponding random copolymer. This topological procedure to achieving miscible blends may be extended to elastomers that have nonuniform distribution of vulcanizable sites on the backbone. This would be the case for an EPDM with a nonuniform, interchain distribution of the diene monomer. The uneven distribution of diene, which is the site for vulcanization in blends of EPDM elastomers, can lead to the formation of two distinct, intermingled vulcanization networks. These can be considered to be blends of miscible elastomers, which differ in their vulcanization.

Multiple notable reviews of elastomer blends exist. The first general treatment of the subject by Hess et al. (1993) reviews the applications, analysis, and the properties of the immiscible elastomer blends. Two related treatments by Roland (1989) and Ngai and Roland (2004) exist and describe the physics of mixing immiscible polymer blends and a more recent account of the analytical methods. Mangaraj (2002) has a more detailed review of elastomer blends. Other reviews by Corish (1978) and McDonel et al. (1978) deal with specific aspects of elastomer blends. A publication by Zhang (2009) on specific EPDM blends with NR/BR for tire sidewall approaches this area from the view of a specific application. Less comprehensive accounts of this area are also described for polyolefin elastomer blends by Ślusarski et al. (2003) and Feldman (2005).

12.2 THERMODYNAMICS AND SOLUBILITY PARAMETERS

Flory and Huggins independently developed the fundamental theory that bears their name to describe the thermodynamics of the mixing of solvents and polymers and binary pairs of polymers (Roland, 1989; Ngai and Roland, 2004; Mangaraj, 2002). Later Flory modified Prigogine's "equation-of-state" concept for polymer blends and introduced the new Flory-Prigogine model (Flory, 1942). Neither of these lattice models, though elegant in concept, are accurate in predicting the miscibility of polymer blends. A part of this inadequacy is because they do not account for the volume change with polymer blending.

Modifications have been suggested to improve these models to describe the thermodynamics of polymer-polymer miscibility (Huggins, 1942; Flory, 1965; McMaster, 1973; Patterson and Robard, 1978).

An alternate approach, which starts from blends and evaluates their interaction parameters, is the use of the Hildebrand Solubility Parameter concept (Hildebrand and Solubility, 1964; Gee, 1942, 1947). This has been used for solvent-solvent miscibility as well as for polymer-solvent mixing because it provides a more intuitive approach to predicting miscibility of polymers with solvents including solutions and gels such as swelling of vulcanized rubbers in organic liquids. Both of the models have similar deficiencies in themselves that they do not account for equation of state effects, and a consideration of specific interactions can be added only as perturbation to these models.

12.2.1 Flory-Huggins Model

The Flory-Huggins lattice model assumes that a polymer chain consists of a number of equivalent segments. The extension to polymer solvent interactions assumes that the polymer solution consists of a three-dimensional lattice and each lattice site is occupied either by a polymer segment or by a solvent molecule. Flory and Huggins calculated the entropy change of mixing as

$$\Delta S = -k(n_1 \ln\varphi_1 + n_2 \ln\varphi_2),$$

where k is Boltzmann's constant; n_1 and n_2 are the number of molecules of polymer and solvent, respectively; and φ_1 and φ_2 are volume fractions, respectively.

Their estimate of the enthalpy change on mixing, taking into account the change in near neighbor interactions and computation of the free energy change on mixing, is

$$\Delta Gm/V = kT\{(\emptyset_1)/V_1 \ln\emptyset_1 + (\emptyset_2)/V_2 \ln\emptyset_2 + (z\Delta W_{12}\emptyset_1\emptyset_2)/VkT\}.$$

Here V is the total volume of the solution, and V_1 and V_2 are molar volumes of solvent and polymer segment, respectively. The term $z\Delta W/kT$ is usually represented by the symbol χ, the Flory Interaction Parameter. Although X is written as a reduced enthalpy term, it is dependent on factors, such as temperature, composition, and molecular weight distribution, and includes the entropy change associated with any volume change on mixing as well as other smaller contributions. X also depends on crosslink density. For a polymer-polymer blend this X is replaced by a mutual interaction term X_{12} when polymer 1 and polymer 2 are blend components. The equation for free energy change of mixing polymer 1 with polymer 2 is given by

$$\Delta Gm/V = kT\{(\emptyset_1)/V_1 \ln\emptyset_1 + (\emptyset_2)/V_2 \ln\emptyset_2 + (X_{12}\emptyset_1\emptyset_2)/V_s\}.$$

Since both V_1 and V_2 (polymer molar volumes) are very large, the term in parentheses is very small. Hence, the sign and the magnitude of ΔGm largely depends on the last term in this equation. V_s, the statistical average volume of the interacting polymer segment, is much larger in a polymer blend than in a polymer solution, because the connectivity of polymer chains leads to the some diminution of neighboring polymer segments from the domain of others. The combinatorial entropy is smaller because of the large chain size. Most segmental contacts are between chains, not within a single chain. The Flory-Huggins model does not provide a method for predicting the value of X_{12}. In working with polymer solutions and polymer blends, X and X_{12} are determined by measuring the free energy change as a function of temperature and composition by osmotic pressure, light scattering viscosity measurements, and neutron scattering. A large database for X and X_{12} is available for polymer-solvent combinations. X_{12} for polymer blends is obtained from the study of phase separation, small angle neutron scattering, and inverse gas liquid chromatography. Very limited data is available for polymer-polymer interaction parameters. Tables 12.1 and 12.2 show some X and X_{12} values both for polymer-solvent and polymer-polymer systems to illustrate the magnitude and dependence on structural factors of the interaction parameters (Barton, 1990). It is instructive that X_{12} for miscible polymer-polymer systems is 5–10 times smaller than X for polymer-solvent systems.

12.2.2 Solubility and Interaction Parameters

In an alternate thermodynamic treatment of polymer blends Flory (1942) had developed an expression for regular solutions of nonelectrolytes. A regular solution is one in which the change in volume on mixing is zero while the

TABLE 12.1 Polymer-Solvent Interaction Parameters (X)

Polymer	Solvent	Interaction Parameter
Polystyrene	Cyclohexane	0.50
Polyvinyl acetate	Methyl ethyl ketone	0.43
Polyvinyl chloride	Tetrahydrofuran	0.14
Polydimethyl siloxane	Cyclohexane	0.40
Cellulose acetate	Acetone	0.45
	Dioxane	0.38
Poly isoprene	Benzene	0.44
	Carbon tetrachloride	0.38
Polyisobutylene	Benzene	0.50

TABLE 12.2 Polymer-Polymer Interaction Parameters (X_{12})

Polymer 1	Polymer 2	Interaction Parameter
PS	PMMA	0.014
PS	BR	0.01
PPO	PS	–0.1
PPO	Poly (o-chlorostyrene)	0.03
PPO	Deutero PS	–0.275

enthalpy of mixing is positive, and the only change in entropy is combinatorial

$$\Delta S = n_1 \ln X_1 + n_2 \ln X_2.$$

Assuming that dissolution requires replacing similar molecules with dissimilar ones, they derived a relationship, between ΔH and the square root of the energy of vaporization, the solubility parameter δ,

$$\Delta H/V = K(\delta_1 - \delta_2)^2 \varphi_1 \varphi_2.$$

Here V is the average molar volume of the two components, K is a constant (nearly = 1), and δ_x and φ_x are the solubility parameters and volume fraction of component x. This treatment affords some inherent experimental advantage over the Flory-Huggins procedure for predicting polymer-solvent and polymer-polymer miscibility. δ for organic liquids can be estimated from physical measurements such as energy of vaporization, boiling point, refractive index, or surface tension. δ for polymers has been estimated by measuring solubility, swelling, or viscosity in a number of solvents. Most importantly, for unknown or experimentally inaccessible materials, both solvents and polymers, δ can be estimated by a group contribution approach from the chemical structure and the molar volume for the molecules (in the case of a solvent) or repeating unit (in the case of a polymer) and existing empirical data on group contributions. These estimates and thus correlations are not unequivocal and often lead to errors in estimation unless the solvents are chosen carefully. Mangaraj (1963) found that the solubility parameter correlation holds well only if the solvents being considered belong to a homologous series, with chemical structures similar to the polymer.

In a further refinement Hansen (1967a,b) and Hansen and Beerbower (1971) have further elaborated the concept and have shown that δ can be described in terms of three components: δd the contribution from dispersion interaction, δp the contribution from polarity, and δh the contribution from hydrogen bonding, and the total solubility parameter δt is the root mean square of the three components:

$$\delta t^2 = \delta d^2 + \delta p^2 + \delta h^2.$$

TABLE 12.3 Solubility Parameter of Elastomer (Mangaraj, 2002)

Elastomer	Total Solubility Parameter		Hansen Solubility Parameters		
	$MPa^{1/2}$	$(cal/cc)^{1/2}$	δd	δp	δh
NBR (wt% AN)					
–18%	18.0	8.9			
–25%	19.0	9.4			
–30%	20.0	9.9			
–39%	21.0	10.4	18.6	8.8	4.2
IIR	16.3	8.0	16.0	2.3	3.3
CR	19.9	9.8	19.4	3.1	2.7
MQ	16.4	8.1	16	–0.04	3.5
SBR	18.1	8.9	17.4	2.9	6.8
EAM	18.6	9.2	16.4	4.3	7.6
EPR/EPDM	17.1	8.4	16.6	8.2	
EVA			17.5	8.6	
Polyisobutylene	16.5	8.1	14.5	2.5	4.7
IR	16.7	8.2	16.6	1.4	–0.8
NR	16.9	8.3			

This concept of solubility parameters is limited to blends with only dispersive enthalpy interactions and combinatorial entropic changes. Quantitatively it does not accurately estimate the very small positive ΔH, which is often required for miscible polymer system. In spite of this limitation, the solubility parameter concept provides a simple capability for estimating values of ΔH and thus understanding and predicting polymer-polymer miscibility. Mangaraj et al. have shown that internal pressure $(\delta E/\delta V)_T$ can be measured directly for most polymers and can be used in the place of δ (Mangaraj et al., 1959). As a guide to miscibility the best that can be done is to match the δ of the candidate polymers as closely as possible. Mangaraj in his various publications has speculated that a difference of 0.1 or more in δ values of the polymer pair can lead to immiscibility (see Table 12.3).

12.2.3 Other Models

In the more common polymer pairs with polarity on the elastomer backbone, chemical moieties in the main chain or side groups may attractively interact with each other to enhance miscibility. Paul and Barlow (1978, 1984) indicates that the existence of exothermic interaction (ΔH –ve or "specific interactions") is responsible for most of the miscible polymer pairs. In these cases, this interactive dissimilarity is the driving force for miscibility. The term "complementary dissimilarity" is often used in reference to this phenomenon. These "complementary dissimilarity" interactions include hydrogen bonding and π-π complexing. Miscibility of polymer pairs such as PVC/NBR, Polymethyl Methacrylate (PMMA)/Polyvinyl Fluoride (PVDF), Polystyrene (PS)/Polyvinyl methyl ether (PVME), EVA/CR, and Styrene Acrylonitrile Copolymer (SAN)/Polycaprolactone (PCL) is due to these effects.

Karasz and McKnight (1977) have developed a model based on segmental interaction parameters in copolymers to quantify these qualitative interactions. Each segment in a copolymer contributes to both a configurational and an enthalpic term due to chemical interaction. A pair of copolymers can be miscible or immiscible, or a single polymer can act as a compatibilizer, or an immiscible blend polymer can act as a compatibilizer. Paul and Barlow have shown that ΔH may be negative in a blend due to dilution of more unfavorable interactions, where at least one component in the blend is a copolymer (Paul and Barlow, 1980).

Prigogine and Mahet authored statistical mechanical theory of solutions for simple molecules based on equations of state concepts. Flory extended this theory to polymer solutions in terms of reduced temperature, pressure, and volume concepts in the Flory Prigogine model. The approach is similar to an equation of state for miscibility and phase separation for small molecules and polymer solution. The Flory Prigogine model has been expanded by McMaster (1973) and later by Patterson and Robard (1978). An advantage of this PVT relationship is the ability to predict polymer-polymer miscibility and phase separation as a function of temperature and pressure. This model also accounts for the role of free volume and volume change on mixing. A model that explicitly accounts for polydispersity in molecular weight, heterogeneity in chemical composition, and stiffness of the main chain has been developed by Koningsveld and Staverman (1968). Sanchez modified the Flory Huggins model to include negative heat and volume change on mixing (Sanchez and Lacombe, 1978).

The effect of polymer-polymer immiscibility is phase separation; that is when the two polymers exist in discrete domains. The small difference in energetics between the miscible and immiscible mixtures for polymers ensures that the phase diagram for binary elastomer blend without specific interactions is complicated by strong pressure and temperature effects. As the temperature is increased, the polymers that are immiscible at lower temperature may become miscible. The temperature for complete miscibility, the upper critical solution temperature (UCST), is lower for blend composition at extremes of the composition rather than in the middle. As the temperature is raised further, often the miscible polymer blends will phase separate into distinct phases, indicating the existence of lower critical solution temperature (LCST). Above this temperature, miscible blends become immiscible and separate into two phases. This advent of immiscibility at higher temperatures is possibly due to a change in the value and sign of ΔH (mixing). The existence of LCST is promoted by negative volume change on mixing. The LCST onset is not affected by molecular weight above a certain critical molecular weight for each polymer pair (Paul, 1981). This confirms that the entropic contribution to LCST behavior is minimal. The significance of phase separation is important to the use of elastomer blends. The blends are masticated in an internal mixer to form the blend. As the previous discussion shows, only the blends made, maintained, and processed until vulcanization at temperatures between the UCST and the LCST will be a miscible system. At other temperatures such blends will be an immiscible mixture leading to inhomogeneous morphology. These immiscible elastomer blends also affect the distribution of compounding ingredients prior to vulcanization and the crosslink density, once the compound is vulcanized.

Schuster, Issel, and Peterseim have obtained accurate solubility parameters of four elastomers, namely NR, HR, EPDM, and *cis*-BR, using inverse gas liquid chromatography (Schuster et al., 1996). The values are consistent with other methods and empirical calculation using group contributions. The estimated Mc at which phase separation would be predicted for all rubbers are by design far below the molecular weights of the common industrial rubbers. Immiscible blends of rubbers have typically coarse-grained morphology and this morphology can be observed by microscopy. The results of this study also show that the critical difference for X between miscible and immiscible blends is about 0.04 at room temperature of an elastomer blend, in line with the speculation by Mangaraj.

In summary, elastomers are miscible where the difference in solubility parameter is small or where specific interaction exists. Otherwise, most polymer blends are immiscible, except at very extreme compositions of the blend.

12.3 PREPARATION

Elastomer blends are prepared by three different techniques, namely latex, solution, and dry blending. Latex blending leads to very fine stable dispersions. Angrove has reviewed latex blending (Angrove, 1967), and Blackley and Charnock have investigated latex blending of NR/BR (Blackley and Charnock, 1973a,b). Solution blending produces coarser dispersions since the low viscosity promotes rapid thermodynamic phase growth by coalescence. The rapid phase growth may be due to the both the low viscosity of the solution as well as the absence of surfactants on the periphery of the polymer latex particles. Keyte and Walters observed this in NR/SBR blends (Keyte and Walters, 1962) when they observed a mixture of SBR encapsulated with NR when the two rubbers were solution blended. Even rapid precipitating with a nonsolvent gave the same type of macro heterogeneity. Blending during solution polymerization reportedly provides better dispersion than solution blending, because the polymer molecules are formed in situ, thereby providing a chance for blending at molecular level. This also helps the incorporation of carbon black. Hybrids of solution and latex blending are also used. In one version the elastomer blend is kept in solution and the black is dispersed in water. The black transfers into the elastomer and produces a master batch. In the reverse process, the black dispersed in a solvent is blended with latex.

Most elastomer blends are prepared by dry blending in an intensive mixer such as a mill, Banbury, or extruder. This efficiently promotes the incorporation of compounding ingredients in one operation. Further blending is facilitated by mastication, which reduces viscosity and provides opportunity for reaction (block or graft polymerization) and covulcanization. Evans and Patridge suggest that incorporation of all the ingredients in one step is desirable (Evans and Patridge, 1963a,b). Chemical assisted blending, where reactive monomers are added to one or both components, provides better blends due to in situ formation of compatibilizer. Angier and Watson have indicated that such graft polymer synthesis occurs in masticated blends of NR, CR, NBR, SBR, and BR (Angier and Watson, 1955, 1960).

12.4 MISCIBLE ELASTOMER BLENDS

12.4.1 Thermodynamics

Miscible blends are most commonly formed from elastomers with similar three-dimensional (Hansen, 1967a,b; Hansen and Beerbower, 1971) solubility parameters. An example of this is blends from copolymer elastomers (e.g., ethylene-propylene or styrene-butadiene copolymers) of slightly different composition, or microstructure. When the forces between the components of the polymer blend are mostly dispersive, miscibility is only achieved in neat polymers with a very close match in Hansen's three-dimensional solubility parameter (Hansen, 1967a,b; Hansen and Beerbower, 1971), such that small combinatorial entropy for high molecular weight elastomers can drive miscibility.

Miscible blends of elastomers differ from corresponding blends of thermoplastics in two important areas. First, the need for elastic properties requires elastomers to be high molecular weight. This reduces both the kinetic rate and the thermodynamic driving force for the interdiffusion and thus formation of a miscible single phase of dissimilar elastomers. Second, elastomers are plasticized in conventional compounding with process oils. The presence of plasticizers leads to both a higher free volume for the blend components and a decrease of the endothermal interactions.

12.4.2 Analysis

(i) Glass Transition

The change of glass transition temperature of an elastomer is an important characteristic of elastomer blends. Above the T_g, the kinetic forces are stronger than the molecular interactions, and diffusion of the elastomer through the bulk of polymer can occur. As a consequence, below T_g elastomers are rigid, while above the T_g, rubbery behavior dominates. At T_g there is a change in modulus of a factor of $10^3/10^4$ within a narrow temperature range encompassing the T_g. A change in activation energy of gas diffusion, specific heat changes, and changes in the coefficient of thermal expansion all occur at the T_g. Single phase blends exhibit a single and unique T_g, generally intermediate between the constituent values. The existence of a single T_g however does not meet the thermodynamic criteria for miscibility. For miscible elastomers, the glass transition temperature as a function of weight fraction can be estimated by various expressions starting with the simple inverse linear form known as the Fox equation (Fox, 1956)

$$1/T_g(\text{blend}) = \{f_1/T_{g1}\} + \{f_2/T_{g2}\},$$

where T_g(blend) is the T_g of the miscible blend and T_{g1} and T_{g2} are the T_g of components 1 and 2, respectively, whose weight fraction in the blend is f_1 and f_2.

(ii) Magnetic Resonance Imaging

Nuclear magnetic resonance (NMR) has been applied to the study of homogeneity in miscible polymer blends and has been reviewed by Cheng (1992) and Roland (1989). When the components of a blend have different T_gs, proton NMR can be used to assess the phase structure of the blend by taking advantage of the rapid decrease of proton-proton coupling with nuclear separation (Albert et al., 1985). For blends containing elastomers of almost identical T_g, proton MAS NMR is applied to blends where one of the components is almost completely deuterated (Vander Hart et al., 1726). Another technique is crosspolarization MAS ^{13}C NMR (Schaefer et al., 1981; Steger et al., 1981). The transfer of spin polarization from protons to the ^{13}C atoms of the deuterated component can occur if these carbons are in proximity (nanometers) to the protons.

(iii) Crystallinity

The formation of miscible rubber blends slows the rate of crystallization (Runt and Martynowicz, 1985; Keith and Padden, 1964) when one of the components is crystallizable. This phenomenon accounts for data that show lower heats of fusion that correlate to the extent of phase homogeneity (Ghijsels, 1977) in elastomer blends. Additionally, the melting behavior of a polymer can be changed in a miscible blend. The stability of the liquid state by formation of a miscible blend reduces the relative thermal stability of the crystalline state and lowers the equilibrium melting point (Nishi and Wang, 1975; Rim and Runt, 1520). This depression in melting point is small for a miscible blend with only dispersive interactions between the components.

Changes in polymer crystallinity have also been employed to study the homogeneity of elastomer blends (Morris, 1967; Sircar and Lamond, 1973).

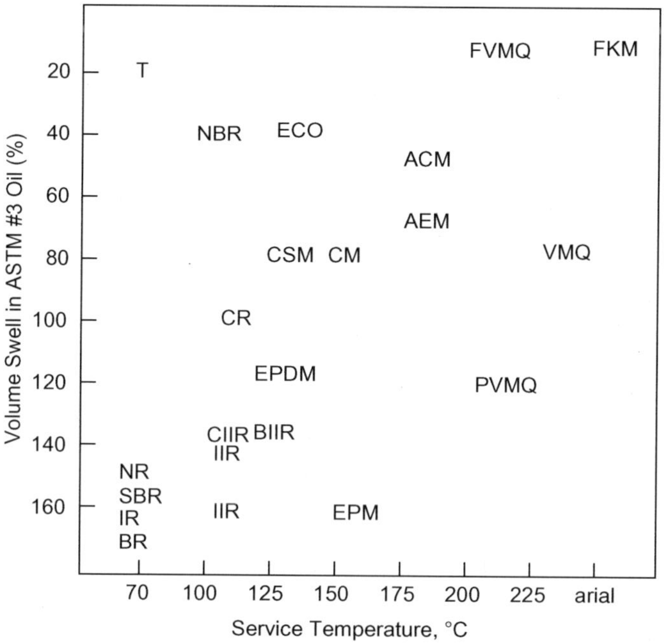

FIGURE 12.1 Elastomer distribution by resistance to aromatic solvents (ordinate) and temperature (abcissae).

Morris (1967) studied the rate of crystallization of *cis*-1,4-BR in blends with SBR. At any given blend composition, the BR crystallization rate diminishes with greater blend homogeneity. Sircar and Lamond (1973) also studied the changes in BR crystallinity in blends with NR, IR, EPM, CIIR, NBR, and CR (Figure 12.1). The nature of the blend component had the greater effect since the more compatible blends (smaller domains) had the greater loss in BR crystallinity.

(iv) Interdiffusion

Interdiffusion between a pair of polymers is a demonstration of their thermodynamic miscibility. The adhesion between contacted rubber sheets parallels the extent of any interdiffusion of the polymer chains (Roland and Bohm, 1985). If the contacted sheets are comprised of immiscible rubbers, no interdiffusion occurs. Natural rubber (NR) and 1,2-polybutadiene (1,2-BR) are miscible even at high molecular weights (Roland, 1988a; Roland, 1987). When NR is brought into contact with 1,2-BR, they interdiffuse spontaneously. When some form of scattering contrast exists between the materials, interdiffusion will enhance the scattering intensity (either X-ray or neutron) measured from the plied sheets. A variety of spectroscopic methods (Klein, 1981;

Klein et al., 1983; Kumugai et al., 1979; Gilmore et al., 1980; Hashimoto et al., 1981; Anderson and Jou, 1987; Summerfield and Ullman, 1987) have been used to detect the interdiffusing species.

(v) Mechanical Properties

Miscible blends should have greater mechanical integrity than a comparable multiphase structure. Miscible rubber blends that react chemically have a densification and a higher cohesive energy density. This may provide improved mechanical properties but has been observed only below the T_g (Kleiner et al., 1979).

12.4.3 Compositional Gradient Copolymers

A significant development (Datta and Kresge, 1988; Datta et al., 1996) in the last decade is the use of miscible blends of compositionally different EPDM. The blends are designed to balance viscoelastic properties such as the rate of extrusion or adhesion with physical properties such as tensile strength by changes in the relaxation characteristics. This is achieved with the components having different average molecular weight or differences in the crystallinity (due to extended ethylene sequences) or both. An example of the components of these blends is shown in Tables 12.4 and 12.5. The blends of component A with one of the B polymers are miscible and are made by mixing hexane solutions of the elastomers. Figures 12.2 and 12.3 show the effect on tensile strength of the compounded but unvulcanized blends from the components in Table 12.4. The blends in Figure 12.2 are different in the molecular weight distribution whereas in Figure 12.3 they are different in composition and crystallinity. This dispersion in molecular weight and crystallinity is apparent in other viscoelastic properties such as peel adhesion. Figures 12.4 and 12.5 show self-adhesion measured by the force needed for the failure of a spliced portion in blends

TABLE 12.4 Ethylene-Propylene Blend Components Differing in Molecular Weight and Crystallinity

Elastomer	Composition C2 (wt%)	Viscosity[a] ML (1 + 4) @125°C
A	60	41
B_1	74	72
B_2	76	247
B_3	78	1900
B_4	68	189
B_5	84	291

[a]*Determined according to ASTM D1646.*

TABLE 12.5 Ethylene-Propylene Blend Components Differing in Crosslink Density

Elastomer	Composition C2 (wt%)	Composition ENB (wt%)	Viscosity[a] ML (1 + 4) @125°C
A_1	57.0	3.2	20
A_2	60.2	2.9	32
A_3	60.3	2.8	41
A_4	59.4	2.6	51
A_5	60.5	3.2	67
B	64	0.9	2100

[a]*determined according to ASTM D1646.*

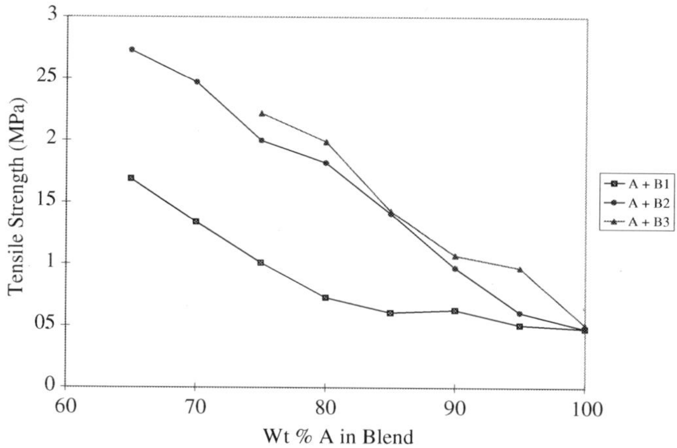

FIGURE 12.2 Variation in tensile strength of unvulcanized, compounded blends of ethylene-propylene copolymer due to differences in molecular weight distribution.

identical to those in Figures 12.2 and 12.3, respectively. Adhesion increases with increasing molecular weight and compositional dispersity of the blend. This is shown by the comparative data for B_1 and B_2 in Figure 12.4, and B_4 and B_2 in Figure 12.5. Small increases in the crystallinity and molecular weight dispersion by blending promote adhesion by slowing down the latter without substantial effect on the former. Further increases in either severely retard the intermingling of chains necessary for the self-adhesion.

Nonuniform vulcanization networks in miscible blends of elastomers have a strong effect on tensile strength and elongation. These networks have an intermolecular distribution in crosslink density and are composed of different concentrations of crosslinkable sites in the components of the blend. Differences in the level of the enchained diene (5-ethylidene-2-norbornene) for EPDM

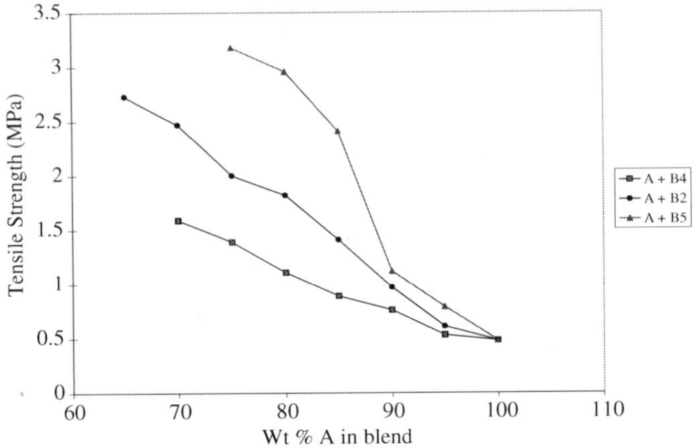

FIGURE 12.3 Variation in tensile strength of unvulcanized, compounded blends of ethylene-propylene copolymer due to differences in composition and crystallinity distribution.

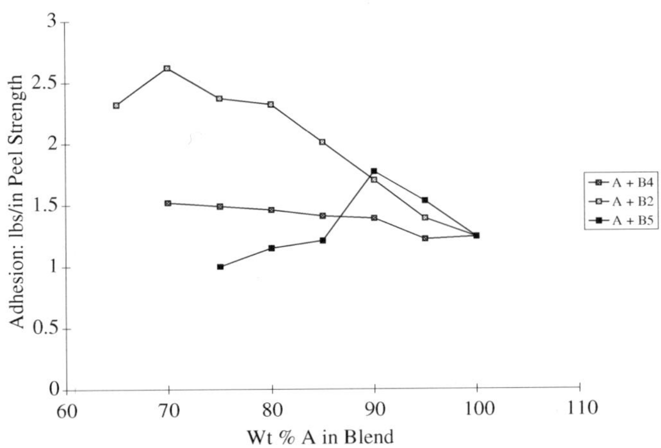

FIGURE 12.4 Variation in peel adhesion of unvulcanized, compounded blends of ethylene-propylene copolymer due to differences in molecular weight distribution.

copolymers (Datta and Kresge, 1988; Datta et al., 1996; Morrar et al., 1996) or differences in the level of the vulcanizable chain end of unsaturation for siloxane polymer define these blends (Zhang and Mark, 1982; Mark and Andrady, 1981). These networks lead to an increase in both the elongation as well as the tensile strength at high elongation compared to vulcanizates of similar viscosity having a uniform network. In particular, nonuniform networks display a nonlinear increase in the tensile modulus at high elongation. This nonlinearity is due to a nonaffine deformation of the network at the high elongation, which continually reallocates the stress during elongation to the lightly crosslinked component of

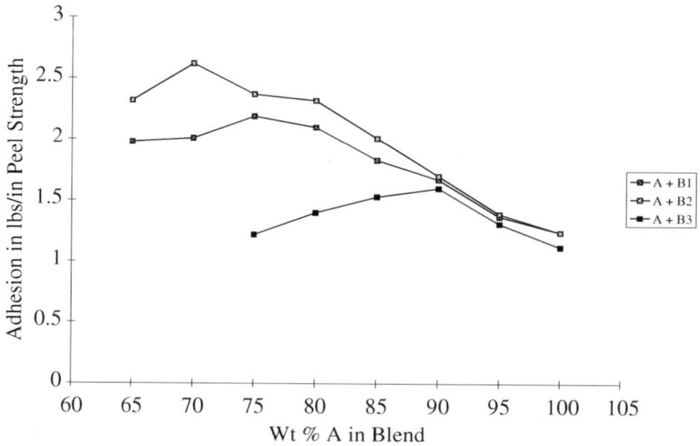

FIGURE 12.5 Variation in peel adhesion of unvulcanized, compounded blends of ethylene-propylene copolymer due to differences in composition and crystallinity distribution.

the blend that is most able to accommodate the strain. Blends of one A and various amounts of polymer B in Table 12.5 were blended in hexane solution, compounded, and vulcanized. Both of the polymers are amorphous, and the A polymers differ in the molecular weight and contain approximately 3% of vulcanizable diene (ENB). The B polymer is much lower in diene and has 0.7% ENB. The tensile strength of the blends derived from all the A with varying amounts of B is shown in Figure 12.6. In all cases where small amounts (<25%) of B are included in the blends, the tensile strength is higher for the blends than for the parent A, even though the extent of vulcanization is lower in the blends than in the blend components A due to reduction in the total amount of diene.

12.4.4 Distinct Polymers

(i) IR–BR Blends

Blends of 1,4-IR with predominantly 1,2-BR are unique since they are miscible, chemically distinct, high-molecular weight homopolymers (Roland, 1988b, 1987; Trask and Roland, 1988, 1989; Roland and Trask, 1989b), even though there is no dipole or specific interaction between the components (Roland, 1987). As the concentration of 1,4-units in the BR increases, there is a decrease in miscibility with IR. The interaction parameter measured for blends of IR with BR of varying 1,2 and 1,4 geometry indicates that the exchange enthalpy between IR and BR becomes more endothermic as the concentration of 1,4-units increases (Trask and Roland, 1988; Roland and Trask, 1989b). Differences in structure are apparent from the thermal expansion coefficients of the components. The difference in thermal expansion coefficients of IR and BR is greatly diminished as the 1,2 content of the latter increases.

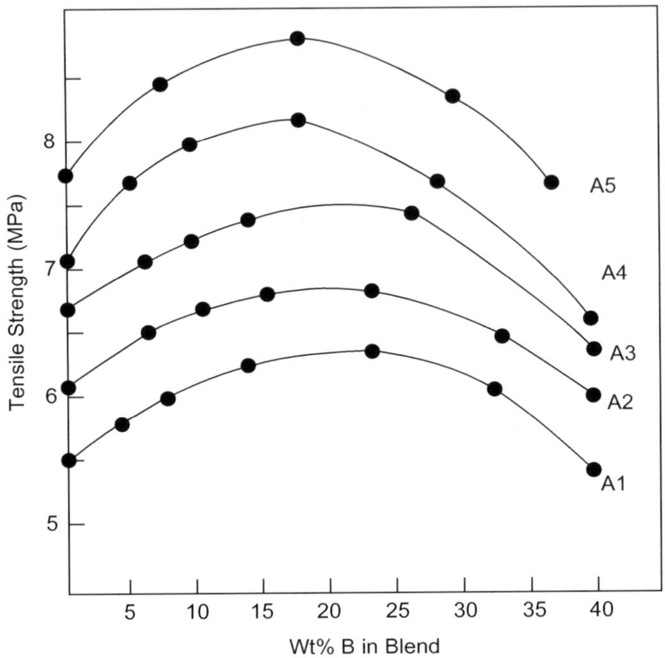

FIGURE 12.6 Tensile strength of blends in Table 12.2, differing in crosslink density.

(ii) Epoxidized PI-CPE Blends

The intramolecular epoxidation of 25 mol% of PI leads to miscible blends with CPE containing 25 wt% chlorine (Gelling, 1987; Margaritis et al., 1987). The origin of the miscibility is the specific interactions involving the oxirane of the epoxy ring with the chlorine.

12.4.5 Reactive Elastomers

The formation of miscible blends of elastomers by mutual chemical reaction of the blend components has been explored by Coleman et al. (1996a,b) and Pehlert et al. (1996, 1997). Chemical reaction in a miscible blend provides the negative free energy needed to compensate for the unlike i-j contacts between dissimilar elastomers. Miscible blends were formed by the use of stabilizing weak hydrogen bonding between a 2,3-dimethyl butadiene-4-vinyl phenol copolymer and carbonyl group of polymers such as EVA and poly alkyl methacrylates. The extent of the H-bonding stabilization was modulated by the concentration of the phenol residues in the unsaturated copolymer. The choice of the stabilizing chemical bond between the elastomers is important. A stable, localized chemical bond would lead to a loss of viscoelastic properties characteristic of an unvulcanized elastomer.

12.5 IMMISCIBLE ELASTOMER BLENDS

12.5.1 Formation

Immiscible blends can be made from any two dissimilar elastomers. They can be made by in situ copolymerization, blending polymer solutions or suspensions, and mechanical mixing. These procedures have been reviewed by Mangaraj (2002) and Corish and Powell (1974). In contrast to blends of other polymers, elastomers can be either "preblended" or "phase mixed" with each other (Corish and Powell, 1974). "Preblended" systems are those where the compounding ingredients (e.g., fillers, plasticizers, and curatives) are added simultaneously to a mixture of the immiscible polymers. In "phase mixed" blends, ingredients are added separately to each of the individual polymers in separate compounding operations. The compounded elastomers are then blended together. Phase mixed blends provide a greater certainty of the initial interphase location of fillers, plasticizers, and curatives than the preblends.

12.5.2 Kinetics of Blend Morphology

The kinetics of the formation of the phase morphology is dominated by two competitive factors. First, the viscoelastic flow of elastomers, the presence of plasticizers, the long resting times (days) between mixing and crosslinking, as well as low shear processing steps (e.g., calendering) are ideal for the development of the equilibrium morphology. However, the sluggish diffusion of high molecular weight elastomers, the presence of particulate fillers, and the ultimate crosslinking generate persistent nonequilibrium morphologies. Tokita (1977) and Avgeropoulos et al. (1976) have studied the mixing of EPDM with the diene rubbers—NR or BR—as a function of relative viscosity.

The detailed morphology of elastomer blends depends on (1) the mixing procedure, (2) the rheology of the blend components, and (3) the interfacial energy. As with other polymer blends, the elastomer of lower viscosity tends to be the continuous phase (Takenaka et al., 1987; Yang et al., 1986). Cocontinuous blend morphology is observed only for elastomers with similar viscosities. The viscoelastic forces developed during the formation of the compounded blend from two rheologically dissimilar elastomers are principally responsible for the morphology. Interfacial tension due to chemical differences between the elastomers is less important (Kumaki and Hashimoto, 1986).

12.5.3 Analysis

(i) Microscopy

Phase contrast light microscopy has been applied extensively to the analyses of unfilled binary elastomer combinations. This method is based on differences in the refractive indices of the polymers and has been reviewed by Kruse (1973). Callan et al. (1965,1971a), and Scott et al. (1969) have shown the versatility of

TABLE 12.6 Average Areas (in μm^2) of the Dispersed Phase in 75:25 Pure Elastomer Blends

Disperse	Matrix 75%							
Phase 25%	NR	CR	BR	SBR	NBR	EPDM	IIR	CIIR
NR	–	45	1.5	1.2	300	1.5	2.0	3.2
CR	35	–	4.0	2.5	1.5	25	20	15
BR	0.7	4.5	–	–	15	2.2	2.1	2.5
SBR	0.5	2.7	–	–	20	2.1	12	10
NBR	400	1.3	17	30	–	250	100	225
EPDM	3.5	75	2.8	2.6	225	–	2.0	1.5
IIR	3.0	15	3.0	4.2	75	1.0		
CIIR	2.2	25	2.3	2.5	85	1.2		

the method for a wide range of binary blends containing NR, SBR, BR, CR, NBR, EPDM, IIR, and CIIR. The results of these experiments are shown in Table 12.6, which lists the measured areas of the disperse phase in more than 50 combinations of 75/25 binary blends mixed in a Banbury containing eight different elastomers. Blends of IIR-CIIR and SBR-BR are excluded since the contrast was low. It can be seen that NBR produced the greatest heterogeneity in all blends except those with CR.

Transmission electron microscopy (TEM) is applicable to both filled and unfilled elastomer blends. However, for most elastomer combinations, there is no contrast between the polymer phases in a TEM. If the polymers differ significantly in unsaturation, osmium tetroxide (OsO_4) or ruthenium tetroxide (RuO_4) staining is the best method for obtaining contrast. The metal oxides selectively oxidize the unsaturation and the location of the metal atoms in the phase with the greater unsaturation provides the electron density contrast for TEM. Contrast for TEM analysis of elastomer blends of diene rubbers is achieved through the sulfur hardening method developed by Roninger, 1933 and utilized by Smith and Andries, 1986. Small rubber specimens are immersed in a molten mixture of sulfur, an accelerator, and zinc stearate. The selective absorption of the zinc salts into the SBR results in phase contrast darker than the BR in a TEM.

Scanning electron microscopy involves simpler specimen preparations than TEM. Both OsO_4 and RuO_4 staining techniques work with SEM and can be applied to both bulk specimens and films. Atomic force microscopy is an extension of scanning tunneling microscope (STM) and has the potential for atomic resolution. In its simplest form, the AFM (Stroup et al., 1993; Stocker et al., 1992; Tsao et al., 1992; Hamada and Kaneko, 1992) acts as a "miniature surface profilometer" and provides topographical images. The potential advantages of these techniques include higher resolution, simplicity of specimen preparation, and greater versatility in varying the mechanisms for achieving image contrast. Digital image analysis has expedited particle size analysis from micrographs (Sax and Ottino, 1073; Nishi et al., 1988; Kruse, 1973; Vesely and Finch, 1988) at resolution limits of a few Angstroms.

Atomic Force Microscopy has been an area of extensive development for elastomer blends. This method is flexible since it may be used for miscible, compatible, or phase separated blends in the presence of fillers and vulcanization agents. BIIR/NR blend (Achalla et al., 2006) shows the versatility of the combined AFM approach to map the incompatible phases of the elastomeric blend. AFM is a tool commonly used to image the surface topography and surface heterogeneity of polymeric materials, organic and inorganic materials, and biological molecules at nanometer scale resolution and under ambient conditions. Previously, transmission electron microscopy (TEM) has been used to visualize the phases of an unfilled polymer blend system. Although this technique can be very informative, the sample preparation is difficult and time consuming.

Recently, a number of scanning probe microscopy techniques have emerged that can provide more direct and revealing information of surfaces. In addition to the imaging at higher resolution, AFM is also capable of providing a general understanding of the local material properties of the heterogeneous samples (Sultan and McGarry, 1973; Kinloch et al., 1983; Mélé et al., 2002; Paredes et al., 2005; De Wolf et al., 2001).

(ii) Glass Transition Temperature

Analytical procedures to differentiate cure levels in blends of elastomers include sol-gel analysis, dynamic mechanical thermal analysis (DMA), differential scanning calorimetry, stress-strain measurements, swollen-state NMR spectroscopy, network visualization microscopy, and isopotential swelling (Mangaraj et al., 1987; Mangaraj, 1987). The most common method of estimating the degree of homogeneity in elastomer blends is by measurement of the temperature of transitions from rubber to glass (Buchdahl and Nielsen, 1955; Morris, 1967; de Decker and Sabatine, 1967; Bohn, 1968; Scheele, 1971; Ramos and Cohen, 1977; Mazich et al., 1986). For elastomers this is usually at subambient temperatures. Glass transition measurements do not provide any information on blend morphology. The observation of distinct transitions corresponding to the respective components of the blend indicates the existence of a multiphase structure. A potential source of error is that vulcanization tends to raise the T_g due to restricted motion of the chain segments; this might be interpreted as miscibility.

(iii) Magnetic Resonance Imaging

The state of cure of the phases in a blend can be determined from changes in the magnitude of the damping peaks (Runt and Martynowicz, 1985) and from freezing-point-depression measurements on swollen networks. Keith and

Padden (1964) have employed the freezing point depression of solvent-swelled NR–BR blends to determine the crosslink densities of the individual polymer phases. The freezing point depression of a solvent in a swollen vulcanizate is dependent on the volume fraction of solvent in the elastomer phase.

Swollen state NMR spectroscopy is the most unequivocal estimate of crosslink densities in different phases. Signals of NMR spectra of polymers are broader than those of simple molecules and the signal width is increased as the polymer is crosslinked due to reduction in chain mobility. Swelling of the vulcanizates permits observation of spectra with sufficient resolution. Hydrogen nuclei of different elastomers that are in different environments are easily observed and are the most convenient tool. Blends of natural rubber (NR) and nitrile rubber (NBR) with 18% acrylonitrile were first investigated in respect of crosslink distribution. The degree of mismatch of the cure level decreased with decrease in acrylonitrile content as did replacement of sulfur with a sulfur donor bis-alkylatophenol disulfide. These studies have shown that the physical properties of the blend vulcanizates are much higher when cure levels are uniform in both elastomer phases. The use of carbon black as filler does not alter the adverse effect. 60:40 blends of NR/EPDM with two different ENB content (1.1% and 10.5%) were vulcanized; the EPDM phase for the first blend (low NB) was hardly crosslinked. The curing of the second EPDM phase was higher compared to the first one, but lower than the NR phase.

NMR imaging of solids has been used to characterize the phase sizes with a spatial resolution of less than 50 μm in immiscible mixtures containing polybutadiene. Butera et al. (1987) have reported improvement over the ^{13}C NMR method by using magic angle spinning (MAS) in polymer blends containing a deuterated and a protonated polymer. More recently, Krejsa and Koenig (1991, 1992), Smith and Koenig (1991), and Sarkar and Komoroski (1992) have used NMR for the determination of vulcanization efficiencies in elastomer blends. Swollen state magnetic resonance has been developed to determine the filler retention and crosslink density of elastomer blends (Wang and Wang, 1997a; Groves, 1998; Loadman and Rubber, 1989).

(iv) Light, X-ray, and Neutron Scattering

Roland et al. (1984) used both SAXS and SANS in conjunction with TEM to assess the size of butadiene domains (5% by weight) in a CR matrix. Roland et al. used different master batches containing deuterated and protonated butadiene, respectively. These were each blended on a two-roll mill to study the coalescence of the polybutadiene phase. Coalescence (measured by SANS) was minimized by low temperature mixing, or by significantly increasing the molecular weight of the disperse or continuous phase to create a greater disparity in viscosity. Some preliminary SANS work on IR blended with deuterated 1,2-BR has been reported (Trask and Roland, 1989; Roland et al., 1984). All scattering techniques are affected by the presence of fillers, which lead to more extensive scattering than the phase morphology. This is a severe limitation to the wide applicability of these techniques to most elastomers.

(v) Freezing Point Depression

Freezing point depression has been used to determine the composition of immiscible phases elastomer blends though the results are sometimes difficult to interpret and correlate to the components of the blend (Wang and Wang, 1997a; Honiball and McGill, 1988).

12.5.4 Interphase Distribution of Filler, Curatives, and Plasticizers

A large number of ingredients including fillers, plasticizers, vulcanizing agents, promoters, and antioxidants are mixed with a rubber to carry out effective vulcanization and to provide physical properties. Further, whereas most thermoplastics are processed much above their melting point, rubber compounds are mixed with polymers as highly viscous. It is, therefore, difficult to get uniform distribution of ingredients in a rubber compound. This nonuniform distribution of compounding ingredients for elastomer blends is modified further because there may be thermodynamic or kinetic partitioning leading to further skewed distribution of one or more ingredients in one component compared to the other, due to the polarity or surface characteristics of the additive. This may influence the morphology and thus the physical properties of the vulcanized rubber blend. This potential of uneven distribution of compounding ingredients in elastomer blends and its net effect on physical properties of the compound is thus important. The uneven distribution of fillers and curatives is accelerated by processing and plasticization but retarded by the ultimate vulcanization. Retention of the favorable properties of a metastable blend, which is often attained only at a select interphase morphology and filler/plasticizer distribution, thus requires careful control of both the processing and the vulcanization procedures.

The compounding ingredients can be either reactive or inert. Vulcanizing ingredients such as sulfur, accelerators, and promoters participate in the vulcanization. Their uneven distribution in the different elastomers in a rubber component is influenced not only by mechanical mixing, but also by their migration during vulcanization. Rubber vulcanization requires several minutes, enough for the reactive ingredients to migrate across phase boundaries. The nonreactive ingredients include fillers, plasticizers, processing aids, and antioxidants.

(i) Curative and Plasticizer Migration in Elastomer Blends

During mixing, curatives are initially located within the continuous phase (Leblanc, 1982). Since the curatives dissolve in the elastomer, curative migration across phase boundaries can occur (Van Amerongen, 1964;

Shershnev, 1982; Huson et al., 1985; Bauer and Crossland, 1988). Owing to the higher solubility of sulfur in elastomers containing diene or styrene groups and the greater affinity of many accelerators for polar rubbers, large differences in crosslink density of the different phases result in vulcanization. Further, increased rate of vulcanization in the diene or styrene containing elastomers can cause depletion of the curatives in this component, leading to even greater curative migration (Le Chatelier's principle). For most elastomer blends, these effects are in concert, and the result is a large cure imbalance (Bhowmick and De, 1980) between the phases. In addition, the boundary layers of the two elastomers, adjacent to the interface, can be different than the bulk since the greatest migration of curatives and plasticizers occurs in the proximity to this region. The transfer of curatives to one phase at the expense of the other creates a morphology where the boundary layers separate a curative-depleted, slightly crosslinked elastomer near a curative-enriched, tightly crosslinked elastomer.

Amerongen, 1964 has an early review on curative migration in heterophasic elastomer blends based on optical and radiochemical analyses. A later, more detailed work by Gardiner (1968, 1969, 1970) used optical analysis to study curative diffusion across the boundaries of elastomer blends consisting of binary combination of polymers of CIIR, IIR, EPDM, CR, SBR, BR, and NR. Gardiner determined that curatives such as sulfur, TMTD, MBTS, and DOTG migrate from compounded rubber to uncompounded ones and between compounded rubbers from those with low unsaturation to the ones with high unsaturation even if the initial concentrations are equal. He also showed that sulfur distribution from NR to SBR is gradual, whereas from NR to Butyl it is abrupt. This is due to large differences of sulfur solubility in NR and IIR. MBTS diffusion is similar but slower due to larger molar mass. This imbalance in sulfur and accelerator concentration leads to differential rates of vulcanization, which can be avoided to some extent by selecting a proper mixing protocol, or by using curatives where the rate of diffusion is independent of rubber polarity. Gardiner measured a diffusion gradient for the concentration change as a function of distance and time. His measurements for the diffusion of accelerator and sulfur from IIR to other elastomers are listed in Table 12.7. Amidon and Gencarelli (1972) have found that the use of long-chain dithiocarbamates, which presumably have less solubility parameter discrepancy with less polar elastomer and slower diffusion, provides more uniform crosslinking in the blends of EPDM and unsaturated rubbers. Similar observations have recently been made for long-chain thiuram disulfides.

(ii) Cure Compatibility

Adequate properties in a vulcanized rubber blend depend on the covulcanization across the phases. Covulcanization is the formation of a single network structure including crosslinked macromolecules of both polymers. The degree of vulcanization is at similar levels in both phases with crosslinking across

TABLE 12.7 Curative Diffusion Coefficients in Elastomer Blends

Curative	From	To	$D \times 10^7 (cm^2/s)$
Accelerator (TDDC)	IIR	BR	12.66
		EPDM	1.09
		CR	1.08
		SBR	0.58
		NR	0.70
Sulfur	IIR	SBR	4.73
		SBR w/50 phr N700 CB	17.2
		NR	2.82

the phase interfaces. Shershnev (1982) and earlier Rehner and Wei (1969) have summarized the experimental requirements for covulcanization of the components of elastomer blends. These are:

Phase blending followed by very short, high-temperature cure cycles (Bhowmick and De, 1980).
Vulcanization agents chemically bound to the elastomer (Baranwal and Son, 1974; Hashimoto et al., 1976; Coran, 1988).
Accelerators similar in solubility parameter to less reactive elastomers (Amidon and Gencarelli, 1972; Mastromatteo et al., 1971; Sumitomo Chemical Company, 1973; Woods and Mars, 1974).
Insoluble vulcanizing systems that cannot migrate.
Nonpolar vulcanizing agents that distribute uniformly and have similar reactivities toward different elastomers (e.g., peroxides and reactive resins).

Gardiner (1968, 1969, 1970) and Woods and Davidson (1976) have improved the covulcanization of EPDM-NR and IIR-NR blends by using an insoluble lead-based curative for the less unsaturated elastomer. The blends with soluble curatives from the same elastomers were much inferior. Coran, 1988 covulcanized EPDM-NR blends by modifying the EPDM with maleic anhydride. The modified EPDM can be crosslinked with zinc oxide to form an ionic crosslink network. The migration of curatives was investigated by Bhowmick and De (1980) in binary blends of NR, BR, and SBR. Cure rates decreased in that order, and the migration of curatives was the greatest for NR-SBR.

Vulcanization of the phase interface of a rubber blend is important to provide strength and dynamic performance. This interfacial covulcanization is difficult in blends of low unsaturation rubbers such as EPDM with high diene elastomers. It can be approached by fast accelerators such as thiuram di- and tetrasulfides. Yoshimura and Fujimoto (1999) and later Corish (1967) showed that initial separate DMA peaks merge on continued vulcanization to give a single peak, characteristic of compatible or a strongly coupled blend. Baranwal and Son (1979) indicate that grafting accelerators to component elastomers improve properties for NR/EPDM blends.

Coran found that the use of maleic anhydride grafted EPDM when blended with NR provides higher modulus and tensile strength than ungrafted blends and he attributed this to the formation of ionic crosslinks formed by ZnO with maleic anhydride (Coran, 1988). In blends with diene rubbers, EPDM modified with a small amount of *N* chlorothio-*N* Methyl p-toluene sulfonamide leads to an increase in the crosslink density in EPDM from 4 to 20 mol/m^3 and the tensile strength of the blend increased (Tinker, 1994). It is likely that MBT (2-mercapro benzothiazole) displaces *N* methyl p-toluene sulfonamide forming a crosslink precursor, which helps in establishing crosslinks in the EPDM phases. Similar crosslink precursors have been formed by the modification EPDM by sulfur donors such as thiodimorpholine, dithiodicaprolactam, and BAPD.

(iii) Interphase Filler Transfer

Fillers, such as carbon black, silica, talc, calcium carbonate, and clay, are used in a rubber compound. Some of them act as reinforcing fillers. The physical properties of many rubber compounds are inadequate without such reinforcing fillers. The distribution of fillers (e.g., carbon black, silica) and various processing aids in heterogeneous elastomer blends can be nonuniform. Interphase transfer of fillers has been observed in blends of both diene and saturated elastomers. This migration is due to the greater solvation between a filler and one of the polymers in the melt-mixed blends. Callan et al. (1971a) showed significant transfer of carbon black from NR to BR and silica from BR to NR in solution and latex blends. Extensive carbon black transfer from a low unsaturation elastomer (IIR and EPDM) to other high unsaturation (e.g., diene) elastomers occurs regardless of the master batch mixing procedure (Callan et al., 1971a). This is seen in blends of EPDM or IIR with NR or SBR. During mechanical mixing, chemisorption of the unsaturated polymer onto the carbon black occurs, thus preventing any subsequent transfer of the carbon black.

In all instances, the carbon black was located almost entirely in the NR phase when added to the preblended elastomers. Transfer of carbon black from EPDM to NBR could be reversed if the EPDM was functionalized with a primary amine (Keith and Padden, 1964). Amines are tenaciously chemisorbed to the surface of carbon black, and the functionalization allows the EPDM phase to retain significant amounts of filler. The carbon black distribution is shown in Figure 12.7a for the blend of EPDM with NBR and is compared to Figure 12.7b for the blend of the amine functionalized EPDM with NBR.

Callan et al. (1971a,b) used the differential swelling technique (*vide infra*) for a ranking of the relative affinity of carbon black to various elastomers.

TABLE 12.8 Effect of Carbon Black on the Distribution in 50:50 Natural-Rubber/SBR Blend Containing 45 Phr Black

Carbon Black	Surface Area	Bound Rubber (%)	Carbon Back loading, phr	
	(CTAB) (m^2/g)		*NR*	*SBR*
N560	43	17.0	41.2	48.8
N347	87	28.0	33.9	56.1
N339	95	30.2	31.1	58.9
N234	120	30.4	30.0	60.0
Vulcan (10H)	136	33.1	27.2	62.8
N326	84	22.5	31.9	68.1

Callan and coworkers studied the distribution of carbon black in 50/50 blends of different rubbers and found that the black affinity decreases in the order of BR, SBR, CR, NBR, NR, EPDM, and IIR. Transfer of black takes place from saturated rubber master batches to those of unsaturated rubbers. These conclusions are in agreement with the findings of Sircar and Lamond (1973). Vonwinkel (1969) showed that carbon black accumulates in the BR from an analysis of 50/50 binary blends. The carbon black retention in each phase was estimated from TEM micrographs of the blend. The ranking of carbon black affinity for different elastomers was in the order SBR > BR, CR, NBR > NR > EPDM > CIIR, IIR. Cotton and Murphy, 1988, in a complementary experiment, have determined the carbon black distribution for seven different carbon blacks in preblended SBR-BR and SBR-NR blends. The data is shown in Table 12.8. The carbon blacks differ in the surface structure and size: they ranged in surface area (CTAB) from 43 to 136 m^2/g. For all cases of SBR-NR blends, the carbon black was preferentially located in the SBR phase.

Ayala et al. (1992) have studied carbon black-rubber interaction for SBR, NR, IIR, and NBR using nitrogen adsorption, dibutyl phthalate absorption by moisture absorption, and adsorption of model hydrocarbons using inverse gas chromatography. In compounds of 45 phr of five different carbon blacks (Nl21, N231, N330, N650, and N472) with these and measured bound (inextractable) rubber content. The mechanical static and dynamic properties of the vulcanized compounds indicate strong interaction with SBR followed by NBR, NR, and HR. Bound rubber content followed this order. The difference in the degree of interaction between carbon black and different rubbers may lead to nonuniform distribution of carbon black in elastomer blends and the resulting change in their mechanical behavior. Polar elastomers contain large amounts of bound rubber, which increases with increase in filler loading.

12.5.5 Analysis of Interphase Transfer

(i) Microscopy

Electron microscopy is a common technique for determining the filler distribution in a heterogeneous elastomer blends. Dias and Galuska (1996) have used time of flight secondary mass spectrometry (TOF-SIMS) to simultaneously determine the morphology as well as curative diffusion in BIMS-diene elastomer blends.

(ii) Differential Swelling

The method of differential swelling of thin section of blends of EPDM and IIR was first utilized by Callan et al. (1951). In blends with EPDM, the IIR phase absorbs more solvent; therefore, the IIR domains are thinner and appear lighter in the TEM. Hess et al. (1965) applied the same differential swelling method to the analysis of carbon black distribution at low filler loadings. Wang and Wang (1997b) have improved the technique by swelling separately with two solvents—one for each of the phases.

(iii) Staining

Staining with volatile reactive metal oxides, OsO_4 and RuO_4, is the preferred method for achieving interphase contrast for TEM analyses. It is applicable to blends of elastomers with different degrees of unsaturation such as NR-EPDM blends.

(iv) Differential Pyrolysis

This method is applicable to blends containing polymers with significantly different thermal degradation temperatures. It has been used for analysis of carbon black distribution in NR-SBR and NR-BR blends (Hess et al., 1985a; Hess and Chirico, 1977).

(v) GC Analysis of Bound Rubber

Bound rubber is the elastomer insoluble in solvents due to chemisorption onto the carbon black during mixing. It is extracted by swelling the unvulcanized polymer in a solvent for an extended period. Any soluble lower molecular weight polymer that is not bound to the carbon black is removed. This method was first used by Callan et al. (1951) for a number of elastomer blends.

(vi) Mechanical Damping

The value of tan δ (at T_g) is lower for a filled elastomer than the pure elastomer (Medalia, 1978a; Fletcher and Gent, 1984). This is due to the increase in dynamic elastic modulus of the filled compound for the higher temperature side of the T_g peak. The effect is governed by filler concentration and loading.

Maiti et al. (1992) have used this lowering of tan δ at T_g to estimate the distribution of filler in an immiscible elastomer blend.

12.5.6 Compatibilization

Since phase separation due to incompatibility is common among elastomers there is usually poor interface and poor adhesion between the components. This leads to poor macroscopic properties. The properties of such blends are often poorer than the individual components. Improved performance characteristics can be accomplished by compatibilizing the blend, either by adding an interfacial agent, called a compatibilizer, or by enhancing the interaction of the two component polymers, chemically or mechanically. The role of the compatibilizer is to: (1) reduce interfacial energy and improve adhesion between phases, (2) achieve finer dispersion during mixing, and (3) stabilize the fine dispersion against agglomeration during processing and throughout the life. The objective of compatibilization is to have a morphology that will allow stress transfer, without interfacial failure from one phase to the other, and allow the product to resist failure under multiple stress. In addition, for elastomer blends compatibilization may be necessary to aid uniform distribution of fillers, curatives, and plasticizers.

Compatibilization of highly incompatible elastomers has only been used to a limited extent (Woods and Mass, 1975; McDonel et al., 1978; Leibler, 1982; Noolandi and Hong, 1982). The amount and composition of the interfacial agents are not designed to affect the bulk properties of either of the phases. Properties of elastomer blends are determined by intensive properties such as cohesive energy, crosslink densities, and chemisorption of fillers of the components that are unaffected by the addition of compatibilizers. However, in binary blends of elastomers with large differences in solubility parameters, as in polyolefin elastomers with polar elastomers, properties are dominated by the large domain size and the lack of interfacial adhesion. These elastomer blends are significantly improved by the addition of a compatibilizer.

Setua and White (1991a,b) used CM (chlorinated polyethylene) as a compatibilizer to improve the homogeneity of binary and ternary blends of CR, NBR, and EPR. NBR-EPM and CR-EPDM blends homogenize more rapidly when small amounts of CM are added. The presence of the compatibilizer leads to reductions in both the time needed for mixing, observed by flow visualization, and the domain size of the dispersed phase, observed by SEM. Arjunan et al. (1997) have used an ethylene acrylic acid copolymer and an EPR-g-acrylate as a compatibilizer for blends of EPDM-CR. The addition of the compatibilizer leads to the reduction in the phase size of the dispersed EPDM phase as well as increase in the tensile tear strength of the blend.

Intensive properties of the blend components that dominate vulcanizate properties of the elastomer are improved if the compatibilizer is the predominant fraction of the elastomer blend. Davison et al. (1982) describe the formation of compatibilized blends of poly(alkyl acrylates) and preformed EPDM-g-acrylate that on vulcanization is resistant to solvents. The acrylate-grafted EPDM are copolymers of methyl-, ethyl-, or *n*-butyl acrylate. These vulcanized blends have excellent tensile strength and modulus, similar to a single elastomer. Huson et al., 1985 were able to generate the graft polymer in situ during compounding. A primary amine copolymerized EPDM blended with NBR to form a graft polymer by the amidine reaction of the amine with the nitrile. The grafting reaction was catalyzed by the presence of the Lewis acidic phophite plasticizer for the NBR phase. The compatibilizer promotes the formation of very small dispersed phase domains. Figure 12.8a and b are micrographs of the dispersion of the EPDM and amine functionalized EPDM, respectively, in the NBR matrix. In this case, the previous micrographs (see Figure 12.7) also show the retention of carbon black filler in the bulk of the EPDM phase. Vulcanization of the amine functional EPDM and NBR blends with nonpolar peroxides that are expected to distribute equally in both phases and lead to blends with excellent solvent and temperature resistance.

Reiss et al. (1967) have shown that for IR/PS, block copolymers provide better compatibilization than graft copolymers, and solubilization of compatibilizer by phases takes place when the molecular weight of the blend components are comparable or smaller than the molecular weight of corresponding block in the compatibilizer. Teysie et al. (1987) had concluded that the structure and the molecular weight of the copolymer control the efficiency of compatibilization and tapered block copolymers are more effective as compatibilizer than linear block copolymers. Gallard et al. (1980) found that addition of styrene-butadiene block copolymers reduced the interfacial tension in PS/BR blends. Coran and Patel (1985) during their investigation on thermoplastic elastomers based on blending of polypropylene with several elastomers, found that parameters such

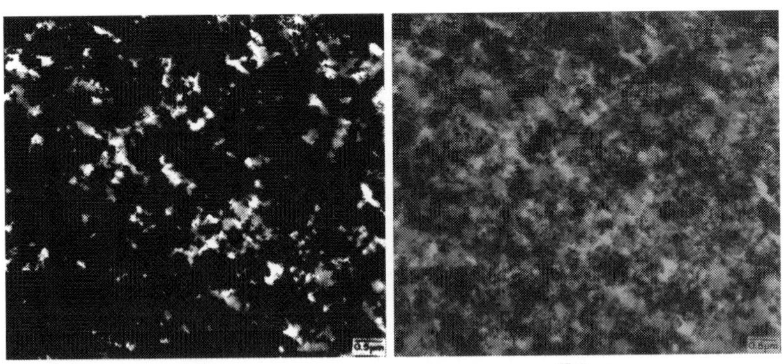

(a) (b)

FIGURE 12.7 Micrographs of a N234 carbon black distribution in 70:30 blend of NBR and EPR. (a) The EPR is a copolymer of 42 mol% ethylene. (b) The EPR has 43 mol% ethylene and 0.9 mol% primary amine functionality.

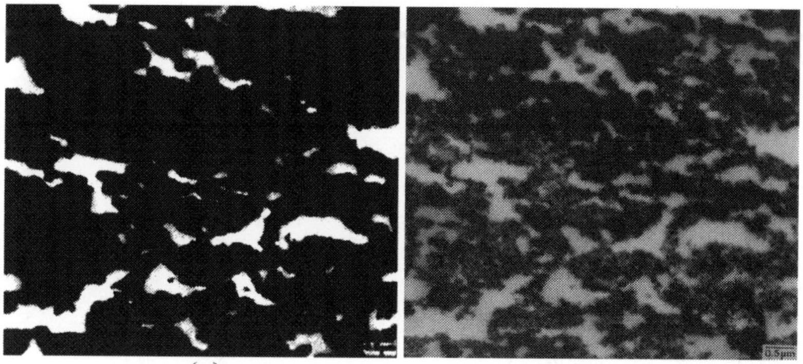
(a) (b)

FIGURE 12.8 Micrographs of a dispersion of 70:30 blend of NBR and EP elastomer. (a) The EP elastomer is a copolymer of 42 mol% ethylene. (b) The EP elastomer has 43 mol% ethylene and 0.9 mol% primary amine functionality.

as critical surface tension for wetting the interface, critical entanglement spacing, crystallinity, and tensile strength of the hard phase determine the mechanical properties of the blend. Leibler et al. (1981) who developed the thermodynamic basis for copolymer compatibilization, suggest that reduction of interfacial tension takes place due to adsorption of the copolymer at the interface, and an asymmetric copolymer is less efficient than a symmetric one.

12.5.7 Properties of Immiscible Blends

While true miscibility may not be required for elastomer properties, adhesion between the immiscible phases is required. Immiscible polymer blends that fulfill this criteria provide a significant opportunity to change the rheological, tensile, and wear properties of elastomer blends compared to miscible blends.

(i) Processing

Blends are often used to improve the processability of rubbers. This improvement may consist of either lowering the viscosity or producing a material less prone to melt fracture during flow. Secondary elastic effects such as die swell can also be affected by blending. Avgeropoulos et al. (1976) showed that the low viscosity phase in a binary blend tends to become continuous. The effect is accelerated under shear as the morphology of the blend responds to the applied stresses. In the vicinity of a wall, the shear rates tend to be highest, and the lower viscosity component will accumulate at the surface (Meinecke and Taftaf, 1988). Incorporation of only a few percent of EPDM to a fluoroelastomer or of PDMS into an SBR (Medalia, 1978b) was found to reduce steady state viscosities. This is due to the lower viscosity component residing at the interface.

Much of the knowledge in this area is either derived from practical experience or anecdotal. Theoretical predictions of the viscosity of elastomer blends (Yang et al., 1986; Meinecke and Taftaf, 1988; Medalia, 1978b; Meinecke and Taftaf, 1988) are of limited use since (1) the inhomogeneous phase morphology of an elastomer blend changes easily in response to applied stress, and (2) the nonuniform distribution of fillers and plasticizers in the phases responds to flow. These structural changes in elastomer blends under shear lead to anomalous rheological properties quite different from the expected average of pure components.

(ii) Modulus

In a heterogeneous blend, the details of the morphology do not generally exert much influence on the stress-strain tensile response. Contrary to expectation that the continuous phase would have more influence, the stress-strain response of unfilled EPDM-BR blends was found to be unaffected by a change in the BR domains from continuous to disperse (Morrisey, 1971).

Carbon black distribution has a profound effect on modulus. Hess et al., 1965 have shown that an increase in the nonuniformity of this filler distribution results in a lower blend modulus. The transfer of a portion of the carbon black from one phase lowers its modulus more than the increase in modulus of the phase with the higher carbon black concentration. This effect is related to the nonlinear dependence of rubber modulus on carbon black loading (Morrisey, 1971).

(iii) Tack and Adhesion

Adhesion is a surface phenomenon due to entangling of chains at the surface. Adhesion of different elastomers before (tack) and after vulcanization (co-cure) can often be obtained only through the blending of the dissimilar elastomers with a single component. However, the composition of the surface and thus the adhesion characteristics can be altered without using a high concentration of a particular elastomer in the blend.

Blends with components that differ in viscosity tend to have the lower viscosity rubber concentrated at the surface during processing. The most common procedure for increasing the tack and co-cure of elastomer blends is to have a single elastomer as the predominant phase in each of the blends. Morrisey (1971) showed that tack of dissimilar elastomers blended with NR improves monotonically with the amount of NR in the blend. Increasing the proportion of a single elastomer in both blends enables the ability of the plied surfaces to fuse together.

(iv) Hysteresis

A principal use of elastomer blends is in sidewalls of automotive tires. The reduction of hysteresis losses ("rolling resistance") is a principal design target.

Lower hysteresis in a single elastomer requires reduced carbon black loading or increased crosslink density. These changes adversely affect other aspects of performance. Blending of elastomers provides a lower hysteresis with few of these adverse effects. The hysteresis of a filled elastomer containing zones of different carbon black concentrations is lower than a uniformly filled elastomer. Blends of dissimilar elastomers, with differing ability to retain carbon black, provide an easy experimental process to obtain this nonuniform carbon black distribution.

(v) Failure

Improved failure properties can result from the blending of elastomers, including performance that exceeds either pure component. An important aspect of the structure of a rubber blend is the nature of the interphase bonding. The mechanical integrity of an interphase crosslinked morphology will usually lead to superior performance. In blends of SBR and chlorobutyl rubber, for example, an increase in fatigue life was obtained by the introduction of interphase crosslinking. Similarly, providing for interfacial coupling improves the tensile strength of EPDM/silicone-rubber blends (Mitchell, 1985; Matthew, 1984; von Hellens, 1989; Cesare, 1989; Hong, 1989; von Hellens et al., 1990; Andrews, 1966; Ambelang et al., 1969; Barrier, 1956; Kinning et al., 1987).

There is improved abrasion resistance associated with a preferential carbon black-BR phase distribution in blends of NR-BR and SBR-BR. The first abrasion studies on the effects of carbon black phase distribution in NR-BR blends were reported by Krakowski and Tinker (1990a,b). Tread wear resistance was found to increase progressively with increasing carbon black in the BR phase, which was determined from TEM analyses. Tse et al. (1998) have shown that blends of dispersed BIMS in BR matrix failure due to fatigue can be retarded if the mean distance between the crosslinks of the BIMS is less than 60 nm.

The incidence of cracking due to ozone attack has been investigated for NR-EPDM blends (Tokita, 1977; Avgeropoulos et al., 1976). Andrews (1966) showed that small zones of EPR in an EPR-NR blend provide a barrier that inhibits ozone crack growth. Ambelang et al. (1969) found the importance of small EPM domain size in EPDM-SBR blends. Matthew (1984) has shown that carbon black improves the ozone resistance of NR-EPDM. An improvement was obtained in the blends with a balanced carbon black phase obtained by phase inversion. This is because (1) there is better reinforcement of the EPDM phase and (2) carbon black in the EPDM expands the volume of that phase. In BIMS-BR blends, the ozone failure can be retarded by reducing the size of the BIMS dispersion (Tse et al., 1998).

The transport properties of polymer blends are of interest both for the practical application of blends and for providing insight into the morphology of the blend. Measurement of the effect of blend composition on permeability in various rubbers has been described (Barrier, 1956). The permeability of elastomer blends depends on the concentration of the continuous phase and the morphology of the dispersed phase. Extended disperse phase structures, particularly lying in a stacked or lamellar configuration, can lead to a reduced permeability because of the more tortuous path that must be taken by penetrants (Kinning et al., 1987).

12.5.8 Applications

(i) Unsaturated Elastomer Blends

The most common blends of unsaturated elastomers are those used in various sections of automotive tires. Table 12.9 lists the important component of tires and the typical blends used for them. Much of the literature of elastomer blends reflects this important application. It is outside the scope of this chapter to discuss each of the applications. We have outlined most of the important principles used in the generation of the blends.

(ii) Saturated and Unsaturated Elastomer Blends

The use of blends of polyolefin elastomers, such as IIR and EPDM, as substantial components in blends of unsaturated elastomers is a rapidly developing area. Mouri and Tonosaki (1997) have compared the properties of EPDM-NR and BIMS-NR blends as sidewall components. In many of the applications, the saturated elastomer is considered a polymeric antioxidant for the diene rubber. It is believed that the higher molecular weight polyolefins are better in these applications due to limited interdiffusion and a more stable morphology. Some of the benefits in tensile properties and abrasion resistance of the blends may be due to the interdiffusion of high molecular chains of dissimilar elastomers across the phase interface. Significant advances have been made in modifying the structure of polyolefin elastomers to increase the compatibility to unsaturated elastomers. Tse et al. (1998) have shown that uncompatibilized blends of saturated elastomers and unsaturated elastomers are possible if the former contains substantial amounts (>12%) of styrene residues. This is expected to be

TABLE 12.9 Elastomer Blends in Automotive Tires (Robeson, 2007)

Component	Passenger Tires	Commercial Tires
Tread	SBR-BR	NR-BR or SBR-BR
Belt	NR	NR
Carcass	NR-SBR-BR	NR-BR
Sidewall	NR-BR or NR-SBR	NR-BR
Liner	NR-SBR-IIR	NR-IIR

an important area of development in the future with the advent of new synthesis procedures for polyolefins.

12.6 CONCLUSION

The formulation and use of elastomer blends is technologically demanding. Miscible blends are widely used but usually not recognized since analytical separation of the vulcanized elastomer is experimentally difficult. Immiscible blends require excellent phase dispersion and interfacial adhesion typical of all polymer blends. In addition, they require control of filler distribution and crosslink density in each component. This is due to the need for mechanical integrity in vulcanized elastomers.

The current design criteria of North American automotive tires require treads to last for 80,000 miles with less than 0.4 in. of wear. Vulcanized roofing membranes require 35 years of outdoor exposure with minimal change in elongation and tensile strength. The technical complexity of analysis and use of elastomer blends has led to secrecy for many of the formulations and uses. In spite of the difficulties of analysis and gaps in understanding, the use of blends containing elastomers continues to be an active and increasing area of research. Part of the impetus is the availability of directed synthesis of many of the older elastomers. These new synthetic tools include new catalysts ("single sited") and process designs for the manufacture of ethylene-based polyolefin elastomers, group transfer polymerization for acrylates, and anionic polymerization for diene elastomers. The availability of elastomers only with a narrow compositional and molecular weight distribution in these syntheses makes the utility of blending to access a broader mixture of composition and molecular weight effects within a single blended elastomer more apparent and useful.

APPENDIX 1: ACRONYMS FOR COMMON ELASTOMERS

ABR	acrylate-butadiene rubber
ACM	copolymer of ethylacrylate and a comonomer (acrylic rubber)
ANM	ethylacrylate-acrylonitrile copolymer (acrylate rubber)
BIMS	brominated isobutylene paramethyl styrene rubber
BR	butadiene rubber (polybutadiene)
BIIR	bromobutyl rubber
CIIR	chlorobutyl rubber
CFM	polychlorotritluoroethylene (fluoro rubber)
CM	chloropolyethylene (previous designation CPE)
CO	epichlorohydrin homopolymer rubber (polychloromethyloxiran)
CR	chloroprene rubber (polychloroprene)
CSM	chlorosulfonylpolyethylene
EAM	ethylene-ethyl acrylate copolymer (e.g., Vamac)
ECO	copolymer of ethylene oxide (oxiran) and chloromethyloxiran
ENM or H-NBR	proposed code for hydrogenated NBR
ENR	epoxidized NR
EPDM	ethylene-propylene-diene terpolymer
EPM	ethylene-propylene copolymer
EVM	ethylene-vinylacetate copolymer (previous code: EVA or EVAC)
FMQ	methyl silicone rubber with fluoro groups (previous designation FSI)
FPM/FPM	rubber having fluoro and fluoroalkyl or fluoroalkoxy substituent group
IIR	isobutylene-isoprene rubber (butyl rubber)
IR	isoprene rubber (synthetic)
MQ (PVMQ)	methyl silicone rubber (with vinyl and phenyl end groups)
NBR	acrylonitrile-butadiene rubber (nitrile rubber)
NR	isoprene rubber (natural rubber)
PUR	generic code for urethane rubbers
Q	generic code of silicone rubbers

SBR	styrene-butadiene rubber
TM	polysulfide rubbers
TOR	trans-polyoctenamer
VMQ	methyl silicone rubber with vinyl groups

REFERENCES

Achalla, P., McCormick, J., Hodge,T., Moreland, C., Esnealt, P., Karim, A., Raghavan, D., 2006. J. Polym. Sci. B: Polym. Phys. 44, 492.
Albert, B., Jerome, R., Teyssie, P., Smyth, G., Boyle, N.G., McBrierty, V.J., 1985. Macromolecules 18, 388.
Alegria, A., Elizetxea, C., Cendoya, I., Colmenero, J., 1995. Macromolecules 28, 8819.
Ambelang, B., Wilson Jr., F.H., Porter, L.E., Turk, D.L., 1969. Rubber Chem. Technol. 42, 1186.
Amerongen, G.J., 1964. Rubber Chem. Technol. 37, 1065.
Amidon, R.W., Gencarelli, R.A., 1972. US Patent 3,674,824.
Anderson, J.E., Jou, J.-H., 1987. Macromolecules 20, 1544.
Andrews, E.H., 1966. J. Polym. Sci. 10, 47.
Angier, D.J., Watson, W F., 1955. J. Polym. Sci. 18, 128.
Angier, D.J., Watson, W F., 1960. Trans. Inst. Rubber Ind. 33, 457.
Angrove, S.N., 1967. Rubber J. 149 (3), 37.
Arjunan, P.A., Kusznir, R.B., Dekmezian, A., 1997. Rubber World, 21.
Avgeropoulos, G.N., Weissert, R.C., Biddison, P.H., Bohm, G.G.A., 1976. Rubber Chem. Technol. 49, 93.
Ayala, J.A., Hess, M., Kistler, F.O., Joyce, G.A., 1992. Rubber Chem. Technol. 64, 19.
Baranwal, K.C., Son, P.N., 1974. Rubber Chem. Technol. 47, 88.
Barrier, J., 1956. Rubber Chem. Technol. 28, 814.
Barton, A.F.M., 1990. CRC Handbook of Polymer Interaction Parameters and Solubility Parameters. CRC Press, Boca Raton.
Bauer, R.F., Crossland, A.H., 1988. Rubber Chem. Technol. 61, 585.
Bhowmick, A.K., De, S.K., 1980. Rubber Chem. Technol. 63, 960.
Blackley, D.C., Charnock, R.C., 1973a. J. Inst. Rubber Ind. 7, 60.
Blackley, D.C., Charnock, R.C., 1973b. J. Inst. Rubber Ind. 7, 113.
Bohn, L., 1968. Rubber Chem. Technol. 41, 495.
Brus, J., Urbanová, M., Kelnar, I., Kotek J., 2006. Macromolecules 39, 5400.
Buchdahl, R., Nielsen, L.E., 1955. J. Polym. Sci. 15, 1.
Butera, R.J., Lando, J.B., Simic-Glavaski, B., 1987. Macromolecules 20, 1724.
Callan, J.E., Topcik, B., Ford, E.P., 1951. Rubber World 151, 60.
Callan, J.E., Topcik, B., Ford, F.P., 1965. Rubber World 161, 60.
Callan, J.E., Hess, W.M., Scott, C.E., 1971a. Rubber Chem. Technol. 44, 814.
Callan, J.E., Hess, W.M., Scott, C.E., 1971b. Rev. Gen. Caoutch. Plast. 48, 155.
Cesare, F.C., 1989. Rubber World 201, 14.
Cheng, H.N.,1992. In: Barth, H.G., Janca, J. (Eds.), Polymer Analysis and Characterization IV. John Wiley and Sons, New York (p. 21).
Coleman, M.M., Pehlert, G.J., Painter, P.C., 1996a. Macromolecules 29, 6820.
Coleman, M.M., Pehlert, G.J., Yang, X., Stallman, J., Painter, P.C., 1996b. Polymer 37, 4753.
Coran, A.Y., 1988. Rubber Chem. Technol. 61, 281.
Coran, A.Y., Patel, R., 1985. Rubber Chem. Technol 58, 1041.
Corish, P.J., 1978. In: Eirich, F.R. (Ed.), Science and Technology of Rubber. Academic Press, New York (Chapter 12).
Corish, P.J., Powell, B.D.W., 1974. Rubber Chem. Technol. 47, 481.
Corish, P.J., 1967. Rubber Chem. Technol. 40, 324.

Cotton, G., Murphy, L.J., 1988. Rubber Chem. Technol. 61, 609.
Datta, S., Kresge, E.N., 1988. US Patent 4,722,971.
Datta, S., Kaufman, L., Ravishankar, P., 1996. US Patent 5,571,868.
Davison, J.A., Nudenberg, W., Rim, Y., 1982. US Patent 4,316,971.
de Decker, H.K., Sabatine, D.J., 1967. Rubber Age 99, 73.
De Wolf, P., Brazel, E., Erickson, A., 2001. Mater. Sci. Semicond. Process 4 (1–3), 71.
Dias, A.J., Galuska, A., 1996. Rubber Chem. Technol. 69, 615.
Dikland, H.G., Van Duin, V., 2003. Rubber Chem. Technol. 76, 495.
Evans, L., Patridge, E.G., 1963a. In: 83rd Meeting ACS/CIC Div Rubber Chem.
Evans, L., Patridge, E.G., 1963b. Rubber Age (NY) 94, 272.
Feldman, D., 2005. J. Macromol. Sci.—Pure Appl. Chem. A42, 587.
Fletcher, W.P., Gent, A.N., 1957. Br. J. Appl. Phys. 8, 1984.
Flory, P.J., 1942. J. Chem. Phys. 10, 51.
Flory, P.J., 1965. J. Am. Chem. Soc. 87, 1883.
Fox, T.G., 1956. Bull. Am. Phys. Soc. 1, 123.
Gallard, P., Sauter, M., Reiss, G., 1980. Makromol Chem. Rapid Commun. 1, 771.
Gardiner, J.R., 1968. Rubber Chem. Technol. 41, 1312.
Gardiner, J.R., 1969. Rubber Chem. Technol. 42, 1058.
Gardiner, J.R., 1970. Rubber Chem. Technol. 43, 370.
Gee, G., 1942. Trans. Faraday Soc. 418.
Gee, G., 1947. Quart. Rev. Chem. Soc. l, 290.
Gelling, I.R., 1987. NR Technol. 18, 21.
German, R., Hank, R., Vaughan, G., 1967. Rubber Chem. Technol. 40, 569.
Ghijsels, A., 1977. Rubber Chem. Technol. 60, 278.
Gilmore, T., Falabella, R., Lawrence, R.L., 1980. Macromolecules 13, 880.
Groves, S.A., 1998. Rubber Chem. Technol. 71, 958.
Hamada, E., Kaneko, R., 1992. Ultramicroscopy 42, 184.
Hansen, C.M., 1967a. The Three Dimensional Solubility Parameter and Solvent Diffusion Coefficients. Danish Technical Press, Copenhagen.
Hansen, C.M., 1967b. J. Paint Technol. 39, 105.
Hansen, C.M., Beerbower, A., 1971. Solubility parameters. in: Encyclopedia of Chemical Technology, second ed. Wiley, New York (Supplement Volume).
Hashimoto, K., Miura, M., Takagi, S., Okamoto, H., 1976. Int. Polym. Sci. Technol. 3, 84.
Hashimoto, T., Tsukahara, Y., Kawai, H., 1981. Macromolecules 14, 708.
Phase blending refers to a practice in the compounding of multiple elastomers where most of one or more of the compounding fillers such as carbon black are mixed with one of the elastomers and the resulting filler concentrate is added to the other elastomers. See Hess, W.M., Chirico, V.E., 1977. Rubber Chem and Tech (50), 301.
Hess, W.M., Chirico, V.E., 1977. Rubber Chem. Technol. 50, 301.
Hess, W.M., Marsh, P.A., Eckert, F.J., 1965. Presented at a Meeting of the Rubber Division. American Chemical Society, Miami Beach, FL.
Hess, W.M., Swor, R.A., Vegvari, P.C., 1985a. Kautsch. Gummi Kunstst. 38, 1114.
Hess, W.M., Herd, C.R., Vergari, P.C., 1993. Rubber Chem Technol. 66, 329.
Hildebrand, H., Scott, R.L., 1964. Solubility of Nonelectrolytes. Dover, NY.
Hong, W., 1989. Rubber Plast. News 19, 14.
Honiball, D., McGill, W.J., 1988. J. Polym. Sci. B: Polym. Phys. 26, 1529.
Huggins, M.L., 1942. J. Am. Chem. Soc. 64, 1712.
Huson, M.G., McGill, W.J., Wiggett, R.D., 1985. Plast. Rubber Process. Appl. 6, 319.
Karasz, F.E., McKnight, W.J., 1977. Polym. Mater. Sci. Eng. 51 (2), 143.
Keith, H.D., Padden, F.J., 1964. J. Appl. Phys. 36, 1286.
Keyte, D.N., Walters, M.H., 1962. Trans. Inst. Rubber Ind. 38, 40
Kinloch, A.J., Shaw, S.J., Tod, D.A., Hunston, D.I., 1983. Polymer 24, 1341.
Kinning, J., Thomas, E.L., Ottino, J.M., 1987. Macromolecules 20, 1129.
Klei, B., Koenig, J.L., 1997. Rubber Chem Technol. 70, 231.
Klein, J., 1981. Philos. Mag. 43, 771.
Kleiner, L.W., Karasz, F.E., MacKnight, W.J., 1979. Polym. Eng. Sci. 19, 519.
Klein, J., Fletcher, D., Fetters, L.J., 1983. Nature (London) 304, 526.
Koningsveld, R., Staverman, A.J., 1968. J. Poly. Sci. A-2 6, 305.

Krakowski, F.J., Tinker, A.J., 1990a. Elastomerics 122, 34.
Krakowski, F.J., Tinker, A.J., 1990b. Elastomerics 122, 24.
Krejsa, M.R., Koenig, J.L., 1991. Rubber Chem. Technol. 64, 635.
Krejsa, M.R., Koenig, J.L., 1992. Rubber Chem. Technol. 65, 956.
Kruse, J.E., 1973. Rubber Chem. Technol. 46, 653.
Kumaki, J., Hashimoto, T., 1986. Macromolecules 19, 763.
Kumugai, Y., Watanabe, K., Miyasaki, T., Hata, J., 1979. J. Chem. Eng. Jpn. 12, 1.
Leblanc, J.L., 1982. Plast. Rubber Process. Appl. 2, 361.
Legrand, A.P., Lecomte, N., Vidal, A., Haidar, B., Papirer, E., 1992. J. Appl. Polym. Sci. 46, 2223.
Leibler, L., 1982. Makromol Chem. Macromol Symp. 16, 1.
Leibler, L., J. Noolandi, et al., 1981. Makromol. Chem. Symp. 16, 17.
Leroy, E., Alegria, A., Colmenero, J., 2002. Macromolecules 35, 5587.
Loadman, M.J.R., Tinker, A.J., 1989. Rubber Chem. Technol. 62, 234.
Maiti, S., De, S.K., Bhowmick, A.K., 1992. Rubber Chem. Technol. 65, 293.
Mangaraj, D., Pfau, J.A., Luttinger, M., 1987. Polym. Eng. Sci. 27, 15.
Mangaraj, D., 1963. Macromol Chem. 65, 29.
Mangaraj, D., 1987. Poster Session. ACS Rubber Division Meeting, Montreal.
Mangaraj, D., 2002. Rubber Chem. Technol. 75, 365
Mangaraj, D., Allen, G., Gee, G., et al., 1959. J. Polym. Sci. 34, 349.
Margaritis, A.G., Kallitsis, J.K., Kalfoglou, N.K., 1987. Polymer 28, 2122.
Mark, J.E., Andrady, A.L., 1981. Rubber Chem. Tech. 54, 366.
Mastromatteo, R.P., Mitchell, J.M., Brett, T.J., 1971. Rubber Chem. Technol. 44, 1065.
Matthew, N.M., 1984. J. Polym. Sci. Polym. Lett. Ed. 22, 135.
Mazich, K.A., Samus, M.A., Kilgoar Jr., P.C., Plummer Jr., H.K., 1986. Rubber Chem. Technol. 59, 623.
McDonel, E.T., Baranwal, K.C., Andries, J.C., 1978. in: Paul, D.R., Newman, S. (Eds.), Polymer Blends, vol. 2. Academic Press, New York, p. 263.
McDonel, E.T., Baranwal, K.C., Andries, J.C., 1978. In: Paul, D.R., Newman, S. (Eds.), Polymer Blends, vol. 11. Academic Press, New York (Chapter 19).
McMaster, L., 1973. Macromol. Sci. 65, 760.
Medalia, A., 1978a. Rubber Chem. Technol. 51, 437.
Medalia, A.I., 1978b. Rubber Chem. Technol. 61, 437.
Meinecke, E.A., Taftaf, M.I., 1988. Rubber Chem. Technol. 37, 1190.
Meinecke, E.A., Taftaf, M.I., 1988. Rubber Chem. Technol. 61, 534.
Mélé, P.S., Brown, D., Puydt, Y., Albérola, D., 2002. Polymer 43 (20), 5577.
Mitchell, J.M., 1985. Rubber Plast. News, 3, 18.
Morrar, F., Ban, L.L., Funk, W.G., Kresge, E.N., Wang, H.C., Datta, S., Keller, R.C., 1996. US Patent 5,428,099.
Morris, M.C., 1967. Rubber Chem. Technol. 40, 341.
Morrisey, R.T., 1971. Rubber Chem. Technol. 44, 1029.
Mouri, H., Tonosaki, Y., 1997. In: 152rd Rubber Division Meeting ACS, Paper 65, Cleveland, OH, October 21–24.
Ngai, K.L., Roland, C.M., 2004. Rubber Chem. Technol. 77, 579.
Nishi, T., Wang, T.T., 1975. Macromolecules 8, 909.
Nishi, T., Hayashi, T., Tanaka, H., 1988. Makromol. Chem. Macromol. Symp. 16, 91.
Noolandi, J., Hong, K.M., 1982. Macromolecules 16, 482.
Paredes, J.I., Gracia, M., Martínez-Alonso, A., Tascón, J.M.D., 2005. J. Colloid Interf. Sci 2005, 288, 19.
Patterson, D., Robard, F., 1978. Macromolecules 11, 690.
Paul, D.R., 1981. Polym. Eng. Sci. 21 (15), 985.
Paul, D.R., Barlow, J.W., 1978. Polym. Eng. Sci. 18. 1225.
Paul, D.R., Barlow, J.W., 1984. Polymer 25, 487.
Paul, D.R., Barlow, J.W., 1980. J. Makromol. Sci. C: J. Macromol. Chem. 18 (1), 109.
Pehlert, G.J., Yang, X., Painter, P.C., Coleman, M.M., 1996. Polymer 37, 4763.
Pehlert, G.J., Painter, P.C., Veytsman, B., Coleman, M.M., 1997. Macromolecules 30, 3671.
Ramos, A.R., Cohen, R.E., 1977. Polym. Eng. Sci. 17, 639.
Rehner Jr., J., Wei, P.E., 1969. Rubber Chem. Technol. 42, 985.
Reiss, G., Kohler, J., Tournut, C., Banderet, A., 1967. Makromol Chem. 101. 58.

Rim, B., Runt, J.P., 1984. Macromolecules 17, 1520.
Robeson, L.M., 2007. Polymer Blends. Hanser, Munich (Chapter 4.2 for an introductory review).
Roland, C.M., 1987. Macromolecules 20, 2557.
Roland, C.M., 1988a. Rubber Chem. Technol. 61, 866.
Roland, C.M., 1988b. J. Polym. Sci. Polym. Phys. Ed. 26, 839.
Roland, C.M., 1989. Rubber Chem. Technol. 62, 456.
Roland, C.M., Bohm, G.G.A., 1985. Macromolecules 18, 1310.
Roland, C.M., Casalini, R., 2007. Macromolecules 40, 3631.
Roland, C.M., Trask, C.A., 1989b. Rubber Chem. Technol. 62, 456.
Roland, C.M., Bohm, G.G.A., 1984. J. Polym. Sci. Polym. Phys. Ed. 22, 79.
Roland, C.M., Santangelo, P.G., Baram, Z., Runt, J., 1994. Macromolecules 27, 5382.
Roninger Jr., F.H., 1933. Ind. Eng. Chem. Anal. Ed. 6, 251.
Runt, J.P., Martynowicz, L.M., 1985. Adv. Chem. Ser. 211, 111.
Sanchez, C., Lacombe, R.H., 1978. Macromolecules 11, 1145
Sarkar, S.N., Komoroski, R.A., 1992. Macromolecules 25, 1420.
Sax, J.E., Ottino, J.M., 1985. Polymer 26, 1073.
Schaefer, J., Sefcik, M.D., Stejskal, E.O., McKay, R.A., 1981. Macromolecules 14, 188.
Scheele, W., 1971. Kautsch. Gummi Kunstst. 24, 387.
Schuster, R.H., Issel, H.M., Peterseirn, V., 1996. Rubber Chem. Technol. 69, 769.
Senake Perera, M.C., Ishiaku, U.S., Mohd, Z.A., 2001. Ishak EPoly J 37, 167.
Setua, D.K., White, J.L., 1991a. Kautsch. Gummi Kunstst. 44, 821.
Setua, D.K., White, J.L., 1991b. Polym. Eng. Sci. 31, 1742.
Shershnev, V.A., 1982. Rubber Chem. Technol. 66, 537.
Sircar, A.K., Lamond, T.G., 1973. Rubber Chem. Technol. 46, 178.
Ślusarski, L., Bieliński D., Włochowicz, A., Ślusarczyk, C., 2003. Polym. Int. 36, 261.
Smith, R.W., Andries, J.C., 1986. Rubber Chem. Technol. 47, 64.
Smith, S.R., Koenig, J.L., 1991. Macromolecules 24, 3496.
Steger, T.R., Schaefer, J., Stejskal, E.O., McKay, R.A., Sefcik, M.D., 1981. Ann. NY Acad. Sci., 371.
Stocker, W., Bickmarm, B., Magonov, S.N., Cantow, H.J., 1992. Ultramicroscopy 42, 1141.
Stroup, E.W., Pungor, A., Hilady, V., Andrade, J.D., 1993. Polym. Prepr. ACS Div. Polym. Chem. 34, 86.
Sultan, J.N., McGarry, F., 1973. Polym. Eng. Sci. 6, 127.
Sumitomo Chemical Company, 1973. British Patent 1,325,064.
Summerfield, G.C., Ullman, R., 1987. Macromolecules 20, 401.
Takenaka, M., Izumitani, T., Hashimoto, T., 1987. Macromolecules 20, 2257.
Teysie, P., Fayat, R., Jerome, R., 1987. Polym. Eng. Sci. 27.328.
Tinker, A.J., 1994. MRPRA Research Disclosure, No. 362, June. International Rubber Forum, ISBN 090036339.
Tokita, N., 1977. Rubber Chem. Technol. 60, 292.
Trask, C.A., Roland, C.M., 1988. Polym. Comm. 29, 332.
Trask, C.A., Roland, C.M., 1989. Macromolecules 22, 256.
Tse, M.F., McElrath, K.O., Wang, H.-C., Wang, Y.F., Tisler, A.L., 1998. In: 153rd Rubber Division Meeting, Paper 24. ACS, Indianapolis, IN, May 5–8.
Van Amerongen, G.J., 1964. Rubber Chem. Technol. 37, 1065.
Vander Hart, D.L., Manders, W.F., Stein, R.S., Herman, W., 1987. Macromolecules 20, 1726.
Vesely, D., Finch, D.S., 1988. Makromol. Chem. Macromol. Symp. 16, 329.
Voelkel, R., 1988. Angew. Chem. 27, 1468.
von Hellens, W., 1989. Rubber Division. American Chemical Society, Detroit, MI (Fall).
von Hellens, W., Edwards, D.C., Lobos, Z.J., 1990. Rubber Plast. News 20, 61.
Vonwinkel, K., 1969. Ned. Rubberind 30 (5), I.
Wang, Y.F., Wang, H.C., 1997a. Rubber Chem. Technol. 70, 650.
Wang, Y.F., Wang, H.C., 1997b. Rubber Chem. Technol. 70, 63.
Wang, M.-J., Wolff, S., 1992a. Rubber Chem. Tech. (65), 890.
Wang, M.-J., Wolff, S., 1992b. Rubber Chem. Tech. (65), 715.
Wang, M.-J., Wolff, S., Donnet, J.-B., 1991a. Rubber Chem. Tech. (64), 714.
Wang, M.-J., Wolff, S., Donnet, J.-B., 1991b. Rubber Chem. Tech. (64), 559.
Woods, M.E., Davidson, J.A., 1976. Rubber Chem. Technol. 49, 112.

Woods, M.E., Mass, T.R., 1974. US Patent 3,830,881.
Woods, M.E., Mass, T.R., 1975. Adv. Chem. Ser. 142, 386.
Yang, H., Shibayama, M., Stein, R.S., 1986. Macromolecules 19, 1667.
Yoshimura, N., Fujimoto, K., 1999. Rubber Chem. Technol. 72, 91.
Zhang, H., 2009. EPDM rubber in blends with NR/BR- elastomers for ozone resistant tyre sidewall applications. PhD Thesis, University of Twente, Enschede, Netherlands.
Zhang, Z.M., Mark, J.E., 1982. Polym. Sci. Polym. Phys. Ed. 20, 473.

Chapter 13

Thermoplastic Elastomers

Brian P. Grady,*[*] Stuart L. Cooper,[†] and Christopher G. Robertson[‡,††]

[*]*School of Chemical, Biological and Materials Engineering, The University of Oklahoma, Norman, OK, USA*
[†]*Department of Chemical and Biomolecular Engineering, The Ohio State University, Columbus, OH, USA*
[‡]*High Institute for Elastomer Industries, Yanbu, Saudi Arabia*
[††]*University of Akron Research Foundation, Akron, OH, USA*

13.1 INTRODUCTION

Thermoplastic elastomers (TPEs) are an extremely fast growing segment of polymer manufacturing. A rate of 6% growth per year is expected until 2015, at which time the total world demand for these materials will reach nearly 6 million mt (World Thermoplastic Elastomers, 2011). The largest use of thermoplastic elastomers is in the production of motor vehicles, and rebounds in the US and European markets for automobiles will account for this growth along with the rising use of TPEs for various applications in developing countries of the world. These high growth areas will be moderated by slower gains in mature markets such as asphalt modification and footwear. A low volume but important new application area for thermoplastic elastomers is in healthcare and medical products. The primary advantage of TPE over conventional rubber is the ease (and therefore low cost) of processing, the wide variety of properties available, and the possibility of recycling and reuse. Besides the obvious environmental benefits of a recyclable raw material, TPE scrap material can be reprocessed. The disadvantage of these materials relative to thermosets is the relatively high cost of raw materials, the general inability to load TPEs with low-cost fillers such as carbon black, and poor chemical and temperature resistance. This last property prevents TPEs from being used in automobile tires.

In order to qualify as a thermoplastic elastomer, a material must have three essential characteristics:

1. The ability to be stretched to moderate elongations and, upon the removal of stress, return to something close to its original shape.

The Science and Technology of Rubber. http://dx.doi.org/10.1016/B978-0-12-394584-6.00013-3

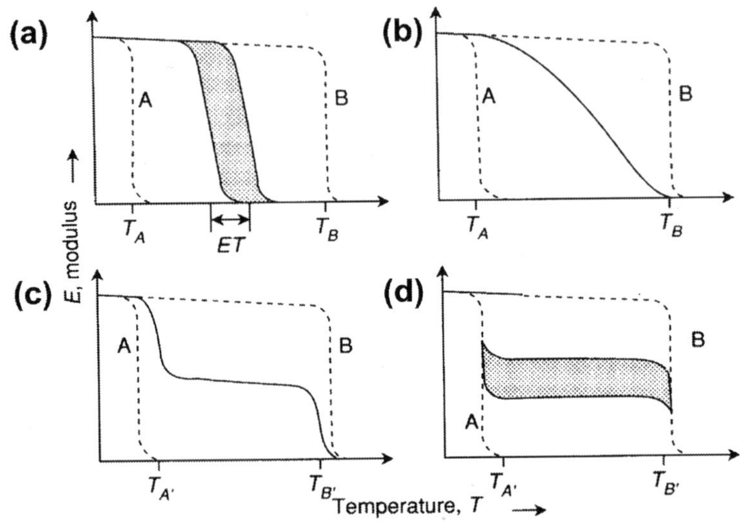

FIGURE 13.1 Diagram of the dependence of modulus on temperature for copolymers. In all the sketches, the dashed lines refer to the behavior for the pure materials. The hatched area shows that the range of the behavior can vary depending on the relative amounts of A and B. (a) Random A–B copolymer. (b) Block copolymer of A and B with extremely short blocks. (c) Segmented block copolymer with imperfect phase separation. (d) Segmented block copolymer with perfect phase separation (from Bonart, 1970).

2. Processable as a melt at elevated temperature.
3. Absence of significant creep.

In nearly all cases, thermoplastic elastomers will be a copolymer (i.e., there will be at least two monomers in the polymer chain). A thermoplastic elastomer will generally have the modulus vs. temperature curve shown in part c of Figure 13.1. The plateau region must include the service temperature of the material. Typically through changes in comonomer composition or identity, the plateau can be shifted upward or downward, giving the manufacturer a great deal of flexibility.

Most TPEs have certain similar structural characteristics. The comonomers typically have long runs, making the material a block copolymer. The comonomers are almost always dissimilar, leading to microphase separation on a nanometer length scale, which means these materials are properly termed nanomaterials (these materials were present long before this term became fashionable!). The driving force for phase separation is always enthalpic and usually one to two orders of magnitude weaker than primary valence bonds. Crystallinity, hydrogen bonding, ionic, and van der Waals driving forces all have been shown to cause microphase separation in these systems.

The two phases in these systems have different properties. One phase, the soft phase, contains a component that is above its glass transition temperature

(T_g) and melting temperature (T_m) so that chains have a high amount of mobility. The other phase, the hard phase, contains chains that are rigidly locked in place, because the service temperature is below either T_m or T_g. The relative amount of the two phases controls the physical properties of the TPE by determining which phase is isolated or continuous. The ability to easily vary these parameters through stoichiometry allows TPEs to be used in the wide variety of applications alluded to earlier.

A large number of structures fall into the category of thermoplastic elastomers. The structures of some common thermoplastic elastomers are shown in Figure 13.2. Shell Development Company developed the first commercially available thermoplastic elastomer in the early 1960s, which became the KRATON (http://www.kraton.com) family of materials. These materials are either poly(styrene-b-butadiene-b-styrene) (SBS), poly(styrene-b-isoprene-b-styrene) (SIS), or poly(styrene-b-ethylenebutylene-b-styrene) (SEBS) triblock copolymers. Phase separation occurs because of the incompatibility between the hard and soft segments. The styrene-rich domains serve as the hard phase since T_g for polystyrene is approximately 100°C. The molecular weight polydispersity is low because these triblocks are typically anionically polymerized. The terminal styrene anchors the polymer, which gives this material the necessary toughness while the flexible soft segment imparts elasticity. Styrenic block copolymers represent the largest volume category of thermoplastic elastomers. Approximately 50% of all thermoplastic elastomers produced are SBS, SIS, or SEBS triblock copolymers.

Another major category of thermoplastic elastomers, accounting for approximately 30% of the thermoplastic elastomer market, is based on polyolefins. The three most important materials that comprise this category are copolymers of ethylene and propylene (EP), copolymers of propylene and higher α-olefins such as 1-butene and 1-octene, and copolymers of ethylene and α-olefins. In the latter two cases, the propylene or ethylene is the major component. Two important differences between the ethylene-rich and propylene-rich materials are flexibility and softening point; the ethylene-rich materials are more flexible but also soften at a temperature roughly 50°C below that of the propylene-rich materials. These types of materials are important enough to be given a very common abbreviation, TPO, which stands for thermoplastic elastomer-olefinic. A specific example of these types of TPEs are the Engage (http://www.dow.com/elastomers/products/engage.htm) family of materials, which are copolymers of ethylene and α-olefins. Metallocene catalysts, mentioned earlier, have allowed better control over the run lengths of the normal EP copolymers. Generally EP copolymers and copolymer blends are of slightly higher cost and show higher performance than the triblock copolymers.

A very important thermoplastic elastomer is comprised of a blend of polypropylene (PP) with an ethylene-propylene-diene (EPDM) terpolymer. This latter material is, of course, a crosslinkable thermoset rubber;

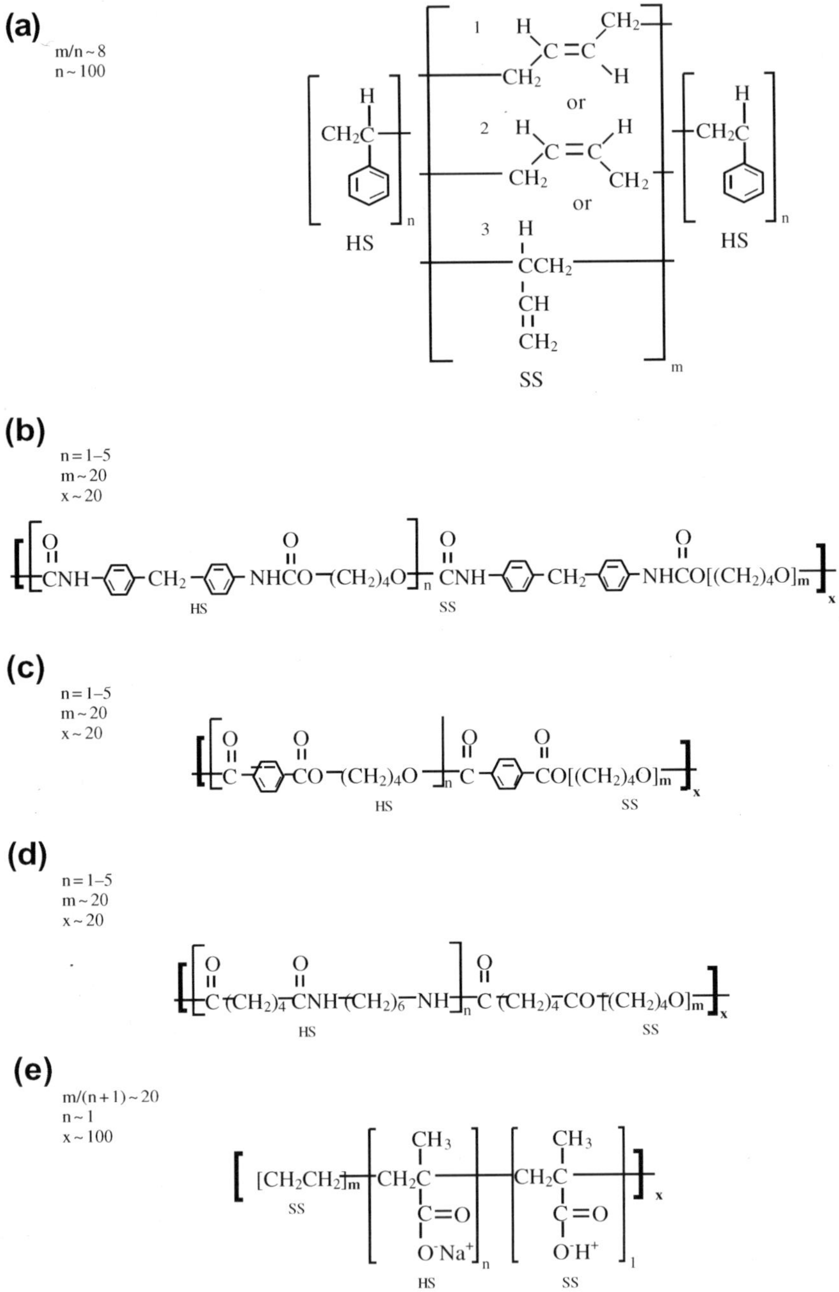

FIGURE 13.2 Structures of commercially important thermoplastic elastomers; HS = hard segment, SS = soft segment. (a) SBS; (b) MDI-BD-PTMO polyurethane; (c) PTMT and PTMO copolyester; (d) nylon 66 and PTMO copolyamide; and (e) random copolymer ionomer E-MAA neutralized with sodium.

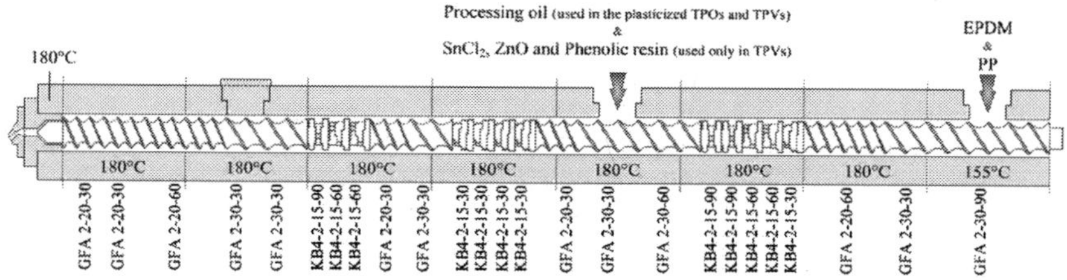

FIGURE 13.3 Screw configuration and temperature profile for a co-rotating twin-screw extruder used to generate TPOs and TPVs from PP and EPDM (from Shahbikian et al., 2012).

however, these materials can be processed as thermoplastics if the crosslinkable component is at low enough concentration to be present as an isolated phase. Melt-processing with crosslinking additives causes the formation of chemical bonds within the isolated rubber phase, a process called dynamic vulcanization. A commercial example of this type of material is Santoprene (http://www.exxonmobilchemical.com/Chem-English/brands/santoprene-thermoplastic-vulcanizate-tpv.aspx) manufactured by ExxonMobil Chemical Company. Thermoplastic elastomers that are formed from crosslinking the minor rubber phase during mixing with a thermoplastic are called thermoplastic vulcanizate (TPV) elastomers. The formation of TPVs in mixing processes is similar to the creation of TPOs from PP and EPDM with the exception that curatives are added during mixing of TPVs as illustrated in Figure 13.3 (Shahbikian et al., 2012). Besides the most common TPV composed of PP and EPDM, other combinations of plastics and crosslinked rubbers have been developed, including higher performance and more costly TPV combinations such as polyester with silicone rubber and polyamide with poly(ethylene acrylate) rubber (Kear, 2003; Babu and Nashkar, 2011).

Polyurethane elastomers are copolymers with a hard segment that contains aromatic rings and a polyether or polyester soft segment. Polyurethane elastomers are part of a family of materials termed segmented block copolymers, which is defined as a material with alternating hard and soft segments that repeat multiple times in a single polymer chain. Microphase separation occurs because of incompatibility between the aromatic rings and the soft segment. In some cases, the hard segment may crystallize as well. Segmented block copolymers have the general formula $(AB)_x$ where a triblock copolymer has the general formula ABA. Polyurethanes are generally manufactured from an aromatic diisocyanate, an oligomeric diol, and a low molecular weight diol. The low molecular weight diol is typically called the chain extender because it links AB segments together. A typical polyurethane based on diphenylmethylene-4,4′ diisocyanate (MDI), poly(tetramethylene oxide) (PTMO), and butanediol (BD) is shown in Figure 13.2. A commercial example of a polyurethane is the MDI-BD family of materials manufactured by the Lubrizol under the commercial name Pellethane (http://www.lubrizol.com/Medical/Products/

Pellethane.html). Polyurethanes are generally expensive and have found uses in high performance structural applications, medical and healthcare products as well as foams. Approximately 15% of the thermoplastic elastomer market is claimed by polyurethanes.

Another class of segmented block copolymers is segmented block copolymers containing an aromatic polyester hard segment and a polyether or polyester soft segment. Hard segment crystallization provides the driving force for phase separation in this system. A copolyester made from poly(tetramethylene terephthalate) and poly(tetramethylene oxide) (4GT-PTMO), which is a member of the Hytrel (http://www2.dupont.com/Plastics/en_US/Products/Hytrel/Hytrel.html) high performance thermoplastic elastomers manufactured by DuPont, is also shown in Figure 13.2. These materials are oil resistant and stable to higher temperatures than other thermoplastic elastomers, which makes these materials more suitable for applications such as automobile engine parts.

Two other thermoplastic elastomers are shown in Figure 13.2. The copolyamide thermoplastic elastomers are comparable to the copolyesters in structure. Crystallization provides the driving force for phase separation in these materials as well. These materials have especially low chemical permeability and can offer good properties at low temperatures. A commercial example of a copolyamide is PEBAX (http://www.pebax.com) marketed by Arkema. Copolyamides compete with polyurethanes and copolyesters for market share. Ionomers are the final material that will be discussed. Ionomers are materials where a small mole fraction of monomers, usually less than 10%, contains an ionic functionality. These materials are not segmented like most of the other materials discussed in this chapter, rather the ionic groups are distributed randomly along the polymer backbone. Incompatibility between the ionic groups and the nonpolar polymer backbone leads to the formation of ionic-rich domains. A commercial example of an ionomer is Surlyn (http://www2.dupont.com/Surlyn/en_US/) manufactured by DuPont, shown in Figure 13.2.

This chapter is organized along the following lines: the synthetic techniques important in the manufacturing of TPEs will be reviewed first. The bulk of the chapter will be concerned with the TPE morphology, since the morphology determines the physical and mechanical properties. This discussion will be followed by specific examples illustrating the effect of structure on the physical properties. The thermodynamics of phase separation, which includes detailed discussions on morphology as well as thermal behavior, will follow. The rheology and processing of these materials will be discussed next. Finally, some applications for thermoplastic elastomers will be highlighted. Emphasis will be given to those topics that are common to all thermoplastic elastomers; however, some discussion specific to commercially important materials will also be included. Also, the discussions are not meant to be exhaustive; for further information, the interested reader can consult some of the references provided throughout this chapter.

13.2 SYNTHESIS OF THERMOPLASTIC ELASTOMERS

13.2.1 Step-Growth Polymerization: Polyurethanes, Polyether-esters, Polyamides

Polyurethanes, copolyesters, and copolyamides are all produced via step-growth polymerization. In step-growth polymerizations relevant to the production of thermoplastic elastomers, a molecule containing two reactive functional groups of one type (e.g., a diisocyanate) reacts with another molecule containing two reactive functional groups of another type (e.g., a diol) to form a polymer. Step-growth polymerizations require extremely high conversions (>99%) to produce high molecular weight product. Generally, TPE properties are only weakly dependent on overall molecular weight, so the breadth of the distribution is usually not very important, although it is important to achieve high molecular weights. Controlling the ratio of functional reactive groups is critical to achieving high molecular weights as the following formula shows in the case of a stoichiometric imbalance between the two reacting functional groups:

$$\bar{n}_n = \frac{1+r}{1+r-2rp}, \tag{13.1}$$

where r is the ratio of the initial imbalance of the functional groups and is defined to be always less than 1, p is the extent of reaction, and n is the number-average degree of polymerization. Even to achieve moderate molecular weights, stoichiometric imbalances of more than a few percent cannot be tolerated.

Before embarking on a short description of the synthesis of polyurethanes, the reader should be aware that polyurethanes are generally divided into three classes: foams, coatings, and TPEs. This chapter concerns only the latter, and the morphological difference between a TPE urethane and others is the fact that the chain is not crosslinked, and the segment lengths are longer. Synthetically, crosslinked materials tend to use water, whereas water must be excluded from a reaction to produce a TPE polyurethane. Books (Uhlig, 1999; Szycher, 1999; Randall and Lee, 2003; Oertel, 1994) on the subject generally cover all three types of polyurethanes, sometimes without a clear distinction between the different uses.

Polyurethanes can be synthesized in solution (Lyman, 1960) or in bulk (Meckel et al., 1996). Solution polymerized polyurethanes generally have more uniform hard and soft segment distributions. Bulk polymerized polyurethanes generally have higher molecular weights, partially caused by side reactions that cause crosslinking (Petrovic and Ferguson, 1991). The majority of industrially produced polyurethanes are made in bulk. Bulk synthesized polyurethanes are reacted at temperatures between 80 and 120°C. The isocyanate-alcohol reaction is highly exothermic, which means that heat must be removed from the reaction mixture so that the temperature will be kept below the degradation temperature of 140°C. Generally, higher temperatures mean more side reactions and crosslinking. To produce totally linear polyurethanes, temperatures under 50°C should be used (Hepburn, 1982). Two methods are used in the bulk, the "one-shot" method and the prepolymer method. In the "one-shot" method, all the ingredients are mixed together; in the prepolymer method, the diisocyanate and oligomeric diol are allowed to react before the chain extender is added. Generally, the oligomeric diol and the diisocyanate are not miscible, so the reaction occurs at the interface between the two components, which can lead to large compositional variations over the course of the reaction. A common solvent for the diols, the diisocyanate, and the polymer is needed for solution polymerization. A relatively polar organic solvent such as N,N-dimethyl acetamide or dimethyl sulfoxide will generally suffice. Usually the analog of the prepolymer method, described earlier for bulk polymerization, is used for solution polymerized polyurethanes. Unlike the bulk reaction, an organotin catalyst is employed in solution polymerization. Organotin compounds are generally not utilized in the bulk method since incomplete catalyst removal will lead to poor hydrolytic stability of the final product.

Another common thermoplastic elastomer produced from a step-growth polymerization are the copolyesters. The general reaction scheme is very similar to the polyurethanes except the hard segment reactants are usually diesters rather than diisocyanates, which means that a small molecule by-product must be removed in order to achieve high conversions. Generally, copolyesters are produced in a melt phase transesterification polymerization described by Witsiepe (1973). In the first step, the oligomeric diol, chain extender, and diester are mixed and allowed to react at elevated temperature (~200°C) in the presence of a titanate catalyst in order to produce prepolymer. The high temperature serves to drive the reaction toward completion as well as remove the low boiling by-product through fractional distillation. Polymers are produced by raising the temperature to 250°C and lowering the pressure to less than 1 mmHg, which causes the second transesterification. The chain extender is removed as a by-product from this second stage. The temperature must be kept below 260°C in order to prevent substantial degradation (Hoeschele, 1974). Reaction completion is monitored by the viscosity of the reaction mixture. Molecular weights of approximately 25,000 g/mol generally result from this procedure.

Different pairs of functional groups can be reacted to form amide linkages, and all of them have been used to produce copolyamides (Nelb and Che, 1996). These include reactions between carboxylic acids and diamines, acid chlorides and diamines, and carboxylic acids and isocyanates. The latter is especially useful for producing copolyamides with aromatic hard segments. The copolyamides are most commonly produced with either ester or amide linkages between the amide hard segment and soft segment. Again, high temperatures are sometimes needed to produce high molecular weight material.

13.2.2 Anionic Polymerization: Styrene-Diene Copolymers

Anionic polymerization is used to produce the styrenic block copolymers and produces a polymer with an extremely narrow block and overall molecular weight distributions. The narrow molecular weight distributions are extremely useful in fundamental studies of polymers, and have led to a great deal of study of anionic polymerization, much more than justified by its commercial importance. In fact, the styrenic-block copolymers are the only polymers produced in large quantities via anionic polymerization. The extremely narrow polydispersity is evident in the following expression for the polydispersity index:

$$\frac{n_w}{n_x} = 1 + \frac{n_x}{(n_x + 1)^2}, \tag{13.2}$$

where n_w is the weight-average degree of polymerization and n_x is the number-average degree of polymerization. This analysis assumes no chain transfer and no chain termination reactions, which are easily achieved using low temperatures and pure ingredients. An extremely fast initiation rate relative to propagation is also assumed.

The triblock copolymers SBS or SIS are produced using a high vacuum process that serves to eliminate both oxygen and water from the reaction mixture (Morton, 1983). It is important to produce a polydiene block with a high 1,4 structure so that the glass transition temperature of the soft phase will be sufficiently low. This is generally accomplished by using a relatively nonpolar solvent (e.g., cyclohexane or toluene), which favors 1,4 formation. SEBS is formed through the hydrogenation of SBS.

Generally three anionic polymerization methods can be used to produce the styrene-diene triblock copolymers: a three-stage process with monofunctional initiator, a two-stage process with monofunctional initiator and a difunctional linking agent, and a two-stage process with a difunctional initiator (Noshay and McGrath, 1977). In all cases, the initiator of choice is an organolithium compound. In the first process, the styrene monomer is anionically polymerized, followed by the butadiene monomer, and then finally more styrene monomer is added to produce a triblock polymer. The butadiene rapidly initiates when added to the styrene blocks; however, a small amount of polar solvent is needed to initiate the final styrene anionic polymerization. In the second process, the styrene and diene blocks are polymerized separately then combined using the difunctional linking agent, which produces symmetric triblocks. The key in this method is to control the stoichiometry exactly so that no diblocks are formed. A small amount of diblock material has an extremely adverse effect on mechanical properties. Finally, a difunctional initiator can be used to initiate the diene block followed by the styrene endblocks. This method suffers from the difficulty in finding an appropriate difunctional initiator.

13.2.3 Catalytic Polymerization

Ethylene-propylene copolymer thermoplastic elastomers and other ethylene-α olefin copolymers are produced from either Ziegler-Natta or metallocene processes. Ziegler-Natta polymerizations are reactions catalyzed by a mixture of alkyl metal halides (e.g., $Al(C_2H_5)_2Cl$) and transition metal salts (e.g., $TiCl_4$). These polymerizations typically produce crystallizable stereospecific products. For example, in propylene-α olefin copolymers, polypropylene crystallites provide the hard phase for the TPE. Long blocks of one component, which are necessary for phase separation, result from the inherent reactivity ratios of the components. In spite of the industrial importance of Ziegler-Natta polymerizations, their kinetics are not well understood. This is a result of the often heterogeneous nature of this reaction as well as the possibility of multiple mechanisms. Blends of EP copolymers with isotactic polypropylene are often made in a two-step reaction process. The first reactor contains only propylene monomer, while the second reactor contains both propylene and ethylene. Other blends are made by mixing the pure components together in the melt. Blending conditions have a large effect on the resulting properties. Compatibilizers, which are often triblock copolymers, can adjust the blend's characteristics.

Metallocene polymerizations have many of the same properties as Ziegler-Natta polymerizations, but as the name implies, the catalyst is very different. Metallocene catalysts have two general properties that lead to stereospecific products: first, the catalyst is rigid, and second, the catalyst is chiral. These properties are obtained by having a metal cation sandwiched between two negatively charged cyclopentadienyl anions. Figure 13.4 shows a representative metallocene catalyst, 1,1′-ethylenedi-η^5-indenylzirconium dichloride. A modification to this approach is to replace one of the cyclopentadienyl anions with a ring containing a nitrogen atom, and then constraining the ring by bonding the nitrogen via a bridging atom, typically silicon. Just as in the Ziegler-Natta polymerization, a compound with labile alkane groups is required; one typical compound used is methyl alumoxane. Unfortunately for the economics of the process, the amount of aluminum required is much higher in the metallocene materials than in the Ziegler-Natta materials. In spite of the higher cost of these materials, metallocene polymerizations are gaining in market share because of the ability to better control the number of defects, which in turn can be used to tune copolymer properties.

13.2.4 Free Radical Polymerization

Free radical copolymerization is used to produce ionomers that are used commercially as thermoplastic elastomers. There are two types of TPE ionomers: copolymers of ethylene and methacrylic acid, and copolymers of ethylene and acrylic acid. The mole fraction of the acid monomer is typically 5% or less. The property difference between the two types of copolymers is

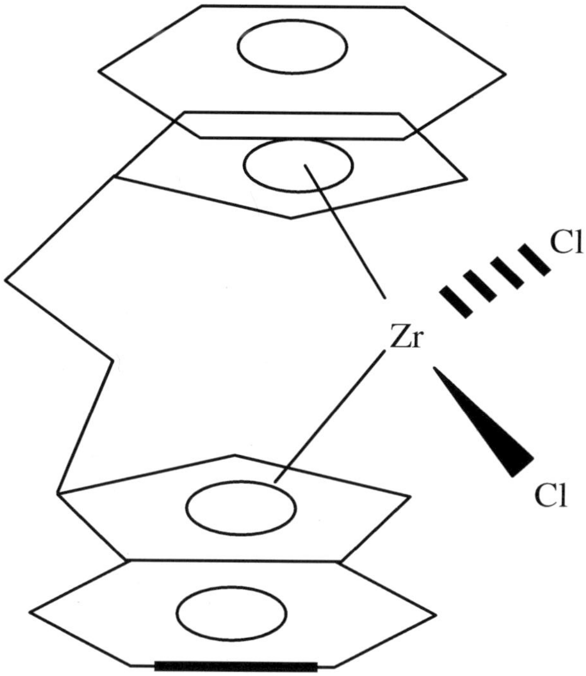

FIGURE 13.4 Structure of 1,1′-ethylenedi-η^5-indenylzirconium dichloride, a common metallocene catalyst. Only one enantiomer is shown; both enantiomers produce isotactic product for this particular catalyst.

small; copolymers of ethylene and methacrylic acid are slightly more resistant to the formation of anhydrides that can crosslink the polymer. The usual method uses high pressure (~1500 atm) and high temperatures (~130°C) similar to the method used for low density polyethylene. In contrast with Ziegler-Natta polymerizations, the kinetics are well known. The neutralizing cation after synthesis is hydrogen; this material does not phase separate into the nanometer-size aggregates required for toughness improvements. Therefore neutralization with an appropriate metal salt occurs industrially by mixing the polymer and the salt together in an extruder or roll mill (Lantman et al., 1989). In commercial materials, not all the hydrogen atoms are replaced by a metal cation, or else a material with too high a viscosity is produced; typical neutralization levels in most ionomers are around 50%.

13.2.5 Molecular Weight and Chain Structure

Determining the molecular weight and the chain microstructure is very difficult for some thermoplastic elastomers. Nearly all molecular weight characterization techniques rely on the ability to dissolve the polymer in a solvent in such a manner that the polymer chains behave individually. Since TPEs are usually composed of two dissimilar components, a good solvent for one of the components may be a poor solvent for the other component, which leads to aggregation of similar components from different chains. Hence the molecular weight is an aggregate molecular weight, rather than a true molecular weight. Even if this problem is overcome, methods that measure molecular weight relative to a standard, such as gel permeation chromatography (GPC), cannot be converted to actual molecular weight because the proportionality constant depends on the copolymer composition.

For polyurethanes and polyesters, end group analysis, either by titration or infrared analysis, can be used to monitor the extent of reaction and the final molecular weight if the reaction stoichiometry is properly controlled. This method only works well if the molecular weight is below about 20,000 g/mol. Membrane osmometry and light scattering can give reliable molecular weights if the appropriate solvent is used; however, a long time is required to make these measurements. Different solvents must be tested so that data can be collected without significant curvature that indicates aggregation. Ultracentrifugation is almost never used because, among other issues, by definition a theta solvent does not exist for thermoplastic elastomers. The molecular weight calculated from viscosity measurements will also depend on the copolymer composition since the radius of gyration (R_g) of the polymer will generally be a function of composition. Fractionation methods also typically fail because the fractionation efficiency depends not only on molecular weight distribution but also on composition distribution. Most often, GPC is used and the molecular weight is usually reported relative to a standard and for a certain solvent.

Molecular weight characterization of polyurethanes using GPC was studied in detail (Lee et al., 1986). Three different molecular weight standardization methods were tested, and it was found that a multidetector method using a refractive index and a UV spectrometer in series provided the most accurate results. The UV spectrometer was used to calculate the variation of the derivative of refractive index with respect to concentration at each point in the chromatogram, which accounts for the effect of changing copolymer composition with molecular weight. The authors found that the normal polystyrene standards used to calibrate the GPC give an upper limit on the actual molecular weight. Nevertheless, the authors were hesitant to call these values absolute. A related issue is the molecular weight of the linear prepolymers if a staged reaction scheme is used; a procedure for determining the molecular weight and soft segment-hard segment distribution of end-capped diisocyanates using GPC and a double detection method was given recently (Aust and Gobec, 2001).

Accurate molecular weights can be easily measured for the triblock copolymers. Because these materials are produced via anionic polymerization, the theoretical molecular weight is usually very close to the actual molecular weight. Depending on the polymerization method, the pure styrene or pure diene block can be removed and the molecular weight measured before making the triblock. Finally, molecular weight standards of polystyrene, polybutadiene, and polyisoprene are commercially available, which means measuring absolute molecular weight of these blocks is easily done using GPC.

The chain microstructure has a very important influence on the properties of TPEs. As mentioned earlier, production of SBS or SIS with a high 1,4 content is necessary. TPO properties also depend quite heavily on any deviations of the microstructure from the ideal head-to-tail, pure isotactic, or syndiotactic microstructure. Properties such as tacticity, *cis*-trans isomerization, and copolymerization content are usually characterized using NMR. Peak positions and peak intensities are used to quantitatively ascertain microstructure to a high degree of accuracy. Copolymer composition can also be determined using NMR. Infrared spectroscopy can also be employed to determine microstructural characteristics in some polymers.

In the segmented block copolymers, the average molecular weight of the hard and soft segments is very important, and the number of studies for the various types of TPE that have investigated this variable is too numerous to list here, although the anionic block copolymers deserve special mention in this regard because of their importance in elucidating fundamental thermodynamic information. In general, the higher the molecular weight of the blocks, the more complete the phase separation. However, it should be noted that complete phase separation is not always desired, since a decrease in toughness will eventually occur as phase separation becomes more complete. A related parameter, which is particularly important in both segmented block copolymers as well as the TPOs, is the distribution (as opposed to the average) of hard and soft segment lengths. In the latter, the hard segments are typically crystalline domains of either propylene or ethylene, and studies of crystallite size distribution are an important topic not just to TPEs. For the step-growth block copolymers, the distribution of soft segments tends to be fairly trivial to characterize, since the fully formed soft segment is typically one of the ingredients fed to the reaction. The hard segment, on the other hand, is typically formed as part of the polymerization, and more sophisticated approaches are required. Two approaches have been applied to study the effect of the distribution of hard segment lengths: the first is to synthesize hard segments with a known length distribution (Eisenbach et al., 1992) and the second is to attempt to measure the hard segment lengths. Of course, the latter is necessary for all commercial materials. A number of experimental methods have been used to determine hard segment length distributions. One method is to chemically cleave the structure at the point where the hard segment and the soft segment join and analyze the residual fractions using GPC or HPLC (Suzuki and Ono 1970a,b; Löcsei et al., 1983; Dombrowski et al., 1986). ^{13}C NMR (Kaji and Murano, 1990) and mass spectroscopy (Yontz and Hsu, 2000) have been used to determine hard segment length in MDI-PTMO polyurethane chains extended with ethylene diamine. Monte Carlo simulations of Markov processes have also been used to derive hard segment molecular weight distributions under ideal and nonideal conditions (Frensdorff, 1971; Sorta and Melis, 1978; Speckhard et al., 1986a,b, 1987a,b).

13.3 MORPHOLOGY OF THERMOPLASTIC ELASTOMERS

13.3.1 General Characteristics

In spite of the wide variety in structure of two-phase thermoplastic elastomers, the number of underlying morphologies in commercially important materials is surprisingly small. If we assume that A is initially the minor component and B the major component, then the five equilibrium morphologies given in increasing A content are (Bates and Fredrickson, 1990):

1. Isolated spheres of A in a continuous matrix of B.
2. Hexagonally packed isolated cylinders of A in a continuous matrix of B.
3. Alternating lamellae of A and B.
4. Hexagonally packed isolated cylinders of B in a matrix of A.
5. Isolated spheres of B in a matrix of A.

Other, more complicated morphologies can be generated with anionically synthesized block copolymers; a complete review of this subject is beyond the scope of this chapter, and the interested reader is referred to reviews (Fredrickson and Bates, 1996; Lodge, 2003) and monographs on the subject (Hamley, 2004, 1998). Transmission electron micrographs are shown in Figures 13.5–13.7 for each of the phases for a SBS block polymer. In general, the morphology in commercial materials is not nearly as well developed as the micrographs imply. Once the general morphology (spheres, cylinders, etc.) has been specified, specific questions remain concerning domain spacing, radii, and arrangement in space. These questions will be dealt with in some detail later in this chapter.

The underlying morphology will have a large effect on the physical properties. The soft phase is usually the continuous phase to maintain the elastomeric behavior of the material. In materials that have crystalline hard segments, such as copolyesters and polyurethanes, both phases are essentially continuous. Thermoplastic elastomers almost never have the soft segment as the isolated phase and the hard segment as the continuous phase. The most common morphology for TPEs is illustrated in Figure 13.8 for styrenic triblock copolymers. The isolated hard phase acts as an intermolecular tie point for the elastomeric soft phase. These are sometimes referred to as "physical crosslinks." Normally the isolated hard domains are between 5 and 30 nm. The ability to crystallize changes the morphology dramatically; the crystalline regions form rectangular thin sheets of material termed lamellae and hence will not be constrained into isolated domains and form more of a continuous structure. As with all polymers, lamellae can organize into spherulites and, a bit surprisingly, the segmented nature of the TPE does not prevent the formation of these large superstructures.

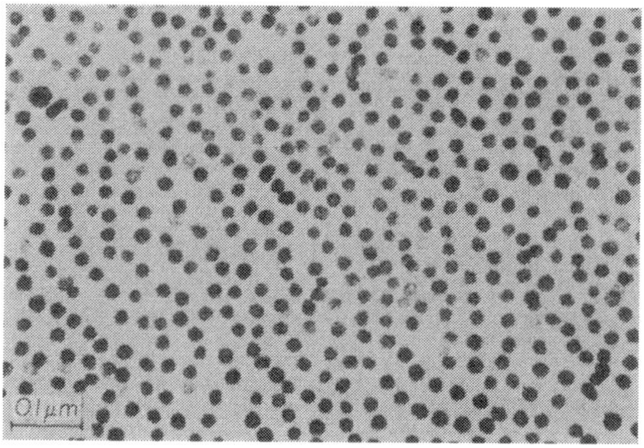

FIGURE 13.5 TEM of spherical butadiene domains in SBS (80% styrene) film cast from toluene. The same patterns were observed in both normal and parallel sections, confirming the periodicity of butadiene domains (from Matsuo, 1968).

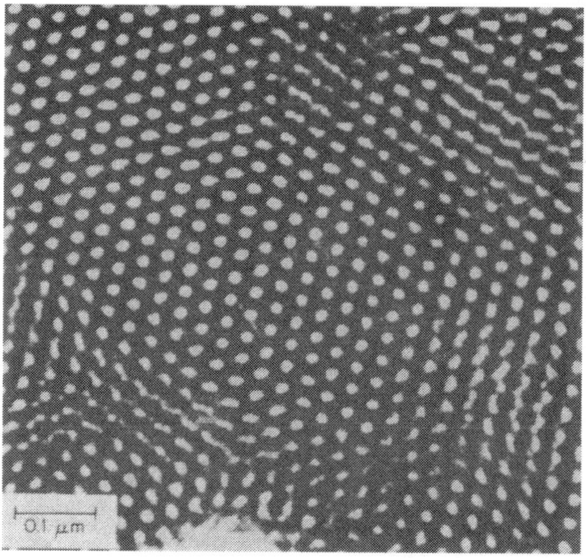

FIGURE 13.6 TEM of cylindrical microdomains in extruded and annealed SBS sample where the micrograph was taken perpendicular to the extrusion direction. An electron micrograph taken parallel to the extrusion direction had a striated structure (from Dlugosz et al., 1970).

The interface between the hard and soft segments is often considered as a separate phase. The influence of the interface on the properties of TPEs is considered to be very important, although not well understood. The higher cost of TPEs relative to commodity polymers is justified in different ways for different materials, but in general a property worth a high premium is toughness

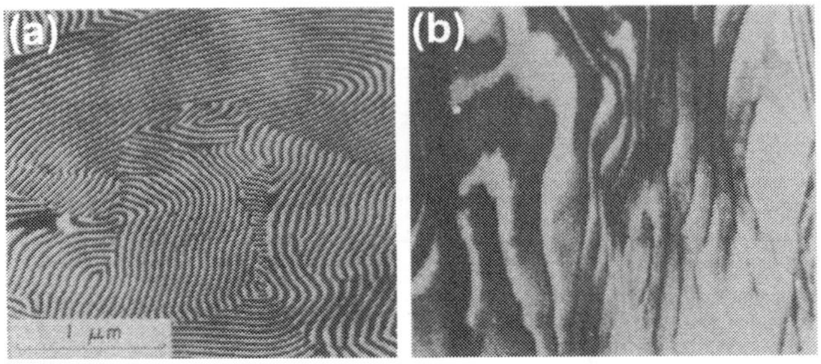

FIGURE 13.7 TEM of lamellar domains in SBS (40% styrene) film cast from cyclohexane. (a) Normal section and (b) parallel section with lamellar layers orienting their surface parallel to film surface.

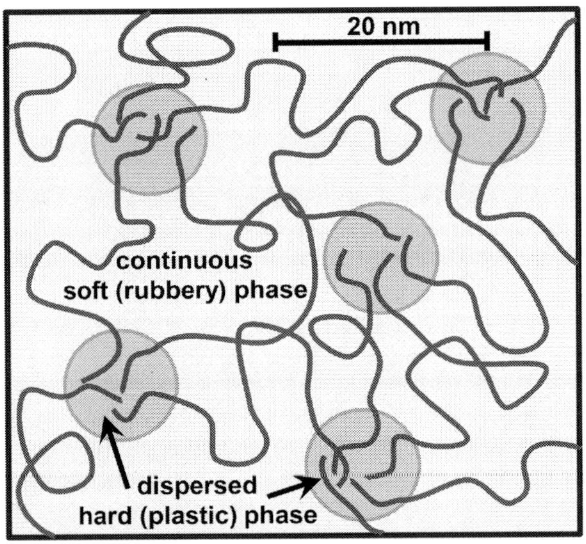

FIGURE 13.8 Schematic illustration of phase separation in a thermoplastic elastomer based on a styrenic triblock copolymer such as SBS. The isolated spherical domains containing the polystyrene end blocks form the hard phase, which acts as both intermolecular tie point ("physical crosslinks") and filler. The continuous phase from the polybutadiene midblock imparts the elastomeric characteristics to this polymer.

(i.e., a great deal of energy is required for failure). Toughness is achieved by creating stress-relieving processes in a material, and in TPEs many of these processes occur at the hard and soft domain interface resulting in the deflection and bifurcation of cracks (Smith, 1974). The adhesion between the interface and the polymer matrix is especially important in developing toughness in EP copolymer blends since failure often occurs in blends because of poor interfacial adhesion.

As implied earlier, kinetic considerations play an extremely important role in the morphology of TPEs. Predicting the precise morphology of any TPE without some knowledge of sample history is impossible. To produce a material that is near the equilibrium morphology, the TPE can be annealed above the dissolution transition of the hard segment but below the order-disorder temperature. Polydispersity in block lengths may also give morphologies that deviate substantially from the underlying morphologies. Finally, many industrially produced parts have oriented microdomains remaining from processing.

The remainder of this section contains an introduction to analytical methods used to characterize the morphology of TPEs. A short description of each method is given, with emphasis on those characteristics relevant to TPEs. Important examples of each method are also discussed, so the reader can understand how each method is used to characterize the morphology of TPEs.

13.3.2 Studies of Morphology

(i) Transmission Electron Microscopy (TEM)

In transmission electron microscopy (TEM), a sample is bombarded with electrons, and the number of electrons that travel through the sample, which is proportional to the sample thickness and the electron density, is measured as a function of position. If the electron densities are not identical for each phase in a TPE, the number of electrons passing through the two phases may be different and hence contrast will be created. TEM can be used on samples with extremely well-ordered morphologies where the length scale of electron density variations is greater than the resolution of the method. If the morphology is not well ordered, then the average number of domains of one type that a single electron passes through does not vary with position. The thickness of the film can be reduced, but a practical limit to thickness reductions exists. Since the electron density difference between atoms that comprise polymers is generally very small, a low molecular weight compound with a high electron density that will preferentially bind or migrate to one of the two phases is almost always added. Typical compounds used for such staining are OsO_4 and RuO_4.

Styrene-diene block copolymers represent the optimum material for TEM. If prepared properly, these materials can be almost 100% phase separated with the staining agent residing almost exclusively in the diene phase. The morphology is generally well ordered, and individual domain sizes tend to be large. Examples of electron micrographs of SBS block copolymers are given in Figures 13.5–13.7. TEM is probably the most important characterization method used for the determination of bulk morphology, so most papers in the field published TEM micrographs of their systems. The review papers and books referenced earlier contain a thorough description of TEM studies of block copolymers.

Copolyesters (Chen et al., 1987) and polyurethanes (Li and Cooper, 1990) have also been imaged. These two materials have not been the focus of electron microscopy studies for many years because of the difficulties in excluding artifacts. In fact, one must cast a very questioning eye on published results of these systems for this very reason. A number of papers by Winey et al. (Laurer and Winey, 1998; Winey et al., 2000; Kirkmeyer et al., 2002) have appeared in the last 5 years describing results of TEM studies of ionomers. The heavy metal atoms are used to provide the necessary contrast, and high-resolution TEM is used to image these systems. A much wider variety of ion aggregate morphologies have been found than previously thought, including spheres and vesicles.

(ii) Infrared and Raman Spectroscopy

In infrared spectroscopy, a beam of infrared light is passed through a sample, and light is absorbed if the frequency of the light is the same as the frequency of a normal mode of vibration. This method is sensitive to molecular bonding between atoms. Beer's law relates the absorbance (A) of a vibration to its absolute concentration:

$$A = \varepsilon t c, \tag{13.3}$$

where ε is the absorption coefficient, t is the path length, and c is the concentration. As indicated previously, infrared spectroscopy can be used to explore chain microstructure. Beyond chemical characterization, infrared spectroscopy can aid in describing interchain interactions such as hydrogen bonding or crystallization.

Hydrogen bonding is the secondary bonding of hydrogen atoms to an atom containing unpaired electrons. Hydrogen bonds generally have strengths on the order of 3 kcal/mol, which is a half order of magnitude less than covalent bond strengths but an order of magnitude greater than simple van der Waals interactions (Coleman et al., 1991). Hydrogen bonding is extremely important in polyurethanes and copolyamides and occurs between the urethane or amide hydrogen and the carbonyl oxygen. Since hydrogen bonding occurs primarily between hard segments, it provides a strong driving force for phase separation and hard segment crystallization.

Hydrogen bonding causes a shift toward lower wave number (lower energy) in the vibration of the bonds involving the hydrogen donating group and the hydrogen accepting group, which indicates that both primary bond strengths have been diminished because of the secondary hydrogen bond. In a typical thermoplastic elastomer that can hydrogen bond, vibrations of functional groups participating in hydrogen bonding are split into bands termed *bonded* and *free*. Table 13.1 gives wave numbers that have been assigned to the bonded and free bands for the C=O and N–H vibrations for a variety of polymers. There is a small but noticeable effect of the polymer type on the exact wave number for the vibration. The two vibrations listed for some C=O stretches are due to the ordered (crystalline) and disordered bonded states.

TABLE 13.1 Band Assignments of Hydrogen Bonded Functional Groups in TPEs

Material	Free (cm^{-1})	Bonded (cm^{-1})	Reference
C=O	1720	1685	Skrovanek et al. (1985), Coleman et al. (1986a)
Polyurethane: hexamethylene diisocyanate/butanediol		1700	
N–H	3440	3320	
C=O	1740	1700	Srichatrapimuk and Cooper (1978)
Polyurethane: 2,6 toluene diisocyanate/butanediol/PTMO			
N–H	3460	3300	
C=O	1733	1703	Seymour et al. (1970), Koberstein et al. (1986)
Polyurethane: MDI/butanediol/PTMO			
N–H	3420	3320	
C=O	1680	1636	Skrovanek et al. (1986), Coleman et al. (1986b)
Polyamide: nylon 11	1645		
N–H	3450	3300	

Since hydrogen bonding is extremely important and infrared studies are relatively inexpensive and simple, a large number of papers have been published in this field. Many studies have tried to relate the amount of hydrogen bonding to the degree of phase separation using Beer's law. Harthcock (1989) explored the carbonyl stretching region in model polyurethanes with monodisperse hard segment lengths. A detailed morphological model that relates the hard segment order to the infrared vibrational frequency was given and is reproduced in Figure 13.9. A similar study was undertaken by Luo et al. for polyurethane ureas (Luo et al., 1996). Changes in hydrogen bonding due to structural changes such as hard segment (Sung and Schneider, 1975) and soft segment type (Seymour et al., 1970) have also been studied. FTIR has aided in the discovery of structure-mechanical property relationships for segmented polyurethanes (Sormana and Meredith, 2004). The thermal behavior of hydrogen bonding has been investigated in some detail (Srichatrapimuk and Cooper, 1978; Koberstein et al., 1986; Luo et al., 1997). A series of papers published by Painter and Coleman (Coleman et al., 1985, 1986a,b; Skrovanek et al., 1985, 1986) on

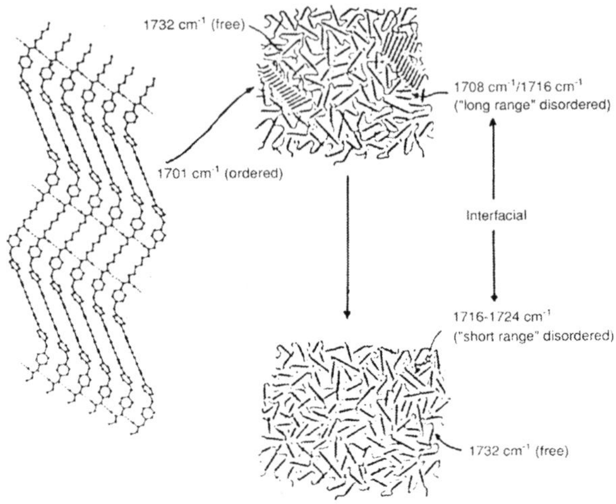

FIGURE 13.9 Morphological model developed by Harthcock that shows the wave number of the C=O stretching vibration absorbed intensity to the local environment (from Harthcock, 1989).

polyamides and polyurethanes concluded that many of the studies involving thermal behavior were based on incorrect reasoning concerning the nature of hydrogen bonds. These authors showed that quantitative analysis of the N–H region was impossible because of the large difference in extinction coefficient between the bonded and free bands along with the large change in the bonded N–H extinction coefficient as a function of temperature. The simple analysis (Srichatrapimuk and Cooper, 1978) normally employed to describe the dependence of the extinction coefficient with temperature could not be used because the vibrational frequency of the N–H bond shifted, which indicated that the hydrogen bond strength changed. Therefore the reader should be very careful interpreting papers that quantitatively analyze the N–H stretching region.

Raman spectroscopy in general gives complementary information to infrared spectroscopy and, because of the historical difficulty in performing these experiments, Raman studies of TPEs have been much less frequent than infrared spectroscopy. Raman experiments have become much simpler in the last 10 years with the development of powerful lasers and CCD detection systems, although the cost of a Raman spectrometer is still much more than an FTIR spectrometer. Raman spectroscopy has two distinct advantages to IR spectroscopy: first, the signal can be measured remotely using fiber optics, which makes it possible to use Raman for process control, and second, laser light can be focused to an approximately 1 μm spot, allowing researchers to image very small cross-sectional areas. The spatial-resolving capabilities of micro-Raman have been used to probe the composition heterogeneity in polyurethanes (Janik et al., 2003).

Another use of infrared spectroscopy employs linearly polarized infrared radiation to determine information about oriented samples, an experiment that

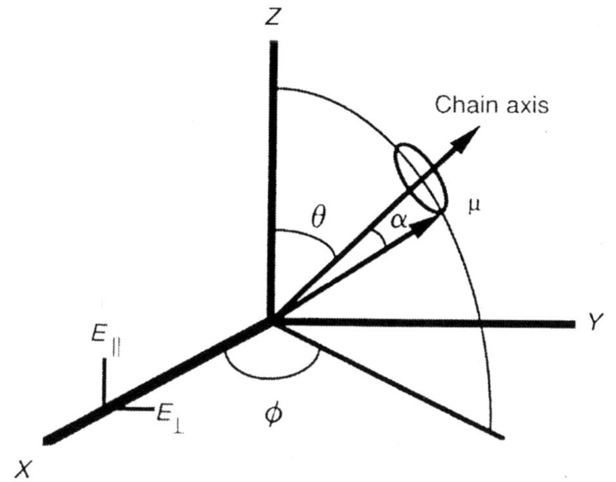

FIGURE 13.10 Geometry for uniaxial extended sample.

has been termed infrared dichroism. The absorbance will be a maximum when the electric field vector and the dipole moment vector are in the same direction, and the absorbance will be zero when the two are perpendicular. Only uniaxial orientation will be considered because this situation is normally found in the literature. R, the dichroic ratio, is defined as follows:

$$R = \frac{A_{\parallel}}{A_{\perp}}, \tag{13.4}$$

where $A_{\parallel}$ is the absorbance when the orientation direction and the polarization direction are parallel to each other and $A_{\perp}$ is the absorbance when the two directions are perpendicular. The relevant geometry is shown in Figure 13.10. Let $f(\theta)$ represent the orientation distribution function (i.e., $f(\theta)d\theta$ is the fraction of chain axis that lies within an angle $d\theta$ of θ). This function can be expanded in terms of the Legendre polynomials:

$$f(\theta) = \frac{1}{2\pi}\sum_{n=0}^{\infty}\frac{2n+1}{2}\langle p_n(\cos\theta)\rangle P_n(\cos\theta). \tag{13.5}$$

Only even number polynomials need to be considered since the assumption of uniaxial symmetry implies that the average values of the odd powers of $\cos\theta$ are zero. From Figure 13.10 after rather extensive mathematical manipulation it can be shown that:

$$\langle P_2(\cos\theta)\rangle = \frac{3\langle\cos^2\theta\rangle - 1}{2} = \frac{(R-1)(R_0+2)}{(R+2)(R_0-1)}. \tag{13.6}$$

R_0 is given by $2\cot^2\alpha$, where α is the angle between the dipole moment vector and the chain axis. $\langle P_2(\cos\theta)\rangle$ is the one-parameter measure of orientation normally given in the literature for uniaxially oriented samples. α is a function of the vibration in question and is known for a number of vibrations useful in studying thermoplastic elastomers (Myers and Cooper, 1994).

Infrared dichroism can be used to follow the orientation of each phase independently if nonoverlapping bands can be found in each domain where α is known. For example, in polyurethanes, the bonded carbonyl band is typically used to monitor hard segment orientation, while the free carbonyl band or CH stretching bands are used to follow soft segment orientation (Seymour et al., 1973; Allegrezza et al., 1974; Wang and Cooper, 1983). The hard segment orientation is almost always less than the soft segment orientation at the same draw ratio. In semicrystalline and more highly ordered polyurethanes, the hard segments are ordered transverse to the stretch direction initially and later become aligned in the elongation direction. Morphologically, at low elongations, the radial arms of the spherulite become oriented in the stretch direction, which means that the chain axes become oriented perpendicular to the stretch direction. At higher elongations, hard segments are physically removed from the arms and align with the elongation. In relaxation experiments, the soft segments tend to relax very quickly to a nearly unperturbed conformation, while hard segments relax much more slowly, especially at high strains. The transverse orientation of the hard segments is reversible, while physically removed hard domains cannot be restored to their previous environment without heating the sample. Dichroism measurements have been made on 4GT-PTMO copolyesters that showed the same negative orientation at low elongations followed by positive orientation at higher elongations (Lilaonitkul et al., 1976). Dichroism measurements at higher temperatures in MDI-BD polyurethanes showed changes in behavior; the onset of positive orientation occurs at a much lower elongation, and the hard segment orientation becomes much greater at a given draw ratio. These results were interpreted as a weakening of hard segment domain cohesion at higher temperatures (Siesler, 1983). The response to elongation is also altered when polyurethanes are hydrolytically degraded (Schoonover et al., 2001) or plasticized (Graff et al., 1999).

(iii) Wide Angle X-ray Scattering (WAXS)

When an electron density difference occurs periodically over a distance that is of the same order of magnitude as the wavelength of X-rays, X-rays will be scattered coherently from a sample. A peak or peaks corresponding to this distance will appear in the scattering pattern if the periodicity occurs enough times. The width and number of the peaks will be proportional to the variation of this repeat distance about its average value, as well as the number of times this periodicity occurs before ending. WAXS measures electron density variations with distances on the order of angstroms, which corresponds to interatomic distances. Therefore, WAXS is utilized to study thermoplastic elastomers with crystalline hard or soft segments.

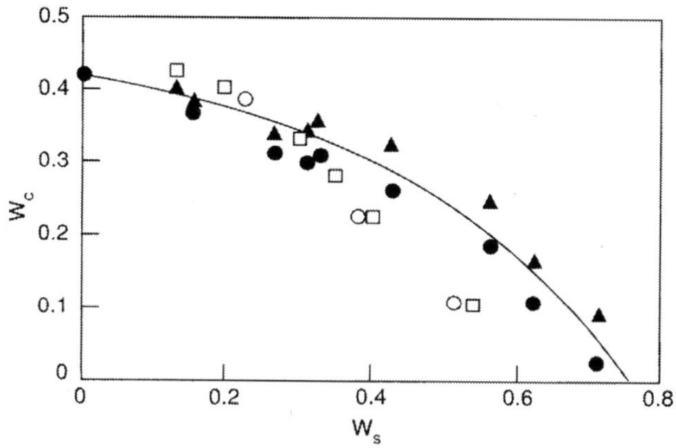

FIGURE 13.11 Weight fraction of crystallites, W_C, vs. weight fraction of soft segments, W_S, for 4GT-PTMO copolyester. ▲ from density measurements, • from WAXS, open symbols from differential scanning calorimetry.

The fundamental relationship that relates the repeat distance of electron density variation and the scattering angle is Bragg's law (Bragg, 1933):

$$n\lambda = 2d \sin\theta, \tag{13.7}$$

where n is an integer, λ is the wavelength of radiation, 2θ is the angle between the incident and exiting radiation, and d is the repeat distance between crystallographic planes. Further details regarding the crystallographic analysis of polymers are quite complicated and beyond the scope of this chapter. The interested reader is referred to the text by Alexander (1985).

Detailed crystallographic studies have been performed on copolyesters, copolyamides, and polyurethanes. Regarding the former, hard segment crystallites of 4GT are identical to 4GT homopolymer crystallites in the quiescent state. A different crystalline form is found in the hard segment when the TPE is extended due to the methylene sequences forming an all-trans configuration (Hall and Pass, 1976; Desborough and Hall, 1977). Using electron microscopy (Cella, 1973) and SAXS (Vallance and Cooper, 1984), the lamellar thickness has been shown to be smaller than the average hard segment length, which means that chain folding must occur. Although the unit cell is insensitive to soft segment fraction or soft segment composition, the overall amount of hard segment crystallites decreases as the soft segment fraction increases, as shown in Figure 13.11. However, the fraction of 4GT units that are crystalline increases as the soft segment fraction increases (Perego et al., 1984). These materials will also show strain-induced crystallization under stretching (Konyukhova et al., 2004).

Copolyamides show many of the same features as the copolyesters in WAXS studies. The crystal structure in the hard segment (nylon) domains is the same as in the homopolymer (Hatfield et al., 1993). Whether chain-folded crystals occur depends on the block length of the hard segment; at short block lengths a fringed micelle model was postulated to occur, while at long block lengths chain folding was present (Yu and Jo, 1994, 1995, 1996). At sufficiently low temperatures (below 0°C), the soft segment will crystallize if the soft segment is PTMO (Sheth et al., 2003). Finally, the PTMO segments show strain-induced crystallization, which is reversible if the sample is heated slightly above room temperature. The crystallization is great enough so that a permanent set will occur (Warner, 1990).

Polyurethanes show many of the same features as the copolyesters and copolyamides with respect to both hard and soft segment crystallinity. One substantial difference between polyurethanes and the previous two materials is that not all commercial polyurethane TPEs show hard segment crystallinity; MDI-based polyurethanes show crystallinity while TDI-based urethanes do not. Two crystalline structures have been found for MDI-BD hard segments in the unoriented state with another distinct form found in the oriented state (Briber and Thomas, 1983, 1985). Other studies have examined the effect of chain extender length on the crystal structure (Blackwell et al., 1982). The differences between polyurethanes with chain-folded and chain-extended hard segment crystallites have been extensively studied by Eisenbach and Stadler (1995).

Copolymers of isotactic propylene (iPP) with α-olefins also exhibit diffraction peaks due to iPP crystallites. As one would predict, increasing the α-olefin content decreases the percent crystallinity. When the copolymer is blended with iPP homopolymer, the copolymer will cocrystallize with the homopolymer, a phenomenon that is rare in polymers. Cocrystallization is believed to substantially contribute to the improved mechanical properties found in the blend (Starkweather, 1980). Large spherulites are not generally found in these blends as opposed to the homopolymer, and the crystal form is monoclinic rather than smectic (Kresge, 1984).

(iv) Small-Angle X-ray Scattering (SAXS)

WAXS measures the scattering of X-rays at distances relatively far away from the primary beam, while SAXS measures the scattering very close to the primary beam. Since the scattering distance is inversely related to the distance in real space, SAXS is sensitive to length scales on the order of nanometers rather than angstroms. SAXS is used to probe the two-phase morphology in TPEs. In other words, WAXS characterizes intraphase morphology, while SAXS characterizes interphase morphology.

Precise collimation systems are needed to make measurements close to the primary beam. In addition, scattering at small angles is usually very weak. Both pinhole and line collimation has been used; block collimation systems require mathematically transforming the data from line collimation to point collimation. During the last 15 years, the development of synchrotron sources and better area detectors has led to a proliferation of scattering experiments that occur in real time.

The interpretation of SAXS curves is more difficult than WAXS curves. In WAXS, the atoms are so small that they can be considered as point scatterers. In SAXS, however, the individual domains are not insignificant on the length scale of X-rays, and this scattering must be considered. Scattering due to individual domains is called *form-factor scattering*, while scattering due to the spatial arrangement of the domains is called *structure factor scattering*. The total scattering can be considered as the product of structure and form-factor scattering if and only if the domains are spherically symmetric. For a spherically symmetric two-phase system with uniform electron densities in each phase and one phase discretely immersed in a sea of the other, the scattered intensity can be written as Guinier and Fournet (1955):

$$\frac{I(q)}{I_e(q)V} = nV_0^2(\rho_1 - \rho_0)^2\phi^2(qR)S(q), \tag{13.8}$$

where $I(q)$ is the scattered intensity at the scattering vector $q(q = 4\pi \sin\theta/\lambda), I_e(q)$ is the scattering of one electron if it was the sample, V is the irradiated volume, n is the number density of discrete domains, $(\rho_1 - \rho_0)$ is the electron density difference between the two phases, $\phi(qR)$ is the form-factor scattering, and $S(q)$ is the structure factor scattering. $\phi(qR)$ has been calculated for a number of common shapes (Pedersen, 1997).

$S(q)$ is generally not a simple function unless the system is very well ordered. If the system is well ordered, then multiple peaks should appear in the SAXS pattern. The relative spacing of these peaks can be used to tentatively identify the domain packing arrangement (see Table 13.2). The only TPEs that show multiple peaks are the anionically synthesized materials, and typically only if laboratory processing procedures are used. A great many studies in the literature have used SAXS to study these types of materials; an excellent listing is found in the review papers given earlier.

In commercially important TPEs, a much more featureless pattern is typically found; the most common SAXS pattern from a TPE is a very broad

TABLE 13.2 Relative Peak Positions in Structure Factor for Common Well-Ordered Morphologies

Arrangement	Relative Peak Positions
Simple cubic packed spheres	$1, \sqrt{2}, \sqrt{3}, \sqrt{4}, \sqrt{5}, \sqrt{6}, \sqrt{8}, \sqrt{9}, \ldots$
Body-centered cubic packed spheres	$1, \sqrt{2}, \sqrt{3}, \sqrt{4}, \sqrt{5}, \sqrt{6}, \sqrt{7}, \sqrt{8}, \ldots$
Face-centered cubic packed spheres	1, 1.155, 1.633, 1.915, 2,...
Diamond packed spheres	1, 1.633, 1.915, 2.309, 3.416,...
Hexagonally packed cylinders	$1, \sqrt{3}, \sqrt{4}, \sqrt{7}, \sqrt{9}, \ldots$
Lamellae	1, 2, 3, 4, 5,...

single peak. Bragg's law can be used to calculate an interdomain spacing; of course this calculation is convoluted with form-factor scattering. Two more quantitative approaches are typically applied to the analysis of the data. One approach is to develop a morphological model, calculate the scattering pattern, and change adjustable morphological parameters until the predicted pattern matches the experimental pattern. The second approach is to Fourier transform the data and calculate a radial distribution function for electron density. A description of these approaches is beyond the scope of this review; the interested reader should examine monographs on the subject (Glatter and Kratky, 1982; Brumberger, 1995).

SAXS can be used to study the interfacial region between the two phases. SAXS gives a one-parameter measure of the interfacial thickness if some concentration profile is assumed. In a two-phase system with sharp interfaces, the scattering at high angles will be given by (after background subtraction):

$$\lim_{q\to\infty}\left[q^4\frac{I(q)}{I_e(q)V}\right] = 2\pi(\rho_1-\rho_0)^2\frac{S}{V}, \tag{13.9}$$

where S is the total interfacial surface area. The presence of a diffuse interface causes the intensity to fall off more rapidly than a q^{-4} dependence predicted above. Ruland (1971) has shown that Eq. (13.9) is modified as shown below:

$$\lim_{q\to\infty}\left[q^4\frac{I(q)}{I_e(q)V}\right] = [H^2(q)]2\pi(\rho_1-\rho_0)^2\frac{S}{V}. \tag{13.10}$$

$H^2(q)$ has been calculated for sigmoidal and linear (Vonk, 1973) concentration gradients. Because of the errors associated with background determination, the use of SAXS to study interface properties should be considered to be relative rather than absolute.

TPE investigations that involve SAXS are numerous. Because the information from SAXS is often ambiguous, the most effective studies are often done in conjunction with a different morphological probe such as electron microscopy or small-angle neutron scattering. Because the number of studies that have used SAXS to study TPEs is far too numerous to list, and in order to give the reader some flavor for SAXS experiments on TPEs, three interesting examples will be highlighted.

Deformed SBS triblock copolymers have been extensively investigated with SAXS (Séguéla and Prud'homme, 1978, 1981, 1988; Pakula et al., 1985). However, patterns were collected after these samples were allowed to relax, which has been shown to substantially affect the morphology (Pakula et al., 1985). The use of synchrotron radiation along with two-dimensional detectors enables intensity measurements while the sample is being drawn. In one such study (Polizzi et al., 1990), short polystyrene cylinders were embedded in a continuous polybutadiene phase. The deformation was found to be affine in the meridional direction until an elongation ratio of 3, which corresponds to the inflection point of the stress-strain curve. Above an elongation ratio of 4, it was shown that the cylinders were aligned with their long axis parallel to the stretch direction. The cylinders were not disrupted up to an elongation ratio of 8.

SAXS determinations of interfacial thickness have been used to show that MDI-BD polyether urethanes exhibit narrower interfaces than polyester urethanes (Ophir and Wilkes, 1980). The difficulty of correct background subtraction has been discussed in detail (Tyagi et al., 1986; Roe, 1982), nevertheless the interfacial thickness assuming a linear gradient profile was approximately twice as large for the polyester vs. the polyether soft segments using two different procedures applied in the same way to the patterns. The absolute magnitude of the numbers must be questioned since other authors have found substantially larger values for similar materials (Bonart and Muller, 1974; Koberstein and Stein, 1983).

Small-angle X-ray scattering has been used to follow morphological changes in copolyesters as a function of temperature (Phillips et al., 1994; Phillips and Cooper, 1995, 1996). In order to slow down the crystallization kinetics, 4GT was replaced by poly(tetramethylene isophthalate) (4GI) as the hard segment. Similar to the 4GT systems, the crystallization rate was found to only weakly influence the morphology of these copolyesters. At a fixed temperature, the Bragg spacing increased with decreasing hard segment concentration, and the long spacing was roughly proportional to the inverse of the undercooling. Annealing at temperatures near the melting point led to morphological reorganization through the melting of imperfect crystallites and recrystallization into more perfect crystallites, which was accompanied by an irreversible increase in the Bragg spacing.

(v) Small-Angle Neutron Scattering (SANS)

The only difference between SAXS and SANS is that the contrast for neutron scattering is a variation in scattering density rather than electron density. Scattering density is a function of the nucleus (not the atomic number!) and varies in a complex way. Because the difference in scattering density between hydrogen and deuterium is large, isotopic substitution is used to create the contrast required for SANS. Perhaps the most famous use of SANS in polymer science was the experimental verification that polymer chains in the bulk assume an unperturbed random coil conformation (Ballard et al., 1973; Cotton et al., 1974).

SANS can be used to look at the same sorts of things that SAXS is used for (i.e., domain size, distance between adjacent domains, interphase sizes, etc.), and a number of studies have used SANS in this manner. However, SANS has a capability unrealizable with SAXS: if a fraction of the chains in the system contain deuterium, with the rest containing hydrogen, or the reverse is true, then it is possible to examine scattering from individual chains. Therefore in two-phase systems with isotopic substitution, scattering will be from two sources: chain scattering, both interchain and single chain, and interphase contrast. By matching the scattering density of the two phases through partial labeling of one or both phases, it is possible to eliminate scattering due to interphase contrast (which gives information similar to SAXS) and study only single-chain scattering. Since the amount of phase mixing is unimportant (assuming no volume change upon mixing) and the compositions of the pure phases are well known in most TPEs, contrast matching is relatively easy to perform. Methods have also been developed for noncontrast matched systems to isolate the single-chain scattering by subtracting the interdomain scattering, using either SAXS or unlabeled SANS patterns (Jahshan and Summerfield, 1980).

For a two-phase system where interphase scattering has been eliminated and only one phase has been partially labeled, the coherent scattering intensity can be written as:

$$I(q) = \left[\frac{(\Delta\beta)^2 m_0}{\rho_m N_A}\right] v_1 v_d (1 - v_d) N S(q), \qquad (13.11)$$

where $\Delta\beta$ is the coherent neutron scattering length density difference between the fully hydrogenous and fully deuterous materials, v_l is the volume fraction of the labeled phase, v_d is the volume fraction of deuterous material in the labeled phase, N_A is Avogadro's number, m_0 and N are the monomer molecular weight and number of repeat units, respectively, and $S(q)$ is the single-chain scattering function. Further review of SANS theory and experimental studies of polymers is found in monographs on the subject (Rennie and Oberthur, 1994; Higgins and Benoit, 1994; Gabrys, 2000).

The effect of temperature and composition on chain conformation has been investigated in MDI-BD-PPO polyurethanes (Naylor et al., 1997), 4GT-PTMO copolyesters, and MDI-BD-PTMO polyurethanes (Miller et al., 1984, 1985; Cooper and Miller, 1985; Cooper et al., 1988). At room temperature, R_g, of the soft segments in the polyurethane, TPEs are approximately 25% larger than in bulk, while for the copolyester the increase is only approximately 10%. The soft segment radius of gyration decreased as the temperature increased above room temperature for all the materials except for a MDI-BD-PTMO material at a 7:6:1 mol ratio. Evidence of phase mixing was found when the temperature reached a high enough value for all materials except the 7:6:1 material as evidenced by an increase in R_g of the soft segment with temperature. SANS measurements of the 4GT segments indicate that substantial chain folding occurs in the copolyesters. The hard segment R_g in the copolyester increased dramatically with temperature, which indicated that the amount of chain folding and/or the degree of phase separation was changing. The hard segment R_g decreased as the temperature was raised in the 7:6:1 polyurethane, which the authors were unable to satisfactorily explain. Measurements of the entire chain dimensions in the copolyesters indicated that the chain initially contracted then expanded as the temperature was raised.

In lamellar styrene-diene diblock copolymers, SANS studies showed that the segment R_g contracts to 70% of the unperturbed value parallel to the interface and expands to 160% of the unperturbed value perpendicular to the interface (Hasegawa et al., 1985, 1987). These values were found for both the styrene and diene blocks. A study of stretching SIS block copolymers having spherical styrene phases showed that the deformation in the direction of stretch was greater than affine, while the deformation perpendicular to the stretch was much less (Richards and Welsh, 1995).

(vi) Nuclear Magnetic Resonance (NMR)

Solid-state NMR has the capability of providing information on a wide variety of characteristics of TPEs, encompassing both static and dynamic properties as well as orientation information. As mentioned previously, NMR can be used to determine chain microstructure information such as tacticity and sequence distributions. NMR has been used in TPEs to investigate spatial interactions between atoms as well as the relative mobility of particular segments. Deuterium labeling significantly expands the capabilities of NMR. Using pulse sequences, relaxation times of segments on different time scales can be probed.

NMR measures change in the spin magnetic moment of nuclei. A strong magnetic field ($\sim 10^5$ Gauss) along one axis, usually taken as the z direction, causes a net population distribution of nuclear spins aligned parallel to the magnetic field. Polarized electromagnetic energy in the radiofrequency region (10^2 MHz) with the magnetic field vector perpendicular to the z direction causes transitions to a higher energy spin state. A voltage that is proportional to the relaxation of nuclei from the higher energy state to a lower energy state is measured as a function of electromagnetic radiation frequency in typical NMR experiments. However, chemical shift rather than frequency is reported where chemical shift is defined as

$$\frac{v(\text{sample}) - v(\text{standard})}{v(\text{standard})}, \qquad (13.12)$$

where v represents the frequency and the standard is a material that contains the same atomic species. Only nuclei with nonzero spin quantum numbers can be studied with this technique. In TPEs, the most common nuclei that fulfill this requirement are ^1H, ^2D, and ^{13}C. In order to remove anisotropic interactions, a technique called magic angle spinning is used. A detailed discussion of the use of

NMR to characterize thermoplastic elastomers is beyond the scope of this chapter. The specifics related to the analysis of various TPEs using NMR can be found elsewhere (Tanaka and Nishi, 1985; Aluas and Filip, 2005; de Ilarduya et al., 2010; Kornfield et al., 1991; Meltzner et al., 1992; Jelinski et al., 1983a,b; Jelinski, 1981; Okamoto et al., 1992a,b, 1993; Sung et al., 1980; Clayden et al., 1998).

13.4 PROPERTIES AND EFFECT OF STRUCTURE

13.4.1 General Characteristics

Table 13.3 lists the properties of some typical thermoplastic elastomers and other common rubbery polymer materials. TPEs generally extend to high elongations without failure and have a high tensile stress at break (i.e., they are extremely tough). As mentioned earlier a thermoplastic elastomer should return to its initial shape after the removal of the stress. The range of extensibilities where a TPE will recover its original shape after stress is removed will generally not be as large as for conventional crosslinked rubbers. At high elongations upon the removal of stress, TPEs will often maintain some residual elongation, termed *permanent set*. As discussed in the section on infrared spectroscopy, the domains can flow under high stress. The toughness of these materials (area under stress-strain curve) is usually many times that of conventional crosslinked rubbers.

The Mullins effect (Mullins, 1950), also called stress softening, occurs in most TPEs. If a TPE sample is strained and then released, then less stress will be required to strain the sample a second time. This effect is generally quantified by the hysteresis energy, which is the difference in the areas under the stress-strain curve for a loading cycle followed by an unloading cycle. The Mullins effect can lead to heat buildup in a material, which is undesired in most applications. However, the ability of a TPE to dissipate energy is related to its strength and toughness.

Fracture of a polymer involves initiation, slow crack growth, and catastrophic crack propagation. The extreme toughness in thermoplastic elastomers is due to the inhibition of catastrophic failure from slow crack growth rather than the prevention of initiation or prevention of slow crack growth. Table 13.4 lists the mechanisms that can strengthen two-phase materials. The hard domains in TPEs act as both filler and intermolecular tie points. Hard domains are effective fillers if the volume fraction exceeds 0.2, their size is less than 100 nm, and the softening temperature is well above the test temperature (Smith, 1986). The filler effect is one key for the extreme toughness inherent in most TPEs. Another key is the interface between the hard and soft segments; in polymer blends (Paul, 1984), it has been hypothesized that the strongest materials result from two polymers on the edge of miscibility. These results suggest that broad interfacial zones lead to improved properties in two-phase block copolymers.

TABLE 13.3 Properties of Typical TPEs Relative to Other Rubbery Polymers

Material	Relative Cost	Tensile Strength at Break (MPa)	Tensile Strain at Break (%)	Service Temperature (°C)	Hardness (Shore A)
Styrene-butadiene rubber	1	15 (reinforced)	500 (reinforced)	High	35–100
Natural rubber	1	30 (reinforced)	500 (reinforced)	High	30–100
Silicone	1.2	5 (reinforced)	150	High	40–100
Polyethylene	1	10	High	−10 to 50	100
Fluorinated elastomer	1.5	10	200	High	50–90
SBS	2	25	800	−20 to 80	50–90
Polypropylene-EPM blend	1.5	20	500	0–110	70
Polyurethane	6	50	600	−20 to 80	50–100
Copolyester	7	40	600	−40 to 150	>100
Ionomer	5	15	500	−20 to 100	50–90

TABLE 13.4 Strengthening Processes in Polymers[43]

Matrix	Dissipation of energy near crack tip
	Strain-induced crystallization
	Orientation processes
Filler particles	Increased dissipation of energy
	Deflection and bifurcation of cracks
	Cavitation
	Deformation of domains

Due to the dual filler and crosslinking nature of the hard domains in TPEs, the molecular deformation process is entirely different than the Gaussian network theories used in the description of conventional rubbers. Chain entanglements, which serve as effective crosslinks, play an important role in governing TPE behavior. The stress-strain results of most TPEs have been described by the empirical Mooney-Rivlin equation:

$$\sigma = \left(\frac{\rho RT}{M_c} + \frac{2C_2}{\lambda}\right)\left(\lambda - \frac{1}{\lambda^2}\right), \tag{13.13}$$

where σ is the stress (force/unit area), R is the universal gas constant, T is the absolute temperature, λ is the extension ratio, and M_c is the average molecular weight between crosslinks. C_2 is an empirical constant that depends on the material. Stress-strain and swelling measurements have shown that M_c is closer to the molecular weight between entanglements than the soft segment length in SBS (Holden et al., 1969). The filler effect is quantified by the Guth-Smallwood equation:

$$\frac{E_F}{E} = (1 + 2.5\phi + 14.1\phi^2), \tag{13.14}$$

where E_F/E is the ratio of the moduli for the filled and unfilled elastomers and ϕ is the volume fraction of filler. Combination of these two equations gives reasonable values for M_c in SBS copolymers; however, objections have been raised to using this analysis (Meier, 1973; Gaylord and Lohse, 1978; Gaylord, 1979). More complicated theories have been proposed to explain the stress-strain behavior of thermoplastic elastomers (Bard and Chung, 1987; Veenstra et al., 2000; Read et al., 2002).

The general tensile behavior of TPEs as the temperature changes is shown in Figure 13.12. TPEs become more rigid as the temperature nears the soft segment T_g, and a discontinuous change in brittleness will occur at this temperature. Normally, the soft segment T_g is never reached in service (i.e., this temperature is far below room temperature). As the temperature rises in a typical TPE, the modulus and strength decrease slightly due to softening of the hard domains. At the hard segment dissolution temperature, the modulus will

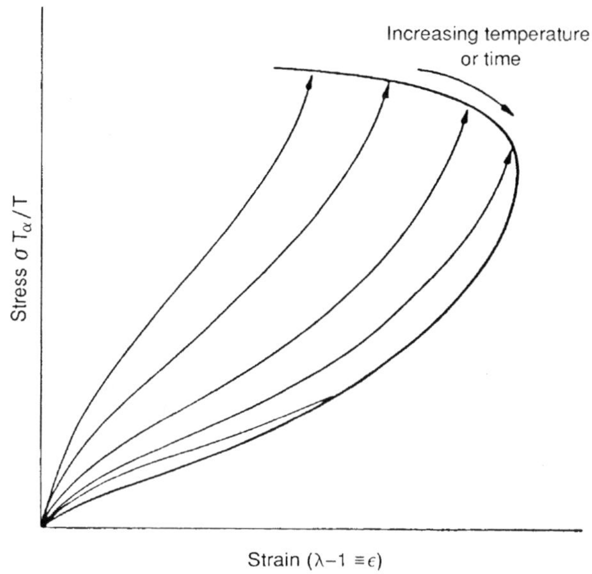

FIGURE 13.12 Effect of temperature and draw rate (inversely related to time) on the stress-strain curve of thermoplastic elastomers. σ is stress, T_{α} is some reference temperature, T is the test temperature, λ is the draw ratio, and ε is the strain (from Petrovic and Ferguson (1991)).

decrease dramatically, and the material can no longer be used as a thermoplastic elastomer. The two-phase structure may persist in the melt however. The precise temperature of the dissolution depends on the nature of the hard block. Plasticizers or other additives may be added to reduce the softening temperature; however, these materials will also tend to disrupt the domain structure.

The chemical resistance of many TPEs is poor compared to that of conventional rubbers. Polyurethanes, copolyesters, and copolyamides are very susceptible to oxidation, especially at elevated temperatures. Antioxidants and other additives are added to commercial products to improve the chemical resistance of these materials. Carbon black can be added to improve stability to UV light if the color of the material is unimportant. Hydrolytic stability is poor for the polyester-based polyurethanes and the copolyamides because the ester linkage can be attacked by water. For all TPEs, certain organic solvents can degrade these materials if one or both of the blocks will dissolve in the particular solvent. The resistance to many common oils and greases is high for the more polar TPEs.

13.4.2 Mechanical Properties

Because the structures of TPEs are diverse, the influence of structure on mechanical properties may not be universal for all materials. Nevertheless,

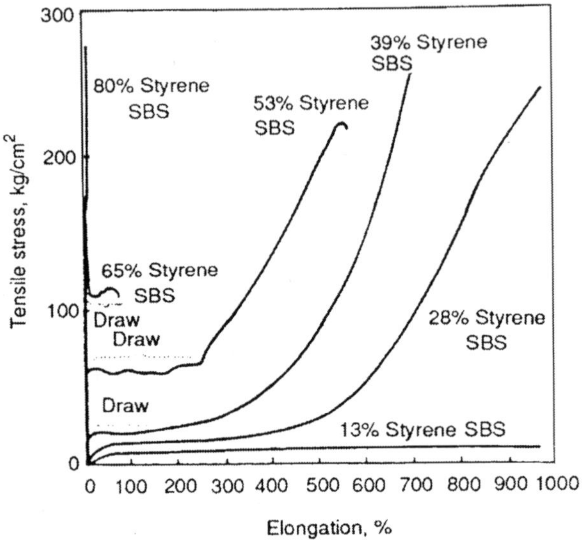

FIGURE 13.13 Tensile properties of SBS as a function of styrene content.

some general characteristics do hold for most TPEs, and these traits will be highlighted in this section.

The relative amount of hard and soft segments influences the mechanical behavior of TPEs. As the hard segment content increases, the material will change from a flexible rubber to a tough, rigid plastic. This is illustrated in Figure 13.13, which shows tensile curves for SBS as a function of styrene content. Similar data has been presented for 4GT-PTMO copolyesters and polyurethanes. TPEs also show a qualitative change in the shape of tensile curves as the hard phase changes from discrete to continuous. The yield point at approximately 40% styrene content has been interpreted as evidence of a continuous polystyrene phase; usually the existence of a yield point in a TPE implies a continuous hard phase. Altering the relative amounts of hard and soft segments in a material can be accomplished either through changing the hard segment length, soft segment length, or both. The effect of segment length has been investigated at constant relative amounts of hard and soft segments. In the styrene-diene triblock copolymers, the tensile properties do not depend on segment lengths as long as the polystyrene is of sufficient molecular weight to phase separate and form strong hard domains (>8000 g/mol) (Holden and Legge, 1996). However, the molecular weight of the blocks will influence the kinetics of phase separation in styrene-diene systems, which can in turn affect the mechanical properties.

In polyurethanes, copolyesters, and copolyamides, the molecular weight of each block is a key factor in determining the mechanical properties of the material. Soft segment block lengths between 1000 and 5000 g/mol have been found to produce materials with optimum properties. Below this value, the materials are unable to adequately phase separate. Lower molecular weights also inhibit the ability of the soft segment to strain crystallize, which provides an important strengthening mechanism. Finally, lower molecular weight soft segments at a constant overall soft segment content require short hard domains, which can severely reduce hard-domain crystallization. Higher soft segment molecular weights introduce practical problems. Since these materials are synthesized industrially in the bulk (without solvent), high molecular weight soft segments result in extremely high viscosities during polymerization, which leads to mixing and pumping difficulties as well as problems in achieving high conversions. Also, higher molecular weight blocks promote phase separation, which will reduce fractional conversion. Since all step-growth polymerizations require high fractional conversions in order to achieve high molecular weights, high molecular weight segments are used with difficulty. The effect of hard segment block length is also important. Longer hard segment blocks generally lead to better phase separation and better properties. However, at the same relative hard segment fraction, longer hard segment blocks contribute to difficulties in synthesis. Some of these difficulties can be overcome with solution polymerization methods, which may or may not lend themselves to commercial scale synthesis.

The overall molecular weight can be altered in TPEs. The overall molecular weight, as long as it is greater than some threshold value, has little effect on the mechanical properties. Two important cases can occur where the overall molecular weight is important. If soft segments in multiblock copolymers are used and have low functionality, then the resulting molecular weight distribution will be quite broad, and the properties will be reduced substantially (Speckhard and Cooper, 1986). Most industrially important soft segments have a functionality near 2, so this effect is usually not important. However, of more importance, especially in polyurethanes, is the possibility of side reactions that can lead to network formation. A small amount of crosslinking can actually improve the properties (Nierzwicki, 1985), which is presumed to occur through strengthening of the hard domain. Large amounts of crosslinking are undesired since crosslinking can inhibit phase separation and increase the brittleness of the TPE.

The effect of hard and soft segment polydispersity has been investigated in different polyurethanes (Eisenbach et al., 1992; Harrell, 1976; Miller et al., 1985). Soft segment polydispersity does not significantly alter the mechanical properties, while materials with monodisperse hard segments have a higher modulus and increased tensile strength. An important underlying assumption of these studies is that the soft segment molecular weight distribution is unaffected by synthesis, which was recently proven to be the case using a novel analytical technique (Mehl et al., 2000). The improved mechanical properties can be attributed to improved phase separation as shown in synchrotron SAXS studies (Blundell et al., 2002); better packing of the hard segments may also be playing a role.

A parameter that can be easily changed, which can have a dramatic effect on the properties, involves the constituents of the TPE. In SBS triblock polymers, this includes changing the center block to isoprene or the endblocks to α-methyl styrene. In the first case, the mechanical properties are unaffected except that in SBS the tensile strength depends on the styrene content, while in SIS the tensile strength does not vary with styrene content. Substitution of α-methyl styrene leads to a tougher polymer, which is at least partially due to the higher glass transition (160 vs. 100°C) of the new end block. Even though T_g for isoprene and butadiene differ substantially, no corresponding effect due to the different glass transition temperature is found in the mechanical properties of the TPEs (Morton, 1996). This difference emphasizes the importance of the hard phase in determining the mechanical properties of the material.

For the segmented block copolymers such as the polyurethanes or copolyesters, the composition of the hard segment or the soft segment can also easily be changed. Changing hard segment type will change the crystallinity of the hard phase. For example, copolyesters can be produced using a mixture of tetramethylene terephthalate and another diacid, such as tetramethylene isophthalate, rather than just pure tetramethylene terephthalate. As shown by Witsiepe (1973), the ultimate properties of the mixed hard segment material are better if the total hard segment content is low, with little effect on hardness or stiffness. The disadvantage of using mixed acids is that crystallization is limited and proceeds much more slowly. In polyurethanes, symmetric diisocyanates produce stronger TPEs. The presence of substituents on the aromatic ring tends to reduce the tensile properties of polyurethanes. In BD-polyester polyurethanes, the following diisocyanates give tensile strengths according to the following: MDI > hexane diisocyanate > isophorone diisocyanate > toluene diisocyanate (TDI) (Pandya et al., 1986).

Soft segment type also plays an important role in the physical properties of multiblock copolymers. Soft segments that strain crystallize produce tougher materials with higher tensile strengths and tear resistance (Morbitzer and Hespe, 1972). Since incorporation into a block copolymer reduces crystallization kinetics and slightly lowers the melting temperature, the unstrained elastomer may not contain any soft segment crystallites. Upon deformation, crystallization may occur, which will cause a large permanent set in the material. An upturn in the tensile curve at high elongations is often taken as evidence of crystallization, but WAXS provides the most direct and conclusive evidence. Soft segments that strain crystallize generally have a melting point slightly above the service temperature, and higher soft segment molecular weights favor strain crystallization. PTMO and polycaprolactone are two common soft segments that can strain crystallize.

Soft segment type also influences the driving force for phase separation. However, improved phase separation does not necessarily lead to improved properties, since both polyether (Srichatrapimuk and Cooper, 1978) and polybutadiene soft segments generally show more complete phase separation than polyesters, yet the polyester-based materials have better mechanical properties. The most likely explanation for this result is poor interfacial adhesion in well-phase-separated systems (Schneider and Matton, 1979).

13.4.3 Thermal and Chemical Properties

The response to changes in thermal or chemical environment is largely the result of the underlying chemical structure of the material. As discussed in the Introduction, lack of resistance to chemical or thermal stimuli limits the use of TPEs, especially as compared to conventional rubbers. Unfortunately it is difficult, if not impossible, to modify thermal and chemical characteristics substantially. The use of small amounts of antioxidants or fillers can improve these properties somewhat, but the overuse of these materials can result in a large change in mechanical behavior.

The sensitivity to hydrolysis is a key issue in many applications. The ester bond in 4GT-PTMO copolymers is sensitive to hydrolysis; however, it is fairly protected since most of the ester is contained in a crystalline structure. The addition of a small amount (1–2%) of a hindered aromatic polycarbodiimide substantially increases the lifetime of this material in the presence of hot water or steam (Brown et al., 1974). Polyurethanes are susceptible to hydrolytic attack, especially those with polyester soft segments. However, polyester soft segment polyurethanes are generally more resistant to oils, organic solvents, and thermal degradation. Ionomers will swell when exposed to water; in fact, a commercial hydrated perfluorosulfonic ionomer (Nafion) is used as a membrane separator in chlor-alkali cells. Styrene-diene copolymers and polyolefin TPEs are insensitive to water.

The ability of a TPE to withstand variations in temperature depends almost entirely on the chemical structure. The maximum service temperature is usually about 20–40°C below the hard segment glass transition or melting temperature. Because of hysteresis, excessive heat buildup can occur during use so that the local temperature of the material can be much higher than the nominal temperature. Changing the maximum service temperature involves changing the structure of the hard block. Using α-methyl styrene in styrene-diene triblock copolymers or ethylene diamine chain extenders in polyurethanes can extend the service temperature in these TPEs substantially. The minimum service temperature is usually about 10°C above the soft segment T_g. Below this temperature, the material will become brittle. If the soft segment can crystallize, then low temperatures can cause crystallization and a corresponding increase in stiffness and brittleness. Using a mixed or copolymer soft segment will eliminate this problem, but at a cost of reducing strain-induced crystallization at higher temperatures.

13.5 THERMODYNAMICS OF PHASE SEPARATION

This section will present the theoretical framework and understanding about the thermodynamics of phase separation in block copolymers. Most theories consider four factors that influence the phase separation of block copolymers: (1) the Flory-Huggins interaction parameter χ, (2) the overall degree of polymerization N, (3) architectural constraints such as the number of blocks and linear vs. starblock polymers, and (4) the weight fraction f of one component. The thermodynamic theories are conveniently divided into three cases: the strong segregation limit (SSL), the weak segregation limit (WSL), and the intermediate segregation region (ISR). In the SSL, the equilibrium state of the material consists of relatively pure phases of A separated from relatively pure phases of B. In the WSL, the two phases are intimately mixed. The ISR is essentially a region that arises because of finite molecular weight; in the case of infinite molecular weight no ISR exists. Whether the ISR should be considered as part of thermodynamic phase space is questionable.

The transition between the WSL and the SSL is termed the order-disorder transition (ODT), which is also called the microphase separation transition (MST). The analogous transition in small molecules is the solid-to-liquid transition. The reader should be aware of the order-order transition (OOT), which is a shift of morphology from one type to another (e.g., spheres to cylinders) that can occur with changes in temperature in anionically synthesized block copolymers at very specific block lengths (Sakarai et al., 2004). If the ODT temperature is below the hard segment T_g or T_m, then the material is one phase above the hard segment dissolution temperature. If the ODT temperature is above the hard segment T_g or T_m, then the material will exist as a two-phase melt. The phase state of a TPE above the dissolution temperature has a substantial effect on the rheological properties since one-phase mixtures have a viscosity much smaller than two-phase mixtures.

A great deal of effort has been spent on developing theories for diblock or triblock copolymers, much of which has been driven by the measurement of phase diagrams, which in turn provide good tests of theories. It should be noted that the rigorous application of these theories to TPEs used in commercial applications is limited, since in nearly all systems some arrest due to slow kinetics occurs. The most rigorous theories were developed for monodisperse block lengths (e.g., anionically polymerized TPEs). In some cases these theories have been extended to multiblock copolymers that have a distribution of block sizes. Further, amorphous systems have had a significantly larger focus on them than crystalline systems.

In the WSL, the chain configuration is unperturbed (e.g., R_g scales as $N^{1/2}$). However, the probability of finding an A or a B segment at a distance r from a particular point does not only depend on f, it also depends on whether the original point sits on an A or B segment. If the original point lies on an A segment, then at short distances the probability is much higher of finding another A segment, and at distances comparable to the chain's radius of gyration, the probability is higher of finding a B segment. This concept is called the correlation hole effect (DeGennes, 1979). At extremely large distances, the probability of finding an A segment reduces to the volume fraction of A segments in the WSL. The chains generally do not assume their unperturbed conformation in the SSL. The probability of finding a segment at a distance r depends strongly on the type of morphology and can also have a directional dependence, whereas in the WSL the probability is isotropic.

The theoretical development of Helfand (1975) and Helfand and Wasserman (1978, 1980, 1982) contains all the necessary ingredients for a complete description of phase separation in the strong segregation limit; in fact, the theory has been modified slightly so that the results are considered to be quantitatively reliable (Whitmore, 2001). Three energetic contributions are included in this theory: (1) confinement entropy loss due to a concentration of AB joints to the interface, (2) conformational entropy loss due to extended chains, and (3) enthalpy due to mixing of A and B segments. An expression is written for a function that is proportional to the probability density that a chain with N segments has one end at r_0 and another at r. The resultant equation is identical to the form for the time-dependent diffusion equation where the differential with respect to time is replaced by a differential with respect to segment. The remainder of the development involves solving this equation with the appropriate boundary conditions. Complete analytical solutions have not been derived, but numerical solutions to the equations have been calculated. This theory was originally developed for diblock copolymers and has been extended to triblock systems.

The interfacial thickness (t) was predicted to be approximately equal to the following:

$$t = 0.816 a \chi^{1/2}, \tag{13.15}$$

where a is the statistical segment length. This equation predicts that the interfacial thickness is independent of molecular weight. In the limit as $N \to \infty$, the domain spacing D was found through numerical analysis to scale as:

$$D \sim a N^{9/14} \chi^{1/7}. \tag{13.16}$$

In this limit, the confinement entropy of a junction is insignificant compared to the stretching entropy of the chain. Numerical predictions of the predicted periodicity have been made for diblock and triblock copolymers for the cylindrical, spherical, and lamellar morphologies. In general, the agreement between the theoretical predictions and the experimentally observed results has been good for cylindrical and lamellar systems, while not as good for spherical systems. Because spheres are isolated, changes in domain size can only occur by transport of segments through an incompatible matrix, which provides a substantial diffusional resistance to changes in morphology. Hence, the morphology tends to not change once it is formed, and the agreement between theory and experiment is poor (Meier, 1996). Numerical procedures were given for predicting the phase diagram including calculation of the underlying morphology (spheres, cylinders, etc.) as a function of composition. The underlying morphology type was predicted to be almost temperature independent.

Numerous studies were completed on block copolymers in the SSL long before a theory of the detail inherent in the Helfand and Wasserman approach existed. In the WSL, experiments have been driven by the mean-field theories originally outlined by Leibler for diblock copolymers with monodisperse blocks (Leibler, 1980). This method uses the random phase approximation (DeGennes, 1970) (RPA) to calculate the free energy in terms of an order parameter $\psi(r)$, which describes the average deviation from the uniform distribution at any point r. The thermodynamic state of the material was found to depend only on χN and f, just as in the Helfand and Wasserman approach. Only three two-phase morphologies are predicted from this model: body-centered-cubic packed (bcc) spheres, hexagonally-closest-packed (hcp) cylinders, and lamellae. Since χ depends on temperature, the morphology can be changed from spheres to cylinders or lamellae simply by changing temperature, in contrast with the Helfand and Wasserman theory, which predicts the underlying morphology is independent of temperature. A critical point (second-order transition) is predicted at $\chi N = 10.495$ for the symmetric diblock copolymer, which also corresponds to the minimum χN that divides the ordered and disordered state. Since the point of demixing for two homopolymers of identical molecular weight is $\chi N = 2$, the joining of two individual chains at the ends means that the individual chains need to be over 2.5 times as long in the diblock copolymer than in the homopolymers to phase separate.

Leibler's theory outlined an experimental method for testing its conclusions. The structure factor for scattering of radiation by the disordered phase was given by:

$$\frac{1}{S(q)} = \frac{F[(qR_g)^2, f]}{N} - 2\chi, \tag{13.17}$$

where F is a dimensionless function that is related to the Debye correlation function of a Gaussian chain. A Lorentzian peak is predicted for the disordered phase. The position of the peak maximum at q^* does not change with temperature. The periodicity of the concentration fluctuations is given by $2\pi/q^*$, which is approximately the radius of gyration of the chain. The height of the peak is a function of temperature because of changes in χ. Generally, χ has been found to depend on temperature according to the following:

$$\chi = \frac{A}{T} + B, \tag{13.18}$$

where A and B are empirically determined constants. According to the above analysis, a plot of 1/peak intensity in SAXS or SANS experiments vs. reciprocal

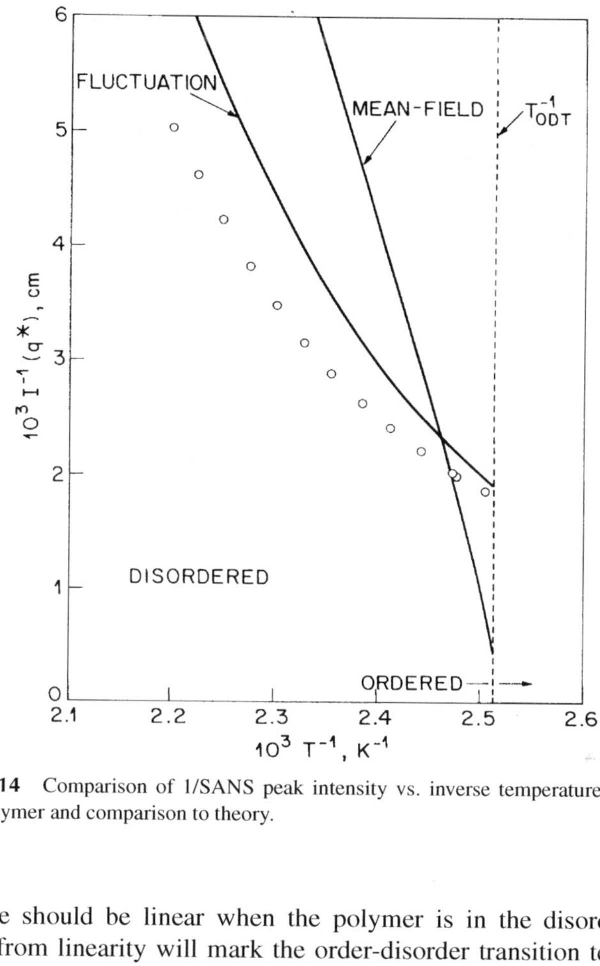

FIGURE 13.14 Comparison of 1/SANS peak intensity vs. inverse temperature for PEP-PEE diblock copolymer and comparison to theory.

temperature should be linear when the polymer is in the disordered state. Deviation from linearity will mark the order-disorder transition temperature, while an extrapolation to zero intensity will allow the calculation of the spinodal decomposition temperature. The ODT according to this theory can also be determined from the temperature where q^* changes as a function of temperature since q^* is predicted to be independent of temperature in the disordered state.

The WSL theory developed by Leibler has been shown to be incorrect because of deviations from the fundamental underlying mean-field assumption. Figure 13.14 shows experimental results for a poly(ethylene-propylene/ethylethylene) (PEP-PEE) diblock copolymer that has been fit to the predictions of the Leibler theory without any adjustable parameters, since the ODT and χ were calculated from rheological measurements (Bates et al., 1990). This mean-field theory does not qualitatively describe the behavior of this material. Other experiments have indicated that the RPA approximation (Stühn and Stickel, 1992) and the Gaussian coil assumption (Bates and Hartney, 1985; Holzer et al., 1991) are inaccurate near the ODT.

In the original analysis, it was understood that this theory should not be applied near the critical point. Mean-field theories ignore concentration fluctuations at distances other than $q = q^*$. Near the critical point, concentration fluctuations on very large length scales become increasingly important. A modification to this theory that includes concentration fluctuations has been developed (Fredrickson and Helfand, 1987). A critical point is not predicted by the fluctuation theory, rather a first-order phase transition is predicted for all compositions. A molecular weight dependence is found for χN, which delineates the ordered from the disordered phase shown below for the symmetric diblock copolymer:

$$\chi N = 10.495 + 41.022N^{-1/3}. \tag{13.19}$$

The minimum χN value corresponding to the disordered phase still occurs at $f = 0.5$. The fluctuation theory predicts an ODT at a slightly lower temperature than the Leibler theory. The lamellar and hcp cylinder phases are directly accessible from the disordered state, which seems to be confirmed by experiment, rather than having to pass through the bcc sphere phase as in the Leibler theory. The rather simple structure factor presented in Eq. (13.17) is retained; however, χ is replaced with an χ_{eff} that depends on temperature, composition, and molecular weight in a complicated manner. At temperatures slightly above the ODT, a partially ordered morphology is predicted. Figure 13.14 also compares the results of the experiment with the fluctuation theory without any adjustable parameters. The qualitative shape of the experimental data is described better by the fluctuation theory; however, quantitative agreement is still not found.

The previously discussed theories were developed for monodisperse diblock copolymers, which are not TPEs. However, Leibler's mean-field theory has been extended to include polydispersity (Leibler and Benoit, 1981) and to include triblock, star, and graft copolymers (Olvera de la Cruz and Sanchez, 1986; Mayes and Olvera de la Cruz, 1989). In the former case, polydispersity corrections tend to lower χN corresponding to the ODT. As would be expected from the analogy between blends and diblocks, triblocks will phase separate at higher χN values than the corresponding diblocks. This theory predicts a monotonic increase in the critical value of χN as the symmetry of the triblock increases, to a maximum of about 18 for the symmetric triblock. Surprisingly, the minimum χN value that separates the order and disordered regions in triblocks does not necessarily correspond to the critical point.

The development of the mean-field theory for triblocks is very similar to the approach followed by Leibler. A second parameter, τ, defines the asymmetry of the triblock. If the block copolymer is labeled ABA, then starting at the center of the B block, τ is the fraction of A units going in one direction along the chain divided by the overall number of A units in the copolymer. By convention, τ is always less than 0.5, so $\tau = 0$ defines a diblock copolymer, while $\tau = 0.5$

is a symmetric triblock. f is defined as the overall fraction of A units in the triblock copolymer. For all τ values, a critical point is predicted at a certain composition and χN. The critical point does not occur at $f = 0.5$ like in the case of symmetric materials, rather the copolymer composition at the critical point is a function of τ. The ordered phase following the ODT is bcc spheres followed by hcp cylinders and finally lamellae except at the critical point. The same experimental methods and analysis that are used for diblock copolymers can also be used for triblocks. q^* seems to be weakly temperature dependent, and the qualitative shape of the scattering curve is different because of an upturn at low q (Koberstein et al., 1990). The difference in the ODT temperature between the diblock PEE-PEP and the triblock PEP-PEE-PEP was 72°C, which is near 61°C predicted from the mean-field theories (Gehlsen et al., 1992).

13.6 THERMOPLASTIC ELASTOMERS AT SURFACES

13.6.1 General Characteristics

The two phases of TPEs also will affect applications that are sensitive to surface or interfacial properties. Generally, the fraction of a component at an interface can be substantially different than the overall bulk fraction. The presence of an interface introduces another thermodynamic consideration that can also alter the morphology that exists in the bulk of the sample. The compositional and morphological variation depends on the processing conditions that were used to generate the surface, as well as the nature of the other surface. The energetic driving force for these variations is a desire to minimize interfacial energy. The primary parameter that characterizes the interfacial characteristics of a material is the surface tension or surface energy.

Although the two terms are often used interchangeably, strictly speaking the two are not identical. The surface tension is defined as:

$$\gamma = \left(\frac{\partial A}{\partial \Omega}\right)_{V,T,N_i}, \tag{13.20}$$

where A is the Helmholtz free energy of the entire system, Ω is the interfacial energy, V is the system volume, T is temperature, and N_i is the number of moles of the ith component. This derivative is defined such that the surrounding medium is a vacuum. A higher surface tension means a stronger opposition to the formation of a larger surface area. The expression for surface energy is identical except the change in the Helmholtz free energy of the surface replaces the Helmholtz free energy of the system. For relatively deformable polymers, the difference between the two expressions is small and is usually ignored (Lelah and Cooper, 1986). Surface tension is experimentally measured either from polymer melts or from contact angle experiments. The former requires extrapolation to the solid, while the latter cannot be measured directly and must be calculated from semiempirical equations. However, the values calculated from the two methods usually agree well (Wu, 1989). The Polymer Handbook lists surface energies for many segments commonly found in TPEs. Most polymers have surface energies in the range 30–45 dyne/cm at room temperature. Exceptions to this simple generalization are fluorine- and silicon-containing polymers, which have significantly lower surface energies.

This section describes analytical methods that are used to characterize the atomic composition or morphology at TPE surfaces. Measurement of surface composition in TPEs can suffer from one major drawback. Since the soft phase is mobile at room temperature, the surface composition can change depending on the medium in which the measurement is being made. Many analytical methods are performed in ultra-high vacuum (UHV), which is not a normal atmosphere for TPE applications. Typically the difference in a TPE surface exposed to UHV or exposed to air is small; however, a significant difference will exist if the surface is exposed to a liquid such as water. Different strategies have been devised to help overcome this problem, and some of these will be discussed in this section. Analytical methods for surface composition also require a very clean surface to give quantitative results. Small amounts of additives normally present in commercial materials may also significantly alter the surface composition (Tyler et al., 1992). Finally, some methods can alter the surface during measurement.

Another extremely important parameter of any surface technique is the depth of penetration (d_p), which is a characteristic distance from the surface that the measurement will probe. Table 13.5 gives approximate depths of penetration for the methods discussed in this section for measurement of TPE surfaces. Since Table 13.5 shows that the depth of penetration varies widely, the change in composition can be monitored from the surface almost continuously to thousands of angstroms. Alternatively, some methods discussed in this section measure the morphology near the surface. Studies of TPE surfaces are less numerous than investigations of bulk properties, which in general speaks to

TABLE 13.5 Depths of Penetration for Surface-Sensitive Analytical Methods

Method	Depth of Penetration
Attenuated total internal reflection	1000–10,000 Å
X-ray photoelectron spectroscopy	10–100 Å
Static secondary ion mass spectrometry	1–10 Å
Dynamic secondary ion mass spectrometry	Varies
Neutron reflectivity	Varies
Atomic force microscopy	Surface

the relative lack of importance of the surface properties of TPEs for most applications. However, exceptions to the latter generalization definitely exist, for example in the medical device area.

13.6.2 Studies of Surfaces

(i) Scanning Electron Microscopy (SEM)

In SEM, electrons are reflected from the surface, and detection is done on the same side of the surface as the source; in TEM, the electrons pass through the sample and are detected on the other side. SEM is done under UHV conditions, generally 10^{-7}–10^{-9} torr. Probably the most common use of SEM in polymer science is to examine fracture surfaces. The morphology of a fracture surface gives details about the fracture mechanism. In TPEs, SEM is not appropriate for determining whether interphase or intraphase fracture occurs, for example; the domains are too small and the large amount of phase mixing may also play a role as well.

In well-phase-separated systems (i.e., anionically synthesized block copolymer cast from solution), information about composition at the surface can be obtained by microtoming perpendicular to the surface and then directing electrons parallel at the just-cut surface. The top (or bottom) of the image contains information about the composition at the free surface. An excellent study on a lamellar SI diblock copolymer with 52% styrene that was cast from toluene (Hasegawa and Hashimoto, 1985) showed that generally the lamellae oriented parallel to the air-polymer interface in agreement with earlier studies (Hashimoto et al., 1977, 1980) and studies of other copolymer systems (Anastasiadis et al., 1990; Chen et al., 1992). A micrograph of SI diblocks with lamellae parallel to the interface is shown in Figure 13.15. The lower surface energy component polyisoprene (dark component in Figure 13.15) was always located at the interface between the polymer and the vacuum. Some micrographs showed perpendicular orientation of the lamellae relative to the surface as also shown in Figure 13.15. However, a thin polyisoprene layer is still at the air-polymer

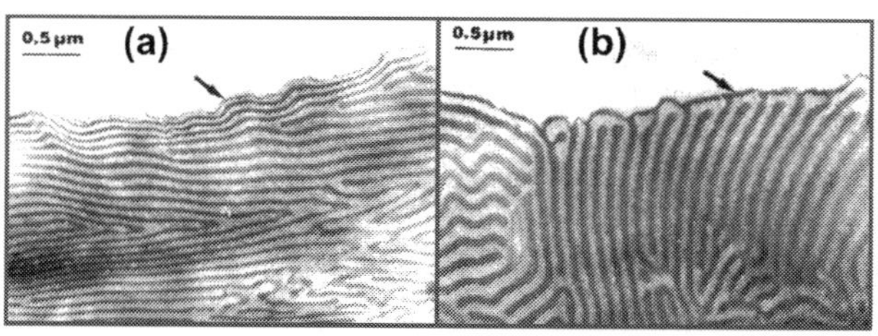

FIGURE 13.15 Electron micrographs of SI diblock copolymer (MW = 524,000 g/mol, 52% styrene) at a free surface. (a) Lamellae oriented parallel to surface (typically observed). (b) Lamellae oriented perpendicular to surface (occasionally observed). The arrows indicate the locations of the free surfaces. (From Hasegawa and Hashimoto (1985); reprinted (adapted) with permission from ACS, copyright 1985.)

interface in this micrograph (i.e., the lower surface energy component was located at the interface). A very similar system, except with a crystallizable block, did not show the same behavior (i.e., the surface composition depended on the orientation of the crystallites (Hasegawa and Hashimoto, 1992)). This type of interaction at a surface can be used in anionically synthesized block copolymers to give very unique morphologies on a surface when very thin films are cast from solution (Segalman et al., 2003; Fasolka and Mayes, 2001).

(ii) Attenuated Total Internal Reflection Infrared Spectroscopy (ATR)

In ATR, a beam of infrared light is totally reflected inside a specially cut infrared transparent material that has a high index of refraction. Typical materials used for ATR prisms are Ge, Si, and ZnSe. Because the index of refraction differs between the polymer and the prism, an evanescent wave penetrates the polymer if it intimately contacts the prism. The infrared radiation will interact with molecular vibrations in the same manner as in conventional infrared spectroscopy. The amplitude of the evanescent wave decays exponentially from the surface, so the depth of penetration is arbitrarily taken as the point where the amplitude decays to 1/e (37%) of its initial value. The depth of penetration depends on the ratio of the refractive indices between the polymer and the prism, the angle of incidence, and frequency of radiation in the following manner (Ishida, 1987):

$$d_p = \frac{1}{2\pi v\left[\sin^2\theta - \left(\frac{n_2}{n_1}\right)^2\right]^{1/2}}, \tag{13.21}$$

where v is the frequency, n_2 and n_1 are the refractive indexes of the polymer and prism, respectively, and θ is the angle of incidence of the infrared wave. The dependence of the depth of penetration is not as simple as the above relationship suggests because the refractive indices also have a frequency dependence. For nonisotropic materials, d_p also depends on the direction of the electric field vector of the radiation relative to the surface of the prism. ATR spectroscopy is a very versatile technique and can theoretically be used in almost any medium. The latter capability has made ATR a very important method for the characterization of TPEs used in the medical device area.

Although ATR has been used to quantify the variation in composition at the surface in TPEs (Sung and Hu, 1980), a related utility is its ability to monitor *in situ* processes such as reaction injection molding (RIM) (Ishida and Scott, 1986) and protein adsorption onto a polyurethane substrate (Pitt and Cooper, 1986). In the latter, the effect of shear rate on the kinetics of protein adsorption and desorption from phosphate-buffered saline (PBS) was studied in a specially designed flow cell. A very thin film of the commercial MDI-ED-PTMO polyurethane Biomer was cast from solution onto a Ge ATR prism. The thickness of the film was less than the penetration depth so the protein concentration could be monitored after the infrared absorption of the polymer and PBS was subtracted. The study found that increasing the wall shear rate does not affect the rate of protein adsorption, but the desorption rate is slowed. In later studies using essentially the same apparatus (Giroux and Cooper, 1991), spectral changes corresponding to protein conformational changes were followed as the protein absorbed to a polyurethane substrate.

(iii) X-Ray Photoelectron Spectroscopy (XPS)
XPS (also termed electron spectroscopy for chemical analysis) takes place in a vacuum and is able to quantify the relative amount of atomic constituents on the surface of a material. With the exception of hydrogen and helium, any chemical element can be identified, and XPS is also sensitive to chemical bonding effects. In XPS, a surface is irradiated with low-energy X-rays (usually aluminum or magnesium K_α), which results in the ejection of core level electrons. Qualitatively, the depth of penetration is limited by inelastic scattering of the ejected photoelectron as it travels toward the surface. Quantitatively, d_p is an extremely complicated function that depends on the energy of X-rays and the structure of the material; typical depths of penetration in a polymer are on the order of a few angstroms. The number of ejected photoelectrons that reach the detector corresponding to a particular distance from the surface decays exponentially as the distance increases. d_p is defined where the number of ejected electrons reaches 37% (1/e) of its initial value and is proportional to $\cos\theta$, where θ is the angle of emission with respect to the sample normal.

The most common use of XPS on TPEs has been to quantify the fractions of each phase at the surface. By varying the emission angle, the composition can be probed at different depths of penetration. XPS studies have consistently shown that the low surface energy component will predominate at a free surface. Even a small difference in surface energy provides a strong driving force for surface enrichment of the lower energy component. The effect of evaporation rate and overall molecular weight was studied in symmetric poly(methyl methacrylate-b-styrene) (PMMA-S) cast from toluene. The surface energy difference between these two components is less than 1 dyne/cm. The amount of polystyrene at the surface was greater for the slowly evaporated film. The fraction of polystyrene at the surface increased as the overall molecular weight increased and reached a constant value of approximately 90% polystyrene at the surface when the overall degree of polymerization was approximately 1000 (Green et al., 1989). A later study on annealed films of these same materials showed that the polystyrene layer completely covered the surface when the degree of polymerization was greater than approximately 2500 (Green et al., 1990). In both studies, the surface excess of polystyrene Φ_{PS} was found to obey the following relationship:

$$\Phi_{PS} = a - bN^{-1/2}, \tag{13.22}$$

where a and b are empirical constants. This functional form agrees with the mean-field theory of block copolymers at surfaces developed by Fredrickson

(1987) and the constants were given a physical interpretation regarding the surface energy differences and the ability of the surface to modify the energetic interactions between the two polymers. The surface excess of polystyrene was also found to decrease with increasing temperature, which agrees with qualitative expectations (Green et al., 1991).

Ultra-high vacuum is required to perform XPS, which is a major disadvantage of this technique. As mentioned earlier, samples can show substantially different surface compositions in UHV than under normal conditions, especially if the material is immersed in a liquid (e.g., blood-contacting applications). Recent techniques that utilize freeze-drying and measurements at temperatures lower than the glass transition of the soft segment have been devised in order to quantify the surface composition at liquid interfaces. In a number of different polyurethanes, the low surface energy component was in excess when spectra were collected under normal conditions while the high-energy component was in excess when the freeze-drying method was used (Kajiyama and Takahara, 1991). This generalization held whether the low energy surface component was the soft or hard segment. This study illustrates surface rearrangement that can occur in any study of polymer surfaces.

(iv) Secondary Ion Mass Spectroscopy (SIMS)

When a surface is bombarded with a beam of high-energy primary ions, atomic collisions between the beam and the solid cause the ejection of secondary ions, which can be characterized using a mass spectrometer. These secondary ions provide information on the atomic and molecular species present at the surface. The ions O_2^+, O^-, and Ar^+ are generally used for SIMS analysis, with the first ion being the most common. SIMS experiments are characterized as either dynamic or static. In static SIMS, the ion beam currents are low (less than 5 nA/cm^2) so that the surface etches very slowly. The depth of penetration for static SIMS is approximately 10 Å. Beam currents are much higher for dynamic SIMS and successive layers are etched away rapidly during the test. This method eventually produces a pit in the sample due to this etching process. Atomic species can be quantitatively characterized as a function of distance from the surface if the relationship between the etching rate can be measured. In SIMS experiments, the sample surface can become charged because most polymers are nonconductive, and this charge must be removed during the experiment. UHV conditions are required for SIMS experiments.

SIMS is extremely sensitive, even to the part per billion range, and is used widely in the semiconductor industry. SIMS and XPS are complementary since they provide very similar information (Sabbatini and Zambonin, 1996), but the depth resolution of static SIMS is substantially higher than XPS. However, the quantitative use of SIMS in polymer science is currently not as well developed as XPS. SIMS has been used to study the surface of polyurethanes (Deslandes et al., 1998; Shard et al., 1995) and PMMA-S lamellar diblocks (Coulon et al., 1989). In the latter study, dynamic SIMS was used to show that the lamellae aligned parallel to the surface after annealing. The thickness of the layer nearest the free surface as well as near the polymer-substrate surface was shown to be approximately half of the thickness of the interior layers. The experimental data were fit with a model that included the lamellar period of the block copolymer, and the results were compared with SAXS results from the same material.

(v) Atomic Force Microscopy

The concept of atomic force microscopy is extremely simple. Essentially a very small probe (tip radius of 10–100 nm in size) is rastered over a surface such that the force between the surface and the probe is maintained constant; for studies of TPEs normally tapping mode is used, which means that a vertical oscillation is superimposed on the rastering. Essentially, the deflection of the probe is measured as a function of position. The interaction between the probe and the surface is not well understood, but the AFM pattern does not seem to be very sensitive to probe characteristics. The resolution of AFM is in the 1 nm range. A primary advantage of AFM is the scanning range, which allows for the investigation of variations in surface roughness and free surface characteristics over a large lateral area.

Different TPEs have been imaged using AFM include polyurethanes (Aneja and Wilkes, 2003; Gisselfalt and Helgee, 2003), polypropylene-based thermoplastic elastomers (Schonherr et al., 2002), copolyamides (Krijgsman et al., 2003), anionically synthesized block copolymers (Annis et al., 1992; Knoll et al., 2001), styrene-isobutylene-styrene triblock copolymers (Puskas et al., 2000), and TPVs and TPOs based on polypropylene and EPDM (Shahbikian et al., 2012). Not only can one visualize topographical differences, the AFM can also distinguish between hard and soft segments, and hence can determine the shape of these regions at the interface and, if the surface is microtomed, within the bulk of the material. The contrast is provided by the difference in modulus of the hard and soft segments; this difference allows AFM to distinguish between the two components. Examples of the use of AFM to characterize phase separation of hard and soft domains in thermoplastic elastomers are shown in Figure 13.16 for poly(styrene-b-isobutylene-b-styrene) (SIBS) TPEs (Puskas et al., 2000) and in Figure 13.17 for TPVs based on PP and EPDM (Shahbikian et al., 2012). Alternatively, the sample can be heated and imaged simultaneously to explore changes in morphology as the system softens (Vasilev et al., 2004). In systems where crystallization is the main driving force toward phase separation, the resulting AFM is qualitatively identical to AFMs from normal semicrystalline polymers. In systems where this is not the case (e.g., polyurethanes), the morphology can be very different. For example, on a system with a monodisperse distribution of hard-domain lengths, such as the spherulitic hard-domain superstructure, the orientation of the chain axis is the same as in normal spherulitic structures (i.e., in a direction tangential to the radius), but no chain folding occurs (Garrett et al., 2001).

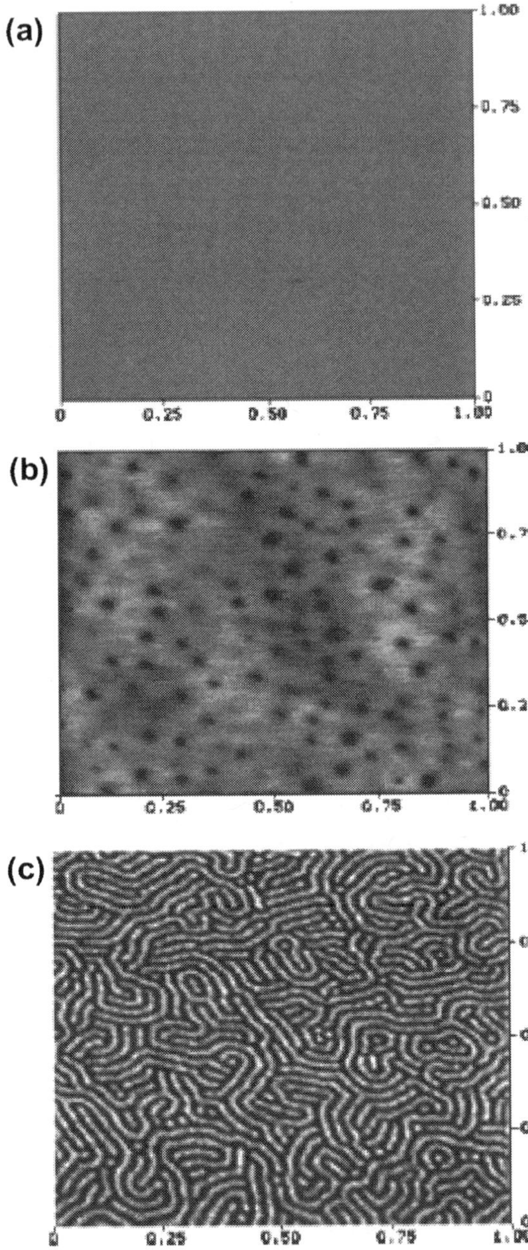

FIGURE 13.16 AFM phase images of poly(styrene-b-isobutylene-b-styrene) (SIBS) triblock copolymers. The images are for: (a) polyisobutylene homopolymer; (b) branched SIBS with 16% styrene content; and (c) linear SIBS with 30% styrene. The darker regions have greater hardness and thus represent the polystyrene domains. (From Puskas et al. (2000); reprinted with permission from John Wiley & Sons, Ltd., copyright 2003.)

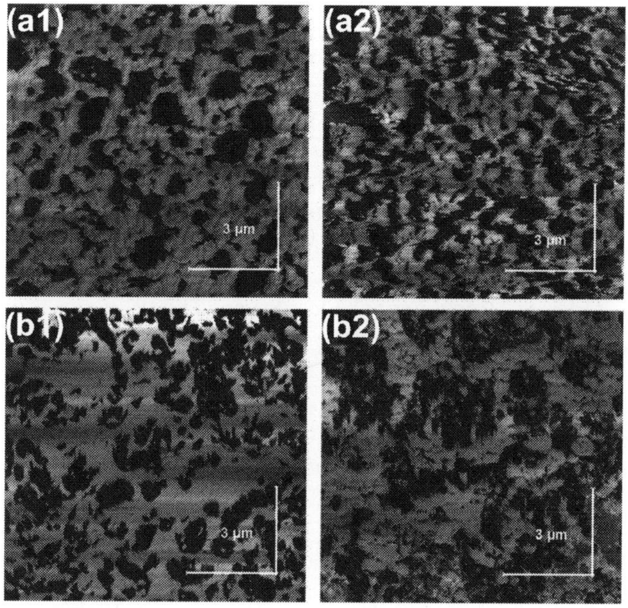

FIGURE 13.17 AFM phase images of PP/EPDM (50 wt.%/50 wt.%) TPVs: (a) Internal mixer; (b) Twin-screw extruder (Column 1: nonplasticized; Column 2: plasticized with paraffinic oil). The dark phase is EPDM and the bright phase is PP. (From Shahbikian et al. (2012); reprinted with permission from John Wiley & Sons, Ltd., copyright 2011.)

13.7 RHEOLOGY AND PROCESSING

The theoretical development and characterization techniques for the rheology of rubbers outlined in other chapters of this book apply equally well to TPEs. However, there are fundamental differences in the response to stress between a conventional rubber and a TPE. Of course the biggest dissimilarity is that a TPE will flow at higher temperatures while a crosslinked rubber will not. Some of the similarities and differences between the two types of materials will be discussed in this section. Also, the processing methods used for TPEs will also be introduced; the processing methods for these materials are no different than for any other thermoplastic, and the interested reader can consult monographs on the subject (Osswald, 1998; Corish, 1991).

At service temperature and frequency, the solid-state rheological properties of a TPE and a conventional rubber are similar, although at large strains the former will generally have some permanent set, while the latter will not. Only by changing the temperature and/or the frequency do substantial differences between the two emerge. If the temperature is lowered or the frequency raised, the qualitative response of the two materials is still very similar. Both materials become more resistant to stress until a temperature is reached that corresponds to the T_g of the soft segment in a TPE or the rubbery matrix in a conventional rubber. Below this temperature, both a conventional rubber and a TPE will behave as a one-phase brittle glassy polymer. As the temperature is raised, the two materials show qualitatively different behavior. In a conventional rubber, the material will continue to behave as an elastomer until the degradation temperature is reached. In a TPE, the hard domains will eventually weaken and flow, which allows the material to be processed as a conventional thermoplastic. The processing characteristics will be intimately related to the ODT temperature since the viscosity is substantially lower above the ODT.

More specifically, the dynamic mechanical properties are significantly different in TPEs vs. conventional elastomers. The Williams-Landel-Ferry (WLF) equation does not generally apply to TPEs since two phases make the material rheologically complex. Dynamic mechanical experiments are used as a macroscopic characterization of phase separation since a more highly phase-separated material will have a larger temperature difference between the soft and hard segment transitions and a flatter plateau modulus. Figure 13.18 (Wang and Cooper, 1983) shows E' and tan δ for a series of polyurethaneureas as a function of increasing hard segment content that demonstrates the forementioned behavior in the more highly phase-separated higher hard segment materials. Figure 13.18 also exhibits the general features common to all dynamic mechanical measurements of TPEs: a soft segment glass transition temperature below room temperature and a hard segment dissolution temperature above room temperature. The normal operating region of a TPE is the flat plateau region between the two transition temperatures. As suggested by the diagram, one can shift the upper transition temperature by increasing phase mixing, which is usually accomplished by reducing the length of the hard segments.

Rheological measurements can also supply a measure of the ODT temperature independent of scattering measurements. In this method, the storage and loss moduli are measured as a function of frequency at different temperatures above and below the MST. One strain level should be used for these experiments since it has been shown that below the ODT, the dynamic mechanical properties are highly strain dependent. The WLF equation is then used to collapse all of the data onto one curve in a modulus vs. reduced frequency plot, if the frequency is high enough. The frequency at which the curves begin to diverge marks the critical frequency (w_c) of the material. The critical frequencies are not necessarily identical for the storage and loss moduli, but generally differ by less than an order of magnitude. After finding w_c, a storage or loss modulus is measured as a function of temperature at constant frequency, which is significantly below the critical frequency. The temperature at which a discontinuity occurs in the modulus marks the ODT temperature, as shown in Figure 13.19. The authors recommend using the storage modulus because the discontinuity is sharper. This method relies on the different behavior of the storage and loss moduli at low frequencies for the disordered and ordered systems. For the disordered system, the behavior is identical to normal terminal

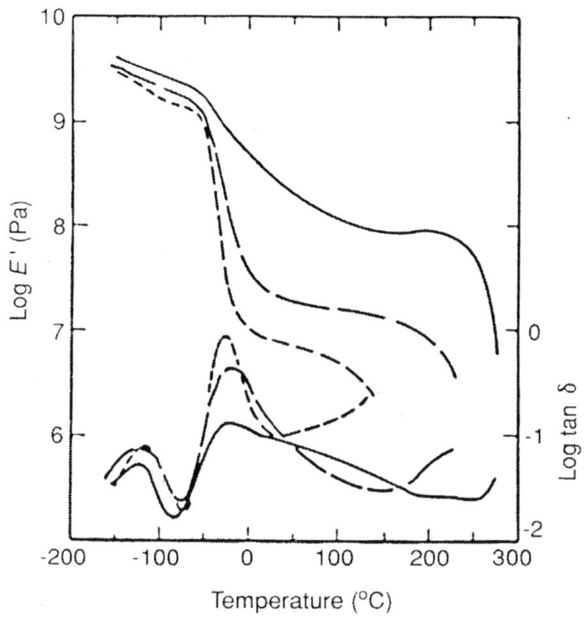

FIGURE 13.18 Storage modulus (E′) and tan $\delta(E''/E')$ vs. temperature for MDI-ED-PTMO (MW = 1000) polyurethaneureas. The molar ratio of the components MDI-ED-PCL is (—) 3:2:1; (– – –) 2:1:1; (- - -) 1.3:0.3:1.

behavior in one-phase materials (i.e., $G' \sim w^2$ and $G'' \sim w$), while in the ordered system, the scaling depends on the morphology and orientation of the material; for a macroscopically unoriented lamellar system, $G' \sim G'' \sim w^{0.5}$.

The viscosity of a TPE depends heavily on whether the system is above or below the ODT temperature. For example, in styrene-diene triblock systems, commercially important materials have ODT temperatures above a reasonable processing temperature; therefore, melt viscosities are usually orders of magnitude higher in the block copolymer than in the homopolymers of the same molecular weight. Typical values at high shear rates are 10^4 poise, which can increase orders of magnitude as the shear rate is decreased.

Generally, TPEs are processed identically to any other thermoplastic (i.e., injection molding, extrusion, calendaring, etc.). Some TPEs, for example ionomers, tend to adhere to surfaces, so special equipment and techniques may be necessary to remove an injection-molded part. The processing temperature must be higher than the hard segment dissolution temperature, but even two-phase melts can generally be handled with conventional equipment. The viscosities of TPEs above the ODT are rather low in comparison to other thermoplastics because of the low overall molecular weights. In fact, in some one-phase materials with low overall molecular weights such as the copolyesters, blow molding cannot be utilized because of low viscosities

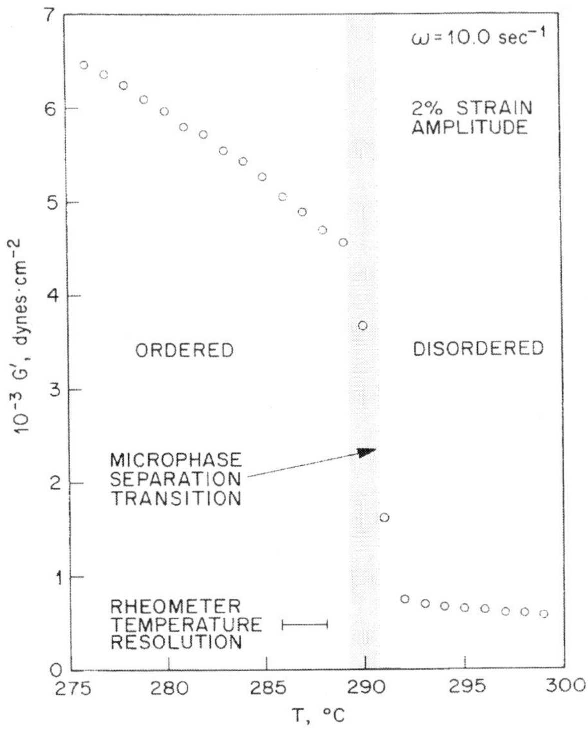

FIGURE 13.19 Temperature dependence of G', the shear storage modulus, at a frequency of 1.6 Hz for the diblock copolymer PEP-PEE. The ODT temperature was calculated as 291 ± 1°C.

(Adams and Hoeschele, 1996). The resultant morphology and properties of the material depend heavily on the processing method.

13.8 APPLICATIONS

The processing advantages of thermoplastic elastomers in comparison to thermoset rubbers were described earlier. Figure 13.20 highlights these differences. Due to the ability to melt TPEs at high temperatures, these materials can be simply shaped into desired products using traditional thermoplastic processing equipment. This benefit is also a detriment; the fact that a TPE can flow at elevated temperatures due to lack of crosslinks in the continuous phase also means that it has a limited temperature range over which it can be used as an elastomeric material. Therefore, the intended use temperature range must be considered in selecting which TPE to use for an application. An illustration of the temperature use window as defined by the temperature dependence of modulus is shown in Figure 13.21. Values of the T_g of the soft phase and T_g or

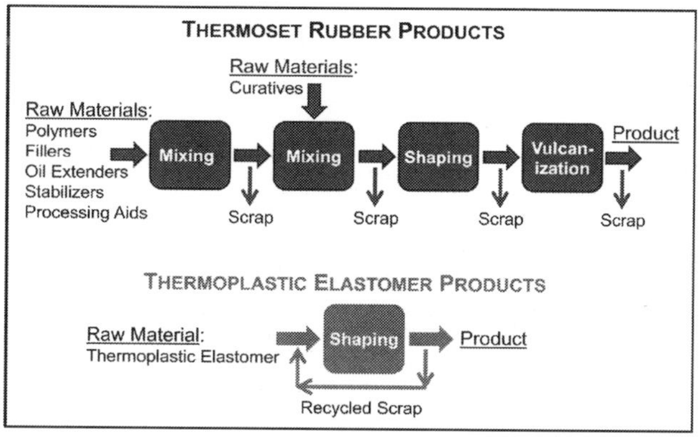

FIGURE 13.20 General contrast between the processing of thermoset rubbers and thermoplastic elastomers.

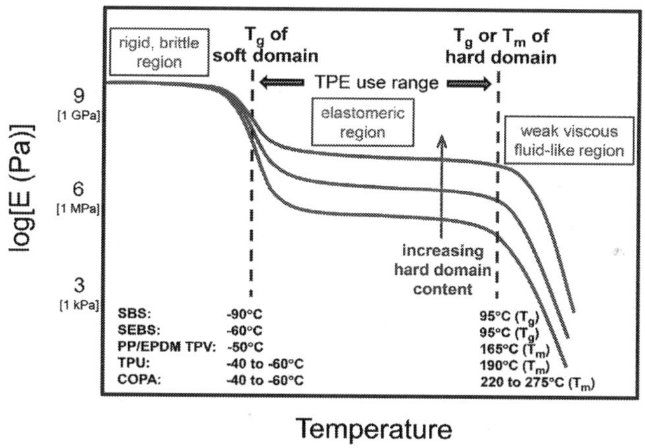

FIGURE 13.21 Schematic illustration of the temperature dependence of tensile modulus (E) for thermoplastic elastomers. The softening points for the various TPEs at the bottom of the figure are from Holden (2010).

T_m of the hard phase of selected thermoplastic elastomers from Holden (2010) are also given in this figure.

Resistance to hydrocarbon oils/solvents and maximum use temperature are key performance parameters for TPEs that dictate in which applications they can be used. As expected, higher performance thermoplastic elastomers are more expensive (Kear, 2003). A qualitative comparison of cost vs. performance is shown for various classes of thermoplastic elastomers in Figure 13.22. Styrenic block copolymers such as SBS and SEBS are less costly than TPVs but have reduced oil resistance and lower maximum use temperatures. The advantage of SEBS over SBS with regard to temperature resistance is not

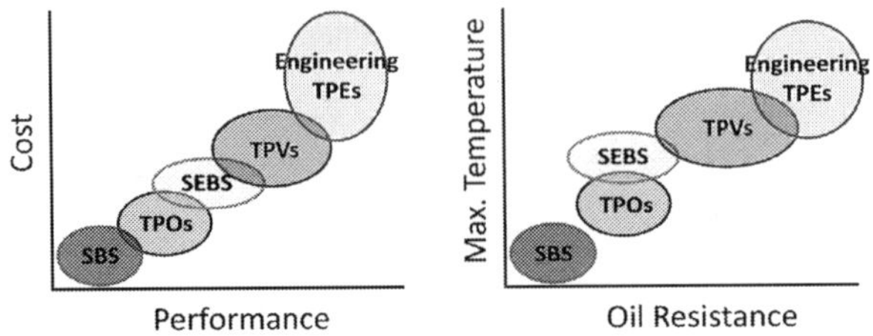

FIGURE 13.22 Qualitative comparison of cost and performance for different classes of TPEs. (Adapted from figures in Kear (2003).)

related to an increased softening temperature of the hard domains, but rather the hydrogenation that is used to make SEBS results in elimination of C=C double bonds in the rubbery phase, thus making the material less susceptible to thermo-oxidative degradation. Engineering TPEs such as thermoplastic urethane (TPU) elastomers, copolyester (COPE) elastomers, and copolyamide (COPA) elastomers are used in demanding applications where both high oil resistance and high service temperatures are required.

As stated previously, styrene-diene triblock copolymers are the most important category of thermoplastic elastomers. Unlike most other TPEs, they can be blended with large quantities of additives without a drastic effect on properties. In almost all applications, the actual triblock copolymer content is less than 50%. Oils are used as a processing aid and do not result in a significant loss of properties if the polystyrene domains are not plasticized. For this reason, naphthalenic oils are preferred. The use of inert fillers such as clays or chalks reduces the cost of the final material. Unlike conventional rubbers, inert fillers do not have a substantial effect on the mechanical properties of TPEs. Thermoplastics such as polyethylene or polypropylene are also used to improve the solvent resistance and can increase the upper service temperature. Polystyrene homopolymer is used as a processing aid, which also increases the hard phase weight fraction and causes the material to stiffen.

KRATON Polymers, the world's largest single producer of styrene-diene block copolymers, produces four types of compounds. SBS compounds are found in shoe soles and as property modifiers for asphalt. In this latter application, the block copolymer is a minor component (~10%) and improves the flexibility at low temperature and increases the softening temperature of the asphalt. SIS materials are employed almost exclusively as adhesives while SEBS triblock copolymers are used as structural materials. The polystyrene content tends to be low in pressure-sensitive adhesives in order to produce a material with more tack, although with hot melt adhesives tack may be undesirable and a harder product is appropriate. The absence of double bonds in SEBS copolymers substantially improves its resistance to UV light and high temperatures, allowing SEBS to be substituted for SIS or SBS in applications where this resistance is necessary. Finally, through postsynthesis processing with maleic anhydride, acid groups are introduced to the SEBS, which improves adhesion to a variety of substrates. These maleated materials are often used in composites (e.g., a polymer containing an inorganic filler).

Applications for TPOs, TPVs, polyurethanes, copolyesters, and copolyamides are substantially different than for the styrenic triblock copolymers. Only small amounts of additives are used in these materials since large amounts of any additive tend to substantially reduce the mechanical properties of these materials. The applications for these TPEs take advantage of their abrasion resistance, tear strength, and toughness along with the improved oil resistance and higher maximum temperature discussed previously. In the automobile industry, TPEs are used in both interior and exterior parts. For example, TPEs are used in boots and bearings for joints and some tubing as well as exterior bumpers and paneling. Other industrial uses include hoses, gears, and cables.

A very visible consumer product group for TPEs has been in sporting goods. Footwear, including ski boots and soccer shoes, often contain a substantial fraction of TPE. Athletic shoe soles are an especially large application for TPEs. Athletic equipment, such as skis and golf ball covers, is another area where TPEs are used.

Although not a large volume market, the medical device area has certainly been a very important user of TPEs. One notable application is the use of polyurethanes in angioplasty devices. Polyurethanes (as well as polyamides and polyesters) comprise the balloon material that is expanded to increase the size of an opening for blood flow within an artery (Lemba et al., 1997). Poly(styrene-b-isobutylene-b-styrene) TPE has established important medical use in drug-eluting coatings for coronary stents (Puskas and Hoerr, 2010; Strickler et al., 2010), and various TPEs have demonstrated potential for biomedical applications in the area of tissue engineering scaffolds (Hong et al., 2010; Lim et al., 2011; Shi et al., 2009). Thermoplastic elastomers clearly represent an exciting and growing class of materials with diverse applications ranging from simple shoe soles to medical products used to treat blocked arteries.

REFERENCES

Adams, R.K., Hoeschele, G.K., 1996. In: Holden, G., Legge, N.R., Quirk, R.P., Schroeder, H.E. (Eds.), Thermoplastic Elastomers, 2nd ed. Hanser Publishers, New York (Chapter 8).
Alexander, L.E., 1985. X-ray Diffraction Methods in Polymer Science. Krieger Publishing Company, Malabar, FL.
Allegrezza Jr., A.E., Seymour, R.W., Ng, H.N., Cooper, S.L., 1974. Polymer 15, 433.
Aluas, M., Filip, C., 2005. Solid State Nucl. Mag. Reson. 27, 165.
Anastasiadis, S.H., Russell, T.P., Satija, S.K., Majkrzak, C.F., 1990. J. Chem. Phys. 92, 5677.
Aneja, A., Wilkes, G.L., 2003. Polymer 44, 7221.
Annis, B.K., Schwark, D.W., Reffner, J.R., Thomas, E.L., Wunderlich, B., 1992. Makromol. Chem. 193, 2589.
Aust, N., Gobec, G., 2001. Macromol. Mater. Eng. 286, 119.
Babu, R.R., Nashkar, K., 2011. Adv. Polym. Sci. 239, 219.

Ballard, D.G., Schelten, J., Wignall, G., 1973. Eur. Polym. J. 9, 965.
Bard, J.K., Chung, C.I., 1987. In: Holden, G., Legge, N.R., Schroeder, H.E. (Eds.), Thermoplastic Elastomers, Section 1. Hanser Publishers, New York (Chapter 12).
Bates, F.S., Fredrickson, G.H., 1990. Ann. Rev. Phys. Chem. 41, 525.
Bates, F.S., Hartney, M.A., 1985. Macromolecules 18, 2478.
Bates, F.S., Rosedale, J.H., Fredrickson, G.H., 1990. J. Chem. Phys. 92, 6255.
Blackwell, J., Nagarajan, M.R., Hoitink, T.B., 1982. Polymer 23, 950.
Blundell, D.J., Eeckhaut, G., Fuller, W., Mahendrasingam, A., Martin, C., 2002. Polymer 43, 5197.
Bonart, R., 1970. Polymer 20, 1389.
Bonart, R., Muller, E.H., 1974. J. Macromol. Sci. Phys. B10, 345.
Bragg, W.L., 1933. The Crystalline State. Macmillan, New York.
Briber, R.M., Thomas, E.L., 1983. J. Macromol. Sci. Phys. B 22, 509.
Briber, R.M., Thomas, E.L., 1985. J. Polym. Sci. Polym. Phys. Ed. 23, 1915.
Brown, M., Hoeschele, G.K., Witsiepe W.K. (to DuPont), 1974. US Patent 3 835 098, September 10.
Brumberger, H. (Ed.), 1995. Modern Aspects of Small-Angle Scattering. Kluwer, New York.
Cella, R.J., 1973. J. Polym. Sci. C42, 727.
Chen, S., Cao, T., Jin, Y., 1987. Polym. Commun. 28, 314.
Chen, X., Gardella Jr., J.A., Kumler, P.L., 1992. Macromolecules 25, 6631.
Clayden, N.J., Nijs, C., Eeckhaut, G., 1998. Macromolecules 31, 7820.
Coleman, M.M., Skrovanek, D.J., Howe, S.E., Painter, P.C., 1985. Macromolecules 18, 302.
Coleman, M.M., Skrovanek, D.J., Painter, P.C., 1986a. Makromol. Chem. Macromol. Symp. 5, 21.
Coleman, M.M., Lee, K.H., Skrovanek, D.J., Painter, P.C., 1986b. Macromolecules 19, 2149.
Coleman, M.M., Graf, J.F., Painter, P.C., 1991. Specific Interactions and the Miscibility of Polymer Blends. Technomic Publishing Company, Lancaster, PA (Chapter 3).
Cooper, S.L., Miller, J.A., 1985. Rubber Chem. Technol. 58, 899.
Cooper, S.L., Miller, J.A., Homan, J.G., 1988. J. Appl. Crystallogr. 21, 692.
Corish, P.J., 1991. Concise Encyclopedia of Polymer Processing and Applications. Elsevier, New York.
Cotton, J.P., Decker, D., Benôit, H., Farnoux, B., Higgins, J., Jannink, G., Ober, R., Picot, C., des Cloiseaux, J., 1974. Macromolecules 7, 862.
Coulon, G., Russell, T.P., Deline, V.R., Green, P.F., 1989. Macromolecules 22, 2581.
DeGennes, P.G., 1970. J. Phys. (Paris) 31, 235.
DeGennes, P.G., 1979. Scaling Concepts in Polymer Physics. Cornell University Press, Ithaca, NY.
de Ilarduya, A.M., Carvalho, E., Alla, A., Munoz-Guerra, S., 2010. Macromolecules 43, 3990.
Desborough, I.J., Hall, I.J., 1977. Polymer 18, 825.
Deslandes, Y., Pleizier, G., Alexander, D., Santerre, P., 1998. Polymer 39, 2361.
Dlugosz, J., Keller, A., Pedemonte, E., 1970. Kolloid Z. Z. Polym. 242, 1125.
Dombrowski, S.A., Goodman, I., Johnson, A.F., 1986. Makromol. Chem. Rapid Commun. 7, 274.
Eisenbach, C.D., Stadler, E., 1995. Colloid Polym. Sci. 273, 352.
Eisenbach, C.D., Heinemann, T., Ribbe, A., Stadler, E., 1992. Angew. Makromol. Chem. 202, 221.
Fasolka, M.J., Mayes, A.M., 2001. Ann. Rev. Mater. Res. 31, 323.
Fredrickson, G.H., 1987. Macromolecules 20, 2535.
Fredrickson, G.H., Bates, F.S., 1996. Ann. Rev. Mater. Res. 26, 501.
Fredrickson, G.H., Helfand, E., 1987. J. Chem. Phys. 87, 697.
Frensdorff, H.K., 1971. Macromolecules 4, 369.
Gabrys, B.J., 2000. Applications of Neutron Scattering to Soft Condensed Matter. Taylor and Francis, New York.
Garrett, J.T., Siedlecki, C.A., Runt, J., 2001. Macromolecules 34, 7066.
Gaylord, R.J., 1979. In: Cooper, S.L., Estes, G.M. (Eds.), Multiphase Polymers. ACS Advances in Chemistry Series, vol. 176, Washington, DC (Chapter 12).
Gaylord, R.J., Lohse, D.J., 1978. Polym. Eng. Sci. 18, 359.
Gehlsen, M.D., Almdal, K., Bates, F.S., 1992. Macromolecules 25, 939.
Giroux, T.A., Cooper, S.L., 1991. J. Colloid Interf. Sci. 146, 179.
Gisselfalt, K., Helgee, B., 2003. Macromol. Mater. Eng. 288, 265.
Glatter, O., Kratky, O. (Eds.), 1982. Small Angle X-ray Scattering. Academic Press, New York.
Graff, D.K., Wang, H.C., Palmer, R.A., Schoonover, J.R., 1999. Macromolecules 32, 7147.
Green, P.F., Christensen, T.M., Russell, T.P., Jérôme, R., 1989. Macromolecules 22, 2189.

Green, P.F., Christensen, T.M., Russell, T.P., Jérôme, R., 1990. J. Chem. Phys. 92, 1478.
Green, P.F., Christensen, T.M., Russell, T.P., 1991. Macromolecules 24, 252.
Guinier, A., Fournet, G., 1955. Small-Angle Scattering of X-rays. John Wiley & Sons, New York.
Hall, I.H., Pass, M.G., 1976. Polymer 17, 807.
Hamley, I.W., 1998. The Physics of Block Copolymers. Oxford University Press, Oxford.
Hamley, I.W. (Ed.), 2004. Developments Block Copolymer Science and Technology. Wiley, West Sussex.
Harrell, L.L., 1976. Macromolecules 2, 607.
Harthcock, M.A., 1989. Polymer 30, 1234.
Hasegawa, H., Hashimoto, T., 1985. Macromolecules 18, 589.
Hasegawa, H., Hashimoto, T., 1992. Polymer 33, 475.
Hasegawa, H., Hashimoto, T., Kawai, H., Lodge, T.P., Amis, E.J., Glinka, C.J., Han, C.C., 1985. Macromolecules 18, 67.
Hasegawa, H., Tanaka, H., Hashimoto, T., Han, C.C., 1987. Macromolecules 20, 2120.
Hashimoto, T., Todo, A., Itoi, H., Kawai, H., 1977. Macromolecules 10, 377.
Hashimoto, T., Shibayama, M., Kawai, H., 1980. Macromolecules 13, 1237.
Hatfield, G.R., Guo, Y., Killinger, W.E., Andrejak, R.A., Roubicek, P.M., 1993. Macromolecules 26, 6350.
Helfand, E., 1975. Macromolecules 8, 552.
Helfand, E., 1982. In: Goodman, I. (Ed.), Developments in Block Copolymers. Applied Science, New York (Chapter 4).
Helfand, E., Wasserman, Z.R., 1976. Macromolecules 9, 879.
Helfand, E., Wasserman, Z.R., 1978. Macromolecules 11, 960.
Helfand, E., Wasserman, Z.R., 1980. Macromolecules 13, 994.
Hepburn, C., 1982. Polyurethane Elastomers. Applied Science Publishers, New York (Chapter 1).
Higgins, J., Benoit, H., 1994. Polymers and Neutron Scattering. Clarendon Press, London.
Hoeschele, G.K., 1974. Chimia 28, 544.
Holden, G., 2010. Thermoplastic Elastomers. In: Encyclopedia of Polymer Science and Technology. John Wiley & Sons, Inc., Hoboken, NJ.
Holden, G., Legge, N.R., 1996. In: Holden, G., Legge, N.R., Quirk, R.P., Schroeder, H.E. (Eds.), Thermoplastic Elastomers, 2nd ed. Hanser Publishers, New York (Chapter 3).
Holden, G., Bishop, E.T., Legge, N.R., 1969. J. Polym. Sci. 26, 37.
Holzer, B., Lehmann, A., Stühn, B., Kowalski, M., 1991. Polymer 32, 1935.
Hong, Y., Guan, J.J., Fujimoto, K.L., Hashizume, R., Pelinescu, A.L., Wagner, W.R., 2010. Biomaterials 31, 4249.
Ishida, H., 1987. Rubber Chem. Technol. 60, 497.
Ishida, H., Scott, C., 1986. J. Polym. Eng. 6, 201.
Jahshan, S.N., Summerfield, G.C., 1980. J. Polym. Sci. Polym. Phys. Ed. 18, 2415.
Janik, H., Palys, B., Petrovic, Z.S., 2003. Macromol. Rapid Commun. 24, 265.
Jelinski, L.W., 1981. Macromolecules 14, 1341.
Jelinski, L.W., Dumais, J.J., Engel, A.K., 1983a. Macromolecules 16, 492.
Jelinski, L.W., Dumais, J.J., Watnick, P.I., Engel, K.A., Sefcik, M.D., 1983b. Macromolecules 16, 409.
Kaji, A., Murano, M., 1990. Polym. J. 22, 1065.
Kajiyama, T., Takahara, A., 1991. J. Biomater. App. 6, 42.
Kear, K.E., 2003. Developments in Thermoplastic Elastomers. Rapra Review Reports, vol. 14, No. (10), Report 166, Rapra Technology Limited, Shawbury, UK.
Kirkmeyer, B.P., Taubert, A., Kim, J.S., Winey, K.I., 2002. Macromolecules 35, 2648.
Knoll, A., Magerle, R., Krausch, G., 2001. Macromolecules 34, 4159.
Koberstein, J.T., Stein, R.S., 1983. J. Polym. Sci. Polym. Phys. Ed. 21, 2181.
Koberstein, J.T., Gancarz, I., Clarke, T.C., 1986. J. Polym. Sci. Part B Polym. Phys. 24, 2487.
Koberstein, J.T., Russell, T.P., Walsh, D.J., Pottick, L., 1990. Macromolecules 23, 877.
Konyukhova, E.V., Neverov, V.M., Chvalun, S.N., Godovsky, Y.K., 2004. Polym. Sci. Ser. A 46, 61.
Kornfield, J.A., Spiess, H.W., Nefzger, H., Hayen, H., Eisenbach, C.D., 1991. Macromolecules 24, 4787.
Kresge, E.N., 1984. J. Appl. Polym. Sci. Appl. Polym. Symp. 39, 37.
Krijgsman, J., Husken, D., Gaymans, R.J., 2003. Polymer 44, 7573.

Lantman, C.W., MacKnight, W.J., Lundberg, R.D., 1989. Ann. Rev. Mater. Sci. 19, 295.
Laurer, J.A., Winey, K.I., 1998. Macromolecules 31, 9106.
Lee, D.C., Speckhard, T.A., Sorenson, A.D., Cooper, S.L., 1986. Macromolecules 19, 2383.
Leibler, L., 1980. Macromolecules 13, 1602.
Leibler, L., Benoit, H., 1981. Polymer 22, 195.
Lelah, M.D., Cooper, S.L., 1986. Polyurethanes in Medicine. CRC Press, Boca Raton, FL.
Lemba, N.M.K., Woodhouse, K.A., Cooper, S.L., 1997. Polyurethanes in Biomedical Applications. CRC Press, Boca Raton, FL.
Li, C., Cooper, S.L., 1990. Polymer 31, 3.
Lilaonitkul, A., West, J., Cooper, S.L., 1976. J. Macromol. Sci. Phys. B12, 563.
Lim, G.T., Puskas, J.E., Reneker, D.H., Jakli, A., Horton, W.E., 2011. Biomacromolecules 12, 1795.
Löcsei, Y., Facosko, O., Pape, R.F., 1983. Polym. Bull. 9, 81.
Lodge, T.P., 2003. Macromol. Chem. Phys. 204, 265.
Luo, N., Wang, D.N., Yang, S.K., 1996. Polymer 37, 3045.
Luo, N., Wang, D.N., Ying, S.K., 1997. Macromolecules 30, 4405.
Lyman, D.J., 1960. J. Polym. Sci. 45, 49.
Matsuo, M., 1968. Jpn. Plast. 2, 6.
Mayes, A.M., Olvera de la Cruz, M., 1989. J. Chem. Phys. 91, 7228.
Meckel, W., Goyert, W., Wieder, W., 1996. In: Holden, G., Legge, N.R., Quirk, R.P., Schroeder, H.E. (Eds.), Thermoplastic Elastomers, second ed. Hanser Publishers, New York (Chapter 2).
Mehl, J.T., Murgasova, R., Dong, X., Hercules, D.M., Nefzger, H., 2000. Anal. Chem. 72, 2490.
Meier, D.J., 1973. Polymer 14, 280, Preprints.
Meier, R.J., 1996. In: Holden, G., Legge, N.R., Quirk, R.P., Schroeder, H.E. (Eds.), Thermoplastic Elastomers, second ed. Hanser Publishers, New York (Chapter 11).
Meltzner, A.D., Spiess, H.W., Eisenbach, C.D., Hayen, H., 1992. Macromolecules 25, 993.
Miller, J.A., Cooper, S.L., Han, C.C., Pruckmayr, G., 1984. Macromolecules 17, 1063.
Miller, J.A., Pruckmayr, G., Epperson, E., Cooper, S.L., 1985. Polymer 25, 1915.
Miller, J.A., Lin, S.B., Hwang, K.K.S., Wu, K.S., Gibson, P.E., Cooper, S.L., 1985. Macromolecules 18, 32.
Morbitzer, L., Hespe, H., 1972. J. Appl. Polym. Sci. 16, 2697.
Morton, M., 1983. Anionic Polymerization: Principles and Practice. Academic Press, New York.
Morton, M., 1996. In: Holden, G., Legge, N.R., Quirk, R.P., Schroeder, H.E. (Eds.), Thermoplastic Elastomers, second ed, Hanser Publishers, New York (Chapter 4).
Mullins, L., 1950. J. Phys. Chem. 54, 239.
Myers, C.W., Cooper, S.L., 1994. J. Appl. Spectrosc. 48, 72.
Naylor, S., Terrill, N.J., Yu, G.E., Tanodekaew, S., Bras, W., King, S.M., Booth, C., Ryan, A.J., 1997. Polym. Int. 44, 371.
Nelb, R.G., Che, A.T., 1996. In: Holden, G., Legge, N.R., Quirk, R.P., Schroeder, H.E. (Eds.), Thermoplastic Elastomers, second ed. Hanser Publishers, New York (Chapter 9).
Nierzwicki, W., 1985. J. Appl. Polym. Sci. 30, 761.
Noshay, A., McGrath, J.E., 1977. Block Copolymers. Academic Press, New York (Chapter 6).
Oertel, G. (Ed.), 1994. Polyurethane Handbook: Chemistry, Raw Materials, Processing, Application, Properties. Hanser Publishers, New York.
Okamoto, D.T., Cooper, S.L., Root, T.W., 1992a. Macromolecules 25, 1068.
Okamoto, D.T., Cooper, S.L., Root, T.W., 1992b. Macromolecules 25, 3301.
Okamoto, D.T., O'Connell, E.M., Cooper, S.L., Root, T.W., 1993. J. Polym. Sci. B-Polym. Phys. 31, 1163.
Olvera de la Cruz, M., Sanchez, I.C., 1986. Macromolecules 19, 2501.
Ophir, Z., Wilkes, G.L., 1980. J. Polym. Sci. Polym. Phys. Ed. 18, 1469.
Osswald, T., 1998. Polymer Processing Fundamentals. Hanser Publishers, New York.
Pakula, T., Saijo, K., Kawai, H., Hashimoto, T., 1985. Macromolecules 18, 1294.
Pakula, T., Saijo, K., Hashimoto, T., 1985. Macromolecules 18, 2037.
Pandya, M.V., Deshpande, D.D., Hundiwale, D.G., 1986. J. Appl. Polym. Sci. 32, 4959.
Paul, D.R., 1984. Polym. Mater. Sci. Eng. 50, 1.
Pedersen, J.S., 1997. Adv. Colloid Interf. Sci. 70, 171.
Perego, G., Cesari, M., Vitali, R., 1984. J. Appl. Polym. Sci. 29, 1157.
Petrovic, Z.S., Ferguson, J., 1991. Prog. Polym. Sci. 16, 695.

Phillips, R.A., Cooper, S.L., 1995. Macromolecules 28, 5734.
Phillips, R.A., Cooper, S.L., 1996. J. Polym. Sci. B-Polym. Phys. 34, 737.
Phillips, R.A., McKenna, J.M., Cooper, S.L., 1994. J. Polym. Sci. B-Polym. Phys. 32, 791.
Pitt, W.G., Cooper, S.L., 1986. Biomaterials 7, 340.
Polizzi, S., Bösecke, P., Stribeck, N., Zachmann, H.G., Zietz, R., Bordeianu, R., 1990. Polymer 31, 639.
Puskas, J.E., Hoerr, R.A., 2010. Macromol. Symp. 326, 291-292.
Puskas, J.E., Chen, Y., Antony, P., Kwon, Y., Kovar, M., Harbottle, R.R., De Jang, K., Norton, P.R., Cadieux, P., Burton, J., Reid, G., Beiko, D., Watterson, J.D., Denstedt, J., 2000. Polym. Adv. Technol. 14, 763.
Randall, D., Lee S. (Eds.), 2003. The Polyurethanes Book. Wiley, New York.
Read, D.J., Teixeira, P.I.C., Duckett, R.A., Sweeney, J., McLeish, T.C.B., 2002. Eur. Phys. J. E8, 15.
Rennie, A.R., Oberthur, R.C., 1994. Practical Small Angle Neutron Scattering. Taylor and Francis, New York.
Richards, R.W., Welsh, G., 1995. Eur. Polym. J. 31, 1197.
Roe, R.J., 1982. J. Appl. Crystallogr. 15, 182.
Ruland, W., 1971. J. Appl. Crystallogr. 4, 70.
Sabbatini, L., Zambonin, P.G., 1996. J. Electron Spectrosc. Rel. Phenom. 81, 285.
Sakarai, S., Okamoto, S., Sakurai, K., 2004. In: Hamley, I.W. (Ed.), Developments in Block Copolymer Science and Technology. Wiley, West Sussex.
Schneider, N.S., Matton, W., 1979. Polym. Eng. Sci. 19, 1122.
Schonherr, H., Wiyatno, W., Pople, J., Frank, C.W., Fuller, G.G., Gast, A.P., Waymouth, R.M., 2002. Macromolecules 35, 2654.
Schoonover, J.R., Thompson, D.G., Osborn, J.C., Orler, E.B., Wrobleski, D.A., Marsh, A.L., Wang, H.C., Palmer, R.A., 2001. Polym. Degred. Stab. 74, 87.
Segalman, R.A., Schaefer, K.E., Fredrickson, G.H., Kramer, E.J., Magonov, S., 2003. Macromolecules 36, 4498.
Séguéla, R., Prud'homme, J., 1978. Macromolecules 11, 1007.
Séguéla, R., Prud'homme, J., 1981. Macromolecules 14, 197.
Séguéla, R., Prud'homme, J., 1988. Macromolecules 21, 635.
Seymour, R.W., Estes, G.M., Cooper, S.L., 1970. Macromolecules 3, 579.
Seymour, R.W., Allegrezza Jr., A.E., Cooper, S.L., 1973. Macromolecules 6, 896.
Shahbikian, S., Carreau, P.J., Heuzey, M.C., Ellul, M.D., Cheng, J., Shirodkar, P., Nadella, H.P., 2012. Polym. Eng. Sci. 52, 309.
Shard, A.G., Davies, M.C., Tendler, S.J.B., Jackson, D.E., Lan, P.N., Schacht, E., Purbrick, M.D., 1995. Polymer 36, 775.
Sheth, J.P., Xu, J., Wilkes, G.L., 2003. Polymer 44, 743.
Shi, R., Chen, D.F., Liu, Q.Y., Wu, Y., Xu, X.C., Zhang, L.Q., Tian, W., 2009. Int. J. Mol. Sci. 10, 4223.
Siesler, H.W., 1983. Polym. Bull. 9, 471.
Skrovanek, D.J., Howe, S.E., Painter, P.C., Coleman, M.M., 1985. Macromolecules 18, 1676.
Skrovanek, D.J., Painter, P.C., Coleman, M.M., 1986. Macromolecules 19, 699.
Smith, T.L., 1974. J. Polym. Sci. Polym. Phys. Ed. 12, 1825.
Smith, T.L., 1986. In: Bever, M.B. (Ed.), Encyclopedia of Materials Science and Engineering. Pergamon Press, New York, pp. 1341.
Sormana, J.L., Meredith, J.C., 2004. Macromolecules 37, 2186.
Sorta, E., Melis, L.A., 1978. Polymer 19, 1153.
Speckhard, T.A., Cooper, S.L., 1986. Rubber Chem. Technol. 59, 405.
Speckhard, T.A., Miller, J.A., Cooper, S.L., 1986a. Macromolecules 19, 1550.
Speckhard, T.A., Miller, J.A., Cooper, S.L., 1986b. Macromolecules 19, 1558.
Speckhard, T.A., Homan, J.G., Miller, J.A., Cooper, S.L., 1987a. Polymer 28, 758.
Speckhard, T.A., Homan, J.G., Miller, J.A., Cooper, S.L., 1987b. Polymer 28, 768.
Srichatrapimuk, V.W., Cooper, S.L., 1978. J. Macromol. Sci. Phys. B15, 267.
Starkweather, H.W., 1980. J. Appl. Polym. Sci. 25, 139.
Strickler, F., Richard, R., McFadden, S., Lindquist, J., Schwarz, M.C., Faust, R., Wilson, G.J., Boden, M., 2010. J. Biomed. Mater. Res. A 92A, 773.
Stühn, B., Stickel, F., 1992. Macromolecules 25, 5306.

Sung, C.S.P., Hu, C.B., 1980. In: Cooper, S.L., Estes, G.M. (Eds.), Multiphase Polymers. Advances in Chemistry Series, vol. 176. American Chemical Society, Washington, DC.
Sung, C.S.P., Schneider, N.S., 1975. Macromolecules 8, 68.
Sung, C.S.P., Hu, C.B., Wu, C.S., 1980. Macromolecules 13, 111.
Suzuki, H., Ono, H., 1970a. Bull. Chem. Soc. Jpn. 43, 682.
Suzuki, H., Ono, H., 1970b. Bull. Chem. Soc. Jpn. 43, 687.
Szycher, M., 1999. Handbook of Polyurethanes. CRC Press, Boca Raton, FL.
Tanaka, H., Nishi, T., 1985. J. Chem. Phys. 82, 4326.
Tyagi, D., McGrath, J.E., Wilkes, G.L., 1986. Polym. Eng. Sci. 26, 1371.
Tyler, B.J., Ratner, B.D., Castner, D.G., Briggs, D., 1992. J. Biomed. Mater. Res. 26, 273.
Uhlig, K., 1999. Discovering Polyurethanes. Hanser, Munich.
Vallance, M.A., Cooper, S.L., 1984. Macromolecules 17, 1208.
Vasilev, C., Heinzelmann, H., Reiter, G., 2004. J. Polym. Sci. B-Polym. Phys. 42, 1312.
Veenstra, H., Verkooijen, P.C.J., van Lent, B.J.J., van Dam, J., de Boer, A.P., Nijhof, A.P.H.J., 2000. Polymer 41, 1817.
Vonk, C.G., 1973. J. Appl. Crystallogr. 6, 81.
Wang, C.B., Cooper, S.L., 1983. Macromolecules 16, 775.
Warner, S., 1990. J. Elastom. Plast. 22, 166.
Whitmore, M.D., 2001. In: Encyclopedia of Materials: Science and Technology. Elsevier Science, Oxford.
Winey, K.I., Laurer, J., Kirkmeyer, B.P., 2000. Macromolecules 33, 507.
Witsiepe, W.K., 1973. ACS Adv. Chem. Ser. 129, 29.
World Thermoplastic Elastomers to 2015—Demand and Sales Forecasts, Market Share, Market Size, Market Leaders, 2011. Study #2803. The Fredonia Group, Inc., Cleveland, OH.
Wu, S., 1989. In: Brandrup, J., Immergut, E.H. (Eds.), Polymer Handbook. John Wiley & Sons, New York.
Yontz, D.J., Hsu, S.L., 2000. Macromolecules 33, 8415.
Yu, Y.C., Jo, W.H., 1994. J. Appl. Polym. Sci. 54, 585.
Yu, Y.C., Jo, W.H., 1995. J. Appl. Polym. Sci. 56, 895.
Yu, Y.C., Jo, W.H., 1997. J. Appl. Polym. Sci. 64, 2155.

Chapter 14

Tire Engineering

Brendan Rodgers and Walter Waddell
ExxonMobil Chemical Company, Houston, TX, USA

14.1 INTRODUCTION

In human history, the wheel is considered one of the most important inventions, because it found use in a wide range of applications such as transportation vehicles, construction equipment, and internal parts of machinery. Like most inventions, the wheel was a development of earlier devices such as rollers, dating to the Bronze Age over 5000 years ago, used to move heavy objects. Wheeled vehicles were recorded in Sumeria in 3500 BCE, Assyria in 3000 BCE, and central Europe toward 1000 BCE. Four-wheeled wagons using a swiveling front axle for steering were recorded in 1500 BCE.

With the introduction of swifter horses from the Asiatic Steppes into Mesopotamia, the cart was adapted for military applications. The spoked wheel was introduced and then given its first "tire," consisting of first a leather and then copper and iron binding to prevent damage to the wooden wheel frame.

The next most important event was probably in 1846 when Thompson was granted a patent for an elastomeric air tube, to be fixed onto a wheel to reduce the power to haul a carriage, make motion easier, and reduce noise. This concept was much refined in the 1880s when the first pneumatic tire was developed for use on tricycles.

The discovery of vulcanization by Charles Goodyear in 1839 and the industrialization of Europe and North America enabled the tire to evolve from a rubberized canvas covering a rubber tube to a complex fabric, steel, and elastomeric composite.

14.2 TIRE TYPES AND PERFORMANCE

In terms of both volume production and consumer awareness, pneumatic tires fall into essentially nine categories, based on vehicle application. There are tires for racing vehicles, passenger vehicles, and light trucks where gross vehicle

The Science and Technology of Rubber. http://dx.doi.org/10.1016/B978-0-12-394584-6.00014-5

weights typically do not exceed 7250 kg. In such tires, significant quantities of fabric are used as a reinforcement. Larger tires such as those for heavy trucks, farm and agricultural vehicles, earthmoving equipment, and large aircraft tend to contain both steel wire and fabric reinforcements. Finally there are a range of specialty tires, which include those used on forklift trucks, light aircraft, light construction equipment, and golf carts.

Regardless of the design or application of the tire, all pneumatic tires must fulfill a fundamental set of functions:

- Provide load-carrying capacity.
- Provide cushioning and dampening.
- Transmit driving and braking torque.
- Provide cornering force.
- Provide dimensional stability.
- Resist abrasion.
- Generate steering response.
- Have low rolling resistance.
- Provide minimum noise and minimum vibration.
- Be durable throughout the expected life span.

Dampening characteristics, elastic properties of rubber, and unique deformability and recovery combine to make tires the only product that satisfies all of these functions.

Essentially three performance parameters govern a tire's functions. These are (1) vehicle mission profile; (2) mechanical properties and performance such as wear resistance and casing durability; and (3) aesthetics, comfort, and behavioral characteristics such as vehicle steering precision.

The mechanical properties of a tire describe the tire's characteristics in response to the application of load, torque, and steering input, resulting in the generation of external forces and deflection. Such mechanical properties are interrelated, and thus a design decision affecting one factor will influence the other factors, either positively or negatively. The result is a complex set of forces acting on a rolling tire on a vehicle (Figure 14.1).

The tire axis system is the center of the tire-road surface contact shown in Figure 14.1. The x axis is the intersection of the wheel plane with the road plane and with the positive direction forward. The z axis is perpendicular to the road plane with a positive direction downward. Thus, the normal force exerted by the tire is positive downward, and the vertical load reaction, which is the road pushing on the tire, is considered negative. The y axis is in the road plane and is directed so as to make the system right-handed and orthogonal. This representation of the tire axis system corresponds closely to the vehicle axis system when the tire is in the right front position of an automobile. In this case, the vehicle would be undergoing a left turn with positive self-aligning torque (M_z) and negative lateral force vector. The forces acting on a tire can

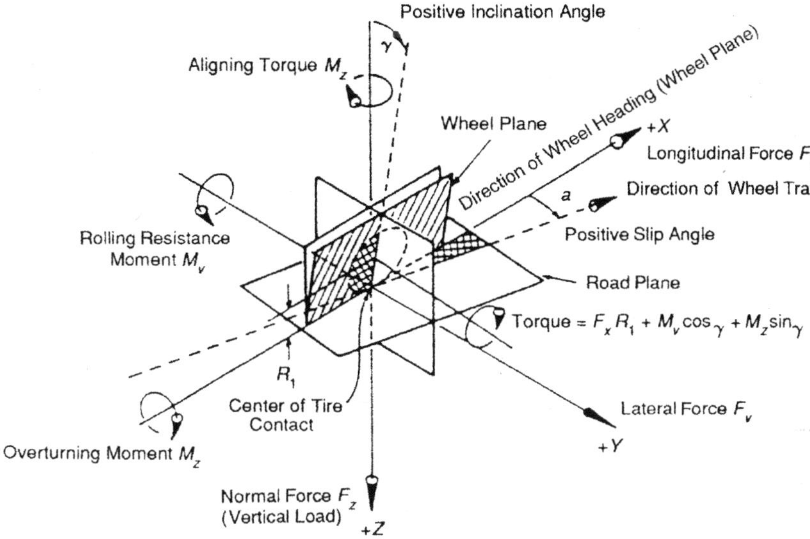

FIGURE 14.1 Tire forces and moments acting at the center of tire contact. (Reprinted with permission. © 1988 from SAE's 34th Buckendale Lecture. Society of Automotive Engineers, Inc.)

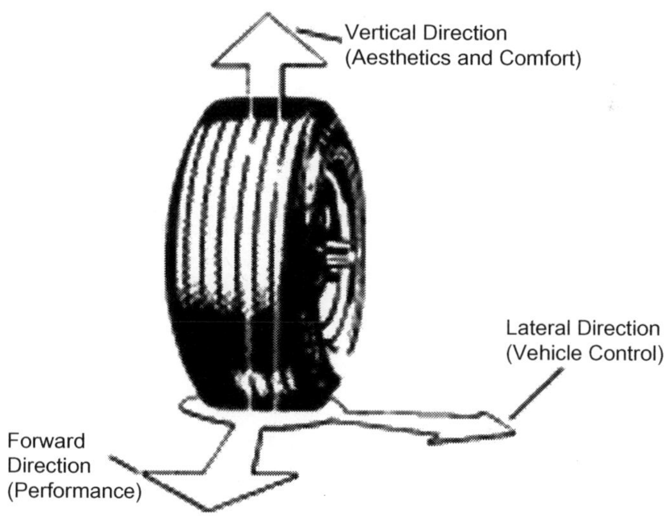

FIGURE 14.2 Tire functions.

thus be broken down into three fundamental vectors: the vertical forces control vehicle aesthetics and comfort, the lateral forces impact vehicle control, and the longitudinal or forward forces control performance such as rolling resistance (Figure 14.2).

14.3 BASIC TIRE DESIGN

A tire is essentially a cord-rubber composite. Tires have plies of reinforcing cords extending transversely from bead to bead, on top of which is a belt located below the tread. The belt cords have low extensibility and are made of steel and fabric depending on the tire application. The belt cords are at a relatively low angle, between 12° and 25°, and serve as restrictions to the 90° casing plies.

14.3.1 Tire Construction

A range of specialized components is found in a tire, which serves to ensure that the product meets its intended design and performance requirements. For example, high-performance passenger tires can have a nylon overlay, also called a cap ply, located over the belt package. Nylon overlays restrict and control tire growth resulting from centrifugal forces created at high speeds. The ply turn up, which describes the manner in which the body ply wraps around the bead wire and turns up the sidewall, anchors the body ply to the bead bundle and further reinforces the lower sidewall region. A toe guard, which is a nylon-reinforced rubber component, protects the bead toe from tearing during mounting and dismounting. Above the toe guard, the chafer further protects the bead area from abrading against the rim flange during use (Figure 14.3).

14.3.2 Tire Components

A tire is an assembly of a series of parts or subassemblies, each of which has a specific function in the service and performance of the product.

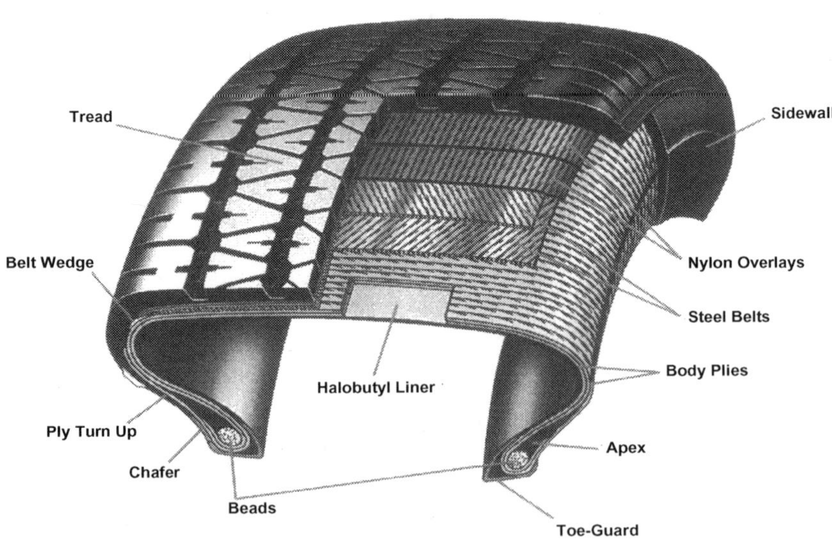

FIGURE 14.3 Cross-section of a high-performance passenger tire.

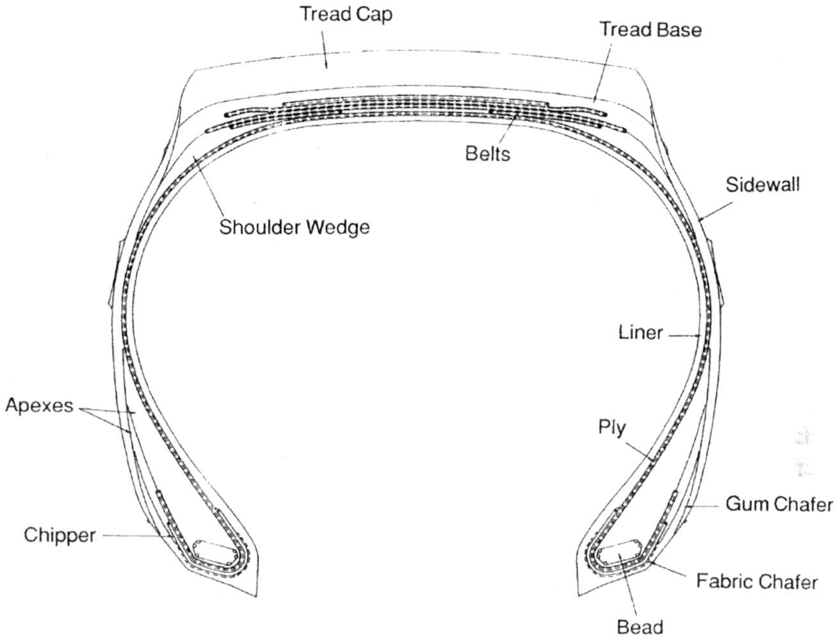

FIGURE 14.4 Components of a radial truck tire.

Figure 14.4 illustrates the key components of a truck tire. The following are the parts of a tire:

- *Tread:* The wear resistance component of the tire in contact with the road. It must also provide traction, wet skid, and good cornering characteristics with minimum noise generation and low heat buildup. Tread components can consist of blends of natural rubber, polybutadiene (BR), and styrene-butadiene rubber (SBR), compounded with carbon black, silica, oils, and vulcanizing chemicals.
- *Tread shoulder:* Upper portion of the sidewall affects tread heat dissipation and tire cornering properties.
- *Tread base:* Also termed the cushion; the rubber compound used to ensure good adhesion between belts and tread, heat dissipation, and low rolling resistance.
- *Sidewall:* Protects the casing from side scuffing, controls vehicle-tire ride characteristics, and assists in tread support. Sidewall compounds consist of natural rubber, SBR, and BR along with carbon black and a series of oils and organic chemicals.
- *Curb guard:* A protrusion of rubber sidewall running circumferentially around the tire to protect it from scuffing on curbs.
- *Beads:* Nonextensible steel wire loops, which anchor the plies and lock the tire onto the wheel assembly so that it will not slip or rock on the rim.

- *Bead area components:* Include the apex or bead filler; the chafer, which protects the wire bead components; the chipper, which protects the lower sidewall; and the flipper, which helps hold the bead in place.
- *Plies:* Textile or steel cords extending from bead to bead and thereby serving as the primary reinforcing material in the tire casing.
- *Belts:* Layers of textile or steel wire lying under the tread and serving to stiffen the casing, thereby allowing improved wear performance and handling response, better damage resistance, and protection of the ply cords from road hazards.
- *Shoulder belt wedge:* High-adhesive rubber compound in the shoulder region between the belts and casing; improves tread wear and durability.
- *Liner:* Butyl rubber or halogenated derivatives of such polymers, which retains the compressed air inside the tire.

14.4 TIRE ENGINEERING

In essence, a tire is a composite of complex elastomer formulations, fibers, textiles, and steel cord. The term *tire structure* defines the number, location, and dimensions of the various components used in a tire's composition. The primary components that govern the performance of a tire are the casing plies, bead construction, belts, sidewall, inner liner, and tread. Chafers, flippers, and overlays, which are strips of rubberized fabric located in the bead and crown area of the tire, are termed *secondary components* because they protect the primary components by minimizing stress concentrations (Ford and Charles, 1988; Davison, 1969).

14.4.1 Tire Nomenclature and Dimensions

Terminology used to describe tire and rim dimensions is explained in Figure 14.5. These dimensions are commonly used throughout the tire industry to describe size, growth, and wheel well clearance factors in addition to computation of variables such as load capacity and revolutions per unit distance traveled (The Tire & Rim Association, Inc. Year Book, 2004; European Tyre and Rim Technical Organization, 2004). In addition, the following definitions should be noted:

- *Aspect ratio:* A numerical term that expresses the relationship between tire section height and cross-section width. An aspect ratio of 70 indicates the tire section is approximately 70% as high as it is wide.
- *Cord angle:* Angle of the cord path to the centerline of the tire; the predominant factor affecting the tire shape or contour. In radial tires the cord angle is 90°.
- *Overall diameter:* Unloaded diameter of a new tire-rim combination.

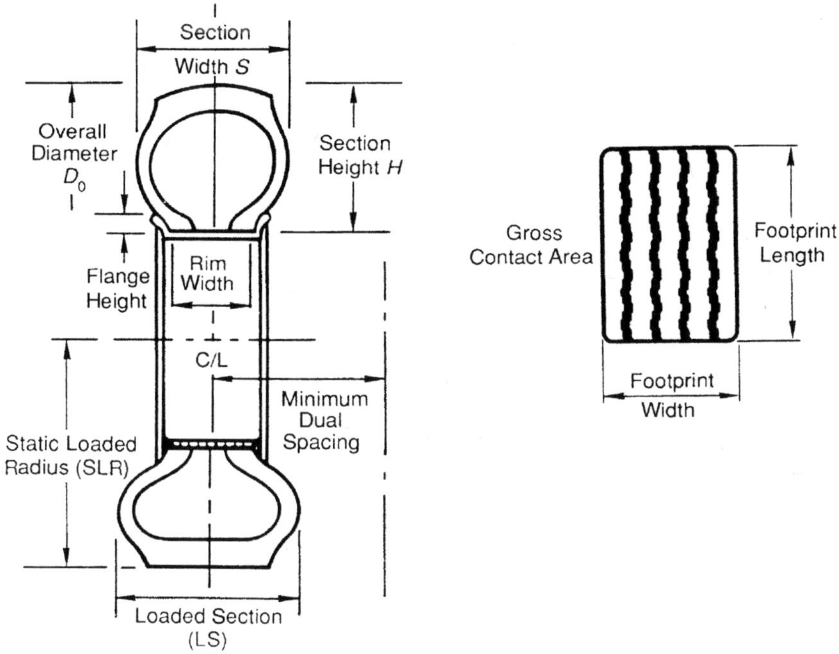

FIGURE 14.5 Tire and rim dimensions.

- *Section width:* Width of a new tire section excluding side ribs, lettering, and decorations.
- *Static loaded radius:* Distance from the road surface to the horizontal centerline of the rim.
- *Minimum dual spacing:* Minimum dimension recommended from rim centerline to rim centerline for optimum performance of a dual-wheel installation.
- *Footprint length:* Length of a loaded footprint.
- *Footprint width:* Width of the loaded footprint.
- *Gross contact area:* Total area under the loaded footprint.
- *Net contact area:* Area of the tread, excluding voids, under the loaded footprint. Also abbreviated as the percent net-to-gross.
- *Asymmetrical:* Tire design in which the tread pattern on one side of the centerline differs from that on the other side.
- *Load rating:* Maximum load a tire is rated to carry for a given usage at a specified inflation. The "load range" is also used to define the load-carrying capability of a tire. Load ranges are specified in Tire & Rim Association tables. Table 14.1 illustrates load values for a single mounted 11R24.5 heavy-duty truck tire.

TABLE 14.1 Tire and Rim Association Load Values for a Single Mounted Radial Medium Truck Tire [The Tire & Rim Association, Inc. Year Book, 2004.]

Load Range Requirement	Ply Rating	Example of Tire Size	Loaded Tire Limit (lbs)	Inflation Pressure
F	12	295/75R22.5	5070	90
G	14	295/75R22.5	5780	100
H	16	295/75R22.5	6610	120
H	16	285/75R24.5	6780	120
J	18	385/65R22.5	9370	120
L	20	425/65R22.5	11,400	120

Three basic tire size designations are used:

- Conventional-size tires used on flat base rims, normally tube-type tires.
- Conventional-size tires used on 15° rims or tubeless.
- Metric sizes also used on 15° rims; tubeless.

In addition, the letter "R" in a size designation indicates "radial," "D" or (—) is bias, and "ML" indicates the tire is for mining and logging applications. The letter "P" denotes passenger vehicle tire, and "LT" is light truck.

For example, the tire size designation "LT 235/75R15" has the following meaning:

- *LT:* Light truck.
- *235*: Approximate section width in millimeters when mounted on the proper rim.
- *75:* Aspect ratio.
- *R:* Radial construction.
- *15:* Nominal rim diameter in inches.

A heavy-duty truck tire would typically have the size designation of 11R24.5 for a conventional tubeless tire; a low-profile metric tire typically could be sized 295/75R22.5. For the conventional tire,

- *11:* Nominal section diameter in inches.
- *R:* Radial construction.
- *24.5:* Rim diameter in inches.

Such conventional-size tires have a standard aspect ratio of 80. The initial step in designing a tire is determination of the required size. Size determination is governed by rim dimensions, wheel well envelope, service load, service speed, and inflation. For a metric heavy-duty truck tire these factors are related

TABLE 14.2 European Tyre and Rim Technical Organization (ETRTO) Load Index Numerical Code for Tire Load-Carrying Capability [European Tyre and Rim Technical Organization, 2004]

Load Index	Load-Carrying Capability (kg)
144	2800
145	2900
146	3000
149	3250
152	3550
155	3875
160	4500

by the equation

$$L = (6.075 \times 10^{-5})K \times P^{0.7} \times S_d^{1.1}(D_r + S_d), \tag{14.1}$$

where L = load at 100 kph (kg); P = pressure (kPa); S_d = dimensional factor, section width adjusted for aspect ratio; D_r = rim diameter; and K = constant dependent on vehicle speed. Using equations such as Eq. (14.1), developed by the Tire & Rim Association, the tire engineer can determine the optimum tire size for a specific application. Normally, load requirements are known so Eq. (14.1) can thus be used to calculate required service pressure. This process is then used in size and load range selection. Tire industry standards are used extensively in the design process (Kovac, 1978). Load specifications are tabulated in Tables 14.1 and 14.2.

14.4.2 Tire Mold Design

Tire mold design initially begins with determination of the inflated dimensions of the required tire size. By use of inflated tire and growth characteristics of the tire, preliminary plyline and mold dimensions are computed (Figure 14.6). Once the mold boundary dimensions, location of the plyline, and tread width and depth are known, the contours of the tread, shoulder, sidewall, and bead components can be established. These dimensions and contours are developed using computer-aided engineering techniques (Figure 14.7).

The primary interest in designing a tire lies in the belt area, bead area, and belt and ply cord tension. As radial tires contain multiple belts, these layups must be viewed as a package. The stiffness of the radial tire belt package is a function of belt wire angles, wire gauge, belt gauge, and compound stiffness. Figure 14.8 illustrates a four-belt configuration of a radial truck tire. As an empirical guide, an increase in the stiffness of the belt package, while maintaining belt width, will improve tread wear performance.

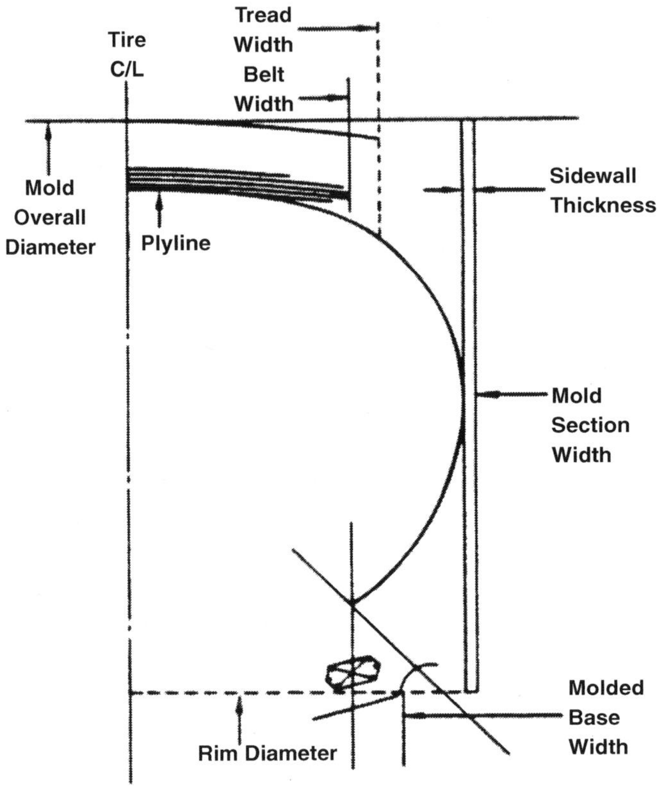

FIGURE 14.6 Plyline boundaries.

The stiffness of a tire belt package can be quantified by determination of the "Gough stiffness" (S), which is a measure of the in-plane bending stiffness of the rigid belt wire, cord, and rubber compound laminate. High radial tire belt rigidity and crown stiffness can significantly improve tire tread wear through provision of a solid foundation for the tread compound (Gough, 1968; Clark, 1982). Elaborating, Gough determined that a simple beam model of tire construction features could be used to predict relative tread wear performance. Using both shearing and bending moments in the computation, an empirical equation was derived defining the stiffness parameter S for a simple laminate

$$S = P/d, \tag{14.2}$$

where load, P, is the force applied to the crown area layup or laminate to give a deflection d.

In a tire, Gough stiffness is a function of the circumferential modulus and shear modulus of the composite.

$$S = (E \times G)/(C_1 \times E) + (C_2 \times G), \tag{14.3}$$

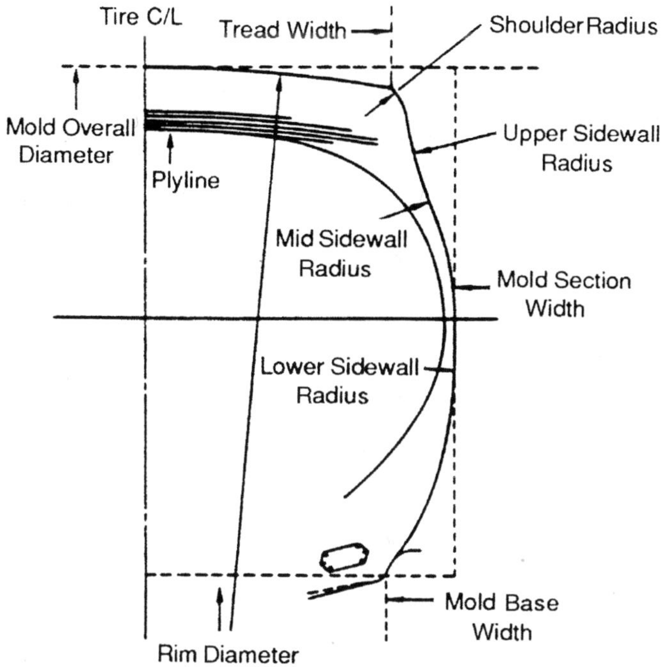

FIGURE 14.7 CADAM-developed mold cavity.

where S = Gough stiffness, E = circumferential modulus of the cord-rubber laminate, G = shear modulus of the cord-rubber laminate, C_1 = constant dependent on the tire size, and C_2 = constant dependent on the tire size. In essence, Eq. (14.3) can be simplified to a model consisting of a simple supported beam of length L with elastic constants E and G deflected distance d by force P as

$$S = PL^3/48EI + 2PL/8AG, \tag{14.4}$$

where A is the in-plane cross-sectional area to which the force P is applied, and I is the moment of inertia of the beam.

Computer analysis of the effect of the belt angle on Gough stiffness of the four-belt layup is illustrated in Figure 14.8b.

For a right/right/left/right belt configuration, the belt angle (θ) of the second and third belts was varied from 10° to 26°. Computation of an optimum belt layup will permit achievement of the required Gough stiffness.

Structural mechanical calculations such as finite-element analysis (FEA) are used to analyze both the inflated and loaded deflected shapes of a tire cross-section and the resulting stress-strain relationships in the belt area. Such studies permit both quantitative analysis and qualitative comparisons of the range of belt configuration options. Figure 14.9 shows a heavy-duty truck tire in the loaded and unloaded states. The density of grids is designed so as to preserve

(a)

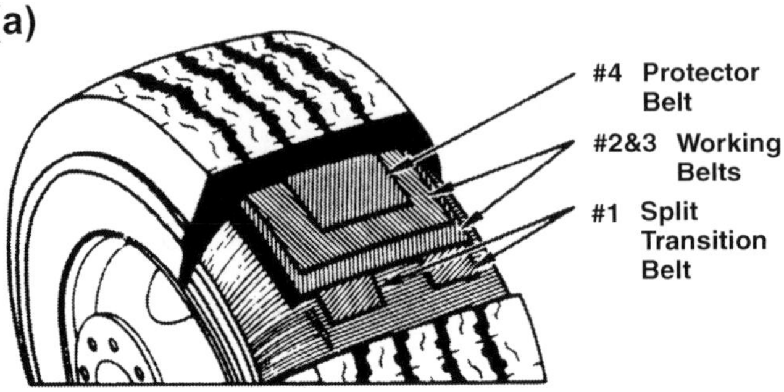

(b)

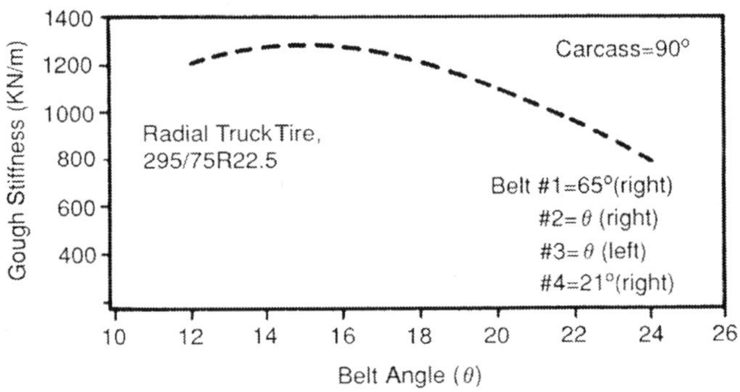

FIGURE 14.8 (a) Typical four-belt layup for a truck tire. (b) Impact of Gough stiffness.

the essential features of the tire cross-section geometry while maintaining the total number of grid points.

In an analysis of the belt package, three conditions can be evaluated that allow computation of the range of strain energy densities: inflated tire condition, loaded tire condition 180° away from the footprint, and loaded tire condition at the center of the footprint. Figure 14.10 illustrates these three conditions. It shows that the strain energy density in megapascals in an inflated tire is similar to that in a load tire 180° away from the footprint. At the center of the footprint in a loaded condition, however, the strain energy density at the belt edge has increased from the range 0.01–0.27 to 0.05–0.51 MPa. These high-strain areas correspond to typical failure-sensitive regions in actual tires.

Similar to the belt area of the tire, the bead region also lends itself to finite-element analysis. Switching grid details to the bead enables analysis of the ply end strains on inflation as well as in a loaded state as the tire makes a complete revolution. By viewing Figure 14.11, it is possible to evaluate strains caused by tire inflation and cyclic deformation via FEA quantitatively, whereas without such tools, tire building and testing are required.

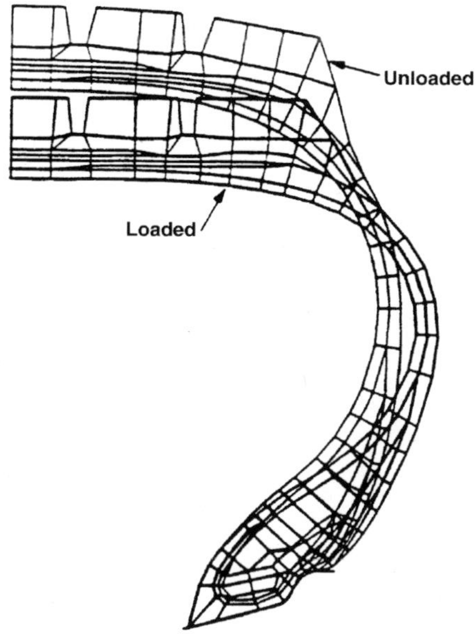

FIGURE 14.9 Finite-element structure of a heavy-duty truck tire.

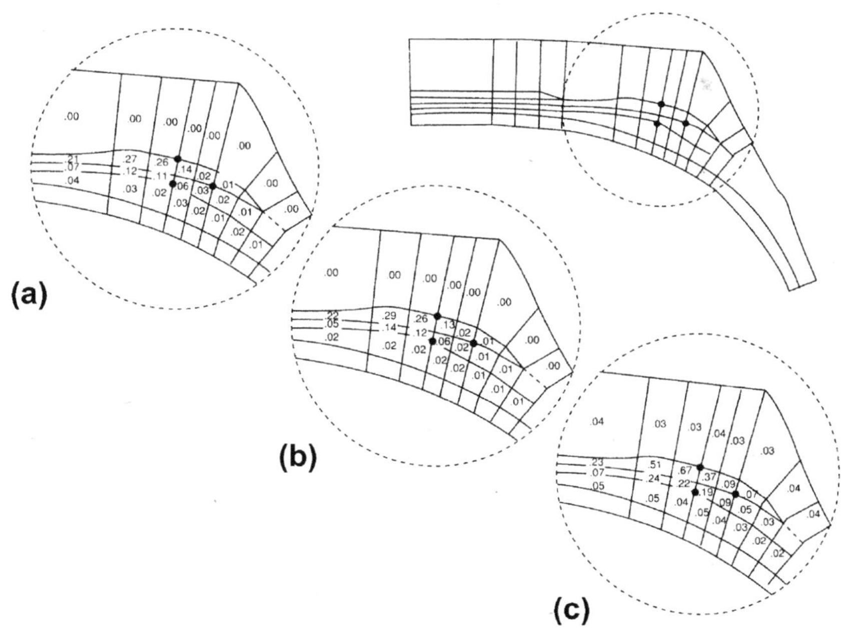

FIGURE 14.10 Finite-element analysis of belt area showing strain energy density (in MPa). (a) Inflation. (b) Loaded 180° away from footprint. (c) Loaded at center of footprint.

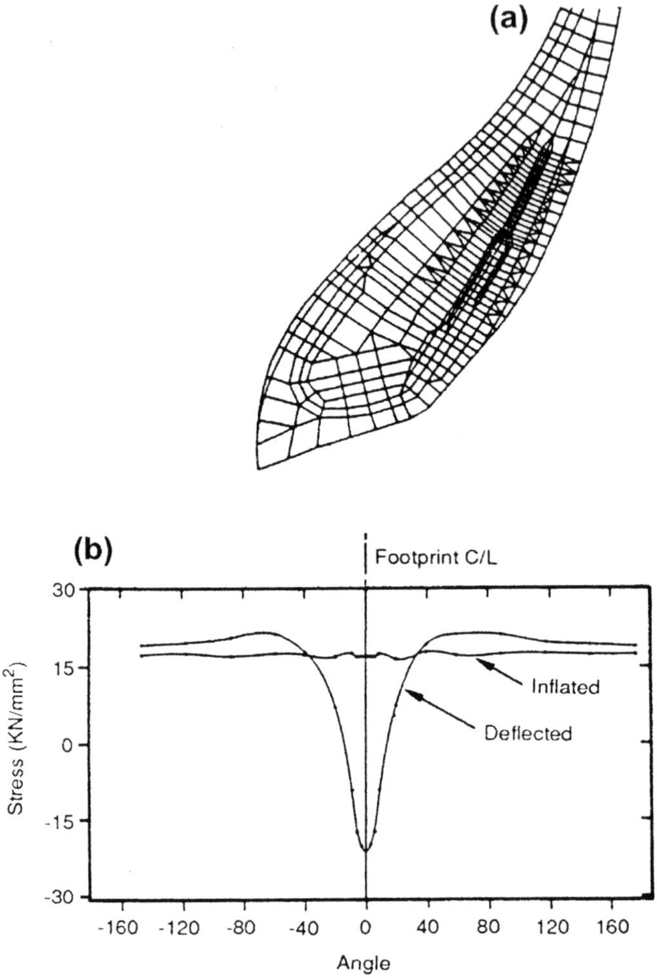

FIGURE 14.11 (a) Finite-element analysis of a bead region. (b) Ply end stress versus rotation from the center of footprint.

14.4.3 Cord Tension

The selection of cord materials for belts and plies in a tire and the associated physical locations lend themselves to further FEA analysis. If one takes an inflated nonbelted tire, the ply cords, whether they are in a single-ply casing or multiple-bias-ply construction, will assume a configuration that minimizes strain within the composite. The resulting cord path is termed the *neutral contour*. Belted tires introduce a restriction to the inflated diameter of the tire, and the neutral contour or plyline of such systems is consequently altered.

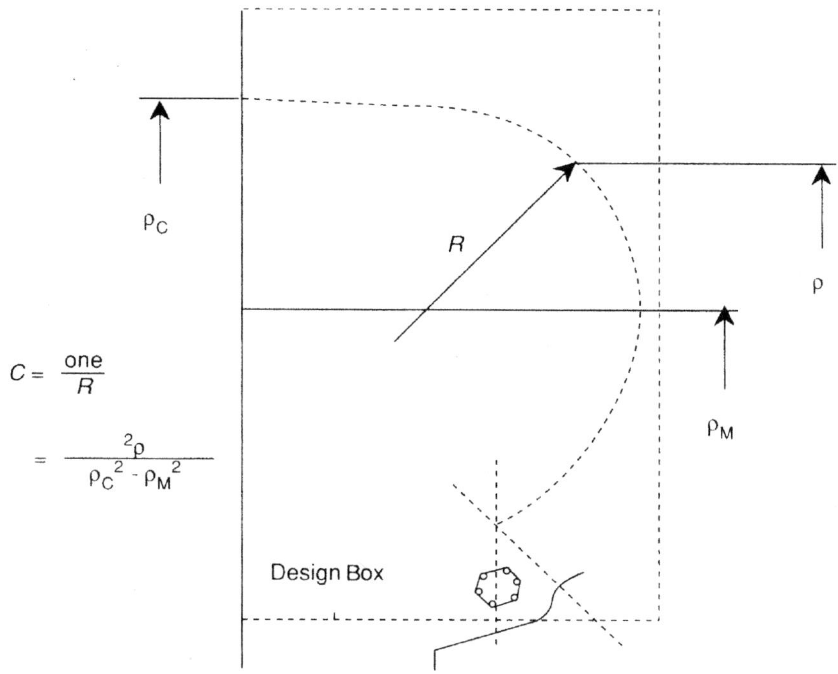

FIGURE 14.12 Plyline definition. Purdy's equation defines ideal inflated or "natural" shape for "thin-film" structure. C = curvature of the plyline, ρ_C = radius from the center of the axle to the center of the plyline or tire centerline, and ρ_M = radius from the center of the axle to the center plyline width.

The plyline is determined as shown in Figure 14.12. The principles of plyline determination were developed by Purdy, a pioneer mathematician who derived the basic mathematical equations for cord path and tire properties (Purdy, 1970).

Typical inflated cord tension plots for a truck tire are shown in Figure 14.13. In an unloaded state the cord tension for the belts tends to be at the tire centerline, and the ply tension is greatest at the point corresponding to the sidewall location; however, on application of a load to the inflated tire and consequent deflection, the cord tensions increase at the belt edges away from the centerline and in the bead zone. As reviewed earlier, these two regions tend to be the failure zones in a tire construction (Figure 14.14).

14.4.4 Tread Design Patterns

The tread pattern influences the ability of the tire to transmit driving forces, braking, and lateral forces while operating on a broad range of highway and off-road surfaces. The design of a tread pattern is essentially the separation or division of a smooth tread into smaller elements or blocks. These elements are arranged in a repetitive pattern of voids, ribs, lugs, slots, and grooves. The pattern

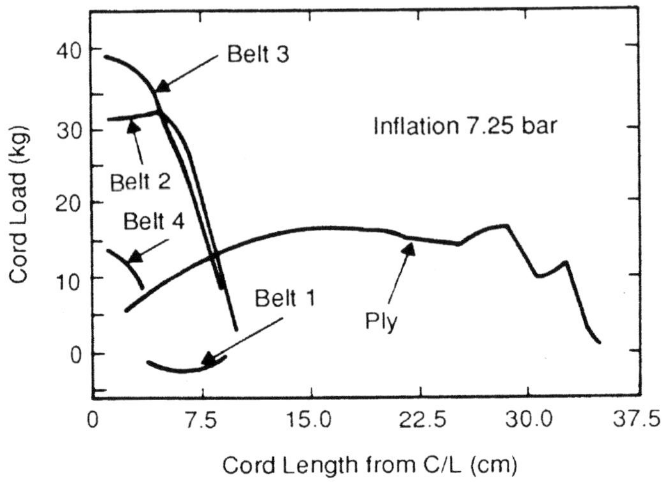

FIGURE 14.13 Inflated tire unloaded.

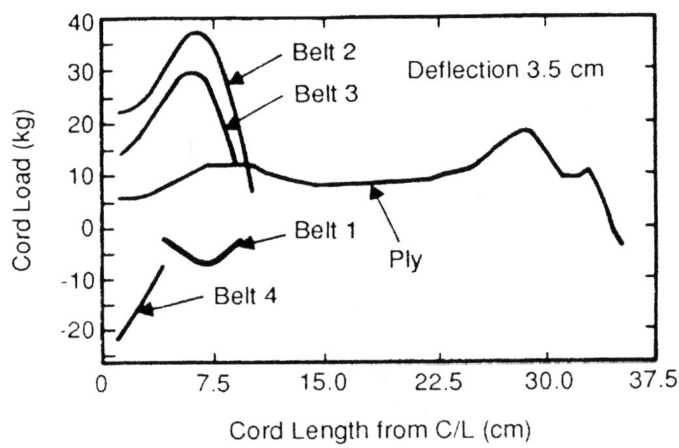

FIGURE 14.14 Inflated tire loaded and deflected.

is first described in terms of length, width, percentage of void of the various elements, and angles. The tread elements are then arranged to give the tread traction characteristics, optimized ratio of net-to-gross contact area, and minimum noise creation (Ford and Charles, 1988; Fitch, 1994; Waddell et al., 2004).

For a truck tire, tread patterns can be classified into five basic types (Figure 14.15):

- Highway rib.
- Highway rib/lug combination, where the rib can be either inboard or outboard in the shoulder area (the lugs will be correspondingly opposite to the ribs, outboard or inboard).

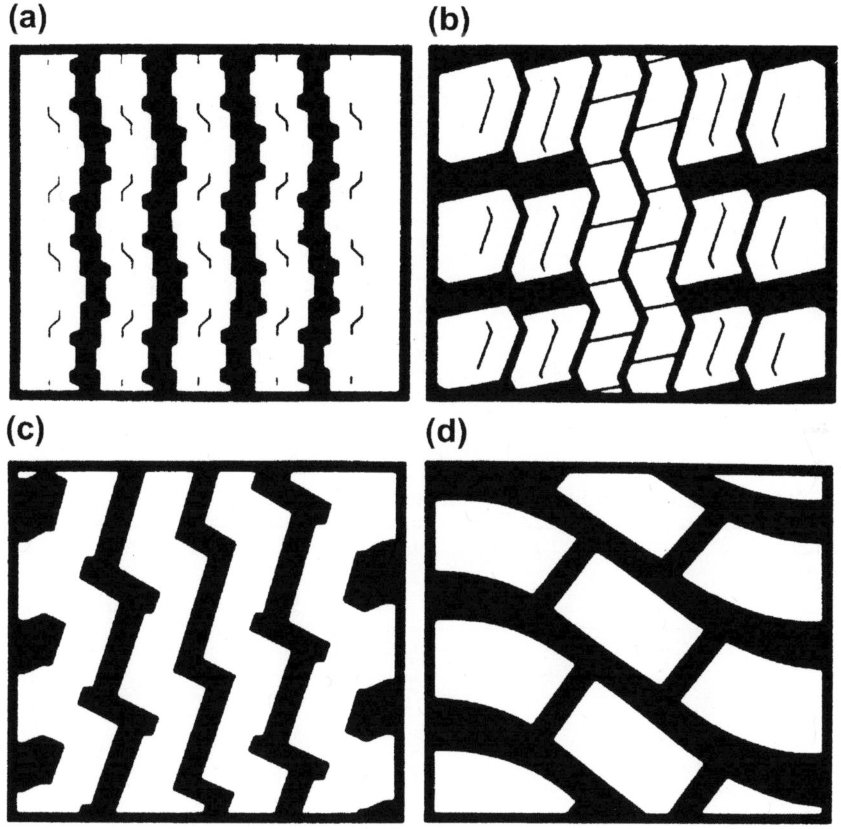

FIGURE 14.15 Basic tread patterns: (a) highway rib; (b) highway rib/lug design; (c) on/off highway (mixed service); (d) off highway.

- Highway lug design.
- On/off highway (special service).
- Off highway.

Rib designs with design elements principally in the circumferential direction are the most common type of tread pattern and show overall good service for all-wheel-position summer service. On heavy trucks, they are used nearly exclusively on steer and trailer axles because of their lateral traction and uniform wear characteristics. Rib/lug combinations tend to find use on all-season tires, which require a balance of good tread wear, traction, and wet skid. On heavy-duty truck drive axles, where forward traction is a prime requirement and where fast tread wear occurs as a result of torque-induced slip, the highway lug design is required. For off-highway service conditions the tread pattern assumes a

TABLE 14.3 Net-to-Gross Contact Area with Nature of Service [Ford and Charles, 1988; Clark, 1982]

Type of Service	% Net to Gross	Relative Tread Depth	Traction Handling	Wear
Highway steer tires	70–75	++	+	+++
Highway drive axle tires	70–80	+++	++	++++
Highway trailer tires	75–85	+	+	++++
On/off road (mixed service)	60–70	+++	+++	++
Off highway	55–65	+++++	+++	++

TABLE 14.4 Automobile Tire Tread Pattern Classes

Category	Design	Application	Net-to-Gross Pavement Contact Area	Vehicle Handling Comfort
1	Central solid rib, outer rib block configuration	High mileage	High	+++++
2	All block	All season	Medium-high	++++
3	Block-rib	Traction	Medium	+++
4	Central groove outboard blocks	Traction, high performance	Low-medium	++
5	Direction —Asymmetric —Symmetrical	High performance	Low-medium	++

staggered joint lateral circumferential direction for both lateral and forward traction. Grooves tend to be larger and deeper, with the rib walls angled to prevent stone retention.

The ratio of net contact to gross tread surface area decreases as wet traction becomes more important. Tables 14.3 and 14.4 illustrate the net-to-gross percentage for various tire tread patterns.

A number of additional terms are used to describe a tread pattern:

- *Groove amplitude:* In staggered groove designs, the distance the groove pattern oscillates about the central direction that the groove follows. It is analogous to a sine wave.

- *Sipes:* Small individual tread voids, generally added to a tread design to improve traction characteristics.
- *Pitch length:* Length of each repeating unit in a tread pattern. Variable pitch lengths in a tread design can be used to minimize noise. Pitch would be analogous to the wavelength of a sine wave.
- *Blade:* A protrusion in a curing mold that forms part of the tread design. The protrusion forms a corresponding depression in the finished tire.
- *Stone ejection rib:* Portion of the tread rib designed to throw off stones with the aid of normal tire flexing. Located up to 75% down in the tread grooves to prevent small stones from locking down at the base of the grooves where they cannot be ejected.

14.5 TIRE MATERIALS

Tire engineering is the study of the stresses created within a tire and includes such factors as straining of components while the tire is being built, the tire stresses while mounted on a wheel of a moving loaded vehicle, quantification of such stresses, and minimization of such stresses through effective distribution of load and proper selection of materials. Consequently, tire technology groups tend to be multidisciplinary teams consisting of mechanical engineers, computer scientists, chemical engineers, chemists, and mathematicians. To understand tire engineering, it is necessary to have knowledge of the function of each of the types of materials used in a tire structure. This could be considered essentially in two aspects: the tire reinforcing system and rubber compounding.

14.5.1 Tire Reinforcement

A tire is a textile-steel-rubber composite; the steel and textile cords reinforce the rubber and are the primary load-carrying structures within the tire. Because of the performance demands of fatigue resistance, tensile strength, durability, and resilience, seven principal materials have been found suitable for tire application: cotton, rayon, nylon, polyester, steel, fiberglass, and aramid; the latter three materials find primary usage in the tire crown or belt region.

The science of tire reinforcement employs a specialized terminology that the tire engineer must understand:

- *Brass weight:* Typically 3.65 gm/kg of cable; brass coat thickness is of the order of 0.3 μm.
- *Breaking strength:* Tensile strength.
- *Cord:* Structure consisting of two or more strands when used as plied yarn or an end product.
- *Denier:* The weight of cord expressed in grams per 9000 m.
- *EPI:* Ends of cord per inch width of fabric.
- *Fibers:* Linear macromolecules orientated along the length of the fiber axis.

- *Filaments:* Smallest continuous element of textile, or steel, in a strand.
- *Filling:* Light threads that run right angles to the warp (also referred to as the "pick") that serve to hold the fabric together.
- *LASE:* Load at a specified elongation or strain.
- *Length of lay:* Axial distance an element or strand requires to make a 360° revolution in a cord.
- *Ply twisting:* Twisting of the tire yarn onto itself the required number of turns per inch; two or more spools of twisted yarn are then twisted again into a cord: for example, if two 840-denier nylon cords are twisted together, an 840/2 nylon cord construction is formed; if three 1300-denier polyester cords are twisted together, they give a 1300/3 cord construction.
- *Rivit:* Distance between cords in a fabric; high rivit typically describes a fabric with a low EPI.
- *Tenacity:* Cord strength, frequently expressed in grams per denier.
- *Tex:* Cord weight expressed in grams per 1000 m.
- *Twist:* Number of turns per unit length in a cord or yarn; direction of twist can be either clockwise ("S" twist) or counterclockwise ("Z" twist); twist imparts durability and fatigue resistance to the cord, though tensile strength can be reduced.
- *Warp:* Cords in a tire fabric that run lengthwise.
- *Weft:* Cords in a fabric running crosswise.
- *Yarn:* Assembly of filaments.

Fibers and steel cord are the primary reinforcement and load-carrying materials in the tire. It is thus appropriate to review the properties of such materials for application in tires.

14.5.2 Steel Cord

Steel wire used in tires is of various configurations, but all are brass-coated wire strands wrapped together to give cords of different characteristics, depending on the application. Steel tire cord is manufactured from high-carbon-steel rod, which is first drawn down to a diameter of approximately 1.2 mm. A brass plating is then added to the wire before a final drawing to 0.15–0.40 mm. These filaments are next stranded to form a cord construction that is designed and optimized for a specific service requirement (Bekaert Corporation, 2004).

Steel tire cord is manufactured from high-quality steel, which is necessary because of the performance demands to which tires are subjected. The composition of a typical steel cord is illustrated in Table 14.5. The key mechanical properties governing a steel cord or wire are its tensile strength, elongation, and bending stiffness. A tire cord construction is normally defined by the structure, the length of lay, and the direction of lay. The full description of a steel cord is given by

$$(N \times F) \times D + (N \times F) \times D + (F \times D), \tag{14.5}$$

TABLE 14.5 Composition of Steel Tire Cord

Element	Composition (%)	Function
Carbon	0.65	Strength
Chromium	0.05	Strength
Copper	0.02	Strength
Manganese	0.60	Deoxidation
Silicon	0.25	Deoxidation
Sulfur	0.03	Machinability

where N = number of strands, F = number of filaments, and D = nominal diameter of filaments (in mm).

An example of a steel cord specification would therefore take the form

$$(1 \times 4) \times 0.175 + (6 \times 4) \times 0.175 + (1 \times 0.15). \quad (14.6)$$

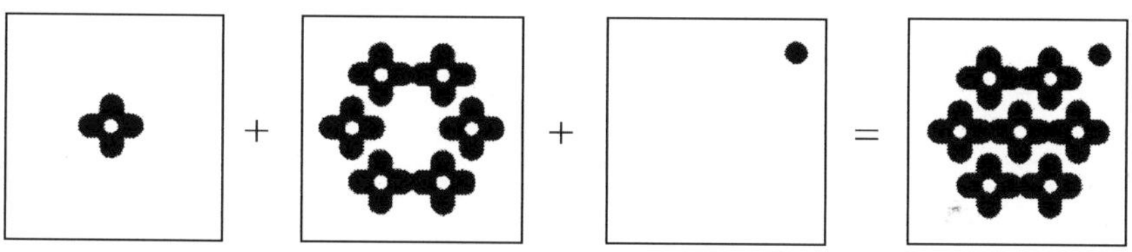

When N or F equals 1, the nomenclature system allows their exclusion. Thus, Eq. (14.6) is reduced to

$$4 \times 0.175 + (6 \times 4) \times 0.175 + 0.15. \quad (14.7)$$

A number of additional conventions are used in defining a steel tire cord:

1. If the diameter D is the same for two or more parts in a sequence, then the diameter is specified only at the end of the sequence.
2. The diameter of the spiral wrap is specified separately.
3. When the innermost strand or wire is identical to the adjacent strand or wire, the definition of the wire can be further simplified by specifying only the sum of the identical components. Then Eq. (14.7) becomes

$$7 \times 4 \times 0.175 + 0.15. \quad (14.8)$$

4. The sequence or order in a wire designation follows the sequence of manufacture. The length of lay and direction of lay for the cord in Eq. (14.8) can be described as illustrated in Figure 14.16.

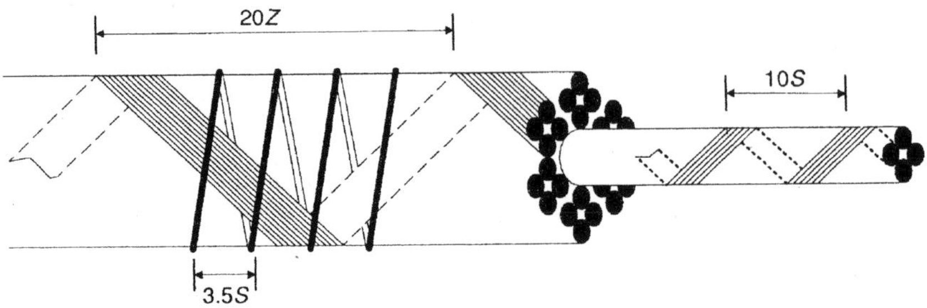

FIGURE 14.16 Length of lay and direction of lay. (Courtesy of Bekaert Corporation.)

TABLE 14.6 Design of a Tire Cord

Design Parameter	Design Factor
Stiffness	Number of strands in cord, gauge
Strength	Gauge, number of strands in cord, steel rod composition
Fatigue	Filament gauge
Elongation	Twist, lay

A number of empirical guidelines govern the construction of a wire for use in tires. For example, if the wire is used in a ply rather than in belts, it will undergo a greater amount of flexing. Hence fatigue performance will be important. If application is in belts, then stiffness becomes a primary design parameter. Thus, key design properties can be specified and are highlighted in Table 14.6.

The thicker a cord is, the stiffer it will be. Thinner cords tend to show better fatigue resistance. Heavier cords tend to find use in the larger tires such as heavy-duty truck tires and tires for earthmoving equipment.

Wire finds application in tire reinforcement in tire belts, heavy-duty tire plies (e.g., in large truck tires), beads, and chippers (which protect the bead from wheel rim damage).

14.5.3 Mechanism of Rubber: Brass Wire Adhesion

The thin coating of brass on the steel cord is the primary adhesive used in steel-to-rubber bonding. The quality of this bonding system built up during vulcanization of, for example, a radial tire will influence the performance of the steel ply or steel belt in the tire and, ultimately, the durability of the product. Though the mechanism of bond formation in rubber-steel cord adhesion is very complex, a brief review of the current understanding of wire to rubber adhesion is presented.

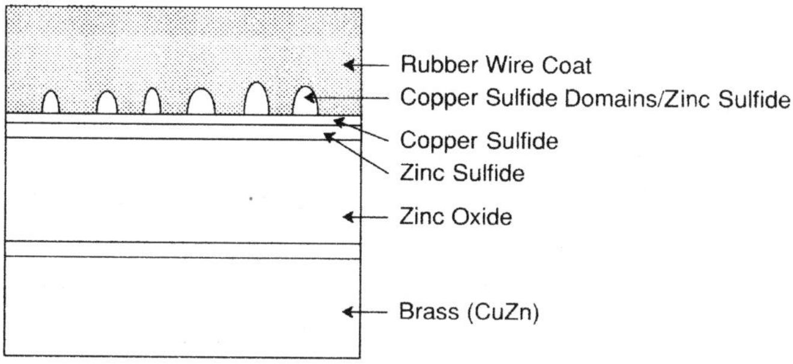

FIGURE 14.17 Interfacial copper sulfide film in rubber-brass bonding (from Gough, 1968.)

Natural rubber typically used in wire coat compounds forms a strong bond with brass as a result of the formation of an interfacial copper sulfide (CuS) film during vulcanization. Copper sulfide domains are created on the surface of the brass film during the vulcanization reaction. Such domains have a high specific surface area and grow within the wire coat compound before the viscous polymer phase is crosslinked into an elastomeric network. Thus the polymer molecules become locked into the crystalline copper sulfide lattice (Figure 14.17). Important factors governing this bonding are formation of copper sulfide, cohesive strength, adhesion to the brass substrate, and rate of secondary corrosion reactions underneath the copper sulfide film.

Zinc sulfide and iron sulfide do not bond because they do not grow rapidly enough during vulcanization, do not form porous domains, and thus cannot interlock with the polymer. Because the primary requirement is the formation of a copper sulfide domain before the initiation of crosslinking, reduction of compound scorch time, consequently, can adversely affect bond formation.

Mechanical stability of the copper sulfide domains is essential to retain long-term durability of the rubber-to-wire adhesion. However, corrosion of the wire-rubber adhesive bond is catalyzed by Zn^{2+} ions that diffuse through the interfacial CuS layer. This will eventually result in an excess of either ZnS or $ZnO/Zn(OH)_2$. Under dry conditions, this process is slow. Nevertheless, Zn^{2+} will migrate to the surface with a consequent drop in mechanical interlocking of the CuS domains and rubber followed by adhesion loss.

Migration of Zn^{2+} ions is a function of the electrical conductivity of the brass coating. Addition of Co^{2+} or Ni^{2+} ions will reduce this conductivity.

Cobalt salts used in the wire coat compounds may accelerate the vulcanization rate of high sulfur compounds, and increase the state of cure or crosslink density. The increase in crosslink density increases the pullout force of the wire in the rubber. More important, cobalt salts form Co^{2+} ions at the interface of the brass surface during vulcanization, and this will affect copper sulfide formation.

Differences in efficiencies between cobalt adhesion promoters are due to the ease with which Co^{2+} ions can be formed. For example, zinc or brass reacts more

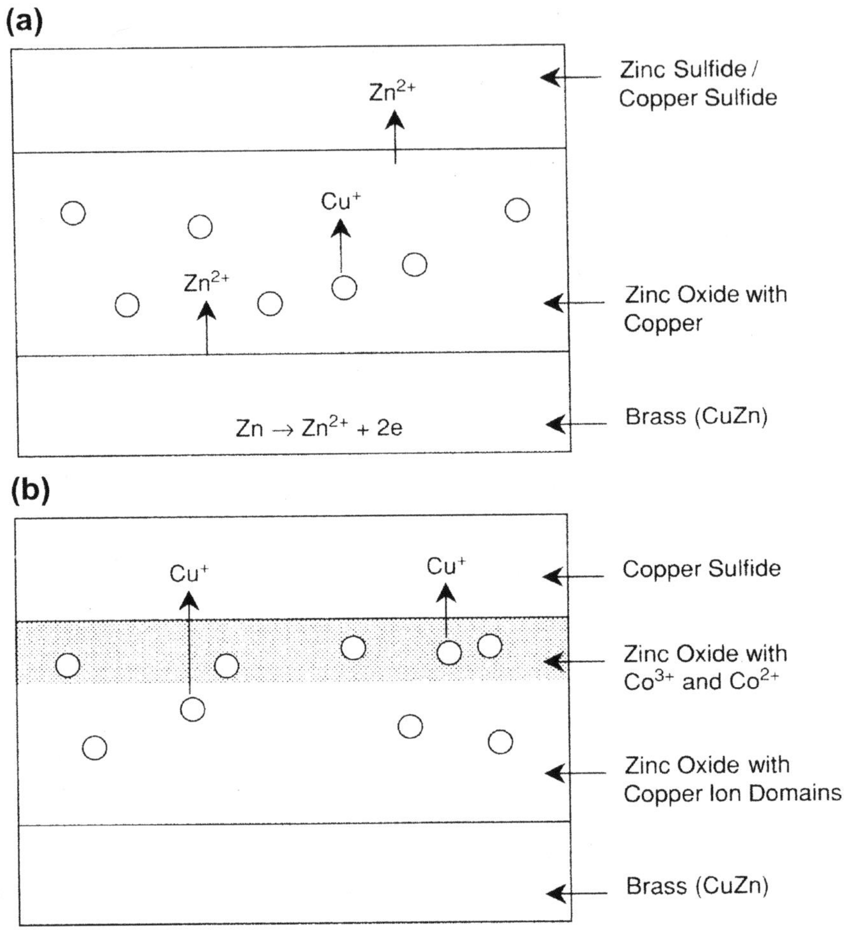

FIGURE 14.18 Copper sulfide formation at low cobalt concentration. (a) Absence of cobalt allows formation of zinc sulfide. (b) Cobalt ions in zinc oxide film hinder zinc ion migration and zinc sulfide formation.

easily with cobalt boron decanoate complexes than with cobalt naphthenate or stearate. The Co^{2+} and Co^{3+} ions are incorporated into the ZnO film before the sulfide film has been built up. Both di- and trivalent cobalt ions reduce the electrical conductivity of the ZnO lattice, thereby reducing the diffusion of Zn^{2+} ions through the semiconducting film.

Diffusion of metallic copper domains to the surface following oxidation by R—S_x is not affected, as Cu^{2+} ions migrate along grain boundaries of the ZnO layer. Thus if a cobalt salt is used, formation of copper sulfide at the cord surface will be accelerated, whereas ZnS generation will be hindered (Figure 14.18). This review is necessarily brief, and the reader is encouraged to consult additional references for further detail on the chemistry of rubber-brass adhesion (Bekaert Corporation, 2004; van Ooij, 1984).

14.5.4 Rayon

The first synthetic fiber for tires was rayon. Cellulose is initially treated with sodium hydroxide to form an alkali cellulose. It is then shredded and allowed to age in air, where it is oxidized and undergoes molecular weight reduction to enable subsequent spinning operations. Treatment with carbon disulfide produces cellulose xanthate, which is then dissolved in sodium hydroxide to form viscose. The material undergoes further hydrolysis and is then fed into spinnerets to produce the fiber. This fiber is passed through a bath of sulfuric acid and sodium sulfate, where the viscose fibers are further coagulated.

Washing, bleaching, and twisting into cords follow. The rayon fibers can be drawn or stretched up to 100% of their original length to enable crystalline orientation to produce a high-tenacity rayon suitable for tires.

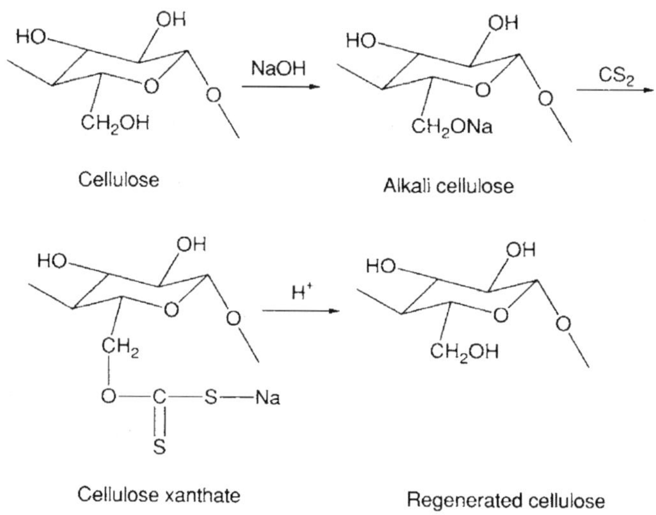

(14.9)

14.5.5 Nylon

Nylon is a polyamide polymer characterized by the presence of amide groups—(CO–NH)—in the main polymer chain. A wide variety of nylon polymers are available but only two have found application in tires: nylon 6,6 and nylon 6.

Nylon 6,6 is produced from a condensation reaction between adipic acid and hexamethylenediamine,

$$\underset{\text{Hexamethylenediamine}}{H_2N{-}(CH_2)_6{-}NH_2} + \underset{\text{Adipic acid}}{HOOC{-}(CH_2)_4{-}COOH}$$

$$\longrightarrow \underset{\text{Nylon 6,6 polymer chain}}{\left(NH{-}(CH_2)_6{-}NH{-}CO{-}(CH_2)_4{-}CO \right)_n} \qquad (14.10)$$

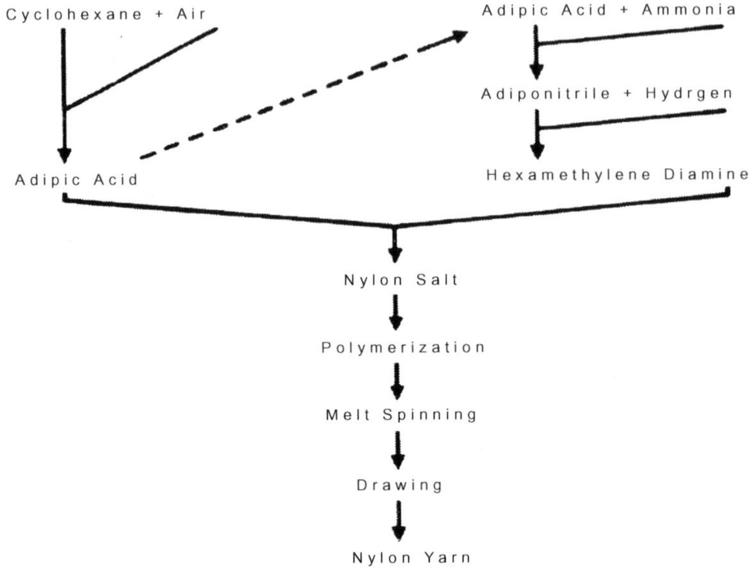

FIGURE 14.19 Nylon 6,6 manufacture.

The "6,6" in the polymer designation denotes the six carbon atoms of hexamethylenediamine and the six carbons of adipic acid. Nylon 6 is produced from caprolactam by a ring-opening polymerization. Since caprolactam contains six carbon atoms and only one monomer is used, the polymer is thus designated as nylon 6.

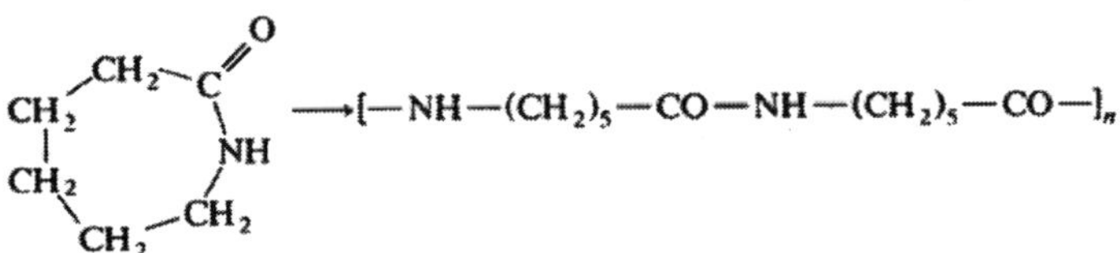

Caprolactam **Nylon 6 polymer chain** (14.11)

After the polymerization stage, the material is passed through a spinneret to form filaments, cooled, and then twisted to form a yarn. This is then drawn by up to 500% to orient the polymer chains, create polymer crystallite zones, and increase tensile strength (Figure 14.19).

14.5.6 Polyester

Like nylon, a range of polyesters is available commercially for use in tires. Polyethylene terephthalate (PET) is the most important. Also, like nylon, polyester is formed by a condensation polymerization but with the monomers ethylene glycol and dimethyl terephthalate.

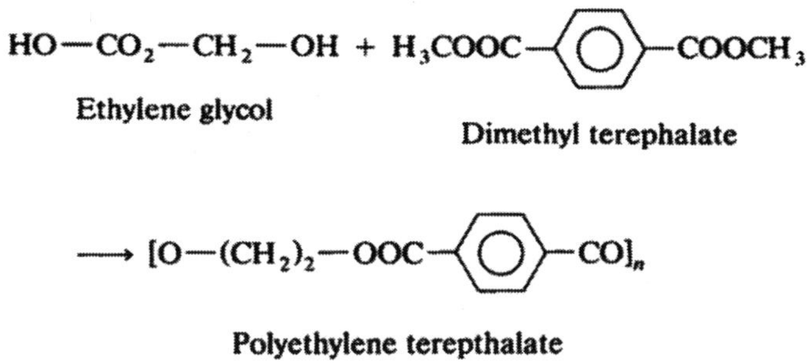

(14.12)

The polymerized material is then extruded through a spinneret to form filaments about 0.025 mm in diameter. These filaments are cooled, spun into a yarn, and drawn to give the required orientation and crystallinity.

14.5.7 Fiberglass

Like steel wire, fiberglass is an inorganic fiber and is essentially a lime-alumina-borosilicate glass (Table 14.7). It is manufactured by blending sand, clay, limestone, and borax, melting the blend at 3000°F, and drawing the filaments of glass through a platinum-rhodium filter. The filaments are then treated with an adhesive precoat and compiled into a yarn with a low twist.

14.5.8 Aramid

Another class of fibers that finds application in tires is the aramids. Kevlar is the trade name of the polymer that has found most extensive use among the aramids. Aramid is like nylon in that it contains the amide bond—(CO–NH)—but is produced by copolymerizing terephthalic acid used in polyester and *p*-phenylenediamine. Hence this aromatic polyamide is termed *aramid*.

TABLE 14.7 Fiberglass Cord Composition

Silicon dioxide	53%
Calcium oxide	21%
Aluminum oxide	15%
Boron oxide	9%
Magnesium oxide	0.30%
Other oxides	1.7%

TABLE 14.8 Trends in Reinforcements

Time	Cord
1900–1956	Cotton
1939–1975	Rayon
1950 to date	Nylon
1965 to date	Polyester
1970 to date	Steel cord
1975–1985	Fiberglass
1980 to date	Aramid
1990 to date	Hybrid cords

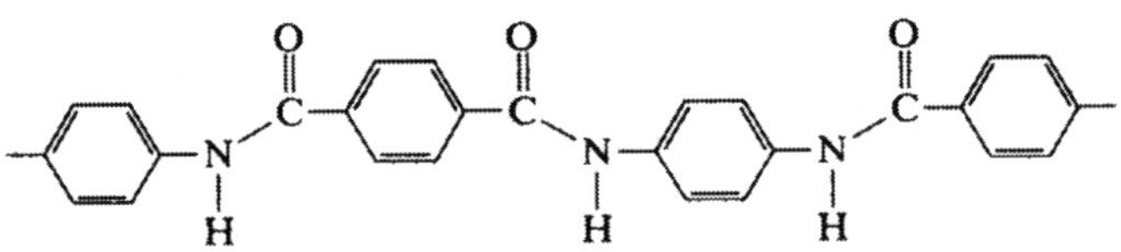

(14.13)

Table 14.8 illustrates the evolution of the range of reinforcement materials that have been used in tires.

14.5.9 Tire Cord Construction

To use the range of fibers for tire applications, the yarns must be twisted and processed into cords. First, yarn is twisted on itself to give a defined number of turns per inch (i.e., ply twisting). Two or more spools of twisted yarn are then twisted into a cord. Generally the direction of twist is opposite that of the yarn; this is termed a *balanced twist*. There are a number of reasons for twist in a tire cord:

1. Twist imparts durability and fatigue resistance to the cord, though tensile strength can be reduced.
2. Without twist, the compressive forces would cause the cord outer filaments to buckle.
3. Increasing twist in a cord further reduces filament buckling by increasing the extensibility of the filament bundle.

Durability reaches a maximum and then begins to decrease with increasing twist. This can be explained by considering the effect of stresses on the cord as the twist increases. As twist increases, the helix angle (the angle between the filament axis and cord axis) increases (Figure 14.20). Thus tension stresses normal to the cord axis result in greater-force parting filaments. The reason for the reduction in strength is also evident in this figure. As cord twist increases,

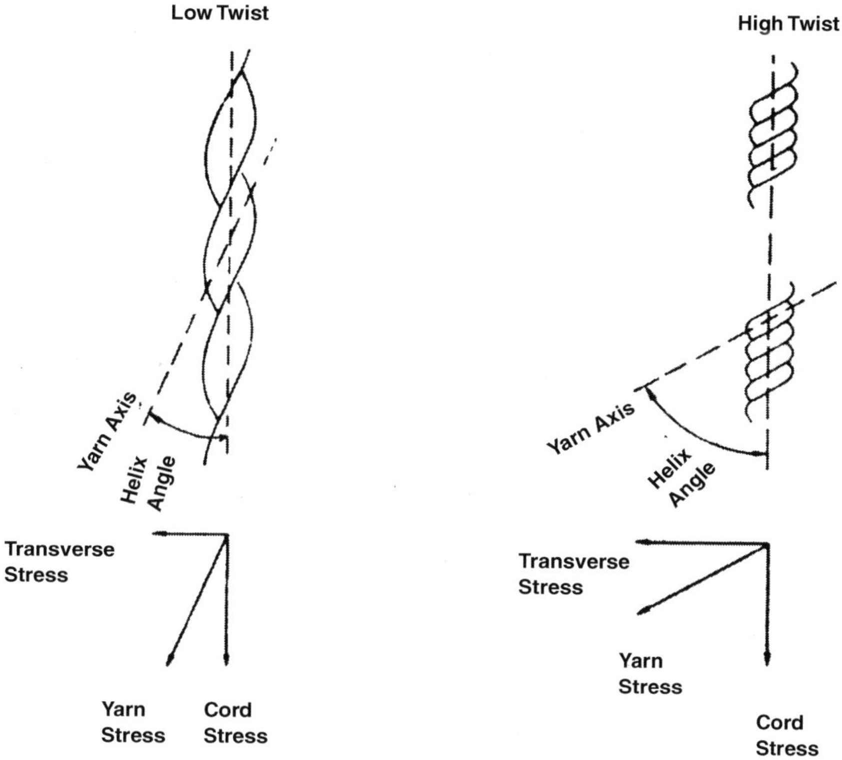

FIGURE 14.20 Cord geometry.

the force in the direction of the yarn axis increases, causing a lower overall breaking strength.

In addition to twist, the cord size may be varied to allow for different applications (Table 14.9). Generally three-ply cords have the best durability. After cable twisting, the cords are woven into a fabric using small fill threads. These threads are also referred to as *picks*. This weaving process introduces an additional construction variable—the number of cords per inch or EPI (ends per inch) that are woven into the fabric. High-end-count fabric gives greater plunger strength or penetration resistance. Low-end-count fabrics have more rivet (distance between cords) and give better separation resistance because of the greater rubber penetration around the cords.

14.5.10 Fabric Processing

The most critical stage in preparing a cord or fabric for use in tires is fabric treatment, which consists of applying an adhesive under controlled conditions

TABLE 14.9 Tire Cord Applications

Fiber Type	Tenacity (Strength) (g·Den)	Ultimate Elongation (%)	Modulus (g·Den)	Relative Durability	Tire Applications
Rayon	5.0	16	50	300	Passenger
Nylon	9.0	19	32	1200	Truck tires, off-road
Polyester	6.5	18	65	400	Passenger, truck, farm
Fiberglass	9.0	4.8	260	5	Passenger
Wire	3.8	2.5	200	3	Truck, off-road
Aramid	20.1	4	350	400	Passenger

of time, temperature, and tension. This process gives the fabric the following properties:

- Adhesion for bonding to rubber.
- Optimization of the physical properties of strength, durability, growth, and elongation of the cord for tire application.
- Stabilization of the fabric.
- Equalization of differences resulting from the source of supply of the fiber.
- Processing consists of passing the fabric through a series of zones, which can be viewed as follows:
- Adhesive application zone or first dip zone.
- First drying zone.
- First heat treatment zone.
- Second dip zone.
- Second drying zone and then heat treatment zone.
- Final cooling zone.

To obtain optimum cord properties of strength, growth, shrinkage, and modulus, specific temperatures and tensions are set at various exposure times within the fabric processing unit. The temperature and tensions in part determine the ratio of crystalline and amorphous areas within the fiber and the orientation of the crystallites, which in turn determine the physical properties of the cord. For example, polyester, when heated, tends to revert to its unoriented form, and the cord shrinks. Stretching the cord in the first heating zone and then allowing the cord to relax in a controlled manner in the second heat treatment zone (i.e., stretch relaxation) will control shrinkage.

Another variable is increase in processing temperature, which can decrease cord tensile strength and modulus but will improve fatigue life. The general relationship between cord properties and processing temperatures and tensions is illustrated in Table 14.10.

TABLE 14.10 Relation of Cord Properties to Processing Tensions and Temperature

Cord Property	Change Effected by First Increase in		Change Effected by Second Increase in	
	Tension Time	Temperature	Tension Time	Temperature
Tensile strength	Slight decrease	Decrease	Slight decrease	Decrease
Loaded at specified elongation (5%)	Decrease	Decrease	Decrease	Decrease
Ultimate elongation	Increase	Increase	Increase	Increase
Shrinkage	Decrease	Decrease	Decrease	Decrease
Rubber coverage	Increase	Increase	Increase	Increase
Fatigue	Decrease	Decrease	Decrease	Decrease
Voids	Decrease	Decrease	Decrease	Decrease

Note that not all cord properties behave similarly with varying processing conditions. It is thus necessary to determine the processing conditions that optimize cord properties for the desired tire end use. When two or more diametrically opposite properties have to be optimized, more complex mathematical methods must be used.

Wire and fiberglass, being high-modulus inorganic belt cords, are not processed like textile cords. Steel cord is brass plated at the foundry and, thus, can be used directly at the calendars. Glass yarn is treated with adhesive dip and then used directly in the weaving operation.

14.5.11 Function of Adhesive

There are three aspects to adhesion of tire cord to the elastomer treatment: molecular, chemical, and mechanical. Molecular bonding is due to absorption of adhesive chemicals from the adhesive dip or elastomer coating onto the fiber surface by diffusion and could be described by hydrogen bonding and van der Waals forces. Chemical bonding is achieved through chemical reactions between the adhesive and the fabric and rubber; that is, crosslinking and resin network formation. Mechanical adhesion is a function of the quality of coverage of the cord by the rubber coat compound. The greater the coverage, the better the adhesion.

The fiber properties of primary importance to adhesion are reactivity, surface characteristics, and finish. Rayon has many reactive hydroxyl groups. Nylon is less reactive but contains highly polar amide linkages. Polyester is relatively inert. Thus an adhesive system must be designed for each type of fiber.

Regardless of the fiber, each adhesive system must conform to a rigid set of requirements:

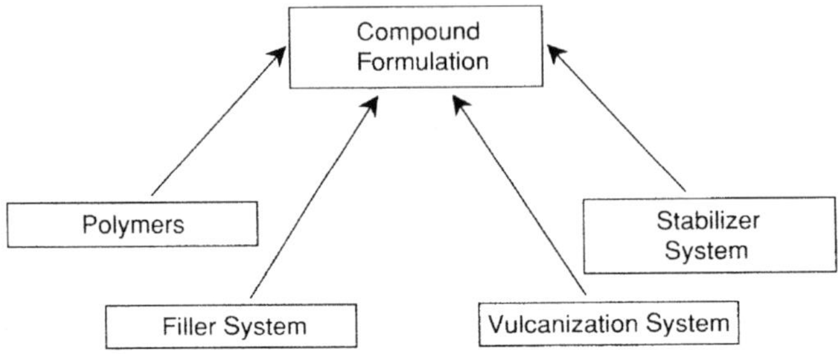

FIGURE 14.21 Basic compound components.

- Rapid rate of adhesion formation.
- Compatibility with many types of compounds.
- No adverse effect on cord properties.
- Heat resistance.
- Aging resistance.
- Good tack.
- Mechanical stability.

The adhesive bond between the rubber and cord is achieved during the tire vulcanization cycle. The rate of adhesive formation should give maximum adhesion at the point of pressure release in the cure cycle.

14.5.12 Rubber Compounding

The principles of compounding were reviewed earlier in this text and cover the fundamental characteristics of polymers, filler systems, and the basics of vulcanization in the context of compound development for tire applications. A compound formulation consists of four basic components: the polymer network, the filler or particulate reinforcing system, the stabilizer system, and the vulcanization system (Figure 14.21). In addition a series of secondary materials such as resins, processing oils, and short fiber reinforcements may be included in a formula (Duddey, 2004; Rodgers and Waddell, 2004; Long, 1985).

Elastomers used in radial tires are basically of four types:

1. Natural rubber.
2. Styrene-butadiene copolymer.
3. Polybutadiene.
4. Butyl rubber (polyisobutylene with approximately 2% isoprene) and halogenated butyl rubber.

Carbon blacks, clays, and silicas constitute the filler or compound reinforcement system. Optimization of these materials in a formulation depends on the application for which the component is intended; for example, tread or sidewall.

The stabilizer system or antioxidant system protects the compounds from aging and oxidation and improves the long-term durability of the tire.

Design of the vulcanization system is probably the most challenging aspect in developing a compound formulation for application in a tire. Knowledge of accelerator activity, reaction kinetics, and nature of the resulting crosslink network is important in constructing such systems.

14.6 TIRE TESTING

Having developed a tire construction for a defined service requirement, the tire engineer must now subject the design to a series of tests (The Tire & Rim Association, Inc. Year Book, 2004; European Tyre and Rim Technical Organization, 2004). This testing falls into two performance categories:

1. Mechanical characteristics of the tire:
   - Load deflection of a vertically loaded mounted tire, load-carrying capacity, and load rating.
   - Steering properties such as aligning torque, cornering characteristics, and tire lateral and tangential stiffness.
   - Traction and wet skid performance.
   - Rolling resistance, which affects vehicle fuel economy.
   - Noise.
2. Durability:
   - Tread wear, which encompasses slow wear rates, fast wear rates, and uniformity of wear.
   - Casing fatigue resistance.
   - Tire heat buildup under loaded dynamic conditions.
   - Chipping, cutting, and tearing resistance of the tread and sidewall.

This testing is broken down into three phases: laboratory testing, proving grounds, and commercial evaluation.

14.6.1 Laboratory Testing

Laboratory testing of tires is preceded by testing of raw materials. For compounds such as the tread, sidewall, wire coat compound, and liner, such tests would include determination of the kinetics of vulcanization, tensile strength, tear strength, resilience, and dynamic properties (e.g., storage modulus and loss modulus). Reinforcements such as the ply cord are similarly tested for tensile strength, but also creep behavior, stability (shrink behavior), and fatigue resistance. Much of the physical testing of compounds in a modern laboratory is done by robotics. Compounds are vulcanized under defined conditions, and samples are cut out depending on the type of testing to be done. Robots then load the samples onto test equipment, and the data generated are collected by computer for the test engineer to access.

Materials that meet the appropriate physical property targets are then used in tire building. Testing of these tires depends on the application for which the tire was designed. It is thus appropriate to introduce another series of definitions:

- *Balance:* Weight distribution around the circumference of a tire; poor balance is due to components having irregular dimensions. Balance can also be affected by irregular component splice widths and poor application of component parts.
- *Conicity:* Tendency of a tire to pull a vehicle to one side or another; it is caused by off-centered or misplaced components during the tire building process.
- *Cornering coefficient:* Lateral force divided by the vertical load at a defined slip angle. Stiffer tread compounds would tend to improve the cornering coefficient.
- *Force variation:* Periodic variation in normal vertical force of a loaded free-rolling tire, which repeats with every revolution.
- *Harmonic:* Periodic or rhythmic force variations occurring in a sinusoidal manner around a tire. One phase is described as the first harmonic. When two phases are noted, it is described as a second harmonic. Lateral force variation first harmonic is typically due to a tread splice. Radial harmonic may be due to irregular placement of the belt layup.
- *Lateral force:* Side force that is exerted by a tire as it rotates under a load.
- *Lateral force coefficient:* The lateral force divided by the vertical load.
- *Lateral force variation:* Change in force from one side of the tire to the other as it rotates under a load. It may cause the tire to wobble and is due to irregular tire component dimensions. Lateral force variation is a summation of the lateral first, second, third, etc., harmonic.
- *Lateral runout:* Difference between the maximum and minimum measurements parallel to the spin axis at the widest point of each tire sidewall when the tire is mounted on a wheel.
- *Radial runout:* Difference between the maximum and minimum measurements on the tread surface and in a plane perpendicular to the spin axis while the tire is mounted on a wheel. It is a measure of the out-of-roundness of the tire. It can also be termed "centerline runout."
- *Radial force:* Force acting on a tire perpendicular to the centerline of rotation or direction of axle. It is caused by heavy tire component splices and will increase radial force. Soft spots in the tire, such as those due to stretched ply cords, cause a decrease in radial force.
- *Radial force variation:* Summation of the radial first, second, third, etc., harmonic. It is the change in radial force as the tire is rotated. Radial force variation will cause the vehicle to have a rough ride (as if on a poor surfaced road).

- *Rolling resistance:* Resistance of a tire to rolling. It has a direct impact on vehicle fuel economy and is influenced most by compound hysteretic properties.
- *Runout:* Differential between the maximum and minimum lateral or radial forces.
- *Self-aligning torque:* Stabilizing reaction to slip angle that helps the tire and vehicle to return to neutral conditions at the completion of a maneuver.
- *Slip angle:* Angle between the vehicle's direction of travel and the direction in which the front wheels are pointing.
- *Speed rating:* Alphabetic ratings that define the design speed capability of the tire. The letter is incorporated into the size description of the tire; for example a 195/75SR14 has a speed rating of "S." Tables of speed ratings and corresponding alphabetic designations are published by the Tire & Rim Association (Table 14.11).
- *Uniformity:* Measure of the tire's ability to run smoothly and vibration free; sometimes measured as tire balance, radial force variation, or lateral force variation.

The laboratory equipment designed to test the aforementioned tire properties is based mostly on a steel flywheel with the appropriate monitoring devices such as transducers and infrared temperature monitors. Data are collected directly into computers for real-time analysis and downloading to the tire engineer's work station. Many of these flywheels are also computer controlled so as to simulate service conditions. For example, aircraft tires can undergo a complete cycle of taxi, takeoff, and landing.

TABLE 14.11 Tire Speed Ratings

Speed Symbol	Maximum Speed (kph)	Typical Application
D	50	Farm tractors
L	120	Commercial truck
M	130	Commercial truck
S	180	Passenger cars and light trucks
T	190	Passenger cars
H	210	Luxury passenger cars
V	240	High-performance cars
Z	270	High-performance sports cars
W	Above 270	Super-high-performance cars

14.6.2 Proving Grounds

The most definitive method of determining the behavior of a tire is to examine its performance when subjected to road testing. Proving ground testing allows all types of tires such as passenger car, truck, earthmover, and farm to be tested under closely monitored conditions. An industry proving ground will generally have the following test tracks and road courses available:

- High-speed tracks, either circular or oval.
- Interstate highway simulation.
- Gravel and unimproved roads.
- Cobblestone.
- Cutting, chipping, and tearing courses.
- Wet and dry skid pads, serpentine and slalom courses for aesthetics, and handling tests.
- Tethered tracks for farm tire durability.
- Glass roads for footprint monitoring.

For example, the 7250-acre Goodyear Proving Grounds in Texas contains a 5-mile high-speed circle, 8 miles of simulated interstate highway, gravel and rock courses for a range of tire type testing, skid pads with spray equipment, and a glass road facility for tire footprint observations and evaluation of water dispersion.

14.6.3 Commercial Evaluation

When a new tire design meets the performance targets identified in the laboratory and proving grounds, commercial evaluation is the next stage in the product development cycle. For truck tires for highway service, a quantity of tires will be placed with a commercial fleet, and their performance will be monitored continuously. Data collected from such tests include tread wear, uniformity of wear, casing durability, and driver assessment. Tires are placed with a range of commercial fleets in both short-haul and long-distance service. The collected data are then entered into computers for detailed analysis, calculation of regression equations, and performance evaluation. Customer input is one of the key parameters in such studies.

14.7 TIRE MANUFACTURING

The manufacture of tires consists of six basic processes:

1. Mixing elastomers, carbon blacks, and chemicals to form the rubber compound.
2. Processing the fabrics and steel cord and coating them with rubber at the calendaring operation.
3. Extruding treads, sidewalls, and other rubber components.

4. Assembling components on the tire building machine.
5. Curing the tire under heat and pressure.
6. Finishing, making the final inspection, and shipping.

The processes involved in the tire manufacturing operation are illustrated in Figure 14.22.

14.7.1 Compound Processing

Polymers are first broken down in an internal mixer where, in addition to the polymer, a peptizer may also be added. This stage is essentially a polymer molecular weight reduction phase. After initial breakdown of the polymer, carbon black, rubber chemicals, and oils can be added to the polymer at intervals to complete the compound formulation. Polymer breakdown and mixing generally occur at a high temperature, up to 180°C.

Compounds may also be mixed on open mills, but this takes considerably more time, batch weights are lower, and it is thus less efficient than use of internal mixers.

Degree of breakdown with both types of equipment is dependent on the "friction ratio" or the difference between the operating speeds of the front and back rolls (or rotors, for internal mixers). In addition, clearance, conditions of the rotor surfaces, pressure, and speed influence breakdown.

The mixing operation is designed to obtain uniform dispersion of all the compounding materials in a formulation. For every batch there is a defined mix period, temperature, mill or mixing speed, and sequence of material addition. Though the general guidelines on compound preparation for both mill mixing and internal mixers are similar, mill mixing has been replaced by internal mixers because of efficiency, automation, quality, and uniformity. For example, internal mixers can be computer controlled, allowing monitoring of power consumption, mix times, and batch drop temperatures and control of temperature gradients through a mixing compound at any point in the mix cycle.

The sequence of addition in an internal mixer is typically (1) polymers and peptizer; (2) plasticizers, most carbon black or silica, and oils; (3) balance of fillers and antioxidants; and (4) vulcanization system components.

Construction of a mix cycle is governed by a set of empirical rules:

1. Keep high-tack resins separate from dry powders.
2. Hold batch drop temperatures above the softening point of hard resins.
3. Contain liquids to prevent leakage.
4. Make use of the shear properties of rubber to accelerate mixing.
5. Avoid scorch and subsequent formation of cured particles and crumb.

These five guidelines are adhered to by reducing internal friction through the use of plasticizers, adequate breakdown of the rubber, holding of the curing system ingredients to the final stage of the mix cycle, and use of master batches

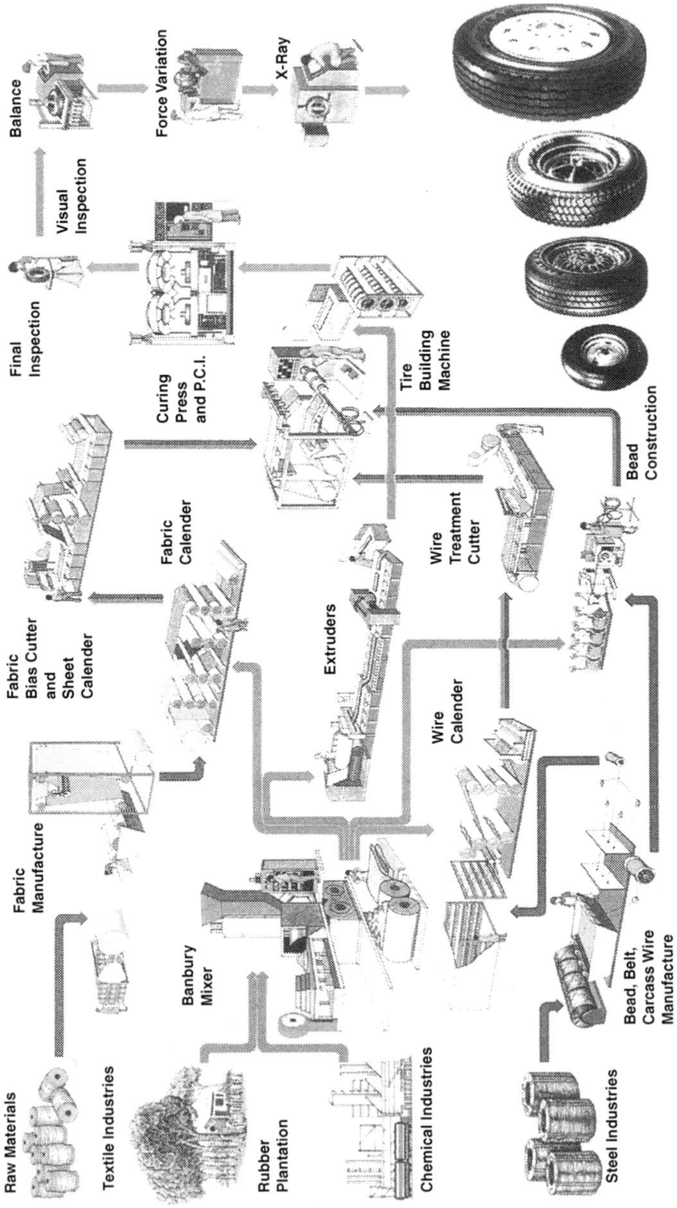

FIGURE 14.22 The tire manufacturing process.

where possible. After mixing, compounds can either be sheeted and water cooled or passed into a pelletizer and then air cooled.

14.7.2 Calendering

Calendering is the forming operation in which the rubber compound is sheeted or spread evenly onto fabric. The calender is a heavy-duty machine equipped with three or more chrome-plated steel rolls that revolve in opposite directions (Figure 14.23). The rolls are heated with steam or circulated water; the gearing allows the rollers to operate at variable speeds like the mill rolls. Fabric or wire is passed through the calender rolls, and compound is applied above and below to fully cover the material.

The amount of compound deposited onto a fabric or steel cord is determined by the distance between rollers and is monitored by Beta Gauges. Each cord is insulated on all sides with rubber. The finished treatment or calendered steel cord sheet is cut to length and angle for tire building. Key compound requirements in calendering operations are minimum shrinkage, optimum tack for component-to-component adhesion, and sufficient scorch resistance to enable long dwell times at high processing temperatures.

14.7.3 Extrusion

Many of the components of a tire, such as tread, sidewall, and apex, are formed by extrusion of the uncured or "green" rubber. Extruders in a tire manufacturing operation are conventional screw-type systems that fall into two categories:

- *Hot feed systems:* Strips of green compound are fed from warm-up mills into an extruder feed box and then into the barrel.
- *Cold feed systems:* Mixed compound is fed directly into the extruder system.

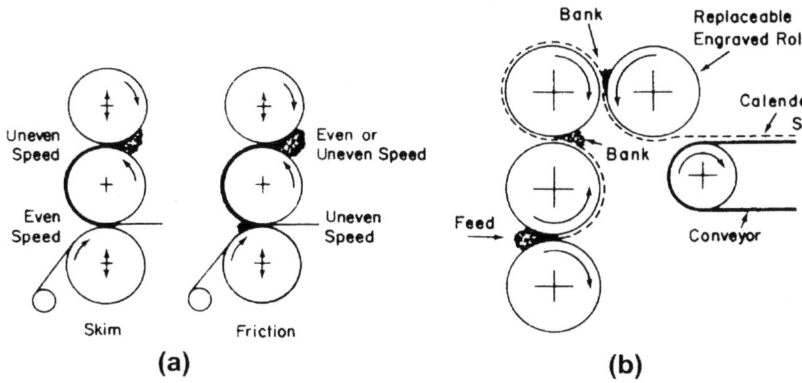

FIGURE 14.23 (a) Applying rubber to fabrics. (b) Profiling by means of a four-roll engraving calender.

Extruders consist of a barrel into which the material is fed. The mixed compound is then pushed forward by a specially designed screw to a filter system, where any foreign material is removed, and then onto a die, where the required compound profile is produced.

The extruder screw consists of three primary zones: feed zone, compression zone, and metering zone. In the feed zone located under the extruder feed box, the screw flights are well spaced to optimize compound flow. The rubber is passed into a compression zone, where the compound is heated through shear; then into a metering zone, where the compound is further heated to reduce viscosity; and finally into the die for profile formation.

Hot feed extruder systems (Figure 14.24) normally have a short barrel and screw with short compression and metering zones because the compound is already hot and has a low viscosity as it enters the feed hopper. Barrel length-to-diameter ratios from 4:1 to 6:1 are typical.

Cold feed extruders (Figure 14.25) have much larger length-to-diameter ratios because of the requirements of reducing green compound plasticity, heat buildup in the compound, and pressure buildup required to produce the extruded profile. The length-to-diameter ratio is typically 24:1. The modern cold feed extruder is also computer controlled, which enables adjustment of the compound temperature profile through the barrel, pressure control, flow rate, and feed rate. This provides accurate control over die swell, extrudate surface quality, and buildability of the extruded product.

Periodic monitoring of screw flight-to-barrel distances is important to ensure minimum back pressure and compound swirl or turbulence in the flow channel.

14.7.4 Tire Building

All tire components (i.e., extruded parts, calendered plies and belts, and beads) are assembled at the tire-building machine. Traditionally, radial tires are built on a flat drum. The initial stage entails application of the inner liner on the cylindrical drum, followed by the ply and any other additional barrier components that the tire engineer has designed into the manufacturing specification. The beads are then positioned and the ply is turned up over them. The structure is transferred to a second-stage machine, where the belts and tread are applied. The building drum is collapsed, and the "green tire" illustrated in Figure 14.26 is transferred to the curing or vulcanization presses.

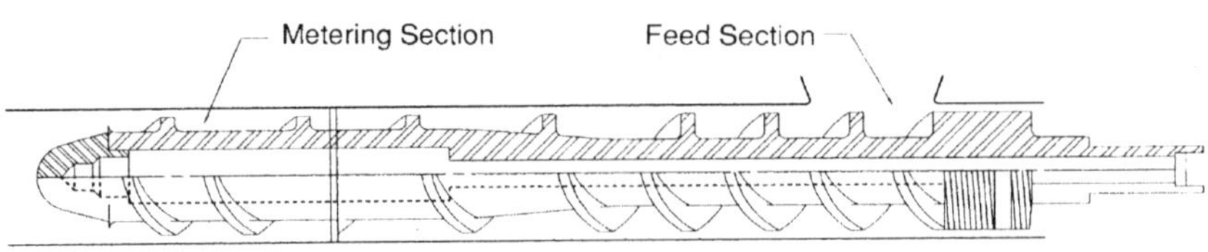

FIGURE 14.24 Hot feed extruder.

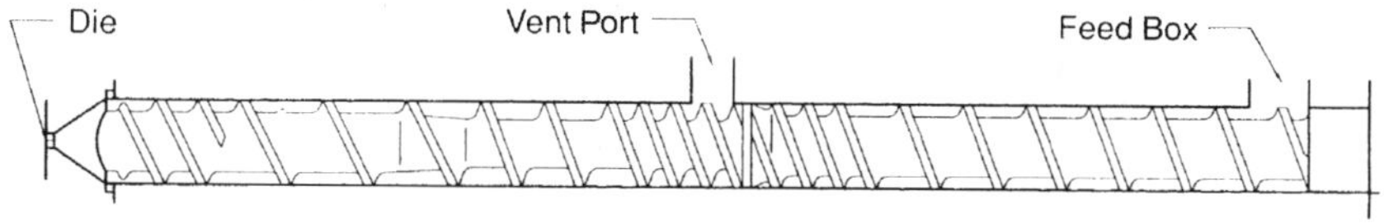

FIGURE 14.25 Cold feed extruder.

FIGURE 14.26 Radial green tire.

During vulcanization, or curing, the green tire is molded into a high-quality engineered product. In the curing press, the compound flows into the mold shape to give a design to the tread and the desired thickness to the sidewall. Good flow depends on the plasticity of the uncured stock. To flow, the compound must resist scorching. Flow must be complete before cure begins, or distortion may result. Proper flow is achieved by effective use of compounding fillers, adjustment of acceleration levels, and use of plasticizers.

14.7.5 Final Tire Inspection

The manufactured tire must meet the following criteria:

- All tire components have been successfully processed through the production facility to the tire-building machine.
- The components have come together satisfactorily at the tire-building machine without hindering the productivity of the tire-building operators.
- The tire meets the quality and performance goals set by the consumer.

Quality assurance is the last stage in the manufacturing cycle. Here the product is checked to ensure customer satisfaction. Quality checks on a tire include:

- Buffing and trimming off mold flash from the tire.
- Visual inspection of each tire for defects.
- X-ray checks on tires for ply cord spacing and belt layup.
- Statistical sampling of tires for durability testing, uniformity, and dynamic balancing; this testing includes many of the development tests reviewed earlier such as conicity, radial runout, and lateral force variation.

After the tire passes the quality checks, it is ready for shipment to the warehouse for distribution.

14.8 SUMMARY

In many ways, a tire is an engineering marvel. Geometrically, a tire is a torus. Mechanically, a tire is a flexible, high-pressure container. Structurally, a tire is a high-performance composite built using elastomers, fibers, steel, and a range of organic and inorganic chemicals.

Tire technology is a complex combination of science and engineering that brings together a variety of disciplines. In the development of a tire, knowledge in the areas of tire geometry, dynamic tire behavior, chemistry of component materials, and technology of composite structures is essential. The result is a broad range of products that satisfy vehicle manufacturers as well as end-consumer needs for optimum performance under a variety of service conditions.

REFERENCES

Bekaert Corporation, 1991. Steel Cord Catalogue. Akron, OH.
Clark, S.K., 1982. Mechanics of Pneumatic Tires. US Department of Transportation, Washington, DC.
Davison, J.A., 1969. Design and Application of Commercial Type Tires. SAE's 15th L. Ray Buckingdale Lecture, SP344.
Duddey, J., 2004. In: Rubber Compounding, Chemistry and Applications. Marcel Dekker Inc., New York.
European Tyre and Rim Technical Organization, 2004. Standards Manual. Brussels, Belgium.
Fitch, J.W., 1994. Motor Truck Engineering Handbook, 4th ed. Society of Automotive Engineers, Warrendale, PA.
Ford, T.L., Charles, F.S., 1988. Heavy Duty Truck Tire Engineering. SAE's 34th L. Ray Buckingdale Lecture, SP729.
Gough, V.E., 1968. Rubber Chem. Technol. 44, 988.
Kovac, F.J., 1978. Tire Technology, 5th ed. The Goodyear Tire & Rubber Company.
Long, H., 1985. Basic Compounding and Processing of Rubber. Rubber Division, American Chemical Society, Lancaster Press, Lancaster, PA.
Purdy, J.E., 1970. Mathematics Underlying the Design of Pneumatic Tires, 2nd ed. Hiney Printing, Akron, OH.
Rodgers, B., Waddell, W.H., 2004. Rubber compounding, In: Kirk-Othmer Encyclopedia of Chemical Technology, 5th ed. John Wiley & Sons, New York.

The Tire & Rim., Association, Inc, 2004. Year Book. Copley, OH.
van Ooij, W.J., 1984. Rubber Chem. Technol. 57, 421.
Waddell, W.H., Rodgers, M.B., Tracey, D.S., 2004. Tire Applications of Elastomers, Part 1: Treads. Rubber Division Meeting, Grand Rapids, MI, Paper H.

Chapter 15

Recycling of Rubbers

Avraam I. Isayev
Department of Polymer Engineering, The University of Akron, Akron, OH 44325-0301, USA

15.1 INTRODUCTION

Manufacturing of tire and other rubber products involves vulcanization process, an irreversible reaction between the elastomer, sulfur, and other chemicals producing crosslinks between the elastomer molecular chains and leading to the formation of a three-dimensional chemical network. The crosslinked elastomers are solid, insoluble, and infusible thermoset materials. This makes the direct reprocessing and recycling of used tires and waste rubbers impossible. Therefore, the environmental problems caused by used tires and other waste rubber products have become serious in recent years. In fact, Goodyear, who invented the sulfur vulcanization process more than 160 years ago (Goodyear, 1844), was also the first who initiated efforts to recycle cured rubber wastes through a grinding method (Goodyear, 1853). Even after so many years of efforts in recycling, the development of a suitable technology to utilize waste rubbers is an important issue facing the rubber industry (Read, 2012).

A large number of tires are scrapped each year. According to a recent survey of the Scrap Tire Management Council of the Rubber Manufacturers Association, approximately 291.8 million scrap tires were generated in the United States alone in 2009 (Table 15.1) (US Scrap Tire Markets, 2009). The market for scrap tires is consuming about 84.9% of that total amount (US Scrap Tire Markets, 2009) while the rest is added to an existing stockpile of an estimated 113.6 million scrap tires located around the United States (Table 15.2) (US Scrap Tire Markets, 2009). The stockpiled tires create serious fire dangers and provide breeding grounds for rodents, snakes, mosquitoes, and other pests, causing health hazards and environmental problems (Snyder, 1998). Moreover, the major use of scrap tires in the United States is to generate the so-called tire-derived energy by burning used tires. However, burning tires may create a danger of air pollution (Hous et al., 1995). About 40.3% of the consumed scrap tires have been burnt in 2009, and only 26.2% of the total consumed amount has been turned into ground tire rubber (GRT), which is the initial material

The Science and Technology of Rubber. http://dx.doi.org/10.1016/B978-0-12-394584-6.00020-0

TABLE 15.1 2009 US Scrap Tire Generation (US Scrap Tire Markets, 2009)

Tire Class	Millions of Tires	Market (%)	Average Weight (lbs)
Light Duty Tires	*259.1*	*88.81*	*22.5*
Passenger tire replacements[a]	189.50	64.94	
Light truck tire replacements[a]	27.80	9.53	
Tire from scrapped vehicles[b]	41.8	14.34	
Commercial Tires	*32.7*	*11.19*	*120*
Medium, wide base, heavy truck replacement tires[a]	13.0	4.45	
Tires from scrapped trucks and buses[b]	19.7	6.74	
Total Scrapped Tires	*291.8*	*100.0*	*33.4*

[a]*Factbook 2010: US Tire Shipment/Activity Report for Statistical Year 2009, Rubber Manufacturers Association.*
[b]*Ward's Motor Vehicle Facts and Figures, 2010. Includes the number of vehicles removed from service in the car/light truck, truck, and bus categories in 2009. Assume four tires scrapped from light-duty vehicles and five tires scrapped from trucks and buses.*

for the tire rubber recycling processes (US Scrap Tire Markets, 2009). Also, the management of other waste rubbers has become a growing problem in the rubber industry since over 150,000 tons or more rubber are scrapped from the production of nontire goods in the form of runners, trim, and pads (Klingensmith and Baranwal, 1998).

Waste tires and rubbers, being made of high-quality rubbers, represent a large potential source of raw material for the rubber industry. The main reasons for the low scale current application of tire and rubber recycling are: more stringent requirements for quality of rubber articles, and hence for that of reclaimed rubber; the substitution of raw rubber by other materials, for example by plastics in some cases; rising cost of reclaim production from tires and rubber waste due to the more stringent regulations for environmental protection; a comparatively high labor input into reclaim production; and as a result of all this, the high cost of reclaimed rubber (Makarov and Drozdovski, 1991). However, the increasing legislation restricting landfills is demanding the search for economical and environmentally sound methods of recycling discarded tires and waste rubbers. Continuous aggressive policies of the automotive industry are aimed to increase the usage of recycled plastic and rubber materials

TABLE 15.2 2009 US Scrap Tire Market Summary (US Scrap Tire Markets, 2009)

Market or Disposition	2005	2007	2009
Tire-derived fuel	2144.64	2484.36	2084.75
Ground rubber	552.51	789.09	1354.17
Land dispose	590.81	593.98	653.38
Used tires	n/a	n/a	371.25
Civil engineering	639.99	561.56	284.92
Reclamation projects	UNK	132.58	130.00
Exported[a]	111.99	102.08	102.10
Baled tires/market	UNK	UNK	27.76
Electric arc furnace	18.88	27.14	27.10
Baled/no market	42.22	9.31	15.57
Agricultural	47.59	7.13	7.10
Punched/stamped	100.51	1.85	1.90
Total to market	3616.11	4105.79	4391.05
Generated[b]	4410.73	4595.72	5170.50
% to market/utilized	82.0%	89.3%	84.9%
% managed (including baled and landfiled tires)	96.3%	102.5%	97.9%

[a]*RMA began tracking tires culled from scrap tire collection entering domestic used passenger and light truck used tire markets in 2009.*
[b]*RMA changed the basis for reporting scrap tire generated annually from state-provided data in 2005–2007 to a calculation of replacement market tires sold and vehicles scrapped in 2009.*

(Scrap Tire News, 1999a,b; Pryweller, 1999). This may serve as an example of the growing industrial demand for such technologies.

The main objective of this chapter is to provide an up-to-date account of the recycling of used tires and waste rubbers, including existing methods and emerging technologies of grinding, reclaiming, and devulcanization, and also the possibility for recycled rubber utilization into products. Rubber devulcanization is a process in which the scrap rubber or vulcanized waste product is converted using mechanical, thermal, or chemical energy into the state in which it can be mixed, processed, and vulcanized again (Franta, 1989). Strictly speaking, devulcanization in sulfur-cured rubber can be defined as the process of cleaving, totally or partially, poly-, di-, and monosulfidic crosslinks formed during the initial vulcanization (Warner, 1994). Devulcanization of peroxide- and resin-cured rubber can be defined as the process of cleaving carbon-carbon or other stronger crosslinks. However, in the present concept, devulcanization is defined as a process that causes the breakup of the chemical network along with the breakup of the macromolecular chains (Rader, 1995).

A number of methods (Pryweller, 1999; Warner, 1994; Rader, 1995; Fix, 1980; Nicholas, 1982; Myhre and MacKillop, 2002) have been applied in an attempt to solve the problem and to find more effective ways of tire rubber recycling and waste rubber utilization. These methods include retreading, reclaiming, grinding, pulverization, microwave and ultrasonic processes, pyrolysis, and incineration. Processes for utilization of recycled rubber are also being developed including the usage of reclaimed rubber to manufacture rubber products and thermoplastic/rubber blends and the usage of GRT to modify asphalt and cement (De et al., 2005).

15.2 RETREADING OF TIRES

Retreading is one method of recycling. Also, it saves energy: It takes about 83 l of oil to manufacture one new truck tire whereas a retread tire requires only about 26 l. The cost of a retread tire can be from 30 to 50% less than a new tire (Myhre and MacKillop, 2002; 2002 Fact Sheet, 2002). Approximately 24.2 million retreaded tires were sold in North America in 2001, with sales totaling more than $2 billion. Mostly medium- and heavy-duty truck tires, off-the-road vehicles, and aircraft tires were retreaded (2002 Fact Sheet, 2002). However, high labor costs and the potential for tougher safety regulations may hurt the retreading business (Myhre and MacKillop, 2002). Accordingly, the Rubber Manufacturing Association estimates that only 15.6 million tire casings were retreaded in the United States in 2006 (Scrap Tire Markets, 2009). These tires were used by commercial aircraft, commercial trucks, school buses, and off-the-road vehicles such as industrial, agricultural, and mining equipment. Three retreading systems are available based on renewed surface including (1) integral retreading that includes renewing the tread and sidewall; (2) semi-integral, which includes renewing the tread and part of the sidewall; (3) only retread according to an adhesion system that includes hot retread in which vulcanization is carried out at molding temperatures of 150–160°C and cold retread in which vulcanization is carried out in autoclaves at temperatures of 98–125°C (Ramos et al., 2011).

15.3 RECYCLING OF RUBBER VULCANIZATES

15.3.1 Reclaiming Technology

Reclaiming is a procedure in which the scrap tire rubber or vulcanized rubber waste is converted, using mechanical and thermal energy and chemicals, into a state in which it can be mixed, processed, and vulcanized again. The principle of the process is devulcanization. In devulcanization, it is assumed that the cleavage of intermolecular bonds of the chemical network, such as carbon-sulfur and/or sulfur-sulfur bonds takes place, with further shortening of the molecular chains (Rader, 1995).

Many different reclaiming processes (Makarov and Drozdovski, 1991; Warner, 1994; Fix, 1980; Nicholas, 1982; Myhre and MacKillop, 2002; Phadke et al. 1983; Siuru 1997; Accetta and Vergnaud 1982; De et al. 1999a) have been applied through the years in an attempt to solve the problem of rubber recycling. Generally ground rubber scrap is, in most cases, the feedstock for the devulcanization step. Warner (1994), Adhikari et al. (2000), and Isayev (2001c) presented reviews of the existing literature that is relevant to various methods of devulcanization.

Reclaiming is the most important process in rubber recycling. Many different reclaiming processes (Klingensmith and Baranwal, 1998; Makarov and Drozdovski, 1991; Rader 1995; Adhikari et al., 2000; Isayev, 2001c; LaGrone, 1986; Bowers et al. 1986; Knorr 1995; Lewis, 1986; Schaefer 1986; Schaefer and Berneking, 1986; Solov'ev et al., 1987; Szilard, 1973) have been used through the years depending on scrap characteristics and economics. Generally, ground rubber scrap is, in most cases, the feedstock for the reclaiming. The pan process, digester process (either wet or dry), and mechanical or reclaimator process are currently the common processes used for reclaiming.

The digester process (Klingensmith and Baranwal, 1998; Makarov and Drozdovski 1991; Warner 1994; Isayev 2001c; LaGrone, 1986; Bowers et al., 1986; Knorr, 1995; Schaefer, 1986; Schaefer and Berneking, 1986; Solov'ev et al., 1987; Szilard, 1973) uses a steam vessel equipped with a paddle agitator for continuous stirring of the crumb rubber while steam is being applied. The wet process may use caustic and water mixed with the rubber crumb while the dry process uses steam only. If necessary, various reclaiming oils may be added to the mixer in the vessel. The dry digester has the advantage of less pollution being generated and was adopted after the Clean Act and Water Act were enacted.

A mechanical or reclaimator process (LaGrone, 1986; Szilard, 1973; Klingensmith, 1991; Leyden 1991) has been used for the continuous reclaiming of whole tire scrap. Fine rubber crumb (typically 30 mesh) mixed with various reclaiming oils is subjected to high temperature with intense mechanical working in a modified extruder for reclaiming the rubber scrap.

Scrap rubber containing natural and synthetic rubbers can be reclaimed by digester process with the use of reclaiming oil having molecular weight between 200 and 1000 Da consisting of benzene, alkyl benzene, and alkylate indanes. The composition of this reclaiming oil and the improved digester process using such reclaiming oil has been patented (Bryson, 1979).

A technology for the devulcanization of sulfur-cured scrap elastomers was developed, (Kohler and O'Neill, 1997a,b) using a material termed "Delink" (Sekhar and Kormer, 1995), and this technique was designated as the Delink process. In this process, 100 parts of 40 mesh or finer rubber crumb is mixed with

2–6 parts of Delink reactant in an open two-roll mixing mill. Delink reactant is a proprietary material and its nature and composition was not disclosed. It should be noted that application of this technology to devulcanize synthetic rubbers is more difficult than natural rubber.

A simple process for reclaiming of rubber with a vegetable product that is a renewable resource material (RRM) was developed (De et al., 1997,2000,1999b). The major constituent of RRM is diallyl disulfide. Other constituents of RRM are different disulfides, monosulfides, polysulfides, and thiol compounds.

Diphenyldisulfide was reported to be an effective chemical agent, aiding devulcanization. Furthermore, devulcanization typically is done under a nitrogen atmosphere. Then, the devulcanized rubber is quenched in liquid nitrogen immediately to suppress the reaction of radicals or other reactive species in the devulcanizate with oxygen or with other reactive species (Rajan et al., 2006, 2005; Jalilvand et al. 2008; Saiwari et al., 2011). Sulfur vulcanized natural rubber (NR) can be completely recycled at a temperature range of 200 to 225°C by using diphenyldisulfide (Knorr, 1994). The efficiency of various disulfides recycling agents for NR and EPDM vulcanizates were reported (Verbruggen, 1999). Whereas complete devulcanization was observed on sulfur-cured NR at 200°C, a decrease of crosslink density by 90% was found when EPDM sulfur vulcanizates with diphenyldisulfide were heated to 275°C in a closed mold for 2 h. At the same time, EPDM cured by peroxide showed a decrease in crosslink density of about 40% under the same conditions.

Another chemical method was also proposed (Hunt et al., 1999), based on the use of 2-butanol solvent as a devulcanizing agent for sulfur-cured rubber under high temperature and pressure. It is claimed that the molecular weight of the rubber is retained and its microstructure is not significantly altered during the devulcanization process. However, the process is extremely slow and requires separation of the devulcanized rubber from the solvent. The process was proven using only a very small amount of rubber.

In addition to use of organic chemicals, rubbers can be devulcanized by means of inorganic compounds. Discarded tires and tire factory waste were devulcanized by desulfurization of suspended rubber vulcanizate crumb (typically 10 ~ 30 mesh) in a solvent such as toluene, naphtha, benzene, or cyclohexane in the presence of sodium (Myers et al., 1997). The alkali metal cleaves mono-, di-, and polysulfidic crosslinks of the swollen and suspended vulcanized rubber crumb at around 300°C in the absence of oxygen. However, this process may not be economical because the process involves swelling of the vulcanized rubber crumb in an organic solvent where the metallic sodium in molten condition should reach the sulfidic crosslink sites in the rubber crumb. Also, solvent may cause pollution and be hazardous. A technology was also proposed to reclaim powder rubbers using an iron oxide phenyl hydrazine based catalyst (Kawabata et al., 1981) and copper(I) chloride-tributylamine catalyst (Kawabata et al., 1979).

Depending on the specification of the finished products, fillers may be added to the devulcanized product before further processing. The devulcanized rubber from each process is then strained and refined as dictated by the specification of the finished product before being powdered, baled, sheeted, or extruded into the finished form.

Chemical reclaiming process is a possible method for devulcanizing the vulcanized network through the use of chemical agents that attack the C–S or S–S bonds. However, this process of devulcanization is very slow and creates further problems with the removal of the solvents and additional waste is generated in the form of sludges. Also, some processes require elaborate chemical processing techniques, therefore, handling and safety has become a concern.

15.3.2 Surface Treatment

This technology (Benko and Beers, 2002a,b,c) utilizes a solvent to treat (devulcanize) the surface of rubber crumb particles of sizes within about 20–325 meshes. It is a variation of earlier disclosed technology (Hunt et al., 1999). The process is carried out at a temperature range between 150°C and 300°C at a pressure of at least 3.4 MPa in the presence of solvent selected from the group consisting alcohols and ketones. Among various solvents, the 2-butanol exhibited the best ability to devulcanize sulfur-cured SBR rubber. Duration of the process is above 20 min.

Reported data on surface devulcanization experiments were obtained by treating small amounts of rubber crumb in the gas chromatography column. The solvent suitable for this process should have a critical temperature in the range of about 200–350°C. The process produces slurry of the surface devulcanized rubber crumb that has to be separated from the solvent. It is claimed that in this process a preferential breakage of S–S and C–S bonds takes place with little breakage of the main chains. The obtained surface modified rubber crumb was subjected to vulcanization as obtained and in blends with virgin rubber. The vucanizates exhibited a good retention of mechanical properties in blends with virgin rubber. However, this process has been tested on a small laboratory scale.

The surface modification of the ground rubber tire (GRT) powder was found to be suitable for enhancement of its adhesion to nitrile rubber (NBR) vulcanizates (Zhang et al., 2009c). By use of the atmospheric pressure dielectric barrier discharge, the hydrophobic surface of the powder transformed to a hydrophilic one as shown by the water contact angle, ATR-FTIR, and X-ray photoelectron spectroscopic (XPS) studies. After atmospheric plasma treatment, the improvement in the tensile strength and tear strength was observed for the modified tire powder-filled NBR vulcanizates. The latter was attributed to the enhanced interfacial interaction between modified GRT and NBR matrix.

Also, ozone, the isotope of oxygen, was applied to the grafting and oxidation of GRT powder (Fan and Lu, 2011b). Compared with conventional chemical grafting methods and irradiation modification techniques, functionalization through ozone treatment has certain advantages including simple equipment, low operation cost, and absence of various noxious chemical additives. Ozone treatment was suggested as an alternative method to modify the surface molecular structure of GRT powder.

15.3.3 Grinding and Pulverization Technology

Use of waste rubber in vulcanized state most often requires reduction of particle size and/or surface area. One widely used method for doing this with scrap tires and rubbers wastes is a grinding process. This method was invented and put forward by Goodyear about 160 years ago (Goodyear, 1853). Presently, there are three methods of grinding waste rubber: ambient grinding, cryogenic grinding, and wet-ambient grinding (Hershaft, 1972, 1977; Ratcliffe, 1972). There are a number of ways to reduce tires (De et al., 2005). The primary reduction of whole tires down to a manageable form is done using the guillotine, the cracker mill, the high impact hammer mill, and the rotary shear shredder. Vulcanized scrap rubber is first reduced to a 5×5 cm^2 or 2.5×2.5 cm^2 chip. Then a magnetic separator and a fiber separator (cyclone) remove all the steel and polyester fragments. This can then be further reduced using ambient ground mill or ground into fine particles while frozen using cryogenic grinding (Makarov and Drozdovski, 1991; De et al., 2005).

A method for obtaining fine-mesh rubber is cooling scrap tires in liquid nitrogen below their glass transition temperature and then pulverizing the brittle material in a hammer mill. Cryogenically ground rubber has a much finer particle size varying from 30 to 100 meshes. But for inexpensive rubbers such as tire rubbers, the process is not economical because of the amount of liquid nitrogen or other cryogen liquids needed to freeze the rubber (LaGrone, 1986). However, the process may be economical for expensive rubbers such as fluorocarbon rubbers. Little or no heat is generated in the process. This results in less degradation of the rubber. In addition, the most significant feature of the process is that almost all fiber or steel is liberated from the rubber resulting in a yield of usable product and little loss of rubber (Klingensmith and Baranwal, 1998).

Because of the high cost of cryogenic size reduction at liquid nitrogen temperature, mechanical size reduction by chopping and grinding is used often. The ambient process often uses a conventional high-powered rubber mill set at close nip. The vulcanized rubber is sheared and ground into small particles. Using this relatively inexpensive method it is common to produce 10–30 meshes material and relatively large crumb. In addition, multiple grinds can be used to further reduce the particle size. Ambient grinding produces an irregularly shaped particle with many small hair-like appendages that attach to the virgin rubber matrix producing an intimate bonded mixture (Szilard, 1973). The lower particle limit for the process is the production of 40 meshes material. The process, however, generates a significant amount of heat. Excess heat can degrade the rubber and if not cooled properly combustion can occur upon storage.

An environmentally friendly approach for partial decrosslinking of postvulcanized fluoroelastomer (FKM) rubber scraps through high-shear mechanical milling was developed for recycling of the FKM rubber (Zhang et al., 2011d). The gel fraction of FKM decreased from its original 97.8% to 79.7% after 32 cycles of milling. The FKM powders, obtained in this process, were compression molded into slabs exhibiting the tensile strength of 6.6 MPa and the elongation of break of 337.1% compared to those for the virgin FKM being 7.9 MPa and 368.7%.

Other suggested recycling processes include mechanical and thermomechanical methods, which only comminute the vulcanizates in rubber and do not devulcanize them. A process using a wet grinding method to achieve a crumb fineness of approximately 200 meshes has been reported (Lynch and LaGrone, 1986). When this product, which had a much higher surface-to-mass ratio, was devulcanized, no chemicals and only minimal heating and mechanical processing were required. Wet or solution process grinding may yield the smallest particle size ranging from 400 to 500 meshes. The advantage of fine particle wet ground rubber is that it allows good processing, producing relatively smooth extrudates and calendered sheets (Lynch and LaGrone, 1986).

The pulverization techniques for rubbers are also being developed based on the concept of polymer pulverization originally proposed for plastics. The process manufactures polymer powder using a twin-screw extruder imposing compressive shear on the polymer at specific temperatures that depend on the polymer (Enikolopian, 1985). Based on this method, the solid-state shear extrusion pulverization method of rubber waste is also proposed (Khait and Torkelson, 1999; Khait, 1994). The obtained rubber particles were fluffy and exhibited unique elongated shape.

This process is further developed to carry out pulverization of rubbers in a single-screw extruder to obtain particles varied in size from 40 to 1700 μm (Bilgili et al., 1999, 2000, 2001a,b). A schematic diagram of the pulverization technique based on a single-screw extruder is shown in Figure 15.1a and b (Bilgili et al., 2000). As indicated in Figure 15.1a, the extruder consists of three zones: feeding (Zone 1), compression (Zone 2), and pulverization (Zone 3). The screw is a square pitched with the compression zone having a uniform taper to create a compression ratio of 5. The water-cooling channel is located in the barrel in order to remove the heat generated by pulverization of rubber. Experimental studies showed that during the pulverization of vulcanized scrap rubber in the extruder, due to friction, a significant amount of heat is generated, leading to partial degradation of rubber (Bilgili et al., 2000). The rubber granulates are fed into the hopper of the extruder, conveyed into compression zone where they are subjected to high compressive shear. Under simultaneous action of this compressive shear and torsion due to the screw rotation, granulates

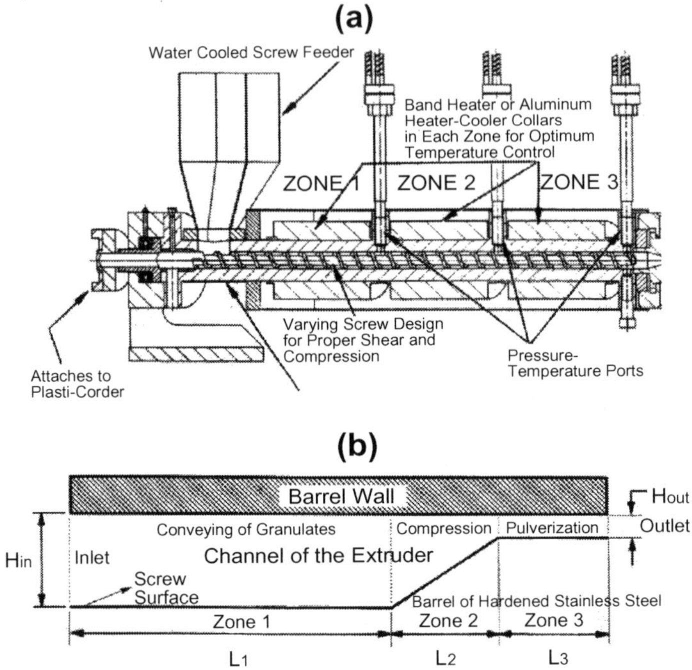

FIGURE 15.1 Schematic diagram of the single-screw extruder for pulverization of rubbers (a) and geometry of the screw channel with variable depth (b) (Bilgili et al., 2000).

are pulverized and emerged from the pulverization zone as rubber powder having smaller particle size. Surface oxidation of the rubber particles and initiation of agglomeration of a fraction of the produced particles may take place. The produced particles exhibit irregular shapes with rough surfaces and have a porous structure. The crosslink density and gel fraction of the particles are reduced in comparison with those of the initial rubber granulates. This would indicate the occurrence of partial devulcanization. Due to this effect, the particles obtained in this process can be molded into products after an exposure to high heat and high pressure for a period of at least 1 h (Arastoopour et al., 1999; Bilgili et al., 2003). Table 15.3 shows the dependence of the elongation at break, ε_b, tensile strength, σ_b, and crosslink density, ν, of compression-molded slabs of the original rubber vulcanizate and the vulcanizates prepared from particles of size in the range of 250–425 μm obtained by solid-state shear extrusion pulverization from discarded by-product of natural rubber (SMR-20) vulcanizates. Approximate composition of the rubber compound was about 54 wt.% of SMR-20, 27 wt.% carbon black (SRF), 11 wt.% aromatic oil, and 8 wt.% vulcanizing ingredients. Molding temperature and pressure were 157°C and 5.11 MPa, respectively. Slab F1, produced without adding sulfur curatives, exhibited the best failure properties among all slabs produced from

TABLE 15.3 Properties of the Slabs of Pulverized Rubber Waste (Bilgili et al., 2003)

	Revulcanizing System		ε_b (%)	σ_b (MPa)	ν (mol/m^3)
Slab Code	Sulfur (phr)	TBBS (phr)			
F1	–	–	360	10.3	50.4
F2	1.0	0.5	350	7.0	73.9
F3	1.0	–	320	8.2	69.5
Original	–	–	470	16.5	66.9

the rubber powder. In sample F1 oil, vulcanizing residue and the sol fraction of the rubber were removed by toluene extraction. This, according to authors, enhanced particle bonding leading to improvement of the failure properties. On the other hand, the slabs F2 and F3 produced by adding sulfur curatives to particles showed inferior failure properties than those of slab F1 due to less particle bonding at increased crosslink density during the revulcanization. Furthermore, the slab F1–F3 showed failure properties inferior to the original slab, indicating the inadequacy of compression molding of rubber particles to achieve the properties of original vulcanizate.

The particles obtained by other grinding processes can also be compression molded into slabs by means of high-pressure high-temperature sintering as shown in Morin et al. (2002) and Tripathy et al. (2002). In particular, in these papers rubber particles of several rubbers, obtained by various grinding methods, were compression molded into slabs with and without an addition of various acids and chemicals. The effect of time, pressure, and temperature on mechanical properties of sintered slabs was studied. In particular, Figure 15.2 shows the effect of molding temperature on mechanical properties of NR/SBR slab compression molded from particles of 80 meshes for 1 h at a pressure of 8.5 MPa (Morin et al., 2002). It clearly shows the importance of the molding temperature. Below approximately 80°C, this process does not work. The highest tensile strength of about 4 MPa was achieved with the sufficiently high elongation at break (about 800%). The mechanism of consolidation of particles in this process is the result of the breakup of bonds into radicals that cross the particle interface and react with other radicals, thus creating a chemical bond across the interface. Authors explained that the inferior properties of the sintered NR rubber particle slabs in comparison with those of the original one were due to the energetics between void propagation and strain-induced crystallization. Less energy is required to generate voids in the sintered slabs than in the original slab, and this does not allow one to achieve a strain-induced crystallization in the sintered slabs.

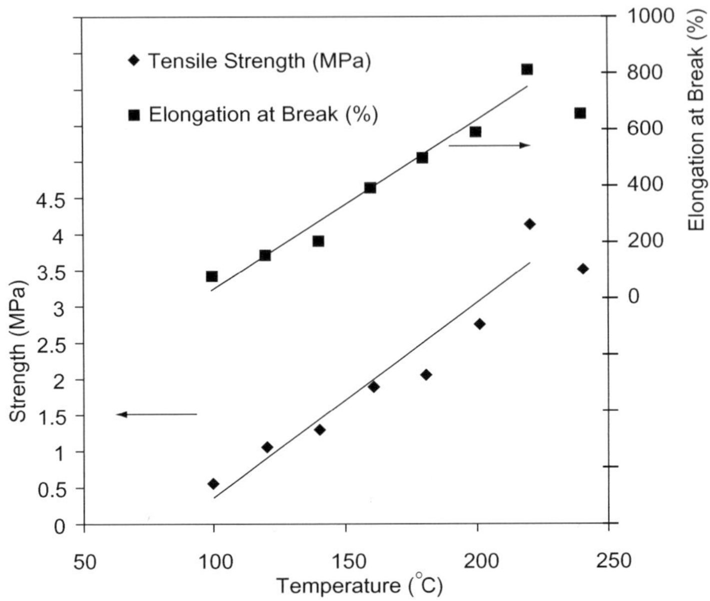

FIGURE 15.2 Effect of the molding temperature on the mechanical properties of NR/SBR slab compression molded from particles of 80 meshes for 1 h at a pressure of 8.5 MPa (Tripathy et al., 2002).

Direct powder compression molding was also utilized to produce large rubber pads from GRT powder without any addition of virgin materials or crosslinking agents. The effect of different powder size distributions on the final performances of the molded parts was investigated. The tensile tests and dynamic mechanical analyses on samples extracted from the pads indicated that the rubber powder distribution strongly affects the mechanical performances of the recycled rubber molded products (Gugliemotti et al., 2012a). Density and strength of rubber moldings and their esthetics were affected by the pressure imposed during compression molding. However, there was a limiting pressure above which the properties were not affected (Gugliemotti et al., 2012b).

15.3.4 Devulcanization Technology

(i) Microwave Method

Microwave technology was also proposed to devulcanize waste rubber (Fix, 1980; Novotny et al., 1978). This process applies the heat very quickly and uniformly on the waste rubber. The method employs the application of a controlled amount of microwave energy to devulcanize a sulfur vulcanized elastomer, containing polar groups or components, to a state in which it could be compounded and revulcanized to useful products, such as hoses, requiring significant physical properties. On the basis of the relative bond energies of carbon-carbon, carbon-sulfur, and sulfur-sulfur bonds, it was presumed that the scission of the sulfur-sulfur and sulfur-carbon crosslinks actually occurred. However, the material to be used in the microwave process must be polar enough to accept energy at a rate sufficient to generate the heat necessary for devulcanization. This method is a batch process and requires expensive equipment.

Thermogravimetry was also employed to study the changes occurring in rubber vulcanizates during devulcanization carried out by microwave treatment (Kleps et al., 2000). The degree of degradation of the polymer chains in response to microwave treatment was obtained, establishing the conditions of devulcanization in order to obtain the best properties of rubber devulcanizates for reuse in rubber processing.

(ii) Ultrasonic Method

Numerous publications are devoted in the past to the study of the effect of ultrasound under static conditions on polymer solutions (Basedow and Ebert, 1987; Price, 1990; Price et al., 1992; Schmid and Rommel, 1939; Schmid, 1940; Jellinek and White, 1951) and on polymer melts during extrusion (Isayev et al., 1987, 1990; Isayev, 1990; Isayev and Mandelbaum, 1991; Peshkovsky et al., 1983; Garcia and Isayev, 1991). Around the 1970s, the effect of vibration on polymer melts was studied by many researchers, as summarized in the review paper by Fridman and Peshkovsky (1990). Significant efforts have also been made to understand the mechanism of the effect of ultrasound on fluids (Suslick, 1989; Suslick et al., 1990) and degradation of polymer in solution (Gooberman, 1960).

The application of ultrasonic waves to the process of devulcanizing rubber is an attractive field of study. Most references indicate that rubber is vulcanized by ultrasound rather than devulcanized. Rubber devulcanization by using ultrasonic energy has been first discussed in Okuda and Hatano (1987). It was a batch process in which a vulcanized rubber was devulcanized at 50 kHz ultrasonic waves after treatment for 20 min under static conditions. The process claimed to break down carbon-sulfur bonds and sulfur-sulfur bonds, but not carbon-carbon bonds. The properties of the revulcanized rubber were found to be very similar to those of the original vulcanizates.

A novel continuous process has been developed for devulcanization of rubbers as a suitable way to recycle used tires and waste rubbers (De et al., 2005; Isayev, 1993, 2001a,b,c; Isayev and Chen, 1994, 1995; Tukachinsky et al., 1996; Levin et al., 1996, 1997a,b; Isayev et al., 1996a,b,c,d, 1997; Yushanov et al., 1996, 1998; Johnston et al., 1997; Diao et al., 1998, 1999; Tapale and Isayev, 1998; Yashin and Isayev, 1999, 2000; Hong and Isayev, 2001a, 2002b,c; Shim and Isayev, 2001, 2003; Yun et al., 2001, 2003; Yun and Isayev, 2003a,b; Shim et al., 2002a,b, 2003; Ghose and Isayev, 2003a; Feng and Isayev, 2004, 2005, 2006; Oh et al., 2004; Feng et al., 2004; Sun and Isayev, 2007, 2008; Massey et al., 2007). This technology is based on the use of the high-power ultrasound.

The ultrasonic waves of certain levels, in the presence of pressure and heat, can quickly break up the three-dimensional network in crosslinked rubber. The process of ultrasonic devulcanization is very fast, simple, efficient, and solvent- and chemical-free. Devulcanization occurs at the order of a second and may lead to the preferential breakage of sulfidic crosslinks in vulcanized rubbers. The process is also suitable for decrosslinking of the peroxide-cured rubbers and plastics. A schematic diagram of the various devulcanization reactors based on single-screw extruders suitable to carry out this process is shown in Figure 15.3. Initially, the so-called coaxial devulcanization reactor (Figure 15.3a) was developed in our laboratory. The reactor consists of a single-screw rubber extruder and an ultrasonic die attachment. A cone-shaped die and the ultrasonic horn have sealed inner cavities for running water for cooling. The shredded rubber is fed into the extruder by a feeder with adjustable output. Thus, the rubber flow rate in the process is controlled by the feed rate. An ultrasonic power supply, an acoustic converter, booster, and a cone-tipped horn are used. The horn vibrates longitudinally with a frequency of 20 kHz and various amplitudes. The ultrasonic unit is mounted onto the extruder flange. The convex tip of the horn matches the concave surface of the die, so that the clearance between the horn and the die is uniform. The clearance is controlled. The rubber flows through the clearance and under the action of ultrasonic waves, propagating perpendicular to the flow direction, it is devulcanized. The die plate and the horn are cooled with tap water.

Later, the barrel (Figure 15.3b) and the grooved barrel (Figure 15.3c) ultrasonic reactors were developed. In the barrel reactor, two ultrasonic water-cooled horns of rectangular cross sections were inserted into the barrel through two ports. Two restrictors made of bronze were placed in the barrel. These restrictors blocked the flow of rubber and forced the rubber to flow through the gap created between the rotating screw and the tip of the horn. In the devulcanization section, the larger diameter provided the converging flow of the rubber to devulcanization zone. The latter may enhance the devulcanization process. In the grooved barrel ultrasonic reactor, two helical channels were made on the barrel surface (grooved barrel). Rubber flows into the helical channel and passes through the gap created between the rotating shaft and the tip of the horns where devulcanization takes place.

Under the license from the University of Akron for the ultrasonic devulcanization technology, NFM Co. of Massillon, Ohio, has built a prototype of the single-screw extruder for ultrasonic devulcanization of tire and rubber products (Boron et al., 1996; Boron, 1999). It was reported that retreaded truck tires containing 15 wt.% and 30 wt.% of ultrasonically devulcanized carbon black-filled SBR had passed the preliminary dynamic endurance test (Boron, 1999).

Extensive studies on the ultrasonic devulcanization of rubbers and some preliminary studies on ultrasonic decrosslinking of crosslinked plastics were

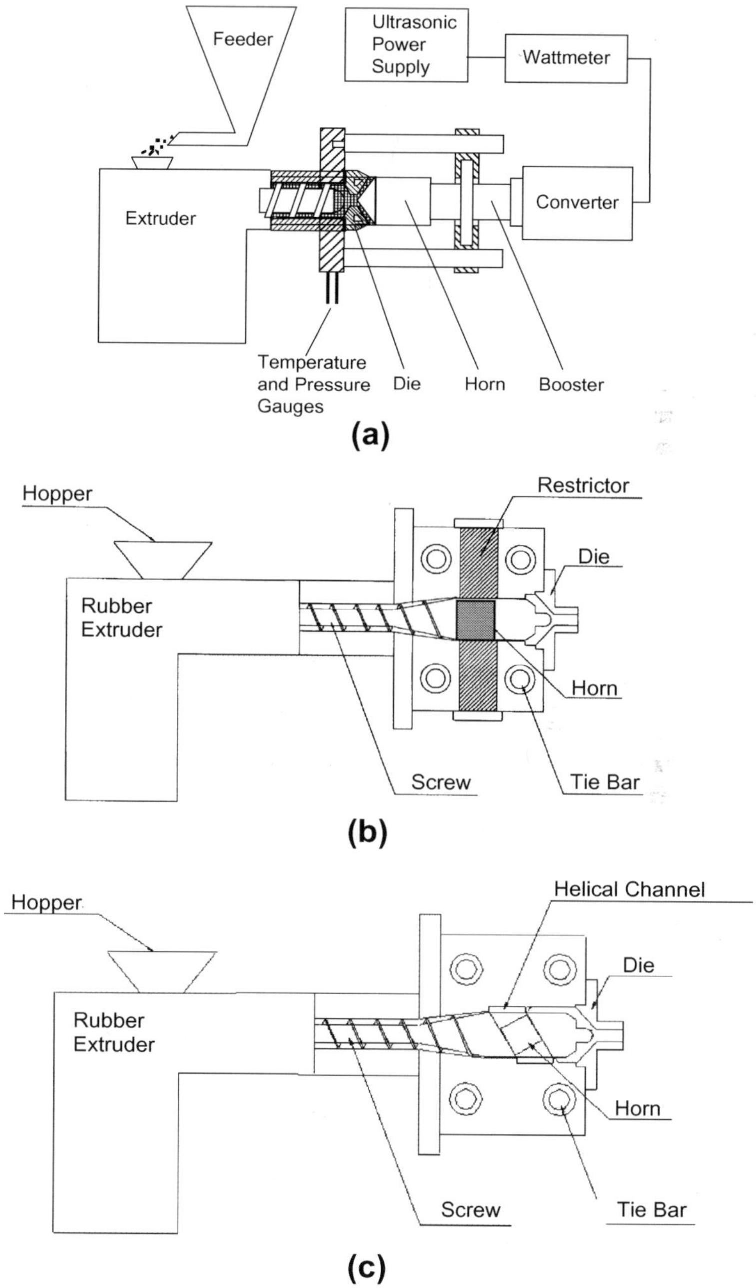

FIGURE 15.3 Schematic diagram of coaxial reactor (a), barrel reactor (b), and grooved barrel reactor (c) built for devulcanization of rubbers.

carried out (Isayev, 2001a,b; Isayev and Chen, 1994, 1995; Tukachinsky et al., 1996; Levin et al., 1996, 1997a,b; Isayev et al., 1996a,b,c,d, 1997; Yushanov et al., 1996, 1998; Johnston et al., 1997; Diao et al., 1998, 1999; Tapale and Isayev, 1998; Yashin and Isayev, 1999, 2000; Hong and Isayev, 2001a, 2002b,c; Shim and Isayev, 2001, 2003; Yun et al., 2001, 2003; Yun and Isayev, 2003a,b; Shim et al., 2002a,b, 2003; Ghose and Isayev, 2003a; Feng and Isayev, 2004, 2005, 2006; Oh et al., 2004; Feng et al., 2004; Sun and Isayev, 2007, 2008; Massey et al., 2007; Boron et al., 1996; Boron, 1999; Gonzalez de Los Santas et al., 1999). It was shown that this continuous process allows one to recycle various types of rubbers and thermosets. As a most desirable consequence, ultrasonically devulcanized rubber becomes soft, therefore enabling this material to be reprocessed, shaped, and revulcanized in very much the same way as the virgin rubber. This technology has been used successfully in the laboratory to devulcanize a ground tire rubber (GRT) (Isayev and Chen, 1994, 1995; Isayev et al., 1996d; Yushanov et al., 1996; Shim and Isayev, 2001; Oh et al., 2004), unfilled and filled NR (Diao et al., 1998; Yashin and Isayev, 2000), guayule rubber (Gonzalez de Los Santas et al., 1999), unfilled and filled SBR (Isayev and Chen, 1995; Isayev et al., 1996b; Johnston et al.,1997; Isayev et al., 1997; Levin et al., 1997b; Diao et al., 1998), unfilled and filled peroxide-cured silicone rubber (Yushanov et al., 1998; Tapale and Isayev, 1998; Hong and Isayev, 2001a; Yun and Isayev, 2003a; Shim et al., 2002a,b; Yun et al., 2003), unfilled and filled EPDM and EPDM roofing membrane (Yun and Isayev, 2003a,b), unfilled polyurethane (Ghose and Isayev, 2003a), unfilled resin-cured butyl rubber (Yun and Isayev, 2003b), used tire curing bladder (Feng and Isayev, 2004), unfilled and filled synthetic isoprene rubber (Sun and Isayev, 2007, 2008), fluoroelastomer, ethylene vinyl acetate foam, and crosslinked polyethylene (Isayev, 1993; Isayev and Chen, 1994). After revulcanization, rubber samples exhibit good mechanical properties, which in some cases are comparable to or exceeding those of virgin vulcanizates (Johnston et al., 1997; Tapale and Isayev, 1998; Hong and Isayev, 2002c).

Ultrasonic devulcanization studies were concerned with finding the effect of processing parameters such as the pressure, power consumption, die gap, temperature, flow rate, and ultrasonic amplitude on devulcanization, structural changes occurring in various rubbers, rheological properties and curing kinetics of devulcanized rubbers, and mechanical properties of revulcanized rubbers and the effect of design of the devulcanization reactor. Figure 15.4 shows the entrance pressure of devulcanization zone versus amplitude of ultrasound at a flow rate of 0.63 g/s, and the entrance pressure of devulcanization zone versus flow rate at the amplitude of 10 μm and clearance of 2 mm during devulcanization of GRT. The entrance pressure of the devulcanization zone was substantially reduced as the amplitude of ultrasound was increased. Ultrasound facilitated the flow of rubber through the gap not only because of reduction of the friction in the presence of ultrasonic waves but also because of the devulcanization taking place as GRT particles entered the devulcanization zone.

The barrel reactor showed a higher pressure in the devulcanization zone than the coaxial reactor and the grooved barrel reactor showed the lowest pressure at low amplitude of ultrasound and a flow rate of 0.63 g/s. The barrel reactor had a converging zone before the devulcanization zone. The GRT flow was essentially blocked by the restrictor of the devulcanization zone at low amplitude of ultrasound. However, at the ultrasound amplitude of 10 μm, the entrance pressure of the devulcanization zone for the coaxial and barrel reactors was almost the same due to a reduction of restrictor effect at high amplitude. The highest flow rate achieved in the barrel and grooved barrel reactors was 6.3 g/s. The devulcanized sample of flow rate of 6.3 g/s for the coaxial reactor could not be obtained due to an overload of the ultrasonic generator. Furthermore, in the grooved barrel reactor, at a flow rate of 6.3 g/s, the gap size needed to be increased to 3.5 mm and ultrasonic amplitude needed to be decreased to 6 μm due to an overload of the ultrasound unit. It was natural that the entrance pressure of devulcanization zone rises with increasing flow rate, as indicated in Figure 15.4 for all three reactors. Nevertheless, at high flow rate the barrel reactor had lower entrance pressure at the devulcanization zone than that of the other reactors at the ultrasound amplitude of 10 μm. The difference in die characteristics (pressure vs. flow rate) among three reactors having the thickness

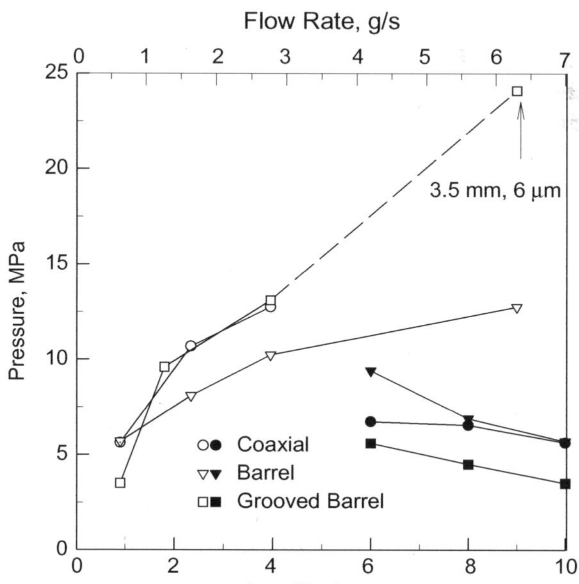

FIGURE 15.4 The entrance pressure of devulcanization zone of different reactors vs. amplitude of ultrasound at a flow rate of 0.63 g/s, and the entrance pressure of devulcanization zone vs. flow rate at the amplitude of 10 μm and clearance of 2 mm during devulcanization of GRT.

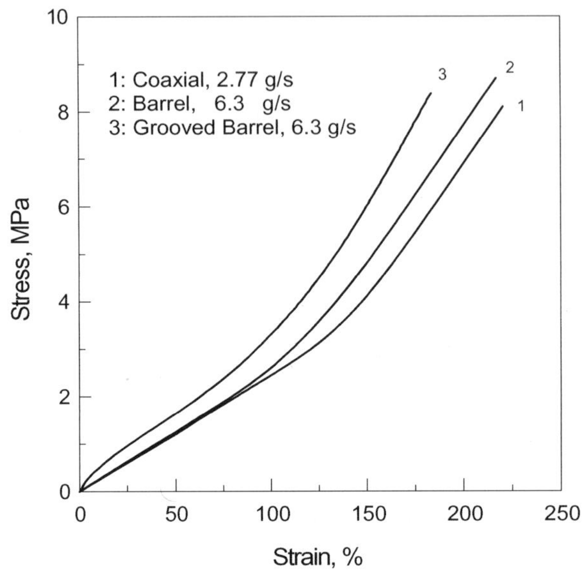

FIGURE 15.5 The stress-strain curves of the vulcanizates prepared from devulcanized GRT produced by the three reactors at the maximum flow rate as shown.

of devulcanization zone of 2 mm was possibly related to the difference in power consumption and the difference in shearing conditions. In the barrel and grooved barrel reactor, the GRT in the devulcanization zone was subjected to a pressure and drag flow while in the coaxial reactor to a pressure flow alone.

The comparison of stress-strain behavior of the vulcanizates prepared from devulcanized GRT produced by the three reactors at the maximum flow rate is shown in Figure 15.5. The revulcanized sample obtained from the barrel reactor having flow rate of 6.3 g/s shows a tensile strength of 8.7 MPa, elongation at break of 217%, and modulus at 100% elongation of 2.6 MPa. In addition, the revulcanized sample obtained from the grooved barrel reactor having flow rate of 6.3 g/s shows the tensile strength of 8.3 MPa, the elongation at break of 184%, and modulus at 100% elongation of 3.3 MPa. The output of the barrel and grooved barrel reactors was higher than that of the coaxial reactor. In addition, the mechanical properties of the sample obtained using the barrel reactor at the higher flow rate, which could not be achieved in the coaxial reactor, were higher. These properties met the higher level of specification made for tire reclaim (Klingensmith and Baranwal, 1998). The samples showing inferior performance were considered overtreated. The overtreatment meant a higher degree of devulcanization along with a significant degradation of the backbone molecular chains. The overtreated samples were usually softer and stickier.

It is believed that the process of ultrasonic devulcanization is based on a phenomenon called cavitation. In this case, acoustic cavitation occurs in a solid body. This is in contrast to cavitation typically known to occur in liquids in the regions subjected to rapidly alternating pressures of high amplitude generated by high-power ultrasonics (Blitz, 1967). During the negative half of the pressure cycle the liquid is subjected to a tensile stress and during the positive half cycle it experiences a compression. Any bubble present in the liquid will thus expand and contract alternately. The bubble can also collapse suddenly during the compression. This sudden collapse is known as cavitation and can result in almost instantaneous release of a comparatively large amount of energy. The magnitude of the energy released in this way depends on the value of the acoustic pressure amplitude and, hence, the acoustic intensity.

Although the presence of bubbles facilitates the onset of cavitation, it can also occur in gas-free liquids when the acoustic pressure amplitude exceeds the hydrostatic pressure in the liquid. For a part of the negative half of the pressure cycle the liquid is in a state of tension. Where this occurs, the forces of cohesion between neighboring molecules are opposed and voids are formed at weak points in the structure of the liquid. These voids grow in size and then collapse in the same way as gas-filled bubbles. Cavitation may be induced in a gas-free liquid by introducing defects, such as impurities, in its lattice structure.

In the case of polymer solutions, it is well known that the irradiation of a solution by ultrasound waves produces cavitation of bubbles (Basedow and Ebert, 1987; Suslick, 1988). The formation and collapse of the bubble plays an important role in the degradation of polymers in solution. Most of the physical and chemical effects caused by ultrasound are usually attributed to cavitation, the growth, and very rapid, explosive collapse of microbubbles as the ultrasound wave propagates through the solution. The intense shock wave radiated from a cavitating bubble at the final stage of the collapse is undoubtedly the cause of the most severe reactions. This shock wave is capable of causing the scission of macromolecules that lie in its path. The degradation arises as a result of the effect of the ultrasound on the solvent.

In any medium, cavities, voids, and density fluctuations exist. It is believed that these induce cavitation, leading to molecular rupture. In solid polymers, the microvoids present intrinsically are responsible for cavitation when they are subjected to a hydrostatic pressure in the manner of an impulse. One of the main causes of microvoid generation in polymer materials is the interatomic bond rupture when they are subjected to mechanical and thermal stresses. Extensive studies showing microvoid formation in stressed polymers have been carried out (Zhurkov et al., 1972).

When applied to rubbers, the cavitation usually corresponds to the effect of formation and unrestricted growth of voids in gas-saturated rubber samples after a sudden depressurization (Gent and Tompkins, 1969; Gent, 1990). In general, this has a broader sense and may be understood as the phenomena related to the formation and dynamics of cavities in continuous media (Isayev et al., 1996c; Yashin and Isayev, 2000). In materials science, for example, it means a fracture mode characterized by formation of internal cavities (Bever, 1986). In acoustics, the cavitation denotes the phenomena related to the dynamics of bubbles in sonically irradiated liquids (Young, 1989).

Structural studies of ultrasonically treated rubber show that the breakup of chemical crosslinks is accompanied by the partial degradation of rubber chain (Isayev and Chen, 1995; Tukachinsky et al., 1996; Levin et al., 1997b; Yushanov et al., 1998). The mechanism of rubber devulcanization under ultrasonic treatment is presently not well understood, unlike the mechanism of the degradation of long-chain polymer in solutions irradiated with ultrasound (Suslick, 1988). Specially, the mechanisms governing the conversion of mechanical ultrasonic energy to chemical energy are not clear. However, it has been shown that devulcanization of rubber under ultrasonic treatment requires local energy concentration, since uniformly distributed ultrasonic energy among all chemical bonds is not capable of rubber devulcanization (Isayev et al., 1995, 1996c,d).

It is well known that some amounts of cavities or small bubbles are present in rubber during any type of rubber processing (Kasner and Meinecke, 1996). The formation of bubbles can be nucleated by precursor cavities of appropriate size (Gent and Tompkins, 1969). The proposed models (Isayev et al., 1996a,c,d; Yashin and Isayev, 1999, 2000) were based upon a mechanism of rubber network breakdown caused by cavitation, which is created by high intensity ultrasonic waves in the presence of pressure and heat. Driven by ultrasound, the cavities pulsate with amplitude depending mostly upon the ratio between ambient and ultrasonic pressures (acoustic cavitation).

It is known that, in contrast to plastics, rubber chains break down only when they are fully stretched (Kinloch and Young, 1983; Kausch, 1987). An ultrasonic field creates high frequency extension-contraction stresses in crosslinked media. Therefore, the effects of rubber viscoelasticity have been incorporated into the description of dynamics of cavitation (Yashin and Isayev, 1999, 2000). The devulcanization of the rubber network can occur primarily around pulsating cavities due to the highest level of strain produced by the power ultrasound (Yashin and Isayev, 2000).

Generally, cleavage in polymer chains results in the production of macroradicals (Tabata and Sohma, 1980a,b), the existence of which has been confirmed spectroscopically by the use of radical scavengers such as diphenyl picrylhydrazyl (DPPH). Obviously, in the absence of scavengers, the macroradicals are free to combine by either disproportionation or combination termination, the former leading to smaller-sized macromolecules and the latter giving a distribution dependent upon the size of the combining fragments (Lorimer, 1990).

It was reported (Isayev et al., 1997) that under some devulcanization conditions the tensile strength of unfilled revulcanized SBR was found to be

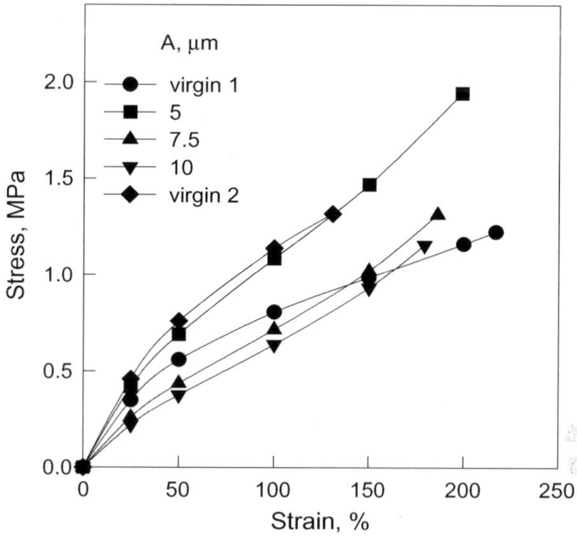

FIGURE 15.6 The stress-strain curves of unfilled virgin vulcanizates and revulcanized SBR obtained from rubbers devulcanized in coaxial reactor at various values of ultrasonic amplitude.

much higher than that of the original vulcanizate with elongation at break being practically intact. In particular, Figure 15.6 shows the stress-strain curves of unfilled virgin vulcanizates and revulcanized SBR obtained from devulcanized rubbers at various values of ultrasonic amplitudes, A. The devulcanized rubbers were obtained by using the coaxial ultrasonic reactor depicted in Figure 15.3a at the barrel temperature of 120°C, the screw speed of 20 rpm, flow rate of 0.63 g/s. The ultrasonic horn diameter was 76.2 mm. In contrast to usual findings that the mechanical properties of reclaimed rubber obtained by using different techniques are inferior to those of virgin vulcanizates, the present data are rather unexpected. It was proposed that the improvement in the mechanical properties of revulcanized SBR was primarily due to the extent of nonaffine deformation of bimodal network, which appears in the process of revulcanization of ultrasonically devulcanized rubber. The superior properties of revulcanized rubbers were also observed in the case of unfilled EPDM (Yun and Isayev, 2003a) and silicone (Diao et al., 1999) rubbers. Unfilled revulcanized NR rubber also shows good properties with the elongation at break remaining similar as in the original NR vulcanizates but with the ultimate strength being about 70% of the original NR. Interestingly, the strain-induced crystallization typical for the original NR vulcanizate is remained intact in revulcanized NR as indicated in Figure 15.7, where upturn of the stress-strain curves is observed in both the original and revulcanized rubbers. These samples were devulcanized in a coaxial reactor at different flow rates and ultrasonic amplitudes and revulcanized with recipe consisting of 2.5 phr ZnO, 0.5 phr stearic acid, and 2 phr sulfur.

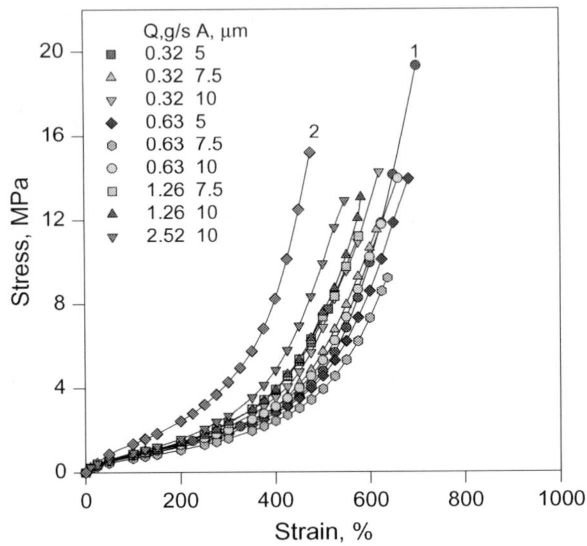

FIGURE 15.7 The stress-strain curves for unfilled NR vulcanizates prepared from ultrasonically devulcanized NR in coaxial reactor at various flow rates and amplitudes at a die gap of 2.54 mm and a barrel temperature of 120°C.

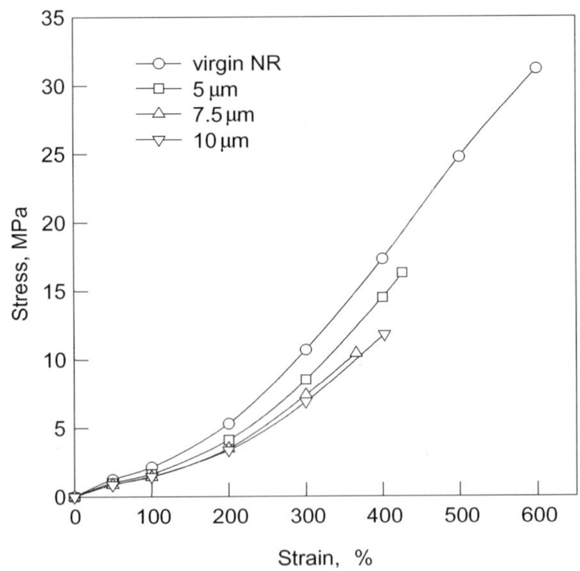

FIGURE 15.8 The stress-strain curves for 35 phr carbon black-filled virgin NR and revulcanized NR devulcanized in coaxial reactor at a barrel temperature of 120°C, a gap of 2.54 mm, a flow rate of 0.63 g/s, and various ultrasonic amplitudes.

It is of interest to establish a role that a filler plays in the devulcanization process. In this regard, Figure 15.8 shows the stress-strain curves for virgin and devulcanized 35 phr carbon black-filled NR vulcanizates. Virgin vulcanizates were cured using 5 phr ZnO, 1 phr stearic acid, 1 phr CBS, and 2 phr sulfur. Revulcanization recipe contained 2.5 phr ZnO, 0.5 phr stearic acid, 0.5 CBS, and 2 phr sulfur. The experiments have shown that upon filling rubbers with carbon black, after devulcanization the mechanical properties of revulcanized rubbers typically deteriorate with the level of deterioration depending on devulcanization conditions. This is clearly evident from Figure 15.8. It was suggested that ultrasonic devulcanization causes a partial deactivation of filler due to the breakup of macromolecular chains attached to the surface of carbon black. In many cases, this effect leads to inferior properties of revulcanized carbon black-filled rubbers. Thus, ultrasonically devulcanized rubber was blended with virgin rubber (Hong and Isayev, 2002a). The blend vulcanizates indicated significantly improved properties.

Also, attempts were made to add a certain amount of a fresh carbon black into the devulcanized rubber. It was shown that the vulcanizates containing a fresh carbon black exhibited better properties than the revulcanized rubber without an addition of fresh carbon black. However, in some cases, even carbon-filled devulcanized rubber shows mechanical properties similar to or better than the original rubber. In particular, this was shown for EPDM roofing membrane containing carbon black and significant amount of oil (Yun and Isayev, 2003b). Apparently, oil plays an important role in devulcanization process. Possibly, the presence of oil prevents a deactivation of the filler that was observed in vulcanizates not containing oil. But in order to prove this hypothesis further experiments are required.

Ultrasonic devulcanization also alters revulcanization kinetics of rubbers. It was shown (Isayev et al., 1996b) that the revulcanization process of devulcanized SBR was essentially different from those of the virgin SBR. The induction period is shorter or absent for revulcanization of the devulcanized SBR. This is also true for other unfilled and carbon black-filled rubbers such as GRT, SBR, NR, EPDM, and BR cured by sulfur containing curative systems, but not for silicone rubber cured by peroxide. It was suggested that a decrease or disappearance of the induction period in case of the sulfur-cured rubbers is due to an interaction between the rubber molecules chemically modified in the course of devulcanization and unmodified rubber molecules resulting in crosslinking. It was shown that approximately 85% of the accelerator remained in the ultrasonically devulcanized SBR rubber (Levin et al., 1997a).

Ultrasonically devulcanized rubbers consist of sol and gel. The gel portion is typically soft and has significantly lower crosslink density than that of the original vulcanizate. Due to the presence of sol the devulcanized rubber can flow and is subjected to shaping. Crosslink density and gel fraction of ultrasonically devulcanized rubbers were found to correlate by a universal master curve (Yushanov et al., 1996, 1998; Diao et al., 1999). This curve is unique for

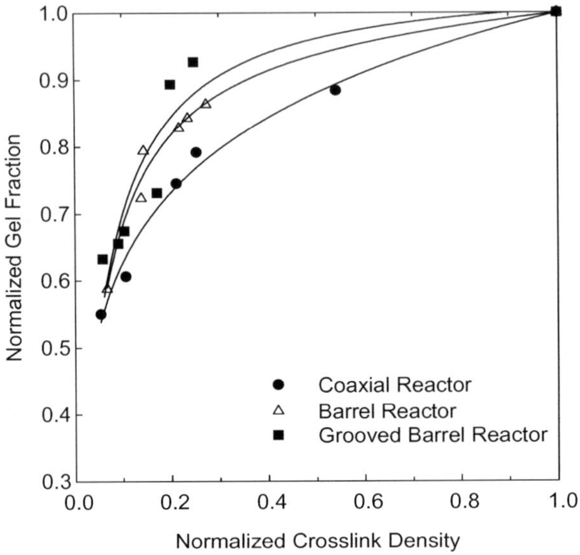

FIGURE 15.9 Normalized gel fraction vs. normalized crosslink density of devulcanized GRT obtained at various devulcanization conditions using the coaxial, barrel, and grooved barrel reactors.

every elastomer due to its unique chemical structure. Figure 15.9 presents the normalized gel fraction as a function of normalized crosslink density of devulcanized GRT obtained from three different reactors. The gel fraction and crosslink density of GRT was 0.82 and 9.9×10^{-2} kmol/m^3, respectively. For each reactor, the dependence of gel fraction on crosslink density was described by a unique master curve that was independent of a processing condition such as flow rate (residence time) and amplitude. The unique correlation between gel fraction and crosslink density obtained in the barrel and grooved barrel reactors was shifted toward lower crosslink density than those obtained in the coaxial reactor, indicating a better efficiency of devulcanization. Possibly, it is considered that additional shearing effect caused by the screw rotation in the barrel and grooved barrel reactors had a positive influence on improving the efficiency of devulcanization.

In search for a more economical process for recycling used tires, the ultrasonic devulcanization technology was further investigated to develop a feasible process to minimize the stockpiles of waste tires. This process induces the cleavage of the chemical networks through the combination of ultrasonic and chemical devulcanization (Kim et al., 2003). According to authors, the products from devulcanization of tires are carbon black and extended oil that can be used in many applications. A claim was also made that the technology can separate sulfur.

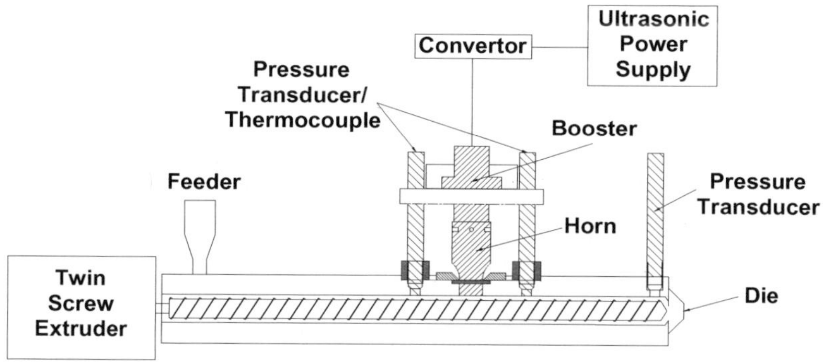

FIGURE 15.10 New ultrasonic corotating twin-screw extruder operating at 40 kHz and various amplitudes.

In addition, a continuous ultrasonic extrusion process was used for recycling of high resiliency of polyurethane foam (Ghose and Isayev, 2004). The foam decrosslinked at various screw speeds and at various amplitudes was blended at different ratios with the virgin polyurethane rubber (PUR) and cured. Blends were also prepared using different proportions of ground samples and virgin rubber. A comparison of the properties between these blends and the blends of the decrosslinked foam was carried out. The results showed that compared to the ground samples, the blends of the decrosslinked samples were easier to mix and exhibited enhanced mechanical properties.

Recently, ultrasonic devulcanization of tire rubbers was carried out in the presence of ozone (Lee et al., 2009a). This extruder was similar to the coaxial ultrasonic extruder depicted in Figure 15.3a. It was shown that by using treatment by ultrasound and ozone combination the mechanical properties of revulcanized rubbers were improved.

Very recently, a new ultrasonic corotating twin-screw devulcanization extruder was developed (Figure 15.10). In this extruder an ultrasonic horn is installed in the barrel. Two screws with a diameter of 16 mm convey waste rubber powder to a devulcanization zone. Devulcanization occurs in the ultrasonic zone, where ultrasonic waves with a frequency of 40 kHz are imposed to the rubber by the horn. The rubber passing through the gap is subjected to longitudinal ultrasonic waves being perpendicular to the flow direction. Some preliminary experiments were carried on micronized GRT and results indicated that indeed ultrasonic devulcanization in this ultrasonic twin-screw extruder is more efficient than in the ultrasonic single-screw extruder. Also, feeding a fine GRT powder, which was difficult in the single-screw extruder, becomes easy in this extruder. Figure 15.11 shows photographs of 40 mesh micronized GRT powder fed into the extruder along with the extrudate of devulcanized rubber, the compounded devulcanized rubber with curatives, and the revulcanized rubber sheet.

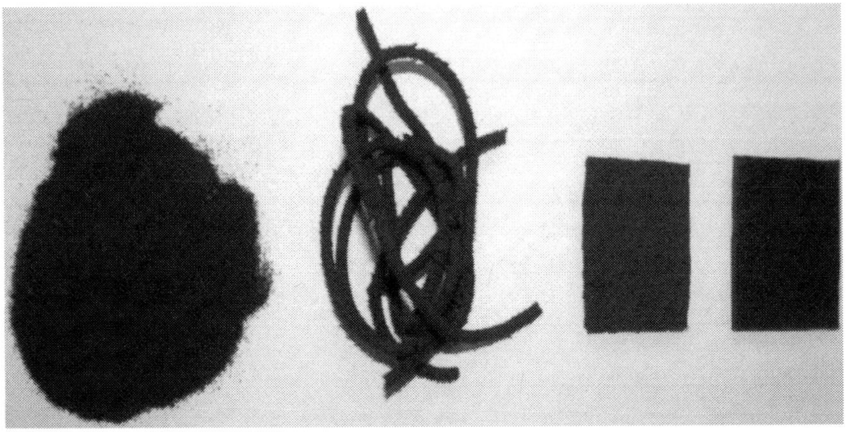

FIGURE 15.11 Photographs of GRT powder, and devulcanized and revulcanized rubbers. Devulcanization conditions in ultrasonic twin-screw extruder are: feed rate of 0.19 g/s, ultrasonic amplitude of 13 μm, frequency of 40 kHz, ultrasonic zone gap of 2.5 mm; temperature at various zones are from the feeding section to die 100/180/180/180/160/150°C.

15.4 USE OF RECYCLED RUBBER

15.4.1 General Remarks

There are certain technical limitations in the devulcanization of rubbers, and vulcanization is, in fact, not truly reversible (Pryweller, 1999). The partial devulcanization of scrap rubber will result in a degradation of physical properties. In many cases, this may limit the amount of substitution levels in high-tech applications such as passenger tires. But it can provide the compounder of less stringent products with an excellent low-cost rubber that can be used as the prime rubber or at very high substitution levels. According to Franta (1989), reclaim cannot be used for tread compounds in tires because every addition may decrease their resistance to wear. However, this statement has not been checked in case rubber devulcanized without an addition of chemicals. Considerable amounts of reclaim are consumed for carcasses of bias ply tires for cars if the compounds are of NR; for carcasses of radial tires no reclaim is added. On the other hand, reclaim is added to compounds for bead wires and it may also be added to sidewalls. Within the framework of direct recycling options a number of applications for GRT outside the rubber industry have been proposed. Such applications include the use as a filler in asphalt for the surface treatment of roads and as a rubberized surface for sport facilities.

The ground scrap rubber can be used as fillers in raw material (Rajalingam and Baker, 1992; Chopey, 1973; Gibala and Hamed, 1994; Gibala et al., 1996; Wolk, 1972; Theodore et al., 1998) and plastic compound (Rajalingam and Baker, 1992; Radusch et al., 1990). However, the problem of compatibility with the matrix and size of the filler, as well as the discontinuity at the interface between the two phases, should be considered. Rubber products containing ground rubber have low tensile properties due to insufficient bonding between the ground rubber and the virgin matrix. However, this bonding can be improved in the case of the addition of devulcanized rubber.

15.4.2 Use in New Tires

The tire is a complicated composite product consisting of tread, undertread, carcass, innerliner, bead, and sidewall. Many different types of rubber and carbon black reinforcement are used in manufacturing tires. Therefore, GRT is a blend of various rubbers and carbon blacks. Accordingly, in using GRT powder and devulcanized GRT in new tire manufacturing, many factors should be considered. Evidently, scrap tire powder can be used as a filler for virgin rubbers and devulcanized GRT can be used in blends with virgin rubbers. This market consumed approximately 1.354×10^6 tons of scrap tire rubber in 2009 (US Scrap Tire Markets, 2009).

Until recently, it was generally understood that only a few percent of ground rubber can be used in new tires. Scrap Tire Management Council reports that 5% of recycled tire rubber is used in an original equipment tire for the Ford Windstar. Although no other information on the amount of devulcanized rubber used in new tires is available in open literature, a possibility exists for the use up to 10 wt.% of recycled tire rubber in new tire compounds (Myhre and MacKillop, 2002). It was reported that actual road tests of truck tire containing 10 wt.% of the devulcanized rubber in the tread exhibited tread wear behavior almost equal to that for standard type with the new rubber compound (Fukumori et al., 2002). The increase in the amount of recycled rubber in tires is growing but it is likely that results will not be available for a number of years.

15.4.3 Rubber-Recycled Rubber Blends

The rubber particles from scrap tires can be incorporated into a virgin rubber as a filler. However, in this case the compatibility with the matrix is a significant issue. Rubber products containing ground rubber typically have lower tensile properties due to insufficient bonding between the ground rubber and the virgin matrix. The effect of GRT particles of different sizes incorporated in a NR compound on its mechanical properties was reported (Naskar et al., 2000). Table 15.4 shows the tensile strength, elongation at break, and tear strength

TABLE 15.4 Properties of GRT-Filled NR Vulcanizates (Naskar et al., 2000)

Properties	Compound									
	Control		A		B		C		D	
	Before Aging	*After Aging*	*Before Aging*	*After Aging*	*Before Aging*	*After Aging*	*Before Aging*	*After Aging*	*Before Aging*	*After Aging*
Tensile strength (MPa)	14.0	8.8 (63)	2.2	1.6 (71)	4.2	2.5 (60)	7.3	3.3 (45)	8.0	2.5 (31)
Elongation at break (%)	1175	770 (66)	430	230 (53)	620	360 (58)	780	410 (53)	860	400 (47)
Tear strength (kN/m)	28.2	20.3 (72)	12.4	9.7 (78)	18.5	10.6 (57)	23.8	11.5 (48)	21.2	9.7 (46)
Sol (%)	2.1	2.2	5.8	6.4	5.2	6.2	4.9	6.0	4.3	6.6
Increase in sol (%) after aging	–	4.8	–	10.3	–	19.2	–	22.4	–	53.5

Values in parentheses show the % retention of properties after aging at 100°C for 36 h.

before and after aging for a virgin vulcanizate and vulcanizates containing GRT particles of various sizes. The vulcanizates A, B, C, and D contained 30 phr of GRT particles of sizes in the range of 650–450 μm, 300–215 μm, 205–160 μm, and 150–100 μm, respectively. Control sample was the virgin NR vulcanizate. Curing recipe contained 6 phr ZnO, 0.5 phr stearic acid, 3.5 phr sulfur, and 0.5 phr MBT. Incorporation of GRT in a NR compound decreased the physical properties of the vulcanizate with the effect being larger in the case of large particles. However, the NR vulcanizate containing GRT exhibited better property retention upon aging. It was shown that smaller particles contain less the amount of rubber but higher the amount of fillers and metals.

Recycled tire rubber in the form of large crumb particles is also used for making prepackaged pour-in-place surfacing product (Scrap Tire News, 2001). GRT is combined with premixed polyurethane to produce a soft, pliable, energy-absorbing rubber surface for playground and other recreational surfaces and intended for placing over compacted gravel, concrete, or asphalt.

In recycling of rubbers it is customary to add various proportions of ground rubbers in the virgin material. Therefore, blends of both filled and unfilled ground and ultrasonically devulcanized rubbers with virgin rubber have been prepared. The mechanical properties of blends of unfilled PUR virgin and devulcanized rubber have been measured (Ghose and Isayev, 2003b). Curatives were added in the blends of devulcanized rubber based on the total rubber, whereas for blends of ground rubber curatives were added based on virgin rubber content. Figure 15.12 shows the tensile properties of the two types of blends of PUR (Ghose and Isayev, 2003b). From a comparison of these two figures it is quite apparent that the blends of the devulcanized samples (Figure 15.12a) have better tensile properties compared to the blends of ground samples (Figure 15.12b). For the latter, only at a ground concentration of 25%, the tensile properties are similar to those of the original, whereas for the former, the properties are superior to those of the original at this concentration and comparable at concentrations of 50%. It may be thought that ultrasonic devulcanization causes a better bonding of the devulcanized rubber to the virgin rubber in the blends than in the case of ground-virgin blends.

Ultrasonically devulcanized and ground CB-filled NR was blended with virgin CB-filled material. Figure 15.13 gives the tensile strength of these NR/NR blends containing 35 phr CB (Hong and Isayev, 2002a). The blends of devulcanized NR and virgin NR have much better tensile properties than blends with fully cured ground rubber. As the proportion of the virgin NR in the blends of devulcanized NR was increased, the mechanical properties progressively increased at or above the rule of mixture. Therefore, the mechanical properties of these kinds of blends can be improved significantly by ultrasonic devulcanization of ground vulcanizates. The modulus at 100% strain of NR/NR blends containing 35 phr CB was measured (Hong and Isayev, 2002a). Blends of devulcanized and virgin NR showed a little drop in the modulus, while the modulus of ground and virgin NR was reduced. It is thought

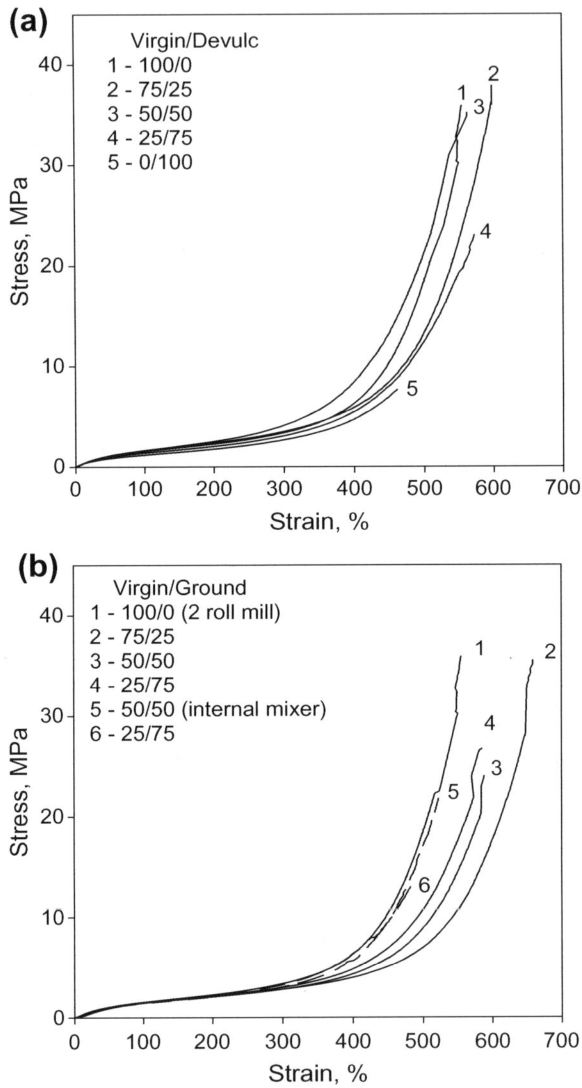

FIGURE 15.12 Stress vs. strain curves of vulcanized blends of virgin and devulcanized sample (a) and ground and virgin sample (b). The condition of devulcanization has a flow rate of 1.26 g/s, a gap size of 3 mm, an amplitude of 7.5 μm, and a barrel temperature of 120°C using the coaxial reactor.

that this was due to the migration of curatives, the blends of ground rubber and virgin rubber prepared using curatives for virgin rubber likely to exhibit reduced modulus. The blends of ground rubber and virgin rubber with curatives added based on total rubber content show higher modulus. However, they indicated lower tensile strength and elongation of break due to excess of curatives.

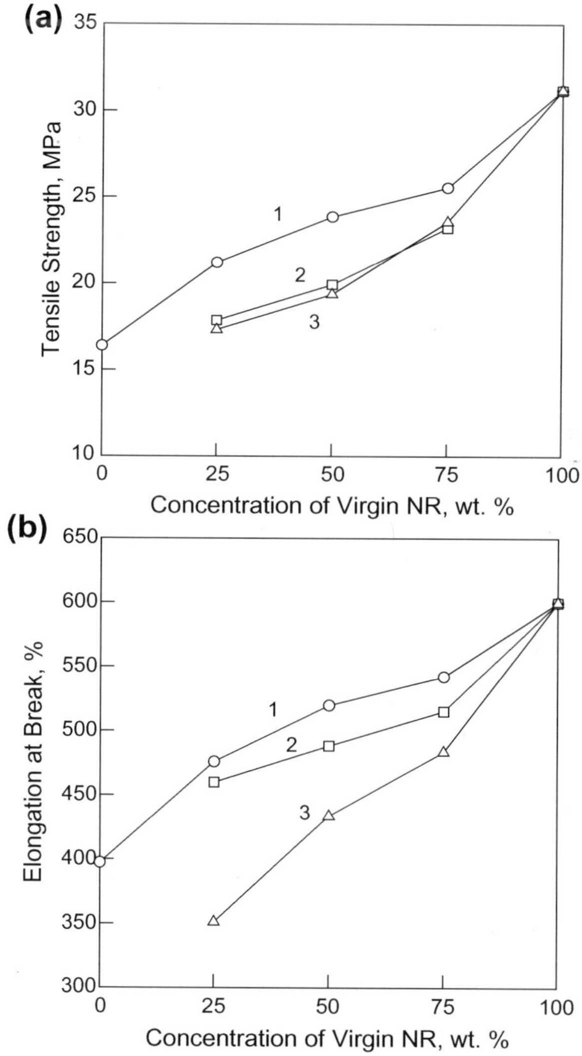

FIGURE 15.13 Tensile strength and elongation at break of 35 phr CB-filled virgin and devulcanized NR blends; (1) curatives were added to total rubber content; (2) blends of ground NR vulcanizates and virgin NR; curatives were added to virgin rubber and (3) blends of ground NR vulcanizates and virgin NR; curatives were added to the total rubber content; NR was devulcanized at a flow rate of 0.63 g/s, a gap size of 2.54 mm, an amplitude of 5 μm, a barrel temperature of 120°C using the coaxial reactor.

Waste rubber tire powder from truck and passenger tires was treated by various concentrations of nitric acid and 30% hydrogen peroxide (Yehia et al., 2004). Surface-modified rubber powder was used as a filler in NR mixes. The incorporation of rubber powder up to 20 phr in combination with a reinforcing carbon black in the NR mixes resulted in vulcanizates that retain at least 50% of their physicomechanical properties before and after aging. It was found that the surface treatment of powder improved the tensile strength and aging resistance of NR vulcanizates in comparison with untreated rubber powder.

Powder of 28–35 mesh of the SBR automotive extruded profile scraps was devulcanized using microwave for periods up to 3 min (Zanchet et al., 2012). The original and microwave devulcanized SBR powder of 80 phr were mixed with the virgin unfilled gum SBR and vulcanized. The absence of carbon black in the recipe of the vulcanizates was aimed to evaluate the effect of SBR scrap as a reinforcing filler on the mechanical properties of mixtures. It was shown that microwave devulcanized SBR for 2 min increased the tensile strength and elongation at the break and decreased the modulus at 100%, due to the improved interfacial adhesion between the devulcanized particles and the virgin rubber matrix. However, the retention of the properties after accelerated aging was less at early time of aging and stabilized at longer aging times due to the formation of a more stable crosslink.

Number of microorganisms exhibiting biological activity toward sulfur was utilized for the desulfurization of GRT (Bredberg et al., 2001; Kanagawa and Mikami, 1989; Romine and Romine, 1998; Sato et al., 2004; Li et al., 2011, 2012). These microorganisms can decrease the sulfur content or convert sulfur to sulfate, increasing the reactive sites on the surface of GRT. Such microbial desulfurization processes break only the surface sulfur crosslinks of GRT, and it is cost effective and environmentally friendly. GRT was subjected to biological desulfurization using various microorganisms isolated from the soil samples of sulfur-rich coal mines. The GRT was mixed with virgin NR in proportion up to 40 phr of powder and vulcanized. After treatment, oxygen content on the surface of waste rubber increased by 30%. Sulfur crosslinks were partially cleaved and converted to sulfate- or oxygen- containing sulfur-based groups. The formation of S–O bonds also took place. A decrease in the contact angle and crosslink density was observed. The latter was due to the migration of sulfur obstructed by the original crosslinks and the lower sulfur content of desulfurized powder. Mechanical properties of vulcanizates were improved due to the enhanced interfacial interaction and homogeneity (Li et al., 2011, 2012).

Solid-state mechanochemical milling at ambient temperature was used to manufacture devulcanized ground fluoroelastomer (FKM) ultrafine powder having particle size of 1 μm or less. This powder was added to acrylic rubber (ACM) to prepare blend compounds and vulcanizates. The ACM/devulcanized FKM blend compounds indicated curing behavior similar to that of the pure ACM. Also, it was shown that the mechanical properties of the ACM-ground FKM blend vulcanizates—their thermal aging and oil resistance—were improved in comparison with ACM. This was due to the good miscibility and superior dispersion of the devulcanized FKM in ACM matrix (Zhang et al., 2012a).

NR/SBR blends were filled with GRT particles of sizes ranging from 0.1 to 1 mm as replacement for soot. The best ratio of soot and rubber powder that would not affect mechanical properties was found (Dekic et al., 2012).

Blends of EPDM rubber with GRT powders of 80 meshes in a proportion of 70/30 with three polyfunctional monomers including diethylene glycol dimethacrylate (2G), trimethylolpropane trimethacrylate (TMPT), and trimethylolpropane triacrylate (A-TMPT) were prepared and cured by the electron-beam irradiation of doses ranging from 25 kGy to 100 kGy (Yasin et al., 2012). The solubility of functional monomers in EPDM/GRT was ranked as A-TMPT > 2G > TMPT. The highest tensile strength was obtained in blend vulcanizates containing A-TMPT.

Virgin EPDM rubber was blended with automotive industry waste EPDM rubber of particle sizes ranging from 0.252 to 0.358 mm at concentrations varying from 5 to 80 phr without and with incorporation of a 5 phr compatibilizer derived from the EPDM rubber modified through reaction with 1-dodecanethiol (Santos et al., 2011). The maximum and minimum torque of cure curve of the blends increased with an increase of concentration of the waste EPDM rubber, but this increase was less in the presence of the compatibilizer. The incorporation of the compatibilizer in the blends generally increased the tensile strength, elongation at break, and tear resistance. This was due to the better interaction between waste EPDM rubber and grafted chains of EPDM matrix and improved dispersion of the waste rubber particle.

Waste NR vulcanizate was devulcanized by benzoyl peroxide of various concentrations by means of chemical and mechanochemical processes (Rooj et al., 2011). The degree of devulcanization was evaluated by the crosslink density and Mooney viscosity. The devulcanized NR was mixed with virgin NR in different proportions and mechanical properties indicated that the waste NR could successfully be reused.

Recently, a number of studies were carried out on blends of NR with waste GRT devulcanized by various techniques (Premachandra et al., 2011; Edirisinghe et al., 2011; Xu et al., 1996; Zhang et al., 2009a; Maridass and Gupta, 2003; Maridass, 2009). Blends of virgin NR with reclaimed GRT of 40 mesh in 85/15 ratio were prepared (Premachandra et al., 2011; Edirisinghe et al., 2011). Environmentally friendly low-cost amino compound of various concentrations at a temperature of 30–70°C in the presence of processing oil using a laboratory two-roll mill was used at two different milling times. It was found that the amino compound acts as a devulcanizing or reclaiming agent for GRT. Prepared blends exhibited a higher viscosity, lower scorch resistance, and lower hysteresis in comparison with the control compounds containing GRT of 40 mesh. Vulcanizates of the blends exhibited the abrasion loss that was comparable to the virgin NR vulcanizate with the tensile strength and elongation at break being lower, but at a level acceptable for tire treads.

The solid-state mechanochemical milling using a pan mill was employed to devulcanize GRT (Xu et al., 1996). A decrease of the gel fraction and crosslink density confirmed the occurrence of partial mechanochemical devulcanization. The partially devulcanized GRT was easier to mix with virgin NR. Blends of virgin NR and GRT particles indicated a lower scorch time and optimum cure time. The latter was ascribed to the presence of more unsaturation in the GRT particles. Also, the blend vulcanizates with partially devulcanized GRT exhibited significantly improved mechanical properties, providing low-cost products with an excellent performance. In particular, at a concentration of a 10 wt.% devulcanized GRT the tensile strength of 23.2 MPa was achieved in comparison with 13.7 MPa for the blend with GRT particles (Zhang et al., 2009a).

The GRT particles of an average size of 190 μm were devulcanized in a counter-rotating twin-screw extruder (Maridass and Gupta, 2003) at rotation speeds varying from 10 to 40 rpm and temperatures from 150 to 200°C at a constant feed rate. The optimal conditions of devulcanization were found to be 30 rpm and 170°C. The devulcanized GRT was mixed with the virgin NR in various proportions and cured. Mechanical properties of blend vulcanizates indicated substantial increase in the tensile strength, elongation at break, and tear strength with concentration of the virgin NR, from 7.8 MPa, 165%, and 21 N/mm for the devulcanized GRT to 16.5 MPa, 830%, and 45 N/mm for the blend containing 60% of the virgin NR (Maridass, 2009).

15.4.4 Thermoplastic-Recycled Rubber Blend

The technology of plastic-rubber blending has emerged as a useful tool in tailoring polymers to the needs of the end users. An exciting development in blending is the introduction of thermoplastic elastomers (TPEs) based on plastic/rubber blends. These TPEs are becoming increasingly important because of their elastomeric properties, easy processability of the blends, and their lower cost.

The blending of waste rubber with thermoplastics is important from the point of view of both disposal of waste and the reduction in the product cost. More attention has been focused on compounding GRT with thermoplastics, which can be subsequently remelted and shaped into a wide range of molded and extruded products (Duhaime and Baker, 1991; Rajalingam et al., 1993). The mechanical properties of such compounds depend upon the concentration of GRT, polymer matrix type, and adhesion between the GRT and polymer matrix, as well as the particle size and their dispersion and interaction between the GRT and matrix (Deanin and Hashemiolya, 1987). Generally, adhesion between the GRT and polymer matrix and size of the GRT particles are the two major factors controlling the mechanical properties of such composites. Also, dynamic vulcanization techniques can be used to improve the properties of the blends of ultrasonically devulcanized GRT and thermoplastics (Luo and Isayev, 1998; Oh and Isayev, 2002; Hong and Isayev, 2001b; Michael et al., 1999).

Dynamic vulcanization is the process of vulcanizing the elastomer during its melt-mixing with the molten plastic (Coran and Patel, 1996).

The use of GRT instead of virgin elastomers, however, results in significant deterioration in the mechanical properties of these composites. It was reported (Hilyward et al., 1983; Phadke et al., 1984; Phadke and De, 1986) that the GRT has a detrimental effect on most of the physical properties of cured rubber, the extent of deterioration increasing with the amount and size of the GRT. There have been many investigations with the aim to improve the adhesion between the GRT and polymer matrix.

The effects of various compatibilizers to promote the adhesion of polyethylene (PE)/GRT blends were studied (Rajalingam and Baker, 1992; Oliphant and Baker, 1993; Pramanik and Baker, 1995a,b). They reported that it was possible to achieve a partial recovery of the same properties through a melt blending process where each component was first conditioned with compatibilizers of similar structure before the actual blending was carried out (Duhaime and Baker, 1991). Among the various compatibilizers, epoxidized natural rubber (Pramanik and Baker, 1995b), ethylene-co-acrylic acid copolymer (Rajalingam and Baker, 1992; Phadke and De, 1986; Oliphant and Baker, 1993), and ethylene-co-glycidyl methacrylate polymer (Duhaime and Baker, 1991, 1993) were found to be effective in improving the impact properties of PE/GRT composites. It was also reported that smaller GRT particle size results in a small increase in the impact property of the composite and has a greater influence on the melt processability of the composites. The percent improvement in the impact energy for linear low-density polyethylene (LLDPE) and the composites prepared from them was greater than that for the corresponding high-density polyethylene (HDPE) composites. It was also suggested that the low polarity and/or low crystallinity of the matrix polymer appeared to favor the compatibility with GRT (Deanin and Hashemiolya, 1987).

Modification of GRT particle surface has been studied to improve the compatibility of GRT and polymer. The polymeric surface modification can be carried out by chemical treatments like chromic acid etching (Briggs, 1980) or thermal oxidation (Briggs, 1978), or by mechanical means. Use of maleic anhydride-grafted and chlorinated GRT, respectively, improved physical properties of GRT/EPDM/acrylated high-density polyethylene and GRT/polyvinyl chloride blends (Naskar et al., 2002a,b). It was found (Stark et al., 1983) that surface treatment of ground rubber with a matrix of unsaturated curable polymer and a curing agent could also improve the performance of the blends. The effect of cryogenically ground rubber (CGR, approximately 250 microns) from old tires on some mechanical properties of an unsaturated polyester resin was investigated (Rodriguez, 1988). Composites made from silane-treated ground rubber showed better mechanical properties than composite made from untreated CGR. However, the particle size of the ground rubber was apparently too large to produce a toughening effect on the filled materials.

High-energy treatment including plasma (Chidambaram and Min, 1994; Campbell and Wise, 1964), corona discharge, and electron-beam radiation (Rajalingam et al., 1993) were used to modify the surface of GRT. The oxidation on the surface of GRT generated by chemical and physical treatment, such as occurs in plasma and autoclave in oxygen atmosphere, was shown to improve adhesion between GRT and polyamide in the blend (Chidambaram and Min, 1994). An epoxy resin compounded with tire rubber particles modified by plasma surface treatment was also studied (Xu et al., 1998). An improvement in mechanical properties of the resulting material over those containing the untreated rubber was observed. The effects of corona discharge treatment of GRT on the impact property of the thermoplastic composite containing the GRT were investigated (Deanin and Hashemiolya, 1987). X-ray photoelectron spectroscopy analysis showed that the corona discharge treatment of GRT increased the oxygen-containing groups on the ground rubber surface. In some composites it has been found that treated GRT marginally improved the impact property of the composites. However, prolonged times of treatment and higher power inputs for corona discharge of GRT reduced the impact strength of the composites.

The phenolic resin cure system and maleic anhydride-grafted PP compatibilizer significantly improved mechanical properties of PP/ultrasonically devulcanized GRT blends prepared by dynamic vulcanization (Luo and Isayev, 1998). Also, a new ultrasonic reactor was built and an improvement in the efficiency of the process and better properties of PP/GRT blends were achieved while carrying ultrasonic treatment during extrusion in this reactor (Oh and Isayev, 2002). In the reactor, two horns were placed in a slit die attached to a plastic extruder. Mechanical properties of PP/GRT, PP/devulcanized GRT (DGRT), and PP/revulcanized GRT (RGRT) that mixed in proportion of 40/60 are shown in Table 15.5. Comparison indicates that the tensile strength, Young's modulus, and elongation at break of the blend prepared by this reactor under certain conditions are higher than those obtained earlier (Luo and Isayev, 1998). Also, properties of PP/RGRT at 10 μm are higher than those of PP/GRT. Evidently, the ultrasonic treatment of PP/GRT blends led to a certain level of compatibilization at the interface between plastic and rubber phases. This was due to mechanochemical reactions induced by ultrasound at the interface. In addition, GRT was blended with ultrasonically devulcanized GRT with HDPE using a Brabender internal mixer and a twin-screw extruder (Oh and Isayev, 2002). These blends were dynamically vulcanized in these mixers. Also, HDPE and GRT blends mixed earlier by using a twin-screw extruder were passed through the ultrasonic devulcanization extruder and subsequently dynamically vulcanized by means of the internal mixer and the twin-screw extruder. The blends mixed by using the twin-screw extruder prior to devulcanization were found to have better tensile properties and impact strength than any other blends. Rheological properties of these blends were also studied.

TABLE 15.5 The Mechanical Properties of PP/GRT, PP/DGRT and PP/RGRT Blends (Oh and Isayev, 2002)

Blend		Tensile Strength (MPa)	Young's Modulus (MPa)	Elongation at Break (%)
PP/GRT		6.7	116	16.6
	2 horns, 5 μm	6.6	98	21.5
	2 horns, 7.5 μm	6.5	100	20.6
	2 horns, 10 μm	5.15	102	7.6
PP/DGRT	1 horn, 5 μm	6.9	110	19.3
	1 horn, 7.5 μm	6.6	104	17.4
	1 horn, 10 μm	7.0	116	21.0
	Earlier work (Kasner and Meinecke, 1996)	5.2	108	20.7
	2 horns, 5 μm	6.7	109	13.2
	2 horns, 7.5 μm	6.7	110	13.3
	2 horns, 10 μm	5.9	111	7.0
PP/RGRT	1 horn, 5 μm	6.4	110	11.8
	1 horn, 7.5 μm	6.6	110	13.3
	1 horn, 10 μm	7.2	122	14.8
	Earlier work (Kasner and Meinecke, 1996)	6.7	110	18.0

Blends of GRT and recycled HDPE from used milk containers were studied and patented (Coran and Howard, 1999; Howard and Coran, 2000). Effects of GRT particle size and concentration on mechanical and rheological properties were determined. The blend systems were optimized by a soft rubber-plastic binder produced from a mixture of HDPE and EPDM, wherein EPDM is dynamically vulcanized during its mixing with the HDPE. It was concluded that the softening of the HDPE binder provides compositions of improved ultimate mechanical properties.

Blends of ABS resin with ground rubber from tire tread were developed (Wu et al., 2007). In particular, two methods were utilized. In the first method, a two-step process was used in which the rubber was first grafted with chlorine molecules and then mixed with ABS to provide a compatibility and interaction with ABS matrix. In the second method, in situ compatibilization of chlorinated rubber and ABS matrix was carried out through the addition of Lewis acid catalyst ($AlCl_3$, $FeCl_3$, $ZnCl_2$) to induce interfacial interactions. The effect of surface treatment was characterized by XPS, indicating the occurrence of oxidation and chlorination on the surface of rubber powder. It was noted that simple mixing of rubber powder leads to tensile properties being inferior of those of the ABS resin. In contrast, a significant improvement in the tensile properties of the composite materials over the untreated rubber powder/ABS sample was observed by a two-step treatment process involving surface chlorination followed by grafting of either amine-terminated butadiene acrylonitrile (ATBN) or triethanolamine zirconium chelate molecules, or via in situ compatibilization of the chlorinated rubber powder and the ABS matrix with the addition of up to 2% of Lewis acid catalysts.

Properties of PP mixed with UV surface-grafted GRT by allylamine in the presence of a benzophenone photoinitiator were investigated (Lee et al., 2009b). A decrease of the dispersive component and an increase of the polar component of the GRT powder indicated an improvement of the surface activity. Blending of unmodified and modified GRT powder with PP and maleic anhydride-grafted PP was carried out. Improved mechanical properties for modified GRT powder/PP-g-MA blends attributed to the increased compatibility due to a chemical reaction between the maleic anhydride-grafted PP with the modified GRT powder. SEM photomicrographs showed the absence of interaction between PP matrix and untreated GRT powder and its presence in the case of the allylamine-grafted GRT powder.

The pretreatments of micronized GRT particles of 400–600 μm were performed by sulfuric acid, trichloroisocyanuric acid (chlorination), and silane gamma-methacryloxypropyltrimethoxysilane (silane A-174) as a coupling agent (Colom et al., 2006). Blends of HDPE with untreated and treated GRT at concentrations up to 40 wt.% were prepared using a two-roll mill at a temperature of 153°C. In all blend compositions, the chlorination led to the tensile strength being inferior to the untreated and other pretreated blends. Among various blends, silane-coupling agent-treated blends showed the highest strength, even being above that of HDPE when concentration of rubber was below 30%. However, the elongation break of all blends was always lower than that of HDPE. The improvement in the tensile strength was due to a chemical interaction of methyl, methylene, and silane with the matrix. In the subsequent study (Colom et al., 2007), three oxidant acids (H_2SO_4, HNO_3, and $HClO_4$) were used for pretreatment of the micronized GRT. The pretreatment with H_2SO_4 or HNO_3 provided blends with HDPE of a greater stiffness due to the improved interaction and the extraction of additives from the rubbers. However, the $HClO_4$ pretreatment did not improve the properties. Blends of recycled HDPE with GRT of various particle sizes were also studied after their pretreatment using various acids (Colom et al., 2009).

Tensile properties of blends were evaluated to determine the effect of particle size and acid pretreatments. It was found that, although pretreatment improved properties of blends, the effect of the particle size was more dominant than that of the acid pretreatments. Blends of HDPE with micronized GRT were also obtained by using wetting additive, low molecular weight unsaturated polycarboxylic acid, and polyethylene wax (Cañavate et al., 2010). PE wax was adsorbed onto the rubber particle surface and protruded into the HDPE matrix, forming fine wax film over the rubber particles. The wetting additive and the wax gave the best performance of the blends at rubber contents of 10%, 20%, and 40%. However, it should also be mentioned that the additives alone also produced an increase in the Young's modulus and tensile strength of the HDPE matrix.

Blends of LDPE with 30 wt.% of GRT of 0.2–0.5 mm without and with content of a 20 wt.% of EVA compatibilizer were prepared in a Brabender twin-screw extruder at a temperature setting of 165–175°C and 10 rpm (Meszaros et al., 2012a). The molded samples were prepared and treated by high-energy electron beam (EB) irradiation up to 200 kGy. The EB irradiation increased the tensile strength and elongation at break of moldings with a slight reduction of the modulus. Cyclic and falling weight impact tests of the moldings showed that the EB irradiation increased the energy absorbed by blends. The modest change in hardness was also observed, proving the occurrence of crosslinking by the EB treatment. The DMTA data confirmed the occurrence of compatibilization between GRT and LDPE in the presence of EVA and by EB treatment. The study was also conducted on recycled LDPE/GRT of a ratio of 40/30 with an addition of virgin LDPE and EVA from 0 to 30 wt.% to the blends (Meszaros et al., 2012b). Rubber-like behavior was improved with increasing EVA content with EB irradiation being very effective for blends containing high amounts of EVA.

The solid-state mechanochemical milling of PP/GRT blends was carried out, leading to the improved dispersion, as confirmed by the fluorescence microscopy and polarized optical microscopy (POM) observations (Zhang et al., 2012b). After 20 cycles of milling, the tensile strength and elongation at break of PP/GRT blends at a ratio of 100/40 were increased by 13.8% and

1200% in comparison to the untreated blend. This increase in properties after mechanochemical milling was due to the presence of smaller and less ordered spherulites of PP as was observed by POM and WAXS analysis.

GRT particles of sizes of lower than 500 μm and between 500–1000 μm were blended with a postconsumer PP and high impact polystyrene (HIPS) at ratios of 90/10 and 70/30 (Montagna and Santana, 2012). The rheological, physical, and mechanical properties of blends were measured. The addition of GRT reduced the mechanical properties with the reduction being dependent on the particle size. The incorporation of GRT particles into the recycled HIPS and PP matrix led to a decrease and an increase of viscosity, respectively. The extent of the effect was dependent on particle size.

TPV nanocomposites of LLDPE/reclaimed rubber with nanoclay and 1 wt.% MA-grafted PE and curative were prepared using a Brabender internal mixer at 170°C (Razmjooei et al., 2012). Contents of the reclaimed rubber, nanoclay, and compatibilizer were varied up to 30, 7, and 21 wt.%, respectively. The blends without the compatibilizer were also prepared. Morphological, thermal, and mechanical properties of the nanoclay-reinforced TPV nanocomposites indicated intercalation and partial exfoliation by the high-shear stress during mixing with the reclaimed rubber. Vulcanization of rubber phase led to an increase of viscosity. The size of rubber particles in TPV was reduced with the addition of nanoclay and compatibilizer.

The physicomechanical behavior and flammability of waste PE/GRT particles of 80 mesh at a ratio of 60/40 containing MA at 2% and antimony trioxide (Sb_2O_3) at contents up to 20 wt.% were studied (Aly et al., 2012). Blends were prepared within a temperature range of 170–175°C in the Brabender internal mixer. Then moldings were gamma irradiated with doses up to 150 kGy. The irradiation of moldings led to an increase of the tensile strength, elongation at break, and modulus within 5–10 wt.% Sb_2O_3 loading. An increase in hardness and resistance to swelling in benzene and volume resistivity of blends was also observed with the increase in concentration of Sb_2O_3. An increase in limited oxygen index from 16 to 20.9 was observed in the blend loaded with 15 wt.% of Sb_2O_3. The SEM micrographs inferred that irradiation facilitated the compatibility among blend components via crosslinking, supported by the presence of the reinforcement. An enhancement in the thermal stability of the blends was observed at 15 wt.% of Sb_2O_3 filler irradiated with a dose of 75 kGy. Slight increase in T_m in the presence of Sb_2O_3 was also observed.

The mechanical and morphological properties of HDPE/flax fiber (FF)/GRT composites and HDPE/FF/SBR were studied at different concentrations of fibers and GRT (Kakroodi et al., 2012). Also, the effect of coupling agent (SEBS) on HDPE-fiber and HDPE-rubber surface adhesion and composite morphology was investigated. An improvement in the adhesion at the presence of the coupling agent was seen, which led to improved tensile, flexural, and impact properties.

A peroxide-induced reactive compounding of PP/GRT blends in a corotating twin-screw extruder was carried out to prepare thermoplastic elastomer alloys (Wiessner et al., 2012). The extrusion experiments showed that the width of the residence time distribution was the key parameter determining the mechanical properties of elastomer alloys, whereas the mean residence time had little effect. This was confirmed by performing both quasi-static and cyclic tensile tests.

Chlorination of waste rubber (WR) of 200–250 μm particles size was carried out by treating with chlorine gas in the presence of 0.5% azobisisobutyronitrile as an initiator (Tan et al., 2012). An 80 phr chlorinated WR with chlorine content of 42% was combined with antimony trioxide in a proportion of 1.5 to 1 and used as a flame retardant. The prepared mixture was compounded with a 100 phr LDPE, 3 phr montmorillonite, 3 phr stearic acid, and 0.8 phr silane coupling agent on a two-roll mill at a temperature of 120°C and compression molded at 130°C. The flame retardancy, thermal properties, limiting oxygen index, and TGA of the neat LDPE and blends were studied. The improved flame retardancy and the thermal properties of the blends were mainly attributed to high char formation and synergistic effect between the chlorinated WR and antimony trioxide at high temperatures. However, the tensile strength and elongation at break of the blends were lower due to a lack of the compatibility with the neat LDPE.

Compounds of ABS/GRT of particle sizes lower than 200 μm, between 200 and 500 μm, and higher than 500 μm, up to concentrations of 70 wt.% were prepared (Mujal-Rosas et al., 2012). The dielectric studies of these compounds at various temperatures and frequencies were conducted. Based on these studies, possible electrical applications were suggested that would satisfy the official regulation requirements. The structure and mechanical behavior of materials were also investigated.

The effect of the vinyl alcohol-phthalic anhydride coupling agent on the tensile properties, morphology, and thermal degradation of LDPE/tire dust composites was studied (Supri and Ismail, 2012). The composites were prepared using Z-blade mixer at rotor speed of 50 rpm and 180°C. The study showed that the addition of the coupling agent led to a better dispersion of the tire dust, higher tensile strength, Young's modulus, mass swell, and better thermal stability, but a lower elongation at break.

Ternary mixtures of PP, EPDM (ethylene-propylene-diene-monomer), and 40 mesh GRT prepared in a corotating twin extruder were injection molded and the response surface methodology was applied to optimize the process (Costa et al., 2010). The tensile strength and impact strength of the mixtures were used as response variables. SEM was also used to investigate the morphology of blends to interpret results. Mechanical properties of ternary blends showed a sharp deterioration with the content of GRT particles due to a poor adhesion between the GRT particle and PP matrix. EPDM was found to improve the interfaces between the particles and PP as was evident from SEM observations, tensile properties, and impact strength. The highest impact strength of more than 80 J/m was achieved at 25% contents of EPDM and GRT particles.

Processing of the foamed PP/50 mesh GRT powder blends was investigated using a single-screw foam extrusion and chemical blowing agent azodicarbonamide (Xin et al., 2009). The regression models were constructed to study the relationships between the void fraction, average cell size, and cell density of foamed blends and the die temperature, screw speed, and concentrations of GRT and blowing agent. In subsequent study, the microcellular compatibilized PP/GRT powder blends were prepared using injection molding process by utilizing a chemical blowing agent to create a new class of materials with unique properties (Zhang et al., 2011a). The addition of the MA-grafted SEBS (SEBS-g-MA) increased the shear viscosity of the unfoamed composites. Cell sizes, cell density, void fraction, and mechanical properties of the foam composites were measured. It was shown that the SEBS-g-MA acted as an effective compatibilizer enhancing mechanical properties. Also, the addition of the compatibilizer increased void fraction and cell density of the composites, while the average cell sizes decreased. A processing temperature range of 180–195°C provided a finer microcellular structure.

Blends of PP/GRT with particle sizes of 30–50 μm were prepared at proportions of 0/100, 75/25, 60/40, and 50/50 in the presence of SEBS-g-MA, as a compatibilizer, with a content of 5 wt.%, based on the weight of rubber, by means of a modular intermeshing twin-screw extruder at a screw speed of 100 rpm, barrel temperatures of 200, 210, 220, and 230°C from the hopper to the die (Xin et al., 2009). Blends were fed into a single-screw extruder with a static mixer and capillary die having a 0.5 mm diameter to conduct foaming using the supercritical CO_2. A 10 mm length nozzle was custom made to generate high and rapid pressure drops with CO_2 pumped to carry out foaming process. The effects of particle, CO_2 contents, and die temperature on cell morphology were investigated. The blend with 25 wt.% of rubber particles demonstrated a smaller cell size, a better cell uniformity, and a higher cell density. The cell morphology deteriorated with increasing the rubber particle content over 40 wt.%. The cell density, void fraction generally increased with CO_2concentration. The lower die temperature led to a smaller cell size, higher cell density, and void fraction.

The GRT powder of 60 mesh was subjected to ozone treatment to create reactive hydroperoxide groups on its surface (Fan and Lu, 2011a). The free radical, produced by heating and by catalyzing with the ferrous ion of hydroperoxidized powder, caused the decomposition of hydroperoxide groups on the powder surface and initiated the graft polymerization of methyl methacrylate (MMA) onto the powder surface. The grafted surface has been characterized by the FTIR, XPS, SEM, EDXS, TGA, and contact angle measurements. The effects of grafting conditions on the degree of grafting were also elucidated. The concentration of hydroperoxide groups and the degree of grafting were increased with ozonization time. With increasing of the polymerization time and temperature, the degree of grafting increased. By altering the grafting conditions, the degree of grafting was controlled. After grafting with MMA, the hydrophilicity of the powder was improved. This approach provided a way to avoid a radical trapping by CB in GRT.

The devulcanization of GRT from the tread of truck tires with the particle size distribution from 40 to 500 μm and decrosslinking of crosslinked PE (XLPE) scrap from the cable insulation were investigated with aim to produce TPV by their blending (Zhang et al., 2011b). This was done by mechanical decrosslinking using repeated milling with subsequent dynamic vulcanization in a Brabender Plasticorder by melt-mixing the components at a temperature of 160°C and a rotor speed of 80 rpm for 10 min. The variation of gel fraction, morphology, and rheological behavior of XLPE/GRT blends during the high-shear mechanical milling were investigated. The dynamically vulcanized blends using sulfur-accelerator system exhibited better mechanical properties than the blends dynamically vulcanized with dicumyl peroxide system. After 20 cycles of milling, the XLPE/GRT 50/50 blends with a tensile strength of 6.0 MPa and an elongation at break of 185.3% were obtained. The tensile strength and elongation at break of the blends increased to 9.1 MPa and 201.2% after dynamic vulcanization and their reprocessability was maintained, According to authors, this approach is cost effective and easy to adopt for future industrial applications.

Recycled EVA/GRT powder blends of three particle sizes of greater than 200 μm, 200–500 μm, and greater than 500 μm with concentrations up to 70 wt.% were prepared by using a Brabender mixer (Mujal-Rosas et al., 2011). The stress-strain behavior showed that upon the addition of smaller particles to the matrix up to 10%, the Young's modulus of the blends increased, while other mechanical properties reduced. At the higher concentration of GRT, all mechanical properties decreased. However, conductivity, permittivity, and dielectric loss factor of blends increased with the powder concentration.

A possibility of obtaining new thermoplastic elastomers by blending of the waste PP and GRT powder with inclusion of bitumen and various compatibilizers, such as SEBS, SEBS-g-MA, and EPDM-g-MA, was considered (Zhang et al., 2009b, 2011c). This was done by studying rheological and mechanical properties, TGA, and morphology of blends. The obtained results indicated that the properties of blends were dependent on the content of bitumen and the kind of compatibilizer utilized. In particular, the presence of bitumen and SEBS-g-MA improved the elongation at break, thermal stability along with the processibility of blends due to the devulcanization and plasticizing effects of bitumen, while SEBS-g-MA, EPDM-g-MA, and SEBS decreased the processibility.

A novel technology combining high-energy electron-induced chemical reactions and melt-mixing process was used to prepare blends of PP random copolymer with GRT particles without and with the presence of polyfunctional monomers (Sritragool et al., 2010). It was shown that to compatibilize PP/GRT blends, specific monomer is required to obtain improved properties.

LDPE and GRT with an average particle size of 0.4 mm were compounded by using an internal mixer in the presence of dicumyl peroxide and sulfur to obtain thermoplastic elastomers (Hrdlicka et al., 2010). The material with a tensile strength of 8.6 MPa and an elongation at break of 260% was obtained.

Blending of PP with 50 wt.% GRT of 80 mesh was carried out using a batch mixer with aim to prepare a TPV (Zhu and Tzoganakis, 2010). To improve the adhesion between the PP matrix and GRT particles PP-derived macromolecules, as interfacial modifiers, were synthesized. Through degradation of PP, terminal double bonds were produced that were subsequently reacted via hydrosilylation with a hydride-terminated PDMS to generate hydrosilylated PP. This hydrosilylation reaction was extended to join the degraded PP and rubber molecules and produce a graft copolymer. Three types of modified PPs with reactive functional groups were tailor-made, and their effects on the mechanical and rheological properties of PP/GRT blends were studied. Using the EDX analysis, it was shown that a layer of interfacial modifiers was formed on the rubber particles in the blend. The presence of this layer reduced the shear viscosity and increased the entry pressure losses and the elongational viscosity of the blends. However, the effect on the tensile properties of blends was not remarkable. The highest tensile strength and elongation at break were 12.7 MPa and 37% with each of this value achieved at different combinations of ingredients.

Blends of PP/GRT of 35 mesh/virgin SBR in a ratio of 30/30/40 were prepared by dynamic vulcanization in the Brabender internal mixer in the presence of dicumyl peroxide of 0.35 and 0.7 phr and N, N′-m-phenylene bismaleimide (BMI) of 0.35 and 2.1 phr (Magioli et al., 2010). The prepared blends showed good reprocessibility and the tensile strength and elongation at break, comparable with several existing TPV systems. The improvement of the properties was attributed to the presence of BMI, as the curing co-agent, acting as a compatibilizer between PP and rubber phases. The decrease of the degree of crystallinity was also found in compatibilized blends.

Blends of the recycled HDPE/GRT powder were obtained by mixing in mills and subsequent sintering (Crespo et al., 2010). The influence of the powder on the thermal and mechanical properties of blends and their degradation was investigated by observation of the deformation and fracture using the SEM to analyze the interaction between the GRT powder and HDPE.

Recycled polyamide was mixed with GRT powder of 80 mesh up to concentrations of 60 wt.% in a Brabender internal mixer and compression molded (Hassan et al., 2010). The prepared moldings were subjected to gamma irradiation for up to absorbed dose of 200 kGy. Properties of the blends as a function of the irradiation dose and GRT concentration were studied. It was shown that due to the weak interfacial adhesion, the mechanical performance of blends reduced by the incorporation of the GRT. The irradiation also decreased the tensile strength and elongation at break, although the SEM studies indicated that irradiation provided a more smooth fracture surface.

Reactive compatibilization of the recycled LDPE or HDPE with GRT was carried out by preparing prefunctionalized components to provide chemical

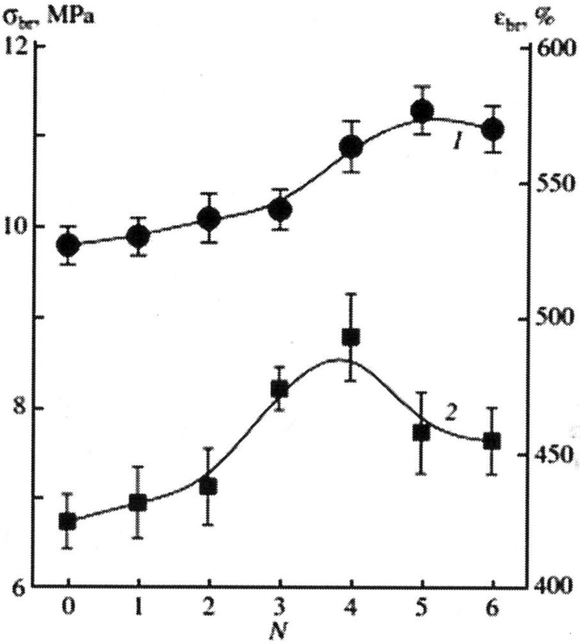

FIGURE 15.14 The tensile strength (curve 1) and elongation at break (curve 2) of TPE as a function of number of processing cycles *N* for TPE samples (Grigoryeva et al., 2009).

interactions (Grigoryeva et al., 2008, 2009). The PE component was functionalized with MA, while the rubber component was functionalized with MA or acrylamide (AA) by chemically or gamma irradiation-induced grafting techniques. The grafting degree and molecular mass distribution of the functionalized polymers were measured by the FTIR and size exclusion chromatography (SEC), respectively. The p-phenylenediamine (PDA) coupling agent and polyamide fiber were used to produce a thermoplastic elastomer (TPE). Dynamically vulcanized TPE based on the synthesized reactive PE, GRT, and EPDM in proportion of 40/25/35 was prepared using a single-screw extruder at a temperature profile of 155, 165, 175°C and 40 rpm. Studies of the phase structure and properties of TPE using DSC, DMTA, and mechanical testing indicated that the high performance TPE with improved mechanical properties can be manufactured. The effect of reprocessing of TPE up to six times on their viscosity and mechanical properties indicated that with an increase of the number of processing cycles the phase separation between the amorphous and crystalline phases in TPE decreases. The latter provided a positive effect on their tensile strength and elongation at break, as shown in Figure 15.14. As can be seen from this figure, remarkable values of the elongation at break were achieved in the prepared TPEs.

Chlorination of GRT is an effective way to make its surface polar and to enhance its compatibility with a polar polymer (Oldfield and Symes, 1983). Chlorination using trichloroisocyanuric acid was employed (Naskar et al., 2002c). The modified GRT becomes compatible with PVC, thus the PVC/chlorinated GRT blends showed better mechanical properties and dielectric constant than those unmodified blends. However, the incomplete chlorination of GRT usually resulted in an unsatisfactory compatibility between PVC and chlorinated GRT to some extent (Tan et al., 2009). An alternative chlorination method was to use chlorine gas to modify GRT before preparing the PVC/chlorinated GRT blends. Mechanical properties, hydrophilicity, swelling behavior, morphology, and thermal stability of the PVC/chlorinated GRT and PVC/GRT blends at proportions of 100/20, 100/40, 100/60, and 100/80 were analyzed. The chlorinated GRT showed a noticeable enhancement in overall properties of blends. At 100/20 and 100/80 blend composition, the tensile strength of PVC/chlorinated GRT was about 18.2 MPa and 20.7 MPa, respectively, while the corresponding value obtained from PVC/GRT blends was only 13.6 MPa and 8 MPa with the tear strength of 110 kN/m and 57.3 against 83 kN/m and 45.1 kN/m, respectively. The PVC/chlorinated GRT blends also exhibited the enhanced resistance to toluene swelling, which indicated that the interfacial adhesion between the PVC matrix and chlorinated GRT was improved. Moreover, their hydrophilicity was also better. The improvement of the polarity of chlorinated GRT also led to the increased thermal stability of the blends.

The waste PP/GRT blends were prepared using a Haake internal mixer at a temperature of 180°C and a rotor speed of 30 rpm for 8 min (Egodage et al., 2009). Blends of nine different compositions with GRT concentrations up to 70 wt.% were prepared. Mixing torque was increased with an increase in the GRT content. The tensile strength, modulus, and tear resistance decreased with the GRT content, while the elongation at break and the impact failure energy increased.

15.4.5 Concrete Modified by Recycled Rubber

Production of rubber-filled concrete compositions is a possible area for a further expansion of usage of GRT (Raghavan et al., 1994, 1998; Goldstein, 1995; Eldin and Senouci, 1992; Eldin and Senouci, 1993a,b; Eldin and Senouci, 1993; Lee et al., 1992). The advantages of using GRT in the cement-concrete structure are an increased crack, freeze-thaw and impact resistance, shock wave absorption, reduced heat conductivity, and increased resistance to acid rain. However, an addition of rubber particles to concrete has shown to reduce the compressive and flexural strengths. Concerning the effect of the size of rubber particles on the compressive strength, the results are contradictory. According to Topcu (1995), Eldin and Senouci (1994) the compressive strength of concrete was lowered upon an addition of the coarse-graded rubber particles while increased with an addition of the fine-graded particles. In contrast, tests (Ali et al., 2000) have shown opposite results. This contradiction possibly can be explained by the difference in rubber source, geometry of particles, and way that particles were prepared.

The influence of the shape of rubber particles on mechanical properties, workability, and chemical stability of rubber-filled cement was studied (Raghavan et al., 1998). The composite containing rubber shred was able to bridge the crack and to prevent catastrophic failure of the specimen, while the composite containing granular rubber particles was unable to bridge the crack. The pull-out test indicated poor interfacial bonding between the granular rubber particles and the matrix. In fact, many studies in this area indicated that the interface between the rubber and cement is weak. Attempts to improve the interface were made by washing the rubber particles (Raghavan et al., 1994; Eldin and Senouci, 1993b; Lee et al., 1993). Some improvement has been achieved by washing the particles with water (Eldin and Senouci, 1993b), water and carbon tetrachloride mixture, and water and latex mixture (Rostami et al., 1993), leading to an enhanced adhesion in rubber-filled cement.

Several surface modifications are also proposed, including the treatment of rubber with sulfuric acid and nitric acid to chemically oxidize rubber and introduce polar groups. Contrary to expectation, the treatment with nitric acid led to a decrease of the strength of the composite (Lee et al., 1993). On the other hand, the treatment with sulfuric acid improved the adhesion of rubber to concrete (Lepore and Tantala, 1997). Using a combination of chemical and surface probing techniques it was shown that the hydrophilicity of the rubber surface is greatly improved by acid or base treatment (Segre and Joekes, 2000; Segre et al., 2002). The rubber surface is typically hydrophobic. It is due to the fact that rubber typically contains zinc stearate that diffuses to the surface and causes hydrophobicity. By acid treatment zinc stearate can be hydrolyzed to stearic acid. Treating the rubber with base, the zinc ions are converted into sodium ions of NaOH. These conversions create the soluble sodium stearate. It was also found (Raghavan, 2000) that addition of rubber particles to the mortar led to a decrease in their compressive and flexural strengths due to pull out of particles. However, the treatment of rubber particles before mixing with bifunctional silane-coupling agent, such as gamma mercapto trimethoxy silane-coupling agent, improved the interface and led to increased ductility (Raghavan, 2000). It is suggested to expand this research to find the effects of the type of coupling agent on the adhesion and the fracture behavior of the rubber-filled cement paste, mortar, and concrete. These materials can be utilized in highway pavement overlays, sidewalks, medians, sound barriers, and other transportation nonstructural uses.

As mentioned earlier, the addition of rubber particles typically leads to the degradation of mechanical properties of the concrete due to a poor adhesion. However, it was shown that the adhesion of GRT of an average size of 120 μm to cement was improved by applying surface plasma treatment to

GRT transforming its hydrophobic surface to hydrophilic one (Cheng et al., 2012). After plasma treatment, peak intensity in the range of 1400–1600 cm^{-1} was increased due to the generation of the carbon-carbon double bonds on the surface of GRT powder. In addition, oxygen-containing polar functional groups on the surface are created. This led to improvements in compactness and impermeability of modified segment due to a decrease of peak pore size from 50 nm to 30 nm. Also, in the triaxial testing plastic and elastic deformations were increased due to the enhanced interfacial interaction and adhesion.

GRT of particle sizes in the range of 300–600 μm was also modified by a partial oxidation by flowing oxygen/nitrogen mixture of the desired ratio through the reactor for 30 min (Chou et al., 2010). Then flow of gases was ceased and the temperature was raised and kept for one hr at a desired temperature of 150, 200, or 250°C to carry out oxidation reaction. It was found that a partial oxidation induced the hydrophilic functional groups, S=O and S—O, on surfaces of the GRT, providing the enhanced hydration of the cement that led to the increase of mechanical properties of the rubberized mortar. The temperature of 250°C, and the oxygen/nitrogen ratio of equal to or less than 0.04 were optimal for the partial oxidation reaction. The addition of the GRT as-received at a concentration of 6 wt.% decreased the compressive strength at 28 days aging from 34.8 MPa to 16.3 MPa, the flexural strength from 6.1 MPa to 1.8 MPa, and the tensile strength from 3.2 MPa to 1.8 MPa. In contrast, the specimens with GRT treated at 250°C had a significant increase in the strength. In particular, at 28 and 56 days aging the compressive strength was 41.2 and 42.1 MPa, respectively, which was greater than that of controlled specimens. Also, the flexural and tensile strength was, respectively, 6.1 and 3.3 MPa at 28 days, and 6.6 and 3.3 MPa at 56 days, and approximately equal to that of controlled specimens.

To improve the mechanical properties of the rubberized concrete, a method of the surface modification of the GRT particles was proposed by using sodium hydroxide (Chou et al., 2007). This method enhances the hydrophilicity of rubber, reducing negative effects of rubber on the hydration of cement. The GRT particles of 30–50 meshes were partially oxidized using hot air/steam in a fluidized bed reactor to produce the hydrophilic groups on the surface of the particles. The concrete containing 5 wt.% modified rubber treated by NaOH solution showed an increase in the compressive, flexural, and tensile strength from 307.3 to 344.6, 54.3 to 55.9, and 37.7 to 38.7 kgf/cm^2, respectively. Oxygen-containing OH groups were formed on the rubber surface by NaOH solution treatment. The latter increased the Hamaker constant of the system, leading to the higher compressive strength. Furthermore, the rubber-modified concrete of the higher toughness was developed by adding to cement the GRT particles that were surface treated by coupling agents. The rubber-modified concrete with the higher strength was prepared by mixing GRT particles with sulfur -concrete mixer and carrying out the partial vulcanization between the rubber and hot sulfur (Xi et al., 2004).

Various properties related to the durability of concrete containing GRT particles including the shrinkage, water absorption by immersion, and capillarity and resistance to carbonation and chloride penetration were studied (Bravo and de Brito, 2012). The effect of rubber aggregates incorporation ratio, size, and grinding process on the production of concrete mixes was investigated. It was found that the durability-related properties of the hardened rubber-modified concretes deteriorate, but they were still acceptable within a limited replacement ratio. However, it was shown that carbonation resistance increased by almost 50%. Also, an increase of the size of the aggregates reduced the durability with only the shrinkage being the exception to this rule. It was also concluded that the concrete durability was not improved by use of fine cryogenically ground GRT in concrete mixes instead of a fine mechanically ground one. A substantial improvement in the chloride penetration resistance was found in the concrete containing the mechanically ground GRT. The utilization of GRT in concrete also provided additional advantages including the good damping characteristics and the good thermal and acoustic performance.

The possible application of cement mortars containing GRT as a flexible interface material was investigated (Nehdi and Khan, 2004). The effects of the water/cement ratio, GRT content, and particle size on the mechanical properties in uniaxial and triaxial compression of the mortars were studied. It was suggested that cement mortars containing GRT particles have a superior ductility to satisfy the requirement for deformations occurring around tunnel linings, pipelines, and other buried infrastructure.

A review of research on the performance of concrete containing GRT particles was recently published (Pacheco-Torgal et al., 2012). In particular, this review discussed the effect of GRT treatment, the size of GRT particles, and the replacement volume on the fresh and hardened properties of concrete. A workability of fresh concrete defining its flowability and the compressive and tensile strength, toughness, elastic modulus, thermal and sound properties, and durability of hardened concretes containing the tire rubber waste was also discussed.

15.4.6 Asphalt Modified by Recycled Rubber

Asphalt can be blended with tire rubber to modify the properties of the asphalt. This is an important market for use of GRT. In 2005, 2007, and 2009, an estimated 120, 100, and 175 million pounds, respectively, were used in the United States (Read, 2012). Mostly, these were consumed in California, Arizona, and Florida. Other states are also recognizing the benefits of the modified asphalt. It was projected that asphalt industry can adsorb up to 40% of scrap tires (Anonymous, 1993).

Utilization of scrap tire rubber in asphalt has advantages in the performance of roads and their longevity. This includes enhanced ductility, crack resistance, skid resistance, and noise reduction. Disadvantages of the rubber-modified asphalt are its cost and a possibility of toxic emissions into the air. Tests indicated that rubber-modified asphalt increases cost of road construction by about 50% in comparison with conventional asphalt (McQuillen et al., 1988). The requirement for an additional step of hot mixing during the processing of a rubber-asphalt mix may possibly cause toxic emissions into the air.

Two processes are used in preparing the rubberized asphalt: dry and wet. In both processes, the GRT particle size ranges from 6.35 mm to 40 mesh. In the wet process, asphalt is blended with GRT particles and then added into the hot mix. In the dry process, GRT is mixed with aggregate and the resulting mix is blended with asphalt.

The use of blending GRT with asphalt has been in existence for quite some time. Depending on the type of tire, the composition of GRT may include different rubbers. These crosslinked rubbers are mostly immiscible in bitumen. The blends show an improvement in basic asphalt properties as well as rubber-like characteristics. The blend is thought of as a dispersion of undissolved swollen rubber particles acting as an elastic aggregate within asphalt, modified by the portion of the rubber particles that have dissolved.

The use of blending GRT with asphalt began in the 1960s by McDonald, who developed and patented a patching material consisted of 25 wt.% scrap and asphalt blended at 375°F for 20 min (Roberts and Lytton, 1987). Macdonald continued his work by expanding it to actual road pavement test sections as a seal coat, and in 1968, Sahmaro Petroleum and Asphalt conducted application of a blend of GRT and asphalt as a binder for hot premix. The hot premix is a mixture of stone aggregate, sand, and the tire-asphalt binder all premixed in a batch or drum-type mixer. This material is then applied as a carpet on top of the road by means of a paving machine followed by a steel roller used to compact material (Renshaw, 1985). Thereafter, many variations of this basic process of hot premix with the blend of GRT and asphalt as the binder have been proposed, and most of them involve replacing stone aggregate with GRT. However, this method does not truly modify the asphalt binder. Thus, the blending process of GRT and asphalt before preparing a mixture is the most efficient in improvement of properties. Typical use levels range from 15 to 30 wt.% (Takallou and Takallou, 1991).

A limited amount of work has been done on the characterization of the blends of GRT and asphalt. Blends are typically mixed at temperatures of 300–400°C for a period of 0.5–2 h. The mix increases in viscosity and has a consistency of slurry with discernable rubber particles spread throughout. At room temperatures the resulting composition is a tough rubbery-elastic-like material. This mixing period is often referred to as reaction or digestion time. It was suggested (Takallou and Takallou, 1991) that the elastic quality of the blend is caused by the mechanical action of the undissolved rubber particles performing as a completely elastic aggregate within asphalt, which is modified by a portion of the rubber particles that have dissolved. According to Roberts and Lytton (1987), the mixing time of 2 h, as opposed to times of 0.5 and 1 h, significantly improves elastic recovery, whereas increasing the mixing time reduces the amount of solid rubber in the mixture and increases both high and low molecular weight fractions of dissolved rubber in the asphalt (Schuller, 1991). This brings up a very interesting point for further investigation as to whether the increase in asphalt soluble rubber fractions that occurs with longer mixing times can be attributed to elastic properties. The addition of GRT at 10, 20, and 30 wt.% levels has significantly increased the softening points and strain recovery over the base asphalt with a viscosity increase of similar magnitude for all three blends (Al-Abdul-Wahhab and Al-Amri, 1991).

Rheological properties as affected by asphalt composition, rubber dissolution, and temperature were studied (Billiter et al., 1997). Rubber content, rubber particle size, and base asphalt composition were found to be the main factors affecting the rheology of asphalt-rubber binder. By controlling these variables the binders with improved cracking resistance and rutting resistance can be produced. Finally, scrap tires, used as a crumb rubber modifier for asphalt, improve paving performance and safety by being an excellent and cost-effective modifier for the highway pavement industry (Rouse, 1995).

Recycled PE/GRT blends were modified by incorporation of bitumen for use in roofing applications and the synergistic effect was obtained (Navarro et al., 2010). The effect of GRT and PE concentration on the rheological and thermal properties and microstructure of the binary PE/GRT blends was evaluated. A reactive polymer with a polyethylene glycol (MDI-PEG prepolymer) was also added to blends to prepare ternary blends. It was shown that SBS copolymer (typical bitumen modifier) or different recycled polymers including EVA, LDPE/EVA blends, and GRT were able to improve binder rheological properties, especially at high in-service temperatures. Bitumen modification by 5 wt.% EVA/LDPE yielded the largest values of the dynamic moduli in the tested temperature range. At low temperature the polymer-modified bitumen that included EVA/SBS and EVA/LDPE blend showed a remarkable increase in the dynamic modulus, while the GRT-modified bitumens showed a reduction in the dynamic modulus. However, the nonreactive polymer-modified bitumens were largely unstable during storage at high temperature.

The influence of highway pavement surface type on tire/pavement noise was studied (Bennert et al., 2006). It was found that the open-graded friction course (OGFC) with the hot mix GRT-modified asphalts attained the lowest noise levels. Also, the GRT-modified OGFC mixes exhibited finer aggregate gradations than the traditionally used OGFC.

Compatibilized blends of LDPE/GRT powder of 60 mesh in proportion of 30/70 were used to modify asphalt (Ouyang et al., 2012). Among various compatibilizers, the blend containing MA-grafted ethylene-octene copolymer (POE-g-MA) showed the highest toughness of 2032.3 MJ/m^3, in comparison with a value of the control blend of 1402.9 MJ/m^3. Therefore, POE-g-MA was selected as an asphalt modifier. Accordingly, GRT/LDPE/POE-g-MA blend exhibited the highest toughness value at POE-g-MA content of 8%.

The softening point of the modified asphalt was higher than 60°C at concentrations above 5% of LDPE/GRT at a ratio of 30/70 and POE-g-MA content of 8% under shearing speed of 3000 rpm for 20 min. In addition, these blends showed good storage stability at high temperatures.

Grafting of GRT of 40 and 300 meshes was carried out by means of bulk polymerization of acrylic acid (AA) at 80–85°C for a period of 1–3 h in nitrogen atmosphere (Kocevski et al., 2012). This grafted GRT was mixed with an asphalt binder. It was found that the presence of the grafted GRT in the asphalt binder increased the failure temperature of the asphalt in comparison with use of unmodified GRT. Rheological properties and SEM studies indicated that AA-grafted GRT improved bonding between the GRT and binder.

The influence of processing variables in the internal mixer on mechanical properties and hot storage stability of the GRT (particle sizes below 0.8 mm)-modified bitumens at a concentration of 12 wt.% in the presence of SBS polymeric modifier was carried out (Gonzalez et al., 2012). The effect of time, mixer gap, and temperature was studied. It was found that the addition of SBS to GRT-modified asphalt improved the elastic properties of asphalt, reduced tan delta, and increased viscosity. The mixing time showed an increase in the primary aging and rubber digestion, especially at times below 2–3 h. The rubber digestion and elasticity of the binder also increased with temperature. The softening point and penetration grade were practically unaffected by the processing temperature. The gap between the rotor and stator of the mixer was found to have little influence on the rubber digestion.

The efforts were undertaken in Nevada to assess and implement the use of GRT- modified binders in the dense graded asphalt mixtures (Hajj et al., 2011). An extensive laboratory performance evaluation culminated in pilot demonstration projects in Nevada and California in 2008. Side-by-side road sections of GRT- and polymer-modified asphalt mixtures were inspected in 2011, revealing no visual distresses. The pavement was maintained in excellent conditions and uniformly along the total length of the road. A study was also initiated in 2010 to evaluate the use of GRT in warm mix asphalt mixes. The effect of two types of warm mix additives on mixture resistance to moisture damage was evaluated using the tensile testing and the dynamic modulus by subjecting samples to multiple freeze-thaw cycles. The potential impact of aggregate residual moisture in the mix due to the insufficient aggregate drying during production was reported.

GRT-modified asphalt with good storage stability was prepared by blending GRT with the SBS polymeric compatibilizer in an extruder (Cheng et al., 2011). The softening point along with photomicrographs was used to determine the storage stability. The long-term storage stability test indicated that the softening point difference of the GRT-modified asphalt was less 2.5°C. The rheological characteristics of the binders, studied using a dynamic shear rheometer (DSR) and a bending beam rheometer (BBR), showed that the GRT-modified asphalts exhibited the better resistance to permanent deformation at high temperature and cracking at low temperature, than those SBS-modified and straight-run asphalts.

The GRT-modified asphalts were prepared by the wet process using continuous blend and terminal blend asphalt rubber (Fontes et al., 2010). The repeated simple shear test at a constant height and the accelerated pavement simulator testing by means of wheel tracking test were carried out. The tests showed a linear correlation between these two tests. The use of asphalt-rubber binder improved significantly the resistance to rutting with the highest resistance exhibited by the mixtures from continuous blend binders and gap-graded aggregate gradation.

The possibility of obtaining penetration grade modified bitumens from vacuum bottom residue (VBR) by adding the waste PE, waste latex, GRT, and heavy vacuum slopes (HVS) was investigated (Mortazavi et al., 2010). These materials were mixed with VBR in different proportions using the high-speed and high-shear double-mixer operating at 8000 rpm and a temperature of 180°C for 90 min. The penetration, softening point, Frass breaking point, ductility, and dynamic properties of the mixtures were measured and the morphology was observed by the optical microscopy. It was found that bitumen with penetration of 40/50, 60/70, and 85/100 grades and improved softening and Frass breaking points can be produced. The modified 60/70 grade bitumen with the improved rheological behavior and decreased probability of cracking at low temperature and rutting at high temperature was obtained in mixtures containing 10% natural bitumen and 5% waste latex with VBR showing homogeneous dispersion. Although GRT particles at low concentrations improved the ductility of the blends, the particle dispersion was not sufficiently fine.

The modification of bitumen with SBS copolymer powder was done using the mechanochemically devulcanized GRT (Zhu et al., 2009). The penetration index, softening point, 5°C ductility, aging behavior, and rheological properties of bitumens modified by 8, 10, and 12 wt.% GRT/SBS mixtures were measured. In comparison with the bitumen modified by incorporation of 5.5 wt.% SBS alone, the majority of properties of the blends were improved, except their penetration. Rheological properties indicated that at high temperatures 10 wt.% GRT/SBS-modified bitumen was better than SBS-modified bitumen. The SEM observation of the fractured surfaces showed that bitumens mixed with the GRT/SBS powder had a better interfacial adhesion with matrix than with SBS alone.

Bitumens modified by nonreactive plastomers and elastomers including SBS, recycled EVA/LDPE blends, and GRT and reactive polymers including MDI-PEG prepolymer were compared (Navarro et al., 2009). The nonreactive and reactive polymers were mixed with bitumen at a processing temperature of 180°C and 90°C, respectively. Modified bitumens were prepared by blending the bitumen and polymer in an open low-shear batch mixer, having a four-blade impeller rotating at a speed of 1200 rpm. The characterization of the thermorheological and thermal behavior and morphology of selected modified bitumens was carried out. It was found that the bitumen modified by SBS or by different recycled polymers improved binder rheological properties at high in-service temperatures. In particular, bitumen containing 5 wt.% EVA/LDPE showed the highest values of the dynamic moduli in the tested temperature range. Moreover, the thermoplastic polymer-modified bitumen (EVA/LDPE, EVA, and SBS) undergoes a remarkable increase in the modulus at low temperature. The bitumen containing GRT showed a reduced modulus. The nonreactive polymer-modified bitumens, due to phase separation or rubber particle settling, were unstable during storage at high temperature, while the bitumens modified by MDI-PEG prepolymer after curing showed increased moduli and stability during storage at 163°C.

The penetration grade, softening point, and ductility of the GRT-modified bitumens were studied as a function of rubber selection, temperature, rotational speed, shearing time, order of rubber, and oil addition (Li et al., 2009). It was established that a shearing time of 40 min, a temperature of 180°C, a rotational speed of 7000 rpm, and a rubber concentration below 25 wt.% with oil content below 4 wt.% were optimal for carrying out the technological process for preparation the GRT-modified asphalts.

Three different bitumen mixes containing limestone with maximum aggregate size of 22 mm were studied (Navarro and Gamez, 2012). The first mix was the reference one manufactured with the conventional bitumen. The second mix was the same bitumen, but it was modified by adding 15 wt.% GRT of the particle size less than 0.8 mm in a refinery by the wet process. In this mix, chemical and elastomeric stabilizers were added to ensure the workability. The third mix was the same bitumen but modified by adding 0.5 wt.% GRT using the dry process. It was determined that mixes containing GRT exhibited a greater stability and resistance to deformation. However, the addition of GRT slightly reduced the tensile strength and cohesion due to a lack of the affinity between the bitumen and rubber. The addition of GRT, especially by the dry process, improved the stiffness of mixes that may be useful in the road construction with thinner pavement layers.

15.4.7 Use of Crumb Rubber in Soil

Patented soil amendment method of using tire crumb can decrease the negative impacts associated with compaction (Malmgren et al., 1991). The resiliency of the turf is not a direct factor of the elastic nature of rubber but rather the result of increased aeration (Logsdon, 1990a,b). Surface hardness characteristics were evaluated (Rogers and Waddington, 1992). Crumb rubber significantly reduced soil hardness, soil shear strength, and water content (Groenevelt and Grunthal, 1998). The building of embankment on soft soil ground using recycled scrap tires showed that the behavior of embankment was significantly improved (Lee et al., 2001).

A possibility of utilization of the rubber-soil mixtures to provide seismic isolation of buildings by using the entire contact surface of the foundation structure is considered (Tsang, 2008). This is in contrast to the conventional system utilizing isolation at certain discrete supporting points. The issues related to nonlinear site response, soil resonance effects, liquefaction, ground settlement, and environmental effects were discussed. The proposed approach provided the ability to reduce both the horizontal and vertical ground accelerations by 60–70% and 80–90%, respectively. The durability of tires also provided protection against termite, fire, and outgassing once they are buried in foundations.

Mixtures of naturally deposited sand, medium plasticity clay, and bentonite with recycled rubber produced composite materials suitable for use in ground improvement (Becker and Vrettos, 2011). Mechanical properties of soils relevant to the application of such composites in geotechnical engineering problems were investigated. The compaction, permeability, compressibility, and flexural and shear strength were determined for typical mixtures along swelling of the bentonite-rubber mixtures.

15.4.8 Products Made from Recycled Rubber

(i) Industrial Products

Compounds containing devulcanized or GRT can be utilized by various manufacturers to make various products. In particular, these compounds are actual or possible major consumers of recycled tire rubber: shoe heels and soles, tubes, conveyor belts, technical rubber moldings, automobile floor mats, mud flaps, livestock stall mattresses, playground and track surfacing, railroad track crossing, lower layers of floor coverings, various molded and extruded profiles, sealing plates, battery boxes, and other hard rubber goods. Since tire rubbers are typically black, they cannot be used for light and colored compounds unless additional measures have been taken to change the color. Obviously, for every such use the recycled tire rubber must undergo extensive testing.

(ii) Absorbents

GRT particles have an ability to adsorb hydrocarbons. However, their adsorption capacity is low in comparison with adsorbent materials currently in use. To improve its adsorption capacity various methods for manufacturing of adsorbents and their various uses were proposed, as discussed in this section. An oil absorptive material of lower cost can be obtained by graft copolymerization through blending of various proportions of GRT of particle size of 100 mesh with 4-tert-butylstyrene (tBS), as a monomer in the presence of divinylbenzene, as a crosslinker, and benzoylperoxide, as an initiator (Wu and Zhou, 2009). Oil absorbency of the grafted blends reached a maximum of 24.0 g/g at a feed ratio GRT/tBS of 60/40 and a divinylbenzene concentration of 1 wt.%.

Active carbon absorbents can also be obtained from the tire tread rubber particles containing good quality CB. This was accomplished by treating the particles of 300 μm size with the activation agents, such as potassium hydroxide and zinc chloride, and then subjecting them to pyrolysis at 500 and 700°C (Nunes et al., 2011). The obtained absorbents exhibited a mesoporous structure containing zinc and sulfur. In contact with water or acid solutions these ingredients were not released and, therefore, the absorbents can be used for waste water treatment. Specifically, they can be used for removal of organic dyes, such as methylene blue dye used in the textile industry; analytical chemistry and medicine; and methyl orange used in the textile, printing, paper manufacturing, and pharmaceutical industries. These dyes have dimensions similar to average pore diameter of the produced carbons.

GRT of particle sizes of 5 mm were carbonized by heating at 773 K in the nitrogen atmosphere for 2 h and subsequent heating to 1223 K at 5 K/min under flowing nitrogen (Mui et al., 2010a,b). Then they were activated by carbon dioxide under isothermal conditions for different time periods (2–16 h) and upon completion of the reaction, cooled in the nitrogen environment. The remaining tire char was subjected to sulfuric acid treatment and filtered. Then, it was repeatedly washed with deionized water to pH of 5.5, and dried in an oven at 378 K for 24 h. The prepared adsorbents exhibited a higher yield, and a higher surface area and volume of micropores comparable with those without acid treatment. The prepared adsorbent can be used for removal of Acid Blue 25 and Acid Yellow 117 dyes from waste water.

The carbonaceous absorbents for adsorption of cadmium were prepared by heat, chemical, and combined heat and chemical treatments of GRT of particle size varying from 1 to 3 mm (Alexandre-Franco et al., 2011). The adsorbent prepared by heat and combined heat and chemical treatments was found to be more effective in the adsorption of cadmium from aqueous solution than that obtained by chemical treatment alone.

Activated carbons for pesticide adsorption were prepared from the char of GRT granules (Betancur et al., 2009). The required surface porosity and desired functional groups on the surface of the activated carbon were achieved by a combination of chemical and physical treatment of tire char prior to physical activation. The adsorption capacity of the carbons was evaluated for pesticide removal from waste water. A cost analysis of the prepared adsorbents indicated that they are more cost effective and efficient than the most commercial adsorbents available for removal of pesticides from waste water.

The thermochemical degradation of GRT particles of 30 mesh in a carbon dioxide environment without pyrolysis to obtain activated carbons with high surface area and porosity was carried out (Quek and Balasubramanian, 2009). In this study, a one-step process was used and an activated carbon with a surface area as high as 414 m^2/g was obtained. Typically, activated carbon from GRT is obtained in a two-step process consisting of a pyrolysis process releasing volatile matter followed by a physical activation process with carbon dioxide and/or steam developing the porosity and surface area. Therefore, a one-step process is possibly advantageous for industrial implementation due to the reduced investment and operating cost.

Scrap tires were pyrolyzed under a nitrogen or carbon dioxide gas environment at various temperatures to produce a char (Manchon-Vizuete et al., 2005). After the completion of pyrolysis, the char was activated by oxygenation at different temperatures. The prepared chars were used to remove copper and lead from aqueous solutions. The optimal temperature for pyrolysis in nitrogen and carbon dioxide atmospheres was 550°C and for activation from 550 to 250°C. Activation of the char by oxygenation significantly improved heavy metal removal efficiencies. It provided much faster removal rates and higher copper removal compared with both pyrolyzed, unactivated chars and commercial activated carbons.

GRT of particle sizes from 1 to 3 mm was treated by applying thermal, chemical, and combined thermal and chemical treatments to prepare carbonaceous adsorbents for removal of mercury in aqueous solution (Gupta et al., 2011). The adsorbents were prepared by heating the rubber at 400 or 900°C for 2 h in the nitrogen atmosphere and then chemically treating with sulfuric acid, nitric acid, or their mixer solutions for 24 h. The heat treatment of the rubber developed mainly the microporosity, particularly the mesoporosity. The chemical treatment provided the creation of macropores. In the combined heat and chemical treatments, the predominant effects on the porous structure were caused by the treatment that provided the first effect. The adsorption capacity of mercury was larger for the adsorbents of higher microporosity.

15.5 PYROLYSIS AND INCINERATION OF RUBBER

15.5.1 Recovery of Hydrocarbon Liquid and Carbon Black

One particular method suitable for recycling used tires is pyrolysis (Moore, 1991; Kaminsky and Sinn, 1996). Pyrolysis is the thermal decomposition of rubbers in the absence of air and oxygen to produce oils and gases for reuse by petrochemical industries. Carbon black and other solid content remaining after pyrolysis can be utilized as fillers. Pyrolysis is typically carried out in boilers, autoclaves, rotary kilns, screw conveyors, and fluidized beds. Also, hydrogenation was performed using the tubing bomb reactor (Mastral et al., 2000). Research activities in tire rubber pyrolysis to recover hydrocarbon liquids and carbon black were quite extensive in the 1960s and 1970s, and led to plant construction for pyrolysis of scrap tires in the 1970s (Kawakami et al., 1980). Since then significant studies were carried out on tire pyrolysis concerning evolution of volatile (Conesa et al., 2000) utilization of oil and carbon black (Roy et al., 1999; Beck and Klingensmith, 1995). However, these attempts proved to be economically unsuccessful due to the low price of crude oil. Also, pyrolysis plants are believed to produce toxic waste as a by-product of operation (Osborn, 1995).

During many years, significant research was carried out and various pyrolysis processes were developed (Kaminsky and Mennerich, 2001a; William, 2002; William and Brindle, 2003). However, despite this progress pyrolysis of scrap tires was made on a limited scale, mainly due to the absence of a wide market for consumption of the oil and the carbon black derived from the pyrolysis process. Encouraging results on pyrolysis of tires were obtained using various technologies, including fixed bed (Berrueco et al., 2005; Islam et al., 2008), fluidized bed (Kaminsky and Mennerich, 2001b; Dai et al., 2001), rotary oven (Li et al., 2004), vacuum moving bed (Pantea et al., 2003; Gupta et al., 2004), and conical spouted bed reactor (Arabiourrutia et al., 2007). In particular, vacuum pyrolysis required a lower flow rate of the inert gas and residence time of volatiles in the reactor leading to a lower energy consumption in the process, simpler devices for volatile condensation, higher liquid yield with better control of its composition, and fuel quality and better quality of carbon black (Lopez et al., 2009). The presence of vacuum also enhanced the volatilization process and internal diffusion of products even at low temperatures with the initial stages of the reaction starting at 485 K under vacuum and at 497 K under atmospheric pressure. Also, the vacuum pyrolysis minimized the secondary reactions of the repolymerization and carbonization of products on the char surface.

The pyrolysis of waste tire powders of sizes of 200 and 600 μm using a capacitively coupled radio-frequency (RF) plasma reactor was carried out to study the reduced pressure RF plasma characteristics and pyrolysis product distribution (Huang and Tang, 2009). In this process the solid conversion varied from 40% to 78.4% over the range of conditions considered. The resulting gaseous product contained a large percentage of hydrogen and carbon monoxide and a small percentage of methane and other light hydrocarbons. The solid conversion and the hydrogen concentration were enhanced at higher power and pressure. The obtained char contained about 85% carbon and may be used as a replacement for the semi-reinforcing carbon black.

Pyrolytic carbon black with a surface area of 81 m^2/g was prepared by pyrolysis of GRT of particle sizes below 420 μm under nitrogen atmosphere (Cataldo, 2005). The resulting carbon black (CBp) and fresh carbon black N339 were compounded with a SBR/NR blend at various ratios of CBp/N339 including 0/74.8, 9.35/65.45, 18.7/56.1, 37.4/37.4, and 74.8/0 phr/phr and various physical properties of compounds, and vulcanizates were measured. The results indicated that the compounds containing CBp cured faster without affecting the scorch safety. This behavior was attributed to the low level of purity of CBp in comparison to a normal furnace black. An incorporation of CBp led to a significant decrease of the tensile strength, tear strength, and hardness of vulcanizates. It was concluded that the CBp cannot replace the furnace carbon black in a formulation at equal part, but may be used as a replacement of N339 carbon black at concentrations of less than 9 phr.

Silica particles were prepared by the pyrolysis-cum-water vapor treatment of waste green tires and used as a filler in fresh SBR rubber blends (Ivanov and Mihaylov, 2011). Pyrolysis was carried out by gradually heating tires between 300°C and 1000°C and continuously purging air, smoke gases, carbon dioxide, and nitrogen with the addition of vapor from 0% to 100%. The final product was removed after a period of 1 min to 48 h and then cooled in an atmosphere being the same or an inert medium without the presence of air. The recovered silica in a combination with the fresh silica, carbon black, and bis (triethoxysilylpropyl) polysulfide (Si 266®) was used to prepare compounds. The size of recovered silica particles was the same as that of the conventional silica. Differences in the mechanical properties (modulus at 300%, tensile strength, abrasion resistance, etc.) of the vulcanizates filled with the recovered silica and with the conventional silica and carbon black in a 2:1 ratio were insignificant. Also, no differences were observed in dependences of the dynamic modulus and the heat build-up on the strain amplitude, as well as in the temperature dependences of tan δ of the vulcanizates. It was concluded that the pyrolysis-cum-water vapor of waste tires is a suitable method for silica recovery.

15.5.2 Tire-Derived Fuel

Tire rubber can be used as energy value. This is the incineration method. It is advocated by the number of the major tire and rubber companies along with the major utility companies. Tire-derived fuel can be in the form of rubber chips containing or not containing the inherent wire. The nominal size of chips is usually about 5–10 cm. The larger sizes, the greater the content of wire and the less likely it is able to be handled and metered. Scrap tires are used as a fuel supplement for coal or gas in kilns for manufacturing Portland cement, lime, and steel. This reduces by 25% the amount of coal consumption by cement industries. Scrap tires are also burned to generate electricity. Tire-derived fuel may reduce sulfur emissions of power plants and may improve the combustion efficiency by adjusting proper stoichiometry in combination with various coals, wood wastes, and household garbage. The data in Table 15.2 indicates that at the end of 2005, 2007, and 2009, the total annual tire-derived fuel consumption from scrap tire rubber was, respectively, 2.145, 2.484, and 2.085 million tons, indicating some decreasing trend. However, it should be noted that burning tires causes valuable rubber materials to be lost. In fact, 60,000 BTU of energy is consumed to make one pound of synthetic tire rubber (William and Ferrer, 1997). In contrast, caloric value recovered by burning is 13,000–16,000 BTU per pound of rubber, which is not much higher than that of burning much cheaper coal. Moreover, burning tires for energy may lead to atmospheric pollution (Mastral et al., 1999).

15.6 CONCLUDING REMARKS

Waste tire and rubber present a problem of international significance. The present work describes some routes available to solve this problem. Numerous technologies are being developed. Among them, in addition to the well-known grinding techniques, are continuous pulverization methods based on a single- or twin-screw extruder that may serve as a possible route to supply rubber powder as a feedstock for various present and future devulcanization and recycling technologies. These include reclaiming, surface treatment, ultrasonic devulcanization, and utilization of rubber particles for making composites with many thermoplastic materials. The latter opens new routes for making thermoplastic elastomers or thermoplastic vulcanizates by combining the GRT with virgin or recycled plastics. Efficient adsorbents and carbon black were created through the pyrolysis process of GRT in combination with various chemical reactions. Also, there were significant activities in these areas, especially during the last decade. Some success has been achieved to further develop ultrasonic technology, which is considered as one of the promising methods to make devulcanized rubbers suitable for making rubber products from 100% recycled rubber and as well as for adding it to virgin rubber, virgin and recycled plastics, asphalt, concrete, and cement. An industrial size of ultrasonic single-screw extruder was developed that is capable of achieving higher output than earlier-developed laboratory-size equipment. This new development will make the process economically acceptable. Also, a laboratory ultrasonic twin-screw ultrasonic extruder for devulcanization of waste tires was developed, which is more efficient than the single-screw extruder and suitable for feeding fine GRT powder. However, the major challenge in industrial implementation of this process still remains due to a limitation in development of a high-power ultrasonic generator capable of continuously operating under high pressure and temperature. Clearly, there is also a lack of scientific understanding of various processes governing recycling of rubbers. Development of the science-based technologies for rubber recycling and novel processes and novel materials from recycled rubbers would significantly reduce the worldwide energy consumption, offer renewable rubbers from scrap tires and rubber waste, and would lead to less pollution of the environment.

ACKNOWLEDGMENT

This work is supported in part by Grant CMMI-1131342 from the National Science Foundation. The author wishes to express his appreciation to many colleagues, former and present graduate students, and associates whose many contributions during the last two decades made this work possible.

REFERENCES

2002 Fact Sheet. Tire Retread Information Bureau.
Accetta, A., Vergnaud, J.M., 1982. Rubber Chem. Technol. 55, 961–966.
Adhikari, B., De, D., Maiti, S., 2000. Prog. Polym. Sci. 25, 909–948.
Al-Abdul-Wahhab, H., Al-Amri, G., 1991. J. Mater. Civil Eng. 3, 189–203.
Alexandre-Franco, M., Fernandez-Gonzalez, C., Alfaro-Dominguez, M., Gomez-Serrano, V., 2011. J. Environ. Manag. 92, 2193.

Ali, N.A., Amos, A.D., Roberts, M., 2000. In: Proceedings of the Internernational Conference on Concrete 2000. University of Dundee, UK, pp. 379–390.
Aly, R.O., Hassan, M.M., Hasanen, J.A., El Sayed, E.F., 2012. J. Appl. Polym. Sci. 124, 4098.
Anonymous, 1993. Biocycle 34, 9.
Arabiourrutia, M., Lopez, G., Elordi, G., Olazar, M., Aguado, R., Bilbao, J., 2007. Chem. Eng. Sci. 62, 5271.
Arastoopour, H., Schocke, D.A., Bernstein, B., Bilgili, E., 1999. US Patent 5,904,885.
Basedow, A.M., Ebert, K., 1987. Adv. Polym. Sci. 22, 83.
Beck, M.R., Klingensmith, W., 1995. ACS Symp. Ser. 609, 254.
Becker, A., Vrettos, C., 2011. Bauingenieur 86, 548.
Benko, D.A., Beers, R.N., 2002a. Inventors. Goodyear Tire and Rubber Company, assignee. US Patent 5,380,269.
Benko, D.A., Beers, R.N., 2002b. Inventors. Goodyear Tire and Rubber Company, assignee. US Patent 6,387,965.
Benko, D.A., Beers, R.N., 2002c. Inventors. Goodyear Tire and Rubber Company, assignee. US Patent 6,462,099.
Bennert, T., Hanson, D., Maher, A., Vitillo, N., 2006. J. Test. Eval. 33 (2), 94.
Berrueco, C., Esperanza, F., Mastral, F.J., Ceamanos, J., García-Bacaicoa, P., 2005. J. Anal. Appl. Phys. 74, 245.
Betancur, M., Martinez, J.D., Murillo, R., 2009. J. Hazard. Mater. 168, 882.
Bever, M.B. (Ed.), 1986. Encyclopedia of Materials Science and Engineering, vol. 4. Pergamon Press, Oxford, pp. 2934.
Bilgili, E., Berstein, B., Arastoopour, H., 1999. AIChE Symp. Ser. 95 (321), 83–89.
Bilgili, E., Arastoopour, H., Bernstein, B., 2000. Rubber Chem. Technol. 73, 340–355.
Bilgili, E., Arastoopour, H., Bernstein, B., 2001a. Powder Technol. 115, 265–276.
Bilgili, E., Arastoopour, H., Bernstein, B., 2001b. Powder Technol. 115, 277–289.
Bilgili, E., Dybek, A., Arastoopour, H., Bernstein, B., 2003. J. Elast. Plastics 35, 235–256.
Billiter, T.C., Davison, R.R., Glover, C.J., Bullin, J.A., 1997. Petrol. Sci. Technol. 15, 205–236.
Blitz, J., 1967. Fundamentals of Ultrasonics, 2nd ed. Butterworth, London (Chapter 8).
Boron, T., Klingensmith, W., Forest, C., Shringarpurey, S., 1999. Paper #136 at the Meeting of the 156th ACS Rubber Division. Orlando, Florida.
Boron, T., Roberson, P., Klingensmith, W., 1996. Tire Technology. International '96, 82–84.
Bowers, B., Barber, D., Allinger, R., 1986. Paper #82 presented at the meeting of the ACS Rubber Division. Atlanta, Georgia, October.
Bravo, M., de Brito, J., 2012. J. Clean. Prod. 25, 42.
Bredberg, K., Persson, J., Christiansson, M., Stenberg, B., Holst, O., 2001. 55, 43.
Briggs, D., 1978. Euro. Polym. J. 14, 1.
Briggs, D., 1980. Surf. Interf. Anal. 2, 107.
Bryson, J.G., 1979. US Patent 4,148,763.
Campbell, R.H., Wise, R.W., 1964. Rubber Chem. Technol. 37, 635.
Cañavate, J., Carillo, F., Casas, P., Colom, X., Sunol, J.J., 2010. J. Compos. Mater. 44, 1233.
Cataldo, F., 2005. Macromol. Mater. Eng. 290, 463
Cheng, G., Shen, B., Zhang, J., 2011. Petrol. Sci. Technol. 29 (2), 192.
Cheng, X., Chen, H., Huang, S., Li, Z., Guo, X., 2012. J. Appl. Polym. Sci. 126, 1837.
Chidambaram, A., Min, K., 1994. SPE ANTEC Tech. Papers 40, 2927–2934.
Chopey, N.P., 1973. Chem. Eng. 80, 54.
Chou, L.H., Lu, C.-K., Chang, J.-R., Lee, M.T., 2007. Waste Manag. Res. 25 (1), 68.
Chou, L.-H., Yang, C.-K., Lee, M.-T., Shu, C.-C., 2010. Compos. B: Eng. 41, 613.
Colom, X., Cañavate, J., Carrillo, F., Velasco, J.I., Pages, P., Mujal, R., Nogues, F., 2006. Eur. Polym. J. 42, 2369.
Colom, X., Cañavate, J., Carrillo, F., 2007. Compos. A 38, 44.
Colom, X., Carrillo, F., Cañavate, J., Sunol, J.J., 2009. J. Appl. Polym. Sci. 112, 1882.
Conesa, J.A., Fullana, A., Font, R., 2000. Energy Fuel 14, 409–418.
Coran, A.Y., Howard, F., 1999. US Patent 5,889,119.
Coran, A.Y., Patel, R.P., 1996. In: Holden, G., Legge, N.R., Quirk, R., Schroeder, H.E. (Eds.), Thermoplastic Elastomers, 2nd ed. Hanser Publishers, New York.
da Costa, H.M., Ramos, V.D., da Silva, W.S., Sirqueira, A.S., 2010. Polym. Test. 29, 572.
Crespo, J.E., Nadal, A., Parres, F., 2010. Materialwiss. Werkstofftech. 41 (5), 293.

Dai, X., Yin, X., Wu, C., Zhang, W., Chen, Y., 2001. Energy 26, 385.
Deanin, R.D., Hashemiolya, S.M., 1987. Polym. Mater. Sci. Eng. 57, 212.
De, D., Adhikari, B., Maiti, S., 1997. J. Polym. Material 14, 333–342.
De, D., Maiti, S., Adhikari, B., 1999a. J. Appl. Polym. Sci. 73, 2951–2958.
De, D., Ghosh, A.K., Maiti, S., Adhikari, B., 1999b. Polym. Recycling 4 (3), 151–161.
De, D., Maiti, S., Adhikari, B., 2000. Kautch. Gummi Kunst. 53, 346–351
De, S.K., Isayev, A.I., Khait, K. (Eds.), 2005. Rubber Recycling. Taylor and Francis, Boca Raton.
Dekic, P.S., Temeljkovski, D.I., Rancic, B., Nusev, S., 2012. J. Sci. Ind. Res. 71, 295.
Diao, B., Isayev, A.I., Levin, V.Y., Kim, S.H., 1998. J. Appl. Polym. Sci. 69, 2691–2696.
Diao, B., Isayev, A.I., Levin, V.Y., 1999. Rubber Chem. Technol. 72, 152–164.
Duhaime, J.R.M., Baker, W.E., 1991. Plast. Rubb. Comp. Proc. Appl. 15 (2), 87–93.
Edirisinghe, D.G., De Silva, M.I.A., Premachandra, J.K., 2011. Prog. Rubber Plast. Recycl. Technol. 27, 161.
Egodage, S.M., Harper, J.F., Walpalage, S., 2009. Progr. Rubber Plast. Recycl. Technol. 25, 213.
Eldin, N.N., Senouci, A.B., 1992. ASCE J. Const. Eng. Mgmt. 118, 561–576.
Eldin, N.N., Senouci, A.B., 1993a. Cem. Concr. Agg. 15, 74.
Eldin, N.N., Senouci, A.B., 1993b. ASCE J. Mater. Civ. Eng. 5, 478.
Eldin, N.N., Senouci, A.B., 1994. Cement Concr. Compos. 16, 287–298.
Enikolopian, N.S., 1985. Pure Appl. Chem. 57, 1707–1711.
Fan, P., Lu, C., 2011a. J. Appl. Polym. Sci. 122, 2262.
Fan, P., Lu, C., 2011b. J Polym. Environ. 19, 943.
Feng, W., Isayev, A.I., 2004. J. Appl. Polym. Sci. 94 (3), 1316.
Feng, W., Isayev, A.I., 2005. J. Mater. Sci. 40, 2883–2889.
Feng, W., Isayev, A.I., 2006. Polym. Eng. Sci. 46, 8.
Feng, W., Isayev, A.I., von Meerwall, E., 2004. Polymer 45 (25), 8459.
Fix, S.R., 1980. Elastomerics 112 (6), 38–40
Fontes, L.P.T.L., Triches, G., Pais, J.C., Pereira, P.A.A., 2010. Constr. Build. Mater. 24, 1193.
Franta, I. (Ed.), 1989. Elastomers and Rubber Compounding Materials. Elsevier, New York.
Fridman, M.L., Peshkovsky, S.L., 1990. Adv. Polym. Sci. 93, 41–79.
Fukumori, K., Matsushita, M., Okamoto, H., Sato, N., Suzuki, Y., Takeuchi, K., 2002. JSAE Rev. 23, 259–264.
Garcia Ramirez, R., Isayev, A.I., 1991. SPE ANTEC Tech. Papers 37, 1084–1087.
Gent, A.N., 1990. Rubber Chem. Technol. 63, G49.
Gent, A.N., Tompkins, D.A., 1969. J. Appl. Phys. 40, 2520.
Ghose, S., Isayev, A.I., 2003a. J. Appl. Polym. Sci. 88, 980–989.
Ghose, S., Isayev, A.I., 2003b. Polym. Eng. Sci. 44, 794–804.
Ghose, S., Isayev, A.I., 2004. J. Cell. Plast. 40, 167.
Gibala, D., Hamed, G.R., 1994. Rubber Chem. Technol. 67, 636–648.
Gibala, D., Laohapisitpanich, K., Thomas, D., Hamed, G.R., 1996. Rubber Chem. Technol. 69, 115–119.
Goldstein, H., 1995. Civ. Eng. 65, 60.
Gonzalez de Los Santas, E.A., Sorieno-Corral, F., Lozano Gonzalez, Ma. J., Cedillo-Garcia, R., 1999. Rubber Chem. Technol. 72, 854.
Gonzalez, V., Martinez-Boza, F.J., Gallegos, C., Perez-Lepe, A., Paez, A., 2012. Fuel Process. Technol. 95, 137.
Gooberman, G., 1960. J. Polym. Sci.,42, 25–33.
Goodyear, C., 1844. Inventor. US Patent 3,633.
Goodyear, C., 1853. Inventor. GB Patent 2,933.
Grigoryeva, O., Fainleib, A., Grenet, J., Saiter, J.M., 2008. Rubber Chem. Technol. 81, 737.
Grigoryeva, O.P., Fainleb, A.M., Shumskii, V.F., Vilenskii, V.A., Kozak, N.V., Babkina, N.V., 2009. Polym. Sci. Ser. A 51, 216.
Groenevelt, P.H., Grunthal, P.E., 1998. Soil Till. Res. 47, 169–172.
Gugliemotti, A., Lucianano, C., Quadrini, F., 2012a. Plast. Rubber Compos. 41, 40.
Gugliemotti, A., Lucianano, C., Quadrini, F., 2012b. Polym. Plast. Technol. Eng. 51, 340.
Gupta, M., Yang, J., Metral, S., Roy, C., 2004. Chem. Eng. Res. Design 82, 34.
Gupta, V.K., Gupta, B., Rastogi, A., Agarwal, S., Nayak, A., 2011. Waste Res. 45, 4047.

Hajj, E.Y., Sebaaly, P.E., Hitti, E., Borroel, C., 2011. Performance Evaluation of Terminal Blend Tire Rubber HMA and WMA Mixtures-Case Studies, vol. 80, p. 665 (in book Asphalt Paving Technology).
Hassan, M.M., Badway, N.A., Gamal, A.M., Elnaggar, M.Y., Hegazy, E.S.A., 2010. Nucl. Instr. Meth. Phys. Res. B 268, 1427.
Hershaft, A.A., 1972. Environ. Sci. Technol. 6, 412–421.
Hershaft, A.A., 1977. Elastomerics 109 (12), 39.
Hilyward, N.C., Tong, S.G., Harrison, K., 1983. Plast. Rubb. Proc. Appl. 3, 315–322.
Hong, C.K., Isayev, A.I., 2001a. J. Appl. Polym. Sci. 79, 2340–2348.
Hong, C.K., Isayev, A.I., 2001b. J. Elastom. Plast. 33, 47–71.
Hong, C.K., Isayev, A.I., 2002a. J. Mater. Sci. 37, 385–388.
Hong, C.K., Isayev, A.I., 2002b. Rubber Chem. Technol. 75, 617–625.
Hong, C.K., Isayev, A.I., 2002c. Rubber Chem. Technol. 75, 133–142.
Hous, P., Bartelds, H., Smit, E., 1995. Rubber World 212, 36–40.
Howard, F., Coran, A.Y., 2000. Paper Presented at the ITEC. Akron, Ohio (September).
Hrdlicka, Z., Kuta, A., Hajek, J., 2010. Polimery 55, 832.
Huang, H., Tang, L., 2009. Energy Conv. Manag. 50, 611.
Hunt. L.K., Kovalak, R.R., 1999. Inventors, The Goodyear Tire and Rubber Company, assignee. US Patent 5,891,926.
A.I. Isayev, 1990. In: Proceeding of the 23rd Israel Conference on Mechanical Engineering, Technion, Haifa, Paper #5.2.3.
Isayev, A.I., 1993. Inventor. The University of Akron, assignee. US Patent 5,258,413.
Isayev, A.I., 2001a. Rubber recycling. In: White, J.R., De, S.K. (Eds.), Rubber Technologist's Handbook. RAPRA Technology Ltd., UK. pp. 511–547.
Isayev, A.I., 2001b. Recycling of elastomers. In: Buschow, K.H.J. (Ed.), Encyclopedia of Materials: Science and Technology, vol. 3, Elsevier, Amsterdam, pp. 2474–2477.
Isayev, A.I., 2001c. In: White, J.R., De, S.K. (Eds.), Rubber Technologist's Handbook. RAPRA, Shawbury, UK, pp. 511–547 (Chapter 15).
Isayev, A.I., Chen, J., 1994. Inventors. The University of Akron, assignee. US Patent 5,284,625.
Isayev, A.I., Chen, J., Tukachinsky, A., 1995. Rubber. Chem. Tech. 68, 267–280.
Isayev, A.I., Mandelbaum, S., 1991. Polym. Eng. Sci. 31, 1051–1056.
Isayev, A.I., Wong, C., Zeng, X., 1987. SPE ANTEC Tech. Papers 33, 207–210.
Isayev, A.I., Wong, C., Zeng, X., 1990. Adv. Polym. Technol. 10 (1), 31–45.
Isayev, A.I., Chen, J., Yushanov, S.P., 1995. In: Shen, S.F., Dawson, P. (Eds.), Simulation of Materials Processing: Theory, Methods and Application. Balkema, Rotterdam, pp. 77–85.
Isayev, A.I., Yushanov, S.P., Schworm, D., Tukachinsky, A., 1996a. Plast. Rubber Compos. Process. Appl. 25, 1–12.
Isayev, A.I., Yushanov, S.P., Kim, S.H., Levin, V.Y., 1996b. Rheol. Acta 35, 616–630.
Isayev, A.I., Yushanov, S.P., Chen, J., 1996c. J. Appl. Polym. Sci. 59, 803–813.
Isayev, A.I., Yushanov, S.P., Chen, J., 1996d. J. Appl. Polym. Sci. 59, 815–824.
Isayev, A.I., Kim, S.H., Levin, V.Y., 1997. Rubber Chem. Technol. 70, 194–201.
Islam, R.M., Haniu, H., Beg, R.A.M., 2008. Fuel 87, 3112.
Ivanov, M., Mihaylov, M., 2011. J. Elast. Plast. 43, 303.
Jalilvand, A.R., Ghasemi, I., Karrabi, M., Azizi, H., 2008. Progress. Rubber Plast. Recycling Technol. 24 (1), 33.
Jellinek, H.H.G., White, G., 1951. J. Polymer Sci. 7, 21–32.
Johnston, S.T., Massey, J., von Meerwall, E., Kim, S.H., Levin, V.Y., Isayev, A.I., 1997. Rubber Chem. Technol. 70, 183–193.
Kakroodi, A.R., Bainier, J., Rodrigue, D., 2012. Int. Polym. Process. 27, 196.
Kaminsky, W., Mennerich, C., 2001a. J. Anal. Appl. Pyrolys. 58, 803–811.
Kaminsky, W., Mennerich, C., 2001b. Anal. Appl. Pyrolys. 58–59, 803.
Kaminsky, W., Sinn, H., 1996. In: Brandrup, J., Bittner, M., Michaeli, W., Menges, G. (Eds.), Recycling and Recovery of Plastics. Hanser Publishers, Munich, pp. 434–443 (Chapter 5.3.1).
Kanagawa, T., Mikami, E., 1989. Appl. Environ. Microbiol. 55, 555.
Kasner, A.I., Meinecke, E.A., 1996. Rubber Chem. Technol. 69, 424–443.
Kausch, H.H., 1987. Polymer Fracture. Springer-Verlag, Berlin.
Kawabata, N., Murakami, T., Yamashita, S., 1979. Nippon Gomu Kyokaishi 52 (12), 768–773.
Kawabata, N., Okuyama, B., Yamashita, S., 1981. J. Appl. Polym. Sci. 26, 1417–1419.

Kawakami, S., Inoue, K., Tanaka, H., Sakai, T., 1980. Pyrolysis process for scrap tires. In: Jones, J.L., Radding, S.B. (Eds.), Thermal Conversion of Solid Wastes and Biomass, Symposium Series 130. ACS Publishers, Washington, DC, pp. 557.
Khait, K., 1994. Paper #24 presented at the meeting of ACS Rubber Division. IL, Chicago.
Khait, K., Torkelson, J.M., 1999. Polym. Plast. Technol. Eng. 38, 445–457.
Kim, D., Shiu, F.J.Y., Yen, T.F., 2003. Energ. Sour. 25, 1099.
Kinloch, A.J., Young, R.J., 1983. Fracture Behavior of Polymers. Applied Science Publishers, London.
Kleps, T., Piaskiewicz, M., Parasiewicz, W., 2000. J. Therm. Anal. Calorim. 60, 271–277.
Klingensmith, W., 1991. Rubber World 203, 16–21.
Klingensmith, W., Baranwal, K., 1998. Rubber World 218 (3), 41–46.
Knorr, K., 1994. Kautchuk Gummi Kunststoffe 47 (1), 54–57.
Knorr, K., 1995. Paper #5 presented at the ACS Rubber Division Meeting. Cleveland, Ohio, October.
Kocevski, S., Yagneswaran, S., Xiao, F., Punith, V.S., Smith Jr, D.W., Amirkhanian, S., 2012. Constr. Build. Mater. 34, 83.
Kohler R., O'Neill, J., 1997. Rubber World 216 (2), 32.
Kohler R., O'Neill, J., 1997. Rubber World 216 (2), 34–36.
LaGrone, B.D., 1986. Conservation and recycling 9, 359–361.
Lee, B.I., Burnett, L., Miller, T., Postage, B., Cuneo, J., 1993. J. Mater. Sci. Lett. 12, 967–968.
Lee, K.M., Cheung, B.K.W., Zhu, G.F., Tang, K., 2001. Soft Soil Engineering. Balkema Publishers, Netherland, 531–536.
Lee, S.H., Hwang, S.H., Kontopoulou, M., Sridhar, V., Zhang, A.X., Xu, D., Kim, J.K., 2009a. J. Appl. Polym. Sci. 112, 3048.
Lee, S.H., Zhang, Z.X., Xu, D., Chung, D., Oh, G.J., Kim, J.K., 2009b. Polym. Eng. Sci. 49, 168.
Lepore, J.A., Tantala, M.W., 1997. In: Proceedings of Concrete Institute of Australia. Concrete 97, 623–627.
Levin, V.Y., Kim, S.H., Isayev, A.I., Massey, J., von Meerwall, E., 1996. Rubber Chem. Technol. 69, 104–114.
Levin, V.Y., Kim, S.H., Isayev, A.I., 1997a. Rubber Chem. Technol. 70, 120–128.
Levin, V.Y., Kim, S.H., Isayev, A.I., 1997b. Rubber Chem. Technol. 70, 641–649.
Lewis, P.M., 1986. NR Technology 17 (4), 57–65.
Leyden, J.J., 1991. Rubber World 203 (6), 28–29.
Li, S.Q., Yao, Q., Chi, Y., Yan, J.H., Cen, K.F., 2004. Ind. Eng. Chem. Res. 43, 5133.
Li, H., Zhao, H., Liao, K., 2009. Petrol. Sci. Technol. 27, 1521.
Li, Y., Zhao, S., Wang, Y., 2011. Polym. Degrad. Stab. 96, 1662.
Li, Y., Zhao, S., Wang, Y., 2012. J. Polym. Res. 19, 9864.
Logsdon, G., 1990a. Biocycle 31, 44–45.
Logsdon, G., 1990b. Biocycle 31, 84–85.
Lopez, G., Aguado, R., Olazar, M., Arabiourrutia, M., Bilbao, J., 2009. Waste Manage. 29, 2649.
Lorimer, J.P., 1990. In: Mason, T.J. (Ed.), Chemistry with Ultrasound. Elsevier, New York (Chapter 4).
Luo, T., Isayev, A.I., 1998. J. Elastom. Plast. 30, 133–160.
Lynch, J., LaGrone, B., 1986. Ultrafine Crumb Rubber. Paper #37 presented at the meeting of the ACS Rubber Division. Atlanta, GA, October.
Magioli, M., Sirqueira, A., Soares, B., 2010. Polym. Test. 29, 840.
Makarov, V.M., Drozdovski, V.F., 1991. Reprocessing of Tyres and Rubber Wastes. Ellis Horwood, New York.
Malmgren, R.C., Parviz, N., Soltanpour, P.N., Cipra, J.E., 1991. US Patent 5,014,562.
Manchon-Vizuete, E., Macias-Garcia, A., Gisbert, A.N., Fernandez-Gonzalez, C., Gomez-Serrano, V., 2005. J. Hazard. Mater. 119, 231.
Maridass, B., 2009. J. Polym. Res. 16, 133.
Maridass, B., Gupta, B.R., 2003. Kautsch. Gummi Kunstst. 56 (5), 17.
Massey, J.L., Parr, J.C., Wagler, T.A., von Meerwall, E., Hong, C.K., Isayev, A.I., 2007. Polym. Int. 56, 860.
Mastral, A.M., Callen, M.S., Murillo, R., Garcia, T., 1999. Environ. Sci. Technol. 33, 4155–4158.
Mastral, A.M., Murillo, R., Callen, M.S., Garcia, T., 2000. Resour. Conserv. Recy. 29, 263–272.
McQuillen Jr., J.L., Takallou, H.B., Hicks, R.G., Esch, D., 1988. ASCE J. Transp. Eng. 114, 259–277.

Meszaros, L., Barany, T., Czvikovszky, T., 2012a. Radiat. Phys. Chem. 81, 1357.
Meszaros, L., Fejos, M., Barany, T., 2012b. J. Appl. Polym. Sci. 125, 512.
Michael, H., Scholz, H., Mennig, G., 1999. Kautch. Gummi Kunst. 52, 510–513.
Montagna, L.S., Santana, R.M.C., 2012. Plast. Rubber Compos. 41, 256.
Moore, M., 1991. Crain's Tire Business 9 (17), 15.
Morin, J.E., Williams, D.E., Farris, R.J., 2002. Rubber Chem. Technol. 75, 955–968.
Mortazavi, S.B., Rasoulzadeh, Y., Yousefi, A.A., Khavanin, A., 2010. Iran. Polym. J. 19 (3), 197.
Mui, E.L.K., Cheung, W.H., Valix, M., McKay, G., , 2010a. J. Coll. Interf. Sci. 347, 290.
Mui, E.L.K., Cheung, W.H., Valix, M., McKay, G., 2010b. Micropor. Mesopor. Mater. 130, 287.
Mujal-Rosas, R., Orrit-Prat, J., Ramis-Juan, X., Marin-Genesca, M., Rahhali, A., 2011. J. Reinf. Plast. Compos. 30, 581.
Mujal-Rosas, R., Orrit-Prat, J., Ramis-Juan, X., Marin-Genesca, M., 2012. AFINIDAD 68 (557), 7.
Myers, R.D., Nicholson, P., MacLeod, J.B., Moir, M.E., 1997. US Patent 5,602,186.
Myhre, M., MacKillop, D.A., 2002. Rubber Chem. Technol. 75, 429–474.
Naskar, A.K., Pramanik, P.K., Mukhopadhyay, R., De, S.K., Bhowmick, A.K., 2000. Rubber Chem. Technol. 73, 902–911.
Naskar, A.K., De, S.K., Bhowmick, A.K., 2002a. J. Appl. Polym. Sci. 84, 370–378.
Naskar, A.K., Bhowmick, A.K., De, S.K., 2002b. J. Appl. Polym. Sci. 84, 622–631.
Naskar, A.K., Khastgir, D., Bhowmick, A.K., De, S.K., 2002c. J. Appl. Polym. Sci. 84, 993.
Navarro, M.F., Gamez, M.C.R., 2012. J. Mater. Civil Eng. 24, 715.
Navarro, F.J., Partal, P., Garcia-Morales, M., Martin-Alfonso, M.J., Martinez-Boza, F., Gallegos, C., Bordado, J.C.M., Diogo, A.C., 2009. J. Ind. Eng. Chem. 15, 458.
Navarro, F.J., Partal, P., Martinez-Boza, F., Gallegos, C., 2010. Polym. Test. 29, 588.
Nehdi, M., Khan, A., 2004. Protective system for buried infrastructure using recycled tire rubber-filled cement mortars. Am. Concr. Inst. Spec. Publ. 219, 99 (in book Recycling Concrete and Other Materials for Sustainable Development).
Nicholas, P.P., 1982. Rubber Chem. Technol. 55, 1499–1515.
Novotny, D.S., Marsh, R.L., Masters, F.C., Tally, D.N., 1978. US Patent 4,104,205.
Nunes, M.R., Perez, G.M., Loguercio, L.F., Alves, E.W., Carreno, N.L.V., Martins, J.L., Garcia, I.T.S., 2011. J. Brazil Chem. Soc. 22, 2027.
Oh, J.S., Isayev, A.I., 2002. Rubber Chem. Technol. 75, 617–625.
Oh, J.S., Isayev, A.I., von Meerwall, E., 2004. Rubber Chem. Technol. 77 (4), 745.
Okuda, M., Hatano, Y., 1987. Japanese Patent Application 62,121,741.
Oldfield, D., Symes, T.E.F., 1983. J. Adhes. 16, 77.
Oliphant, K., Baker, W.E., 1993. Polym. Eng. Sci. 33, 166–174.
Osborn, J.D., 1995. Rubber World, 34.
Ouyang, C., Gao, Q., Shi, Y., Shan, X., 2012. J. Appl. Polym. Sci. 123, 485.
Pacheco-Torgal, F., Ding, Y., Jalali, S., 2012. Const. Build. Mater. 30, 714.
Pantea, D., Darmstadt, H., Kaliaguine, S., Roy, C., 2003. J. Anal. Appl. Pyrol. 67, 55.
Peshkovsky, S.L., Friedman, M.L., Tukachinsky, A.I., Vinogradov, G.V., Enikolopian, N.S., 1983. Polym. Compos. 4, 126–134.
Phadke, A.A., De, S.K., 1986. Polym. Eng. Sci. 26, 1079–1087.
Phadke, A.A., Bhattacharya, A.K., Chakraborty, S.K., De, S.K., 1983. Rubber Chem. Technol. 56, 726–736.
Phadke, A.A., Chakraborty, S.K., De, S.K., 1984. Rubber Chem. Technol. 57, 19–33.
Pramanik, P.K., Baker, W.E., 1995a. J. Elast. Plast. 27, 253–267.
Pramanik, P.K., Baker, W.E., 1995b. Plast. Rubb. Comp. Proc. Appl. 24 (4), 229–237.
Premachandra, J.K., Edirisinghe, D.G., De Silva, M.I.A., 2011. Prog. Rubber Plast. Recycl. Technol. 27, 31.
Price, G.J., 1990. In: Mason, T.J. (Ed.), Advances in Sonochemistry, vol. 1. JAI Press Ltd., Greenwich, CT, pp. 231–287.
Price, G.J., Norris, D.J., West, P.J., 1992. Macromolecules 25, 6447–6454.
Pryweller, J., 1999. Eur. Rubber J. 181 (9), 17.
Quek, A., Balasubramanian, R., 2009. J. Air Waste Manag. Assoc. 59, 747.
Rader, C.P. (Ed.), 1995. Plastic, Rubber and Paper Recycling, American Chemical Society, Washington DC.
Radusch, H.J., Luepke, T., Poltersdorf, S., Laemmer, E., 1990. Kautschuk Gummi 43, 767–769.
Raghavan, D., 2000. J. Appl. Polym. Sci. 77, 934–942.

Raghavan, D., Tratt, K., Wool, R.P., 1994. Mater. Res. Soc. Symp. 344, 177–188.
Raghavan, D., Huynh, H., Ferraris, C.F., 1998. J. Mater. Sci. 33, 1745–1752.
Rajalingam, P., Baker, W.E., 1992. Rubber Chem. Technol. 65, 908–916.
Rajalingam, P., Sharpe, J., Baker, W.E., 1993. Rubber Chem. Technol. 66, 664–677.
Rajan, V.V., Dierkes, W.K., Noordermeer, J.W.M., Joseph, R., 2005. Rubber Chem. Technol. 78, 855.
Rajan, V.V., Dierkes, W.K., Joseph, R., Noordermeer, J.W.M., 2006. J. Appl. Polym. Sci. 102, 4194.
Ramos, G., Alguacil, F.J., Lopez, F.A., 2011. Revista de Metalurgia 47, 227–284.
Ratcliffe, A., 1972. Chem. Eng. 79 (7), 62.
Razmjooei, F., Naderi, G., Bakhshandeh, G., 2012. J. Appl. Polym. Sci. 124, 4864.
Read, K., 2012. Tire Technol. Int. October, 42.
Renshaw, R.H., 1985. Rubber in Roads. Plastics and Rubber Institute, South African Section, S. Africa, pp. 1.
Roberts, F.L., Lytton, R.L., 1987. Transport. Res. Rec. 115, 216–225.
Rodriguez, E.L., 1988. Polym. Eng. Sci. 28, 1455–1461.
Rogers III, J.N., Waddington, D.V., 1992. Agron. J. 84, 203–209.
Romine, R.A., Romine, M.F., 1998. Polym. Degrad. Stab. 59, 353.
Rooj, S., Basak, G.C., Maji, P.K., Bhowmick, A.K., 2011. J. Polym. Environ. 19, 382.
Rostami, H., Lepore, J., Silverstram, T., Zandi, I., 1993. In: Proceedings of the Internernational Conference on Concrete 2000. Dundee, UK, pp. 391–399.
Rouse, M., 1995. Rubber World 23.
Roy, C., Chaala, A., Darmstadt, H., 1999. J. Anal. Appl. Pyrol. 51, 201–221.
Saiwari, S., Dierkes, W.K., Noordermeer, J.W.M., 2011. Paper # 82, presented at a meeting of the Rubber Division. Oct, ACS, Cleveland, Ohio.
Santos, G.C., Carmo, D.M., Rezende, C.G.F., Zattera, A.J., Oliveira, M.G., Oliveira, P.J., 2011. J. Appl. Polym. Sci. 122, 948.
Sato, S., Honda, Y., Kuwahara, M., Kishimoto, H., Yagi, N., Muraoka, K., Watanabe, T., 2004. Biomacromolecules 5, 511.
Schaefer, R., 1986. Paper #79 presented at the meeting of ACS Rubber Division. Atlanta, Georgia, October.
Schaefer, R., Berneking, R., 1986. Paper #80 presented at the meeting of ACS Rubber Division. Atlanta, Georgia, October.
Schmid, G., 1940. Physik. Z. 41, 326–337.
Schmid, G., Rommel, O., 1939. Z. Elektrochem 45, 659–661.
Schuller, I.S., 1991. PhD Dissertation, Texas A&M University.
Scrap Tire Markets in the United States, 2009. 9th Biannual Report, May.
Scrap Tire News, 1999a. 13 (2), 6.
Scrap Tire News, 1999b. 13 (6), 7.
Scrap Tire News, 2001. 15 (10) 1–3.
Segre, N., Joekes, I., 2000. Cement Concr. Res. 30, 1421–1425.
Segre, N., Monteiro, P.J.M., Sposito, G., 2002. J. Coll. Interf. Sci. 248, 521–523.
Sekhar, B.C., Kormer, V.A., 1995. European Patent Application, EP 0690091A1.
Shim, S.E., Isayev, A.I., 2001. Rubber Chem. Technol. 74, 303–316.
Shim, S.E., Isayev, A.I., 2003. J. Appl. Polym. Sci. 88, 2630–2638.
Shim, S.E., Ghose, S., Isayev, A.I., 2002a. Polymer 43, 5535–5543.
Shim, S.E., Parr, J.C., von Meerwall, E., Isayev, A.I., 2002b. J. Phys. Chem. B 106, 12072–12078.
Shim, S.E., Isayev, A.I., von Meerwall, E., 2003. J. Polym. Sci. B: Polym. Phys. 41, 454–465.
Shutov, F., Ivanov, G., Arastoopour, H., Volfson, S., 1992. Polym. Mater. Sci. Eng. 67, 404.
Siuru, B., 1997. Scrap Tire News 11, December, 14.
Snyder, R.H., 1998. Scrap Tyres: Disposal and Reuse. Society of Automotive Engineers, Inc., Warrendale, PA.
Solov'ev, E.M., Pavlov, V.B., Enikolopov, N.S., 1987. Int. Polym. Sci. Technol. 14, 10–12.
Sritragool, K., Michael, H., Gehde, M., Gohs, U., Heinrich, G., 2010. Kautch. Gummi Kunst. 63, 554.
Stark, F.J. Jr, Leigton, A., Wagner, D., 1983. Rubber World 188 (12), 36–51.
Sun, X., Isayev, A.I., 2007. J. Mater. Sci. 42 (17), 7520.
Sun, X., Isayev, A.I., 2008. Rubber Chem. Technol. 81, 38.
Supri, A.G., Ismail, H., 2012. Polym. Plast. Technol. Eng. 51, 549.

Suslick, K.S., 1988. Ultrasound: Its Chemical. Physical and Biological Effects, VCH, New York.
Suslick, K.S., 1989. Scientific American 260, 80.
Suslick, K.S., Doctycz, S.J., Flint, E.B., 1990. Ultrasonics 28, 280–290.
Szilard, J.A., 1973. Reclaiming Rubber and Other Polymers. Noyes Data Corporation, London.
Tabata, M., Sohma, J., 1980a. Chem. Phys. Lett. 73, 178–180.
Tabata, M., Sohma, J., 1980b. Eur. Polym. J. 16, 589–595.
Takallou, H.B., Takallou, M.B., 1991. Elastomerics 123 (7), 19.
Tan, K., Li, C., Meng, H., Wang, Z., 2009. Polym. Test. 28, 2.
Tan, K.L., Chen, X.-S., Yan, B.-S., Lu, Y.-Z., Li, C.-X., 2012. J. Appl. Polym. Sci. 123, 3495.
Tapale, M., Isayev, A.I., 1998. J. Appl. Polym. Sci. 70, 2007–2019
Theodore, A.N., Pett, R.A., Jackson, D., 1998. Rubber World 218 (2), 23–25.
Topcu, I.B., 1995. Cement Concr. Res. 25, 304–310.
Tripathy, A.R., Morin, J.E., Williams, D.E., Eyles, S.J., Farris, R.J., 2002. Macromol. 35, 4616–4627.
Tsang, H.-H., 2008. Earthquake Eng. Struct. Dynam. 37 (2), 283.
Tukachinsky, A., Schworm, D., Isayev, A.I., 1996. Rubber. Chem. Tech. 69, 92–103.
US Scrap Tire Markets, 2009. Scrap Tire Management Council, Rubber Manufactures Association.
Verbruggen, M.A.L., 1999. L van der Does, J.W.M. Noordermeer, M. van Duin and H.J. Manuel. Rubber Chem. Technol. 72, 731–740.
Warner, W.C., 1994. Rubber Chem. Technol. 67, 559–566.
Wiessner, S., Wagenknecht, U., Heinrich, G., Michael, H., Mennig, G., 2012. Kautch. Gummi Kunst. 65 (4), 28.
William, P.T., 2002. Chem. Rev. 12, 17–19.
William, P.T., Brindle, A.J., 2003. J. Anal. Appl. Pyrolys. 67, 143–164.
William, P.T., Ferrer, F., 1997. Resour. Conserv. Recycling 19, 221.
Wolk, R.H., 1972. Rubber Age 104, 103.
Wu, B., Zhou, M.H., 2009. Waste Manage. 29, 355.
Wu, D.Y., Bateman, S., Partlett, M., 2007. Compos. Sci. Technol. 67, 1909.
Xi, Y.P., Li, Y., Xie, Z.H., Lee, J.S., 2004. In: Proceedings of the International Workshop on Sustainable Developement, Beijing, China, May. Concr. Technol. 45–54.
Xin, Z.X., Zhang, Z.X., Zhang, B.S., Pal, K., Deng, X., Lee, S.H., Kim, J.K., 2009. J. Compos. Mater. 43, 3003.
Xin, Z.X., Zhang, Z.X., Pal, K., Kim, K.J., Kang, D.J., Kim, J.K., Bang, D.S., 2009. J. Cell. Plast. 45, 499.
Xu, X., Wang, Q., Kong, X.A., Zhang, X.D., Huang, J.G., 1996. Plast. Rubber Compos. Process. Appl. 25, 152.
Xu, Z., Losure, N.S., Gardner, S.D., 1998. J. Adv. Mater. 30 (2), 11–18.
Yashin, V.V., Isayev, A.I., 1999. Rubber Chem. Technol. 72, 741–757.
Yashin, V.V., Isayev, A.I., 2000. Rubber Chem. Technol. 73, 325–339.
Yasin, T., Khan, S., Nho, Y.C., Ahmad, R., 2012. Radiat. Phys. Chem. 81, 421.
Yehia, A.A., Mull, M.A., Ismail, M.N., Hefny, Y.A., Abdel-Bary, E.M., 2004. J. Appl. Polym. Sci. 93, 30.
Young, F.R., Cavitation. McGraw-Hill Co., London, 1989 (Chapter 2).
Yun, J., Isayev, A.I., 2003a. Rubber Chem. Technol. 76, 253–270.
Yun, J., Isayev, A.I., 2003b. Polym. Eng. Sci. 43, 809–821.
Yun, J., Oh, J.S., Isayev, A.I., 2001. Rubber Chem. Technol. 74, 317–330.
Yun, J., Isayev, A.I., Kim, S.H., Tapale, M., 2003. J. Appl. Polym. Sci. 88, 434–441.
Yushanov, S.P., Isayev, A.I., Levin, V.Y., 1996. J. Polym. Sci. B: Polym. Phys. 34, 2409–2418.
Yushanov, S.P., Isayev, A.I., Kim, S.H., 1998a. Rubber Chem. Technol. 71, 168–190.
Zanchet, A., Carli, L.N., Giovanela, M., Brandalise, R.N., Crespo, J.S., 2012. Mater. Design 39, 437.
Zhang, X.X., Lu, C.H., Liang, M., 2009a. J. Polym. Res. 16, 411.
Zhang, S.L., Xin, Z.X.A., Zhang, Z.X., Kim, J.K., 2009b. Waste Manag. 29, 1480.
Zhang, X.X., Zhu, X.Q., Liang, M., Lu, C.H., 2009c. J. Appl. Polym. Sci. 114, 1118.
Zhang, Z.-X., Fan, J.-L., Pal, K., Kim, J.K., Xin, Z.-X., 2011a. J. Vinyl Add. Technol. 17, 254.
Zhang, X.X., Lu, C.H., Liang, M., 2011b. J. Appl. Polym. Sci. 122, 2110.
Zhang, S.L., Zhang, Z.X., Kim, J.K., 2011c. J. Macromol. Sci. Phys. 50, 762.
Zhang, X.X., Lu, C.H., Zheng, Q., Liang, M., 2011d. Polym. Adv. Technol. 22, 2104.

Zhang, X.X., Li, H., Tian, D., He, X., Lu, C.H., 2012a. Mater. Res. Innov. 16, 143.
Zhang, X., Zhou, Z., He, X., Li, J., Lu, C., 2012b. Int. Polym. Process. 27, 427.
Zhu, S.-H., Tzoganakis, C., 2010. J. Appl. Polym. Sci. 118, 1051.
Zhu, X.-Q., Lu, C.-H., Liang, M., 2009. J. Mater. Civil Eng. 21, 699.
Zhurkov, S.N., Zakrevskii, V.A., Korsukov, V.E., Kuksenko, V.S., 1972. Soviet Phys. Solid State 13, 1680.

Index

C

D

E

F

G

H

I

N

R

S

T

W

X

Y

Z

Zeitfracht Medien GmbH
Ferdinand-Jühlke-Straße 7,
99095 - DE, Erfurt
produktsicherheit@zeitfracht.de

FSC
www.fsc.org
MIX
Papier | Fördert
gute Waldnutzung
FSC® C083411